► ENGINEERING MECHANICS
►
► STATICS

PWS SERIES IN ENGINEERING

Anderson, *Thermodynamics*
Askeland, *The Science and Engineering of Materials, Third Edition*
Borse, *FORTRAN 77 and Numerical Methods for Engineers, Second Edition*
Bolluyt / Stewart / Oladipupo, *Modeling for Design Using SilverScreen*
Clements, *68000 Family Assembly Language*
Clements, *Microprocessor Systems Design, Second Edition*
Clements, *Principles of Computer Hardware, Second Edition*
Das, *Principles of Foundation Engineering, Third Edition*
Das, *Principles of Geotechnical Engineering, Third Edition*
Das, *Principles of Soil Dynamics*
Duff / Ross, *Freehand Sketching for Engineering Design*
El-Wakil / Askeland, *Materials Science and Engineering Lab Manual*
Fleischer, *Introduction to Engineering Economy*
Gere / Timoshenko, *Mechanics of Materials, Third Edition*
Glover / Sarma, *Power System Analysis and Design, Second Edition*
Janna, *Design of Fluid Thermal Systems*
Janna, *Introduction to Fluid Mechanics, Third Edition*
Kassimali, *Structural Analysis*
Keedy, *An Introduction to CAD Using CADKEY 5 and 6, Third Edition*
Keedy / Teske, *Engineering Design Using CADKEY 5 and 6*
Knight, *The Finite Element Method in Mechanical Design*
Knight, *A Finite Element Method Primer for Mechanical Design*
Logan, *A First Course in the Finite Element Method, Second Edition*
McDonald, *Continuum Mechanics*
McGill / King, *Engineering Mechanics: Statics, Third Edition*
McGill / King, *Engineering Mechanics: An Introduction to Dynamics, Third Edition*
McGill / King, *Engineering Mechanics: Statics and An Introduction to Dynamics, Third Edition*
Meissner, *Fortran 90*
Raines, *Software for Mechanics of Materials*
Ray, *Environmental Engineering*
Reed-Hil / Abbaschian, *Physical Metallurgy Principles, Third Edition*
Reynolds, *Unit Operations and Processes in Environmental Engineering*
Russ, *CD-ROM for Materials Science*
Schmidt / Wong, *Fundamentals of Surveying, Third Edition*
Segui, *Fundamentals of Structural Steel Design*
Segui, *LRFD Steel Design*
Shen / Kong, *Applied Electromagnetism, Second Edition*
Sule, *Manufacturing Facilities, Second Edition*
Vardeman, *Statistics for Engineering Problem Solving*
Weinman, *VAX FORTRAN, Second Edition*
Weinman, *FORTRAN for Scientists and Engineers*
Wempner, *Mechanics of Solids*
Wolff, *Spreadsheet Applications in Geotechnical Engineering*
Zirkel / Berlinger, *Understanding FORTRAN 77 and 90*

Third Edition

▶ ENGINEERING MECHANICS

▶

▶ STATICS

DAVID J. McGILL AND WILTON W. KING
Georgia Institute of Technology

PWS Publishing Company
Boston

I(T)P An International Thomson Publishing Company

Boston • Albany • Bonn • Cincinnati • Detroit • London • Madrid
Melbourne • Mexico City • New York • Paris • San Francisco
Singapore • Tokyo • Toronto • Washington

PWS PUBLISHING COMPANY
20 Park Plaza, Boston, Massachusetts 02116-4324

 This book is printed on recycled, acid-free paper

International Thomson Publishing
The trademark ITP is used under license.

For more information, contact:

PWS Publishing Company
20 Park Plaza
Boston, MA 02116

International Thomson Publishing Europe
Berkshire House I68-I73
High Holborn
London WC1V7AA
England

Thomas Nelson Australia
102 Dodds Street
South Melbourne, 3205
Victoria, Australia

Nelson Canada
1120 Birchmount Road
Scarborough, Ontario
Canada M1K 5G4

International Thomson Editores
Campos Eliseos 385, Piso 7
Col. Polanco
11560 Mexico D.F., Mexico

International Thomson Publishing GmbH
Konigswinterer Strasse 418
53227 Bonn, Germany

International Thomson Publishing Asia
221 Henderson Road
#05-10 Henderson Building
Singapore 0315

International Thomson Publishing Japan
Hirakawacho Kyowa Building, 31
2-2-1 Hirakawacho
Chiyoda-ku, Tokyo 102
Japan

Library of Congress Cataloging-in-Publication Data

McGill, David J., 1939–
    Engineering mechanics, statics / David J. McGill
and Wilton W. King.
   — 3rd ed.
     p. cm.
    Includes index
    ISBN 0-534-93393-9
    1. Mechanics, Applied 2. Statics I. King,
Wilton W., . II. Title
TA350.M385 1994
620.1'03 — dc20
               94-33845
               CIP

Printed and bound in the United States of America
95 96 97 98 99 — 10 9 8 7 6 5 4 3 2

Sponsoring Editor: Jonathan Plant
Editorial Assistant: Cynthia Harris
Developmental Editor: Mary Thomas
Production Coordinator: Kirby Lozyniak
Marketing Manager: Nathan Wilbur
Manufacturing Coordinator: Marcia Locke
Production: York Production Services

Cover Designer: Julie Gecha
Interior Designer: York Production Services
Compositor: Progressive Information Technologies
Cover Photo: © Grant Faint. Courtesy of the Image Bank
Cover Printer: John P. Pow Company, Inc.
Text Printer: Quebecor Printing, Hawkins

TO THE MEMORY OF ROBERT W. SHREEVES,
FRIEND AND COLLEAGUE

CONTENTS

►
►
► PREFACE

Statics is the first book in a two-volume set on basic mechanics. It is a text for standard courses in statics as found in most colleges of engineering. This text includes more material than is normally covered in such a course because we have attempted to include most traditional special applications from which schools and instructors make selections to augment the core subject matter.

In the writing of this text we have followed one basic guideline—to write the book in the same way we teach the course. To this end, we have written many explanatory footnotes and included frequent questions interspersed throughout the chapters. The answer to each question is provided on the same page (instead of at the end of the chapter as in previous editions). These questions are the same kind as the ones we ask in class; to make the most of them, treat them as serious homework as you read, and look at the answers only after you have your own answer in mind. The questions are intended to encourage thinking about tricky points and to emphasize the basic principles of the subject.

In addition to the text questions, a set of approximately one dozen review questions and answers are included at the end of each chapter. These true-false questions are designed for both classroom discussion and for student review. Homework problems of varying degrees of difficulty appear at the end of every major section. There are over 1,300 of these exercises, and the answers to the odd-numbered ones constitute Appendix F in the back of the book.

There are a number of reasons (besides carelessness) why it may be difficult to get the correct answer to a homework problem in statics or dynamics on the first try. The problem may require an unusual amount of thinking and insight; it may contain tedious calculations; or it may challenge the student's advanced mathematics skills. We have placed an asterisk beside especially difficult problems falling into one or more of these categories.

Statics is characterized by only two basic equations, but these equations are applied in a wide variety of circumstances. Thus it is imperative that students develop a feel for realistically modeling an engineering situation. Consequently, we have included a large number of actual engineering problems among the examples and exercises. Being aware of the assumptions and accompanying limitations of the model and of the solution method can be developed only by sweating over many problems outside the classroom. Only in this way can students develop the insight and creativity needed to solve engineering problems.

Some examples and problems are presented in SI (Système International) metric units, whereas others use traditional United States engineering system units. Whereas the United States is slowly and painfully converting to SI units, our consulting activities make it clear that much engineering work is still being performed using traditional units. Most United States engineers still tend to think in pounds instead of newtons and in feet instead of meters. We believe students will become much better engineers, scientists, and scholars if they are thoroughly familiar with both systems.

In Chapter 1 we introduce engineering mechanics and its primitives, and we set forth the basic laws of statics. This chapter also covers units and dimensions as well as techniques of problem solving and the importance of accuracy.

In this edition, we have split the concepts of forces and moments into separate chapters. Chapter 2, new to the third edition, introduces the reader to the force, to its vector representation, and to a group of equilibrium problems that can be solved without the need for summing moments.

In Chapter 3, the moment of a force is defined and covered in detail, after which the pair of equilibrium equations for the finite-sized body are presented. The reader is then prepared for the second half of the chapter, which deals with equipollence and with resultants of discrete and distributed force systems.

The heart of the book is Chapter 4, in which we analyze equilibrium problems. The chapter begins with the free-body diagram—crucial to successful analysis of problems in statics and dynamics. We then examine the equilibrium of a single body and expand that study to interacting bodies and to parts of a structure.

In Chapter 5 we extend our study to structures of four common types: trusses, frames, beams, and cables. In preparation for later courses in strength of materials or deformable bodies, we include a section on shear and moment diagrams. The studies of Chapter 5 differ from most of those in Chapter 4 in that the bodies are routinely "cut" (on paper) in order to determine their important internal force distributions.

Though friction forces may sometimes act on the bodies studied in Chapter 4, the special nature of these forces was not elaborated on there. This detailed study is done in Chapter 6, which deals exclusively with Coulomb (or dry friction). This chapter also includes fundamental problems and applications of dry friction along with special applications such as the friction on a flexible flat belt wrapped around a cylindrical surface.

Chapters 7 and 8, although not statics per se, treat topics often covered in statics courses. Chapter 7 includes the topics of centroids (of lines, areas, and volumes) and of centers of mass. Chapter 8 follows with a study of inertia properties of areas—a necessary background for studies of the strength and deflection of beams in courses on the mechanics of deformable solids. To this end we include a closing section on Mohr's circle for principal axes and principal moments of inertia of areas. Mohr's circle is also useful in studies of stress and strain as well as in studies of moments of inertia of masses (the latter of which is covered in our dynamics volume).

Finally, Chapter 9 includes two special topics in statics. The first is the principle of virtual work, a very powerful method in mechanics and an elegant alternative to the equations of equilibrium. The second is "fluid statics," or the statics of submerged bodies subjected to hydrostatic fluid pressure.

We thank Meghan Root for cheerfully typing all our third edition changes. And for their many useful suggestions, we are grateful to our third edition reviewers:

William Bickford
Arizona State University

Vincent WoSang Lee
University of Southern California

Donald E. Carlson
University of Illinois at Urbana

Joseph Longuski
Purdue University

Robert L. Collins
University of Louisville

Robert G. Oakberg
Montana State University

John Dickerson
University of South Carolina

Joseph E. Parnarelli
University of Nebraska

John F. Ely
North Carolina State University

Mario P. Rivera
Union College

Laurence Jacobs
Georgia Institute of Technology

Wallace S. Venable
West Virginia University

Seymour Lampert
University of Southern California

Carl Vilmann
Michigan Technological University

We also thank our colleagues at *Georgia Institute of Technology:* Larry Jacobs, Charles Ueng, Wan-Lee Yin, Don Berghaus, Jianmin Qu, Al Ferri, Dewey Hodges, Manohar Kamat, and Alan Larson, for helpful comments since the second edition.

We are grateful to the following professors, who each responded to a questionnaire we personally sent out in 1991: Don Carlson, *University of Illinois;* Patrick MacDonald and John Ely, *North Carolina State University;* Vincent Lee, *University of Southern California;* Charles Krousgrill, *Purdue University;* Samuel Sutcliffe, *Tufts University;* Larry Malvern and Martin Eisenberg, *University of Florida;* John Dickerson, *University of South Caro-*

lina; Bill Bickford, *Arizona State University;* James Wilson, *Duke University;* Mario Rivera, *Union College;* and Larry Jacobs, *Georgia Institute of Technology.* Their comments were also invaluable.

Special appreciation is expressed to our insightful editor, Jonathan Plant, and to the following individuals involved in the smooth production of this third edition: Mary Thomas and Kirby Lozyniak of *PWS Publishing Company,* and Tamra Winters of *York Production Services.*

David J. McGill
Wilton W. King

We are pleased to introduce to this edition a new set of model-based problems. These problems, presented in a full-color insert bound into the book, introduce students to the process of building three-dimensional models from commonly found objects in order to observe as well as calculate mechanical behavior. Many students beginning their engineering education lack a hands-on, intuitive feel for this behavior, and these specially designed problems can help build confidence in their observational and analytical abilities.

We wish to acknowledge the following contributors to the model-based problems insert:

David J. McGill and Wilton W. King for their initial conception and presentation of the model-based problem idea in *Dynamics Model Problems* written to accompany *Engineering Mechanics: An Introduction to Dynamics,* Third Edition.

David Barnett, *Stanford University,* Mario P. Rivera, *Union College;* Robert G. Oakberg, *Montana State University;* John F. Ely, *North Carolina State University;* Carl Vilmann, *Michigan Technological University;* Robert L. Collins, *University of Louisville;* Nicholas P. Jones, *Johns Hopkins University* and William B. Bickford, *Arizona State University* for their evaluations of McGill and King's *Dynamics Model Problems.*

We thank Mario P. Rivera, Robert G. Oakberg, and John F. Ely, for developing additional model problems for the insert. And a very special thanks to Michael K. Wells, *Montana State University,* for developing and editing the final text of the insert, and for providing an introduction and additional problems.

PWS Publishing Company

► **Engineering Mechanics**

►

► **Statics**

1 ▶ INTRODUCTION

1

1.1   Engineering Mechanics

Two things that are basic to understanding the physical world and universe in which we live are (a) the motions of bodies and (b) their mechanical interactions. Engineering mechanics provides the basic principles by which these motions and interactions are described, related, and predicted.

There are many diverse applications of mechanics, which begin in most undergraduate engineering curricula with studies of statics, dynamics, mechanics of materials, and fluid mechanics. Applications of the principles learned in these studies have led to solutions of such problems as:

1. The invention and continuing refinement of the bicycle, the automobile, the airplane, the rocket, and machines for manufacturing processes.
2. The description of the motions of the planets and of artificial satellites.
3. The description of the flows of fluids that allow motion and flight to occur.
4. The determination of the stresses (intensities of forces) produced in machines and structures under load.
5. The control of undesirable vibrations that would otherwise cause discomfort in vehicles and buildings.

In solving problems such as these, mathematical models are created and analyzed. It will be important for students to learn to bridge the gap between problems of the real world and the mathematical models used to describe them. This, too, is part of mechanics — being able to visualize the actual problem and then to come up with a realistic and workable model of it. Proficiency will come only from the experience of comparing the predictions of mathematical models with observations of the physical world for large numbers of problems. The reader will find that there are not a great number of basic ideas and principles in mechanics, but they provide powerful tools for engineering analysis if they are thoroughly understood.

In the first part of this introductory mechanics text, we shall be considering bodies at rest in an inertial (or Newtonian) reference frame; a body in this situation is said to be in equilibrium. Statics is the study of the equilibrium interactions (forces) of a body with its surroundings. In another study, called dynamics, we explore the relation between motions and forces, especially in circumstances in which the body may be idealized as rigid.

1.2   The Primitives

There are several concepts that are *primitives* in the study of mechanics.

   Space   We shall be using ordinary Euclidean three-dimensional geometry to describe the positions of points on the bodies in which we are interested, and, by extension, the regions occupied by these bodies. The coordinate axes used in locating the points will be locked into a reference frame, which is itself no more or less than a rigid body (one for which the distance between any two points is constant).

Time   Time will be measured in the usual way. It is, of course, the measure used to identify the chronology of events. Time will not really enter the picture in statics; it becomes important when the bodies are no longer at rest, but are instead moving in the reference frame.

Force   Force is the action of one body upon another, most easily visualized as a push or pull. A force acting on a body tends to accelerate it in the direction of the force.

Mass   The resistance of a body to motion is measured by its mass and by the distribution of that mass. Mass per unit volume, called density, is a fundamental material property. Mass is a factor in the gravitational attraction of one body to another. It is this manifestation of mass that we shall encounter in statics.

1.3   Basic Laws

When Isaac Newton first set down the basic laws or principles upon which mechanics has come to be based, he wrote them for a particle. This is a piece of material sufficiently small that we need not distinguish its material points as to locations (or velocities or accelerations). Therefore, we could actually consider the Earth and Moon as particles for some applications such as the analysis of celestial orbits (as Newton did).

Newton published a treatise called *The Principia* in 1687, in which certain principles governing the motion of a particle were developed. These have come to be known as **Newton's Laws of Motion** and are commonly expressed today as follows:

1. In the absence of external forces, a particle has constant velocity (which means it either remains at rest or travels in a straight line at constant speed).

2. If a force acts on a particle, it will be accelerated in the direction of the force, with an acceleration magnitude proportional to that of the force.

3. The two forces exerted on a pair of particles by each other are equal in magnitude, opposite in direction, and collinear along the line joining the two particles.

We must recognize that the laws will not apply when velocities approach the speed of light, when relativistic effects become important. Neither will Newton's Laws apply at a spatial scale smaller than that of atoms. It is also important to understand that what we are really doing is hypothesizing the existence of certain special frames of reference in which the laws are valid. These frames are called Newtonian, or inertial.

This poses a chicken-and-egg problem where one tries to reason which comes first — the inertial frame or the three laws. It is true that the laws hold only in inertial frames, but also that inertial frames are those in which the laws hold, so that neither is of any value without the other. To establish that a frame is inertial requires numerous comparisons of the predictions of the laws of motion with experimental observations. Such comparisons have failed to provide any contradiction of the assertion that a frame containing the mass center of the solar system and having fixed orientation relative to the "fixed" stars is inertial. For this reason many writers refer to this frame of reference as "fixed" or "absolute." While the earth, which moves and turns relative to this standard, is not an inertial frame, it closely enough approximates one for the analysis of most earth-bound engineering problems.

An important extension of Newton's Laws was made in the 18th century by the Swiss mathematician Leonhard Euler. The extension was the postulation of two vector laws of motion for the finite-sized body. These laws **(Euler's Laws),** again valid only in inertial frames, are expressible as:

1. The resultant of the external forces on a body is at all times equal to the time derivative of its momentum.
2. The resultant moment of these external forces about a fixed point is equal to the time derivative of the body's moment of momentum about that point.

Euler's Laws allow us to study the motions (or the special case in which the motions vanish) of bodies, whether or not they are particles. The first law yields the motion of the mass center, and the second leads to the orientational, or rotational, motion of a rigid body. It can be shown (see Appendix E) that an "action-reaction" principle (equivalent to Newton's Third Law) follows from these two laws of Euler.

Another contribution by Isaac Newton which is of monumental importance in mechanics is his **Law of Gravitation,** which expresses the gravitational attraction between two particles in terms of their masses (m_1 and m_2) and the distance (r) between them. The magnitude (F) of the force on either particle is given by

$$F = \frac{Gm_1 m_2}{r^2}$$

where G is the universal gravitation constant. For a small body (particle) being attracted by the earth, the force is given approximately by an equation of the same form,

$$F = \frac{GMm}{r^2}$$

where now M is the mass of the earth, m is the mass of the particle, and r the distance from the particle to the center of the earth. If the particle is near the earth's surface, r is approximately the radius, r_e, of the earth and to good approximation

$$F = \left(\frac{GM}{r_e^2}\right) m = mg$$

The symbol g is called the strength of the gravitational field or the gravitational acceleration, since this is the free-fall acceleration of a body near the surface of the earth. Although g varies slightly from place to place on the earth, we shall, unless otherwise noted, use the nominal values of 32.2 lb/slug (or ft/sec^2) and 9.81 N/kg (or m/s^2). The force, mg, that the earth exerts on the body is called the weight of the body.

1.4 Units and Dimensions

The numerical value assigned to a physical entity expresses the relationship of that entity to certain standards of measurement called **units.** There is currently an international set of standards called the International System (SI) of Units. This is a descendant of the MKS metric system. In the SI system the unit of time is the **second** (s), the unit of length is the **meter** (m), and the unit of mass is the **kilogram** (kg). These independent (or *basic*) units are defined by physical entities or phenomena: the second is defined by the period of a radiation occurring in atomic physics, and the meter is defined by the wavelength of a different radiation. One kilogram is defined to be the mass of a certain piece of material that is stored in France. Any other SI units we shall need are *derived* from these three basic units. The unit of force, the **newton** (N), is derived by way of Newton's Second Law, so that, for example, one newton is the force required to give a mass of one kilogram an acceleration of one meter per second per second, or 1 N = 1 kg · m/s^2.

Until very recently almost all engineers in the United States have used a different system (sometimes called the British gravitational or U.S. system) in which the basic units are the **second** (sec) for time, the **foot** (ft) for length, and the **pound** (lb) for force. The pound is the weight, at a standard gravitational condition (location) of a certain body of material that is stored in the United States. In this system the unit of mass is derived and is the **slug,** one slug being the mass that is accelerated one foot per second per second by a force of one pound, or 1 slug = 1 lb-sec^2/ft. For the foreseeable future, United States engineers will find it desirable to be as comfortable as possible with both the U.S. and SI systems; for that reason we have used both sets of units in examples and problems throughout this book.

We next give a brief discussion of unit conversion. The conversion of units is very quickly and efficiently accomplished by multiplying by equivalent fractions until the desired units are achieved. For example, suppose we wish to know how many newton · meters (N · m) of torque are equivalent to 1 lb-ft; since we know there to be 3.281 ft per m and

4.448 N per lb,

$$1 \text{ lb-ft} = 1 \text{ lb-ft} \left(\frac{1 \text{ m}}{3.281 \text{ ft}} \right) \left(\frac{4.448 \text{ N}}{1 \text{ lb}} \right) = 1.356 \text{ N} \cdot \text{m}$$

Note that if the undesired units don't cancel, the fraction is erroneously upside-down. For a second example, let us find how many slugs of mass there are in a kilogram:

$$1 \text{ kg} = 1 \frac{\text{N} \cdot \text{s}^2}{\text{m}} \left(\frac{1 \text{ lb}}{4.448 \text{ N}} \right) \left(\frac{1 \text{ m}}{3.281 \text{ ft}} \right)$$

$$= 0.06852 \frac{\text{lb-sec}^2}{\text{ft}} \quad \text{or} \quad 0.06852 \text{ slug}$$

Inversely, 1 slug = 14.59 kg. A table of units and conversion factors may be found in Appendix B.

It is a source of some confusion that sometimes there is used a unit of mass called the pound, or pound mass, which is the mass whose weight is one pound of force at standard gravitational conditions. Also, the term kilogram has sometimes been used for a unit of force, particularly in Europe. Grocery shoppers in the U.S. are exposed to this confusion by the fact that packages are marked as to weight (or is it mass?) both in pounds and in kilograms. Throughout this book, without exception, *the pound is a unit of force and the kilogram is a unit of mass.*

The reader is no doubt already aware of the care that must be exercised in numerical calculations using different units. For example, if two lengths are to be summed in which one length is 2 feet and the other is 6 inches, the simple sum of the measures, 2 + 6 = 8, does not provide a measure of the desired length. It is also true that we may not add or equate the numerical measures of different types of entities; thus it makes no sense to attempt to add a mass to a length. These are said to have different dimensions, a **dimension** being the name assigned to the *kind* of measurement standard involved, as contrasted with the choice of a particular measurement standard (unit). In science and engineering we attempt to develop equations expressing the relationships among various physical entities in a physical phenomenon. We express these equations in symbolic form so that they are valid regardless of the particular choice of system of units; nonetheless, they must be *dimensionally consistent.*

To aid in verification of dimensional consistency, we assign some common symbols for basic dimensions: L for length, M for mass, F for force, and T for time. Just as there are derived units of measure, there are derived dimensions; thus the dimension of velocity or speed is L/T and the dimension of acceleration is L/T^2. In SI units, force is derived from L, M, and T; we have, dimensionally, $F = ML/T^2$. In U.S. units, mass is derived from L, F, and T; hence, dimensionally, $M = FT^2/L$. Some things are dimensionless. An example of this is the radian measure of an angle. Since the measure is defined by the ratio of two lengths, the numerical value is thus independent of the choice of unit of length. Arguments of transcendental functions must always be dimensionless.

To check an expression for dimensional consistency, we replace each symbol for a physical quantity by the symbol (or symbols) for its dimension. We likewise replace any dimensionless quantity by unity. The dimension symbols in each separate term of an equation must combine to yield the same dimension for each term. The following examples illustrate this process:

1. The distance, d, of a runner from the finish line of a race has been derived to be (for an interval of constant acceleration)

$$d = d_0 - v_0 t - \frac{1}{2} a t^2$$

where t is time, d_0 is the distance at $t = 0$, v_0 is the speed at $t = 0$, and a is the constant acceleration. Substituting the dimension symbols in each term

$$L = L - \frac{L}{T}(T) - \frac{L}{T^2}(T^2)$$

where the equality sign and the minus signs serve only the purpose of identifying the terms under consideration. Since each term has the dimension of length (L) the equation is dimensionally consistent.

2. A square plate is supported by a pair of ropes; suppose that a student deduces that the force, P, exerted by one rope is

$$P = mg(2\ell + 3\ell^2)$$

where ℓ is the length of a side of the plate, m is the mass of the plate, and g is the acceleration of gravity. If, as is intended here, every length appearing in the problem is a multiple (or fraction) of ℓ, then a student must immediately conclude that the analysis is in error since the dimension of 2ℓ is L and the dimension of $3\ell^2$ is L^2; thus they cannot be added.

A second student analyzing the problem concludes that

$$P = \frac{1}{2} mg\ell$$

This student also must conclude that the analysis is in error since the dimension of P is F while the dimension of $mg\ell$ is FL.

A third student analyzing this problem concludes that

$$P = \frac{1}{2} mg$$

This solution may be in error, but at least it satisfies the requirement of dimensional consistency.

3. Analyzing the dynamics of a rotating plate with edge lengths a and b, a student finds the angular speed, ω (the dimension is $1/T$, and typical units are rad/sec), at a certain instant to be

$$\omega = 5g/(a + b^2)$$

which cannot be true since the denominator is dimensionally inconsistent (adding an L to an L^2).

A second student obtains

$$\omega = 5g/(a + b)$$

Noting that g, the acceleration of gravity, has dimension L/T^2, we test the dimensional consistency of the result by writing

$$\frac{1}{T} \overset{?}{=} \frac{L/T^2}{L} = \frac{1}{T^2}$$

which demonstrates that this result is not dimensionally consistent either.

A third student obtains

$$\omega = 5\sqrt{g/(a + b)}$$

which is dimensionally consistent since the dimension of $\sqrt{g/(a + b)}$ is $\sqrt{1/T^2} = 1/T$.

4. A student's analysis of vibrations of an airplane wing yields the displacement, u (its dimension is L), of a certain point to be

$$u = Ae^{-\alpha t} \sin \beta t$$

where t is time. For this equation to be dimensionally consistent, (αt) and (βt) must be dimensionless; therefore, α and β must each have the dimension $1/T$. Moreover, the dimension of A must be length (L).

These examples illustrate a compelling reason for expressing the solutions to problems in terms of symbols so that *any system of units can be used*. When that is done it is relatively easy to check the dimensional consistency of the proposed expressions. With a solution in terms of symbols, we can also examine limiting cases of the parameters to check the solution itself. Sometimes we can even undertake to optimize a solution quantity with respect to one or more of the parameters.

1.5 Problem Solving and Accuracy of Solutions

In Chapters 2 and 3 we shall undertake a study of the two vectors of prime importance in statics: forces and moments. These vectors will be used to develop the concept of the resultant of a force/couple system. We shall then be ready in Chapter 4 to solve general equilibrium problems. At that time we shall give a detailed discussion of problem solving, emphasizing one of the most useful concepts in mechanics — the free-body diagram. Until we reach that point in our study, however, it is important that the student/reader do the following with the problems in the first three chapters:

1. Read the problem carefully, digest the physical meaning, and list the "givens" and the "requireds."

2. Sketch any diagrams that might be helpful.
3. Carry out the calculations, using only as many digits as the least accurate number in the given data.*
4. Look over your answers. See if they make sense, and draw and state all the conclusions you can from them.

In the examples, unless stated otherwise, we shall retain three significant digits (unless one or more digits are lost through additions or subtractions; for example, $90.2 - 90.1 = 0.1$). If, say, a length ℓ is given in the data to be 2 ft, it will be assumed throughout the example that ℓ is actually 2.00 ft.

In the next two chapters (indeed, throughout most of the rest of the book), we shall be using vectors to represent the three entities commonly known as force, moment, and position. In Appendix A we offer a review of vectors, and we encourage all student readers to glance through this appendix at this time and to study any unfamiliar topics.

---

* For instance, if $g = 32.2$ ft/sec^2 or 9.81 m/s^2 is used in a calculation, it is ridiculous to give an answer to four significant digits.

PROBLEMS ▶ Chapter 1

1.1 Describe a physical problem in which we already know the configuration (location) of a body at rest, and are interested in knowing the forces that keep it there.

1.2 Describe a physical problem in which we know at least one of the forces acting on a body at rest and are interested in knowing its configuration.

1.3 Explain why velocity and energy are not primitives in the study of mechanics.

1.4 A dyne is one gram · centimeter/s^2. How many dynes are there in one pound?

1.5 How many kilometers are there in one mile?

1.6 What is the weight in newtons of a 2500-pound automobile?

1.7 The Btu (British thermal unit) is a unit of energy used in thermodynamic calculations. There are 778 ft-lb in one Btu. How many joules are there in one Btu? (One joule $= 1$ N · m of energy.)

1.8 Determine which of the terms in the following equation is dimensionally inconsistent with all the others:

$$mg \cos \theta - N = \frac{mv^2}{r} + \frac{mr^2}{t^2}$$

where $m =$ mass, $g =$ gravitational acceleration, $N =$ force, $v =$ velocity, $r =$ radius, and $t =$ time.

1.9 With the same symbols as in the preceding problem, is the equation $v = \sqrt{2gr}$ dimensionally consistent?

1.10 Suppose that a certain (fictitious) quantity has dimension $L^2 M^3 / T^4$, and that one quix $= 1$ m^2 · kg^3/s^4. The corresponding unit in U.S. units is a quax. How many quix in a quax?

1.11 Determine the units of the universal gravitation constant G, using the fact that the gravity force is expressible as Gm_1m_2 / r^2. Roughly calculate the value of G using your own weight and mass, and the fact that for the earth, (a) radius ≈ 3960 miles, and (b) average specific gravity $= 5.51$. (The specific gravity of a material is the ratio of its density to that of water.)

1.12 In studying the flow of a fluid around an object (such as an airplane wing), there arises a quantity called the drag coefficient, which is defined as $C_F = F / (\frac{1}{2}\rho v^2 L^2)$. In this equation, F is the force on the object in the direction of the flow, ρ is the mass density of the fluid, v is its velocity, and L is a characteristic length of the object. Show that C_F is a dimensionless parameter.

1.13 The unit of stress in the SI system of units is the Pascal (Pa). If a material has an elastic modulus of 30×10^6 psi (pounds per square inch), what is its elastic modulus in (a) Pascals? (b) kilopascals (kPa)? (c) megapascals (MPa)? (d) gigapascals (GPa)? Note: 1 kPa $= 1000$ Pa, 1 MPa $= 10^6$ Pa, 1 GPa $= 10^9$ Pa.

1.14 A spring has a modulus of 30 lb/ft. What is its modulus in N/m?

1.15 If the measure of a quantity is known to three significant figures, what is the maximum percentage of uncertainty?

1.16 A rectangular parallelepiped has sides of lengths 2.00 m, 3.00 m, and 4.00 m. Another has sides 2.02 m, 3.03 m, and 4.04 m. Find the difference in their volumes. To how many significant figures is the difference known?

* 1.17 Each edge of a cube is increased in length by 0.002%. What is the percent increase in volume of the cube? Notice the number of significant figures required to calculate the change in volume if it is done by calculating a numerical value for the new volume and then subtracting the old. Repeat the problem for an increase of 2×10^{-6} %. Try to find a way to avoid this "small difference of large numbers" problem.

---

* Asterisks identify the more difficult problems.

2

Forces and Particle Equilibrium

11

2.1   Introduction

To understand the subject of statics, one must master two fundamental concepts: (a) forces, and (b) moments of forces (i.e., turning effects of forces). In this chapter, we will study the first of these, the force. The reader will see that a force is just what one thinks it is — a "push" or "pull," but that with the help of vectors we can actually represent forces in a simple mathematical form.

Having learned how to express forces as vectors in Section 2.2, we will then (in Section 2.3) learn to use the force vectors in equilibrium equations to solve some simple engineering problems involving a single particle. In the process, we will also encounter a critical concept called the "free-body diagram," which is a sketch of the object to be analyzed, showing clearly all externally-applied forces acting upon it.

Finally, in Section 2.4, we expand our study of forces to include the case of more than one particle in equilibrium.

2.2   Forces

A **force** is a mechanical action exerted by one physical body on another. Very simply it is what we perceive as a "push" or a "pull." A force has *magnitude* which is given numerical value in a system of units as described in Section 1.4. Most commonly the unit would be the pound (lb) or the newton (N). A force also has *direction*, which we can describe using the tools of geometry such as angles.

It is usually natural to think of the action of a force on a body to be distributed over a surface or a volume. Push on a table with your finger and the action is distributed over the fingerprint. Sometimes, however, the action is sufficiently localized that it makes sense to characterize its place of application by a single point on the body; this is called the **point of application** of the force. For the time being we shall adopt this viewpoint, leaving the details of the "fingerprint" until later.

The **line of action** of a force is the line in space that passes through the point of application and has the same direction as the force. This concept is illustrated in Figure 2.1 where the force exerted on a ball by a bat is shown. We shall see that the line of action of a force plays a central role in mechanics.

Forces as Vectors   A **vector*** is the mathematical entity by which a force is represented. The magnitude-and-direction qualities of a vector can be displayed graphically by an "arrow" — that is, a directed line segment whose length is proportional to the magnitude of the vector. A defining property of vectors is that they satisfy the **parallelogram law of addi-**

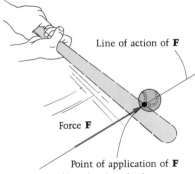

Line of action of **F**

Force **F**

Point of application of **F**

Figure 2.1   Line of action of a force.

_____

* See Appendix A for a more formal treatment of vector algebra.

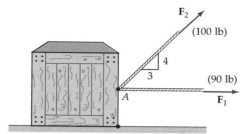

Figure 2.2

tion. To illustrate this law, first consider Figure 2.2 where is shown a crate and the vectors representing two forces exerted on it by cables. Note that we use bold type to denote vectors. Figure 2.3(a) shows the application of the parallelogram law to form F_1 and F_2, their sum being labeled F_3. It's possible to find the magnitude of F_3 by laying out the parallelogram with a ruler and then measuring the diagonal. Or we can observe that the 4-to-3 slope of the line of action of F_2 means that it makes with the horizontal and vertical the angles of a 3-4-5 triangle. Thus in Figure 2.3(a)

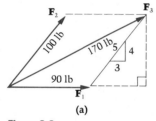
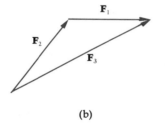
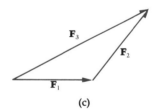

(a)   (b)   (c)

Figure 2.3

we can see a right triangle whose base is of "length"

$$90 + \frac{3}{5}(100) = 150 \text{ lb}$$

and whose "height" is

$$\frac{4}{5}(100) = 80 \text{ lb}$$

The magnitude of F_3 is then the "hypotenuse" of this triangle and is given by

$$\sqrt{(150)^2 + (80)^2} = 170 \text{ lb}.$$

Equivalent head-to-tail representations of the sum of the forces are shown in Figure 2.3(b) and (c).

The legitimacy of representing forces by vectors rests upon the experimental evidence that the effect of two forces simultaneously applied (and having a common point of application) is the same as the effect that arises from a single force, related to the first two by the parallelogram law. However, it is also important to point out that the *mathematical* operation of summing the two vectors representing the forces does not depend in any way upon the forces having a common point of application; only their magnitudes and directions come into play.

In our first example we further illustrate the addition of forces by the parallelogram law.

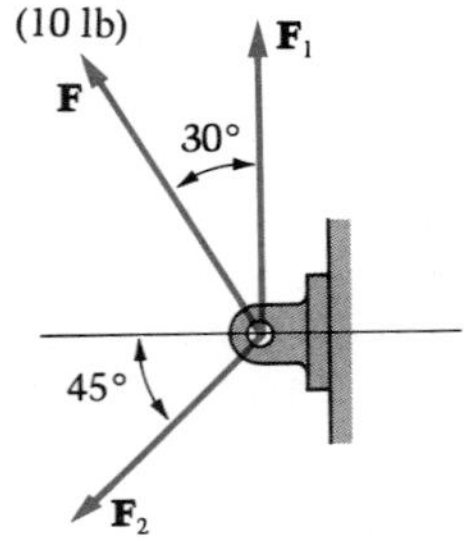

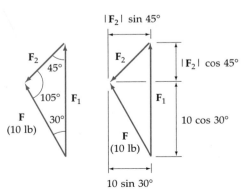

Figure E2.1a

(10 lb) F
F_1
30°
45°
F_2

EXAMPLE 2.1

Find the magnitudes of forces F_1 and F_2 having the directions shown in Figure E2.1a so that their sum is the force F, whose magnitude is 10 lb. All three forces lie in the plane of the paper.

Solution

The head-to-tail version of the parallelogram law of addition is shown in the sketch.

One way to solve this problem is by the Law of Sines; that is, using the triangle in Figure E2.1b,

$$\frac{\sin 45°}{10} = \frac{\sin 30°}{|F_2|} = \frac{\sin 105°}{|F_1|}$$

where $|F_1|$ and $|F_2|$ denote the magnitudes of forces F_1 and F_2. Thus

$$|F_2| = \frac{10(0.5)}{0.707} = 7.07 \text{ lb}$$

$$|F_1| = \frac{10}{0.707}(0.966) = 13.7 \text{ lb}$$

$|F_2| \sin 45°$

F_2
45°
105° F_1
F
(10 lb) 30°

F_2
F_1
F
(10 lb)

$|F_2| \cos 45°$

$10 \cos 30°$

$10 \sin 30°$

Figure E2.1b

An alternative approach is to observe that:

1. The horizontal projection of $|\mathbf{F}_2|$ must be equal to the horizontal projection of 10; that is,

$$|\mathbf{F}_2| \sin 45° = 10 \sin 30°$$

$$|\mathbf{F}_2| = \frac{10(0.5)}{0.707} = 7.07 \text{ lb}$$

2. $|\mathbf{F}_1|$ is the sum of the vertical projections of 10 and $|\mathbf{F}_2|$; that is,

$$|\mathbf{F}_1| = 10 \cos 30° + |\mathbf{F}_2| \cos 45°$$

$$= 10(0.866) + (7.07)(0.707)$$

$$= 8.66 + 5$$

$$= 13.7 \text{ lb}$$

This second approach is closely associated with the concept of orthogonal components of a force, which will be discussed after this example.

A third approach that could be used for this problem is graphic. That is, we could use a scale, a straightedge, and a protractor to draw the "force triangle" shown. The student is encouraged to do this and then to think about the effects of measurement errors on the accuracy of a solution by this method.

Unit Vectors and Orthogonal (Mutually Perpendicular) Components   The directionality of a vector is easily communicated by the graphical means so far shown in this section when we are working in a plane (two-dimensional space). This is often awkward in three-dimensional space and, besides, it's useful to have some formal tools which can be used for mathematical manipulations. **Unit vectors** are the *direction indicators* of vector algebra. Such a vector, as the name suggests, has a *magnitude of unity* and it is *dimensionless.* It is a way of labeling some preassigned direction relative, of course, to some physical reference body. Suppose we let x, y, and z be mutually perpendicular axes, or reference directions, and we let $\hat{\mathbf{i}}, \hat{\mathbf{j}}$, and $\hat{\mathbf{k}}$ be dimensionless unit vectors* parallel, respectively, to those directions as shown in Figure 2.4. Two applications of the parallelogram law (using the shaded plane first) allow us to decompose the force $\mathbf{F}$ into three mutually perpendicular parts written $F_x\hat{\mathbf{i}}, F_y\hat{\mathbf{j}}$, and $F_z\hat{\mathbf{k}}$ so that, as suggested by Figure 2.4,

$$\mathbf{F} = F_x\hat{\mathbf{i}} + F_y\hat{\mathbf{j}} + F_z\hat{\mathbf{k}} \tag{2.1}$$

$F_x\hat{\mathbf{i}}, F_y\hat{\mathbf{j}}$, and $F_z\hat{\mathbf{k}}$ are called orthogonal (or rectangular) **vector components of F**, and F_x, F_y, and F_z are called the corresponding **scalar compo-**

---

* In this book a caret, or "hat," over a bold lower-case letter signifies that the vector is a unit vector. All unit vectors that we use are dimensionless. Throughout the book, the unit vectors $(\hat{\mathbf{i}}, \hat{\mathbf{j}}, \hat{\mathbf{k}})$ are always parallel, respectively, to the assigned directions of (x, y, z).

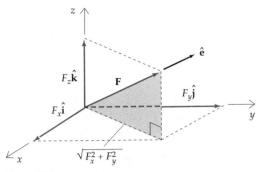

Figure 2.4

nents.* Referring to Figure 2.4, we see that the magnitude, $|\mathbf{F}|$, of $\mathbf{F}$ is given by

$$|\mathbf{F}| = \sqrt{(\sqrt{F_x^2 + F_y^2})^2 + F_z^2} = \sqrt{F_x^2 + F_y^2 + F_z^2} \qquad (2.2)$$

where $\sqrt{F_x^2 + F_y^2}$ is itself the magnitude of the component of $\mathbf{F}$ in the xy plane. (Thus the components of a force need not be associated with coordinate directions.) We can speak of the component in a plane, or normal to a plane, or along a skewed line, and so on.

Sometimes we shall need to write a force as the product of its magnitude $|\mathbf{F}|$ and a unit vector $\hat{\mathbf{e}}$ (as shown in Figure 2.4) in its direction:

$$\mathbf{F} = |\mathbf{F}|\hat{\mathbf{e}} \qquad (2.3)$$

Both Equations (2.1) and (2.3) are very important in the study of statics. It is also important to realize that the scalar components of $\hat{\mathbf{e}}$ are the cosines of the angles (or direction cosines) that $\mathbf{F}$ makes with the positive x, y, and z axes:

$$\mathbf{F} = |\mathbf{F}|(e_x\hat{\mathbf{i}} + e_y\hat{\mathbf{j}} + e_z\hat{\mathbf{k}})$$
$$= |\mathbf{F}|[(\cos\theta_x)\hat{\mathbf{i}} + (\cos\theta_y)\hat{\mathbf{j}} + (\cos\theta_z)\hat{\mathbf{k}}]$$

A comment about notation as it relates to figures is in order here. Sometimes the figures show an arrow labeled with a bold letter denoting a vector. The purpose of this is to display a vector pictorially, usually to depict some general relationship. At other times the figures show an arrow labeled with a scalar. In these instances we are communicating that *the vector in question is expressed by the scalar multiplying a unit vector in the direction of the arrow*. The examples that follow illustrate the use of this "code."

---

* Sometimes in this book we refer to "components" without an adjective; in such instances it should be clear from the context which orthogonal components, scalar or vector, are intended.

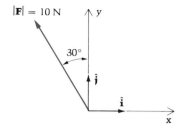

$|\mathbf{F}| = 10$ N

Figure E2.2a

EXAMPLE 2.2

A force **F** of magnitude 10 newtons (N) is depicted in Figure E2.2a. Express the vector in component form using the reference directions x, y, and z.

Solution

$$\theta_x = 90° + 30° = 120°$$
$$\theta_y = 30°$$
$$\theta_z = 90°$$

Therefore,

$$\cos \theta_x = \cos 120° = -0.5$$
$$\cos \theta_y = \cos 30° = 0.866$$
$$\cos \theta_z = \cos 90° = 0$$

and so

$$F_x = 10(-0.5) = -5 \text{ N}$$
$$F_y = 10(0.866) = 8.66 \text{ N}$$
$$F_z = 10(0) = 0$$

The force is therefore expressible as:

$$\mathbf{F} = F_x\hat{\mathbf{i}} + F_y\hat{\mathbf{j}} + F_z\hat{\mathbf{k}}$$
$$= -5\hat{\mathbf{i}} + 8.66\hat{\mathbf{j}} \text{ N}$$

The same result may be obtained by decomposing **F** as shown in Figure E2.2b. Thus we see that, because a unit vector to the left is $(-\hat{\mathbf{i}})$ and a unit vector upward is $\hat{\mathbf{j}}$,

$$\mathbf{F} = 10(0.5)(-\hat{\mathbf{i}}) + 10(0.866)\hat{\mathbf{j}}$$
$$= -5\hat{\mathbf{i}} + 8.66\hat{\mathbf{j}} \text{ N}$$

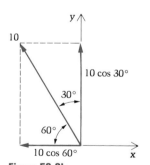

Figure E2.2b

Consequently, by picking off the coefficients of $\hat{\mathbf{i}}$, $\hat{\mathbf{j}}$, and $\hat{\mathbf{k}}$,

$$F_x = -5 \text{ N}$$
$$F_y = 8.66 \text{ N}$$
$$F_z = 0 \text{ N}$$

are seen again to be the scalar components.

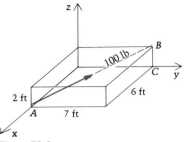

Figure E2.3

EXAMPLE 2.3

Find the components of the force of 100 lb having as its line of action the diagonal of the rectangular solid as shown in Figure E2.3.

Solution

A vector in the direction of the force is the vector from A to B*:

$$\mathbf{r}_{AB} = -6\hat{\mathbf{i}} + 7\hat{\mathbf{j}} + 2\hat{\mathbf{k}} \text{ ft}$$

The unit vector parallel to $\mathbf{r}_{AB}$ is $\mathbf{r}_{AB}/|\mathbf{r}_{AB}|$, or $\hat{\mathbf{e}}_{AB}$:

$$\hat{\mathbf{e}}_{AB} = \frac{-6\hat{\mathbf{i}} + 7\hat{\mathbf{j}} + 2\hat{\mathbf{k}}}{\sqrt{6^2 + 7^2 + 2^2}} = -0.636\hat{\mathbf{i}} + 0.742\hat{\mathbf{j}} + 0.212\hat{\mathbf{k}}$$

The reader should note that $\hat{\mathbf{e}}_{AB}$ has unit magnitude and is dimensionless. Now, writing the force as a vector in the form of its magnitude times the unit vector in its direction, we have

$$\mathbf{F} = 100(-0.636\hat{\mathbf{i}} + 0.742\hat{\mathbf{j}} + 0.212\hat{\mathbf{k}}) \text{ lb}$$
$$\mathbf{F} = -63.6\hat{\mathbf{i}} + 74.2\hat{\mathbf{j}} + 21.2\hat{\mathbf{k}} \text{ lb}$$

and so the scalar components of $\mathbf{F} = F_x\hat{\mathbf{i}} + F_y\hat{\mathbf{j}} + F_z\hat{\mathbf{k}}$ are

$$F_x = -63.6 \text{ lb}$$
$$F_y = 74.2 \text{ lb}$$
$$F_z = 21.2 \text{ lb}$$

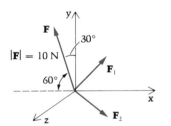

Figure E2.4

EXAMPLE 2.4

Express the force of Example 2.2, $\mathbf{F} = -5\hat{\mathbf{i}} + 8.66\hat{\mathbf{j}}$ N, as the sum of one force making equal angles with x, y, and z and a second force whose direction is in the xz plane. (See Figure E2.4.)

Solution

The first force, $\mathbf{F}_1$, may be expressed as

$$\mathbf{F}_1 = F_1\hat{\mathbf{e}}_1$$

where $\hat{\mathbf{e}}_1$ is a unit vector in the preassigned direction.

To find $\hat{\mathbf{e}}_1$, we write a vector in the direction making equal angles with x, y, and z, such as $\hat{\mathbf{i}} + \hat{\mathbf{j}} + \hat{\mathbf{k}}$, and then divide it by its magnitude:

$$\hat{\mathbf{e}}_1 = \frac{\hat{\mathbf{i}} + \hat{\mathbf{j}} + \hat{\mathbf{k}}}{\sqrt{1^2 + 1^2 + 1^2}}$$

$$\hat{\mathbf{e}}_1 = \frac{1}{\sqrt{3}}\hat{\mathbf{i}} + \frac{1}{\sqrt{3}}\hat{\mathbf{j}} + \frac{1}{\sqrt{3}}\hat{\mathbf{k}}$$

Now the scalar F_1 may be positive or negative, but, since $\hat{\mathbf{e}}_1$ is a unit vector, the absolute value of F_1 is the magnitude, $|\mathbf{F}_1|$, of $\mathbf{F}_1$.

---

* To get a vector from point A to point B, we merely start at A and write down "what we have to do" to get to B; thus, in this case, we travel $-6\hat{\mathbf{i}}$ to get to the origin plus $7\hat{\mathbf{j}}$ to get to C, plus $2\mathbf{k}$ to move finally up to B.

The only thing we know about the second force is that it has no component in the y direction. Therefore, without any loss of generality we may express $\mathbf{F}_2$ as

$$\mathbf{F}_2 = F_{2x}\hat{\mathbf{i}} + F_{2z}\hat{\mathbf{k}}$$

Now, setting the sum of $\mathbf{F}_1$ and $\mathbf{F}_2$ equal to $\mathbf{F}$:

$$\mathbf{F}_1 + \mathbf{F}_2 = -5\hat{\mathbf{i}} + 8.66\hat{\mathbf{j}}$$

Therefore

$$F_1(0.577\hat{\mathbf{i}} + 0.577\hat{\mathbf{j}} + 0.577\hat{\mathbf{k}}) + F_{2x}\hat{\mathbf{i}} + F_{2z}\hat{\mathbf{k}} = -5\hat{\mathbf{i}} + 8.66\hat{\mathbf{j}}$$

Equating the respective coefficients of $\hat{\mathbf{i}}$, $\hat{\mathbf{j}}$, and $\hat{\mathbf{k}}$, we have

$$\hat{\mathbf{i}}: \quad 0.577\, F_1 + F_{2x} = -5$$
$$\hat{\mathbf{j}}: \quad 0.577\, F_1 = 8.66$$
$$\hat{\mathbf{k}}: \quad 0.577\, F_1 + F_{2z} = 0$$

from which

$$F_1 = 8.66/0.577 = 15.0 \text{ N}$$
$$F_{2z} = -0.577\, F_1 = -8.66 \text{ N}$$
$$F_{2x} = -5 - 0.577\, F_1 = -5 - 8.66 = -13.7 \text{ N}$$

Therefore, the required forces are

$$\mathbf{F}_1 = 8.66\hat{\mathbf{i}} + 8.66\hat{\mathbf{j}} + 8.66\hat{\mathbf{k}} \text{ N}$$
$$\mathbf{F}_2 = -13.7\hat{\mathbf{i}} - 8.66\hat{\mathbf{k}} \text{ N}$$

Throughout this book, we have inserted questions for the reader to think about. The answer to each may be found below the Example or at the bottom of the page. The first question follows:

Question 2.1   Are there sources of force other than those from direct pushes or pulls — that is, those involving physical contact?

Dot Product to Find Components   As the reader is perhaps aware, the **dot product** (or **scalar product**) of two vectors can be used to find the orthogonal component of one of them in the direction of the other. The dot product of two vectors $\mathbf{F}$ and $\mathbf{Q}$ is defined by

$$\mathbf{F} \cdot \mathbf{Q} = |\mathbf{F}||\mathbf{Q}|\cos\theta = \mathbf{Q} \cdot \mathbf{F} \qquad (2.4)$$

where θ is the angle between $\mathbf{F}$ and $\mathbf{Q}$ in their plane. Thus if $\mathbf{F}$ represents a force and we wish to find its component (see Figure 2.5) in the direction of $\mathbf{Q}$, we just dot $\mathbf{F}$ with the unit vector $\hat{\mathbf{u}}$ in the direction of $\mathbf{Q}$, which is

Figure 2.5

$\hat{\mathbf{u}} = \mathbf{Q}/|\mathbf{Q}|$, and obtain

$$\mathbf{F} \cdot \hat{\mathbf{u}} = |\mathbf{F}|(1)\cos\theta$$

Answer 2.1   Yes. Gravity and electromagnetic forces are two such examples.

which is the desired projection. Therefore the vector rectangular component of $\mathbf{F}$ in the direction of $\mathbf{Q}$ is

$$(\mathbf{F} \cdot \hat{\mathbf{u}})\hat{\mathbf{u}} = \mathbf{F}_Q \qquad (2.5)$$

We now develop a useful expression for the dot product of two vectors when they are expressed in component form. We have:

$$
\begin{aligned}
\mathbf{F} \cdot \mathbf{Q} &= (F_x\hat{\mathbf{i}} + F_y\hat{\mathbf{i}} + F_z\hat{\mathbf{k}}) \cdot (Q_x\hat{\mathbf{i}} + Q_y\hat{\mathbf{i}} + Q_z\hat{\mathbf{k}}) \\
&= F_xQ_x(\hat{\mathbf{i}} \cdot \hat{\mathbf{i}}) \quad + F_xQ_y(\hat{\mathbf{i}} \cdot \hat{\mathbf{j}}) + F_xQ_z(\hat{\mathbf{i}} \cdot \hat{\mathbf{k}}) \\
&\quad + F_yQ_x(\hat{\mathbf{j}} \cdot \hat{\mathbf{i}}) + F_yQ_y(\hat{\mathbf{j}} \cdot \hat{\mathbf{j}}) + F_yQ_z(\hat{\mathbf{j}} \cdot \hat{\mathbf{k}}) \\
&\quad + F_zQ_x(\hat{\mathbf{k}} \cdot \hat{\mathbf{i}}) + F_zQ_y(\hat{\mathbf{k}} \cdot \hat{\mathbf{j}}) + F_zQ_z(\hat{\mathbf{k}} \cdot \hat{\mathbf{k}})
\end{aligned}
$$

But Equation (2.4) gives

$$\hat{\mathbf{i}} \cdot \hat{\mathbf{i}} = \hat{\mathbf{j}} \cdot \hat{\mathbf{j}} = \hat{\mathbf{k}} \cdot \hat{\mathbf{k}} = (1)(1) \cos 0° = 1$$

and

$$\hat{\mathbf{i}} \cdot \hat{\mathbf{j}} = \hat{\mathbf{j}} \cdot \hat{\mathbf{k}} = \hat{\mathbf{i}} \cdot \hat{\mathbf{k}} = (1)(1) \cos 90° = 0$$

Therefore

$$\mathbf{F} \cdot \mathbf{Q} = F_xQ_x + F_yQ_y + F_zQ_z$$

We shall now use the dot product in an example to find a component of a force.

EXAMPLE 2.5

Given the forces

$$\mathbf{F}_1 = 2\hat{\mathbf{i}} + 3\hat{\mathbf{j}} - 4\hat{\mathbf{k}} \text{ N}$$

and

$$\mathbf{F}_2 = \hat{\mathbf{i}} - 2\hat{\mathbf{j}} + 5\hat{\mathbf{k}} \text{ N}$$

find the component of $\mathbf{F}_1 + \mathbf{F}_2$ in the direction of the line through the points whose rectangular coordinates are (0, 6, 5,) and (4, 0, 2) m as shown in Figure E2.5.

Solution

$$
\begin{aligned}
\mathbf{F}_1 + \mathbf{F}_2 &= (2 + 1)\hat{\mathbf{i}} + (3 - 2)\hat{\mathbf{j}} + (-4 + 5)\hat{\mathbf{k}} \\
&= 3\hat{\mathbf{i}} + \hat{\mathbf{j}} + \hat{\mathbf{k}} \text{ N}
\end{aligned}
$$

We next construct the directed line segment (vector whose dimension is length) from point A to point B.

$$\mathbf{r}_{AB} = (4 - 0)\hat{\mathbf{i}} + (0 - 6)\hat{\mathbf{j}} + (2 - 5)\hat{\mathbf{k}} = 4\hat{\mathbf{i}} - 6\hat{\mathbf{j}} - 3\hat{\mathbf{k}} \text{ m}$$

A unit vector $\hat{\mathbf{e}}_{AB}$ in the direction of the line is then

$$\hat{\mathbf{e}}_{AB} = \frac{\mathbf{r}_{AB}}{|\mathbf{r}_{AB}|} = \frac{4\hat{\mathbf{i}} - 6\hat{\mathbf{j}} - 3\hat{\mathbf{k}}}{\sqrt{(4)^2 + (6)^2 + (3)^2}} = \frac{1}{\sqrt{61}}(4\hat{\mathbf{i}} - 6\hat{\mathbf{j}} - 3\hat{\mathbf{k}})$$

$$= 0.512\hat{\mathbf{i}} - 0.768\hat{\mathbf{j}} - 0.384\hat{\mathbf{k}}$$

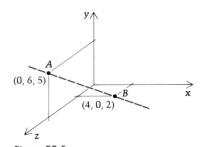

Figure E2.5

Therefore the scalar component of $\mathbf{F}_1 + \mathbf{F}_2$ associated with the direction of $\hat{\mathbf{e}}_{AB}$ is

$$(\mathbf{F}_1 + \mathbf{F}_2) \cdot \hat{\mathbf{e}}_{AB} = 3(0.512) + (1)(-0.768) + 1(-0.384) = 0.384 \text{ N}$$

The vector component along the line AB is

$$0.384 \, \hat{\mathbf{e}}_{AB} = 0.384(0.512\hat{\mathbf{i}} - 0.768\hat{\mathbf{j}} - 0.384\hat{\mathbf{k}})$$
$$= 0.197\hat{\mathbf{i}} - 0.295\hat{\mathbf{j}} - 0.147\hat{\mathbf{k}} \text{ N}$$

Had we begun the analysis by forming

$$\hat{\mathbf{e}}_{BA} = \frac{\mathbf{r}_{BA}}{|\mathbf{r}_{BA}|} = -\hat{\mathbf{e}}_{AB}$$

we would have found the *scalar* component, $(\mathbf{F}_1 + \mathbf{F}_2) \cdot \hat{\mathbf{e}}_{BA}$, to be -0.384 N, but the *vector* component along AB is of course the same, because

$$[(\mathbf{F}_1 + \mathbf{F}_2) \cdot \hat{\mathbf{e}}_{AB}]\hat{\mathbf{e}}_{AB} = [(\mathbf{F}_1 + \mathbf{F}_2) \cdot \hat{\mathbf{e}}_{BA}]\hat{\mathbf{e}}_{BA}$$

PROBLEMS ▶ Section 2.2

2.1 Which force has the largest magnitude?

$$\mathbf{F}_1 = 2\hat{\mathbf{i}} + 3\hat{\mathbf{j}} + 6\hat{\mathbf{k}} \text{ N}$$
$$\mathbf{F}_2 = 9\hat{\mathbf{j}} \text{ N}$$
$$\mathbf{F}_3 = 3\hat{\mathbf{i}} - 7\hat{\mathbf{j}} + \sqrt{7}\hat{\mathbf{k}} \text{ N}$$

2.2 If $\mathbf{F}_1 = 5\hat{\mathbf{i}} + 6\hat{\mathbf{j}}$ lb and $\mathbf{F}_2 = 2\hat{\mathbf{i}} - 3\hat{\mathbf{j}} - 4\hat{\mathbf{k}}$ lb, find $\mathbf{F}_3$ so that the sum of the three forces is zero.

2.3 Prove that the sum of the magnitudes of two forces $\mathbf{F}_1$ and $\mathbf{F}_2$ is greater than or equal to the magnitude of their sum.

2.4 Express the 238-N force $\mathbf{F}$ in Figure P2.4 as a vector. Write it (a) as a magnitude times a unit vector in its direction; (b) in terms of its components parallel to $\hat{\mathbf{i}}$ and $\hat{\mathbf{j}}$.

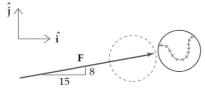

Figure P2.4

2.5 The orthogonal components of a certain force are

    40 N in the positive x direction
    50 N in the positive y direction
    60 N in the negative z direction

  a. What is the magnitude of the force?
  b. What are its direction cosines?

2.6 A force has a magnitude of 100 lb and direction cosines $l = 0.7$, $m = -0.2$, $n = ?$ relative to an xyz frame of reference. Determine the orthogonal components of the force.

2.7 What is the unit vector in the direction of the force $6000\hat{\mathbf{i}} - 6000\hat{\mathbf{j}} + 7000\hat{\mathbf{k}}$ lb?

2.8 Find a force along $\hat{\mathbf{e}} = 0.8\hat{\mathbf{i}} - 0.6\hat{\mathbf{j}}$ and another force normal to $\hat{\mathbf{e}}$ that add up to the force $\mathbf{F} = 5\hat{\mathbf{i}} - 10\hat{\mathbf{j}} + 3\hat{\mathbf{k}}$ N.

2.9 Determine the component of the force in Problem 2.6 along a line having the direction cosines $(-0.3, 0.1, 0.9487)$.

2.10 Obtain the dot product of the two vectors $\mathbf{F} = 10\hat{\mathbf{i}} + 6\hat{\mathbf{j}} - 3\hat{\mathbf{k}}$ lb and $\mathbf{B} = 6\hat{\mathbf{i}} - 2\hat{\mathbf{j}}$ ft.

2.11 Given the vectors $\mathbf{A} = 2\hat{\mathbf{i}} - 4\hat{\mathbf{j}}$ lb, $\mathbf{B} = 3\hat{\mathbf{j}} - 48\hat{\mathbf{k}}$ lb, and $\mathbf{C} = 3\hat{\mathbf{i}}$ (dimensionless), determine $\mathbf{C}(\mathbf{A} \cdot \mathbf{C}) + \mathbf{B}$.

2.12 Express the 500-N force making equal angles with x, y, and z (see Figure P2.12): (a) as a magnitude multiplied by a unit vector; (b) in terms of its orthogonal vector components.

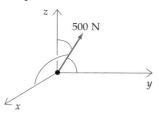

Figure P2.12

2.13 Find a unit vector in the xy plane that is perpendicular to the force $3\hat{i} - 4\hat{j} + 12\hat{k}$ N.

2.14 Find the direction cosines of the force $\mathbf{F} = 30\hat{i} + 40\hat{j} - 120\hat{k}$ lb, and use them to determine the angles the force makes with the coordinate axes (x, y, z).

2.15 For what value of F_y are the vectors $\mathbf{F}_1 = 3\hat{i} + F_y\hat{j} + 15\hat{k}$ N and $\mathbf{F}_2 = 7\hat{i} - 2\hat{j} + 3\hat{k}$ N orthogonal?

2.16 A force is given by $\mathbf{F} = 20\hat{i} - 60\hat{j} + 90\hat{k}$ N. Find its magnitude and the angles it forms with the coordinate axes.

2.17 Given the forces $\mathbf{F}_1 = 6\hat{i} + 10\hat{j} + 16\hat{k}$ lb, $\mathbf{F}_2 = 2\hat{i} - 3\hat{j}$ lb, and $\mathbf{F}_3 = $ a third force in the xy plane at an inclination of $45°$ to both the negative y and positive x axes. The magnitude of $\mathbf{F}_3$ is 25 lb. Find (a) $\mathbf{F}_1 + \mathbf{F}_2 + \mathbf{F}_3$ and (b) $\mathbf{F}_1 - 2\mathbf{F}_2 + 3\mathbf{F}_3$.

2.18 If $\mathbf{F}_1 = 2\hat{i} + 4\hat{j}$ kN (kilonewton), $\mathbf{F}_2 = \hat{i} - 2\hat{k}$ kN, $\mathbf{F}_3 = \hat{i} + \hat{j} - 7\hat{k}$ kN, and $\mathbf{F}_4 = 2\hat{i} - 9\hat{j} + 3\hat{k}$ kN, determine scalars a, b, and c such that $\mathbf{F}_4 = a\mathbf{F}_1 + b\mathbf{F}_2 + c\mathbf{F}_3$.

2.19 If the tension in the guy wire AB in Figure P2.19 is 10 kN, find the tension in the other guy wire BC if the sum of the two tension forces exerted on the column at B is known to be vertical. Then find the sum of the two forces.

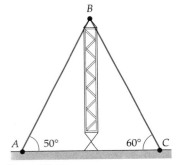

Figure P2.19

2.20 Find the angle between the forces $\mathbf{F}_1 = 2\hat{i} + \hat{j} - \hat{k}$ N and $\mathbf{F}_2 = 5\hat{i} - 6\hat{j} + 8\hat{k}$ lb.

2.21 Express the 20 kN force acting on the beam in Figure P2.21 in terms of its axial (parallel to the beam's axis, x) and transverse (normal to the beam's axis, y) components.

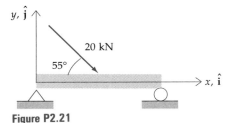

Figure P2.21

2.22 Show that if a, b, and c are nonvanishing scalars, and if $a\mathbf{F}_1 + b\mathbf{F}_2 + c\mathbf{F}_3 = 0$, then the three forces $\mathbf{F}_1$, $\mathbf{F}_2$, and $\mathbf{F}_3$ have lines of action in parallel planes.

2.23 Prove that $(\mathbf{A} \cdot \mathbf{B})^2$ is never greater than $|\mathbf{A}|^2 |\mathbf{B}|^2$.

2.24 The cable BCA in Figure P2.24 passes smoothly through a hole at the end of the strut at C, and ties to the ground at A and to the end of the pole at B. If the tension in the cable is 800 lb, what is the force exerted by the cable onto the pole at B?

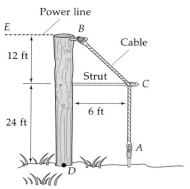

Figure P2.24

2.25 Resolve the force $\mathbf{F}$ into a part perpendicular to AB and a part parallel to BC. (See Figure P2.25.)

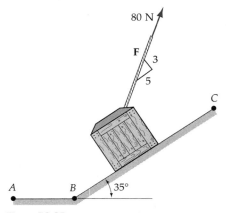

Figure P2.25

2.26 What are the orthogonal components of the 100-N force shown in Figure P2.26? What are the direction cosines associated with this force?

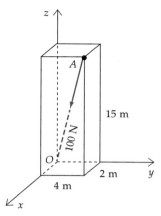

Figure P2.26

2.27 Express the force in Figure P2.27 (a) as a magnitude times a unit vector and (b) in terms of its components.

2.28 The x component of the force **P** in Figure P2.28 is 140 N to the left. Find **P**.

2.29 Show that the component of the downward force W in Figure P2.29 that is:

 a. perpendicular to the inclined plane is $W \cos \theta$ (toward the plane) and

 b. parallel to the plane is $W \sin \theta$ (down the plane). You will use this result many times in the study of engineering mechanics.

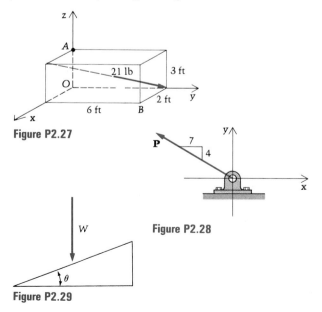

Figure P2.27

Figure P2.28

Figure P2.29

2.30 The girl in Figure P2.30 slowly pushes the lawnmower up the incline by exerting a 30-lb force **F** parallel to the handle as shown. Find the component of **F** (a) parallel to the incline; (b) normal to the incline; (c) parallel to the direction of gravity; (d) perpendicular to the direction of gravity.

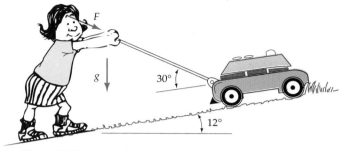

Figure P2.30

2.31 Resolve the 250-N force shown in Figure P2.31 into parts acting along the members PR and QR.

2.32 The x and z components of the force **F** in Figure P2.32 are known to be 100 N and -30 N, respectively. What is the force **F**, and what are its direction cosines?

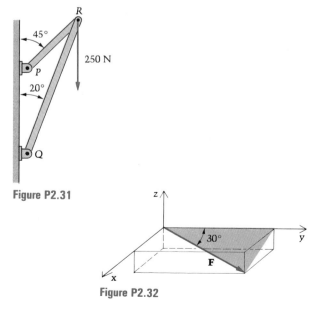

Figure P2.31

Figure P2.32

2.33 Resolve the 170-lb force, **F**, in Figure P2.33 into three parts — one of which is parallel to OQ, another parallel to *OP*, and the third parallel to the *y* axis. Are these the components of **F** in these directions?

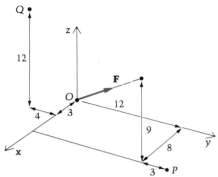

Figure P2.33

2.34 Two forces **P** and **Q** with respective magnitudes 100 and 200 N are applied to the upper corner of the crate in Figure P2.34. The sum of the two forces is a horizontal force to the right of magnitude 250 N. Find the angles that **P** and **Q** each make with their sum — that is, with the horizontal line through A.

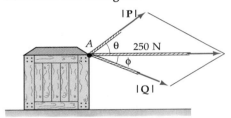

Figure P2.34

2.35 Find two 80-lb forces whose sum is the force $40\hat{i}$ lb.

2.36 Given that $\mathbf{F}_1 = \hat{i}$, $\mathbf{F}_2 = \hat{j}$, $\mathbf{F}_3 = 3\hat{i} - 4\hat{j} + 5\hat{k}$, $\mathbf{F}_4 = 6\hat{i} - 4\hat{j}$ lb, find a vector that is simultaneously in the plane of $\mathbf{F}_1$ and $\mathbf{F}_2$ and in the plane of $\mathbf{F}_3$ and $\mathbf{F}_4$.

2.37 Find all unit vectors that are perpendicular

to each of the forces $\mathbf{F}_1 = \hat{i} + 2\hat{j} + 3\hat{k}$ N and $\mathbf{F}_2 = 8\hat{i} - 9\hat{j} - 12\hat{k}$ N.

2.38 Determine a unit vector in the plane of the forces $\hat{i} + \hat{j}$ lb and $\hat{j} + \hat{k}$ lb and simultaneously perpendicular to the force $\hat{i} + \hat{j} + \hat{k}$ lb.

2.39 (a) The *x*-component of each of the four forces in Figure P2.39(a) is 10 lb. Which of the forces has the largest magnitude?

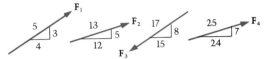

Figure P2.39(a)

(b) The vertical component of each of the six forces in Figure P2.39(b) is 1 N. Which force has the smallest magnitude?

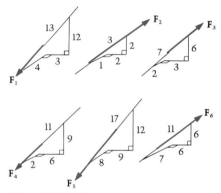

Figure P2.39(b)

2.40 Express the force of Examples 2.2 and 2.4, $\mathbf{F} = -5\hat{i} + 8.66\hat{j}$ N, as the sum of one force making equal angles with *x*, *y*, and *-z*, and a second force whose direction is in the *yz* plane.

2.3 Equilibrium of a Particle

When none of the particles (bits of material) making up a body is accelerating relative to an inertial frame of reference, Euler's laws (Section 1.3) take on particularly simple forms involving only the external forces acting on the body. The first of these two laws of **equilibrium** requires that the external forces exert no net thrust on the body. In equation form this is

$$\Sigma \mathbf{F} = 0 \qquad (2.6)$$

The second law is a statement that there be no net turning effect exerted by the external forces. To delve more deeply into this we need the concept of moment of a force, which is to come in the next chapter. For the remainder of this chapter we shall concern ourselves with simpler problems for which only Equation (2.6) is needed. These are problems for which the lines of action of the different external forces all intersect at a single point so that there can be no net turning effect and thus the actual size of the body is irrelevant. We often say that these are problems of *particle equilibrium,* a particle being a piece of material so small that we need not distinguish its different points.

In most engineering problems some of the external forces are known (or prescribed) before any analysis is carried out; we usually refer to these as **loads.** The external forces exerted by attached or supporting bodies are called **reactions;** usually we can think of these as forces that constrain the body against motion that the loads tend to produce.

The Free-Body Diagram   The **free-body diagram** (FBD) is an extremely important and useful concept for the analysis of problems in mechanics. It is a figure, usually sketched, depicting and hence identifying precisely the body under consideration. On the FBD (Figure 2.6) we show, by arrows, all of the external forces which act on that body. Thus we have a catalog, graphically displayed, of all the forces that contribute to the equilibrium equations. Of particular importance is the fact that the free-body diagram provides us a way to express what we know about reactions (for example, that a certain reaction force has a known line of action) before applying the equations of equilibrium.

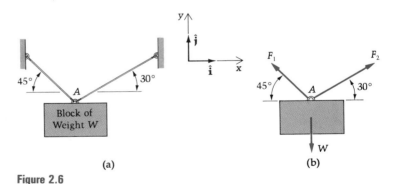

(a)   (b)

Figure 2.6

To illustrate the concept of a free-body diagram, consider the block shown in Figure 2.6a. It is supported (held in equilibrium) by two ropes or cables that counter the effect of gravity. The free-body diagram, Figure 2.6b, shows that three external forces act on the block. One of these is the weight W, which we show to have a line of action downward through the center of gravity of the block. That this single force can represent the cumulative effect of the distributed action of gravity will be shown in the next chapter. The other two forces, labeled F_1 and F_2, along with associated arrows represent the forces exerted *on* the block by the two cables. The directions of these forces are perceived to be along the (assumed) straight cables. The *arrow code* tells us that we are describing the force exerted on the block by the left cable by the expression

$$F_1(-\cos 45°\hat{\mathbf{i}} + \sin 45°\hat{\mathbf{j}})$$

Similarly the free-body diagram is conveying that the analytical description of the force exerted on the block by the right cable is to be

$$F_2\,(\cos 30°\hat{\mathbf{i}} + \sin 30°\hat{\mathbf{j}})$$

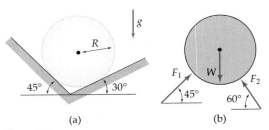

(a)   (b)

Figure 2.7

    As a second illustration we consider the uniform sphere of weight W, supported by smooth (frictionless) surfaces as shown in Figure 2.7a. Because the contacting surfaces are smooth, the forces exerted on the sphere by the planes must be perpendicular to the surface. Thus we reason that the representations of those forces by F_1 and F_2 and the arrow codes are as depicted on the free-body diagram, Figure 2.7b.

    Later on, in Chapter 4, when preparing to analyze general equilibrium problems involving the moment equation mentioned in Section 1.3 and non-concurrent force systems, we will dedicate an entire section of the book (4.2) to the free-body diagram. For now, we simply "sketch the particle" with all its forces intersecting at a point. Since all we are doing in this chapter is summing forces, displaying distances on the FBD is unimportant here.

Analysis of Equilibrium   Let us now return to the block of Figure 2.6 and substitute into the equilibrium equation, $\Sigma\mathbf{F} = \mathbf{0}$, to obtain

$$F_1\,(-\cos 45°\hat{\mathbf{i}} + \sin 45°\hat{\mathbf{j}}) + F_2\,(\cos 30°\hat{\mathbf{i}} + \sin 30°\hat{\mathbf{j}}) + W(-\hat{\mathbf{j}}) = 0$$

or, because the $\hat{\mathbf{i}}$ and $\hat{\mathbf{j}}$ coefficients must separately add to zero, we get two equations in two unknowns:

$$-F_1 \cos 45° + F_2 \cos 30° = 0$$

and

$$F_1 \sin 45° + F_2 \sin 30° - W = 0$$

from which

$$F_1 = 0.897W$$

and

$$F_2 = 0.732W$$

We observe that there were only two independent scalar equations embodied in $\Sigma\mathbf{F} = 0$ because all the external forces have lines of action in a single plane and thus none of the forces has a component in the direction (z) perpendicular to that plane. In the language of Chapter 3, our force system here is coplanar and concurrent. Our two scalar equations express the vanishing of the sums of components in the x and y directions, respectively. We shall often write them using the notation

$$\Sigma F_x = 0 \quad \text{and} \quad \Sigma F_y = 0.$$

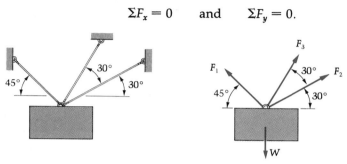

Figure 2.8

Before leaving this illustrative problem let us see what would happen if there were a third cable helping to support the block as shown in Figure 2.8. The force system is still coplanar so that there still will be only two independent equations, now relating F_1, F_2, and F_3. (The symbol $\xrightarrow{+}$ is a reminder that the unit vector in that direction has been suppressed from each term in the equation to follow. Some call it the "positive direction" for the force summation.)

$$\xrightarrow{+} \quad \Sigma F_x = 0$$
$$-F_1 \cos 45° + F_2 \cos 30° + F_3 \cos 60° = 0$$
$$+\uparrow \quad \Sigma F_y = 0$$
$$F_1 \sin 45° + F_2 \sin 30° + F_3 \sin 60° - W = 0$$

With three unknowns and only two equations the problem is now **statically indeterminate.** We need more information to determine F_1, F_2, and F_3. Problems such as this are solved in engineering courses variously titled Mechanics of Solids, Mechanics of Materials, Mechanics of Deformable Bodies, or Strength of Materials, where the equilibrium equations are supplemented by information describing the manner in which the cables stretch. Most of the problems in this book are statically determinate, but it is important to realize that the equations of equilibrium are necessary ingredients in the analysis of statically indeterminate problems as well. It's just that statics alone is insufficient to solve such problems.

There follow two more examples of problems of equilibrium. The student should note the central role played by the free-body diagram in each of these analyses. Throughout the book, whenever the weights of bodies subjected to other loads are not given as data in examples and in problems, it is to be understood that these weights may be neglected in comparison with the other loads.

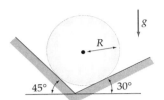

Figure E2.6a

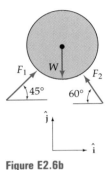

Figure E2.6b

EXAMPLE 2.6

A uniform sphere of weight W is supported by smooth (frictionless) plane surfaces as shown in Figure E2.6a. The plane of the page is vertical. Find the forces exerted by the supporting surfaces on the sphere.

Solution

The free-body diagram (Figure E2.6b) shows that each unknown force has a line of action through the sphere's center, which is also its mass center. This is an example of a body held in equilibrium by three forces (see Problem 4.141 for a general discussion).

The force-equation of equilibrium, $\Sigma\mathbf{F} = \mathbf{0}$, yields

$$F_1 (\cos 45°\hat{\mathbf{i}} + \sin 45°\hat{\mathbf{j}}) + F_2(-\cos 60°\hat{\mathbf{i}} + \sin 60°\hat{\mathbf{j}}) + W(-\hat{\mathbf{j}}) = \mathbf{0}$$

Separating the coefficients of $\hat{\mathbf{i}}$ and $\hat{\mathbf{j}}$, we get

$$\hat{\mathbf{i}} \text{ coefficients:} \quad \frac{\sqrt{2}}{2}F_1 - \frac{1}{2}F_2 = 0 \qquad (1)$$

which we could obtain directly by $\Sigma F_x = 0$, and

$$\hat{\mathbf{j}} \text{ coefficients:} \quad \frac{\sqrt{2}}{2}F_1 + \frac{\sqrt{3}}{2}F_2 = W \qquad (2)$$

which we could get from $\Sigma F_y = 0$. (To do this, we would just sum up the vertical components of all forces and omit the unit vector $\hat{\mathbf{j}}$ as we go.) The solution to Equations (1) and (2) is

$$F_1 = 0.518W$$
$$F_2 = 0.732W$$

EXAMPLE 2.7

The sandbag weighing 2500 N is held up by three cables as shown in Figure E2.7a. Each of the cables is tied to the ceiling 2 m above point O. The cable OC makes equal angles with the positive coordinate directions shown. Find the tension in each cable.

Solution

Let $\hat{\mathbf{i}}$, $\hat{\mathbf{j}}$, and $\hat{\mathbf{k}}$ be unit vectors along x, y, and z. To express the three forces exerted on the connection by the cables vertically, we first construct unit vectors along OA, OB, and OC. So

$$\hat{\mathbf{e}}_A = \frac{(-1.5\hat{\mathbf{i}} - 2.5\hat{\mathbf{j}} + 2\hat{\mathbf{k}})}{\sqrt{(1.5)^2 + (2.5)^2 + (2)^2}} = -0.424\hat{\mathbf{i}} - 0.707\hat{\mathbf{j}} + 0.566\hat{\mathbf{k}}$$

$$\hat{\mathbf{e}}_B = \frac{2\hat{\mathbf{j}} + 2\hat{\mathbf{k}}}{\sqrt{(2)^2 + (2)^2}} = 0.707\hat{\mathbf{j}} + 0.707\hat{\mathbf{k}}$$

$$\hat{\mathbf{e}}_C = \frac{2\hat{\mathbf{i}} + 2\hat{\mathbf{j}} + 2\hat{\mathbf{k}}}{\sqrt{(2)^2 + (2)^2 + (2)^2}} = 0.577\hat{\mathbf{i}} + 0.577\hat{\mathbf{j}} + 0.577\hat{\mathbf{k}}$$

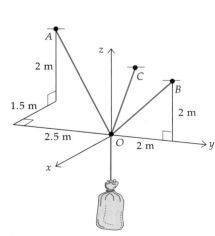

Figure E2.7a

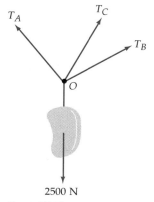

Figure E2.7b

Using these unit vectors, along with the scalars T_A, T_B, and T_C (see the FBD, Figure E2.7b), the equilibrium equation becomes

$$\Sigma \mathbf{F} = \mathbf{0}$$

$$T_A \hat{\mathbf{e}}_A + T_B \hat{\mathbf{e}}_B + T_C \hat{\mathbf{e}}_C + 2500\,(-\hat{\mathbf{k}}) = \mathbf{0}$$

Lifting off the coefficients of $\hat{\mathbf{i}}$, or equivalently writing $\Sigma F_x = 0$, we obtain

$$-0.424\,T_A + 0.577\,T_C = 0$$

and from $\Sigma F_y = 0$,

$$-0.707\,T_A + 0.707\,T_B + 0.577\,T_C = 0$$

and from $\Sigma F_z = 0$,

$$0.566\,T_A + 0.707\,T_B + 0.577\,T_C - 2500 = 0$$

The result of solving these three equations simultaneously is

$$T_A = 1970 \text{ N}$$
$$T_B = 787 \text{ N}$$

and
$$T_C = 1450 \text{ N}$$

PROBLEMS ▶ Section 2.3

2.41 A large block of wood with an equilateral triangle cross section of side s as shown in Figure P2.41 is to be lifted by a sling of length $4s$. Without writing any equations, explain why the sling tension will be greater in configuration (b) than in (a).

2.42 A boat is in the middle of a stream whose current flows from right to left (see Figure P2.42). If the forces F_1 and F_2, exerted on the ropes shown, are holding the boat in equilibrium against a force due to the current of 80 lb, what are the values of F_1 and F_2?

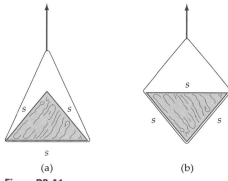

(a)   (b)

Figure P2.41

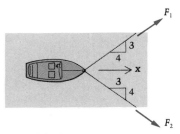

Figure P2.42

2.43   A 50-lb traffic light sags 1 ft in the center of a cable as shown in Figure P2.43. Determine the tension in the cable to which it is clamped.

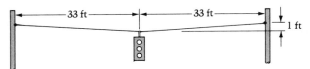

Figure P2.43

2.44   (a) A cable is being used to lift the 2000-N beam in Figure P2.44. Find the force in the lengths AB and BC of cable which are tied at B, in terms of the angle θ. (b) If the cable breaks when the tension exceeds 5000 N, what is the smallest angle θ that can be used?

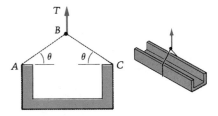

Figure P2.44

* 2.45   Find α such that the tension in cable AC is a minimum. (See Figure P2.45.)

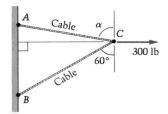

Figure P2.45

2.46   A smooth ball weighing 10 lb is supported by a cable and rests against a wall as shown in Figure P2.46. (a) Find the tension T in the cable and the normal force N exerted on the ball by the wall, as functions of the distance H. (b) Explain the limiting case results for T and N as H gets very large.

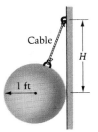

Figure P2.46

2.47   The 7-lb lamp in Figure P2.47 is suspended as indicated from a wall and a ceiling. Find the tensions in the two chains.

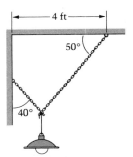

Figure P2.47

2.48   In the preceding problem, suppose the chains are replaced by a continuous 7-ft cord. (See Figure P2.48.) If it supports the lamp by passing through a smooth eye-hook so that the tension is the same on both sides, find this tension.

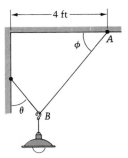

Figure P2.48

2.49 In Figure P2.49 force P is applied to a small wheel that is free to move on cable ACB. For a cable tension of 500 lb, find P and α.

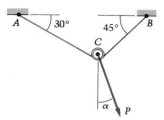

Figure P2.49

2.50 In Figure P2.50 a 1500-N weight is attached to a small, light pulley that can roll on the cable ABC. The pulley and weight are held in the position shown by a second cable DE, which is parallel to the portion BC of the main cable. Find the tension in cable ABC and the tension in cable DE.

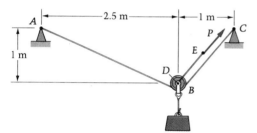

Figure P2.50

2.51 In Figure P2.51 the cable ABC is 10 feet long and flexible. A small pulley rides on the cable and supports a weight $W = 50$ lb. Find the tension T in the cable.

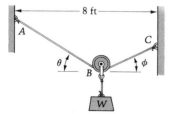

Figure P2.51

2.52 The man in Figure P2.52 is slowly pulling a drum over a circular hill. The drum weighs 60 N, and the hill is smooth. In the given position, find the tension in the rope (which does not vary along the rope if the hill is smooth).

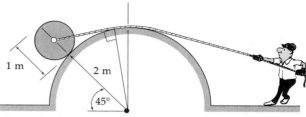

Figure P2.52

2.53 Find the forces exerted by the smooth planes on the 100-kg cylinder $\mathcal{C}$ shown in Figure P2.53. (The dotted cylinder $\mathcal{D}$ is absent from this problem.)

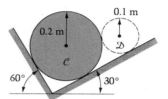

Figure P2.53

2.54 The man in Figure P2.54 is holding the 200-N drum in equilibrium at the point $(x, y) = (2, 1)$ m, with a force parallel to the plane at the point of contact. If friction is negligible, find the force exerted by the man.

2.55 The cylinder of weight W is in equilibrium between the two smooth planes. (See Figure P2.55.) Find the reactions N_1 and N_2 of the planes on the cylinder. Check your results by showing that $N_1 \to W$ and $N_2 \to 0$ as $\theta_1 \to 0$.

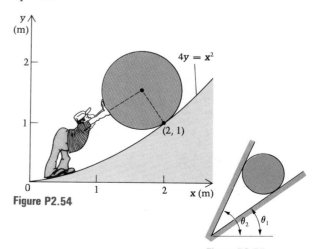

Figure P2.54

Figure P2.55

2.56 Three smooth cylinders A, B, and C, each of weight W, are arranged as shown in Figure P2.56. Find the forces exerted onto C by A and B.

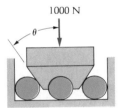

Figure P2.56

2.57 A shaft that carries a thrust of 1000 N terminates in a conical bearing as shown in Figure P2.57. If the angle θ is 25°, find the normal force that each of four equally spaced ball bearings exerts on the conical surface. Assume symmetry.

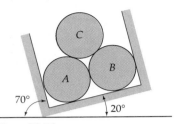

Figure P2.57

2.58 The 500-N reinforced concrete slab in Figure P2.58 is being slowly lowered by a winch at the end of the cable 𝒞. The cables 𝒜, ℬ, and 𝒟 are each attached to the slab and to the hook. Find the forces in each of the four cables if the distance from the upper surface of the slab to the hook is 2 m.

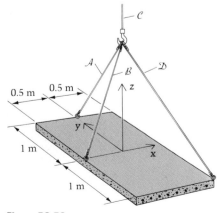

Figure P2.58

2.59 In Figure P2.59 the weight W is 600 N, and it is supported by cables AD, BD, and CD. Find the tension in each cable. (Points A, B, and C are all in the xz plane.)

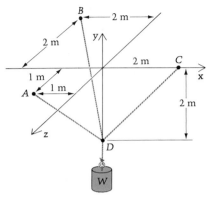

Figure P2.59

2.60 An equilateral triangular plate weighs 500 N. Three equal-length ropes (which break at a tension of 1500 N) are tied to the corners of the plate and to each other as shown in Figure P2.60, and hold up the horizontal plate. Find the shortest value of ℓ such that the ropes don't break.

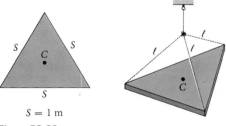

Figure P2.60

2.61 The weight in Figure P2.61 is to be suspended by 60 ft of cable in any possible number of equally spaced lengths (one 60-ft length, two 30-ft lengths, etc.). The cables are to be symmetrically attached to the 20-ft diameter ring as suggested in the figure for three lengths. Show that the load in the cables is less for four lengths than for one, two, three, or five, and that for six, the load in each cable is theoretically infinite. In each case, assume all the cable forces to be equal by symmetry, noting that for four or more cables, the problem is actually statically indeterminate.

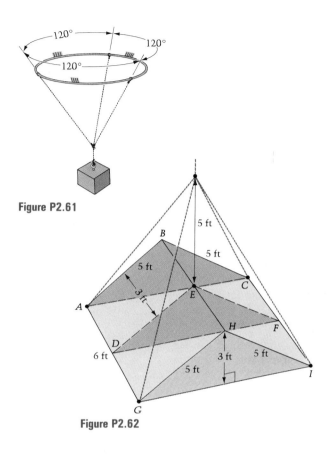

Figure P2.61

Figure P2.62

plates. Calculate the tension in each of the four cables, assuming them to be equal by symmetry. Could you work the problem without this assumption?

2.63 A particle located at the origin [$(x, y, z) = (0, 0, 0)$ m] is suddenly acted on by the three forces $\mathbf{F}_1$, $\mathbf{F}_2$, and $\mathbf{F}_3$, of magnitudes 14, 6, and 10 newtons, respectively. (See Figure P2.63.) $\mathbf{F}_1$ acts on the line from (0, 0, 0) to (3, 6, 2) m; $\mathbf{F}_2$ acts on the line from (0, 0, 0) to (3, 6, −6) m; and $\mathbf{F}_3$ acts on the line from (4, 3, 0) to (0, 0, 0) m. Is the particle still in equilibrium after the forces are applied? If not, what additional force $\mathbf{F}_4$ will keep it at the origin?

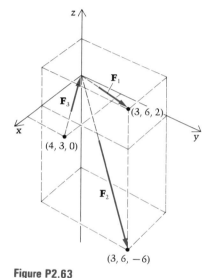

Figure P2.63

2.62 A small roof made of two identical rectangular plates (*ABHG* and *BCIH*), each weighing 60 lb, is supported symmetrically as shown in Figure P2.62. The plates are braced by three triangular plates (*ABC, DEF, GHI*) of the same density and thickness as the rectangular

2.4 Equilibrium of a System of Particles

Sometimes it is useful in an analysis to separate a "body" into constituent parts, and it is necessary to do this if we wish to determine the forces of interaction between those parts. The key concept in such an analysis is the **action-reaction principle** which states that the force exerted on a first body by a second body is equal in magnitude but opposite in direction to the force exerted on the second by the first.

To illustrate these concepts consider the two cylinders of Figure 2.9. They are supported by the smooth floor and the two vertical smooth walls. In Figure 2.10 are free-body diagrams of each of the cylinders and

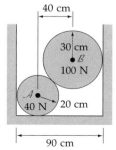

Figure 2.9

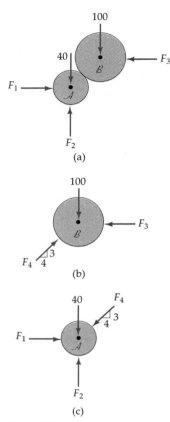

(a)

(b)

(c)

Figure 2.10

of the two-cylinder combination. The free-body diagram of the larger cylinder (Figure 2.10(b)) shows that the force exerted on it by the smaller one is given by

$$F_4 \left(\frac{4}{5} \hat{\mathbf{i}} + \frac{3}{5} \hat{\mathbf{j}} \right)$$

Execution of the action-reaction principle is seen in the free-body diagram, Figure 2.10(c), of the small cylinder where F_4 is associated with an oppositely directed unit vector. That is, the force exerted by the larger cylinder on the smaller must be

$$F_4 \left(-\frac{4}{5} \hat{\mathbf{i}} - \frac{3}{5} \hat{\mathbf{j}} \right)$$

The student should note that, by using the same scalar F_4 along with the arrow codes in the two free-body diagrams, we easily force satisfaction of the action-reaction principle.

Question 2.2   Why does F_4 not appear on the free-body diagram of the two-cylinder composite?

Turning to the free-body diagram of the larger cylinder and using the equilibrium equations, we obtain

$$+\uparrow \quad \Sigma F_y = 0$$

$$\frac{3}{5} F_4 - 100 = 0 \Rightarrow F_4 = 167 \text{ N}$$

and

$$\xrightarrow{+} \quad \Sigma F_x = 0$$

$$\frac{4}{5} F_4 - F_3 = 0$$

$$F_3 = \frac{4}{5} (167) = 133 \text{ N}$$

And then from the free-body diagram of the smaller cylinder, the equilibrium equations yield

$$\xrightarrow{+} \quad \Sigma F_x = 0$$

$$F_1 - \frac{4}{5} F_4 = 0 \Rightarrow F_1 = 133 \text{ N}$$

and

$$+\uparrow \quad \Sigma F_y = 0$$

$$F_2 - 40 - \frac{3}{5} F_4 = 0$$

$$F_2 = 140 \text{ N}$$

Answer 2.2   For the two-cylinder composite body, F_4 is an *internal* force. Only *external* forces are drawn on the free-body diagram of a body, because only they affect its equilibrium equations.

The student should note carefully that these forces, as they relate to the free-body diagram of the two-cylinder composite body, cause the equation $\Sigma\mathbf{F} = \mathbf{0}$ to be satisfied for that body.

An important class of multibody problems in which we may ignore the moments (turning effects) of forces is that of systems of pulleys. The following example illustrates application of equilibrium analysis to such a system. We must use in the analysis that, for a pulley in equilibrium with frictionless bearings, the tension in a rope or belt going around the pulley is everywhere the same. That fact will be rigorously proved in Chapter 4; for now, suffice it to say that if the tensions differed, a frictionless pulley would be turning.

EXAMPLE 2.8

Determine the force the man in Figure E2.8a must exert to hold the blocks in equilibrium.

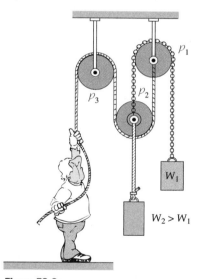

Figure E2.8a

Solution

In a problem involving frictionless (at the axle) pulleys in equilibrium, first remember that the tension in the rope is the same going on as it is coming off. The tensions would differ, of course, if either (a) the pulley were accelerating angularly, or (b) there were friction between the pulley and its axle.

Let us now start with the load W_1 and proceed to draw free-body diagrams of the various bodies. Normally the pulleys' weights are neglected, and we shall do so here.

The bodies are sketched below in proximity to one another to facilitate glancing from one to a neighboring one. The dashed lines are merely reminders of how the ropes or chain are connected to the bodies.

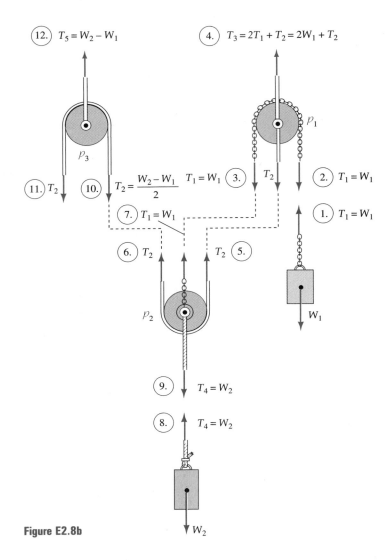

Figure E2.8b

Steps in the solution are correspondingly numbered in the various FBDs of Figure E2.8b; the reader should refer to these FBDs while carefully reading through the steps that follow this figure.

Step: Reason for result on figure:

1. $\Sigma F_y = 0$ on FBD of W_1.
2.   Same straight portion of chain as in FBD of W_1, so same tension in it.
3.   Same chain as on other side of this pulley.
4. $\Sigma F_y = 0$ on FBD of P_1; note T_2 remains unknown to this point.
5.   Same rope as in FBD of P_1, so same tension in it.
6.   Same rope as on other side of this pulley.
7.   Same chain as in FBD of P_1, so same tension in it.
8. $\Sigma F_y = 0$ on FBD of W_2.
9.   Same rope as in FBD of W_2, so same tension in it.

10. $\Sigma F_y = 0$ on FBD of P_2 yields

$$2T_2 + W_1 - W_2 = 0$$

so that

$$T_2 = \frac{W_2 - W_1}{2}$$

Then on FBD of P_3, same rope as in FBD of P_2.

11. Same rope as on other side of this pulley. Thus the man's pulling force on the rope is $(W_2 - W_1)/2$.

12. For completeness, $\Sigma F_y = 0$ on P_3 gives $T_5 - 2T_2 = 0$,

or $T_5 = 2\left(\dfrac{W_2 - W_1}{2}\right) = W_2 - W_1$.

In the preceding example, it is interesting to show the results on the FBDs of the three separated pulleys (see Figure 2.11) . . . . .

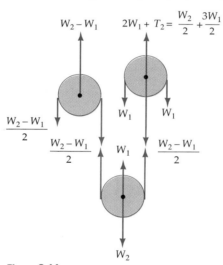

Figure 2.11

. . . . . and then to put the ropes and chains back together to form Figure 2.12. From this combined FBD we obtain a nice check on the solution:

$$+\uparrow \quad \Sigma F_y = \underbrace{(W_2 - W_1)}_{\substack{\text{rope} \\ \text{above} \\ P_3}} + \underbrace{\left(\frac{W_2}{2} + \frac{3}{2} W_1\right)}_{\text{rope above } P_1} - \underbrace{\left(\frac{W_2 - W_1}{2}\right)}_{\text{man's force}} - \underbrace{W_2}_{\substack{\text{weight} \\ \text{of } W_2}} - \underbrace{W_1}_{\substack{\text{weight} \\ \text{of } W_1}}$$

or

$$\Sigma F_y = 0 \checkmark$$

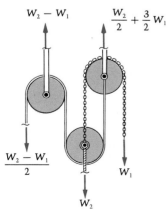

Figure 2.12

The reader should note that if $W_1 = W_2$, then no force is required to hold the blocks in equilibrium. This is because, in that case, the tensions on either side of P_1 are equal to $W_1(= W_2)$, while the tensions on either side of P_3 and P_2 vanish.

Question 2.3   If the man begins to pull on the rope with a force slightly greater than the equilibrium value of $(W_2 - W_1)/2$, what happens to the blocks?

Question 2.4   How much does the man have to weigh to stay on the ground?

Answer 2.3   The center of P_2 moves upward. W_2 moves up and W_1 moves down.
Answer 2.4   At least $(W_2 - W_1)/2$, or else he would need to pull down with more than his weight.

PROBLEMS ▶ Section 2.4

2.64 In Figure P2.64 the force of attraction between a pair of particles is 52 lb. What forces would have to be applied (if any), and where, for the system to be in equilibrium? (No other forces act on the particles.)

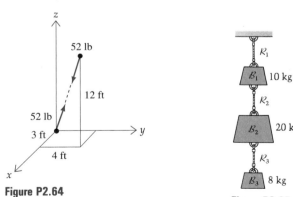

Figure P2.64

Figure P2.65

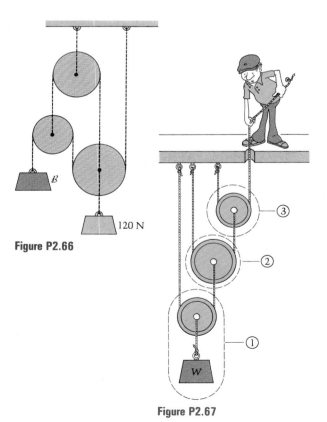

Figure P2.66

Figure P2.67

2.65 Using free-body diagrams, find the forces in the ropes R_1, R_2, and R_3. (See Figure P2.65.)

2.66 Find the weight of B for equilibrium of the system shown in Figure P2.66.

2.67 Find, by successively drawing the free-body diagrams suggested by 1, 2, and 3 in Figure P2.67, the force that the man must exert to hold the weight in equilibrium.

2.68 The weight in Figure P2.68 is 1400 N. How much force P does the woman have to exert on the rope to lift the weight? How heavy does she have to be to stay on the ground?

2.69 (a) Show that the force F that the man must exert in order to lift the engine of W lb using the chain hoist is

$$F = \left(\frac{R - r}{2R} \right) W$$

(b) If $R = 10$ in., $r = 8.5$ in., and $W = 400$ lb, how many pounds are required? (See Figure P2.69.)

2.70 In the block and tackle shown in Figure P2.70, a single rope passes back and forth over pulleys that are free to rotate within the blocks about axes ll and mm in the figure. The particular block and tackle shown has two pulleys in each block.

a. If the man in the preceding problem uses the block and tackle as indicated to raise the engine, how much force must he exert this time?

b. What is the ratio of r to R in the preceding problem for which the force he must exert is the same as it is with the block and tackle?

Assume all rope segments to be vertical.

2.71 In Figure P2.71, what is the force P needed to hold the weight W in equilibrium? Assume that all rope segments are vertical.

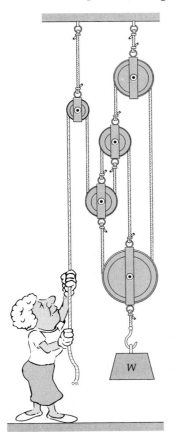

Figure P2.68

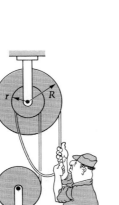

Figure P2.69

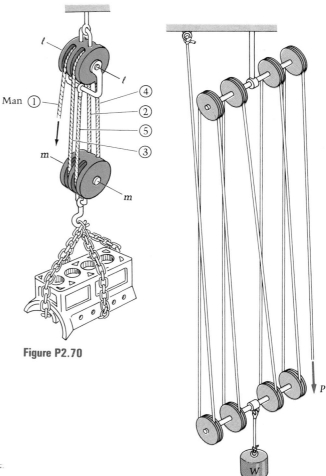

Figure P2.70

Figure P2.71

Figure P2.72

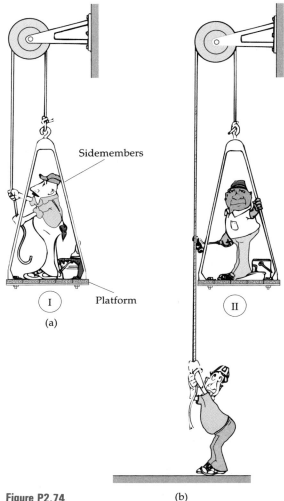

Sidemembers

Platform

I

(a)

II

(b)

Figure P2.74

2.72 In Figure P2.72, what force must the man exert in order to raise the crate of weight W?

2.73 Repeat the preceding problem if the weights of the pulleys A, B, C, and D are, respectively, $W/2$, $W/8$, $W/16$, and $W/4$.

2.74 The two painters in Figure P2.74 are slowly lifted on scaffolds. The first man lifts himself (case a). The second man is lifted by a colleague on the ground (case b). Each scaffold weighs 40 lb. Each painter weighs 180 lb. For each case:

a. Draw free-body diagrams of the painter and the scaffold.

b. Determine the magnitudes and the directions of all forces on the painter and on the scaffold.

Assume that the pulley is small and frictionless.

2.75 The 900-lb platforms in Figure P2.75 are supported by the light cable and pulley system as shown in the three configurations. Find the tension in the cable over pulley A

and the tension in the cable over pulley B for each configuration. Assume mass center locations so that the platforms remain horizontal.

2.76 The five ropes in Figure P2.76 can each take 1500 N without breaking. How heavy can W be without breaking any?

2.77 The mass of the man in Figure P2.77 is 70 kg, and the mass of the scaffold on which he is sitting is 10 kg. The pulleys and ropes are light. Find the tension in the cable that the man is holding, and also the force he exerts directly on the scaffold.

2.78 Find the relationship between the load W and the force P for equilibrium of the differential winch shown in Figure P2.78. The rope is wrapped around the different-sized cylinders in opposite directions.

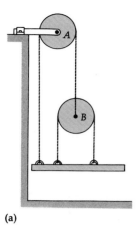

(a)

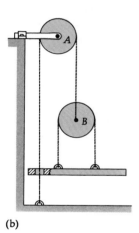

(b)

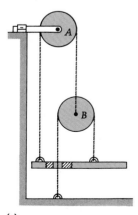

(c)
Figure P2.75

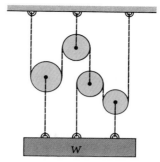

Figure P2.76

Scaffold

Figure P2.77

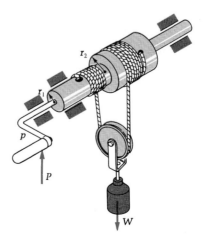

Figure P2.78

2.79 The collar in Figure P2.79 can slide along the rod without friction. The spring, which is attached to the collar and to the ceiling, exerts a force $k\delta$ proportional to its stretch δ, where k is the modulus of the spring. If $k = 50$ lb/in, $W = 50$ lb and the collar weighs 20 lb, find how much the spring is stretched in the given position.

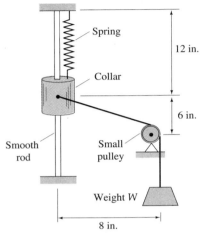

Figure P2.79

2.80 Two identical pieces of pipe rest against an incline and a vertical wall as shown in Figure P2.80. Each pipe has weight W, and all surfaces are smooth.

   a. Draw free-body diagrams of the two pipes.

   b. Determine the magnitude and show the direction of each force acting on each pipe.

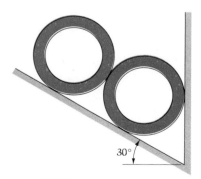

Figure P2.80

2.81 Two 50-lb traffic lights cause a cable sag of 8 in. as shown in Figure P2.81. Find the tensions in the three sections of the cable to which they are clamped.

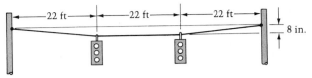

Figure P2.81

2.82 The two weights in Figure P2.82 are supported with six light, flexible, inextensible cords. Find the forces in the cords.

2.83 Find the weight of block P if the system is in equilibrium and W has a mass of 40 kg. (See Figure P2.83.)

2.84 The rope in Figure P2.84 has length l. It is attached at one end to a pin at A and at the other to mass m after passing under the free, small pulley at D and over the fixed, small pulley at B. The mass M is suspended from the free pulley. Find the height H for equilibrium of the system.

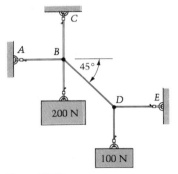

Figure P2.82

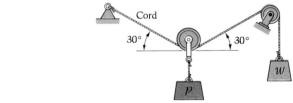

Figure P2.83

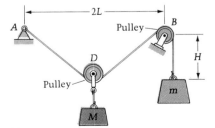

Figure P2.84

2.85 The man in Figure P2.85 weighs 800 N. He pulls down on the rope, raising the 250-N weight. He finds that the higher it goes, the more he must pull to raise it further. Explain this, and calculate and plot the rope tension T as a function of θ. What is the value of the tension, and the angle θ, when the man can lift it no further? Neglect the sizes and weights of the pulleys.

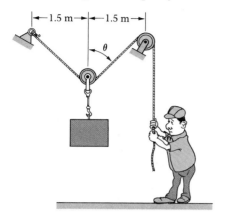

Figure P2.85

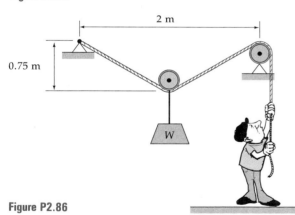

Figure P2.86

2.86 The 460-N man is holding the 360-N weight W in equilibrium as shown in Figure P2.86. (a) What is the tension in the rope? (b) How much higher can he raise the weight?

2.87 In the preceding problem, if in the figure there is a slack length of 0.5 m of rope below the man's hands and if he can reach 0.2 m higher than the position in the figure, what is the minimum rope tension possible?

2.88 Five identical smooth 5-kg cylinders are at rest on a 30° incline as shown in Figure P2.88. Find the normal force exerted by B on C at A. Repeat the problem if there are 100 cylinders instead of five.

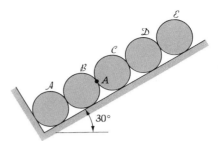

Figure P2.88

2.89 Cylinder A (mass 15 kg) rests on cylinder B (mass 20 kg) as shown in Figure P2.89. Find all forces acting on cylinder B.

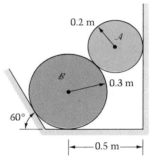

Figure P2.89

2.90 In Figure P2.90 the cylinders A (weight 50 N) and B (weight 150 N) are assumed to be smooth, and they rest on smooth planes oriented at right angles as shown. Find the angle Ψ between the horizontal line xx and the line joining the centers of the cylinders.

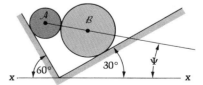

Figure P2.90

2.91 In Figure P2.91 two identical cylinders C_1 and C_2, each of radius r and weight P, are tied together by a cord. They support a third cylinder C_3 of radius R and weight Q. There is no friction, and the tension in the cord is just sufficient to make the contact force between C_1 and C_2 zero. Find:

  a. The tension in the cord
  b. The force exerted by the ground on C_1
  c. The normal force between C_1 and C_3

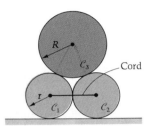

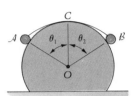

Figure P2.91

Figure P2.92

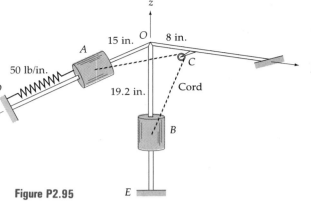

Figure P2.95

* **2.92** Two small balls A and B have masses m and $2m$, respectively. (See Figure P2.92.) They rest on a smooth circular cylinder with a horizontal axis and with radius R. They are connected by a thread of length $2R$. Find the angles θ_1 and θ_2 between the radii and the vertical line OC for equilibrium, as well as the tension in the thread and the forces exerted by A and B on the cylinder. Assume that the balls are very small and that the tension is constant.

* **2.93** Three identical spheres are at rest at the bottom of the spherical bowl shown in Figure P2.93. If a fourth sphere is placed on top, what is the largest ratio R/r for equilibrium if there is no friction?

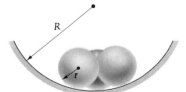

Figure P2.93

* **2.94** Repeat the previous problem if there are four spheres in the bowl and a fifth is placed on top.

2.95 Collars A and B in Figure P2.95 may slide smoothly on the rods OD and OE. Collar B weighs 20 pounds, and is connected to A by means of an inextensible cord which wraps around a pulley at C of negligible dimensions. The spring has a modulus (see Problem 2.79) of 50 lb /inch. How much is it stretched?

* **2.96** Three identical spheres that each weigh 10 N rest on a horizontal plane touching each other. They are tied together by a cord wrapped around their equatorial planes. A fourth 10-N sphere $\mathcal{D}$ is placed atop the others as shown in Figure P2.96(a). Neglecting friction, find the smallest cord tension needed to hold the spheres together. [Refer also to Figure P2.96(b).]

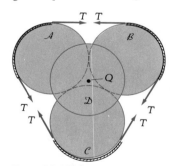

Figure P2.96(a)

Geometrical hints:

In horizontal plane through centers of $A,B,C.$

In vertical plane through centers of A and $\mathcal{D}$.

Q is in the plane of the three centers, equidistant from each.
Figure P2.96(b)

* **2.97** Four identical marbles are stacked as shown in Figure P2.97 with the three horizontal forces P maintaining equilibrium. There is no friction to be considered anywhere. Find the reactive forces between upper and lower marbles and give the minimum value of P for which equilibrium can exist.

Top view

Side view

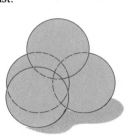

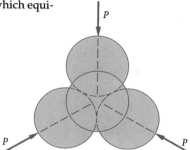

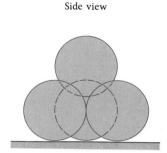

Figure P2.97

SUMMARY ▶ Chapter 2

In Chapter 2, we learned to express a force as a vector. Sometimes we do this in terms of the components (F_x, F_y, F_z) of the force:

$$\mathbf{F} = F_x\hat{\mathbf{i}} + F_y\hat{\mathbf{j}} + F_z\hat{\mathbf{k}},$$

while at other times we write the force as its magnitude $|\mathbf{F}|$ multiplied by a unit vector $(\hat{\mathbf{e}})$ in its direction:

$$\mathbf{F} = |\mathbf{F}|\hat{\mathbf{e}}$$

The two representations above are related by

$$\mathbf{F} = \underbrace{\sqrt{F_x^2 + F_y^2 + F_z^2}}_{|\mathbf{F}|} \underbrace{\left(\frac{F_x\hat{\mathbf{i}} + F_y\hat{\mathbf{j}} + F_z\hat{\mathbf{k}}}{\sqrt{F_x^2 + F_y^2 + F_z^2}}\right)}_{\hat{\mathbf{e}} = e_x\hat{\mathbf{i}} + e_y\hat{\mathbf{j}} + e_z\hat{\mathbf{k}}}$$

where $F_z = 0$ if the force lies in the xy plane. The components (e_x, e_y, e_z) of the unit vector $\hat{\mathbf{e}}$ are also the cosines of the angles $(\theta_x, \theta_y, \theta_z)$ between the line of action of the force and the three coordinate axes (x, y, z).

To find the orthogonal component of a force $\mathbf{F}$ in the direction of a line L, we write a unit vector $\hat{\mathbf{u}}$ along the line, then form the dot product $\mathbf{F} \cdot \hat{\mathbf{u}}$. To write a unit vector along a line, we write *any* vector along the line, then divide by its magnitude. For example, if a 20-N force is directed from the point with coordinates $(1, 3, 5)$ m to the point $(4, 5, -6)$ m, then its vector representation is

$$\mathbf{F} = 20\frac{(4 - 1)\hat{\mathbf{i}} + (5 - 3)\hat{\mathbf{j}} + (-6 - 5)\hat{\mathbf{k}}}{\sqrt{3^2 + 2^2 + (-11)^2}} \text{ N}$$

which may be simplified into component form as:

$$\mathbf{F} = 5.18\hat{\mathbf{i}} + 3.46\hat{\mathbf{j}} - 19.0\hat{\mathbf{k}} \text{ N}$$

To analyze a particle in equilibrium, we first draw a free-body diagram of it, which is a sketch of the particle including all the external forces being exerted upon it. Then, with the help of the free-body diagram, the equilibrium equation $\Sigma \mathbf{F} = \mathbf{0}$ is written, which has the scalar component equations (if rectangular coordinates are used):

$$\Sigma F_x = 0 \qquad \Sigma F_y = 0 \qquad \Sigma F_z = 0$$

These equations may be written for each particle in a system if more than one constitute the system being analyzed.

REVIEW QUESTIONS ▶ Chapter 2

True or False?

1. A force on a body has to result from direct contact with another body.
2. Forces have magnitudes and directions but are not vectors because they do not obey the parallelogram law of addition.
3. It can be proved mathematically that forces are vectors.
4. Unit vectors have dimension of length.
5. The dot product of two vectors is a scalar.
6. The cross product of two vectors is a vector.
7. In particle equilibrium problems, only the equation $\Sigma \mathbf{F} = \mathbf{0}$ is needed.
8. Only the external forces acting on a body being analyzed for equilibrium are drawn on its free-body diagram.
9. If a frictionless pulley is in equilibrium, the tensions on either side, in a rope wrapped around it, are equal.

Answers: 1. F   2. F   3. F   4. F   5. T   6. T   7. T   8. T   9. T

3

THE MOMENT OF A FORCE; RESULTANTS

3.1   Introduction

In the preceding chapter, we learned that the external forces acting on a body in equilibrium sum to zero. For the relatively small subset of equilibrium problems examined in Chapter 2, that equation, $\Sigma \mathbf{F} = \mathbf{0}$, was all that was needed to complete the solution.

Most of the time, however, we will need a complementary, independent equation to complete the solution to statics problems. This second equation is that the moments, about an arbitrary point P, of all the forces acting on the body also add to zero. For this reason we shall spend a chapter learning a number of things about moments of forces.

In Section 3.2, we will begin by examining the moment of a force about a point, using three different definitions: a "common sense" formula, a vector representation, and a theorem which allows us to sum the moments of the *components* of a force and thereby obtain the moment of the entire force.

Sometimes we need to find the moment of a force about a line through a certain point, instead of about the point itself. We learn to do this in Section 3.3, where we also develop the physical interpretation that the moment of a force $\mathbf{F}$ about line ℓ is the turning effect, about ℓ, of the part of $\mathbf{F}$ that is perpendicular to ℓ.

Section 3.4 contains a study of the concept of the couple, which is a pair of non-collinear, equal-magnitude, oppositely directed forces. A couple will be seen to have a turning effect but no resultant force, and it has the same moment about every point of space. The couple is an important concept in the study of moments of forces.

In Section 3.5, we will present the other equilibrium equation (the "moment equation," $\Sigma \mathbf{M}_P = \mathbf{0}$) as a companion to the "force equation," $\Sigma \mathbf{F} = \mathbf{0}$, that we studied in Chapter 2. We then develop the relationship between the sum of the moments about two points ($\Sigma \mathbf{M}_P$ and $\Sigma \mathbf{M}_Q$) which leads in Section 3.6 to the concept of equipollent systems of forces, meaning they have equal power, or strength; more precisely, equipollent force systems make identical contributions to the equations of equilibrium (and also to the equations of motion in a later study of Dynamics).

Any system of forces and couples can be replaced at any point P by an equipollent system consisting of a force and couple there, which is called a resultant of the original system. This is proved and illustrated in Section 3.7, and followed in Section 3.8 by a further reduction to the simplest resultant. This resultant is just a simple force in the cases of concurrent, coplanar, and parallel force systems. For more complicated three-dimensional force systems, the simplest resultant is a collinear force and couple, which for obvious reasons is called a "screwdriver."

In the last section, 3.9, we examine distributed force systems. We shall find that the resultant of a continuously distributed system of parallel forces is the area beneath the loading curve, located at the centroid of this area.

3.2 Moment of a Force About a Point

Common Sense Definition

The moment of a force is a measure of the tendency of the force to turn a body to which the force is applied. The moment of a force *about a point* (or *with respect to a point*) is defined to be a vector whose magnitude is the product of (a) the magnitude of the force and (b) the perpendicular distance between the point and the line of action of the force. The vector is perpendicular to the plane defined by the point and the line of action of the force.

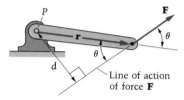

Figure 3.1

> **Question 3.1**   Is there a case when the plane of P and F is not defined? If so, what is the moment?

The direction is assigned by the "right-hand rule": If the fingers of the right-hand curve in the direction of the perceived sense of the turning effect, the thumb will point in the direction of the moment. Thus, if we have the situation shown in Figure 3.1 and if we let $\mathbf{M}_P$ stand for the moment of $\mathbf{F}$ about P, then

$$\mathbf{M}_P = |\mathbf{F}|\,d\hat{\mathbf{n}} \tag{3.1}$$

In this case, the page is the plane of $\mathbf{r}$ and $\mathbf{F}$, and $\hat{\mathbf{n}}$ is a unit vector pointing out of the page toward the reader because we envision the turning effect to be counterclockwise in Figure 3.1. The magnitudes of the two sides of Equation (3.1) are of course equal:

$$|\mathbf{M}_P| = |\mathbf{F}|\,d \tag{3.2}$$

and this scalar equation allows us to find $|\mathbf{M}_P|$, $|\mathbf{F}|$, or d if we know the other two. Thus, for example, the perpendicular distance "d" from P to the line of action of the force is the magnitude of the moment $|\mathbf{M}_P|$ divided by the magnitude of the force $|\mathbf{F}|$.

Note that except for the case in Question 3.1, P, $\mathbf{r}$, and $\mathbf{F}$ always form a plane, so the above equations (3.1, 3.2) are always valid, whether the vectors are easy to depict (as in Figure 3.1) or not.

Vector Representation

Another way of representing $|\mathbf{M}_P|$ follows from the fact that, if $\mathbf{r}$ is the directed line segment from P to *any* point on the line of action of $\mathbf{F}$, then $d = |\mathbf{r}|\sin\theta$ as shown in Figure 3.1. Thus

$$\mathbf{M}_P = [|\mathbf{F}||\mathbf{r}|\sin\theta]\hat{\mathbf{n}}$$

This result can be expressed in terms of the cross (or vector) product of the

---

Answer 3.1   If the point lies on the line of action of the force, then, of course, the plane isn't defined. But then the distance is zero, so the moment is zero.

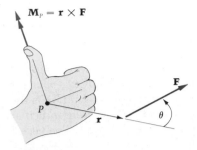

$$\mathbf{M}_P = \mathbf{r} \times \mathbf{F}$$

Figure 3.2

two vectors **r** and **F**, which is defined by

$$\mathbf{r} \times \mathbf{F} = [\,|\mathbf{r}|\,|\mathbf{F}|\sin\theta]\hat{\mathbf{n}}$$

where $\hat{\mathbf{n}}$ is the unit vector determined by the right-hand rule as shown in Figure 3.2, that is, the right thumb points in the direction of $\hat{\mathbf{n}}$ (and hence of $\mathbf{r} \times \mathbf{F}$) if the fingers are turned from **r** into **F**.* Thus

$$\mathbf{M}_P = \mathbf{r} \times \mathbf{F} \tag{3.3}$$

Note from Figure 3.2 that we shall depict moment vectors by using double-headed arrows.

If we express **r** and **F** in component form, then

$$
\begin{aligned}
\mathbf{r} \times \mathbf{F} = &(r_x\hat{\mathbf{i}} + r_y\hat{\mathbf{j}} + r_z\hat{\mathbf{k}}) \times (F_x\hat{\mathbf{i}} + F_y\hat{\mathbf{j}} + F_z\hat{\mathbf{k}}) \\
= &\; r_xF_x(\hat{\mathbf{i}} \times \hat{\mathbf{i}}) + r_xF_y(\hat{\mathbf{i}} \times \hat{\mathbf{j}}) + r_xF_z(\hat{\mathbf{i}} \times \hat{\mathbf{k}}) \\
&+ r_yF_x(\hat{\mathbf{j}} \times \hat{\mathbf{i}}) + r_yF_y(\hat{\mathbf{j}} \times \hat{\mathbf{j}}) + r_yF_z(\hat{\mathbf{j}} \times \hat{\mathbf{k}}) \\
&+ r_zF_x(\hat{\mathbf{k}} \times \hat{\mathbf{i}}) + r_zF_y(\hat{\mathbf{k}} \times \hat{\mathbf{j}}) + r_zF_z(\hat{\mathbf{k}} \times \hat{\mathbf{k}})
\end{aligned}
$$

But by the definition of the cross product,

$$\hat{\mathbf{i}} \times \hat{\mathbf{i}} = 1(1)\sin 0\ \hat{\mathbf{n}} = 0$$

and similarly

$$\hat{\mathbf{j}} \times \hat{\mathbf{j}} = \hat{\mathbf{k}} \times \hat{\mathbf{k}} = 0$$

If we specify $\hat{\mathbf{i}}, \hat{\mathbf{j}},$ and $\hat{\mathbf{k}}$ to constitute a right-handed system (as will be the case throughout this book), then

$$\hat{\mathbf{i}} \times \hat{\mathbf{j}} = (1)(1)\sin 90°\ \hat{\mathbf{k}} = \hat{\mathbf{k}}$$
$$\hat{\mathbf{j}} \times \hat{\mathbf{k}} = \hat{\mathbf{i}}$$
$$\hat{\mathbf{k}} \times \hat{\mathbf{i}} = \hat{\mathbf{j}}$$

and similarly

$$\hat{\mathbf{j}} \times \hat{\mathbf{i}} = -\hat{\mathbf{k}}$$
$$\hat{\mathbf{k}} \times \hat{\mathbf{j}} = -\hat{\mathbf{i}}$$
$$\hat{\mathbf{i}} \times \hat{\mathbf{k}} = -\hat{\mathbf{j}}$$

Therefore

$$
\begin{aligned}
\mathbf{r} \times \mathbf{F} = &\; r_xF_y(\hat{\mathbf{k}}) + r_xF_z(-\hat{\mathbf{j}}) \\
&+ r_yF_x(-\hat{\mathbf{k}}) + r_yF_z(\hat{\mathbf{i}}) \\
&+ r_zF_x(\hat{\mathbf{j}}) + r_zF_y(-\hat{\mathbf{i}}) \\
= &\;(r_yF_z - r_zF_y)\hat{\mathbf{i}} + (r_zF_x - r_xF_z)\hat{\mathbf{j}} \\
&+ (r_xF_y - r_yF_x)\hat{\mathbf{k}}
\end{aligned}
$$

---

* Through the smaller ($<180°$) of the two angles between **r** and **F** in their plane.

The reader can easily verify that this can be put in the form of a determinant

$$\mathbf{r} \times \mathbf{F} = \begin{vmatrix} \hat{\mathbf{i}} & \hat{\mathbf{j}} & \hat{\mathbf{k}} \\ r_x & r_y & r_z \\ F_x & F_y & F_z \end{vmatrix}$$

The cross-product method of finding $\mathbf{M}_P$ is particularly useful in a situation in which the plane containing P and the line of action of $\mathbf{F}$ is not a natural reference plane for the problem under investigation. In such a circumstance the determination of d and $\hat{\mathbf{n}}$ by nonvector methods of analytic geometry becomes a difficult task. Equation (3.3) effectively reduces this task to a single straightforward operation.

> **Question 3.2**   Why does $\mathbf{r} \times \mathbf{F}$ yield the same result for $\mathbf{M}_P$ regardless of which point on the line of action of $\mathbf{F}$ is intersected by $\mathbf{r}$?
>
> **Question 3.3**   Does it matter whether $\mathbf{M}_P$ is computed as $\mathbf{r} \times \mathbf{F}$ or $\mathbf{F} \times \mathbf{r}$?

Varignon's Theorem

We now proceed to prove a very important theorem concerning the moments of forces. Suppose that we have the two forces of Figure 3.3 again acting on the crate at point A, and suppose that we are now interested in the moment about the lower right-hand corner point, P, of the sum, $\mathbf{F}$, of $\mathbf{F}_1$ and $\mathbf{F}_2$.

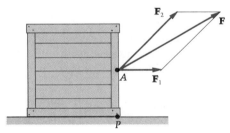

Figure 3.3

Then

$$\mathbf{M}_P = \mathbf{r} \times \mathbf{F}$$
$$= \mathbf{r} \times (\mathbf{F}_1 + \mathbf{F}_2)$$

Answer 3.2   In Figure 3.1 we see that $|\mathbf{r}| \sin \theta$ is a constant, making $|\mathbf{M}_P|$ the same for all intersection points. This figure also shows by the right-hand rule that the *direction* of the cross product is also independent of the intersection point. Thus $\mathbf{M}_P$ is the same regardless of which point on $\mathbf{F}$ is intersected by $\mathbf{r}$ — that is, regardless of which $\mathbf{r}$ is used to form it.

Answer 3.3   Yes! $\mathbf{r} \times \mathbf{F} = -(\mathbf{F} \times \mathbf{r})$, so if $\mathbf{F} \times \mathbf{r}$ is used, the moment will be in the wrong direction.

and by the distributive property of the cross product,

$$\mathbf{M}_P = \mathbf{r} \times \mathbf{F}_1 + \mathbf{r} \times \mathbf{F}_2 = |\mathbf{F}_1| d_1 \hat{\mathbf{n}}_1 + |\mathbf{F}_2| d_2 \hat{\mathbf{n}}_2$$

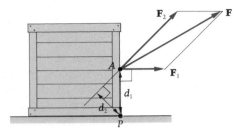

Figure 3.4

where d_1 and d_2 are shown in Figure 3.4 and where $\hat{\mathbf{n}}_1$ and $\hat{\mathbf{n}}_2$ are the same unit vector, directed into the paper. Thus we see that the moment of $\mathbf{F}$ about P is equal to the sum of the moments of $\mathbf{F}_1$ and $\mathbf{F}_2$ about P. By extension we can say in general that for any number n of forces acting at a common point A,

$$\mathbf{M}_P = \mathbf{r} \times \mathbf{F}$$
$$= \mathbf{r} \times (\mathbf{F}_1 + \mathbf{F}_2 + \cdots + \mathbf{F}_n)$$
$$= \mathbf{r} \times \mathbf{F}_1 + \mathbf{r} \times \mathbf{F}_2 + \cdots + \mathbf{r} \times \mathbf{F}_n$$
$$\mathbf{M}_P = |\mathbf{F}_1| d_1 \hat{\mathbf{n}}_1 + |\mathbf{F}_2| d_2 \hat{\mathbf{n}}_2 + \cdots + |\mathbf{F}_n| d_n \hat{\mathbf{n}}_n \qquad (3.4)$$

Therefore the moment about a point P of the sum of n forces acting at a point A is the sum of the moments of the separate forces (more briefly, "the moment of the sum is the sum of the moments"). This statement, which perhaps seems obvious, is often known as *Varignon's Theorem*. It is of practical value especially when we can decompose a force like $\mathbf{F}$ into parts whose perpendicular distances from P are easily determined.

Question 3.4   Does the development of Equation 3.4 require that $\mathbf{F}_1$, $\mathbf{F}_2, \ldots, \mathbf{F}_n$ all lie in the same plane?

In the example to follow, we will use each of the above approaches [Equations (3.1), (3.3), and (3.4)] to calculate the moment of a force about a point.

Answer 3.4   No. This was never required in the derivation, and thus it isn't necessary.

EXAMPLE 3.1

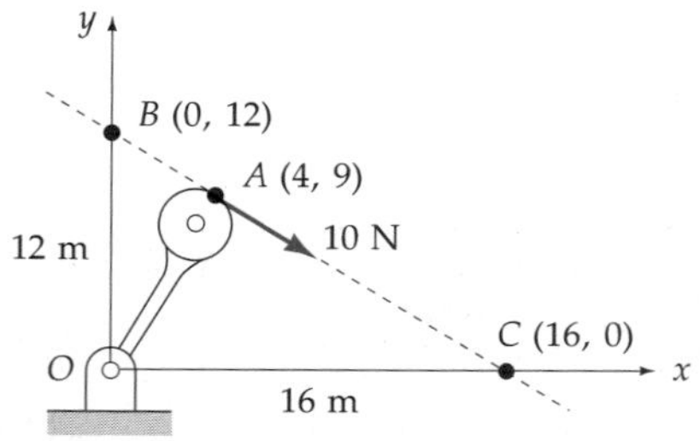

Calculate the moment of the 10-N force with respect to point O, the origin of the rectangular coordinate system shown in Figure E3.1a. Use all three approaches suggested by Equation (3.1), (3.3), and (3.4).

Figure E3.1a

Solution

To use Equation (3.1), we need the perpendicular distance d between O and the line of action of the force. It is calculated below to be 9.6 meters by using the similar (shaded) triangles, in Figure E3.1b.

$$\frac{16}{d} = \frac{5}{3} \Rightarrow d = \frac{48}{5} = 9.6 \text{ m}$$

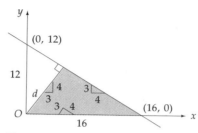

Figure E3.1b

Therefore, Equation (3.1) yields

$$\mathbf{M}_O = |\mathbf{F}| \, d\hat{\mathbf{n}} = (10)(9.6)(-\hat{\mathbf{k}}) = -96\hat{\mathbf{k}} \text{ N} \cdot \text{m} \text{ or } 96 \circlearrowleft \text{N} \cdot \text{m}$$

We emphasize again that this approach is of little value in three-dimensional situations in which the perpendicular distance d is not easily calculated and the unit vector is not obvious.

To use Equation (3.3), we note that the force is $\mathbf{F} = 8\hat{\mathbf{i}} - 6\hat{\mathbf{j}}$ N and the directed line segment from O to point A on the line of action of $\mathbf{F}$ is $\mathbf{r}_{OA} = 4\hat{\mathbf{i}} + 9\hat{\mathbf{j}}$ m. So, by Equation (3.3),

$$\mathbf{M}_O = \mathbf{r}_{OA} \times \mathbf{F} = (4\hat{\mathbf{i}} + 9\hat{\mathbf{j}}) \times (8\hat{\mathbf{i}} - 6\hat{\mathbf{j}})$$
$$= (4)(-6)(\hat{\mathbf{i}} \times \hat{\mathbf{j}}) + (9)(8)(\hat{\mathbf{j}} \times \hat{\mathbf{i}})$$
$$= -24\hat{\mathbf{k}} + 72(-\hat{\mathbf{k}}) = -96\hat{\mathbf{k}} \text{ N} \cdot \text{m} \qquad \text{(as above)}$$

In arriving at this result, we have recalled that the cross-product of a vector with itself is zero, i.e.,

$$\hat{\mathbf{i}} \times \hat{\mathbf{i}} = \hat{\mathbf{j}} \times \hat{\mathbf{j}} = \hat{\mathbf{k}} \times \hat{\mathbf{k}} = \mathbf{0}.$$

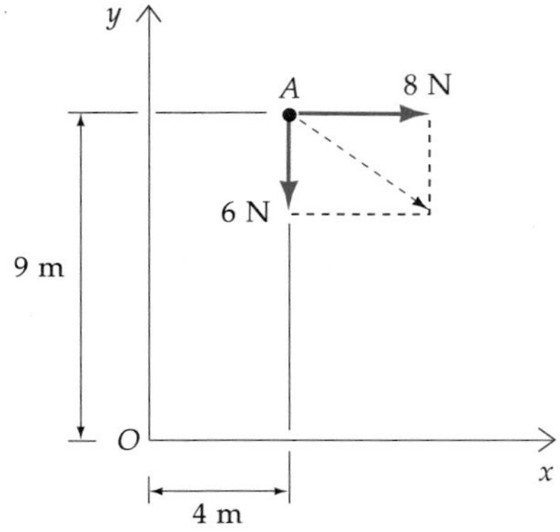

To use Equation (3.4), we shall decompose the force $\mathbf{F}$ into the 8-N and 6-N forces at A as shown in Figure E3.1c. We then revert to the "force times perpendicular distance" definition to calculate separately the moments of these component forces with respect to O.* For the 8-N force we have

$$\mathbf{M}_O' = 8(9)(-\hat{\mathbf{k}}) = -72\hat{\mathbf{k}} \text{ N} \cdot \text{m}$$

and for the 6-N force we have

$$\mathbf{M}_O'' = 6(4)(-\hat{\mathbf{k}}) = -24\hat{\mathbf{k}} \text{ N} \cdot \text{m}$$

Therefore the sum is $\mathbf{M}_O = -96\hat{\mathbf{k}} \text{ N} \cdot \text{m}$, the same result we obtained in the first two parts of the example. Note how the unit vector is attached using the

Figure E3.1c

---

* This is the practical importance of Varignon's Theorem.

right-hand rule. Also, it is important to observe the correspondence of $\mathbf{M}'_O$ and $\mathbf{M}''_O$ with the two terms in the cross-product calculation above. We have obtained the result without the formality of taking the cross product.

Before leaving this example, note that because point B is also on the line of action of $\mathbf{F}$, we could alternatively compute $\mathbf{M}_O$ by using $\mathbf{r}_{OB}$ instead of $\mathbf{r}_{OA}$:

$$\mathbf{M}_O = \mathbf{r}_{OB} \times \mathbf{F} = 12\hat{\mathbf{j}} \times (8\hat{\mathbf{i}} - 6\hat{\mathbf{j}}) = 12(8)(\hat{\mathbf{j}} \times \hat{\mathbf{i}})$$
$$= 96(-\hat{\mathbf{k}}) = -96\hat{\mathbf{k}} \ \text{N} \cdot \text{m}$$

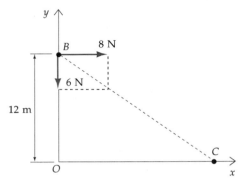

Figure E3.1d

The equivalent informal calculation follows from decomposing $\mathbf{F}$ into 8-N and 6-N components with lines of action intersecting at B (see Figure E3.1d). In this case the 6-N force produces no moment about O since its line of action passes through O, and the 8-N force produces a moment of $8(12)(-\hat{\mathbf{k}})$ so that (for the fifth time!):

$$\mathbf{M}_O = 0 + (-96\hat{\mathbf{k}}) = -96\hat{\mathbf{k}} \ \text{N} \cdot \text{m}$$

This illustrates the fact that it matters not where $\mathbf{F}$ is decomposed on its line of action in order to obtain a moment. In Problem 3.4, the student will be asked to recompute $\mathbf{M}_O$ by breaking the 10-N force into components at point C.

In the following example, the plane of $\mathbf{r}$ and $\mathbf{F}$ is skewed with respect to the xy, yz, and xz planes, and we shall use the cross product to advantage since finding the value of d needed in Equation (3.1) would actually require the same vector operation, the cross product.

EXAMPLE 3.2

A position vector from point P to point Q on the line of action of a force $\mathbf{F}$ is given by

$$\mathbf{r}_{PQ} = 2\hat{\mathbf{i}} + 3\hat{\mathbf{j}} - 4\hat{\mathbf{k}} \ \text{ft}$$

The force $\mathbf{F}$ is

$$\mathbf{F} = \hat{\mathbf{i}} - 2\hat{\mathbf{j}} + 5\hat{\mathbf{k}} \ \text{lb}$$

Find the moment of $\mathbf{F}$ about point P.

Solution

By Equation (3.3),

$$\mathbf{M}_P = \mathbf{r}_{PQ} \times \mathbf{F} = (2\hat{\mathbf{i}} + 3\hat{\mathbf{j}} - 4\hat{\mathbf{k}}) \times (\hat{\mathbf{i}} - 2\hat{\mathbf{j}} + 5\hat{\mathbf{k}})$$

$$= 2(1)(\hat{\mathbf{i}} \times \hat{\mathbf{i}}) + 2(-2)(\hat{\mathbf{i}} \times \hat{\mathbf{j}}) + 2(5)(\hat{\mathbf{i}} \times \hat{\mathbf{k}}) + 3(1)(\hat{\mathbf{j}} \times \hat{\mathbf{i}})$$

$$+ 3(-2)(\hat{\mathbf{j}} \times \hat{\mathbf{j}}) + 3(5)(\hat{\mathbf{j}} \times \hat{\mathbf{k}}) + (-4)(1)(\hat{\mathbf{k}} \times \hat{\mathbf{i}})$$

$$+ (-4)(-2)(\hat{\mathbf{k}} \times \hat{\mathbf{j}}) + (-4)(5)(\hat{\mathbf{k}} \times \hat{\mathbf{k}})$$

$$= (2)(0) + (-4)(\hat{\mathbf{k}}) + (10)(-\hat{\mathbf{j}}) + (3)(-\hat{\mathbf{k}}) + (-6)(0)$$

$$+ (15)(\hat{\mathbf{i}}) + (-4)(-\hat{\mathbf{j}}) + 8(-\hat{\mathbf{i}}) + (-20)(0)$$

$$= 7\hat{\mathbf{i}} - 14\hat{\mathbf{j}} - 7\hat{\mathbf{k}} \text{ lb-ft}$$

Alternatively, we could use the determinant method of computing the cross product:

$$\mathbf{M}_P = \mathbf{r}_{PQ} \times \mathbf{F} = \begin{vmatrix} \hat{\mathbf{i}} & \hat{\mathbf{j}} & \hat{\mathbf{k}} \\ 2 & 3 & -4 \\ 1 & -2 & 5 \end{vmatrix}$$

$$= \hat{\mathbf{i}} \begin{vmatrix} 3 & -4 \\ -2 & 5 \end{vmatrix} - \hat{\mathbf{j}} \begin{vmatrix} 2 & -4 \\ 1 & 5 \end{vmatrix} + \hat{\mathbf{k}} \begin{vmatrix} 2 & 3 \\ 1 & -2 \end{vmatrix}$$

$$= \hat{\mathbf{i}}(15 - 8) - \hat{\mathbf{j}}(10 + 4) + \hat{\mathbf{k}}(-4 - 3)$$

$$= 7\hat{\mathbf{i}} - 14\hat{\mathbf{j}} - 7\hat{\mathbf{k}} \text{ lb-ft}$$

(as before, but obtained much more quickly)

Sometimes we have to compute the position vector before we can take the moment, as in the next example:

EXAMPLE 3.3

A force, $\mathbf{F} = 3\hat{\mathbf{i}} - 5\hat{\mathbf{j}} + \hat{\mathbf{k}}$ lb, has a line of action through point A of Figure E3.3 with coordinates (0, 3, 4) ft. (a) Find the moment of $\mathbf{F}$ about point B whose coordinates are (4, 1, 2) ft. (b) Find the distance from point B to the line of action of $\mathbf{F}$.

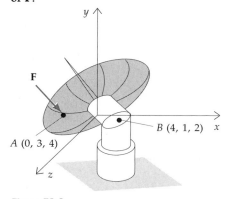

Figure E3.3

Solution

(a) The moment about B is given by

$$\mathbf{M}_B = \mathbf{r}_{BA} \times \mathbf{F}$$

where

$$\mathbf{r}_{BA} = (0 - 4)\hat{\mathbf{i}} + (3 - 1)\hat{\mathbf{j}} + (4 - 2)\hat{\mathbf{k}} = -4\hat{\mathbf{i}} + 2\hat{\mathbf{j}} + 2\hat{\mathbf{k}} \text{ ft}$$

Thus

$$\begin{aligned}\mathbf{M}_B &= (-4\hat{\mathbf{i}} + 2\hat{\mathbf{j}} + 2\hat{\mathbf{k}}) \times (3\hat{\mathbf{i}} - 5\hat{\mathbf{j}} + \hat{\mathbf{k}}) \\ &= (-4)(-5)\hat{\mathbf{k}} + (-4)(1)(-\hat{\mathbf{j}}) + (2)(3)(-\hat{\mathbf{k}}) + 2(1)\hat{\mathbf{i}} \\ &\quad + 2(3)\hat{\mathbf{j}} + 2(-5)(-\hat{\mathbf{i}}) \\ &= 12\hat{\mathbf{i}} + 10\hat{\mathbf{j}} + 14\hat{\mathbf{k}} \text{ lb-ft}\end{aligned}$$

(b) The distance from B to the line of action of F is found by Equation (3.2):

$$d = \frac{|\mathbf{M}_B|}{|\mathbf{F}|}$$

Computing the magnitudes of $\mathbf{M}_B$ and $\mathbf{F}$, we find:

$$|\mathbf{M}_B| = \sqrt{(12)^2 + (10)^2 + (14)^2} = \sqrt{440} = 21.0 \text{ lb-ft}$$
$$|\mathbf{F}| = \sqrt{(3)^2 + (-5)^2 + (1)^2} = \sqrt{35} = 5.92 \text{ lb}$$

Therefore

$$d = \frac{21.0}{5.92} = 3.55 \text{ ft}$$

PROBLEMS ▶ Section 3.2

3.1 Find the moment of the 12-N force about the origin in Figure P3.1.

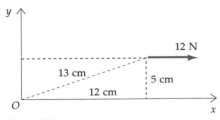

Figure P3.1

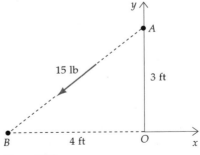

Figure P3.2

3.2 Find the moment of the force in Figure P3.2 about the origin using Equation (a) 3.1; (b) 3.3.

3.3 In the preceding problem, break the force into its components at (a) A and (b) B, and in each case find the moment of the force about the origin using Varignon's Theorem.

3.4 In Example 3.1, compute $\mathbf{M}_O$ by using the fact that C is also on the line of action of F. Do this in two ways:

a. vectorially, with $\mathbf{r}_{OC} \times \mathbf{F}$;
b. by decomposing $\mathbf{F}$ into its x- and y-components at C and then summing the separate moments of the components.

3.5 A force $\mathbf{F}$ with magnitude 1000 lb is exerted on the tooth of the sector gear shown in Figure P3.5 by a tooth of another gear that is not shown. The force makes a 20° angle with the normal to the radius drawn to the tooth from point O, as shown in the small figure. Find the moment of $\mathbf{F}$ about O, and the perpendicular distance from O to the line of action of the force.

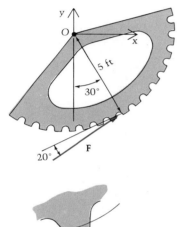
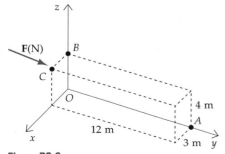

Figure P3.5

3.6 The force $\mathbf{F}$ in Figure P3.6 acts along the line CA and has a magnitude of 520 N.

a. Write the unit vector $\hat{\mathbf{u}}$ that has the direction CA, and use it to form the vector force $\mathbf{F}$.

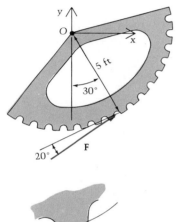

Figure P3.6

b. Calculate the moments about B of the three components of $\mathbf{F}$ acting at C by multiplying these forces by their perpendicular distances from B. Assign the correct unit vector to each using the right-hand rule and add them to form $\mathbf{M}_B$.
c. Calculate the moment of $\mathbf{F}$ about B using $\mathbf{r}_{BC} \times \mathbf{F}$, and compare it with $\mathbf{M}_B$ from part (b).
d. Calculate the moment of $\mathbf{F}$ about B using $\mathbf{r}_{BA} \times \mathbf{F}$, and note that the result is still the same because A also lies on the line of action of $\mathbf{F}$.

3.7 Find the sum of the moments about point A of the five applied forces. (See Figure P3.7.)

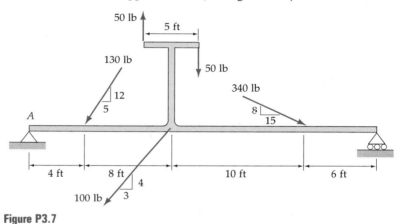

Figure P3.7

3.8 In Figure P3.8, find:

a. the moment of the 50-lb force about point P having the coordinates $(x, y, z) = (2, 3, 5)$ ft;
b. the perpendicular distance from P to the line of action of the force.

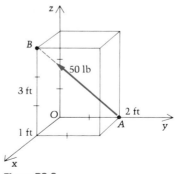

Figure P3.8

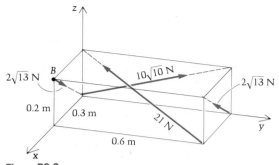

Figure P3.9

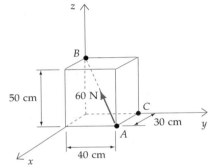

Figure P3.10

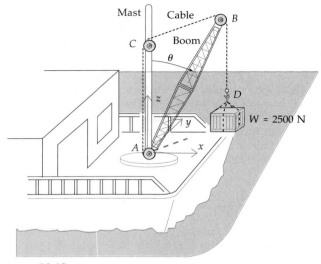

Figure P3.12

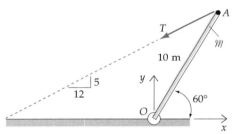

Figure P3.13

3.9 Find (see Figure P3.9):

a. The sum of the forces

b. The sum of their moments about point B.

3.10 Determine the moment of the 60-N force in Figure P3.10 with respect to corner C using

a. $\mathbf{r} \times \mathbf{F}$;

b. Varignon's Theorem after resolving the force into its components at A.

3.11 Find the moment about the origin due to the force $\mathbf{F} = -30\hat{\mathbf{i}} + 40\hat{\mathbf{k}}$ lb, which acts at the point $(x, y, z) = (20, 10, 0)$ ft.

3.12 In terms of angle θ, find the moment of the weight W about the base A of the mast in Figure P3.12 if the intersection line (dashed) of the deck plane (xy) with the plane of the mast, boom, and cable forms angles with x and y of 30° and 60°, respectively. The length of the boom is 10 m.

3.13 In raising the heavy mast $\mathcal{m}$ in Figure P3.13, the tension T in the cable is supplying a moment about O of magnitude 5000 N · m. Find the tension in the cable.

3.14 The force of 140 lb acts at A with a line of action directed toward B. (See Figure P3.14.) Find the moment of this force about O and the shortest distance from O to the line of action of the force.

✱ 3.15 A force of given magnitude P lb has a line of action through A, and is to act on the rod, as shown. Find (see Figure P3.15) the angle θ for which the moment of the force about B is a maximum.

---

* Asterisks identify the more difficult problems.

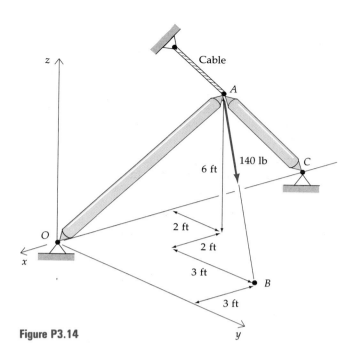

Figure P3.14

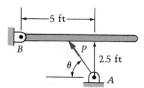

Figure P3.15

3.3 Moment of a Force About a Line

Definition of M_l

In statics we are often interested in knowing the moment of a force about a line rather than about a point. As suggested by Figure 3.5, we shall define the moment of a force $\mathbf{F}$ about a line l to be the projection along l of the moment of $\mathbf{F}$ about any point P lying on l. The moment $\mathbf{M}_l$ is thus defined to be

$$\mathbf{M}_l = (\mathbf{M}_P \cdot \hat{\mathbf{u}})\hat{\mathbf{u}} \tag{3.5}$$

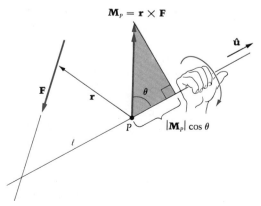

Figure 3.5

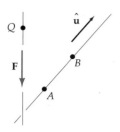

Figure 3.6

or

$$\mathbf{M}_\ell = [(\mathbf{r} \times \mathbf{F}) \cdot \hat{\mathbf{u}}]\hat{\mathbf{u}} \qquad (3.6)$$

where $\hat{\mathbf{u}}$ is the unit vector in the direction of ℓ.

> **Question 3.5** How would the calculated $\mathbf{M}_\ell$ be different if $-\hat{\mathbf{u}}$ had been chosen as the reference unit vector instead of $\hat{\mathbf{u}}$ in Equation 3.6?

Our definition of $\mathbf{M}_\ell$ (Equation 3.5) suggests that the same value is obtained regardless of the location of P on the line ℓ. In fact we can prove, if A and B are points on ℓ (see Figure 3.6), that

$$\mathbf{M}_\ell = (\mathbf{M}_B \cdot \hat{\mathbf{u}})\hat{\mathbf{u}} = (\mathbf{M}_A \cdot \hat{\mathbf{u}})\hat{\mathbf{u}}$$

To do this, we use the fact that if Q is any point on the line of action of $\mathbf{F}$, then

$$\mathbf{M}_A = \mathbf{r}_{AQ} \times \mathbf{F}$$

But

$$\mathbf{r}_{AQ} = \mathbf{r}_{AB} + \mathbf{r}_{BQ}$$

so that

$$\mathbf{M}_A = \mathbf{r}_{AB} \times \mathbf{F} + \mathbf{r}_{BQ} \times \mathbf{F}$$

Thus

$$(\mathbf{M}_A \cdot \hat{\mathbf{u}})\hat{\mathbf{u}} = [(\mathbf{r}_{AB} \times \mathbf{F}) \cdot \hat{\mathbf{u}}]\hat{\mathbf{u}} + [(\mathbf{r}_{BQ} \times \mathbf{F}) \cdot \hat{\mathbf{u}}]\hat{\mathbf{u}}$$

But $\mathbf{r}_{AB} \times \mathbf{F}$ is perpendicular to ℓ, that is, to $\hat{\mathbf{u}}$, so that

$$(\mathbf{M}_A \cdot \hat{\mathbf{u}})\hat{\mathbf{u}} = 0 + [(\mathbf{r}_{BQ} \times \mathbf{F}) \cdot \hat{\mathbf{u}}]\hat{\mathbf{u}}$$

$$(\mathbf{M}_A \cdot \hat{\mathbf{u}})\hat{\mathbf{u}} = (\mathbf{M}_B \cdot \hat{\mathbf{u}})\hat{\mathbf{u}} \qquad (3.7)$$

and thus $\mathbf{M}_\ell$ is independent of the point on ℓ which is chosen for the computation.

Physical Interpretation of $\mathbf{M}_\ell$

We now provide a physical interpretation of $\mathbf{M}_\ell$ as the turning effect, about ℓ, of the part of $\mathbf{F}$ that is perpendicular to ℓ. To do this we first identify, as in Figure 3.7, the plane defined by the line of action of $\mathbf{F}$ and any line intersecting it and parallel to ℓ. Let d be the distance between this plane and ℓ, as shown in Figure 3.7.

> **Question 3.6** If $\mathbf{F}$ is itself already parallel to ℓ, then this plane does not exist. Why? What is $\mathbf{M}_\ell$ in this case?

Answer 3.5 It would be the same.

Answer 3.6 If $\mathbf{F} \parallel \ell$, then the only line intersecting $\mathbf{F}$ and parallel to ℓ is coincident with $\mathbf{F}$; thus not a plane, but just a line, is defined. In that case, $\mathbf{r} \times \mathbf{F}$ is $\perp \hat{\mathbf{u}}$, so $\mathbf{M}_\ell = 0$ by Equation (3.6).

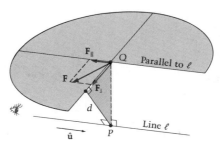

Figure 3.7

Next we construct a plane perpendicular to ℓ. (Again see Figure 3.7.) We shall let P be its intersection with ℓ, and Q its intersection with the line of action of **F**. Furthermore, at Q we decompose **F** into components $\mathbf{F}_\parallel$ and $\mathbf{F}_\perp$, respectively, parallel and perpendicular to ℓ as shown.*

Taking the moment of **F** about point P, we obtain

$$\mathbf{M}_P = \mathbf{r}_{PQ} \times \mathbf{F} = \mathbf{r}_{PQ} \times (\mathbf{F}_\parallel + \mathbf{F}_\perp)$$
$$= \mathbf{r}_{PQ} \times \mathbf{F}_\parallel + \mathbf{r}_{PQ} \times \mathbf{F}_\perp \qquad (3.8)$$

Our last step is to substitute $\mathbf{M}_P$ from Equation (3.8) into Equation (3.5) for $\mathbf{M}_\ell$:

$$\mathbf{M}_\ell = (\mathbf{M}_P \cdot \hat{\mathbf{u}})\hat{\mathbf{u}}$$
$$= [(\mathbf{r}_{PQ} \times \mathbf{F}_\parallel + \mathbf{r}_{PQ} \times \mathbf{F}_\perp) \cdot \hat{\mathbf{u}}]\hat{\mathbf{u}} \qquad (3.9)$$

Because $\mathbf{r}_{PQ} \times \mathbf{F}_\parallel$ is perpendicular to $\mathbf{F}_\parallel$ and thus also to ℓ and $\hat{\mathbf{u}}$, the dot product $(\mathbf{r}_{PQ} \times \mathbf{F}_\parallel) \cdot \hat{\mathbf{u}}$ is zero. This means that, as expected, the component of **F** parallel to line ℓ produces no moment about ℓ.

Since $\mathbf{r}_{PQ}$ and $\mathbf{F}_\perp$ are each perpendicular to ℓ, then $\mathbf{r}_{PQ} \times \mathbf{F}_\perp$ is in the direction of ℓ, or, in other words, proportional to $\hat{\mathbf{u}}$. Referring to Figure 3.8, we see that

$$\mathbf{M}_\ell = \mathbf{r}_{PQ} \times \mathbf{F}_\perp = |\mathbf{F}_\perp| d\,\hat{\mathbf{u}} \qquad (3.10)$$

Finally, the moment $\mathbf{M}_\ell$ of a force **F** about a line ℓ can be summarized by Figure 3.9. Only the component $\mathbf{F}_\perp$ of **F** that is normal to ℓ has a turning effect about ℓ. The moment $\mathbf{M}_\ell$ has for its magnitude the product of $|\mathbf{F}_\perp|$ and the distance d between $\mathbf{F}_\perp$ and ℓ. The direction of $\mathbf{M}_\ell$ is given by the right thumb when the fingers of the right hand curl about ℓ in the direction of the turning effect of $\mathbf{F}_\perp$ about ℓ.†

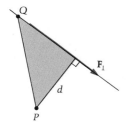

Figure 3.8 View along ℓ ($\hat{\mathbf{u}}$ is into the paper).

---

* If **F** is itself *already* perpendicular to ℓ, i.e., if $\mathbf{F} = \mathbf{F}_\perp$, then Q is not a unique point. In this case we take the plane perpendicular to ℓ to be the one *containing* the entire line of action of **F**, and Q to be any point on that line of action. For this case, in what follows, $\mathbf{F}_\parallel$ will of course be zero.

† We note that one point on the line of action of **F** in Figure 3.7 (nearer to the "eye" in the figure) will be closest to line ℓ; it will be the distance d from ℓ.

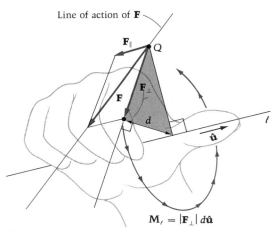

Line of action of **F**

$$\mathbf{M}_l = |\mathbf{F}_\perp|\, d\hat{\mathbf{u}}$$

Figure 3.9

Question 3.7   Could the magnitude of the moment of a force about a line ever be greater than the magnitude of the moment of the force about any point on the line?

Question 3.8   Does the distance d in Equation (3.10) depend on the choice of point P on line l?

Question 3.9   Suppose the moment of a force **F** about a point P has been computed using $\mathbf{M}_P = \mathbf{r} \times \mathbf{F}$, where **r** is a vector from P to some point on the line of action of **F**. About some *line* through P, the moment of **F** will be the largest in magnitude. What is the unit vector in the direction of this line?

Answer 3.7   No! $\mathbf{M}_l$ is the same for all points on l. But both $\mathbf{F}_\parallel$ and $\mathbf{F}_\perp$ make other contributions to the moment of **F** about points lying on l— contributions perpendicular to $\mathbf{M}_l$— so that the magnitude of moments of **F** about points on l is greater than or equal to $|\mathbf{M}_l|$.

Answer 3.8   No. Since both $\mathbf{M}_l$ and $\mathbf{F}_\perp$ are independent of the choice of P, Equation (3.10) shows that d is, too. The distance d is simply the distance indicated between the line l and the plane containing **F** in Figure 3.7.

Answer 3.9 $(\mathbf{r} \times \mathbf{F})/|\mathbf{r} \times \mathbf{F}|$, for then the line is in the direction of $\mathbf{M}_P$ and none of it is "lost" in the dot product of Equation (3.6).

EXAMPLE 3.4

The chain in Figure E3.4 carries 20 lb of tension as it holds open the trap door. Find the moment of the chain force exerted at A, about the x-axis.

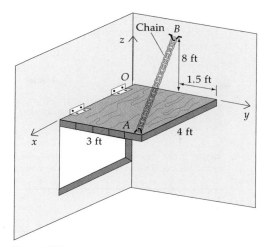

Figure E3.4

Solution

The force in the chain may be expressed as a vector (for use in Equation 3.6) by multiplying its magnitude by the unit vector in its direction:

$$\mathbf{F} = 20 \left(\frac{\mathbf{r}_{AB}}{|\mathbf{r}_{AB}|} \right) \text{lb}$$

The points A and B have the respective coordinates (4, 3, 0) and (0, 1.5, 8) ft; therefore,

$$\mathbf{F} = 20 \left(\frac{(0 - 4)\hat{\mathbf{i}} + (1.5 - 3)\hat{\mathbf{j}} + (8 - 0)\hat{\mathbf{k}}}{\sqrt{(-4)^2 + (-1.5)^2 + 8^2}} \right)$$

$$= 20 \left(\frac{-4\hat{\mathbf{i}} - 1.5\hat{\mathbf{j}} + 8\hat{\mathbf{k}}}{9.07} \right)$$

$$= -8.82\hat{\mathbf{i}} - 3.31\hat{\mathbf{j}} + 17.6\hat{\mathbf{k}} \text{ lb} \tag{1}$$

The origin O is a convenient point on the line (x-axis) about which we desire the moment of $\mathbf{F}$. Therefore, using Equation 3.6,

$$\mathbf{M}_{x\text{-}axis} = \overbrace{[(\mathbf{r}_{OA} \times \mathbf{F})}^{\mathbf{M}_O} \cdot \hat{\mathbf{i}}]\hat{\mathbf{i}}$$

$$= [(4\hat{\mathbf{i}} + 3\hat{\mathbf{j}}) \times (-8.82\hat{\mathbf{i}} - 3.31\hat{\mathbf{j}} + 17.6\hat{\mathbf{k}}) \cdot \hat{\mathbf{i}}]\hat{\mathbf{i}}$$

We note here that only the $\hat{\mathbf{i}}$-coefficient of the cross product will survive the subsequent dot product with $\hat{\mathbf{i}}$, so we only need to compute that component. Of the six vectors arising from the cross product, only $\hat{\mathbf{j}} \times \hat{\mathbf{k}}$ results in an "$\hat{\mathbf{i}}$"; therefore,

$$\mathbf{M}_{x\text{-}axis} = [(3\hat{\mathbf{j}} \times 17.6\hat{\mathbf{k}}) \cdot \hat{\mathbf{i}}]\hat{\mathbf{i}}$$

$$= 52.8\hat{\mathbf{i}} \text{ lb-ft}$$

Except possibly for a small amount of friction in the hinges, only the above (x) component of the total moment about O of the tension in the chain is holding the door open (i.e., opposing the moment of its weight). The rest of the moment $\mathbf{M}_O$ (its y and z components) is "wasted" and is only causing stress in the hinge connections.

Question 3.10   How could the chain be reattached so as to minimize the hinge reactions yet still hold the door open?

Answer 3.10   Connect the chain to the center of the free 4-ft edge and run it straight up to the ceiling. This may or may not be convenient, however.

EXAMPLE 3.5

At the instant shown in Figure E3.5, the wrench lies in a horizontal (xy) plane. Two 20-lb forces are applied to a ratchet wrench, down (into the paper) at A and up (out of the paper) at B, in an effort to loosen the bolt B. Find the sum of the moments of the two forces about the axis (x) of the bolt.

Solution

We shall find $\mathbf{M}_Q$ as the sum of the two $\mathbf{r} \times \mathbf{F}$'s each dotted with the unit vector along the line ($\hat{\mathbf{u}} = \hat{\mathbf{i}}$ here) as in Equation (3.5).

The vector from Q to B is seen to be

$$\mathbf{r}_{QB} = \frac{8}{\sqrt{2}}\,\hat{\mathbf{i}} + \frac{8}{\sqrt{2}}\,\hat{\mathbf{j}} \text{ in.}$$

So, for the force at B,

$$\mathbf{M}_x = [(\mathbf{r}_{QB} \times \mathbf{F}_B) \cdot \hat{\mathbf{i}}]\hat{\mathbf{i}}$$

$$= \left\{ \left[\left(\frac{8}{\sqrt{2}}\,\hat{\mathbf{i}} + \frac{8}{\sqrt{2}}\,\hat{\mathbf{j}} \right) \times (20\hat{\mathbf{k}}) \right] \cdot \hat{\mathbf{i}} \right\} \hat{\mathbf{i}}$$

$$= \left[\left(\frac{-160}{\sqrt{2}}\,\hat{\mathbf{j}} + \frac{160}{\sqrt{2}}\,\hat{\mathbf{i}} \right) \cdot \hat{\mathbf{i}} \right] \hat{\mathbf{i}} = \frac{160}{\sqrt{2}}\,\hat{\mathbf{i}} \text{ lb-in.}$$

Next, the vector from Q to A is:

$$\mathbf{r}_{QA} = \mathbf{r}_{QB} + \mathbf{r}_{BA} = \left(\frac{8}{\sqrt{2}}\,\hat{\mathbf{i}} + \frac{8}{\sqrt{2}}\,\hat{\mathbf{j}} \right) + \left(\frac{5}{\sqrt{2}}\,\hat{\mathbf{i}} - \frac{5}{\sqrt{2}}\,\hat{\mathbf{j}} \right)$$

$$= \frac{13}{\sqrt{2}}\,\hat{\mathbf{i}} + \frac{3}{\sqrt{2}}\,\hat{\mathbf{j}} \text{ in.}$$

so that, for the force at A,

$$\mathbf{M}_x = [(\mathbf{r}_{QA} \times \mathbf{F}_A) \cdot \hat{\mathbf{i}}]\hat{\mathbf{i}}$$

$$= \left\{ \left[\left(\frac{13}{\sqrt{2}}\,\hat{\mathbf{i}} + \frac{3}{\sqrt{2}}\,\hat{\mathbf{j}} \right) \times (-20\hat{\mathbf{k}}) \right] \cdot \hat{\mathbf{i}} \right\} \hat{\mathbf{i}}$$

$$= \left[\left(\frac{260}{\sqrt{2}}\,\hat{\mathbf{j}} - \frac{60}{\sqrt{2}}\,\hat{\mathbf{i}} \right) \cdot \hat{\mathbf{i}} \right] \hat{\mathbf{i}}$$

$$= \frac{-60}{\sqrt{2}}\,\hat{\mathbf{i}} \text{ lb-in.}$$

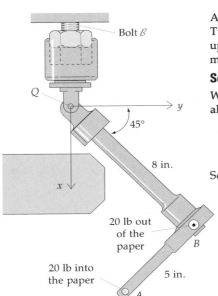

Bolt B

Q

y

$45°$

8 in.

x

20 lb out of the paper

B

20 lb into the paper

5 in.

A

Figure E3.5

The net effect of the two forces toward loosening the bolt is the sum of their separate contributions to $\mathbf{M}_x$:

$$\mathbf{M}_x = \frac{160 - 60}{\sqrt{2}} \,\hat{\mathbf{i}} = 70.7\hat{\mathbf{i}} \text{ lb-in.}$$

The two noncollinear forces in the preceding example, being equal in magnitude but opposite in direction, form what is called a *couple*. We shall study the properties of a couple in the next section of this chapter and will then rework the example using the most powerful of those properties.

EXAMPLE 3.6

Referring to Figure E3.6, find the moment of the 8-N force about the line *CD*.

Solution

To obtain $\mathbf{M}_{CD}$ we need the moment with respect to some point on the line such as C or D. Using C,

$$
\begin{aligned}
\mathbf{M}_C &= \mathbf{r}_{CA} \times \mathbf{F} \\
&= [(0 - 0)\hat{\mathbf{i}} + (6 - 3)\hat{\mathbf{j}} + (5 - (-4))\hat{\mathbf{k}}] \times 8\hat{\mathbf{e}}_{AB} \\
&= (3\hat{\mathbf{j}} + 9\hat{\mathbf{k}}) \times 8\left[\frac{-4\hat{\mathbf{i}} - 6\hat{\mathbf{j}} - 3\hat{\mathbf{k}}}{\sqrt{4^2 + 6^2 + 3^2}}\right] \\
&= (3\hat{\mathbf{j}} + 9\hat{\mathbf{k}}) \times (-4.10\hat{\mathbf{i}} - 6.14\hat{\mathbf{j}} - 3.07\hat{\mathbf{k}}) \\
&= 46.1\hat{\mathbf{i}} - 36.9\hat{\mathbf{j}} + 12.3\hat{\mathbf{k}} \text{ N} \cdot \text{m}
\end{aligned}
$$

The unit vector directed from C toward D is

$$
\begin{aligned}
\hat{\mathbf{e}}_{CD} &= \frac{\mathbf{r}_{CD}}{|\mathbf{r}_{CD}|} = \frac{1}{\sqrt{56}}(4\hat{\mathbf{i}} - 2\hat{\mathbf{j}} + 6\hat{\mathbf{k}}) \\
&= 0.535\hat{\mathbf{i}} - 0.267\hat{\mathbf{j}} + 0.802\hat{\mathbf{k}}
\end{aligned}
$$

By Equation (3.5),

$$
\begin{aligned}
\mathbf{M}_{CD} &= (\mathbf{M}_C \cdot \hat{\mathbf{e}}_{CD})\hat{\mathbf{e}}_{CD} \\
&= [(46.1)(0.535) + (-36.9)(-0.267) + (12.3)(0.802)]\hat{\mathbf{e}}_{CD} \\
&= 44.4\hat{\mathbf{e}}_{CD} \text{ N} \cdot \text{m} \\
&= 23.8\hat{\mathbf{i}} + 11.9\hat{\mathbf{j}} + 35.6\hat{\mathbf{k}} \text{ N} \cdot \text{m}
\end{aligned}
$$

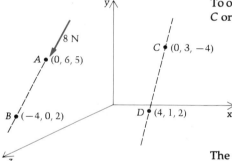

Figure E3.6

In the preceding example, the same result could have been obtained by first computing $\mathbf{M}_D$ instead of $\mathbf{M}_C$:

$$
\begin{aligned}
\mathbf{M}_D &= \mathbf{r}_{DA} \times \mathbf{F} \\
&= (-4\hat{\mathbf{i}} + 5\hat{\mathbf{j}} + 3\hat{\mathbf{k}}) \times (-4.10\hat{\mathbf{i}} - 6.14\hat{\mathbf{j}} - 3.07\hat{\mathbf{k}}) \\
&= 3.07\hat{\mathbf{i}} - 24.6\hat{\mathbf{j}} + 45.1\hat{\mathbf{k}} \text{ N} \cdot \text{m}
\end{aligned}
$$

so that

$$\mathbf{M}_{CD} = (\mathbf{M}_D \cdot \hat{\mathbf{e}}_{CD})\hat{\mathbf{e}}_{CD}$$

$$= [(3.07)(0.535) + (-24.6)(-0.267) + (45.1)(0.802)]\hat{\mathbf{e}}_{CD}$$

$$= 44.4\hat{\mathbf{e}}_{CD} \text{ N} \cdot \text{m} \qquad \text{(as before)}$$

One of the problems at the end of the section (Problem 3.17) is to rework this example using $\mathbf{r}_{CB}$ instead of $\mathbf{r}_{CA}$ (and $\mathbf{r}_{DB}$ instead of $\mathbf{r}_{DA}$).

Let us also give an illustration of the use of Equation 3.10, in conjunction with the preceding example, to find the shortest distance between the line of action of $\mathbf{F}$ and the line CD. First, the component of $\mathbf{F}$ parallel to CD is:

$$F_{\parallel} = \mathbf{F} \cdot \hat{\mathbf{e}}_{CD} = (-4.10\hat{\mathbf{i}} - 6.14\hat{\mathbf{j}} - 3.07\hat{\mathbf{k}})$$

$$\cdot (0.535\hat{\mathbf{i}} - 0.267\hat{\mathbf{j}} + 0.802\hat{\mathbf{k}})$$

$$= (-4.10)(0.535) + (-6.14)(-0.267)$$

$$+ (-3.07)(0.802)$$

$$= -3.02 \text{ N}$$

so that

$$|\mathbf{F}_{\perp}| = \sqrt{|\mathbf{F}|^2 - F_{\parallel}^2}$$

$$= \sqrt{8^2 - 3.02^2}$$

$$= 7.41 \text{ N}$$

Now using Equation (3.10) and the result of Example 3.6, we can obtain the desired shortest distance between the line of action of $\mathbf{F}$, and the line CD:

$$\mathbf{M}_l = \mathbf{M}_{CD} = |\mathbf{F}_{\perp}| \, d\hat{\mathbf{u}}$$

so that, equating the magnitudes of both sides,

$$44.4 = 7.41 \, d$$

$$d = 5.99 \text{ m}$$

EXAMPLE 3.7

Using Equation 3.10 and the result of Example 3.4, find the shortest distance between the x-axis and the line of action of the force in the chain.

Solution

Applying Equation (3.10),

$$|\mathbf{M}_{x\text{-}axis}| = |\mathbf{F}_{\perp}| \, d$$

or

$$d = \frac{52.8}{|\mathbf{F}_{\perp}|}$$

Now the part of **F** which is perpendicular to the line of interest (the x-axis) is in this case simply the vector sum of its y- and z-components (See Eq. (1) of Example 3.4):

$$\mathbf{F}_\perp = -3.31\hat{\mathbf{j}} + 17.6\hat{\mathbf{k}} \text{ lb}$$

or

$$|\mathbf{F}_\perp| = \sqrt{3.31^2 + 17.6^2}$$
$$= 17.9 \text{ lb}$$

Therefore

$$d = \frac{52.8}{17.9} = 2.95 \text{ ft}$$

We can check the result of the preceding example by looking at Figure 3.10 below, which is a view looking from the positive x-axis:

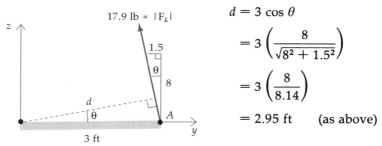

$$d = 3 \cos\theta$$
$$= 3\left(\frac{8}{\sqrt{8^2 + 1.5^2}}\right)$$
$$= 3\left(\frac{8}{8.14}\right)$$
$$= 2.95 \text{ ft} \qquad \text{(as above)}$$

Figure 3.10

Using "Force Times Perpendicular Distance" in Certain 3-D Applications to Find M_ℓ

In finding the moment of a force **F** about a point P, the formal use of the cross product is not always necessary. To see an alternative approach that uses what we have learned about moments with respect to lines, we express the moment, $\mathbf{M}_P$, of **F** about P in component form. Let Q again be some point on the line of action of **F** as shown in Figure 3.11 and note that we have positioned the axes x, y, and z to intersect at P. Expressing both $\mathbf{r}_{PQ}$ and **F** in component form as

$$\mathbf{r}_{PQ} = r_x\hat{\mathbf{i}} + r_y\hat{\mathbf{j}} + r_z\hat{\mathbf{k}}$$
$$\mathbf{F} = F_x\hat{\mathbf{i}} + F_y\hat{\mathbf{j}} + F_z\hat{\mathbf{k}}$$

then, as we noted in Section 3.2,

$$\mathbf{M}_P = \mathbf{r}_{PQ} \times \mathbf{F}$$
$$= (r_x\hat{\mathbf{i}} + r_y\hat{\mathbf{j}} + r_z\hat{\mathbf{k}}) \times (F_x\hat{\mathbf{i}} + F_y\hat{\mathbf{j}} + F_z\hat{\mathbf{k}})$$
$$= (r_yF_z - r_zF_y)\hat{\mathbf{i}} + (r_zF_x - r_xF_z)\hat{\mathbf{j}} + (r_xF_y - r_yF_x)\hat{\mathbf{k}} \qquad (3.11)$$

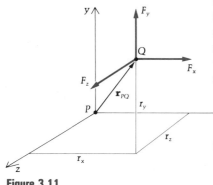

Figure 3.11

Since the x component of this vector is $\mathbf{M}_P \cdot \hat{\mathbf{i}}$, then from the foregoing analysis and discussion we recognize $(r_y F_z - r_z F_y)\hat{\mathbf{i}}$ as the moment of $\mathbf{F}$ about the x axis through P. Similarly, the other components of $\mathbf{M}_P$ are the moments of $\mathbf{F}$ about the y and z axes. The term $r_y F_z \hat{\mathbf{i}}$ is clearly the moment of $F_z \hat{\mathbf{k}}$ about the x axis since r_y (or, more precisely, its magnitude) is the distance from the x axis to the line of action of $F_z \hat{\mathbf{k}}$. Likewise, $-r_z F_y \hat{\mathbf{i}}$ is the moment of $F_y \hat{\mathbf{j}}$ about the x axis. Obviously the force $F_x \hat{\mathbf{i}}$ produces no moment about the x axis, since its line of action is parallel to this axis.

Thus, we see from Equation (3.11) that the moment about a point may be constructed by finding the moments about three mutually perpendicular axes through the point, and each of these may be computed by decomposing the force into components parallel to the axes and summing the moments of the individual components about these axes.

The above process allows us to utilize the fundamental concept of "magnitude of force times perpendicular distance," and hence even in three dimensions we can avoid, if we wish, the formal (vector product) calculation. When we do this we must, of course, manually attach the correct unit vector (including the proper sign) with the help of the right-hand rule. We are not necessarily advocating the evasion of vector products here, but rather we are calling attention to the more physical interpretation that can be given to the terms in Equation (3.11). The following example illustrates these ideas.

EXAMPLE 3.8

Find the moment of the force $\mathbf{F}$ in Figure E3.8a about point Q, using the ideas expressed by Equation (3.11).

Solution

The force $\mathbf{F}$ has the components:

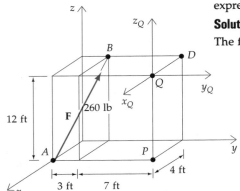

Figure E3.8a

$$F_x = \frac{-4}{\sqrt{3^2 + 4^2 + 12^2}}\, 260 = \left(\frac{-4}{13}\right) 260 = -80 \text{ lb}$$

$$F_y = \frac{3}{13}\, 260 = 60 \text{ lb}$$

$$F_z = \frac{12}{13}\, 260 = 240 \text{ lb}$$

We shall obtain $\mathbf{M}_Q$ by adding the separate moments of these components of $\mathbf{F}$ at A about the axes $(x_Q,\ y_Q,\ z_Q)$ through Q.

The component $F_x \hat{\mathbf{i}}$, or $-80\hat{\mathbf{i}}$ lb, of $\mathbf{F}$ at A produces moments about lines through Q parallel to y and to z. These are $80(10)$ lb-ft about axis z_Q and $80(12)$ lb-ft about y_Q, as shown in Figures E3.8b,c on the next page. Thus the contribution of $F_x \hat{\mathbf{i}}$ to $\mathbf{M}_Q$ is $960\hat{\mathbf{j}} - 800\hat{\mathbf{k}}$ lb-ft. The reader should note how the unit vector is attached by using the right-hand rule. If the fingers curl in the direction of the turning effect about an axis through Q, the thumb will aim in the direction of the moment and the unit vector in this direction is then written down.

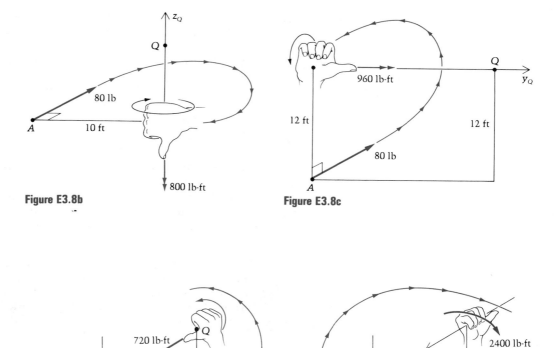

Figure E3.8b

Figure E3.8c

Figure E3.8d

Figure E3.8e

Next, the component $F_y\hat{\mathbf{j}}$, or $60\hat{\mathbf{j}}$ lb, of $\mathbf{F}$ at A produces a moment about x_Q but not about y_Q or z_Q.

> **Question 3.11**   Why not y_Q or z_Q?

The contribution is seen in Figure 3.8d to be $60(12)$ lb-ft about axis x_Q. Therefore, the contribution of $F_y\hat{\mathbf{j}}$ to $\mathbf{M}_Q$ is $720\hat{\mathbf{i}}$ lb-ft.

Finally, the third component $F_z\hat{\mathbf{k}}$, or $240\hat{\mathbf{k}}$ lb, of $\mathbf{F}$ at A also produces a moment about just one of the axes (x_Q, y_Q, z_Q) through point Q, and this axis is again x_Q. The moment contribution is seen in Figure E3.8e to be $240(10)$ lb-ft about x_Q, and so $F_z\hat{\mathbf{k}}$ adds $-2400\hat{\mathbf{i}}$ lb-ft to $\mathbf{M}_Q$. The total moment of $\mathbf{F}$ about Q is then

$$\mathbf{M}_Q = (960\hat{\mathbf{j}} - 800\hat{\mathbf{k}}) + (720\hat{\mathbf{i}}) + (-2400\hat{\mathbf{i}}) \text{ lb-ft}$$
$$= -1680\hat{\mathbf{i}} + 960\hat{\mathbf{j}} - 800\hat{\mathbf{k}} \text{ lb-ft}$$

Answer 3.11 $F_y\hat{\mathbf{j}}$ at A is parallel to y_Q and it intersects z_Q; thus, no moment about either of these lines!

The reader is encouraged to check this answer by finding and adding the moments about Q of F_x, F_y, and F_z placed at B instead of A and also by computing either $\mathbf{r}_{QA} \times \mathbf{F}$ or $\mathbf{r}_{QB} \times \mathbf{F}$. Comparing the two approaches, you will probably observe some loss of physical feeling for the moment when using the cross product.

The idea of "force times perpendicular distance" symbolized by Equation (3.11) and utilized in the preceding example can also be used to compute the moment about a *line* through a point. If in Example 3.8 we had been seeking the moment about axis y_Q, then only the second of the four figures would have been of interest to us because $\mathbf{F}_z$ passes through axis y_Q, and $\mathbf{F}_y$ is parallel to it; thus neither of these components contributes to the moment of $\mathbf{F}$ about axis y_Q. The answer would then have been simply the y-component of $\mathbf{M}_Q$, namely $\mathbf{M}_{y_Q} = 960\hat{\mathbf{j}}$ lb-ft.

> **Question 3.12**   If in Example 3.8 we were using "force components (at A) times perpendicular distances" to find the moment about the line through point D parallel to the y-axis, how many products would be needed? Is there a better point than A to decompose the force in order to find this moment?

Answer 3.12   Two, because this time $\mathbf{F}_z$ doesn't pass through the line. Yes, point B, which lies *on* the line of interest so that by inspection $\mathbf{M}_D \cdot \hat{\mathbf{j}}$ vanishes.

EXAMPLE 3.9

Rework Example 3.4, using the ideas expressed by Equation (3.11) in conjunction with the preceding discussion. We are seeking the moment of the 20-lb chain force about the x-axis (see Figure E3.9a).

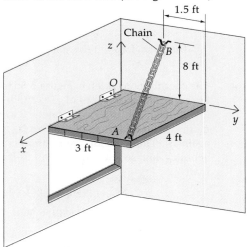

Figure E3.9a

Solution

We are again seeking the moment of the 20-lb chain force **F** at A, about the x-axis. The components of **F** at A were computed in Example 3.4 and are shown in Figure E3.9b. We see immediately that neither the x- nor y-components exert any moment (turning effect) about the x-axis. This is because the x-component of **F** is parallel to the line, and the y-component intersects it. Thus the moment about the x-axis is simply

$$\mathbf{M}_{x\text{-}axis} = (17.6)3\hat{\mathbf{i}} = 52.8\hat{\mathbf{i}} \text{ lb-ft,}$$

as before.

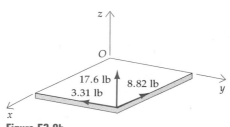

Figure E3.9b

PROBLEMS ► **Section 3.3**

3.16 The 200-N force in Figure P3.16 lies in a plane parallel to xz. Find the moment of this force about the axis of the vertical shaft (z).

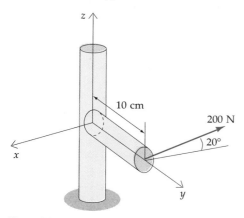

Figure P3.16

3.17 Recompute the moment $\mathbf{M}_{CD}$ in Example 3.6, using both of the suggestions following that example, namely, using $\mathbf{r}_{CB}$ instead of $\mathbf{r}_{CA}$ and $\mathbf{r}_{DB}$ instead of $\mathbf{r}_{DA}$.

3.18 Find the moment of the 21-lb force in Figure P3.18 about the line OD.

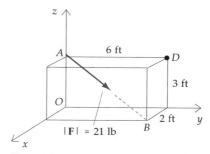

Figure P3.18

3.19 In Example 3.8, find the moment of **F** about line PQ using:

   a. $[(\mathbf{r}_{PA} \times \mathbf{F}) \cdot \hat{\mathbf{k}}]\hat{\mathbf{k}}$
   b. $[(\mathbf{r}_{PB} \times \mathbf{F}) \cdot \hat{\mathbf{k}}]\hat{\mathbf{k}}$
   c. $[(\mathbf{r}_{QA} \times \mathbf{F}) \cdot \hat{\mathbf{k}}]\hat{\mathbf{k}}$
   d. $[(\mathbf{r}_{QB} \times \mathbf{F}) \cdot \hat{\mathbf{k}}]\hat{\mathbf{k}}$

The answer should each time be $-800\hat{\mathbf{k}}$ lb-ft, which was the z-component of $\mathbf{M}_Q$ in Example 3.8.

3.20 Find the moment of the force in Problem 3.11 about the line that passes through points $y = 12$ ft on the y axis and $z = 5$ ft on the z axis.

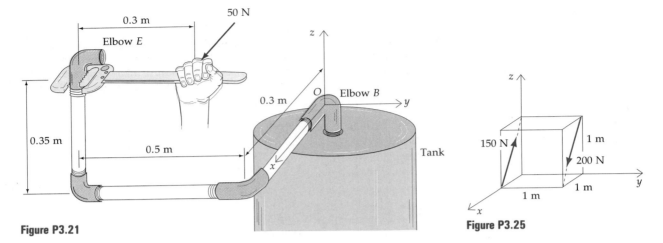

Figure P3.21

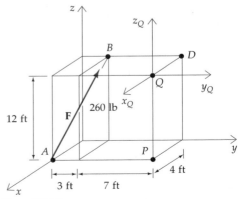

Figure P3.25

3.21 A do-it-yourselfer is (in a most unwise fashion) tightening an elbow onto a length of pipe. (See Figure P3.21.) Find the moment of the 50-N force about the z-axis through point O, and note from the answer that the "plumber" is actually exerting a force that is tending to unscrew the elbow B from the top of the tank.

3.22 Using the result of Problem 3.10, find the moment of the 60-N force about the y-axis (noting that it is a line through point C). (See Figure P3.10.)

3.23 In Problem 3.8, determine the shortest distance between the 50-lb force and the line through point P that makes equal angles with the coordinate axes.

3.24 With no writing, make a rough estimate of the moment of the 200-N force **F** about the line CE in Figure P3.24. Then calculate the moment and compare it with your estimate.

3.25 In Figure P3.25 find the moment of the two forces about:

a. The point $(x, y, z) = (1, 2, 3)$ m

b. The line defined by the intersection of the plane $x = 1$ m with the plane $z = 3$ m.

3.26 Find the moment of the 2000-N force in Figure P3.26 about the diagonal line AB.

Figure P3.26

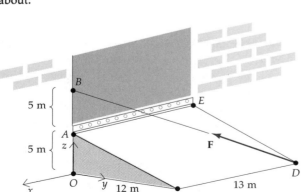

Figure P3.24

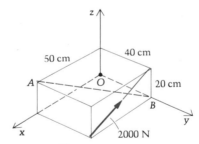

Figure P3.27

3.27 Find the moment of the force **F** in Example 3.8 about the line through point D parallel to the z-axis. (See Figure P3.27 on the preceding page.)

3.28 Using the result of Example 3.8, find the shortest distance between the force **F** and

   a. point Q;
   b. axis z_Q.

3.29 Determine the moment of the 180-lb force in Figure P3.29 about the line AC.

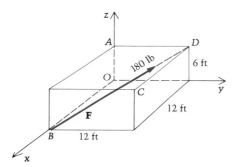

Figure P3.29

3.31 In Figure P3.31 determine the moment of the 280-lb force **F** with respect to:

   a. Point A
   b. Line OA
   c. Point B
   d. Line BR.

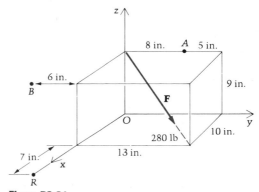

Figure P3.31

3.30 Find the moment of the force in Figure P3.30 about point C. Then find the moment about the line that passes through C and:

   a. Point A
   b. Point B
   c. Point D.

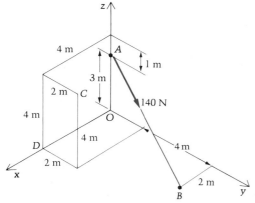

Figure P3.30

3.32 A particle located at $\mathbf{r}_{OP} = -\hat{\mathbf{i}} + 3\hat{\mathbf{j}} - 8\hat{\mathbf{k}}$ m is acted upon by the two forces

$$\mathbf{F}_1 = -2\hat{\mathbf{i}} + 3\hat{\mathbf{j}} - \hat{\mathbf{k}} \text{ N and } \mathbf{F}_2 = 7\hat{\mathbf{i}} + \hat{\mathbf{j}} - \hat{\mathbf{k}} \text{ N}$$

Find the moment of the sum of these forces about the z axis.

3.33 Find the moment of the 21-lb force **F** in Figure P3.33 about line AB by (a) using $(\mathbf{r}_{AP} \times \mathbf{F}) \cdot \hat{\mathbf{u}}_{AB}$, and (b) resolving **F** into its components at P and finding the moments of each about the line with the help of the right-hand rule.

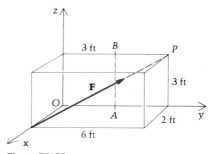

Figure P3.33

3.34 Find the moment of the force in Figure P3.34 with respect to:

   a. Line BC

   b. Point A

   c. Line BA.

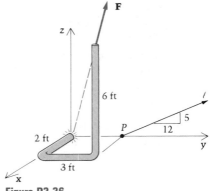

Figure P3.36

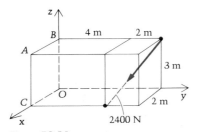

Figure P3.34

3.35 Find (in Figure P3.35):

   a. The direction cosines associated with the 39-lb force

   b. The 39-lb force expressed in terms of the unit vectors $\hat{\mathbf{i}}$, $\hat{\mathbf{j}}$, and $\hat{\mathbf{k}}$

   c. The moment of the 39-lb force about point A

   d. The moment of the 39-lb force about a line from A to B

   e. The moment of the 39-lb force about a line from A to D.

3.37 The force $\mathbf{F}$ of magnitude 10 lb in Figure P3.37 acts through P in the direction of the unit vector $\hat{\mathbf{e}}_F = 0.8\hat{\mathbf{i}} + 0.6\hat{\mathbf{j}}$. Determine:

   a. The moment of $\mathbf{F}$ about the origin O

   b. The moment of $\mathbf{F}$ about the z axis

   c. The direction cosines of the line through O about which $\mathbf{F}$ has the largest moment, and the value of this moment.

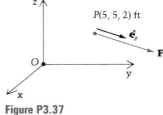

Figure P3.37

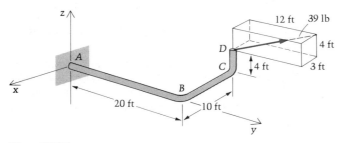

Figure P3.35

3.36 A bent bar is rigidly attached to a wall at the point $(0, 0, 0)$. A force with magnitude $|\mathbf{F}| = 7$ lb acts at its free end with a line of action passing through the origin, as shown in Figure P3.36. Find:

   a. The moment of $\mathbf{F}$ about point P

   b. The moment about the line ℓ passing through P with a slope of $5/12$ in the yz plane as shown.

3.38 All coordinates in Figure P3.38 are given in inches. Find the moment of the 340-lb force with respect to:

   a. Point A

   b. Line AB.

   c. At any point P on the line of action of the force, it may be resolved into components $\mathbf{F}_{\parallel}$ and $\mathbf{F}_{\perp}$, parallel and perpendicular to line AB. Give the shortest distance between $\mathbf{F}_{\perp}$ and the line AB (which is independent of the choice of P).

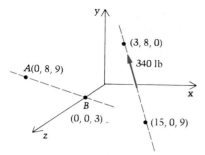

Figure P3.38

* **3.39** Find the moment of the force in Figure P3.39 about

    a. the point A;

    b. the line defined by:

$$\frac{x-1}{2} = \frac{y-1}{2} = \frac{z-0}{-4}$$

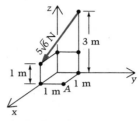

Figure P3.39

3.4 The Couple

Definition and Moment of a Couple

A pair of forces, equal in magnitude but having opposite directions and different (but, of course, parallel) lines of action, constitutes a **couple**. The pair of forces exerts no net thrust on a body, but there is obviously a turning effect. Hence the moment of the couple about a point is of great importance. Referring to Figure 3.12, we let A and B be points on the lines of action of a "couple" of forces $-\mathbf{F}$ and $\mathbf{F}$, respectively. By the moment of the couple we mean the sum of the moments of the two forces. Thus the moment $\mathbf{M}_P$ about point P is

$$\mathbf{M}_P = \mathbf{r}_{PB} \times \mathbf{F} + \mathbf{r}_{PA} \times (-\mathbf{F})$$
$$= (\mathbf{r}_{PB} - \mathbf{r}_{PA}) \times \mathbf{F} = \mathbf{r}_{AB} \times \mathbf{F} \quad\quad (3.12)$$

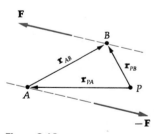

Figure 3.12

We see from Equation (3.12) that it makes no difference where A and B lie on the respective lines of action, since the cross product produces a vector whose magnitude is that of $\mathbf{F}$ multiplied by the distance d between the lines of action of the two forces. Moreover, recalling the direction associated with the cross product, we see from Figure 3.13 that $\mathbf{M}_P$ is perpendicular to the plane defined by the lines of action of the two forces and in the direction with which we would associate, by the right-hand rule, the sense of turning of the two forces. In Figure 3.13 the fingers show the turning effect of $\mathbf{F}$ and $-\mathbf{F}$ in their plane while the thumb points in the direction of the (axis of turning of the) couple.

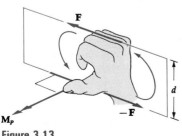

Figure 3.13

Question 3.13   In the above derivation, is point P required to lie in the plane containing the lines of action of $\mathbf{F}$ and $-\mathbf{F}$?

The Most Important Property of a Couple

If we equate the magnitudes of the two sides of Equation (3.12), we obtain a very useful result:

$$|\mathbf{M}_P| = |\mathbf{F}| d \qquad (3.13)$$

where, again, d is the distance between the lines of action of $\mathbf{F}$ and $-\mathbf{F}$ (see Figure 3.13). This simple scalar equation states that the magnitude of the moment of a couple (or its *strength*) is the magnitude of either force times the perpendicular distance between them.

In vector form,

$$\mathbf{M}_P = |\mathbf{F}| d\hat{\mathbf{u}} \qquad (3.14)$$

where $\hat{\mathbf{u}}$ is the unit vector in the direction of $\mathbf{M}_P$ (normal to the plane of the two forces, along the "right thumb" of Figure 3.13).

In Equation 3.12, all reference to point P has been lost; therefore, $\mathbf{M}_P$ does not depend in any way upon the location of point P. Thus, *the moment of a couple about every point is the same.* This is an extremely powerful property. For example, the moment of the couple in Figure 3.14 is 20 N · m counterclockwise or, in vector terms, directed out of the page. If the page is the xy-plane as shown, the moment of the couple is $20\hat{\mathbf{k}}$ N · m, which is its moment about the point $(0, 0, 0)$, or the point $(1, 2, 3)$, or *any other point of space!*

Furthermore, as far as the moment of a couple is concerned, its forces and the distance between them may be altered as desired, so long as the product $|\mathbf{F}| d$ and the couple's direction (right thumb in Figure 3.13) remain the same. The various couples in Figure 3.15 below thus have the same moment as did the couple of Figure 3.14:

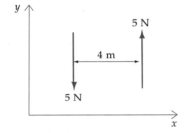

Figure 3.14

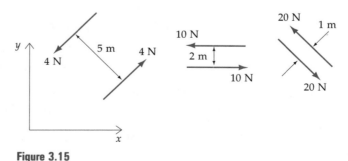

Figure 3.15

Answer 3.13   No. Nothing in the derivation of Equation (3.12) placed any restriction whatsoever on the location of point P.

The next four examples illustrate how to obtain the moment of a couple.

EXAMPLE 3.10

A mechanic applies two forces, each one of magnitude 20 lb as shown in Figure E3.10a, to a lug wrench in the process of changing a tire. Find the moment of the couple comprising the equal-magnitude, oppositely directed forces.

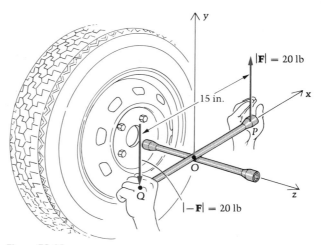

Figure E3.10a

Solution

The moment of the couple is calculated as the sum of the moments of **F** and −**F** about *any* point. Let us use point Q (at the mechanic's left hand (see Figure E3.10b):

$$\mathbf{C} = \mathbf{r}_{QQ} \times (-\mathbf{F}) + \mathbf{r}_{QP} \times \mathbf{F}$$
$$= 0 + 15\hat{\mathbf{i}} \times 20\hat{\mathbf{j}}$$
$$= 300\hat{\mathbf{k}} \text{ lb-in.}$$

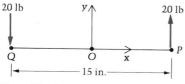

Figure E3.10b

Alternatively, using point O,

$$\mathbf{C} = -7.5\hat{\mathbf{i}} \times (-20\hat{\mathbf{j}}) + 7.5\hat{\mathbf{i}} \times 20\hat{\mathbf{j}}$$
$$= +150\hat{\mathbf{k}} + 150\hat{\mathbf{k}}$$
$$= 300\hat{\mathbf{k}} \text{ lb-in.} \quad \text{(as before)}$$

Regardless of the point chosen, **C**, as we have proved, always will be the same vector. We could also simply multiply either force times the perpendicular distance between them, obtaining:

$$\mathbf{C} = 20(15)\hat{\mathbf{k}} = 300\hat{\mathbf{k}} \text{ lb-in.} \quad \text{(once again)}$$

where we attach the unit vector $\hat{\mathbf{k}}$ in accordance with the direction in which the forces are perceived to turn about an axis normal to their plane.

EXAMPLE 3.11

In Example 3.5, the two 20-lb forces perpendicular to the paper, down at A and up at B, are now seen to form a couple (see Figure E3.11a). Use the property that a couple has the same moment about any point of space to rework the previous example, i.e., to find the moment of the two forces about the axis (x) of the bolt.

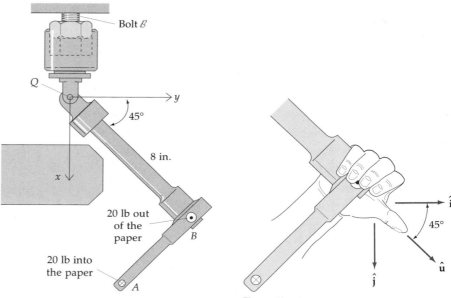

Figure E3.11a   Figure E3.11b

Solution

Using Equation (3.14),

$$\mathbf{C} = |\mathbf{F}| \, d\hat{\mathbf{u}} = 20(5)\left(\frac{\hat{\mathbf{i}} + \hat{\mathbf{j}}}{\sqrt{2}}\right)$$

$$= \frac{100}{\sqrt{2}}\,(\hat{\mathbf{i}} + \hat{\mathbf{j}})$$

where Figure E3.11b gives the unit vector $\hat{\mathbf{u}}$ by the right-hand rule.

The vector $\mathbf{C}$ is the moment of the couple about any point, in particular about Q. Thus

$$\mathbf{M}_{x\text{-}axis} = (\mathbf{M}_Q \cdot \hat{\mathbf{i}})\,\hat{\mathbf{i}} = \left[\frac{100}{\sqrt{2}}\,(\hat{\mathbf{i}} + \hat{\mathbf{j}}) \cdot \hat{\mathbf{i}}\right]\hat{\mathbf{i}} = \frac{100}{\sqrt{2}}\,\hat{\mathbf{i}} = 70.7\hat{\mathbf{i}} \text{ lb-in,}$$

a *much* simpler solution than was possible before!

EXAMPLE 3.12

The two forces shown in Figure E3.12 have magnitudes of 50-N and are oppositely directed. Find the moment of the couple that they constitute.

Solution

As in the previous example, we compute the moment of the couple by adding the moments of the two forces. We can add them about any point, and we choose the origin:

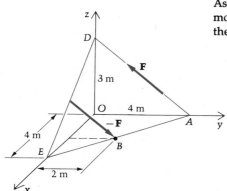

Figure E3.12

$$C = \mathbf{r}_{OA} \times \mathbf{F} + \mathbf{r}_{OB} \times (-\mathbf{F})$$

$$= (\mathbf{r}_{OA} - \mathbf{r}_{OB}) \times \mathbf{F}$$

$$= [4\hat{j} - (2\hat{i} + 2\hat{j})] \times 50\left(\frac{-4\hat{j} + 3\hat{k}}{5}\right)$$

$$= (-2\hat{i} + 2\hat{j}) \times 50(-0.8\hat{j} + 0.6\hat{k})$$

$$= 50(1.2\hat{i} + 1.2\hat{j} + 1.6\hat{k})$$

$$= 60\hat{i} + 60\hat{j} + 80\hat{k} \text{ N} \cdot \text{m}$$

As a check, let us form the moment of the couple, $\mathbf{C}$, by adding the moments of $\mathbf{F}$ and $-\mathbf{F}$ about A:

$$C = \mathbf{r}_{AA} \times \mathbf{F} + \mathbf{r}_{AB} \times (-\mathbf{F})$$

$$= 0 + (2\hat{i} - 2\hat{j}) \times 50\left(\frac{4\hat{j} - 3\hat{k}}{5}\right)$$

$$= \tfrac{50}{5}(8\hat{k} + 6\hat{j} + 0 + 6\hat{i})$$

$$= 60\hat{i} + 60\hat{j} + 80\hat{k} \text{ N} \cdot \text{m}$$

as before. We note again that the couple has the same moment about *all* points.

Working the above example using $|\mathbf{F}| \, d\hat{u}$ (Equation 3.14) is *much* harder (see Problem 3.56). Usually, when a couple's forces are not parallel to coordinate axes or do not lie in a coordinate plane (and the ones above are neither!), the use of vectors to compute its moment is easier.

EXAMPLE 3.13

One of the two forces that constitute a couple is $\mathbf{F} = 3\hat{i} - 4\hat{j} + 5\hat{k}$ lb, having a line of action that passes through the point A at (0, 6, 5) ft. The other force has a line of action that passes through point B at (−4, 0, 2) ft. Find the moment of the couple, and the distance between the lines of action of the forces.

Solution

The moment, $\mathbf{C}$, of the couple is the sum of the moments of its two forces about any point. Choosing B as the point,

$$C = \mathbf{r}_{BA} \times \mathbf{F} + \mathbf{r}_{BB} \times (-\mathbf{F})$$

$$= \mathbf{r}_{BA} \times \mathbf{F}$$

because A is on the line of action of $\mathbf{F}$ and B is on the line of action of the companion force $-\mathbf{F}$. The position vector $\mathbf{r}_{BA}$ is computed as

$$\mathbf{r}_{BA} = [0 - (-4)]\hat{\mathbf{i}} + (6 - 0)\hat{\mathbf{j}} + (5 - 2)\hat{\mathbf{k}}$$
$$= 4\hat{\mathbf{i}} + 6\hat{\mathbf{j}} + 3\hat{\mathbf{k}} \text{ ft}$$

Therefore

$$\mathbf{C} = (4\hat{\mathbf{i}} + 6\hat{\mathbf{j}} + 3\hat{\mathbf{k}}) \times (3\hat{\mathbf{i}} - 4\hat{\mathbf{j}} + 5\hat{\mathbf{k}})$$
$$= 4(-4)\hat{\mathbf{k}} + 4(5)(-\hat{\mathbf{j}}) + 6(3)(-\hat{\mathbf{k}}) + 6(5)\hat{\mathbf{i}} + 3(3)\hat{\mathbf{j}} + 3(-4)(-\hat{\mathbf{i}})$$
$$= 42\hat{\mathbf{i}} - 11\hat{\mathbf{j}} - 34\hat{\mathbf{k}} \text{ lb-ft}$$

The distance between the lines of action is, using Equation (3.13),

$$d = \frac{|\mathbf{C}|}{|\mathbf{F}|}$$

$$= \frac{\sqrt{(42)^2 + (11)^2 + (34)^2}}{\sqrt{(3)^2 + (4)^2 + (5)^2}} = \frac{\sqrt{3041}}{\sqrt{50}}$$

$$= 7.80 \text{ ft}$$

In the remainder of Chapter 3, we shall extend our study to include systems of forces and couples, in preparation for solving equilibrium problems in Chapters 4, 5, and 6 which will involve moments and thus be much more difficult than those of Chapter 2. We shall first present the full equations of equilibrium of a body, which will help us both motivationally and practically. Motivationally, the equilibrium equations show the importance of mastering force and moment relationships. Practically, the equilibrium equations will be used in our definition of equipollent force systems. Only after these ideas are mastered will we be fully prepared to solve problems involving the equilibrium of bodies.

PROBLEMS ▶ Section 3.4

Write the vector expressions for the moments of the couples in the following three problems (the forces are in the planes labeled in Figures P3.40–P3.42).

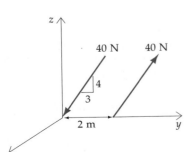

Figure P3.40

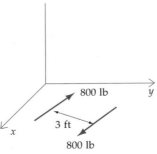

Figure P3.41

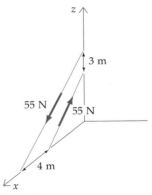

Figure P3.42

3.43 The radius of each of the two pulleys in Figure P3.43 is 1 ft. Determine the resultant moment of the two pulley tension forces about (a) point A and (b) any other point.

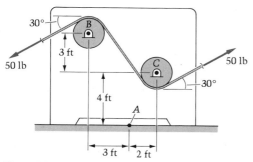

Figure P3.43

3.44 Friction is causing a set of uniformly distributed tangential forces of intensity 300 lb/in. to act on the circular ring of radius 16 in. shown in Figure P3.44. Determine the moment of these forces about the center, O, of the ring. What is the moment about A?

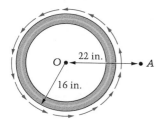

Figure P3.44

3.45 Determine the moment of the couple in Figure P3.45 about (a) point O, (b) point A, (c) the y axis, and (d) the line in the yz plane defined by $z = 4$ ft.

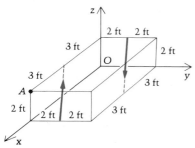

Each force has magnitude 265 lb

Figure P3.45

3.46 Find the moment of the couple about line AB in Figure P3.46. The lines of action of the forces are both in the yz plane.

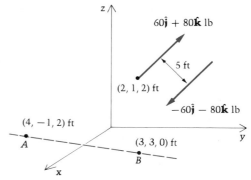

Figure P3.46

3.47 Write the moment of the couple formed by the two 30-lb forces shown in Figure P3.47. What is the moment of the couple about the point $(x, y, z) = (1, 5, -8)$?

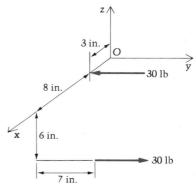

Figure P3.47

3.48 After the couple was defined and analyzed, it was found to have (1) the same moment about every point, and (2) no net thrust (or force). Show that the set of three forces shown in Figure P3.48 possesses these same two properties.

Figure P3.48

In each of the following four problems, show that the system of forces has no net thrust (or force). In Section 3.5 we shall see that whenever this happens, the moment of the forces about all points is the same. Find the moment about A and B in each problem and observe that they are equal.

3.49

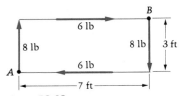

Figure P3.49

3.50

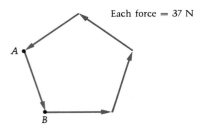

Pentagon, side 2 m
Figure P3.50

3.51

Hexagon, side 2 ft

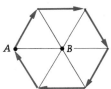

Each force = 5 lb
Figure P3.51

3.52   Sixteen forces of 14 lb each at 22.5° spacing on a 6-ft diameter circle. See Figure P3.52.

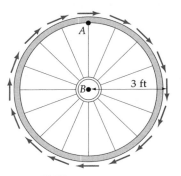

Figure P3.52

3.53   The two parallel forces in Figure P3.53 each have magnitude $10\sqrt{3}$ N. Find the (vector) moment of this system of forces about:

 a. point A;
 b. point B, which lies at $(10, -6, 12)$ m;
 c. line AB.

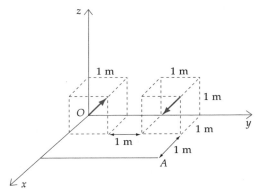

Figure P3.53

3.54   Force **F** in Figure P3.54 has magnitude $|\mathbf{F}| = 15$ lb, and couple **M** has magnitude $4\sqrt{41}$ lb-in. and is normal to plane ACD. Find the moment of the system of **F** and **M** with respect to point D. Then determine the moment of the system about line CD.

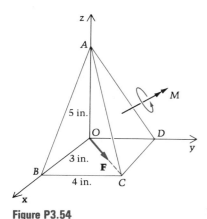

Figure P3.54

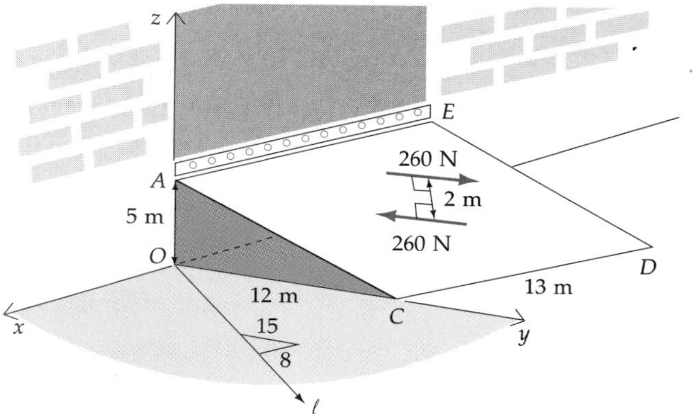

Figure P3.55

3.55 The two 260-N forces lie in the inclined plane *AEDC* shown in Figure P3.55. Find the moment of the couple, which they constitute, about the indicated line *l* in the *xy* plane.

* **3.56** Find the moment of the couple in Example 3.12 by means of $\mathbf{M} = |\mathbf{F}| \, d\hat{\mathbf{u}}$, where $\hat{\mathbf{u}}$ is the unit vector in the direction of the couple (normal to the plane *ADE*). Hint: Crossing $\mathbf{r}_{EA}$ into $\mathbf{r}_{ED}$ gives a vector in that direction; then, dividing the result by its magnitude yields $\hat{\mathbf{u}}$. Finding the distance *d* is also tricky.

3.5 Laws of Equilibrium: Relationship Between Sums of Moments

The Equilibrium Equations

The purposes of this brief section are (a) to provide motivation for the topics covered in the remainder of the chapter, and (b) to establish a useful relationship having to do with the moments that a system of forces produce about different points in space. In particular, we need to clearly identify those characteristics of a force system that are essential to the analysis of problems in statics. In our view, the easiest way to do this is by displaying the equations of equilibrium; we are of course already familiar with the first of these from our studies in the preceding chapter.

When none of the particles making up a body is accelerating relative to an inertial frame of reference, Euler's laws (Section 1.3) are reduced to the statements (1) that the sum of the external forces vanishes, and (2) that the sum of their moments about any point also vanishes. A body that is stationary in an inertial frame clearly falls into this category. Letting the uppercase Greek letter sigma (Σ) indicate the process of summation, the two laws, called the **equilibrium equations,** are written symbolically as

$$\Sigma \mathbf{F} = \mathbf{0} \tag{3.15}$$

$$\Sigma \mathbf{M}_P = \mathbf{0} \tag{3.16}$$

where P denotes the point with respect to which the moments are calculated.*

In Chapter 2, we only needed Equation (3.15). That is because we were then solving only problems for which the "moment equation" (3.16) was not required. All we needed to know about forces in Chapter 2 was how to express them as vectors and sum them. A second characteristic of a force now comes into play with the second equilibrium equation (3.16), and that is the location of its line of action. As we know from our study of moments of forces earlier in this chapter, however, the specific point of application (along the line of action) is not important in statics.

Question 3.14   Why not?

Question 3.15   What about a couple acting on the body: How is it to be incorporated into the equations?

Relationship Between Sums of Moments

We shall not actually use the equilibrium Equations (3.15) and (3.16) together in problems until Chapter 4. But it is important to recognize the operations we shall then have to perform on a system of forces†; namely, we must sum the forces as we did in Chapter 2, and now we must in addition sum the moments of the forces. These very same operations are required in dynamics, where we analyze the motion of an accelerating body. Of immediate importance to us is the fact that, for a given system of forces, the sums of moments with respect to two different points are related in a particularly simple way. To establish this relationship, let the body (Figure 3.16) be acted upon by a number (N_F) of forces and a number (N_C) of couples. Let $\mathbf{F}_i$ be the ith force, for which the point of application is A_i (which actually could be *any* point on the line of action of $\mathbf{F}_i$). Also, let $\mathbf{C}_j$ be the moment of the jth couple. Recall from Section 3.4 that the moment of a couple about every point is the same. To distinguish, in figures such as Figure 3.16, couples from single forces we use

---

* Somewhat surprisingly the word "equilibrium," having a connotation of balance, does not have a universally accepted technical definition in the literature of mechanics. Some writers associate it with a force system, one that satisfies Equations (3.15) and (3.16); other writers give it the kinematic definition of stationarity of the body in an inertial frame. Still other writers define it by "stationarity" of the body in some, not necessarily inertial, frame of reference. The important point, as we study statics, is that the equilibrium equations are necessary conditions for a body to be stationary in an inertial frame of reference. In this text, whenever we refer to a "body in equilibrium" the reader should take that to mean that the body is stationary in an inertial frame.

† We use the phrases "system of forces" and "force system" to denote any collection of forces and/or couples of interest to us.

Answer 3.14   As we have seen, the moment of a force is the same regardless of its point of application on the line of action.

Answer 3.15   The couple may be ignored in $\Sigma \mathbf{F} = 0$ because its equal-magnitude, oppositely directed forces cancel. To the left-hand side of $\Sigma \mathbf{M}_P = 0$, we add the moments about P of the two forces constituting the couple (or add the moment of the couple directly, if we know it). Since the moment of a couple is the same about any point, any convenient point may be used to find the moment.

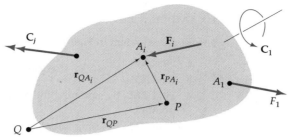

Figure 3.16

double-headed arrows, or sometimes curved arrows indicating sense of turning, to denote moments of couples.

Letting $\Sigma \mathbf{M}_Q$ denote the sum of the moments about Q of all the forces and couples in the system, we have

$$\Sigma \mathbf{M}_Q = \sum_{i=1}^{N_F} \mathbf{r}_{QA_i} \times \mathbf{F}_i + \sum_{j=1}^{N_C} \mathbf{C}_j \qquad (3.17)$$

But

$$\mathbf{r}_{QA_i} = \mathbf{r}_{QP} + \mathbf{r}_{PA_i}$$

so that, making the substitution in Equation (3.17),

$$\Sigma \mathbf{M}_Q = \sum_{i=1}^{N_F} (\mathbf{r}_{QP} + \mathbf{r}_{PA_i}) \times \mathbf{F}_i + \sum_{j=1}^{N_C} \mathbf{C}_j$$

$$= \sum_{i=1}^{N_F} \mathbf{r}_{QP} \times \mathbf{F}_i + \underbrace{\sum_{i=1}^{N_F} \mathbf{r}_{PA_i} \times \mathbf{F}_i + \sum_{j=1}^{N_C} \mathbf{C}_j}$$

$$\Sigma \mathbf{M}_Q = \underbrace{\mathbf{r}_{QP} \times \Sigma \mathbf{F}} \quad + \quad \Sigma \mathbf{M}_P \qquad (3.18)$$

Equation (3.18) is very important because it tells us that the sum of the moments about one point is completely determined by the sum of the forces, the sum of the moments about a second point, and the relative locations of the two points. In particular we note that if Equations (3.15) and (3.16) both hold for a force system, then $\Sigma \mathbf{M}_Q$ is also zero for *any point* Q. We shall use Equation (3.18) a number of times in the remainder of this chapter.

> **Question 3.16**   In view of Equation (3.18), would there be a system of forces for which the moment is (a) the same at all points? (b) zero at all points?
>
> **Question 3.17**   Why is Equation (3.18) also valid when the body on which the forces and/or couples act is accelerating (i.e., in dynamics)?

Answer 3.16   (a) Yes, if and only if $\Sigma \mathbf{F} = 0$. (b) Yes, if and only if $\Sigma \mathbf{F} = 0$ *and* $\Sigma \mathbf{M}_P = 0$ at some one point P.

Answer 3.17   Because we didn't use the equilibrium equations (3.15, 3.16) in developing it. Equation (3.18) is just a useful result valid for any force system.

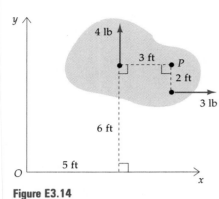

Figure E3.14

EXAMPLE 3.14

For the force system in Figure E3.14,

  a. Find $\Sigma \mathbf{F}$;

  b. Find $\Sigma \mathbf{M}_P$;

  c. Find $\Sigma \mathbf{M}_O$ by summing the moments of the two forces with respect to point O;

  d. Find $\Sigma \mathbf{M}_O$ by using Equation (3.18) together with the results from (a, b) above.

Solution

  a. Adding the two forces,

$$\Sigma \mathbf{F} = 3\hat{\mathbf{i}} + 4\hat{\mathbf{j}} \text{ lb}$$

  b.
$$\Sigma \mathbf{M}_P = (3 \text{ lb})(2 \text{ ft})\hat{\mathbf{k}} + (4 \text{ lb})(3 \text{ ft})(-\hat{\mathbf{k}})$$
$$= -6\hat{\mathbf{k}} \text{ lb-ft}$$

or, with vectors,

$$\Sigma \mathbf{M}_P = -2\hat{\mathbf{j}} \times 3\hat{\mathbf{i}} + (-3\hat{\mathbf{i}}) \times 4\hat{\mathbf{j}}$$
$$= 6\hat{\mathbf{k}} - 12\hat{\mathbf{k}} = -6\hat{\mathbf{k}} \text{ lb-ft}$$

  c.
$$\Sigma \mathbf{M}_O = (4 \text{ lb})(5 \text{ ft})\hat{\mathbf{k}} + (3 \text{ lb})(4 \text{ ft})(-\hat{\mathbf{k}})$$
$$= 8\hat{\mathbf{k}} \text{ lb-ft}$$

  d.
$$\Sigma \mathbf{M}_O = \Sigma \mathbf{M}_P + \mathbf{r}_{OP} \times \Sigma \mathbf{F}$$
$$= -6\hat{\mathbf{k}} + (8\hat{\mathbf{i}} + 6\hat{\mathbf{j}}) \times (3\hat{\mathbf{i}} + 4\hat{\mathbf{j}})$$
$$= -6\hat{\mathbf{k}} + 32\hat{\mathbf{k}} - 18\hat{\mathbf{k}}$$
$$= -8\hat{\mathbf{k}} \text{ lb-ft, as above.}$$

PROBLEMS ▶ Section 3.5

3.57 To the system of two forces in Example 3.14, add a third force of $5\hat{\mathbf{j}}$ lb along the line $x = 2$ ft. Answer the same four questions in the example for this three-force system.

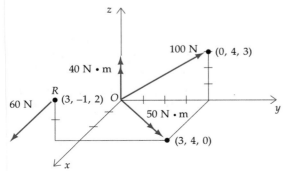

Figure P3.58

3.58 For the system shown in Figure P3.58, find:

  a. $\Sigma \mathbf{F}$;

  b. $\Sigma \mathbf{M}_R$ by summing the couples and the moments of forces about R;

  c. $\Sigma \mathbf{M}_O$ by summing the couples and the moments of forces about O;

  d. $\Sigma \mathbf{M}_O$ by using (a), (b), and Equation (3.18). The result must of course agree with that of (c).

3.59 A body $\mathcal{B}$ is subjected to five forces acting at points of an inertial reference frame given in the table on the next page.

  a. Find $\Sigma \mathbf{F}$ and $\Sigma \mathbf{M}_P$, where P is the origin.

  b. Show that $\mathcal{B}$ is not in equilibrium.

  c. Add forces at $(1, 0, 0)$ m and $(4, 0, 0)$ m that would allow $\mathcal{B}$ to be in equilibrium.

Force (N)	Position Vector to Point of Application (m)
1. $2\hat{\imath} + 3\hat{\jmath} - 5\hat{k}$	$-\hat{\imath} + \hat{\jmath}$
2. $7\hat{\imath} - 2\hat{k}$	$\hat{\jmath} - \hat{k}$
3. $7\hat{\imath} - 2\hat{k}$	$\hat{\imath} - \hat{\jmath}$
4. $-\hat{\imath} - 3\hat{\jmath} + 5\hat{k}$	$\hat{\imath} - \hat{k}$
5. $-14\hat{\imath} + 4\hat{k}$	$2\hat{\jmath} - \hat{k}$

3.60 Find the moments of each of the five forces in the preceding problem about the point Q at (x, y, z) $= (1, 2, 3)$ m. Add them to obtain $\Sigma \mathbf{M}_Q$. Then illustrate Equation (3.18) by obtaining the same answer for $\Sigma \mathbf{M}_Q$ by using $\Sigma \mathbf{F}$ and $\Sigma \mathbf{M}_P$ (from the preceding problem) in this equation.

3.6 Equipollence of Force Systems

The Meaning of Equipollence

Two different systems of forces are said to be **equipollent** (meaning "of equal power, or strength") if they make the same contributions to the equations of equilibrium, (3.15) and (3.16). They will also make the same contributions to the corresponding equations of motion for a body that is *not* in equilibrium. Equipollence, then is a special kind of equivalence in which two force systems exert the same net push (or pull) on a body and also exert the same net turning action (moment).*

In general, a body will not respond in the same way to a force system S_1 as it will to an equipollent force system S_2, the exception being a rigid body whose responses to two such systems are indistinguishable. However, we are *not restricting our discussion to rigid bodies* because it is only the contributions to equilibrium equations that concern us here.

A simple example of equipollent forces is shown in Figure 3.17 where we can compare the processes of pushing and towing an automobile. The effects of the two 500-lb forces on the bumpers of the automobile are of course quite different. However, if the forces have the same line of action, they are equipollent because the equations of equilibrium (or motion, in the case of dynamics) relating the external forces acting on the automobile would not distinguish between them.

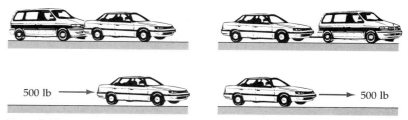

500 lb   500 lb

Figure 3.17

---

* Some authors use "statically equivalent," or "rigid-body equivalent," or simply "equivalent" to describe this relationship. We are using the less familiar word equipollent in order to emphasize that the relationship is restricted neither to statics nor to rigid bodies, and also to call attention to the special nature of the equivalence.

The Two Conditions for Equipollence

It follows from the definition that the two conditions for equipollence of systems S_1 and S_2 are:

1. *Force Condition:* $(\Sigma F)_1 = (\Sigma F)_2$ (3.19)

2. *Moment Condition:* $(\Sigma M_P)_1 = (\Sigma M_P)_2$ (3.20)

in which P is some common point in S_1 and S_2.

Question 3.18 Could a system consisting of a single force be equipollent to a system consisting only of a couple?

It is important to note that satisfaction of the two equipollence conditions, namely the force condition and the moment condition at a single point P, guarantees the satisfaction of the moment condition at *every* point. To show that this is the case, we let Q be any point and recall the important result, Equation (3.18):

$$(\Sigma M_Q)_2 = \mathbf{r}_{QP} \times (\Sigma F)_2 + (\Sigma M_P)_2$$

where we have applied the result to the force system of S_2. Next, by the equipollence of S_1 and S_2, we may replace $(\Sigma F)_2$ by $(\Sigma F)_1$ and $(\Sigma M_P)_2$ by $(\Sigma M_P)_1$ and get

$$(\Sigma M_Q)_2 = \underbrace{\mathbf{r}_{QP} \times (\Sigma F)_1 + (\Sigma M_P)_1}$$
$$= (\Sigma M_Q)_1 \quad \text{(by Equation (3.18) again)}$$

Therefore if the force summations are equal and the moment summations are equal about one point P, then they are also equal about *every other point*. Thus to determine whether two systems are equipollent, we need only to compare ΣF, and then ΣM at any one point.

Question 3.19 Given three force systems S_1, S_2, and S_3, with S_1 equipollent to both S_2 and S_3, determine whether S_2 and S_3 are necessarily equipollent.

We now illustrate the equipollence of force systems by means of some examples. In the first one, a force and couple system equipollent to a system of two forces and a couple is found at a pre-assigned point.

Answer 3.18   No, because in this case ΣF could not possibly be the same. Recall that $\Sigma F = 0$ for a couple! (This assumes the force and couple are not both zero, of course.)
Answer 3.19   Yes, they are.

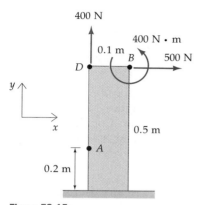

400 N

400 N • m

0.1 m

B

D

500 N

y

x

0.5 m

A

0.2 m

Figure E3.15a

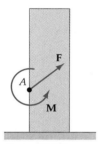

F

A

M

Figure E3.15b

EXAMPLE 3.15

Find a force and couple at point A which together form a system S_2 equipollent to the system S_1 of Figure E3.15a.

Solution

We wish to determine $\mathbf{F}$ and $\mathbf{M}$ in Figure E3.15b such that our equipollence conditions 1 and 2 (Equations 3.19, 20) hold.

$$\text{Condition 1:} \qquad (\Sigma\mathbf{F})_1 = (\Sigma\mathbf{F})_2$$

$$500\hat{\mathbf{i}} + 400\hat{\mathbf{j}} = \mathbf{F}$$

Next we proceed to determine the couple $\mathbf{M}$: we shall use A as the moment center (any point could be used):

$$\text{Condition 2:} \qquad (\Sigma\mathbf{M}_A)_1 = (\Sigma\mathbf{M}_A)_2$$

$$400\hat{\mathbf{k}} + 500(0.5 - 0.2)(-\hat{\mathbf{k}}) = \mathbf{M}$$

$$\mathbf{M} = 250\hat{\mathbf{k}} \ \text{N} \cdot \text{m}$$

Question 3.20   Why was A a good choice for the moment center here?

The equipollent system at A may therefore be drawn as shown in Figure E3.15c (with the force expressed in terms of its components), or in Figure E3.15d.

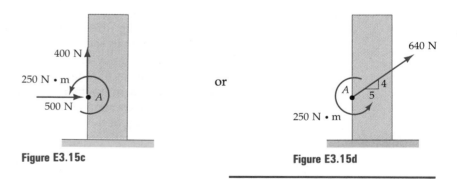

400 N

250 N • m

A

500 N

Figure E3.15c

or

640 N

4

A

5

250 N • m

Figure E3.15d

It is interesting to note that the preceding example could alternately be worked in the following manner:

Answer 3.20   In system S_1, the 400-N force passes through A; and in system S_2, force $\mathbf{F}$ passes through A. These forces thus produce no moment about A, making for simpler moment calculations.

1. For any force (the 400-N force in this case) which passes through A, simply slide it along its line of action and place it at A. Also place and combine at A all couples in the system. See Figure 3.18.

2. "Add and subtract" copies of the remaining forces (just the 500-N force in this case) at the point A of interest. Note that this does not change $\Sigma\mathbf{F}$ or $\Sigma\mathbf{M}$ at any point since it's merely adding two cancelling, collinear forces. See Figure 3.19.

3. Recognize and replace each "equal-but-opposite" pair like the two 500-N forces enclosed by the dashed line in Figure 3.20 as a couple with moment $= (500)(0.3) \circlearrowleft \mathrm{N} \cdot \mathrm{m} = 150 \circlearrowleft \mathrm{N} \cdot \mathrm{m}$. Couples have the same moment about any point, so we may draw it at A and combine it with the couple already there.

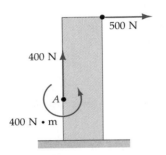

Figure 3.18

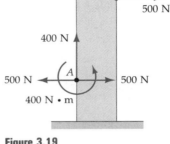

Figure 3.19

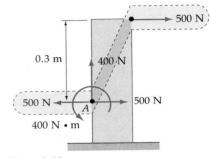

Figure 3.20

4. What now remains (see Figure 3.21) is of course the same result as we obtained more formally in the preceding example. The above approach, however, is often handy for simple two-dimensional systems as we shall see again in section 3.8.

The equipollent system S_2 will always have the same effect on the body, if it is rigid, as did the original system S_1: the same force tending to move it, and the same moment (about a common point) tending to turn it.

In our second example, we require the equipollent system to comprise forces acting at more than one pre-assigned point:

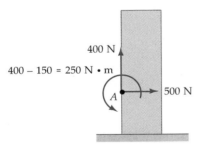

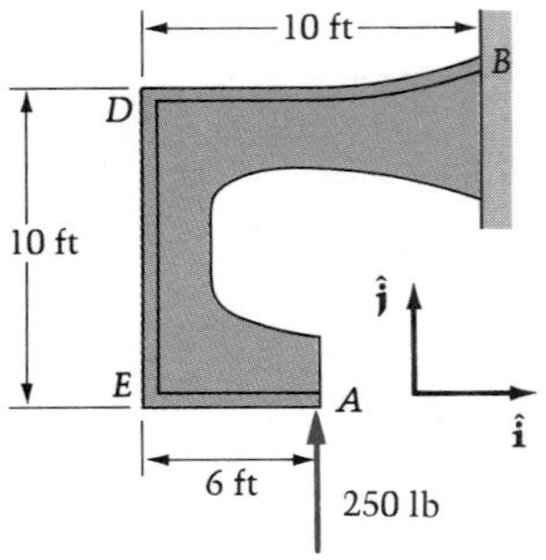

Figure 3.21

EXAMPLE 3.16

Let the system in Figure E3.16a, consisting only of the 250-lb force at A, be called S_1. Determine a force system S_2 that is equipollent to S_1 and that consists of a vertical force acting at B and a pair of horizontal forces acting at D and E.

Solution

Since the original system S_1 contains no horizontal forces, the two horizontal forces at D and E must be equal in magnitude and opposite in direction; that is, they have to form a couple. But let us proceed as if we had not realized this, and

Figure E3.16a

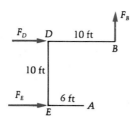

System S_2

Figure E3.16b

sketch the sought system S_2 shown in Figure E3.16b. Our force condition, which is necessary for equipollence of S_1 and S_2, gives

$$(\Sigma \mathbf{F})_1 = (\Sigma \mathbf{F})_2$$
$$250\hat{\mathbf{j}} = F_D\hat{\mathbf{i}} + F_E\hat{\mathbf{i}} + F_B\hat{\mathbf{j}}$$

Thus, from the $\hat{\mathbf{i}}$ coefficients, we find that the forces at D and E are equal in magnitude and opposite in direction:

$$F_D = -F_E \tag{1}$$

The $\hat{\mathbf{j}}$ coefficients give

$$250 = F_B \tag{2}$$

To find the value of F_D, we next ensure that our moment condition is also satisfied; we choose point A as our moment center (*any* point could be used!)

$$(\Sigma \mathbf{M}_A)_1 = (\Sigma \mathbf{M}_A)_2$$
$$\mathbf{0} = F_B(4)\hat{\mathbf{k}} + F_D(10)(-\hat{\mathbf{k}})$$

But $F_B = 250$ lb from Equation (2):

$$F_D = \frac{250(4)}{10} = 100 \text{ lb}$$

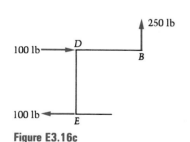

Figure E3.16c

Thus the final system S_2 appears as shown in Figure E3.16c.

As a check, we note that the moment condition is also satisfied at, say, B:

$$(\Sigma \mathbf{M}_B)_1 = (\Sigma \mathbf{M}_B)_2$$
$$250(10 - 6)(-\hat{\mathbf{k}}) = 100(10)(-\hat{\mathbf{k}})$$

or

$$-1000\hat{\mathbf{k}} = -1000\hat{\mathbf{k}} \text{ lb-ft}$$

In our third example, the original system S_1 consists of a simple couple:

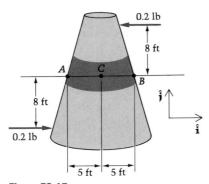

Figure E3.17a

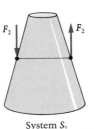

System S_2

Figure E3.17b

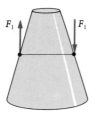

Figure E3.17c

EXAMPLE 3.17

Find a pair of vertical forces, F_1 at A and F_2 at B, that together form a force system S_2 equipollent to the pair of 0.2-lb forces acting on the spacecraft in Figure E3.17a.

Solution

If we call the original force system in the figure S_1, then using Condition 1 — that is, the force condition — we have

$$(\Sigma F)_1 = (\Sigma F)_2$$

$$-0.2\hat{i} + 0.2\hat{i} = F_1 + F_2$$

$$0 = F_1 + F_2$$

so that $F_1 = -F_2$. If $F_2 = F_2\hat{j}$, the system S_2 shown in Figure E3.17b results.

Next, we use Condition 2, or the moment condition, with point C selected as the moment center:

$$(\Sigma M_C)_1 = (\Sigma M_C)_2$$

$$8(0.2)\hat{k} + 8(0.2)\hat{k} = F_2(5)\hat{k} + F_2(5)\hat{k}$$

$$F_2 = \frac{3.2}{10} = 0.32 \text{ lb}$$

The reader has probably noticed that the original 0.2-lb forces form a couple; therefore, the two forces in S_2 must do the same, as we have found. Note that if the forces had been drawn as shown in Figure E3.17c, then the moment condition would have yielded

$$8(0.2)\hat{k} + 8(0.2)\hat{k} = F_1(5)(-\hat{k}) + F_1(5)(-\hat{k})$$

$$F_1 = -0.32 \text{ lb}$$

which of course corresponds to the same pair of forces as the F_2's in the earlier figure (E3.17b) of system S_2.

Note also that if we had not required the forces at A and B to be vertical, then there would have been lots of correct answers, examples of which are shown in the figures below:

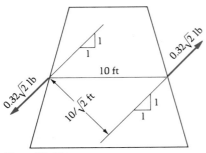

Figure E3.17d

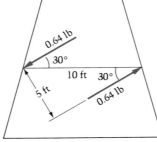

Figure E3.17e

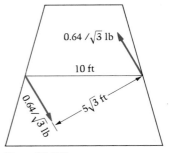

Figure E3.17f

The reader is encouraged to demonstrate that each of these systems is equipollent to S_1.

Question 3.21   Could the forces at A and B constituting system S_2 be (a) horizontal? (b) nonparallel?

In the next example in this section, we shall investigate a series of different force systems and determine which of them are equipollent.

Answer 3.21   (a) No, for they must form a couple. If they were horizontal, they would be collinear and their moment would be zero about points on their common line of action. (b) No, then they still couldn't form a couple. This time neither the force condition *nor* the moment condition can be satisfied.

EXAMPLE 3.18

Consider a triangular plate under the six loading conditions shown in Figures E3.18a–f (SI units throughout). Determine which of the loadings, or force systems, are equipollent.

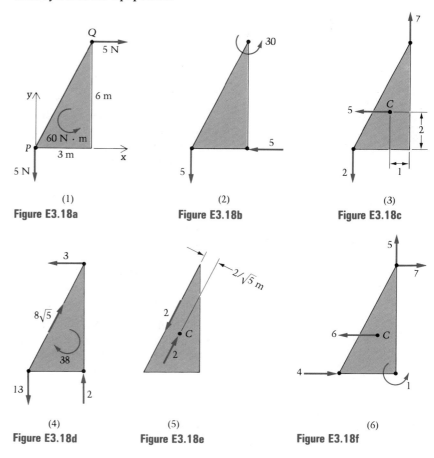

(1)	(2)	(3)
Figure E3.18a	Figure E3.18b	Figure E3.18c
(4)	(5)	(6)
Figure E3.18d	Figure E3.18e	Figure E3.18f

Solution

Since we must compute ΣF and ΣM_P for each system and then compare these results, it would be useful to construct a table. We select the common moment center P to be the lower left-hand corner of the plate because forces pass through this point in each case, and thus their $r \times F$ is zero, making our calculations easier. The resulting table follows.

Force System	ΣF (N)	ΣM_P (N · m)
1	$5\hat{i} - 5\hat{j}$	$30\hat{k}$
2	$-5\hat{i} - 5\hat{j}$	$30\hat{k}$
3	$-5\hat{i} + 5\hat{j}$	$31\hat{k}$
4	$5\hat{i} + 5\hat{j}$	$-14\hat{k}$
5	0	$1.79\hat{k}$
6	$5\hat{i} + 5\hat{j}$	$-14\hat{k}$

Notes on the entries of the six systems above are now listed:

1. Here, the value of ΣF is just the sum of the two forces, $5\hat{i}$ and $-5\hat{j}$. The moment about P is due to the $60\hat{k}$ N · m couple plus the moment of the horizontal force, $r_{PQ} \times 5\hat{i} = (3\hat{i} + 6\hat{j}) \times 5\hat{i} = 30\hat{k}$ N · m. Note that the latter is merely 5 newtons times its perpendicular distance (6 m) from P, with the $-\hat{k}$ attached since the force turns clockwise around P.

2. Both forces pass through P, leaving only the couple to contribute to ΣM_P. Remember that couples have the same moment about all points! Note that although the moment condition is satisfied in comparing systems 1 and 2, the force condition is not, and it takes *both* for equipollence!

3. In the vertical direction, the force is the combination of 7 N up and 2 N down, or $5\hat{j}$ N. Rotationally, the 7-N force and 5-N force each have moments around P, their "lever arms" being 3 m and 2 m, respectively.

4. In this case, the sum of the forces is calculated as

$$\Sigma F = 8\sqrt{5}\left(\frac{\hat{i} + 2\hat{j}}{\sqrt{5}}\right) - 3\hat{i} + 2\hat{j} - 13\hat{j} = (5\hat{i} + 5\hat{j}) \text{ N}$$

The sum of moments is computed vectorially as follows:

$$\Sigma M_P = (3\hat{i} + 6\hat{j}) \times (3\hat{i}) + 3\hat{i} \times 2\hat{j} - 38\hat{k} = -14\hat{k} \text{ N} \cdot \text{m}$$

Alternatively, using forces times lever arms and attaching the unit vector with correct sign,

$$\Sigma M_P \text{ also} = [3(6) + 2(3) - 38]\hat{k} = -14\hat{k} \text{ N} \cdot \text{m}$$

5. In this one, we have just a pure couple since the two forces cancel each other, making $\Sigma F = 0$. The magnitude of the couple (whose unit vector and direction is seen to be $+\hat{k}$) is given by either force times the distance d between them:

$$\Sigma\Sigma M_P = 2\left(\frac{2}{\sqrt{5}}\right)\hat{k} = 1.79\hat{k} \text{ N} \cdot \text{m}$$

Note that by inspection of the table through force system (5), there are no equipollent loadings thus far.

6. The sum of forces is $\Sigma \mathbf{F} = (4 - 6 + 7)\hat{\mathbf{i}} + 5\hat{\mathbf{j}} = 5\hat{\mathbf{i}} + 5\hat{\mathbf{j}}$ N. And $\Sigma \mathbf{M}_P$
$= 2\hat{\mathbf{j}} \times (-6\hat{\mathbf{i}}) + 6\hat{\mathbf{j}} \times 7\hat{\mathbf{i}} + 3\hat{\mathbf{i}} \times 5\hat{\mathbf{j}} + 1\hat{\mathbf{k}} = (12 - 42 + 15 + 1)\hat{\mathbf{k}}$
$= -14\hat{\mathbf{k}}$ N $\cdot$ m, where we have used $2\hat{\mathbf{j}}$ and $6\hat{\mathbf{j}}$ m for the position vectors to points
on the lines of action of the 6- and 7-N horizontal forces, respectively.

We see that $(\Sigma \mathbf{F})_4 = (\Sigma \mathbf{F})_6$ *and* $(\Sigma \mathbf{M}_P)_4 = (\Sigma \mathbf{M}_P)_6$, so that, in conclusion, the
only equipollent systems among the six are numbers 4 and 6.

In our last example, we shall examine a three-dimensional force
system; such a system is one in which (having expressed each couple as a
pair of equal-but-opposite forces) not all the forces in the system can lie in
a single plane. Note that according to this definition, all previous examples in this section have been *two*-dimensional.

EXAMPLE 3.19

Find a system S_2 comprising a force at P and a couple that is equipollent to the
300-N force and 250-N $\cdot$ m couple in Figure E3.19. The couple's turning effect
lies in the plane *PCBA*, so its unit vector is $\hat{\mathbf{k}}$.

Solution

Condition (1) yields:

$$(\Sigma \mathbf{F})_2 = (\Sigma \mathbf{F})_1 = 300 \left(\frac{3\hat{\mathbf{i}} - 4\hat{\mathbf{k}}}{5} \right) = 180\hat{\mathbf{i}} - 240\hat{\mathbf{k}} \text{ N}$$

To form system S_2, we place this force at P and then proceed to sum moments
there to determine the accompanying couple. Condition (2) yields, with point P
as the moment center:

$$(\Sigma \mathbf{M}_P)_1 = (\Sigma \mathbf{M}_P)_2$$

$$\underbrace{250\hat{\mathbf{k}}}_{\substack{\text{moment of} \\ \text{couple}}} + \underbrace{\mathbf{r}_{PA} \times \mathbf{F}}_{\substack{\text{moment of} \\ \text{force about } P}} = (\Sigma \mathbf{M}_P)_2$$

so that

$$(\Sigma \mathbf{M}_P)_2 = 250\hat{\mathbf{k}} + (-0.3\hat{\mathbf{i}}) \times (180\hat{\mathbf{i}} - 240\hat{\mathbf{k}})$$
$$= -72\hat{\mathbf{j}} + 250\hat{\mathbf{k}} \text{ N} \cdot \text{m},$$

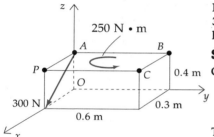

Figure E3.19

which is the couple accompanying the force at P; together they form the
system S_2 equipollent to the original system.

The reader should remember just two main things about equipollent
systems in working the problems which follow: (1) Make the forces add
up to the same vector in the two systems; (2) About some common point
in the two systems, make the moments also add up to the same vector.

PROBLEMS ▶ Section 3.6

3.61 Replace the 2000-lb force in Figure P3.61 by an equipollent system consisting of a horizontal force through A, a vertical force along line CB, and a couple.

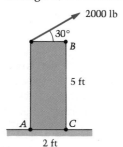

Figure P3.61

3.62 Find (and illustrate with a sketch) a pair of forces which (a) is equipollent to the couple in the figure, and (b) has one force equal to $50\hat{i}$ N through the point $(x, y) = (10, 20)$ cm. (See Figure P3.62.)

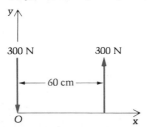

Figure P3.62

3.63 Repeat the preceding problem with condition (b) changed to a force of $180\hat{j}$ N through $(x, y) = (30, 40)$ cm.

3.64 Replace the 250-lb force in Figure P3.64 by an equipollent system consisting of a force at Q and a couple. Then show that *both* systems have the same moment about point B.

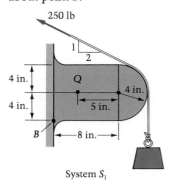

Figure P3.64

3.65 Find a system S_2 consisting of a force at A and a couple, which is equipollent to the system of two forces in Figure P3.65. Use the procedure of Examples 3.15–3.17.

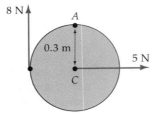

Figure P3.65

3.66 Repeat the preceding problem using the ideas which were illustrated following Example 3.15. Add and subtract two pairs of forces at A, both $\pm5\hat{i}$ N and $\pm8\hat{j}$ N. Then identify two couples and add them, leaving $5\hat{i} + 8\hat{j}$ N at point A, plus the combined couple.

3.67 By filling in the chart [Figure P3.67(b)], determine which of the six beam loadings in Figure P3.67(a) are equipollent.

3.68 Continue Example 3.18 by determining which of the force systems in Figure P3.68 are equipollent either to any of the loadings in that example or to one another. How many different comparisons of two systems at a time have now been made?

3.69 A bending moment of 20,000 lb-ft acts on a large plate bolted to the ground with eight equally spaced bolts as shown in Figure P3.69. Determine an equipollent system of eight vertical forces, each acting at one of the bolts and having a magnitude proportional to the distance of the bolt from line ℓ.

3.70 In the preceding problem, re-determine the eight forces if the bolt pattern is aligned with the moment vector as shown in Figure P3.70.

3.71 A body is acted on by two forces: one with components (10, 20, 30) N at a point having coordinates (3, 2, 1) meters and the other with components (30, 20, 10) N at a point (1, 2, 3) m. Find the equipollent system consisting of a force at the point (1, 1, 1) m and a couple.

3.72 In Figure P3.72 the magnitude of the moments of couples $\mathbf{M}_1$, $\mathbf{M}_2$, $\mathbf{M}_3$ are each $= \mathbf{M}_0 =$ constant. The forces act in planes ABC, ACD, and $BODC$, respectively, so that the moments of the couples are normal to these planes. Find a single couple that is equipollent to these three couples.

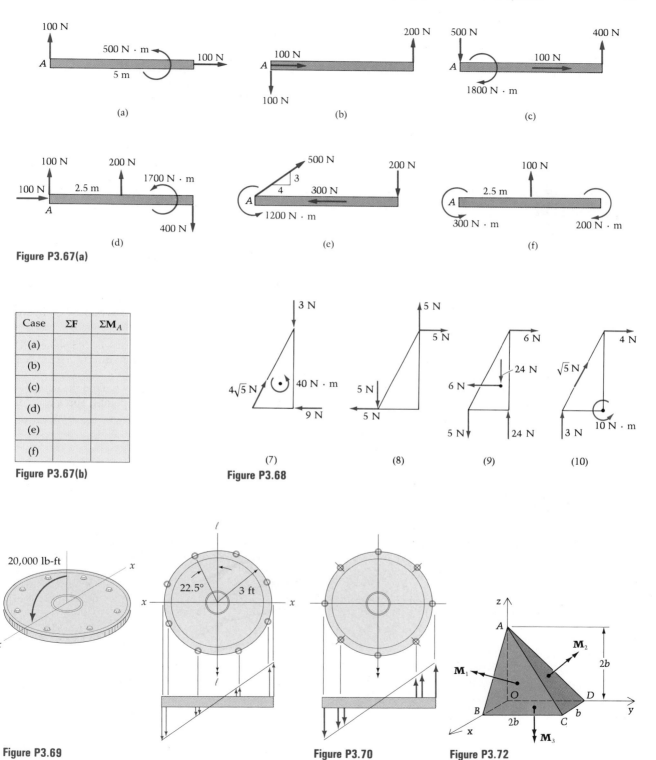

Figure P3.67(a)

Figure P3.67(b)

Figure P3.68

Figure P3.69

Figure P3.70

Figure P3.72

3.73 Find a force at E and an accompanying couple that together form a system equipollent to the one shown in Figure P3.73.

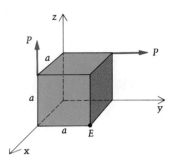

Figure P3.73

3.74 The couple's forces lie in the shaded plane, and there are two other applied forces in the system shown in Figure P3.74. Find an equipollent system consisting of a force at the origin O and a couple.

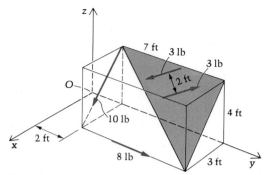

Figure P3.74

3.75 Find the force and couple at the origin equipollent to the five forces and the couple shown in Figure P3.75. Then repeat the problem considering the two vertical 30-N forces as a couple, and the two horizontal 40-N forces as another couple. That is, repeat the problem with the system treated as one force and three couples. Compare the results.

3.76 A couple of moment $\mathbf{C}$ lies in the shaded plane ABC and has a magnitude of $\sqrt{13}$ lb-ft and a direction indicated by the right-hand rule. (See Figure P3.76.) A force $\mathbf{F}$ of magnitude $2\sqrt{13}$ lb also acts as shown. Find the equipollent system consisting of a force at point C and a couple.

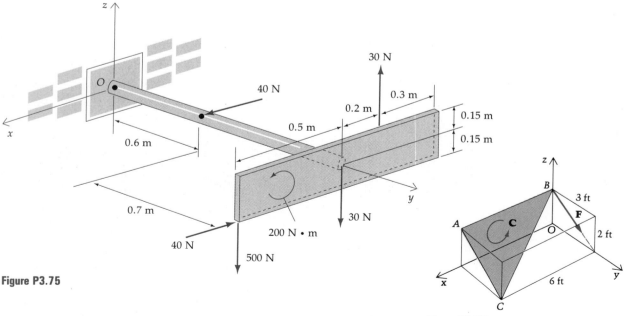

Figure P3.75

Figure P3.76

3.7 The Force-and-Couple Resultant of a System of Forces

The Definition of a Resultant

The definition of equipollent force systems correctly suggests that we may replace any system, no matter how complicated, by a force at any point P and a couple (see Figure 3.22).

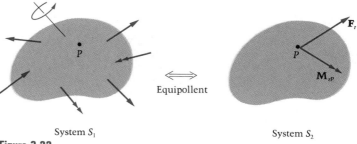

Equipollent

System S_1 System S_2

Figure 3.22

We have already seen that for the system S_2 to be equipollent to S_1 we need only to ensure that:

1. $\mathbf{F}_r$, the force at P in S_2, is the sum of all the forces acting in system S_1.

2. $\mathbf{M}_{rP}$, the moment of the couple in S_2, is the sum of the moments of all the couples in S_1 plus the sum of the moments about P of all the forces in S_1. Note that, in system S_2, $\mathbf{F}_r$ produces no moment about P.

The force-couple pair, $\mathbf{F}_r$ and $\mathbf{M}_{rP}$, is called a **resultant*** of the system S_1. For simplicity we are using the phrase "force and couple at point P" to describe the resultant, but it is important to realize that there is no reason to assign a location-point subscript to the couple since the moment of a couple about every point is the same. And while the force $\mathbf{F}_r$ does not depend upon the choice of reference point P, the couple $\mathbf{M}_{rP}$ depends upon the choice of the *line of action* of $\mathbf{F}_r$. Thus, by "force and couple at P" we mean *the* resultant (of S_1) when the line of action of $\mathbf{F}_r$ is chosen to pass through point P.

> **Question 3.22** The implication of the preceding paragraph is that for all other points on the line of action of $\mathbf{F}_r$ through P, the force and couple resultant is the same as it is at P. Why is this indeed so?

---

* Hence the subscript "r" for resultant, on each member of the pair.

Answer 3.22 Because (1) $\mathbf{F}_r$ is always the same from point to point, and (2) if Q is any other point on the line,

$$\mathbf{M}_{rQ} = \mathbf{M}_{rP} + \mathbf{r}_{QP} \times \mathbf{F}_r = \mathbf{M}_{rP}$$

since $\mathbf{r}_{QP}$ is coincident with the line of action of $\mathbf{F}_r$.

We consider now a number of examples of replacement of a system of forces and/or couples by an equipollent system of a force and a couple (that is, a resultant) at a preselected point.

EXAMPLE 3.20

Determine the resultant, at the pin P, of the two belt tensions shown in Figure E3.20a. The pulley has a radius $R = 0.6$ m.

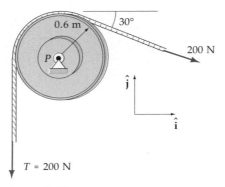

Figure E3.20a

Solution

Using the force condition, the force part of the resultant is calculated as follows:

$$(\Sigma \mathbf{F})_1 = (\Sigma \mathbf{F})_2$$

$$200(\cos 30°)\hat{\mathbf{i}} - [200 + 200(\sin 30°)]\hat{\mathbf{j}} = \mathbf{F}_r$$

or

$$\mathbf{F}_r = 173\hat{\mathbf{i}} - 300\hat{\mathbf{j}} \text{ N}$$

Next, the moment condition tells us that the couple part of the resultant at P vanishes this time:

$$(\Sigma \mathbf{M}_P)_1 = (\Sigma \mathbf{M}_P)_2$$

$$200(0.6)\hat{\mathbf{k}} + 200(0.6)(-\hat{\mathbf{k}}) = \mathbf{0} = \mathbf{M}_{rP}$$

Figure E3.20b shows what has happened in going from the original system (S_1) to the resultant at P (S_2).

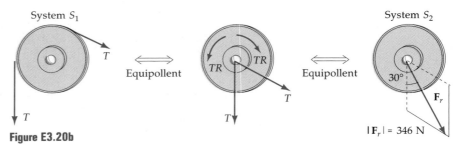

Figure E3.20b

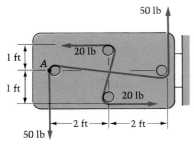

Figure E3.21a

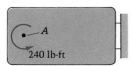

Figure E3.21b

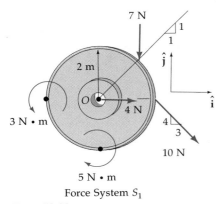

Force System S_1

Figure E3.22a

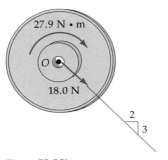

Figure E3.22b

In the next example, the original system consists only of couples.

EXAMPLE 3.21

Determine the resultant at point A of the four forces shown in Figure E3.21a.

Solution

Since the forces are seen to form a pair of couples, the force $\mathbf{F}_r$ of the resultant at A is zero. The moment of the couple is

$$\mathbf{M}_{rA} = \overset{S_1}{\sum} \mathbf{M}_A = 50(4)(\hat{\mathbf{k}}) + 20(2)\hat{\mathbf{k}} = 240\hat{\mathbf{k}} \text{ lb-ft}$$

The resultant is shown in Figure E3.21b.

> **Question 3.23** What is the resultant at the point 2 ft to the right of A?

In the next example, the original system comprises three forces in a plane and two couples with turning effects in that same plane:

Answer 3.23 At all points, the resultant in this problem is the same: a couple of $240\hat{\mathbf{k}}$ lb-ft, unaccompanied by a force.

EXAMPLE 3.22

Replace the force and couple system in Figure E3.22a by its resultant (an equipollent force and couple) at O.

Solution

The resultant force is

$$\mathbf{F}_r = [4 + \tfrac{3}{5}(10)]\hat{\mathbf{i}} + [-7 - \tfrac{4}{5}(10)]\hat{\mathbf{j}} = 10\hat{\mathbf{i}} - 15\hat{\mathbf{j}} \text{ N}$$

The resultant moment about O in S_1 is

$$\mathbf{M}_{rO} = \underbrace{(3\hat{\mathbf{k}} - 5\hat{\mathbf{k}})}_{\substack{\text{moments of} \\ \text{couples}}} + \underbrace{2\left(\frac{\hat{\mathbf{i}} + \hat{\mathbf{j}}}{\sqrt{2}}\right) \times (-7\hat{\mathbf{j}}) + 2\hat{\mathbf{i}} \times (6\hat{\mathbf{i}} - 8\hat{\mathbf{j}})}_{\text{moments of forces}}$$

$$= \left[-2 - \frac{14}{\sqrt{2}} - 16\right]\hat{\mathbf{k}} = -27.9\hat{\mathbf{k}} \text{ N} \cdot \text{m}$$

The equipollent system is shown in Figure E3.22b.

In each of the next three examples, we have a three-dimensional system.

EXAMPLE 3.23

Find the force at the origin and couple that are equipollent to the system S_1 in Figure E3.23.

Solution

The force is

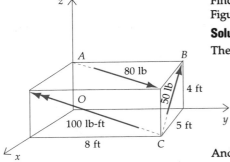

Figure E3.23

$$\mathbf{F}_r = \Sigma\mathbf{F} = 80\left(\frac{5\hat{\mathbf{i}} + 8\hat{\mathbf{j}}}{\sqrt{89}}\right) + 50\left(\frac{-5\hat{\mathbf{i}} + 4\hat{\mathbf{k}}}{\sqrt{41}}\right)\text{lb}$$

$$= (42.4 - 39.0)\hat{\mathbf{i}} + 67.8\hat{\mathbf{j}} + 31.2\hat{\mathbf{k}} \text{ lb}$$

$$= 3.4\hat{\mathbf{i}} + 67.8\hat{\mathbf{j}} + 31.2\hat{\mathbf{k}} \text{ lb}$$

And the moment of the couple is

$$\mathbf{M}_{rO} = \Sigma\mathbf{M}_O = 100\left(\frac{-8\hat{\mathbf{j}} + 4\hat{\mathbf{k}}}{\sqrt{80}}\right) + \overbrace{(4\hat{\mathbf{k}})}^{\mathbf{r}_{OA}}\times 80\left(\frac{5\hat{\mathbf{i}} + 8\hat{\mathbf{j}}}{\sqrt{89}}\right) + \overbrace{(8\hat{\mathbf{j}} + 4\hat{\mathbf{k}})}^{\mathbf{r}_{OB}}$$

$$\times 50\left(\frac{-5\hat{\mathbf{i}} + 4\hat{\mathbf{k}}}{\sqrt{41}}\right)$$

$$= \left[\frac{-80(32)}{\sqrt{89}} + \frac{50(32)}{\sqrt{41}}\right]\hat{\mathbf{i}} + \left[\frac{-800}{\sqrt{80}} + \frac{80(20)}{\sqrt{89}} - \frac{50(20)}{\sqrt{41}}\right]\hat{\mathbf{j}}$$

$$+ \left[\frac{400}{\sqrt{80}} + \frac{50(40)}{\sqrt{41}}\right]\hat{\mathbf{k}}$$

$$= -21.5\hat{\mathbf{i}} - 76.0\hat{\mathbf{j}} + 357\hat{\mathbf{k}} \text{ lb-ft}$$

Together, $\mathbf{F}_r$ and $\mathbf{M}_{rO}$ form the resultant at point O of the original system S_1.

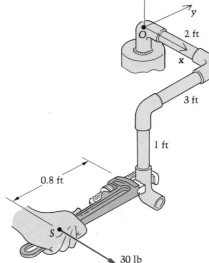

Figure E3.24a

EXAMPLE 3.24

In Figure E3.24a, a person exerts a force of 30 lb in the x direction to turn the elbow onto the threaded pipe. Determine the force and couple resultant at the origin O, where the pipe is screwed into another elbow above a tank.

Solution

The force is simply $\mathbf{F}_r = 30\hat{\mathbf{i}}$ lb. The moment at O is

$$\mathbf{M}_{rO} = \mathbf{r}_{OS} \times 30\hat{\mathbf{i}}$$

$$= (2\hat{\mathbf{i}} - 3.8\hat{\mathbf{j}} - \hat{\mathbf{k}}) \times 30\hat{\mathbf{i}} = -30\hat{\mathbf{j}} + 114\hat{\mathbf{k}} \text{ lb-ft}$$

Before moving to another example, we note that the force and moment are undesirable concerning stress and deflection (and maybe leaks!) in the pipes. This can be avoided by the wise use of a second pipe wrench (see Figure E3.24b). Now, we see that for the two forces, $\mathbf{F}_r = 0$. Furthermore, the z component of $\mathbf{M}_{rO}$ is

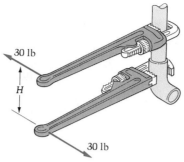

Figure E3.24b

eliminated. The couple at O formed by the two 30-lb forces is simply $30H(-\hat{\jmath})$, which, if H is, say, 2 in., amounts to only $-5\hat{\jmath}$ lb-ft. This illustrates the advantage that may accrue from using two forces to produce a desired moment, which here is $24\hat{k}$ lb-ft on the elbow.

EXAMPLE 3.25

For the system shown in Figure E3.25, find the force and couple resultant at (a) the origin and (b) point A.

Solution

The force resultant is the same at all points, and is

$$\mathbf{F}_r = 10\left(\frac{3\hat{\imath} - 4\hat{k}}{5}\right) + 10\left(\frac{-3\hat{\imath} + 4\hat{\jmath}}{5}\right)$$

$$= (6\hat{\imath} - 8\hat{k}) + (-6\hat{\imath} + 8\hat{\jmath})$$

$$= 8\hat{\jmath} - 8\hat{k} \text{ N}$$

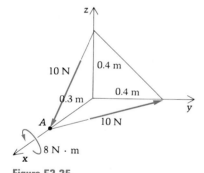

Figure E3.25

Obtaining the resultant moment at the origin, and noting that $0.3\hat{\imath}$ m is a vector from O to a point on the line of action of both forces,

$$\mathbf{M}_{rO} = 0.3\hat{\imath} \times (6\hat{\imath} - 8\hat{k}) + 0.3\hat{\imath} \times (-6\hat{\imath} + 8\hat{\jmath}) + 8\hat{\imath}$$

$$= 8\hat{\imath} + 2.4\hat{\jmath} + 2.4\hat{k} \text{ N} \cdot \text{m}$$

Thus at the origin, the original system is equipollent to

$$\mathbf{F}_r = 8\hat{\jmath} - 8\hat{k} \text{ N}$$

$$\mathbf{M}_{rO} = 8\hat{\imath} + 2.4\hat{\jmath} + 2.4\hat{k} \text{ N} \cdot \text{m}$$

At point A, the moment is simply the couple of $8\hat{\imath}$ N $\cdot$ m, since both forces pass through A. Therefore, at A the system equipollent to the given one is

$$\underbrace{\mathbf{F}_r = 8\hat{\jmath} - 8\hat{k} \text{ N}}_{\substack{\text{(The force never changes} \\ \text{from point to point!)}}} \quad \text{and} \quad \mathbf{M}_{rA} = 8\hat{\imath} \text{ N} \cdot \text{m}$$

We should note in this example that the two systems at A and O are of course not only each equipollent to the given system, but also to each other. To show this, note that their forces are each $8\hat{j} - 8\hat{k}$ N, and that the moment of the system at A, about O, is

$$\underbrace{\mathbf{M}_{rO} = \mathbf{M}_{rA}}_{\substack{\text{couple at } A \\ \text{(same moment} \\ \text{everywhere!)}}} + \underbrace{\mathbf{r}_{OA} \times \mathbf{F}_r}_{\substack{\text{moment about } O \text{ of} \\ \text{the force at } A}}$$

$$= 8\hat{i} + 0.3\hat{i} \times (8\hat{j} - 8\hat{k})$$

$$= 8\hat{i} + 2.4\hat{j} + 2.4\hat{k} \text{ N} \cdot \text{m}$$

This is indeed what we had previously obtained for $\mathbf{M}_{rO}$.

Question 3.24   In terms of the $\mathbf{F}_r$ and $\mathbf{M}_{rP}$ "resultant" notation, what do the equilibrium equations (see Section 3.5) of a body look like?

Before closing this section, we remark that if the force ($\mathbf{F}_r$) and couple ($\mathbf{M}_{rP}$) resultant at point P has been computed for some system S_1 of forces and couples, then by Varignon's theorem the moment of these forces and couples about a line ℓ through P is simply

$$\mathbf{M}_\ell = (\mathbf{M}_{rP} \cdot \hat{u}_\ell)\hat{u}_\ell \qquad (3.21)$$

where $\hat{u}_\ell$ is a unit vector along ℓ.

As an illustration of Equation (3.21), the moment of the forces and couple of Example 3.25 about the y-axis (through O) is (with $\hat{u}_\ell = \hat{j}$):

$$\mathbf{M}_{y\text{-axis}} = (\mathbf{M}_{rO} \cdot \hat{j})\hat{j} = [(8\hat{i} + 2.4\hat{j} + 2.4\hat{k}) \cdot \hat{j}]\hat{j}$$

$$= 2.4\hat{j} \text{ N} \cdot \text{m}$$

Answer 3.24 $\Sigma \mathbf{F} = 0$ becomes $\mathbf{F}_r = 0$, and $\Sigma \mathbf{M}_P = 0$ becomes $\mathbf{M}_{rP} = 0$. Of course, in the context of equilibrium, the force-couple system we are considering is the set of *all* the external forces and couples exerted on the body.

PROBLEMS ▶ Section 3.7

3.77 Four truss members carrying the indicated forces have their center lines all intersecting at point P of the shaded gusset plate shown in Figure P3.77. Find the resultant of the four forces at P.

3.78 The resultant of the three-force system shown in Figure P3.78 is a single force of 300 lb pointing up along the y axis. Find the force $\mathbf{F}$ and the angle θ it forms with the x axis.

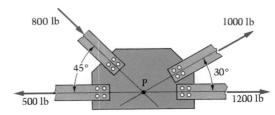

Figure P3.77

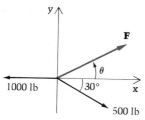

Figure P3.78

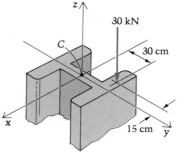

Figure P3.82

3.79   Find the resultant of the three concurrent forces at point C within the equilaterial triangle in Figure P3.79.

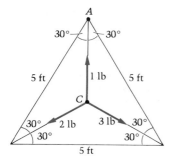

Figure P3.79

3.80   Referring to the preceding problem, find the force and couple at vertex A that are equipollent to the given system (that is, find the resultant at A).

3.81   Find the magnitude of the couple $\mathbf{C}$ (whose axis is in the xy plane), and its orientation angle θ, for which the three couples have a zero resultant (see Figure P3.81).

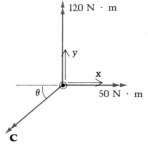

Figure P3.81

3.82   The 30-kN load is eccentrically applied to the column as shown in Figure P3.82. Determine the force and couple at C that are together equipollent to the 30-kN load.

3.83   Determine the force and couple at O that constitute the resultant there of the three forces, the 40 lb-ft twisting couple, and the two bending couples acting on the end of the cantilever beam shown in Figure P3.83.

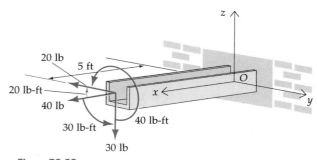

Figure P3.83

3.84   For the system shown in Figure P3.84, find the resultant at point B.

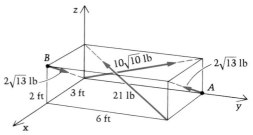

Figure P3.84

3.85 a.   Repeat the preceding problem for A instead of B.
  b.   Demonstrate that $\mathbf{M}_{rA} = \mathbf{M}_{rB} + \mathbf{r}_{AB} \times \mathbf{F}_r$, which is, of course, generally true.

3.86 Determine the resultant at A of the forces applied to the bar shown in Figure P3.86. Then, find the moment at the origin and demonstrate that $\mathbf{M}_{rA} = \mathbf{M}_{rO} + \mathbf{r}_{AO} \times \mathbf{F}_r$. Finally, use Equation (3.21) to find the moment of the forces in the figure about the line through A that forms equal angles with the coordinate axes.

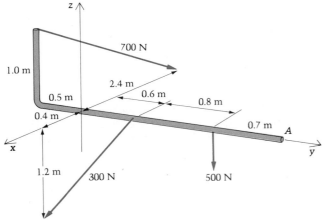

Figure P3.86

3.87 Check the answer to Example 3.23 by finding the resultant at C and then computing $\mathbf{M}_{rO} + \mathbf{r}_{CO} \times \mathbf{F}_r$, and comparing the result with $\mathbf{M}_{rC}$. The two results should of course be equal.

3.88 Determine the resultant of the four forces and the two couples that act on the shaft shown in Figure P3.88, expressed as a force and couple at the origin O.

3.89 In Figure 3.89, couple $\mathbf{C}$ lies in plane OBG, and has magnitude $2\sqrt{34}$ lb-ft. Find:

  a. the moment of $\mathbf{F}_1$ about O;

  b. the moment of $\mathbf{F}_2$ about O;

  c. the moment of $\mathbf{C}$ about O;

  d. the moment of the resultant (of $\mathbf{F}_1$, $\mathbf{F}_2$, and $\mathbf{C}$) about line OG;

  e. the moment of the resultant (of $\mathbf{F}_1$, $\mathbf{F}_2$, and $\mathbf{C}$) about line OE.

  f. Adding the answers to a, b, and c, then subtracting those of d and e, gives a vector normal to plane EOG. Without any calculations, explain why this must be so.

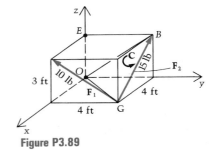

Figure P3.89

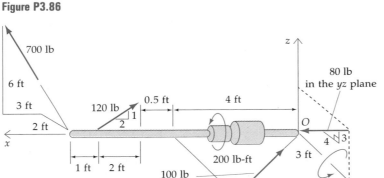

Figure P3.88

3.8 The Simplest Resultant of a Force System

The Single-Force Resultant

In Example 3.16, we constructed a force and couple that were equipollent to a single force. There are situations in which this procedure is desirable. Often, however, our motivation is to produce the simplest resultant — that is, where possible, simply a force or simply a couple.* Thus, in the

---

* Some authors use the term "resultant" to denote what we are calling "simplest resultant." Our choice of language is motivated by what the student will encounter in the analysis of stresses in deformable solids. There, the most useful form of resultant is often a force-couple pair with a preassigned reference point, and, even if further reduction to a single force is possible, it is not useful.

example mentioned above, the original force itself is already the simplest resultant. Similarly, the system of forces in Example 3.17 had $\mathbf{F}_r = \mathbf{0}$, and so the original *couple* was already the simplest resultant.

In this section we are going to think about the conditions under which we may reduce an arbitrary force system to an equipollent one consisting only of a force. Figure 3.23 depicts three systems; the first one being an "original system" S_1 comprising any number of forces and couples.

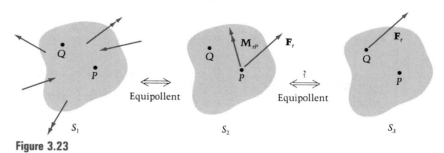

Figure 3.23

The second system, S_2, is equipollent to S_1 and consists of a force $\mathbf{F}_r$ and couple $\mathbf{M}_{rP}$ at some point P. This system can always be found, because, by the two conditions for equipollence,

$$\mathbf{F}_r = (\Sigma\mathbf{F})_1$$
$$\mathbf{M}_{rP} = (\Sigma\mathbf{M}_P)_1$$

The third system S_3, however, may or may not exist. It is required to be equipollent to S_2 (and hence also to S_1) and consist only of a "force alone" at some point Q. If this system is to exist, the force in S_3 must be the same as the sum of the forces in S_2 (and S_1):

$$\mathbf{F}_r = (\Sigma\mathbf{F})_1$$

which of course is already true.

The second condition that must be fulfilled for S_3 to exist is that the sum of the moments about P in S_3 be the same as that sum in S_2 (and in S_1); in S_2 it was called $\mathbf{M}_{rP}$:

$$(\Sigma\mathbf{M}_P)_3 = \mathbf{M}_{rP}$$

By using Equation (3.18),

$$(\Sigma\mathbf{M}_P)_3 = (\Sigma\mathbf{M}_Q)_3 + \mathbf{r}_{PQ} \times \mathbf{F}_r$$

But we want $(\Sigma\mathbf{M}_Q)_3$, or $\mathbf{M}_{rQ}$, to be zero. Thus, replacing $(\Sigma\mathbf{M}_P)_3$ with the equivalent notation $\mathbf{M}_{rP}$,

$$\mathbf{M}_{rP} = \mathbf{r}_{PQ} \times \mathbf{F}_r \qquad (3.22)$$

Now, Equation (3.22) can be satisfied (or is meaningful) only if $\mathbf{M}_{rP}$ is orthogonal, or perpendicular, to $\mathbf{F}_r$. This is because a cross product is always perpendicular to each of the vectors making up the product, so that the cross product, $\mathbf{M}_{rP}$, is perpendicular both to $\mathbf{r}_{PQ}$ and to $\mathbf{F}_r$. Whenever this is the case, Equation (3.22) can be solved for vectors $\mathbf{r}_{PQ}$ identifying points on the line of action of the single force resultant $\mathbf{F}_r$ in system S_3.

We note that if in system S_1 we happen to have $\mathbf{F}_r = 0$ with $\mathbf{M}_{rP} \neq 0$, then by Equation (3.18) we have $\mathbf{M}_{rQ} = \mathbf{M}_{rP}$ and the resultant is a couple, period! And that if in system S_1 we have $\mathbf{F}_r \neq 0$ with $\mathbf{M}_{rP} = 0$, then we *already have* the location of the force-alone resultant — its line of action passes through P.*

We next explore three special force systems for which a single-force resultant exists.

Special Force Systems that are Equipollent to a Single Force

1. Concurrent Force Systems A concurrent force system is one for which all of the lines of action intersect at a single point. If we call that point A as in Figure 3.24, then $\mathbf{M}_{rA} = 0$ because none of the forces produces a moment about A. Thus the resultant is simply $\mathbf{F}_r = \Sigma\mathbf{F}$ with its line of action passing through A.

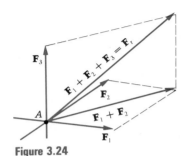

Figure 3.24

EXAMPLE 3.26

Find the single-force resultant, or "force-alone" resultant, of the system of forces shown in Figure E3.26.

$$\mathbf{F}_1 = 2\hat{\jmath} \text{ N}$$
$$\mathbf{F}_2 = \hat{\imath} + 2\hat{\jmath} + 5\hat{k} \text{ N}$$
$$\mathbf{F}_3 = -4\hat{\jmath} + 7\hat{k} \text{ N}$$
$$\mathbf{F}_4 = 6\hat{\imath} \text{ N}$$

Solution

Because the four forces are concurrent at Q, there is a single force resultant that is simply their sum $\mathbf{F}_r$, acting at Q (or any other point on the line of $\mathbf{F}_r$ through Q):

$$\mathbf{F} = \mathbf{F}_1 + \mathbf{F}_2 + \mathbf{F}_3 + \mathbf{F}_4$$
$$= 2\hat{\jmath} + (\hat{\imath} + 2\hat{\jmath} + 5\hat{k}) + (-4\hat{\jmath} + 7\hat{k}) + 6\hat{\imath}$$
$$= 7\hat{\imath} + 12\hat{k} \text{ N}$$

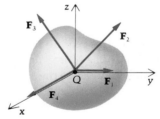

Figure E3.26

Note in Example 3.26 that at any point *off* the line of $\mathbf{F}_r$ through Q, the equipollent system will consist not only of the force $\mathbf{F}_r$, but also of a non-zero couple. For example, at the point B with coordinates (1, 1, 1),

---

* And if $\mathbf{F}_r$ and $\mathbf{M}_{rP}$ are *both* zero in S_1, then the sum of the forces and the sum of the moments about any point vanish for all systems equipollent to S_1. This in fact is what happens with the force system acting on a body in equilibrium.

we would still have

$$\mathbf{F}_r = 7\hat{\mathbf{i}} + 12\hat{\mathbf{k}}\ \text{N}$$

(the resultant force does not change from one equipollent system to another) accompanied by

$$\begin{aligned}\mathbf{M}_{rB} &= \overset{\mathbf{0}}{\cancel{\mathbf{M}_{rQ}}} + \mathbf{r}_{BQ} \times \mathbf{F}_r \\ &= (-\hat{\mathbf{i}} - \hat{\mathbf{j}} - \hat{\mathbf{k}}) \times (7\hat{\mathbf{i}} + 12\hat{\mathbf{k}}) \\ &= -12\hat{\mathbf{i}} + 5\hat{\mathbf{j}} + 7\hat{\mathbf{k}}\ \text{N} \cdot \text{m}\end{aligned}$$

which of course is not zero.

2. Coplanar Force Systems In this type of system, all of the forces have lines of action that lie in the same plane (say xy), while all the vectors representing moments of couples are normal to this plane (see Figure 3.25).

Figure 3.25

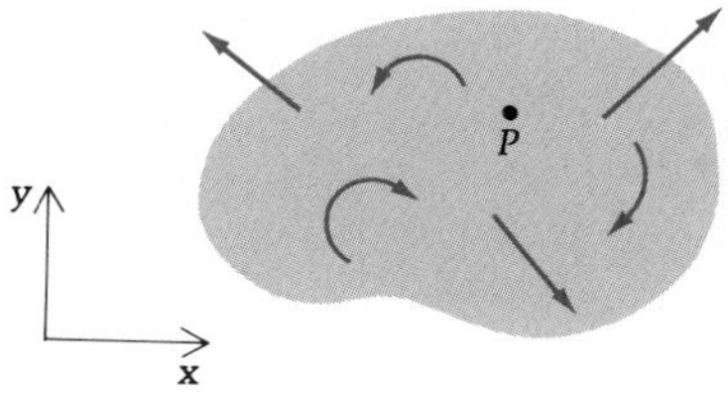

In this case if we sketch the equipollent system at P (that is, the resultant at P) in the xy plane (see Figure 3.26), we will necessarily have

$$\mathbf{F}_r = \mathbf{F}_{rx}\hat{\mathbf{i}} + \mathbf{F}_{ry}\hat{\mathbf{j}}$$
$$\mathbf{M}_{rP} = M_{rP}\hat{\mathbf{k}}$$

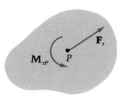

Figure 3.26

Question 3.25 (a) Why does $\mathbf{F}_r$ have no z component? (b) Why does $\mathbf{M}_{rP}$ have *only* a z component?

Now if we let P be the origin for rectangular coordinates and let x, y, and z be coordinates of a point through which the line of action of a "force-alone" resultant would pass, then by Equation (3.22),

$$M_{rP}\hat{\mathbf{k}} = (x\hat{\mathbf{i}} + y\hat{\mathbf{j}} + z\hat{\mathbf{k}}) \times (F_{rx}\hat{\mathbf{i}} + F_{ry}\hat{\mathbf{j}})$$

or

$$M_{rP}\hat{\mathbf{k}} = -zF_{ry}\hat{\mathbf{i}} + zF_{rx}\hat{\mathbf{j}} + (xF_{ry} - yF_{rx})\hat{\mathbf{k}}$$

From either the $\hat{\mathbf{i}}$- or $\hat{\mathbf{j}}$-coefficients of this equation, we see that $z = 0$. That is, as we would have anticipated, the line of action of the force-alone resultant lies in the xy plane (the plane of the forces in the original system). But also, from the $\hat{\mathbf{k}}$-coefficients, we find

$$xF_{ry} - yF_{rx} = M_{rP} \qquad \text{or} \qquad y = \frac{F_{ry}}{F_{rx}}x - \frac{M_{rP}}{F_{rx}} \tag{3.23}$$

which provides the equation of the line of action of the force-alone resultant.

Answer 3.25 (a) Because all forces in the system lie in the xy plane. (b) $\mathbf{M}_{rP}$ is made up of (a) couples of the form $C\hat{\mathbf{k}}$, and/or (b) moments of forces of form $(x\hat{\mathbf{i}} + y\hat{\mathbf{j}}) \times (F_x\hat{\mathbf{i}} + F_y\hat{\mathbf{j}})$. The latter is $(xF_y - yF_x)\hat{\mathbf{k}}$, thus only a z component.

By comparing Equation (3.23) with the familiar form of the equation of a straight line, $y = mx + b$, we see that the slope of the line is $m = F_{ry}/F_{rx}$, as it must be.

Question 3.26   Why must it?

Similarly, the y-intercept is $b = -M_{rP}/F_{rx}$. These results are depicted below in Figure 3.27:

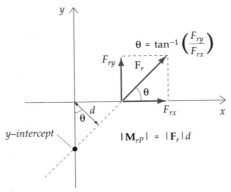

Figure 3.27

Note from the figure that:

$$m = \text{slope} = \tan\theta = \frac{F_{ry}}{F_{rx}}$$

and

$$b = y\text{-intercept} = \frac{-d}{\cos\theta} = -\frac{|\,\mathbf{M}_{rP}\,|\,/\,|\,\mathbf{F}_r\,|}{F_{rx}/\,|\,\mathbf{F}_r\,|} = -\frac{M_{rP}}{F_{rx}},$$

so that Equation (3.23) has been verified.

Before proceeding to examples of coplanar systems, we illustrate a slightly different view* of this process of finding the line of action of the force-alone resultant. Since $\mathbf{M}_{rP}$ and $\mathbf{F}_r$ are perpendicular, we may replace $\mathbf{M}_{rP}$ by an equipollent pair of forces $\mathbf{F}_r$ and $-\mathbf{F}_r$, as shown in the second frame of Figure 3.28, where we let the $-\mathbf{F}_r$ part of this pair have a line of action through our reference point P. Now the "canceling" of the $\mathbf{F}_r$ and $-\mathbf{F}_r$ pair through P leaves the third frame of the figure as the final result. Clearly this reasoning may be used any time $\mathbf{M}_{rP}$ and $\mathbf{F}_r$ are perpendicular.[†] This illustrates the fact that as long as $\mathbf{F}_r \perp \mathbf{M}_{rP}$ to start with, we are guaranteed a force-alone resultant for a coplanar force system.

Answer 3.26   Because the single force resultant is always $\mathbf{F}_r$.
* Which was previewed on page 90.
† See footnote on page 111.

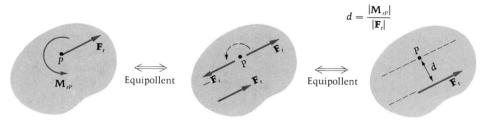

$$d = \frac{|\mathbf{M}_{rP}|}{|\mathbf{F}_r|}$$

Figure 3.28

EXAMPLE 3.27

In Example 3.22, the original system of forces and couples was shown to be equipollent to the single force and couple at O shown in Figure E3.27a. Now, further reduce this coplanar system to a single force.

Solution

We have $\mathbf{M}_{rO} \perp \mathbf{F}_r$. Therefore, the equation

$$\mathbf{r}_{OA} \times \mathbf{F}_r = \mathbf{M}_{rO}$$

will identify all points A on the line of action of the single force, $\mathbf{F}_r$, in an equipollent system consisting only of this force. Thus,

$$(x\hat{\mathbf{i}} + y\hat{\mathbf{j}}) \times (10\hat{\mathbf{i}} - 15\hat{\mathbf{j}}) = -27.9\hat{\mathbf{k}}$$

from which

$$-15x - 10y = -27.9$$

or

$$y = -1.5x + 2.79 \text{ m} \tag{1}$$

The solution, then, is the force $\mathbf{F}_r$ placed along the line defined by Equation (1), and shown below in Figure E3.27b:

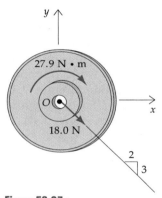

Figure E3.27a

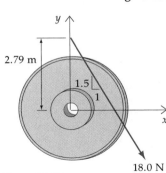

Figure E3.27b

---

† This is true whether or not the original system, to which $\mathbf{F}_r$ and $\mathbf{M}_{rP}$ are equipollent, is coplanar! We will see later in Example 3.31 an example of a non-coplanar system (which is also not concurrent or parallel) in which $\mathbf{F}_r \perp \mathbf{M}_{rP}$ so that a force-alone resultant exists.

In the previous example, we could have proceeded in a different manner: Realizing that we know $\mathbf{F}_r$, all that is really required to establish the line of action is the location of a single point on the line. Thus, we could have simply sought a specific point such as the intersection of the line with the x axis. Setting $y = 0$ and letting $x = a$ at that intersection, we can write

$$a\hat{\mathbf{i}} \times \mathbf{F}_r = \mathbf{M}_{rO}$$

or

$$a\hat{\mathbf{i}} \times (10\hat{\mathbf{i}} - 15\hat{\mathbf{j}}) = -27.9\,\hat{\mathbf{k}}\,\text{N} \cdot \text{m}$$

Thus

$$15a = 27.9$$

$$a = 1.86\ \text{m}$$

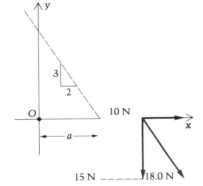

Figure 3.29

as seen in Figure 3.29.

The same result is obtained with less formality, and perhaps in a manner that facilitates physical insight, by referring again to Figure 3.29 where $\mathbf{F}_r$ has been decomposed into its horizontal and vertical components. Since the horizontal part (10 N) has a line of action through O, it produces no moment about O. The clockwise moment (15a) of the vertical part must then be equal to the clockwise 27.9 N · m. Thus

$$15a = 27.9$$

$$a = 1.86\ \text{m}$$

(as before, and also obtainable from Equation (1) of Example 3.27 when y is set to zero).

Yet a fourth approach is to replace $\mathbf{M}_{rO}$ by two forces $\mathbf{F}_r$ and $-\mathbf{F}_r$ (as was suggested by Figure 3.28):

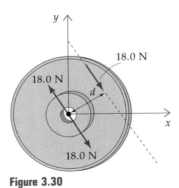

Figure 3.30

Since $\mathbf{M}_{rO}$ and $\mathbf{F}_r$ are given, the distance d in Figure 3.30 has to have the value

$$d = \frac{|\mathbf{M}_{rO}|}{|\mathbf{F}_r|} = \frac{27.9}{18.0} = 1.55\ \text{m}$$

and the two 18.0 N forces at the origin cancel, leaving a force-alone resultant matching that of Figure E3.27b in the previous example. The

y-intercept of that figure is obtainable from Figure 3.30 and the similar triangles in Figure 3.31 below:

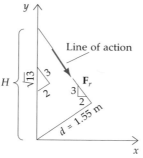

$$\frac{d}{H} = \frac{1.55}{H} = \frac{2}{\sqrt{13}}$$

or

$$H = 2.79 \text{ m, as before.}$$

Figure 3.31

EXAMPLE 3.28

Find the single force that is equipollent to the system shown in Figure E3.28a.

Solution

The resultant force is

$$\mathbf{F}_r = 4\hat{\mathbf{i}} + (3 - 1 + 1)\hat{\mathbf{j}} = 4\hat{\mathbf{i}} + 3\hat{\mathbf{j}} \text{ N}$$

(Note that the 1-N forces form a couple and thus have no resultant force.) At the origin O, the resultant moment is

$$\mathbf{M}_{rO} = \underbrace{5\hat{\mathbf{i}} \times (-1\hat{\mathbf{j}}) + 10\hat{\mathbf{k}}}_{\text{couples}} + \underbrace{4\hat{\mathbf{j}} \times 4\hat{\mathbf{i}}}_{\substack{\text{moment of the 4-N} \\ \text{force about } O}}$$

$$= -11\hat{\mathbf{k}} \text{ N} \cdot \text{m}$$

In a case such as this where it is easy to identify the perpendicular distances from O to the lines of action of various forces, the moments can be calculated with less formality by using the "force times perpendicular distance" method in conjunction with the right-hand rule. With this approach

$$\mathbf{M}_{rO} = 1(5)(-\hat{\mathbf{k}}) + 10\hat{\mathbf{k}} + 4(4)(-\hat{\mathbf{k}})$$

$$= -11\hat{\mathbf{k}} \text{ N} \cdot \text{m} \quad \text{or} \quad 11 \circlearrowright \text{N} \cdot \text{m}$$

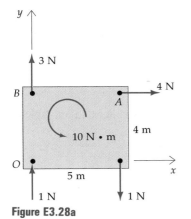

Figure E3.28a

Question 3.27 Why isn't the 3-N force included in the $\mathbf{M}_{rO}$ calculation?

Question 3.28 Why doesn't the vector $\mathbf{r}$, in the $\mathbf{r} \times \mathbf{F}$ calculation for the 4-N force, extend all the way to the application point A of the force?

The equipollent system at O thus appears as shown in Figure E3.28b.

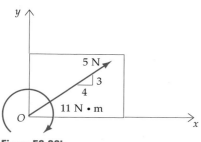

Figure E3.28b

Answer 3.27 It passes through O, and therefore has zero moment about O.

Answer 3.28 As we have seen in Section 3.2, the vector $\mathbf{r} \times \mathbf{F}$ may intersect $\mathbf{F}$ at any point on its line of action and the moment will be the same; in this case $\mathbf{r} = 4\hat{\mathbf{j}}$ intersects the line of action of the force at point B.

The force-alone resultant must now be positioned so that it produces a clockwise 11-N · m moment about point O. This means (see Figure E3.28c below) that this resultant intersects the y-axis *above* point O:

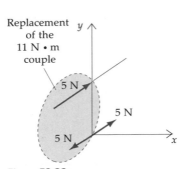

Figure E3.28c

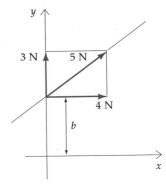

Figure E3.28d

By decomposing the 5-N resultant force into horizontal and vertical components where the line of action of $\mathbf{F}_r$ intersects the y axis (see Figure E3.28d above), we can easily determine that intersection by:

$$4b = 11 \text{ N} \cdot \text{m}$$

$$b = \tfrac{11}{4} = 2.75 \text{ m}$$

This pins down the line of action of the force-alone resultant and completes the example.

___

An alternative approach to the previous example would be to use the more formal vector algebra approach to determine the force-alone resultant and thereby obtain the equation of the line of action of the force, at the possible expense of a bit of physical understanding:

$$\mathbf{r}_{OA} \times \mathbf{F}_r = \mathbf{M}_{rO}$$

where A is any point on the line of action of the force. Therefore,

$$(x\hat{\mathbf{i}} + y\hat{\mathbf{j}}) \times (4\hat{\mathbf{i}} + 3\hat{\mathbf{j}}) = -11\hat{\mathbf{k}}$$

or

$$3x - 4y = -11$$

Question 3.29   Why hasn't a $z\hat{\mathbf{k}}$ been included as part of the vector $\mathbf{r}_{OA}$ above?

Thus

$$y = \tfrac{3}{4}x + \tfrac{11}{4} \text{ m}$$

___

Answer 3.29   We have shown in general that the force-alone resultant for a coplanar force system lies in that same plane.

or

$$y = 0.75x + 2.75 \text{ m}$$

Note from Figure E3.28d that the point where the single-force resultant intercepts the y-axis, namely $(x, y) = (0, 2.75)$, obviously lies on this line.

3. Parallel Force Systems   In this system all of the lines of action of the various forces are parallel; we can let the common direction of these forces be z (Figure 3.32).

An intermediate equipollent system of one force and one couple at the origin has

$$\mathbf{F}_r = \Sigma\mathbf{F}_i = \Sigma F_i\hat{\mathbf{k}} = F_r\hat{\mathbf{k}}$$

where F_r could be negative here, together with the couple

$$\mathbf{M}_{rO} = \Sigma[(x_i\hat{\mathbf{i}} + y_i\hat{\mathbf{j}}) \times F_i\hat{\mathbf{k}}]$$

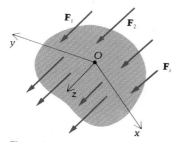

Figure 3.32

in which (x_i, y_i) locates the point where $\mathbf{F}_i$ pierces the xy plane. Performing the cross products,

$$\mathbf{M}_{rO} = \Sigma[F_i(-x_i\hat{\mathbf{j}} + y_i\hat{\mathbf{i}})]$$

and we see that $\mathbf{M}_{rO} \perp \mathbf{F}_r$. This means that the simplest resultant is once again a single force $\mathbf{F}_r$,* with no accompanying couple. To find where this resultant acts, we once again use Equation (3.22):

$$(x\hat{\mathbf{i}} + y\hat{\mathbf{j}} + z\hat{\mathbf{k}}) \times F_r\hat{\mathbf{k}} = \mathbf{M}_{rO} = \Sigma[F_i(y_i\hat{\mathbf{i}} - x_i\hat{\mathbf{j}})]$$

where (x, y, z) are the coordinates of a point on the line of action of $\mathbf{F}_r$. Matching the coefficients of $\hat{\mathbf{i}}$, $\hat{\mathbf{j}}$, and $\hat{\mathbf{k}}$, we obtain

$$\hat{\mathbf{i}} \text{ coefficients:} \quad y = \frac{\Sigma(F_iy_i)}{F_r} \tag{3.24a}$$

$$\hat{\mathbf{j}} \text{ coefficients:} \quad x = \frac{\Sigma(F_ix_i)}{F_r} \tag{3.24b}$$

$$\hat{\mathbf{k}} \text{ coefficients:} \quad 0 = 0 \quad \text{(means } z \text{ can have any value)}$$

Thus the line parallel to the z axis with x and y given by Equations (3.24a,b) is the line of action. At any point of this line, a system equipollent to the original one of Figure 3.32 is given quite simply by Figure 3.33.

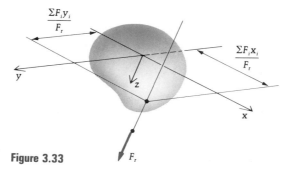

Figure 3.33

---

* Unless, of course, $\mathbf{F}_r = 0$ in which case the simplest resultant is the couple $\mathbf{M}_{rO}$.

We shall now consider two examples of parallel force systems:

EXAMPLE 3.29

Find the single-force resultant for the system of parallel forces acting on the plate in Figure E3.29.

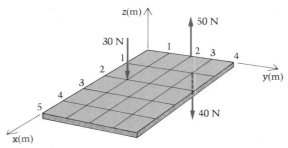

Figure E3.29

Solution

Let us call the given system S_1, and the system we are seeking, S_2. The force in S_2 is found to be

$$\mathbf{F}_r = \overset{S_1}{\sum}\mathbf{F} = -30\hat{\mathbf{k}} - 40\hat{\mathbf{k}} + 50\hat{\mathbf{k}}$$
$$= -20\hat{\mathbf{k}} \text{ N}$$

If we let x and y be the coordinates locating the intersection of the line of action of the force-alone resultant with the surface of the plate, then

$$(x\hat{\mathbf{i}} + y\hat{\mathbf{j}}) \times \mathbf{F}_r = \mathbf{M}_{rO}$$

or

$$(x\hat{\mathbf{i}} + y\hat{\mathbf{j}}) \times (-20\hat{\mathbf{k}}) = 2\hat{\mathbf{j}} \times 50\hat{\mathbf{k}} + (\hat{\mathbf{i}} + 3\hat{\mathbf{j}})$$
$$\times (-40\hat{\mathbf{k}}) + (2\hat{\mathbf{i}} + \hat{\mathbf{j}}) \times (-30\hat{\mathbf{k}})$$

or

$$20x\hat{\mathbf{j}} - 20y\hat{\mathbf{i}} = 100\hat{\mathbf{i}} + (40\hat{\mathbf{j}} - 120\hat{\mathbf{i}}) + (60\hat{\mathbf{j}} - 30\hat{\mathbf{i}})$$

Thus, collecting like terms,

$$\hat{\mathbf{i}}: \quad -20y = 100 - 120 - 30 = -50$$
$$y = 2.5 \text{ m}$$

and

$$\hat{\mathbf{j}}: \quad 20x = 40 + 60 = 100$$
$$x = 5 \text{ m}$$

These, of course, are the x and y coordinates of *every point* on the line of action of the force $\mathbf{F}_r$, since in this case the line of action parallels the z axis.

It is important to realize that the $\hat{\mathbf{i}}$ and $\hat{\mathbf{j}}$ parts of $\mathbf{M}_{rO}$ in the preceding example are the moments about the x and y axes, respectively. This fact may be used to locate the line of action of the force-alone resultant without recourse to the formalities of vector algebra. The force-alone resultant must produce the same moments about these axes as do the forces in the original system. And since all of the forces are perpendicular to the axes in question, the "force times perpendicular distance" method easily may be used to calculate the moment. Referring to Figure 3.34, then, with the 20-N resultant force shown dashed,

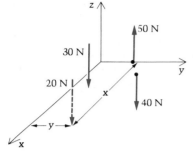

Figure 3.34

$$\text{Moment about } x = -20y = -30(1) + 50(2) - 40(3) = -50$$

$$y = 2.5 \text{ m} \quad \text{(as before)}$$

and

$$\text{Moment about } y = 20x = 30(2) + 40(1) = 100$$

$$x = 5 \text{ m} \quad \text{(as before)}$$

EXAMPLE 3.30

Determine the force-alone resultant for the system of six parallel forces acting on the beam as shown in Figure E3.30.

Solution

The required resultant consists of a single force computed as

$$\mathbf{F}_r = \Sigma\mathbf{F} = (-10 - 20 - 30 - 40 - 50 - 60)\hat{\mathbf{k}}$$

$$= -210\hat{\mathbf{k}} \text{ lb}$$

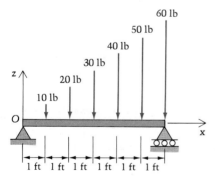

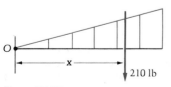

Figure E3.30

Its location is given by the value of x for which the moment of $\mathbf{F}_r$ about, say, O is the same as the moment of the six forces about O:

$$-210x = -10(1) - 20(2) - 30(3) - 40(4) - 50(5) - 60(6)$$

$$x = \frac{910}{210}$$

$$= 4.33 \text{ ft}$$

The y coordinate of the force-alone resultant vanishes this time because all the forces lie in the xz plane.

In this example, we have a set of forces, finite in number, that happen to increase proportionately. In Section 3.9 we shall consider what happens when a loading becomes *continuously distributed* across a line or an area, instead of acting at a few discrete points as above.

The Simplest Resultant of a General Force-and-Couple System: A Collinear Force and Couple ("Screwdriver")*

We close this section with two examples. In the first, the force and couple parts of the resultant at P, $\mathbf{F}_r$ and $\mathbf{M}_{rP}$, happen to be perpendicular. Thus, even though the original system is neither concurrent, coplanar, nor parallel, a force-alone resultant can be found.

In the second example, $\mathbf{F}_r$ and $\mathbf{M}_{rP}$ are not perpendicular, so a force-alone resultant does not exist. However, this example will show how we can always reduce such a system to *a force together with a parallel couple*. Since this is the mechanical action required to advance a screw with a screwdriver (see Figure 3.35), we call this simplest resultant the "equipollent screwdriver" for the system.† As the reader will see, the reduction is accomplished by applying Equation (3.22) to *that part of $\mathbf{M}_{rP}$ that is perpendicular to* $\mathbf{F}_r$.

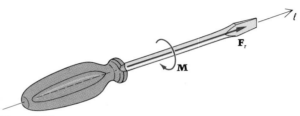

Figure 3.35

EXAMPLE 3.31

Reduce the system of Example 3.25 (see Figure E3.31a) to a force-alone resultant, if possible.

Solution

We have in the earlier example already reduced the system to its resultant at the origin O:

$$\mathbf{F}_r = 8\hat{\mathbf{j}} - 8\hat{\mathbf{k}} \text{ N} \qquad \text{and} \qquad \mathbf{M}_{rO} = 8\hat{\mathbf{i}} + 2.4\hat{\mathbf{j}} + 2.4\hat{\mathbf{k}} \text{ N} \cdot \text{m}$$

We note that:

$$\mathbf{M}_{rO} \cdot \mathbf{F}_r = (8\hat{\mathbf{i}} + 2.4\hat{\mathbf{j}} + 2.4\hat{\mathbf{k}}) \cdot (8\hat{\mathbf{j}} - 8\hat{\mathbf{k}})$$

$$= 2.4(8) + 2.4(-8) = 0$$

This zero result means that the resultant moment at O is in fact perpendicular to the resultant force. Therefore, three-dimensional though it is, a force-alone resultant can be found in this example. We proceed then toward determining its

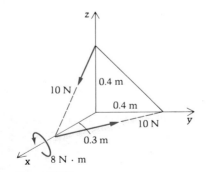

Figure E3.31a

---

* This material has been included for completeness, but nothing essential to the study of statics is missed if this part of the section is omitted.
† In most texts this is called a "wrench," but the action of a screwdriver is more descriptive.

line of action. If A is a point on this line, then we may write

$$\mathbf{r}_{OA} = x\hat{\mathbf{i}} + y\hat{\mathbf{j}} + z\hat{\mathbf{k}}$$

and then from Equation (3.22),

$$\mathbf{r}_{OA} \times \mathbf{F}_r = \mathbf{M}_{rO}$$

we get

$$(x\hat{\mathbf{i}} + y\hat{\mathbf{j}} + z\hat{\mathbf{k}}) \times (8\hat{\mathbf{j}} - 8\hat{\mathbf{k}}) = 8\hat{\mathbf{i}} + 2.4\hat{\mathbf{j}} + 2.4\hat{\mathbf{k}}$$

$$(-8y - 8z)\hat{\mathbf{i}} + (8x)\hat{\mathbf{j}} + (8x)\hat{\mathbf{k}} = 8\hat{\mathbf{i}} + 2.4\hat{\mathbf{j}} + 2.4\hat{\mathbf{k}}$$

so that

$\hat{\mathbf{i}}$ coefficients:	$-8y - 8z = 8$	(1)
$\hat{\mathbf{j}}$ coefficients:	$8x = 2.4$	(2)
$\hat{\mathbf{k}}$ coefficients:	$8x = 2.4$	(3)

Equations (2) and (3) each give the result that $x = 0.3$; this means that the line lies in the plane parallel to yz at this value of x. In this plane, its equation is given by (1):

$$y + z = -1$$

The simplest resultant is shown in Figure E3.31b.

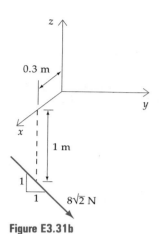

Figure E3.31b

EXAMPLE 3.32

The three forces in Figure E3.32a each have a magnitude of 10 lb. Find the screwdriver equipollent to this system.

Solution

The force resultant is

$$\mathbf{F}_r = 10\hat{\mathbf{i}} + 10\hat{\mathbf{j}} + 10\hat{\mathbf{k}} = 10\sqrt{3}\left(\frac{\hat{\mathbf{i}} + \hat{\mathbf{j}} + \hat{\mathbf{k}}}{\sqrt{3}}\right) \text{ lb}$$

Already, then, we know that (a) the force of the "screwdriver" is $10\sqrt{3}$ lb and (b) the orientation of its line in space is given by the unit vector $(\hat{\mathbf{i}} + \hat{\mathbf{j}} + \hat{\mathbf{k}})/\sqrt{3}$, which means in this case that the axis of the screwdriver makes equal angles (each $= \cos^{-1}(1/\sqrt{3}) = 54.7°$) with the coordinate axes.

At the origin, an equipollent system to the three given forces is $\mathbf{F}_r$ accompanied by the resultant couple there:

$$\mathbf{M}_{rO} = 3\hat{\mathbf{k}} \times 10\hat{\mathbf{i}} + (6\hat{\mathbf{i}} + 2\hat{\mathbf{j}}) \times 10\hat{\mathbf{k}}$$

$$= 20\hat{\mathbf{i}} - 30\hat{\mathbf{j}} \text{ lb-ft}$$

Here we pause to note that since

$$\mathbf{M}_{rO} \cdot \mathbf{F}_r = 20(10) - 30(10) + 0(10) \neq 0,$$

there is no possibility of a force-alone resultant this time. The non-zero dot product means there is a component of $\mathbf{M}_{rO}$ lying *along* the line of action of $\mathbf{F}_r$ so that the reduction illustrated by Figure 3.23 is not possible.

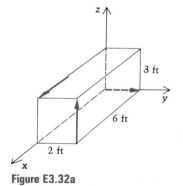

Figure E3.32a

The "parallel component" of $\mathbf{M}_{rO}$ (the part parallel to $\mathbf{F}_r$) is

$$M_{rO\parallel} = \mathbf{M}_{rO} \cdot \left(\frac{\mathbf{F}_r}{|\mathbf{F}_r|}\right) = (20\hat{\mathbf{i}} - 30\hat{\mathbf{j}}) \cdot \left(\frac{\hat{\mathbf{i}} + \hat{\mathbf{j}} + \hat{\mathbf{k}}}{\sqrt{3}}\right)$$

$$= \frac{20 - 30}{\sqrt{3}} = \frac{-10}{\sqrt{3}} \text{ lb-ft}$$

Therefore,

$$\mathbf{M}_{rO\parallel} = \frac{-10}{\sqrt{3}} \frac{\mathbf{F}_r}{|\mathbf{F}_r|} = \frac{-10}{\sqrt{3}}\left(\frac{\hat{\mathbf{i}} + \hat{\mathbf{j}} + \hat{\mathbf{k}}}{\sqrt{3}}\right) = \frac{-10}{3}(\hat{\mathbf{i}} + \hat{\mathbf{j}} + \hat{\mathbf{k}}) \text{ lb-ft}$$

This moment, $\mathbf{M}_{rO\parallel}$, is the couple of the screwdriver. Note that this time it is *opposite* in direction to $\mathbf{F}_r$, as if a wood screw were being *unscrewed* instead of advanced.

The "perpendicular component" of $\mathbf{M}_{rO}$ is the other part of $\mathbf{M}_{rO}$, this time *normal* to $\mathbf{F}_r$:

$$\mathbf{M}_{rO\perp} = \mathbf{M}_{rO} - \mathbf{M}_{rO\parallel}$$

$$= (20\hat{\mathbf{i}} - 30\hat{\mathbf{j}}) - \left[\frac{-10}{3}(\hat{\mathbf{i}} + \hat{\mathbf{j}} + \hat{\mathbf{k}})\right]$$

$$= \tfrac{70}{3}\hat{\mathbf{i}} - \tfrac{80}{3}\hat{\mathbf{j}} + \tfrac{10}{3}\hat{\mathbf{k}} \text{ lb-ft}$$

Note as a check that, by inspection, $\mathbf{F}_r \cdot \mathbf{M}_{rO\perp} = 0$.

Finally, if A is any point on the line of action of the screwdriver, located by the position vector

$$\mathbf{r}_{OA} = x\hat{\mathbf{i}} + y\hat{\mathbf{j}} + z\hat{\mathbf{k}}$$

then the condition

$$\mathbf{r}_{OA} \times \mathbf{F}_r = \mathbf{M}_{rO\perp}$$

gives the equation of the axis of the equipollent screwdriver:

$$(x\hat{\mathbf{i}} + y\hat{\mathbf{j}} + z\hat{\mathbf{k}}) \times (10\hat{\mathbf{i}} + 10\hat{\mathbf{j}} + 10\hat{\mathbf{k}}) = \tfrac{70}{3}\hat{\mathbf{i}} - \tfrac{80}{3}\hat{\mathbf{j}} + \tfrac{10}{3}\hat{\mathbf{k}}$$

from which we obtain

$\hat{\mathbf{i}}$ coefficients: $10y - 10z = \tfrac{70}{3}$ (1)

$\hat{\mathbf{j}}$ coefficients: $10z - 10x = -\tfrac{80}{3}$ (2)

$\hat{\mathbf{k}}$ coefficients: $10x - 10y = \tfrac{10}{3}$ (3)

Note that by adding Equations (1) and (2), and multiplying the result by -1, we obtain Equation (3). This means that only two of the equations are independent and necessary to define the axis of the screwdriver. Indeed, a pair of equations such as (1) and (2) constitute the general form of a line in three-dimensional space. Before leaving this example, let us find two points on the line, and sketch the final answer:

(1) $y - z = \tfrac{7}{3}$

(3) $x - y = \tfrac{1}{3}$

Point 1:   If $z = 0$, $y = \tfrac{7}{3}$ If $y = \tfrac{7}{3}$, $x = \tfrac{8}{3}$

Point 2:   If $y = 0$, $x = \tfrac{1}{3}$ If $y = 0$, $z = -\tfrac{7}{3}$

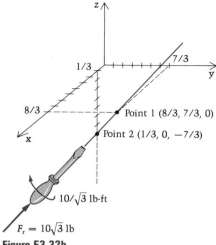

Figure E3.32b

The screwdriver may thus be sketched as shown in Figure E3.32b.

Question 3.30   Are the direction numbers of the screwdriver axis, obtained from the coordinates of points 1 and 2 in Example 3.32, in agreement with direction numbers obtained from $\mathbf{F}_r$ itself?

Answer 3.30   Yes, as they must be. Those of the lines are $[\frac{8}{3} - \frac{1}{3}, \frac{7}{3} - 0, 0 - (-\frac{7}{3})]$ $= (\frac{7}{3}, \frac{7}{3}, \frac{7}{3})$, and these are proportional to the direction numbers of the line of action of $\mathbf{F}_r$, which are $(1/\sqrt{3}, 1/\sqrt{3}, 1/\sqrt{3})$.

PROBLEMS ▶ Section 3.8

3.90   Five members of a truss are exerting the indicated loads on the pin at point O as shown in Figure P3.90. Find the single-force resultant of the five forces.

Find the force-alone resultants of the concurrent force systems in Figures P3.91–P3.96.

3.91

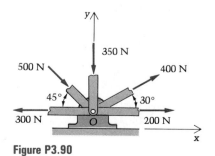

Figure P3.90

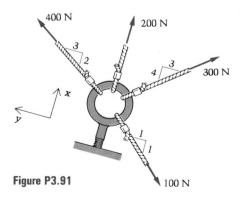

Figure P3.91

3.92

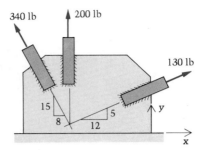

Figure P3.92

3.93

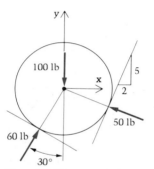

Figure P3.93

3.94

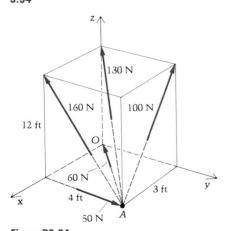

Figure P3.94

3.95

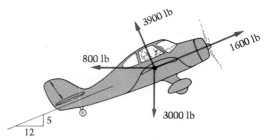

Figure P3.95

3.96

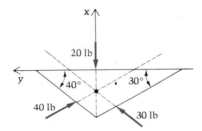

Figure P3.96

3.97 Find the force-alone resultant of the five forces shown in Figure P3.97. Locate the intersection of the line of action of the resultant with line *BC*.

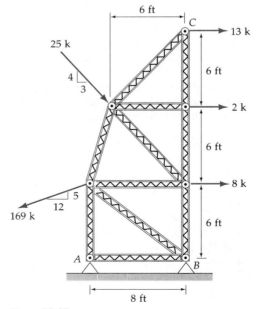

Figure P3.97

Find the simplest resultant for each of the coplanar force systems shown in Figures P3.98–P3.103. Give the equation of the line of action in each case, and determine where it crosses the x axis.

3.98

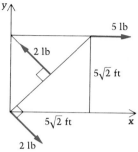

Figure P3.98

3.99

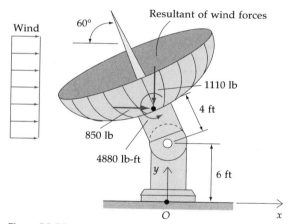

Figure P3.99

3.100

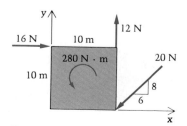

Figure P3.100

3.101

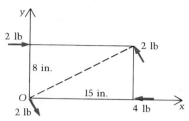

Figure P3.101

3.102

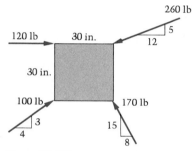

Figure P3.102

3.103

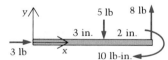

Figure P3.103

3.104 The forces have vertical lines of action and act through a horizontal unit grid as shown in Figure P3.104. Find and locate the single force that is equipollent to the given force system.

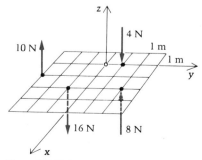

Figure P3.104

Find the simplest resultant for each of the parallel force systems shown in Figures P3.105–P3.109. Locate the line of action of each in the coordinate system given.

3.105

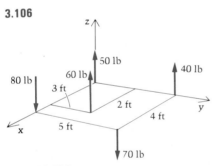

Figure P3.105

3.106

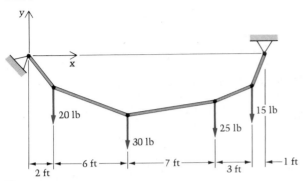

Figure P3.106

3.107

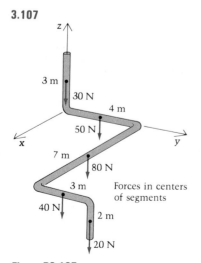

Figure P3.107

3.108

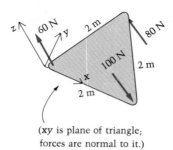

(**xy** is plane of triangle; forces are normal to it.)

Figure P3.108

3.109

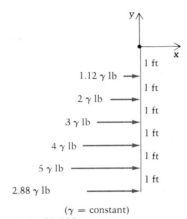

(γ = constant)

Figure P3.109

3.110 Replace the force system shown in Figure P3.110 by:

    a. A force through O and a couple

    b. A single force

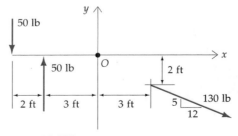

Figure P3.110

3.111 Is it possible to reduce the force system shown in Figure P3.111 to a single force? If so, what is the shortest distance from P to the line of action of this single force?

Figure P3.111

3.112 Suppose we take a complicated system of forces and couples and reduce them to an equipollent system, at P, of force $\mathbf{F}_r$ and moment $\mathbf{M}_{rP}$. When can the system be further reduced to (a) a single force? (b) a single couple? Hint: Think about the figure of the preceding problem.

3.113 Replace the two forces acting on the bar in Figure P3.113 by a single force. Give its magnitude, direction and line of action.

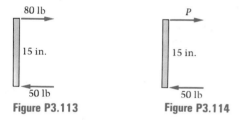

Figure P3.113   **Figure P3.114**

3.114 Find the force P so that the force-alone resultant of the two forces acts through the center of the bar in Figure P3.114.

3.115 With reference to Figure P3.115, find the following:

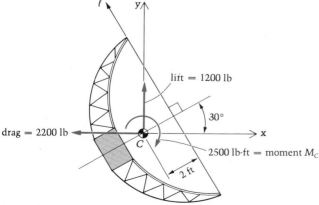

Figure P3.115

a. The force-alone resultant of the lift and drag forces and moment $\mathbf{M}_C$, which represent the resultant at C of the wind forces on the parabolic antenna dish.

b. The point where the resultant from part (a) crosses the line ℓ in the figure. (This line is the intersection of the rim plane and the xy plane.)

3.116 Figure P3.116 shows a force-couple system. The magnitudes are:

$$|\mathbf{F}_1| = 3\sqrt{13}\ \text{lb}$$
$$|\mathbf{F}_2| = 2\sqrt{40}\ \text{lb}$$
$$|\mathbf{C}_1| = 21\ \text{lb-ft}$$
$$|\mathbf{C}_2| = 9\ \text{lb-ft}$$

Give an equipollent system consisting of a single force and couple through point A (2, 0, 0). Can your system be further reduced to a single force or to a single couple?

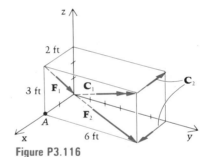

Figure P3.116

3.117 Replace the force system of Figure P3.117 by an equipollent system consisting of (a) a force through O and a couple; (b) a single force.

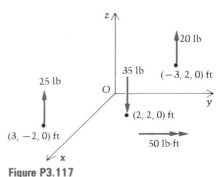

Figure P3.117

3.118 Show that if the elements of a force-couple resultant at one point are perpendicular, then the elements at any other point are also perpendicular.

3.119 Three forces of magnitude P, $2P$, and $3P$ act on a block as shown in Figure P3.119.

a. Find a equipollent system consisting of a force and a couple at point A.

b. Give the relation between a, b, and c so that the system may be reduced to a single force.

c. Can the system be reduced to a single couple? Why or why not?

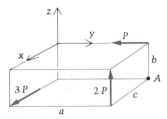

Figure P3.119

3.120 If possible, replace the system of forces in Figure P3.120 by a single equipollent force. Otherwise replace the system by a force through O and a couple.

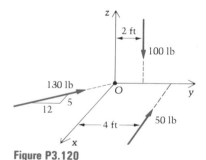

Figure P3.120

3.121 With reference to Figure P3.121, and in terms of l, find the following:

a. The force and couple at the origin O that are equipollent to the three forces and three couples shown acting on the bent bar.

b. The value of l for which the system can be further reduced to a single force.

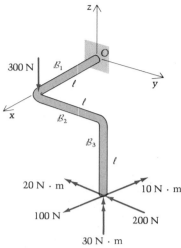

Figure P3.121

3.122 In Figure P3.122 a system of four forces and a couple is shown. Replace this system by an equipollent system consisting of a single couple and a single force whose line of action passes through point A, located at coordinates $(2, 6, 0)$. Further reduce this new system to a single force, if possible.

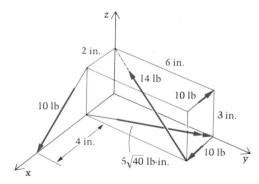

Figure P3.122

* 3.123 A circular plate of radius R supports three vertical loads as shown in Figure P3.123 on the next page. Determine the magnitude and point of application of the smallest additional vertical force that must be applied onto the surface of the plate if the four loads are to be equipollent to: (a) zero (that is, to a system with $\mathbf{F}_r = \mathbf{0}$ and $\mathbf{M}_{rO} = \mathbf{0}$); (b) a force through the center of the plate. What is this force?

3.124 Equation (3.23) in the text must obviously be rederived if $F_{rx} = 0$. Obtain the correct equation of the line for this special case.

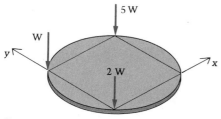

Figure P3.123

3.125 Return to Problem 3.74 and find the screwdriver equipollent to the forces and couples shown again in Figure P3.125. At what point does it pierce the ground (xy plane)?

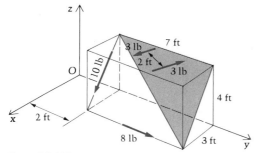

Figure P3.125

3.126 Return to Example 3.18 and find the screwdriver equipollent for each of the six force systems. Why can none of them be a single force through P?

3.127 Find the force and couple that must be added at point A to the force system shown at the left in Figure P3.127 so that it will be equipollent to the force system at the right. What is the screwdriver equipollent to the original system on the left?

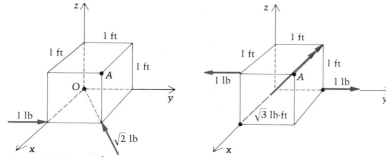

Figure P3.127

In Problems 3.128–3.139, the procedure outlined in the text is to be followed in establishing the "equipollent screwdriver" for the given system — that is, the resultant consisting of a collinear force and couple. Give, in addition to the force and couple, a complete description of the line of action of the screwdriver.

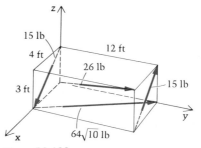

Figure P3.128

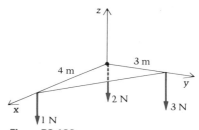

Figure P3.129

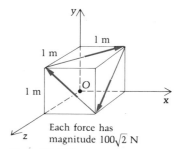

Each force has magnitude $100\sqrt{2}$ N

Figure P3.130

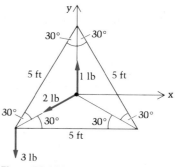

Figure P3.131

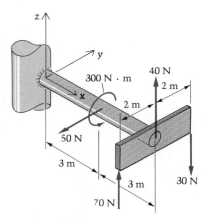

Figure P3.132

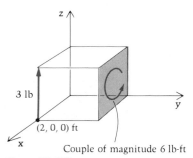

Figure P3.133

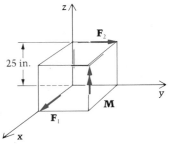

Figure P3.134 $F_1 = 3$ lb, $F_2 = 4$ lb, and $M = 50$ lb-in.

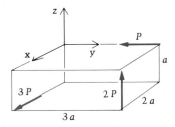

Figure P3.135

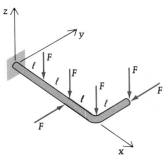

Figure P3.136

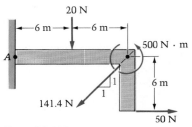

Figure P3.137

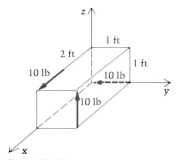

Figure P3.138

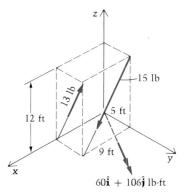

Figure P3.139

3.140 Even though the force system of Examples 3.25 and 3.31 was not concurrent, coplanar, or parallel, there *was* a force-alone resultant. From the force and couple resultant found in Example 3.25 at the point of intersection A of the two 10-N forces, note that $\mathbf{F}_r \cdot \mathbf{M}_{rA} = 0$. Now state why *any* change in the magnitude of just *one* of the two forces would make a single-force resultant impossible.

3.141 Explain the general process of going from the force and couple resultant of Figure P3.141(a) to the equipollent screwdriver of Figure P3.141(d). Tell what was done in each step.

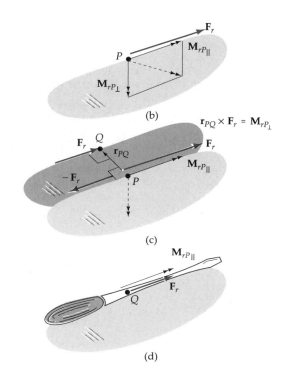

(b)

(c)

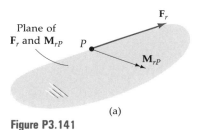

(a)

Figure P3.141

(d)

3.9   Distributed Force Systems

Probably the most important reason for studying the resultants of force systems is to have at our fingertips the resultants of frequently occurring distributed force systems. In engineering mechanics we usually are dealing with bodies on a large scale (macroscopic), where material is perceived (or modeled) to be continuously distributed in space. This is in contrast to a microscopic view where we might be distinguishing individual atoms and the spaces between them. We perceive mechanical actions to be exerted on bodies either by direct contact or by the action of a "field" such as gravity or electromagnetism. In the first instance (for example, pressing a finger against this book), it is natural to view the force exerted as the net effect of something distributed over the surface area of contact. In the second case (for example, the gravitational force exerted by the earth on the book), it is natural to view the force as the net effect of "weights" of individual particles, or elements of mass, which are distributed through the volume of the body. Thus, the mechanical actions that naturally arise in engineering mechanics may be classified as either **surface forces** or **body forces**.

Forces Distributed Along a Straight Line

We begin our calculations of resultants of distributed force systems by considering the simplest case, in which the distribution is over a *line*. This, of course, doesn't fit the classifications of the preceding paragraph, but it commonly arises as a system that is equipollent to one of those classifications. In particular we consider the frequently occurring case, illustrated in Figure 3.36, in which the force system is distributed over a straight line and the mechanical actions are all parallel to one another and perpendicular to the line.

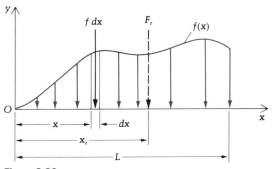

Figure 3.36

Letting $\mathbf{f} = -f\hat{\mathbf{j}}$ be the distributed force intensity — that is, force per unit of length along the line — then an elemental, or infinitesimal, force in the system at location x is $(-f\,dx)\hat{\mathbf{j}}$. To find the resultant force $\mathbf{F}_r$ we, as always, must simply add up all the forces in the system. Here the process of summation becomes integration so that

$$\mathbf{F}_r = \int_0^L \mathbf{f}\,dx = \int_0^L (-f\hat{\mathbf{j}})\,dx$$

$$= -\left(\int_0^L f\,dx\right)\hat{\mathbf{j}}$$

where the unit vector $-\hat{\mathbf{j}}$ is a constant, both in magnitude *and* direction. The couple part of the resultant at O is obtained by adding up (integrating) the moments of the elemental forces in the distributed system so that

$$\mathbf{M}_{rO} = \int_0^L x\hat{\mathbf{i}} \times (-f\hat{\mathbf{j}}\,dx) = -\hat{\mathbf{k}} \int_0^L xf\,dx$$

Thus if $\int_0^L f\,dx \neq 0$, there is a force-alone resultant (as we may have anticipated since our distributed system is both parallel *and* coplanar). Letting x_r locate the line of action of this force-alone resultant, we can find x_r by ensuring that the moment condition for equipollence is satisfied:

$$x_r\hat{\mathbf{i}} \times \mathbf{F}_r = \mathbf{M}_{rO}$$

or

$$x_r \int_0^L f \, dx = \int_0^L xf \, dx$$

or

$$x_r = \frac{\displaystyle\int_0^L xf \, dx}{\displaystyle\int_0^L f \, dx} \qquad (3.25)$$

It is important to observe the similarity of Equation (3.25) to the second of Equations (3.24), where parallel discrete forces were being studied. It is also of interest to observe that $\int_0^L f \, dx$ may be interpreted as the "area" under the force intensity curve or loading curve, when $f(x)$ is graphed.

We shall see in Chapter 7 that x_r locates the x coordinate of the "centroid" of the area beneath the loading curve of Figure 3.31. The denominator of Equation (3.25) measures the "total" (or magnitude of the resultant) of the distributed force.

In summary, then, the force-alone resultant of a parallel distributed line loading has (a) a magnitude and sense given by the signed area beneath the loading curve, (b) a direction parallel to that of the parallel distributed forces, and (c) a line of action given by Equation (3.25). We shall now consider a series of examples in which we compute resultants of distributed line loadings.

EXAMPLE 3.33

Find the resultant of the uniformly distributed weight of the sacks of cement stacked on the dock. There are 12 piles as shown, with 8 bags per pile, and each 94-lb bag is 20 in. wide by 2.5 ft long as shown in Figures E3.33a, b.

20 in. →| |←

Figure 3.33a

2.5 ft

20 in. →| |←

|— 20 ft —|

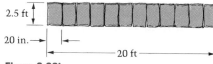

Figure 3.33b

Solution

The total weight of the cement is

$$94 \frac{lb}{bag}(8 \times 12) \text{ bags} = 9020 \text{ lb} \qquad \text{(to three digits)}$$

This of course is the magnitude of the resultant $\mathbf{F}_r$. By intuition, its line of action is downward, 10 feet from either end of the cement. But let us obtain the result by ensuring that the moments of the two systems shown in Figure E3.33c are the same about the left end A of the loading.

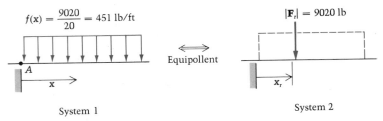

$$f(x) = \frac{9020}{20} = 451 \text{ lb/ft} \qquad\qquad |\mathbf{F}_r| = 9020 \text{ lb}$$

Equipollent

System 1   System 2

Figure 3.33c

$$(M_{rA})_1 = \int_0^L x f(x)\, dx = \int_0^{20} \frac{9020}{20} x\, dx = 451 \left.\frac{x^2}{2}\right|_0^{20} = 90{,}200 \text{ lb-ft}$$

This must equal the moment $(M_{rA})_2$ of the force-alone resultant:

$$9020\, x_r = 90{,}200$$

$$x_r = 10 \text{ ft (as expected)}$$

Note that in the solution we did not need the 2.5-ft dimension. Since the loading distribution does not vary in the direction perpendicular to the page, we have idealized the weight as a line load acting in the central plane of the bags. More will be said about this later in the section.

EXAMPLE 3.34

A concrete ramp has the weight distribution shown in Figure E3.34a, where p is the weight per unit length at the right end of the ramp. Find the resultant of the triangularly distributed weight.

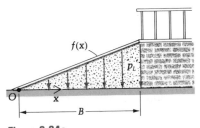

Figure 3.34a

Solution

The distributed loading has the equation (see Figure E3.34b):

$$f(x) = \frac{p_L}{B} x$$

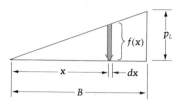

Figure 3.34b

The differential force dF at x is given as suggested in the figure by

$$df = f(x)\, dx = \frac{p_L}{B} x\, dx$$

so that the resultant $\mathbf{F}_r$ has magnitude

$$\int_0^B \frac{p_L}{B} x\, dx = \frac{p_L B}{2}$$

This is 1/2 the base times the height of the triangle, or the area beneath the loading curve, as we have previously shown. We next locate the line of action of the resultant by finding the value of x_r in Figure E3.34c for which the moments of the two systems are the same about some common point, say, O:

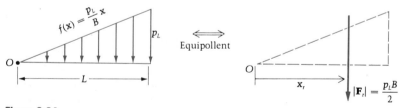

Figure 3.34c

We obtain

$$|\mathbf{F}_r|\, x_r = \int_0^B x f(x)\, dx = \int_0^B x\, \frac{p_L}{B} x\, dx$$

or

$$\frac{p_L B}{2} x_r = \frac{p_L}{B} \frac{x^3}{3} \Big|_0^B = \frac{p_L B^2}{3}$$

$$x_r = \frac{2}{3} B$$

Therefore, whenever we encounter a triangularly distributed loading curve, as in the example to follow, we will know the magnitude of its resultant is its area, and that its line of action is 2/3 the distance from the vertex to the opposite side.

Question 3.31   How would you handle a case in which some of the distributed loading was upward and some downward?

We now address the fact that loadings such as we have just seen in Examples 3.33 and 3.34 are actually (that is, physically) distributed over *surfaces*. Force systems distributed over surfaces occur with great frequency in engineering mechanics. We shall treat here the special case in which each elemental force is perpendicular to the surface — that is, simple pressure. To begin, let us suppose that the surface is plane (flat) and rectangular in shape and that the pressure $p(x)$ varies only with the coordinate x, measured along one edge of the surface (Figure 3.37(a)).

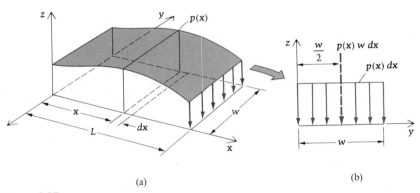

(a)   (b)

Figure 3.37

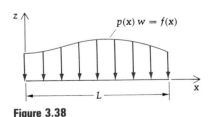

Figure 3.38

Along a strip of infinitesimal width dx and length w we have a set of infinitesimal forces $p(x)\, dx\, dy$, which can be viewed as a line-distributed force system, the intensity at each point being $p(x)\, dx$. The resultant of the strip of infinitesimal width is then $p(x)w\, dx$ [see Figure 3.37(b)] with line of action at $y = w/2$. Since this is true in *each* such strip, the pressure loading is equipollent to a line-distributed system along the line $y = (w/2)$ with $p(x)w = f(x)$ being the intensity as shown in Figure 3.38.* It is in this fashion that line-distributed force systems

---

Answer 3.31   In Equation (3.25), if f is negative over a part (or parts) of the interval $(0, L)$, it simply contributes a negative result for that portion of the integrals. Note that if $\int_0^L f\, dx = 0$, then the resultant will be (at most) simply a couple.

* The same type of equipollent line loading along $y = (w/2)$ occurs when $p = p(x, y)$ if the pressure is symmetric about $y = (w/2)$ — that is, if $p(x, y)$ is an even function of $(y - w/2)$. In that more general case, $f(x) = \int_0^w p(x, y)\, dy$.

arise in mechanics. When they are encountered in examples, the reader should always bear in mind that they are the result of the process we have just been through. This was the case in the two previous examples and will be again in the one to follow, in which we shall discuss its resolution into a line load.

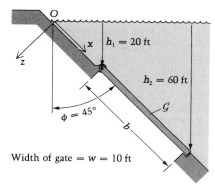

Figure E3.35a

EXAMPLE 3.35

Water is held back by the submerged rectangular gate $\mathcal{G}$ in Figure E3.35a. Find the value of the single force (and its location) that is equipollent to the fluid pressure forces acting on $\mathcal{G}$.

Solution

As the reader may recall from previous studies, the pressure p in a fluid at rest:

1. is equal in all directions at a point (Pascal's Law)
2. is constant through the fluid in each horizontal plane
3. causes a force that is normal to every differential area of surface on which it acts
4. is equal to $\rho g h$, where ρ is the mass density of the fluid (taken to be constant here), and h is the depth below the free surface

For a flat gate, the distributed loading caused by the water pressure therefore forms a parallel force system. Multiplying the pressure by the constant width w of the gate, we obtain the distributed line load as discussed in the preceding text and as illustrated in Figure E3.35b.

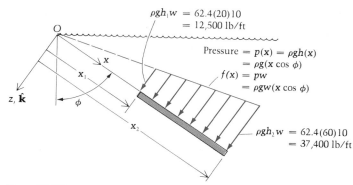

Figure E3.35b

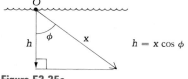

Figure E3.35c

In Figure E3.35b we have used the fact that h and the "slanted coordinate" x are related quite simply, as shown in Figure E3.35c. The single force that is equipollent to the system of parallel distributed forces caused by the water pressure is calculable from

$$F_r = \int f(x)\, dx = \int_{20\sqrt{2}}^{60\sqrt{2}} \underbrace{62.4(10)\frac{1}{\sqrt{2}}}_{\rho g w\, \cos\phi}\, x\, dx$$

$$= 441\left.\frac{x^2}{2}\right|_{20\sqrt{2}}^{60\sqrt{2}}$$

$$= 1.41 \times 10^6\ \text{lb*}$$

The location of this resultant follows from equating the moment about O of the single-force resultant to that of the distributed loading:

$$\underbrace{F_r}_{1.41(10^6)\ \text{lb}}\ x_r = \int xf(x)\, dx = \int_{20\sqrt{2}}^{60\sqrt{2}} 441x^2\, dx = 441\left.\frac{x^3}{3}\right|_{20\sqrt{2}}^{60\sqrt{2}}$$

$$x_t = \frac{86.5 \times 10^6}{1.41 \times 10^6}$$

$$= 61.3\ \text{ft}$$

The single-force resultant is shown in Figure E3.35d at the left.

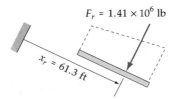

$F_r = 1.41 \times 10^6$ lb

$x_r = 61.3$ ft

Figure E3.35d

Dividing a Distributed Force System into Composite Parts

We next illustrate an approach which can be used to avoid integrations whenever the loading can be divided into several individually familiar "composite parts":

EXAMPLE 3.36

Find the single-force resultant of the distributed loading acting on the beam shown in Figure E3.36a.

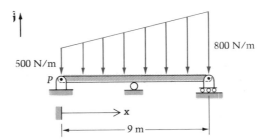

800 N/m

500 N/m

P

x

9 m

Figure E3.36a

---

* We shall show in Chapter 9 that this resultant equals the product of the pressure at the centroid of the gate and its area. Thus, as a check, $F_r = [(62.4)40][(60 - 20)\sqrt{2}(10)]$ $= 1.41 \times 10^6$ lb.

Solution

We shall consider the loading as the sum of the two distributions illustrated in Figures E3.36b,c. Note that we are familiar with each of the loadings (1) and (2)

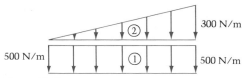

Figure E3.36b

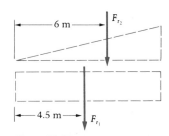

Figure E3.36c

from Examples (3.33) and (3.34). Therefore, the given loading is equipollent to the following pair of single forces:

$$F_{r2} = \tfrac{1}{2}(9)(300) = 1350 \text{ N} \qquad (\text{6m from left end of beam as shown})$$

$$F_{r1} = (9)(500) = 4500 \text{ N} \qquad (\text{4.5m from left end of beam as shown})$$

The final step is to find the single force $\mathbf{F}_r$ that is equipollent to this two-force system. Its value is obtained from the force condition:

$$\mathbf{F}_r = -1350\hat{\mathbf{j}} - 4500\hat{\mathbf{j}} = -5850\hat{\mathbf{j}} \text{ N}$$

And if the x coordinate of its vertical line of action is called x_r, we find the value of x_r from the moment condition

$$(5850)x_r = 4500(4.5) + 1350(6)$$

$$x_r = 4.85 \text{ m}$$

Therefore, our solution is shown in Figure E3.36d below.

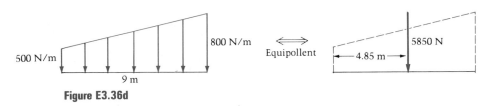

Figure E3.36d

We will see the method of composite parts again in Chapter 7 when we study a concept called centroids.

Let us now check the above results for $\mathbf{F}_r$ and its location by integrating the distributed load. Equating moments about the pin P we have

$$\underbrace{\left(\int_0^9 f(x)\, dx \right)}_{\substack{\text{magnitude} \\ \text{of force-alone} \\ \text{resultant}}} \underbrace{(x_r)}_{\substack{\text{location} \\ \text{of force-alone} \\ \text{resultant}}} = \underbrace{\int_0^9 xf(x)\, dx}_{\substack{\text{moment of} \\ \text{distributed load} \\ \text{about } P}}$$

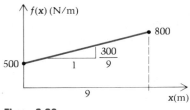

Figure 3.39

The loading curve in Figure 3.39 has the equation

$$f(x) = \frac{300}{9} x + 500$$

so that

$$\left[\int_0^9 \left(\frac{300x}{9} + 500 \right) dx \right] x_r = \int_0^9 \left(\frac{300}{9} x^2 + 500x \right) dx$$

$$\left[\frac{300}{9} \frac{x^2}{2} \Big|_0^9 + 500x \Big|_0^9 \right] x_r = \frac{300}{9} \frac{x^3}{3} \Big|_0^9 + 500 \frac{x^2}{2} \Big|_0^9$$

$$\underbrace{= |\mathbf{F}_r| = 5850 \text{ N}}_{} \qquad \underbrace{28400 \text{ N} \cdot \text{m}}_{}$$

or

$$x_r = 4.85 \text{ m} \qquad \text{(as before)}$$

EXAMPLE 3.37

Find the value of the single-force resultant (and its location) in Example 3.35 using the idea of composite parts.

Solution

The loading comprises the rectangular and triangular parts of Figure E3.37 below:

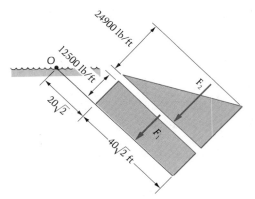

Figure E3.37

We have for the individual resultants:

$$F_1 = (40\sqrt{2}) \, 12{,}500 = 707{,}000 \text{ lb}$$
$$F_2 = \tfrac{1}{2}(40\sqrt{2}) \, 24{,}900 = 704{,}000 \text{ lb}$$

The single force that is equipollent to these two forces (and thus also to the original system of forces caused by the water) is obtained by the force condition for equipollence:

$$F_r = 707{,}000 + 704{,}000$$

$$= 1.41(10^6) \text{ lb} \quad \text{(as before)}$$

The location of F_r is obtained by satisfying the moment condition for equipollence:

$$M_{rO} = F_r x_r = F_1\left(20\sqrt{2} + \frac{40\sqrt{2}}{2}\right) + F_2[20\sqrt{2} + \tfrac{2}{3}(40\sqrt{2})]$$

$$1.41(10^6)x_r = 40.0(10^6) + 46.5(10^6)$$

$$x_r = 61.3 \text{ ft} \quad \text{(as before)}$$

As a practical matter, in a situation such as that of Example 3.36 or 3.37 we very seldom have an interest in determining a single-force resultant. What we really want to do is avoid, if possible, explicitly evaluating the integrals that arise when we evaluate the contributions of the distributed forces to the equilibrium equations for the beam. This was achieved in the two examples when we decomposed the distributed loading into two parts and were able to recognize the force-alone equipollent of each part. Our objective is then satisfied as suggested by Figure 3.40 below for Example 3.36. The next example will further illustrate this point.

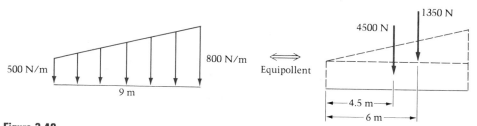

Figure 3.40

EXAMPLE 3.38

Replace the distributed loading on the cantilever beam in Figure E3.38a by an equipollent set of forces.

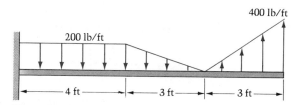

Figure E3.38a

Solution

By viewing the distributed loading as a uniform part plus two linearly varying parts, and using the results of Examples 3.33 and 3.34, we obtain the results shown in Figure E3.38b. These three forces will collectively make the same contributions to the equilibrium equations for the whole beam as will the original distributed loading.

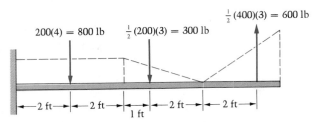

Figure E3.38b

Forces Distributed Over Surfaces

We now broaden the scope of our study by again considering pressure on a flat surface, S, but letting the surface boundary have any shape and allowing the pressure to vary arbitrarily with both of the coordinates x and y in the plane of the surface (Figure 3.41).

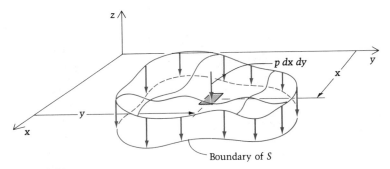

Figure 3.41

The resultant force $\mathbf{F}_r$ is given by

$$\mathbf{F}_r = (-\hat{\mathbf{k}}) \int_A p(x, y) \, dA = -P\hat{\mathbf{k}}$$

where A is the area of the surface S. The resultant couple at the origin is

$$\mathbf{M}_{rO} = \int_A (x\hat{\mathbf{i}} + y\hat{\mathbf{j}}) \times [p(x, y) \, dA(-\hat{\mathbf{k}})]$$

$$= -\hat{\mathbf{i}} \int_A yp(x, y) \, dA + \hat{\mathbf{j}} \int_A xp(x, y) \, dA$$

We pause at this point to note that sometimes there is negative pressure over a portion of the surface so that $P = \int\int p \, dx \, dy$ might vanish and *the* resultant is then the couple $\mathbf{M}_{rO}$. This occurs particularly in the mechanics of solids where we are concerned with an internal surface separating two portions of a body. In that application, "negative pressure" is called **tensile normal stress** and ordinary or positive pressure is called **compressive normal stress.** The resultant couple is usually then referred to as a **bending moment.**

When $P \neq 0$ there is, of course, a force-alone resultant, and if x_r and y_r locate its line of action,

$$x_r = \frac{\int xp \, dA}{\int p \, dA}$$

$$y_r = \frac{\int yp \, dA}{\int p \, dA}$$

In the special case of $p = $ constant,

$$x_r = \frac{\int x \, dA}{\int dA}$$

$$y_r = \frac{\int y \, dA}{\int dA} \tag{3.26}$$

where A is the area of surface S. We shall see in Chapter 7 that x_r and y_r locate the *centroid* of the area of the plane surface in this case.

With regard to pressure on curved surfaces, there are a few special properties of resultants we shall have occasion to use:

1. For uniform pressure the component of the resultant in a given direction is the pressure times the projection of the surface area onto the plane perpendicular to that direction. To show that this is the case, we refer to Figure 3.42, in which dA is an element of area on the surface and $\hat{\mathbf{n}}$ is a unit vector normal to the surface. The elemental force is $(p \, dA)\hat{\mathbf{n}}$ and the component in the x direction is $(p \, dA)\hat{\mathbf{n}} \cdot \hat{\mathbf{i}}$, so that

$$F_{rx} = \int_S (\hat{\mathbf{n}} \cdot \hat{\mathbf{i}})p \, dA$$

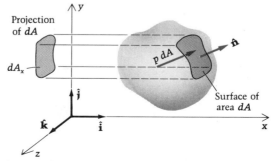

Figure 3.42

In the integration, $\hat{\mathbf{n}}$ varies since its direction is always normal to the surface. However, $\hat{\mathbf{n}} \cdot \hat{\mathbf{i}}$ is the cosine of the angle between the normal to the surface and x (which is perpendicular to the yz plane). Thus $(\hat{\mathbf{n}} \cdot \hat{\mathbf{i}}) \, dA$ is the projection, dA_x, of that element of the surface onto the yz plane. Since p is uniform,

$$F_{rx} = p \int \hat{\mathbf{n}} \cdot \hat{\mathbf{i}} \, dA = p \int dA_x$$
$$= pA_x$$

where A_x is the projection of the surface area A onto the yz plane.

2. The resultant of pressure on a spherical surface has a line of action through the center of the sphere (see Figure 3.43). This follows from the fact that each elemental force has a line of action through that point. That is, we have here a concurrent distributed force system.

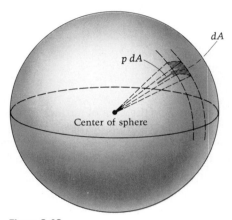

Figure 3.43

3. Pressure on a circular cylindrical surface produces no resultant moment about the cylinder axis (see Figure 3.44). This follows from the fact that each elemental force has a line of action that intersects that axis.

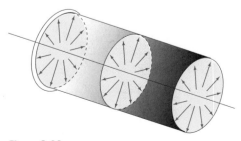

Figure 3.44

Question 3.32   Do the above results (2) and (3) concerning pressure on spherical and cylindrical surfaces require that the spheres and cylinders be complete (that is, whole)?

Question 3.33   Does the pressure in (1), (2), and (3) above have to be constant for the results to be valid?

Question 3.34   Would the moment about the axis of a cone, due to pressure on a portion of a cone's surface, be zero?

Forces Distributed Throughout a Volume; Gravity

We now turn to the most common example of a parallel force system distributed over a volume. Suppose a body $\mathcal{B}$ can be treated as if the gravitational attractions of the earth on all of the particles of $\mathcal{B}$ are parallel (vertical). The magnitude of the force exerted by the earth on an element of mass in the body is $g(dm)$, where g is the strength of the gravitational field (or the acceleration of gravity), assumed constant over the body, and $dm = \rho\, dV$ is the mass of an elemental volume dV, where the density (mass per unit of volume) is ρ (see Figure 3.45).

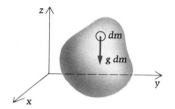

Figure 3.45

For this (assumed) parallel force system, the resultant force (weight) is vertical (downward) with magnitude

$$F_r = \int_{\mathcal{B}} g\, dm = g \int_{\mathcal{B}} dm = mg$$

where m is the mass of the body. In order to produce the same moments about the x and y axes as does the distributed system, we must have

$$x_r F_r = \int xg\, dm$$
$$= g \int x\, dm \tag{3.27}$$

and similarly,

$$y_r F_r = \int yg\, dm$$
$$= g \int y\, dm \tag{3.28}$$

where x_r and y_r are coordinates of points on the line of action of the force-alone resultant.

Since $F_r = mg$, Equations (3.27) and (3.28) simplify to

$$x_r = \frac{\int x\, dm}{m}$$

$$y_r = \frac{\int y\, dm}{m} \tag{3.29}$$

Answer 3.32   No. It is the fact that all the forces intersect the center of the sphere (or the axis of the cylinder) that makes the moment zero.

Answer 3.33   1: Yes; 2: No; 3: No.

Answer 3.34   Yes. Even though the forces caused by the pressure intersect the cone's axis nonperpendicularly, they still produce no moment about this line (see Figure 3.46).

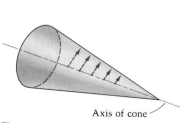

Axis of cone

Figure 3.46

The coordinates given by Equations (3.29) define a line on which the **center of mass** C of the body is located. A third equation,

$$z_r = \frac{\int z \, dm}{m}$$

pins down the actual point C. We usually write the coordinates of C as $(\bar{x}, \bar{y}, \bar{z})$ rather than (x_r, y_r, z_r), incidentally.

Thus we see that, when it is practical to assume that the gravity forces acting on a body are parallel and g is constant, the earth exerts a force (called the weight) with its line of action passing through the center of mass of the body. This will be the case in all the exercises in this book. In this case, the center of mass is sometimes called the **center of gravity.***

We note that if in Equations (3.29) we replace the differential mass dm by $\rho \, dV$, where ρ is the mass density and dV is the differential volume enclosing dm, then we obtain three coordinate equations, the first of which is

$$x_r = \frac{\int x \, \rho \, dV}{\int \rho \, dV} \tag{3.30}$$

in which the denominator is the mass m of the body:

$$m = \int dm = \int \rho \, dV$$

We see from Equation (2.30) that if the density ρ of the body is constant, it cancels, leaving

$$x_r = \frac{\int x \, dV}{\int dV} = \frac{\int x \, dV}{V} \tag{3.31}$$

and similarly

$$y_r = \frac{\int y \, dV}{V} \quad \text{and} \quad z_r = \frac{\int z \, dV}{V} \tag{3.32}$$

---

* The terms "center of mass" and "center of gravity," of course, refer to different physical concepts — the former having to do with the mass distribution of a body and the latter having to do with the resultant of distributed gravitational attractions. In the literature of mechanics "center of gravity" is used in two ways. One usage refers to the location of the equivalent particle (same mass as the body) that would cause this particle to be subjected to the same force by an attracting particle as would be the actual body in question; this center of gravity has a location which depends upon, among other things, the orientation of the body. The second and more common usage refers to the location of the point through which the resultant weight passes when the gravity field is uniform and parallel. This center of gravity is independent of the orientation of the body, and, as we have seen, it has the same location as the mass center of the body. Thus, in this case there is no reason to distinguish center of gravity from the center of mass except, perhaps, to remind the reader of the physical phenomenon motivating location of the point. However, it is important to realize that the solutions of some engineering problems require recognition of the fact that a gravitational field is not uniform and parallel. An example of this arises in the attitude control of an earth satellite. The resultant of the earth's gravitational attraction on the satellite is a force through the mass center and a couple whose small moment is called the "gravity gradient torque," and this is a very important factor in establishing control of the satellite.

The coordinates given by Equations (3.31) and (3.32) define the **centroid of volume** of the body. We see that, for a body having constant density, this centroid and the center of mass coincide.

When positions of non-obvious mass centers are needed in the equilibrium problems in Chapters 4–6, these mass centers will be indicated in the various figures.

We have presented in this section a number of special cases of distributed force systems, particularly those for which the force-alone resultant, or perhaps only its line of action, is easily recognized. It is for these cases that the concept of a resultant is most useful. For the sake of completeness, however, we set down here the method of handling the general case. Suppose **f** is the intensity of a distributed force system (force per unit of length or area or volume), and suppose we let dQ be an element of length or area or volume as the case may be. Let **r** be the directed line segment from point P to the point of application of an elemental force $\mathbf{f}\, dQ$. Then the resultant at P is

$$\mathbf{F}_r = \int \mathbf{f}\, dQ \qquad \text{and} \qquad \mathbf{M}_{rP} = \int \mathbf{r} \times \mathbf{f}\, dQ$$

This is nothing more than the obvious extension of our work with discrete systems; here, though, we are integrating instead of summing. Once $\mathbf{F}_r$ and $\mathbf{M}_{rP}$ are calculated, any further reduction follows in the same manner as for a system of discrete forces.

PROBLEMS ▶ Section 3.9

3.142 Specify and locate the single force that is equipollent to the concentrated and distributed loads acting on the shaded plate in Figure P3.142.

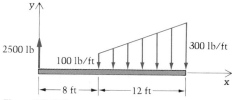

Figure P3.142

3.143 A cantilever beam is loaded as shown in Figure P3.143. Replace the distributed line load by an equipollent system at the wall.

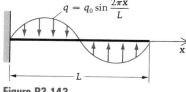

$q = q_0 \sin \dfrac{2\pi x}{L}$

Figure P3.143

3.144 Pressure acts on a rectangular solid as shown in Figure P3.144. Find the single force that is equipollent to the given loading (magnitude *and* line of action).

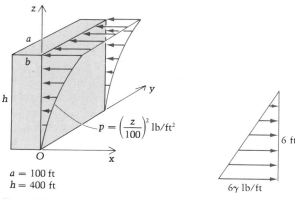

$p = \left(\dfrac{z}{100}\right)^2 \ \text{lb/ft}^2$

$a = 100$ ft
$h = 400$ ft

Figure P3.144

6 ft

6γ lb/ft

Figure P3.145

3.145 Compare the resultant of Problem 3.109 with that of the distributed load shown in Figure P3.145. Tell why the force is the same but the line of action is slightly different.

3.146 Wind velocity varies with height due to a number of factors, among which are the earth's angular velocity, wind path curvature pressure gradient, air density, latitude, and viscosity. In coastal areas, if the wind speed at 30 feet (called v_{30}) is under 60 mph, then it is assumed that at height z ft (for $z \le 600$ ft), the wind velocity is

$$v_z = v_{30} \left(\frac{z}{30} \right)^{0.3}$$

Use this equation to compute the total wind force on the windward sides of two buildings of equal areas facing the wind, one with height $h = 100$ ft and width $w = 50$ ft, the other with $h = 50$ ft and $w = 100$ ft. Take the value of v_{30} to be 20 ft/sec and the dynamic pressure to be $(1/2) \rho v_z^2 C_D$, where ρ = wind density (use 0.0024 slug/ft³) and C_D = drag coefficient (use 1.4).

3.147 A concrete ramp leading up to a hospital emergency room door is 1 m wide and has the length and height shown in Figure P3.147. If the concrete weighs 22,600 N/m³, determine the ramp's weight and its line of action.

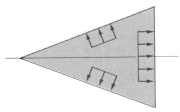

Figure P3.147

3.148 With reference to Figure P3.148,

a. Show that a uniform pressure p within a conically shaped tank with a closed, flat base produces no resultant moment about the axis of the cone.

b. Show further that the resultant of this distributed force system is zero.

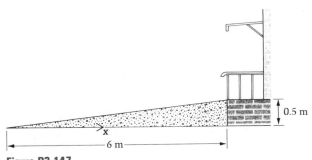

Figure P3.148

3.149 Replace the distributed loading shown in Figure P3.149 by a single force.

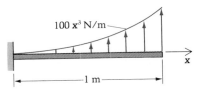

Figure P3.149

3.150 The distributed load on the arch shown in Figure P3.150 varies according to $q = 300 \cos \theta$ N/m. Find the resultant of this load, expressed as a force at Q and a couple.

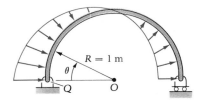

Figure P3.150

3.151 Repeat the preceding problem if $q = 300 \cos 2\theta$ (see Figure P3.151).

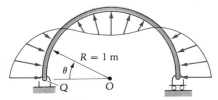

Figure P3.151

* **3.152** Repeat the preceding problem if $q = 300 \cos 3\theta$ (see Figure P3.152).

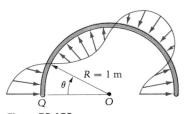

Figure P3.152

For each of the following five problems, replace the distributed loading on the beam by an equipollent set of forces.

3.153

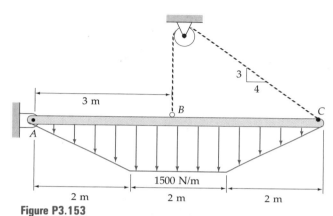

Figure P3.153

3.154

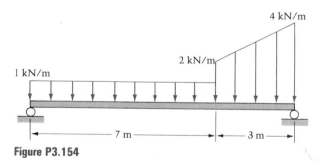

Figure P3.154

3.155

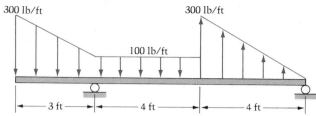

Figure P3.155

3.156

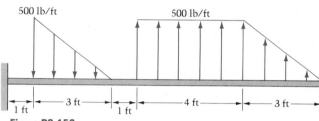

Figure P3.156

*3.157

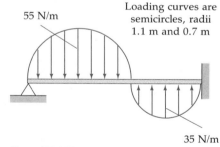

Figure P3.157

COMPUTER PROBLEMS ▶ Chapter 3

3.158 Write a computer program that will read the components of the three-dimensional vectors r_{AB}, F and L, where B is a point on the line of action of force F, r_{AB} is a position vector from point A to B, and L has as components a set of direction numbers of a line l through A (see Figure P3.158). The program is to calculate and print (a) the moment of F about A; (b) the moment of F about line l; and (c) the distance from A to the line of action of F. Run the program for these data: $r_{AB} = 2\hat{i} + 3\hat{j} - 6\hat{k}$ m; $F = \hat{i} - 2\hat{j} + 2\hat{k}$ N; and $L = 3\hat{i} + 4\hat{j} - 12\hat{k}$.

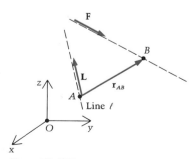

Figure P3.158

3.159 Write a program that will read any number of forces and couples in a coplanar force system, along with a point on the line of action of each of the forces. The program is then to compute and print the single-force resultant and the equation of its line of action. Run the program for these forces and couples:

$$\mathbf{F}_1 = 100\hat{\mathbf{i}} - 200\hat{\mathbf{j}} \text{ lb, passing through } (x, y)$$
$$= (1, 3) \text{ ft};$$

$$\mathbf{F}_2 = 50\hat{\mathbf{i}} - 340\hat{\mathbf{j}} \text{ lb, passing through } (x, y)$$
$$= (2, -5) \text{ ft};$$

$$\mathbf{F}_3 = 250\hat{\mathbf{j}} \text{ lb, lying along the line } x = -7 \text{ ft};$$

$$\mathbf{F}_4 = 300 \text{ lb, lying along the positive } x \text{ axis};$$

$$\mathbf{C}_1 = 400\hat{\mathbf{k}} \text{ lb-ft};$$

$$\mathbf{C}_2 = -730 \; \hat{\mathbf{k}} \text{ lb-ft}.$$

3.160 Write a program that will read any number of forces parallel to the z axis (with $+\hat{\mathbf{k}}$ suppressed), along with the (x, y) coordinates of the points where each force pierces the plane $z = 0$. The program is then to compute and print the single-force resultant and *its* piercing point in the plane $z = 0$. Run the program for this data:

$$\mathbf{F}_1 = 200 \text{ N through the origin};$$

$$\mathbf{F}_2 = 1000 \text{ N through } (x, y) = (10, -20) \text{ m};$$

$$\mathbf{F}_3 = -580 \text{ N through } (x, y) = (-5, 15) \text{ m};$$

$$\mathbf{F}_4 = -900 \text{ N through } (x, y) = (-10, -2) \text{ m}.$$

SUMMARY ► **Chapter 3**

In Section 3.2, we learned that the moment $\mathbf{M}_P$ of a force $\mathbf{F}$ about a point P has a magnitude equal to that of the force multiplied by the distance from the point to the line of action of the force. The direction of the moment vector is that of the right thumb when the right hand's fingers turn in the direction of the turning effect of $\mathbf{F}$ about P.

We also learned that a vector representation of $\mathbf{M}_P$ is $\mathbf{r} \times \mathbf{F}$, where $\mathbf{r}$ is a vector from P to *any point* on the line of action of $\mathbf{F}$. This cross product makes moment calculations much easier in cases of complicated geometry, particularly in three dimensions.

The final important topic in Section 3.2 was Varignon's theorem, which states that the moment about P, of the sum of a set of forces acting at a point A, is the same as the sum of the moments of the separate forces about P. This allows us to decompose a force into its components at A when the perpendicular distances along coordinate directions from A to P are easily seen or determined.

Section 3.3 was concerned with computing the moment of a force $\mathbf{F}$ about a line ℓ. This moment, $\mathbf{M}_\ell$, was the projection along ℓ of the moment of $\mathbf{F}$ about any point P lying on ℓ, i.e.,

$$\mathbf{M}_\ell = (\mathbf{M}_P \cdot \hat{\mathbf{u}}) \, \hat{\mathbf{u}}$$

where $\hat{\mathbf{u}}$ is the unit vector in the direction of ℓ. We proved that $\mathbf{M}_\ell$ does not depend on the point P along ℓ used to compute it.

We proved and then displayed (in Figure 3.9) a physical interpretation of $\mathbf{M}_\ell$ as the turning effect, about ℓ, of the component of $\mathbf{F}$ that is perpendicular to ℓ. And finally, we illustrated the power of Varignon's theorem in finding the moment of a force about a line.

An interesting concept called a couple was studied in Section 3.4. Defined as a pair of equal-magnitude, oppositely directed, and non-collinear forces, the couple has no resultant force, but a moment *which is the same about all points of space.* If A lies on $-\mathbf{F}$ and B lies on $\mathbf{F}$, where $\mathbf{F}$ and $-\mathbf{F}$ constitute a couple, then this moment is simply $\mathbf{r}_{AB} \times \mathbf{F}$. But it is also $|\mathbf{F}|d$ in magnitude, where d is the distance between the lines of action of $\mathbf{F}$ and $-\mathbf{F}$, and directed according to the right-hand rule as illustrated in Figure 3.13.

The equations of equilibrium, $\Sigma\mathbf{F} = \mathbf{0}$ and $\Sigma\mathbf{M}_P = \mathbf{0}$, were presented in Section 3.5 for motivational purposes. We noted that there is more to the equilibrium of bodies than merely setting the summation of external forces equal to zero as we did in Chapter 2. For problems in which all the forces acting on a body are not concurrent, we must also set the sum of moments about a point P equal to zero.

In Section 3.5 we also proved that if Q is some other point besides P, then it is always true that

$$\Sigma\mathbf{M}_Q = \mathbf{r}_{QP} \times (\Sigma\mathbf{F}) + \Sigma\mathbf{M}_P$$

and we noted from this equation that if $\Sigma\mathbf{F} = \mathbf{0}$ and $\Sigma\mathbf{M}_P = \mathbf{0}$, then $\Sigma\mathbf{M}_Q$ is automatically also zero. This means that nothing is to be gained by a second moment equation once $\Sigma\mathbf{F} = \mathbf{0}$ and $\Sigma\mathbf{M}_P = \mathbf{0}$ have been used.

The equipollence (equal power) of force systems was defined and studied in Section 3.6. Two equipollent systems will make the same contributions to the equations of equilibrium (and also to the equations of motion in later studies in dynamics). Two systems are called equipollent if two things happen:

1. $(\Sigma\mathbf{F})$ in system 1 $= (\Sigma\mathbf{F})$ in system 2

and

2. $(\Sigma\mathbf{M}_P)$ in system 1 $= (\Sigma\mathbf{M}_P)$ in system 2 (where P is some common point in systems 1 and 2)

We showed that if conditions 1 and 2 above hold for two force systems, then the moment sums are also equal for *any other* point Q. Thus in checking for equipollence, the sum of the moments need only be compared at one point.

For any system of forces (and/or couples), we learned in Section 3.7 how to replace it by a force at any pre-selected point P and a couple. This force and couple pair, $\mathbf{F}_r$ and $\mathbf{M}_{rP}$, is called a resultant of the original system. The force, $\mathbf{F}_r$, is the summation $(\Sigma\mathbf{F})$ of all the forces in the original system, placed at P. The couple, $\mathbf{M}_{rP}$, is the sum of the moments about P of the forces in the original system, plus the sum of the moment vectors of the couples. Thus, regardless of how complicated a force-couple system may be, it can always be reduced to an equipollent force at any desired point and a couple.

We learned in Section 3.8 that in several cases of interest, the force and couple resultant, with $\mathbf{F}_r \neq \mathbf{0}$, can be further reduced to an equipollent system consisting of a single force. These cases are (a) concurrent force systems; (b) coplanar force systems; and (c) parallel force systems. Any other systems for which $\mathbf{F}_r$ is perpendicular to $\mathbf{M}_{rP}$ will also have a "force-alone" resultant. When $\mathbf{F}_r$ is *not* normal to $\mathbf{M}_{rP}$, the simplest resultant is a collinear force and couple, called a "screwdriver."

The final section in the chapter (3.9) dealt with an important subject: distributed force systems. When the distributed forces are all parallel and act on a body along a line normal to the forces, then the magnitude and sense of the force-alone resultant are given by the signed area beneath the loading curve. Its direction is that of the parallel distributed forces, and its line of action is given by Equation 3.25. (It will be seen in Chapter 7 that this resultant passes through a point called the centroid of the signed area beneath the loading curve.) These results were seen to facilitate problems in which distributed loadings due, for example, to weight or water pressure were involved.

Pressure varying in two dimensions over a flat surface was also considered, as were uniform pressure on curved surfaces and variable pressure on spherical and cylindrical surfaces.

Finally, the force of gravity was considered, and the meanings of, and differences between, the center of mass and the center of gravity were discussed.

REVIEW QUESTIONS ▶ Chapter 3

True or False?

1. The magnitude of the moment of a force about a point is the magnitude of the force multiplied by the distance from its line of action to the point.
2. A force has zero moment about any point on its line of action.
3. A force has no moment about a line parallel to its line of action.
4. The magnitude of the moment of a force about a line equals the product of the component of the force perpendicular to the line and the distance between this component and the line.
5. The moment about a point P of the sum of a set of concurrent forces, placed at the point of concurrency, equals the sum of the moments about P of the separate forces.
6. A couple has the same moment about any point.
7. If we know the moment of a couple about a point A, then we can find the moment of the couple about a specified line BC even if A does not lie on BC.
8. The sum of the moments, about any point, of the external forces acting on a body at rest in a noninertial frame is always zero.
9. Two force and couple systems are equipollent if they have either the same resultant force or if the moments about some point P are the same for both systems.
10. It is possible for a system consisting of a single couple to be equipollent to a system of one force and 17 couples.
11. Given any system of forces and couples, and any point P, the system may be reduced to an equipollent system comprising a force at P and a couple (where either or both might be zero).

12. The couple part of the resultant of a concurrent force system will vanish at all points.

13. If a system of forces is coplanar, there is a point Q in the plane where the resultant moment vanishes, and the moment also vanishes for any point on the line through Q normal to the plane of forces.

14. If a system of forces which don't sum to zero is parallel, then there is a line l at every point of which the resultant moment is zero, and both the line l and the resultant force are parallel to each of the forces.

15. If a system S_1 of forces and couples has a "force-alone" equipollent system S_2, then S_1 is either a concurrent, coplanar, or parallel force system.

16. The resultant of the distributed loading shown in Figure 3.47 is a zero force and a zero couple.

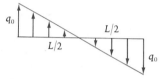

Figure 3.47

17. The simplest resultant of a general system of forces and couples is a "screwdriver," that is, a collinear force and couple along a specific axis in space.

18. Let a general system of forces and couples be resolved into a resultant force at point P and an accompanying couple. If the couple is normal to the force, then the equipollent screwdriver consists of a single force.

Answers: 1. T 2. T 3. T 4. T 5. T 6. T 7. T 8. F 9. F 10. F 11. T 12. F 13. F 14. T 15. F 16. F 17. T 18. T

4

ANALYSIS OF GENERAL EQUILIBRIUM PROBLEMS

4.1 Introduction

We are now in position to solve some general equilibrium problems. By "general," we mean problems which require us to use *both* of Euler's laws to effect a solution; i.e., we must invoke not only the "force equation" from Chapter 2:

$$\Sigma \mathbf{F} = \mathbf{0} \tag{4.1}$$

but also the "moment equation" mentioned briefly back in Chapter 1:

$$\Sigma \mathbf{M}_P = \mathbf{0} \tag{4.2}$$

where P is any point.

With Equation (4.2) now in our box of tools, we are no longer restricted to examining the relatively small number of equilibrium problems in which the forces acting on the body are concurrent. For a point-mass, of course, they must be so,* but most everyday, finite-sized bodies in the state of equilibrium are acted on by forces that do not all meet at one point. Examples are parked cars, desks, buildings — the list is endless.

So, when the external forces acting on a body are not concurrent, we must sum and set to zero the moments of forces as well as the forces themselves. This of course is the reason we spent Chapter 3 learning all about the moment of a force. Without such knowledge, we could not form Equation (4.2) from free-body diagrams of the bodies we wish to analyze.

This chapter is actually the heart of the book; the three chapters that preceded it are preparation *for* it, while the two that immediately follow will be special applications *of* it. By the end of this chapter the reader should be able to write and solve the equations relating the external forces acting on any body in equilibrium. Now the "body" might be a single indentifiable physical object, such as a table. But it can also be a collection of objects taken *together* to form a "combined body," such as a diver standing still on a diving board, or a table with a pitcher of water on it.

The equations of equilibrium [Equations (4.1) and (4.2)] relate *all* of the *external* forces acting on a body at rest in an inertial frame of reference. In most engineering problems some of these external forces are known (or prescribed) before any analysis is carried out; we usually refer to these as **loads.** The external forces exerted by attached or supporting bodies are called **reactions;** usually we can think of these as forces that constrain the body against motion the loads tend to produce. It is by the equations of equilibrium that we try to find these reactions. We emphasize, however, that a force "by any other name" is still a force as far as the equations of equilibrium are concerned; whether we think of a force as a cause (load) or as an effect (reaction) makes no difference in the equa-

---

*Because for a "point-mass," there is only one point on which they may act!

tions of equilibrium. If a body could be equipped with sensors to measure all of the forces acting on it, the sensors would not be able to distinguish applied loads from constraint reactions.

To illustrate one of the difficulties encountered in statics, we consider the problem of determining the forces (reactions) exerted by the supports of a diving board when a diver (whose weight is the load) stands on the end of the board as in Figure 4.1.

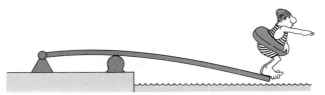

Figure 4.1

The configuration shown is that to which the equations of equilibrium must be brought to bear; the distances from the supports to the diver are important because of the moment equations. However, the diver's distance from each of the supports depends upon how much the board is bent — the greater the bending, or sag, of the board, the smaller the distance of the diver from each of the supports. Clearly, the bending of the board depends upon the weight of the diver and the stiffness of the board. Thus we might be tempted to conclude that we can do nothing useful until we study the geometry changes that occur when a body deforms. Fortunately, for many engineering problems the picture is not quite as bleak as the one we have painted. Frequently the deformations arising from the application of loads are small enough that gross changes in geometry can be ignored. A rigid body is the idealization in which no deformation at all occurs, a rigid body being one in which the distances between all possible pairs of points are unchanged when the body undergoes a change in configuration. That is, no portion of the body can change in shape or size. No real body is rigid, but it may be near-rigid in the sense just described — that is, for small deformations. If that can be assumed to be the case with the diving board, we have the situation shown in Figure 4.2:

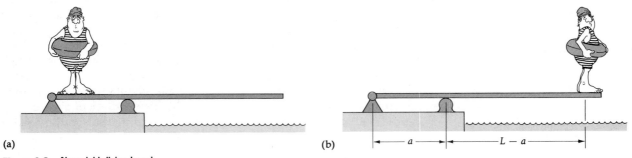

(a)

(b)

Figure 4.2   Near-rigid diving board.

Here the geometry with which we have to deal is, for all practical purposes, independent of the weight of the diver. The assumption of near-rigidity is to be understood throughout this book with few exceptions that will be obvious. However, as students proceed into further studies of mechanics (mechanics of deformable solids, mechanics of fluids), it is important for them to realize that the equations of equilibrium are valid for *any* body at rest in an inertial frame, regardless of the degree of deformability and regardless of the phase (solid, liquid, or gas).

In engineering mechanics, as in other areas of science and applied mathematics, we must be alert to the possibility of posing a meaningless, or silly, problem. For example, suppose we remove the interior support from the diving board and inquire as to the hinge reaction when the board is in equilibrium in the configuration of Figure 4.3. The difficulty here is that this configuration cannot be an equilibrium configuration. The equations of equilibrium will tell us that, but, of course, we don't need them here. The impossibility should be obvious from the fact that the board is free to rotate as a rigid body about the hinge. We shall usually be immune to the possibility of this situation if the body is constrained in such a way that a rigid-body change in configuration is prohibited. Such would be the case, for example, if one point of the body were fixed and rotation about each of three distinct nonplanar axes through the point were prohibited.

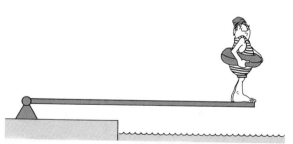

Figure 4.3   Equilibrium impossible.

Section 4.2 will be concerned with a much more complete treatment of the free-body diagram (FBD) than was possible in Chapter 2. We will learn, for example, to remove the supports in figures such as 4.1 and 4.2, thereby isolating for analysis bodies such as the diving board and/or diver. *We cannot overemphasize the importance of mastering the FBD concept.*

As we have said, we applied Equation (4.1) to a number of simple, particle-like problems in Chapter 2. The next step in analyzing equilibrium problems, now that we are knowledgeable about moments from Chapter 3, is to apply *both* Equations (4.1) and (4.2) to a single finite-sized* body (or to two or more physical objects left intact and considered as a single body). This will be the subject of Section 4.3.

Finally, in Section 4.4, we learn to separate bodies joined by simple connections such as pins, rollers, cables and the like, and to draw FBD's and write equilibrium equations for the separate bodies as well as the "combined body." In that process, we shall discover that not all the resulting equations are independent.

We will defer to Chapter 5 a last step in this graduated approach to studying equilibrium problems: that of imagining bodies to be sliced completely in two, so as to expose and then solve for the forces with*in* the body known as *internal forces*.

4.2   The Free-Body Diagram

In Chapter 2, we spoke briefly of the free-body diagram (FBD) when studying the equilibrium of a particle. In that context, the FBD actually needed be nothing more than a sketch of the forces acting on the particle, all interacting at the one point of concurrency.

As important as it was in equilibrium problems of particles, the FBD will be seen to be even *more* important in general equilibrium problems of finite-sized bodies, on which the forces need not be concurrent and on which couples may also act. This is because the FBD now gives not just a sketch of all the external forces and their directions for use in $\Sigma \mathbf{F} = \mathbf{0}$, but also illustrates the various points of application of these forces along with appropriate distances for use in $\Sigma \mathbf{M}_P = \mathbf{0}$. For this reason, it is now *vital* that the student sketch the actual body as more than just a point-mass.

Because there is this additional importance now attached to the free-body diagram, we devote this short but all-important section to its use. And to make the section self-contained, we shall not refer further to the brief FBD coverage in Section 2.2.

What the Free-Body Diagram Is

The **free-body diagram** is an extremely important and useful concept for the analysis of problems in mechanics. It is a figure, usually sketched, depicting (and hence identifying precisely) the body under consideration. On the figure we show, by arrows, all of the external forces, and moments of couples, that act on the body. Thus, we have a catalog, graphically displayed, of all the forces that contribute to the equilibrium equations (or the equations of motion if the problem is one of dynamics). Of particular importance is the fact that the free-body diagram provides us a way to express what we know about reactions (for example, that a

---

* "Finite-sized" means that the body cannot be analyzed as a particle unless all of the external forces are concurrent. The dimensions of the body and various points of application of the forces now become important because moment equations must also be written.

certain reaction force has a known line of action) before applying the equations of equilibrium. This is best illustrated by example.

Let us return to the diver and the diving board of the preceding section. We want to emphasize that a **body** is whatever collection of material we chose to focus on; we shall choose here to let the body be the diver and the diving board, taken together. The free-body diagram is

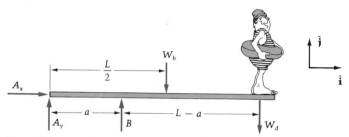

Figure 4.4   Free-body diagram of diver and board.

then shown in Figure 4.4.* The meaning of, and reasoning behind, the symbols appearing on the diagram are as follows:

1. W_d is the weight of the diver; thus the resultant force exerted by the earth on the diver by way of gravity is W_d times a unit vector downward. The line of action of this resultant is through the mass center of the diver. Similarly, W_b is the weight of the board, and the line of action of its resultant passes through the mass center of the board. In each case the letter and the arrow identify the vector description of the force, $W_d(-\hat{\mathbf{j}})$ or $W_b(-\hat{\mathbf{j}})$, and the location of its line of action. It is this information that will be needed in the equations of equilibrium.

2. The board is connected at its left end to the supporting structure by a hinge; basically, a cylindrical bar or pin fits into holes in brackets at the end of the board. In the absence of any significant friction we perceive that the pin exerts only pressure on the cylindrical surface of the hole. We saw in Section 2.9 that the resultant of this pressure will be a force with a line of action through the center of the hole. Because we don't know in advance the direction of this resultant, we express it by its unknown components A_x and A_y. That is, the force exerted on the left end is $A_x\hat{\mathbf{i}} + A_y\hat{\mathbf{j}}$. The letters and arrows at the left end constitute a code for how we have chosen to express the force, each arrow denoting a unit vector in the direction of the arrow.

3. The interior support is perceived to exert pressure perpendicular to the board. While we don't know precisely the location of the line of action of the resultant, if this pressure is distributed over a small region then we know the line of action close enough for engineering purposes.

---

* One need not be an artist to draw good-free body diagrams. The roughest of sketches will suffice as long as the body is clearly identified.

Because the pressure acts perpendicular to the board here, we know the direction of the force-alone resultant (it is vertical). Thus the force may be expressed in terms of a single unknown scalar B; that is, the diagram is communicating that the force may be expressed as $B\hat{\mathbf{j}}$ without any loss of generality.

> **Question 4.1**   Why does the force exerted on the board by the diver's feet not appear on this free-body diagram?

It is important to realize that each of the forces appearing on the free-body diagram (Figure 4.4) is in fact the resultant of a distributed force system. We recall that the resultant of a force system embodies all of the characteristics of that system that show up in the equations of equilibrium. Thus, for example, when we write the equations of equilibrium for the diver-plus-board body, the only information we can hope to determine about the hinge pressure is its resultant, and that is precisely what is depicted on the free-body diagram. Shortly we shall look at several other kinds of mechanical connections between bodies and identify the nature of the resultants at these connections.

Separating Two Bodies; The Action-Reaction Principle

The hinge reaction in the diver-plus-board problem alternatively could be determined by an analysis in which the board alone is the body under consideration. The free-body diagram of the board is shown in Figure 4.5, where we see that the external forces are the same as before except that the weight of the diver does not appear. This is because the free-body diagram does not include him, and thus we have a new external force, P, that is the resultant of the pressure exerted on the board by the feet of the diver. The free-body diagram of the diver communicates (by the arrow) that the same scalar P multiplies an upward unit vector to express the force exerted *on* him *by* the board; that is, the force (vector) exerted by the board on the diver is the negative of the force (vector)

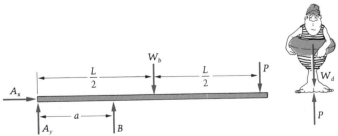

Figure 4.5   Free-body diagrams of board and of diver.

Answer 4.1   The body is the composite of the diver and board. This force is not external to that body; it is an internal force (interaction between parts of the body).

exerted by the diver on the board. This is an example of the **action-reaction principle,** which states that, when two bodies interact mechanically, the resultant exerted by the first body on the second body is the (vector) negative of that exerted by the second body on the first. This is often loosely referred to as Newton's Third Law. The principle is almost self-evident when we consider two bodies in intimate contact so that the interaction between the bodies is that of force systems distributed over the surface of contact. At each point on the common surface the mechanical action (force per unit area) of the first body on the second is the negative of what is exerted on the first body by the second. Because the force intensities are opposites, the resultants likewise must be the negatives of one another.

Identifying Resultants Transmitted by Common Types of Connections

Before we take up a variety of examples of free-body diagrams, it will be helpful to identify the resultants that are transmitted at different types of connections between bodies. To do this we follow the procedure we used on the diving-board hinge; that is, we study the particular "distributed-over-a-surface" force system involved and deduce what is the most general form of the resultant of that distributed system. This process could be tedious, but there is a shortcut that fortunately is in harmony with the intuition of most engineering students. The reasoning that leads to the correct conclusions is based upon the rigid-body motion that would be inhibited, or impeded, if each body attached to the one under consideration were to be fixed in the frame of reference. If there are several attachments, then to determine the resultant exerted by one of them we first imagine the others to be removed. We then determine how the single attachment hinders a rigid motion of the body. In particular, if we desire the resultant at a point A, then the force part of the resultant will have a component in any direction in which a displacement of A is inhibited. That is, if the connection resists a movement of A in a given direction, then the resultant force at A must have a component in that direction. The couple part of the resultant at A will have a component in the direction of any line through A about which the rotation of the body is inhibited. Results of applying this reasoning process are presented in Table 4.1 for several types of connections.

The first item (I) in Table 4.1 is a reminder that, when there is no special freedom associated with a connection, the resultant of the distributed forces of interaction is a general force and couple. In this case we know nothing about the direction of either, and hence each must be described by three unknown components. Under "Plane Counterpart" are displayed the symbols used to denote the connections when we have reason to be concerned only with forces having lines of action in a given plane together with moments perpendicular to that plane. (In our discussion of the diver and the diving board we tacitly assumed that to be the case.) We shall discuss this assumption in more detail later in one of the example problems. Unless otherwise indicated, items II-IX in Table 4.1

Table 4.1 Selected Connections and the Corresponding Unknown Components of Resultants on Body $\mathcal{B}$ (shaded)

Type of Connection	Reaction on $\mathcal{B}$ (Shaded)	Plane Counterpart	
I. General interaction (no freedom) (a)	(b)	(c)	(d)
II. Hinge or clevis pin	 ↑ Same	Pin ↑ Same	
III. Rod in sleeve (hinge without thrust support)		Same as II	

Table 4.1   Continued

Type of Connection	Reaction on $\mathcal{B}$ (Shaded)	Plane Counterpart	
IV. Pin in slot, or roller, or line contact along smooth surface			
V. Ball-and-socket, or self-aligning bearing, or local contact with rough surface			
VI. Ball bearing		Same as II	

Table 4.1   Continued

Type of Connection	Reaction on $\mathcal{B}$ (Shaded)	Plane Counterpart
VII. Roller bearing 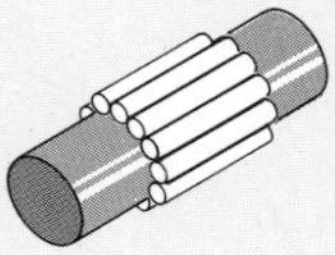	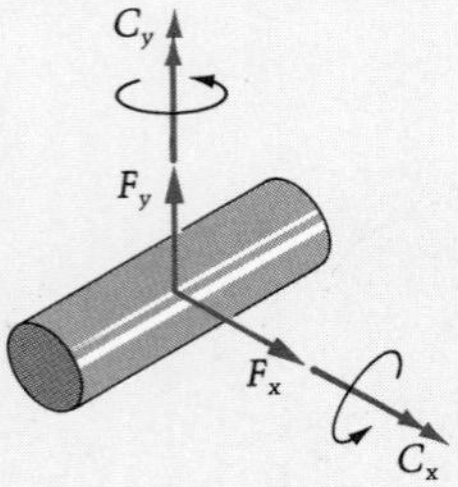	Same as II
VIII. Cable (rope, wire) 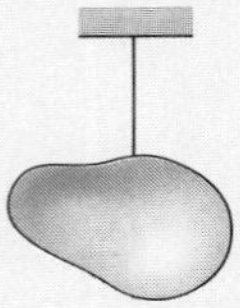	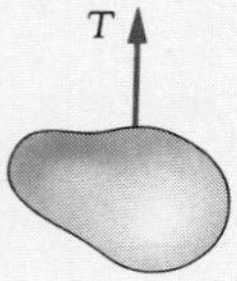	Same as in the figure at the left
IX. Clevis pinned to collar supporting smooth bar 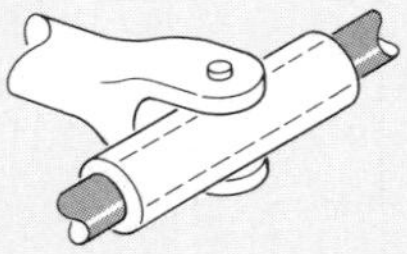	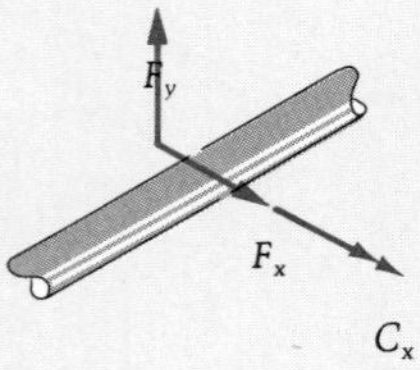	Same as II

are based on the assumption of negligible friction at contact surfaces. *The student should study them* to gain familiarity with the reasoning process and with the symbols that often will be used in the figures that depict exercise problems. A brief description of each follows:

II. An ordinary door hinge is the most common example of this connection. In the absence of friction, rotation is free to take place about the axis of the hinge; thus the only vanishing component of the resultant at a point on the axis is the component of the couple along the axis. Of course if the hinge is "rusty," friction will produce a component of couple in that direction, too, and the resultant will revert to that of I.

III. If there is no resistance to sliding along the hinge axis, the component of force in that direction also vanishes. An example of this connection is a cylindrical pin attached to one body and snugly fit into a cylindrical cavity in a second body if the pin is free to slide without friction along the cavity.

IV. If the pin of item III is inserted in a slot, there is only one component of force and one component of couple. That is, the pin is free to slide in two directions (in the plane of the slot) and to turn about an axis that is perpendicular to the plane of the slot. If a cylindrical roller is inserted between two (necessarily parallel) surfaces, the same kind of resultant is generated provided the contact is along a region sufficiently narrow to be approximated as line contact. In that case we could reason that the resultant should be a force alone. However, not knowing in advance where along the contact line the resultant will act, we express the resultant as a force with preassigned location on this contact line and a companion couple.

V. A ball-and-socket connection is a spherical ball on one body snugly fit into a spherical cavity in the other. No rotation is inhibited, but the attachment point (center of ball or socket) cannot move in any direction. In self-aligning bearings, the bearing housing is supported in this manner. The same resultant is also transmitted when one body is in local ("point") contact with a surface of a second body, there being friction at the interface.

VI. The balls in a ball bearing will exert what are essentially radial point loads on a circumferential line on the surface of a shaft. These produce no moments about diameters of the shaft that intersect that circumferential line. Consequently, the resultant is a force (with components F_x and F_y) having a line of action through the center of the shaft. All of this assumes that the bearing provides no thrust support — that is, resistance to motion of the shaft along its axis. The figure is of course simplified in that the bearing races are not shown.

VII. The rollers in a roller bearing exert loads essentially distributed over longitudinal (axial) lines on the surface of the shaft. Consequently, there is resistance to turning about diameters. Thus the possibility of a couple perpendicular to the axis of the shaft exists in addition to the resultant of F_x and F_y.

VIII. A flexible cable or wire exerts a tensile force in the direction of tangency to the cable at the attachment point. Unless otherwise indicated we shall assume that cables are sufficiently taut that the centerline will be a straight line joining its ends.

IX. The last item is a composite of the clevis pin (II) and rod-in sleeve (III). Note that freedom for rotation about the pin axis eliminates one component of the couple shown in III. That vanishing component is along the axis of the pin.

We now illustrate the construction of free-body diagrams through several examples. As we do this, we keep in mind that the equilibrium equations are

$$\Sigma \mathbf{F} = 0 \quad \text{or} \quad \mathbf{F}_r = 0$$
$$\Sigma \mathbf{M}_P = 0 \quad \text{or} \quad \mathbf{M}_{rP} = 0$$

where P is any point, and the subscript "r" denotes resultant (of the external force system). The free-body diagram will provide:

1. A catalog of all the external forces and/or couples acting on the body.
2. A graphical display of how much we know (directions, locations) about unknown reactions.
3. The geometric dimensions needed for establishing moments of the forces.

Thus, we shall find that the free-body diagram will include all of the information to be incorporated in the equations of equilibrium.

In many of the examples and subsequent exercise problems, information is given only for two spatial dimensions. It is reasonable for the student to be uneasy about the fact that we have not set down any criteria by which to decide when we can ignore considerations of the third dimension. Universally applicable criteria are not easy to establish, and this, like other issues of mathematical modeling in mechanics, requires experience. Working a large number of three-dimensional problems will provide students some of the experience with which to supplement their raw intuition.

EXAMPLE 4.1

Draw the free body diagram of the homogeneous block of Figure 2.6 if the supporting cables are vertical and attached to its upper corners as shown in Figure E4.1a, instead of meeting at point A.

Solution

We replace the cables by forces that act along their length, as shown in Figure E4.1b. The force of gravity is drawn as a weight W, acting downward through the center of gravity. Note that when the forces are not concurrent, it will become important to label the dimensions of bodies and precise locations of forces.

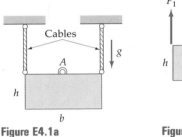

Figure E4.1a   **Figure E4.1b**

In the extended Problem 4.141, we will outline a proof of the fact that if a body is in equilibrium under the action of three forces, then these forces are necessarily coplanar and either (a) concurrent or (b) parallel. The block B supported as in Figure 2.6 is an example of (a), while the same body, supported as in the preceding example, is an illustration of (b). It is interesting that any body in equilibrium under the action of only three forces is never more complicated than one of these two types!

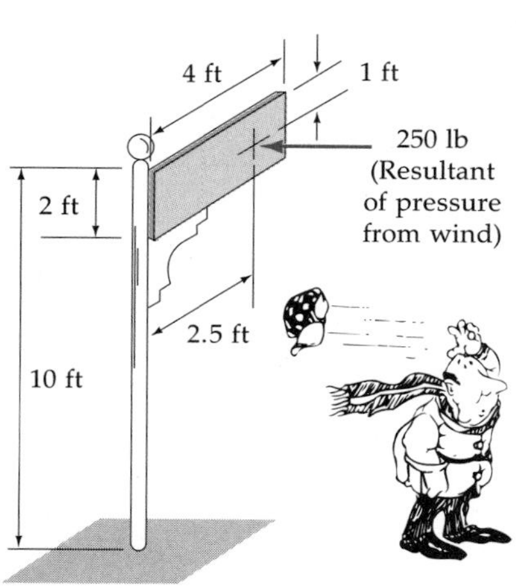

EXAMPLE 4.2

Sketch the free-body diagram of the small advertising sign and supporting post if there is a steady wind load producing the resultant shown in Figure E4.2a. The sign itself is sheet metal with a weight of 200 lb, and the post weighs 500 lb.

Solution

When we separate the sign and post from the ground we see that the external forces on the body are:

a. The resultant of the pressure from the wind.

b. The weights of the sign and the post.

c. The reaction of the ground on the base of the post.

Thus the free-body diagram is as shown in Figure E4.2b.

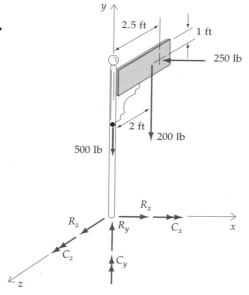

Figure E4.2a

Figure E4.2b

Note that we have tacitly assumed that the mass center of the post is on its centerline and that the sign itself is of constant thickness and constant density so that its mass center is "in the middle." It is important to realize that the connection of the post to the ground provides resistance to displacement in every direction and resistance to rotation about any axis through the base. Thus the force and couple there will *each* have three unknown components.

Question 4.2 Why has it been unnecessary to specify the elevation of the mass center of the post on the free-body diagram?

Answer 4.2 It is only the *line of action* of the force that is important.

EXAMPLE 4.3

Sketch the free-body diagram for the pliers of Figure E4.3a and for each of its parts.

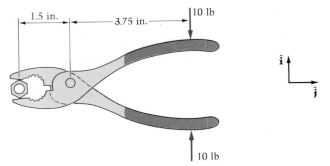

Figure E4.3a

Solution

The only external forces acting on the pliers are the hand-applied 10-pound forces and the reactions of the nut being gripped. The free-body diagram is shown in Figure E.4.3b.

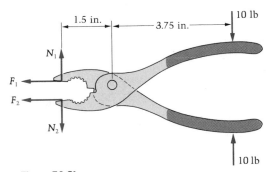

Figure E4.3b

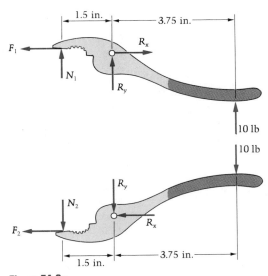

Figure E4.3c

Removing the pin that holds the two parts together, the free-body diagrams of the individual parts are shown in Figure E4.3c. Note that the action transmitted at the pin is just a force since the pin provides no resistance to relative rotation of the parts. Note further that the action-reaction principle has been satisfied automatically in depicting the pin forces.

Question 4.3   How have we guaranteed satisfaction of the action-reaction principle here?

Answer 4.3   The arrow code on the upper free-body diagram communicates the decision to express the force exerted by the lower part on the upper part as $R_x\hat{\mathbf{i}} + R_y\hat{\mathbf{j}}$. Similarly the arrow code on the lower free-body diagram communicates that the force exerted by the upper part on the lower part is $R_x(-\hat{\mathbf{i}}) + R_y(-\hat{\mathbf{j}}) = -(R_x\hat{\mathbf{i}} + R_y\hat{\mathbf{j}})$.

PROBLEMS ▶ Section 4.2

Draw a free-body diagram of body $\mathcal{A}$ in each of the following eight problems.

4.1   Body $\mathcal{A}$ in Figure P4.1 is a uniform 40-lb rod, 6 ft long.

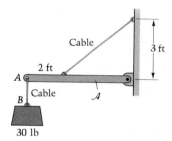

Figure P4.1

4.2   Body $\mathcal{A}$ in Figure P4.2 is a crowbar of negligible weight. Assume the force of the man's hand is directed along his arm, and that the force of the nail is along the axis of the exposed part of the nail.

Figure P4.2

4.3   Body $\mathcal{A}$ in Figure P4.3 is the 200-N uniform cylinder in equilibrium on the rough plane. The cable is parallel to the plane.

4.4   Body $\mathcal{A}$ in Figure P4.4 is the 30-kg ladder, together with a 70-kg painter.

Figure P4.3

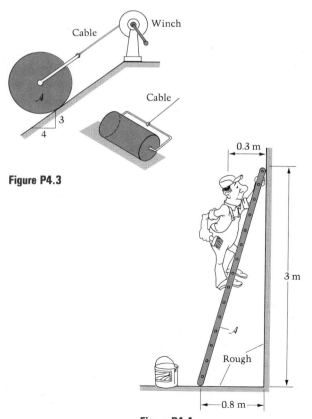

Figure P4.4

4.5   Body $\mathcal{A}$ in Figure P4.5 is the boom of the crane, weighing 1000 lb.

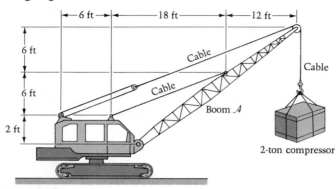

Figure P4.5

4.6   Body $\mathcal{A}$ in Figure P4.6 is a 50-lb door, supported by two hinges, each capable of exerting thrust in the direction of the hinge axis as well as a lateral force and couple.

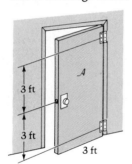

Figure P4.6

4.7   Body $\mathcal{A}$ in Figure P4.7 is a 100-N solid door. Each hinge can exert forces in all three directions and couples about both lateral axes (parallel to y and z, in this case).

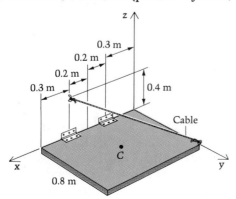

Figure P4.7

4.8   Body $\mathcal{A}$ in Figure P4.8 is the 180-N "corner bar," supported by three smooth eyebolts.

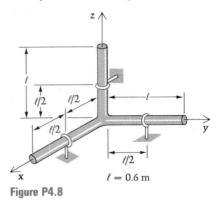

Figure P4.8

4.9   The block W in Figure P4.9 weighs 200 N and the uniform bar *ABD* weighs 300 N. Draw free-body diagrams of (a) W and (b) *ABD* including the roller at *A* as shown. The turning effect of the 1000-N couple applied to *ABD* is in the plane of the page.

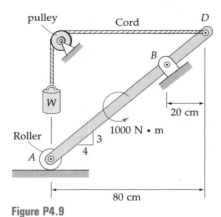

Figure P4.9

4.10   Draw a free-body diagram of the slotted bar *ABC* of Figure P4.10. The smooth pin bearing against the slot is fixed to the other bar, *DBE,* and the weights of the members may be neglected.

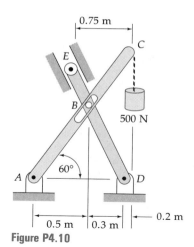

0.75 m

500 N

60°

A   D

0.5 m   0.3 m   0.2 m

Figure P4.10

4.11 Draw free-body diagrams of the three bars *CE*, *BEF*, and *ABCD* in Figure P4.11. Their weights may be neglected in comparison with the 600-lb load applied.

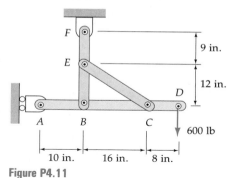

9 in.

12 in.

600 lb

10 in.   16 in.   8 in.

Figure P4.11

4.3 Fundamental Applications of the Equilibrium Equations

Number of Independent Equations

The most common problem in statics is the determination of unknown forces and couples which together with prescribed forces and couples hold a body in equilibrium. As we have seen in our discussion of free-body diagrams, the unknown forces and couples will be expressed in terms of the least number of unknown scalars (components) consistent with what is known about directions. There arises then the question: will the equilibrium equations provide a sufficient number of algebraic equations relating these scalars so that they may be found? At first glance we might be tempted to answer with an unqualified yes, since $\Sigma\mathbf{M}_P = \mathbf{0}$ for every point P so that there is no limit to the number of moment equations of equilibrium.

There is, however, a fallacy in this reasoning, because these additional moment equations are not independent. Let us first illustrate this for a planar equilibrium problem, in which all the forces (and turning effects of couples, if any) lie in a plane as in Figure 4.6(a). We recall that any system of forces and couples may be replaced by an equipollent system consisting of a force at a preselected point plus a couple. Thus the system at Point A in Figure 4.6(b), with the force broken into its x- and y-components, makes the same contribution to the equilibrium equations as does the original, actual system (a):

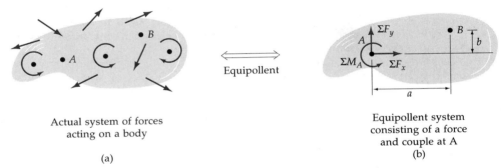

Actual system of forces
acting on a body

(a)

Equipollent system
consisting of a force
and couple at A

(b)

Figure 4.6

Suppose now that we have written the three equilibrium equations $\Sigma F_x = 0$, $\Sigma F_y = 0$, and $\Sigma M_A = 0$, for the body. If we now additionally write $\Sigma M_B = 0$, *this will not be an independent equation*, because it is seen below to be *already satisfied*:

$$\Sigma M_B = 0$$

$$\overbrace{\Sigma M_A + (\Sigma F_x)b + (\Sigma F_y)(-a)} = 0$$

$$0 \;+\; 0 \;\;+\; 0 \qquad = 0$$

$$0 = 0$$

Therefore, no new information is gained by writing additional moment equations once three independent equilibrium equations have been expressed. Problems 4.24–26 at the end of the section will deal with the question of when two (or three) moment equations could be used *if* we do not use one (or either) of the force equations.

A more formal vector proof of the above argument goes as follows for general (including three-dimensional) equilibrium problems: In Section 3.5 we found that for any system of forces:

$$\Sigma \mathbf{M}_Q = \Sigma \mathbf{M}_P + \mathbf{r}_{QP} \times (\Sigma \mathbf{F})$$

Thus, if $\Sigma \mathbf{M}_P = \mathbf{0}$ for some point P, and $\Sigma \mathbf{F} = \mathbf{0}$, then $\Sigma \mathbf{M}_Q$ *automatically* vanishes for *every* point Q. Therefore, the equations of equilibrium are the *two* independent vector equations

$$\Sigma \mathbf{F} = \mathbf{0}$$

$$\Sigma \mathbf{M}_P = \mathbf{0}$$

or in the notation of resultants

$$\mathbf{F}_r = \mathbf{0}$$

$$\mathbf{M}_{rP} = \mathbf{0}$$

Question 4.4 These equations pertain when the body is in equilibrium relative to what frames of reference?

Static Indeterminacy

The vanishing of a vector is guaranteed by the vanishing of three distinct (usually taken to be mutually perpendicular) components. Thus the two vector equations of equilibrium are equivalent to *six* component (scalar) equations. If we have more than six unknown scalars, then we cannot have enough independent equations, and the problem may fall into a category known as **static indeterminacy,** mentioned previously in Section 2.3. Just because we have twice as many vector equations of equilibrium (two instead of one) as we did for the particle (in Chapter 2) does not mean that we cannot still run into the problem of indeterminacy. For example, suppose we alter the problem of Example 2.6 and say that the surfaces of contact are *not* smooth, which means there is the possibility of friction forces as well as normal forces exerted by the planes on the sphere. The free-body diagram then becomes Figure 4.7 below:

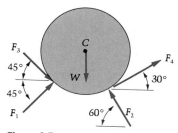

Figure 4.7

where F_3 and F_4 represent the friction forces.

The moment-equation of equilibrium $\Sigma \mathbf{M_C} = \mathbf{0}$ now yields

$$\overset{\curvearrowright}{+} \qquad F_4 R + F_3 R = 0$$

or

$$F_4 + F_3 = 0$$

From $\Sigma \mathbf{F} = \mathbf{0}$ we have, as before, two scalar equations, but now in the four unknowns F_1, F_2, F_3, and F_4. These two equations plus the above result of balancing moments, $F_4 + F_3 = 0$, constitute three equations in four unknowns and hence the problem is **statically indeterminate.**

We note that if just one surface is smooth, this condition renders the problem statically determinate because the remaining friction force vanishes (from the moment equation), and F_1 and F_2 revert to the same answers we found back in Example 2.6.

Answer 4.4 Inertial frames of reference.

Whenever we encounter indeterminacy, additional information, usually having to do with the deformability of the body, is needed to find the unknown forces.

Now in many situations the equilibrium equations provide *fewer* than six independent scalar equations, yet the problems need not be indeterminate. We have seen examples of this in Chapter 2. One of these was the block of Figure 2.6, supported by two taut cables and reproduced below as Figure 4.8:

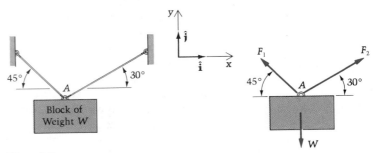

Figure 4.8

First we observe from the free-body diagram that all the external forces have lines of action passing through the same point (A);* in Chapter 3 we called this a concurrent force system. Since $\Sigma M_A = 0$ regardless of the values of F_1 and F_2, the moment equation provides no information about these forces. More generally, we may state that whenever the external forces on a body constitute a concurrent system, there is a moment equation of equilibrium that is identically satisfied, and the greatest number of scalar equations relating the forces will be three.

In the above problem, the z-component of $\Sigma F = 0$ also yields $0 = 0$, i.e., no information, so in fact only two equations were available. We found in Section 2.3 that they were sufficient to solve for the cable tensions, however.

Generally, then, a body in equilibrium will have between one and six useful equations available for determining forces, depending on the physical configuration in which the body is placed. We shall not attempt to catalog all the various circumstances in which one or more of the component-equations of equilibrium is identically satisfied, but we shall point out some commonly occurring situations as they arise in the examples.

Types of Problems in This Section: Bodies Left Intact

The examples and problems in this section will feature finite-sized bodies which are either (a) a single physical object, or (b) two or more physical objects in contact and not needing to be separated in order to effect a

---

* The block must be "hanging" in such a way that the mass center is directly below A; otherwise, $\Sigma M_A \neq 0$.

solution to the equilibrium equations. In other words, this section comprises problems involving objects or groups of objects that are *left intact*. In Section 4.4 to follow, we will find it necessary to go a step further and separate the contacting objects in order to obtain solutions. We will then draw FBDs and write equilibrium equations for the separated objects, treating each in turn as the body being analyzed.

Finally, while we intentionally have not made a clear separation of discussions of two- and three-dimensional problems, the examples that follow are ordered so that the two-dimensional ones come first. In these, where we are dealing with a planar force system, it is important to realize that three of the component equations are satisfied identically. If the xy plane is the plane of the force system, $\Sigma F_z \equiv 0$ and each force produces, with respect to a point in the plane, a moment perpendicular to the plane. Thus, if P is a point in this plane, then $(\Sigma M_P)_x \equiv 0$ and $(\Sigma M_P)_y \equiv 0$. For this restricted class of problems, then, the component equations not automatically satisfied are

$$\Sigma F_x = 0$$
$$\Sigma F_y = 0$$
$$(\Sigma M_P)_z = 0$$

In the future, we shall shorten $(\Sigma M_A)_z$ to simply ΣM_A when the problem is a "plane" one; when this is the case, only "z moments" are normally written.

There follows a series of examples of equilibrium problems that can be solved by leaving the body intact. In each one, as in Chapter 2, the student should note carefully the great importance of the free-body diagram in formulating the equations of equilibrium. The first example is concerned with the equilibrium of a single physical object, a beam. Note that throughout the book, where weights of bodies subjected to other loads are not given as data in examples and in problems, it is to be understood that these weights may be neglected in comparison with the other loads. That is the case with the beam of this example, which also deals with a distributed load.

EXAMPLE 4.4

Find the reactions on the ends of the simply supported beam shown in Figure E4.4a.

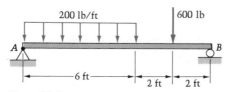

Figure E4.4a

Solution

From Section 3.9, we recognize the resultant of the distributed load to be the area beneath the loading curve, or 1200 lb. Its line of action is at the center of the

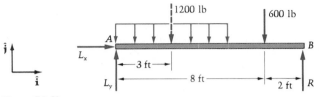

Figure E4.4b

loaded segment as indicated on the free-body diagram (Figure E4.4b) of the beam. Using the equilibrium equations,

$$\Sigma \mathbf{F} = \mathbf{0}$$

yields

$$L_x \hat{\mathbf{i}} + L_y \hat{\mathbf{j}} + 1200(-\hat{\mathbf{j}}) + 600(-\hat{\mathbf{j}}) + R\hat{\mathbf{j}} = \mathbf{0}$$

which has the following component equations:

$$L_x = 0 \tag{1}$$

and

$$L_y - 1200 - 600 + R = 0 \tag{2}$$

Also,

$$\Sigma \mathbf{M}_A = \mathbf{0}$$

yields

$$3(1200)(-\hat{\mathbf{k}}) + 8(600)(-\hat{\mathbf{k}}) + 10R\hat{\mathbf{k}} = \mathbf{0}$$

from which

$$-3600 - 4800 + 10R = 0$$

$$R = 840 \text{ lb} \tag{3}$$

Substituting R into Equation (1),

$$L_y = 1800 - R = 1800 - 840$$

$$L_y = 960 \text{ lb}$$

In the preceding example, note that the roller reaction at B was found with a single equation, by summing moments about the point of intersection (A) of the other two unknowns. We will often find this idea useful.

Scalar Vs. Vector Approach

We wish now to note a slightly different approach to the preceding example. Instead of writing the vector equations $\Sigma \mathbf{F} = \mathbf{0}$ and $\Sigma \mathbf{M}_A = \mathbf{0}$ and then picking off the coefficients of the unit vectors $\hat{\mathbf{i}}, \hat{\mathbf{j}}$, and $\hat{\mathbf{k}}$ to form the scalar equilibrium equations (1, 2, 3), we could have written, a bit more quickly,

$$\xrightarrow{+} \quad \Sigma F_x = 0 = L_x$$

$$+\uparrow \quad \Sigma F_y = 0 = L_y - 1200 - 600 + R$$

$$\curvearrowright + \quad \Sigma M_A = 0 = -3(1200) - 8(600) + 10R$$

This scalar approach has the advantage of omitting the unit vectors and proceeding immediately to the algebraic equations. One must be especially careful here to affix the correct sign on the various terms; it is also helpful to display a symbol such as $\xrightarrow{+}$ to the left of an equation as a reminder of the direction of the unit vector that is being suppressed.

One more point is worth mentioning in conjunction with the preceding example: It is often worthwhile to use another moment equation of equilibrium to check the numerical results. Let's see if $\Sigma M_B = 0$ is satisfied by the values we have calculated for L_x, L_y, and R:

$$\Sigma M_B = 0$$

$$2(600) + 7(1200) - 10L_y = 0$$

$$1200 + 8400 - 10(960) = 0$$

$$9600 - 9600 = 0$$

This check correctly suggests that we could have solved this problem using the three scalar equations: $\Sigma F_x = 0$, $\Sigma M_A = 0$, and $\Sigma M_B = 0$. As mentioned earlier, Problems 4.24–4.26 are concerned with the possibilities of using two and three moment equations when the force system (loads and reactions) is coplanar.

For an example of the equilibrium of a body comprising more than one physical object in contact, let us revisit the diver and diving board of Sections 4.1 and 4.2:

EXAMPLE 4.5

If $a = 5$ ft, $L = 14$ ft, and the respective weights of the diver and the board are 200 lb and 90 lb, find the reactions onto the uniform board in Figure E4.5a at the pin A and the roller B.

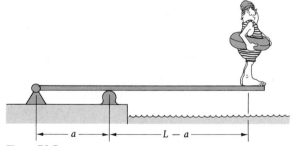

Figure E4.5a

Solution

The free-body diagram of the body (diver plus board) is shown in Figure E4.5b. Note that we must allow for the possibility of two force components (A_x and A_y) from the pin, but only a vertical component (B) from the roller:

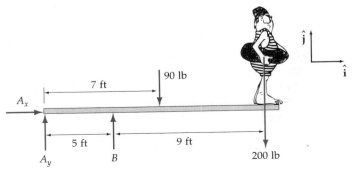

Figure E4.5b

We see from the FBD that we have three unknowns (A_x, A_y, B), and we know we can write three independent equations and solve for them. The first equation expresses the fact that the x-components of the external forces must sum to zero:

$$\xrightarrow{+} \quad \Sigma F_x = 0 = A_x$$

Thus A_x, being the only force in the x-direction acting on the body, must in this case vanish.

> **Question 4.5**   Will the force A_x remain zero while the diver dives off the board?

Next we use the FBD to write the "y-equation" of equilibrium:

$$+\uparrow \quad \Sigma F_y = 0 = A_y + B - 90 - 200$$

or

$$A_y + B = 290 \tag{1}$$

It is clear that we cannot solve for A_y and B without invoking the remaining equation — the "moment equation" of equilibrium:

$$\curvearrowleft_+ \quad \Sigma M_A = 0 = B(5 \text{ ft}) - (90 \text{ lb})(7 \text{ ft}) - (200 \text{ lb})(14 \text{ ft})$$

$$B = \frac{630 + 2800}{5} = \frac{3430}{5} = 686 \text{ lb}$$

Equation (1) then yields:

$$A_y = 290 - B$$
$$= 290 - 686$$
$$= -396 \text{ lb}$$

Answer 4.5   No.

The minus sign tells us that the pin at A is pushing *down* on the board. Although we wrote the "y-equation" in scalar form, we were actually representing the force as $A_y \hat{\jmath}$, and we obtained $A_y = -396$. Thus the force in vector form is $-396\hat{\jmath}$, or $396 (-\hat{\jmath})$, or $396 \downarrow$ lb.

As an overall check on our answers, let us sum the moments at C:

$$\Sigma M_C = 396(7) - 686(7 - 5) - 200(7)$$

$$= 2772 - 1372 - 1400$$

$$= 0 \quad \checkmark$$

Just because the sum of the moments about C is zero does not *guarantee* that we haven't made a mistake, but it does make it highly unlikely.

In the next example, three bars are pinned together to form a body known as a *frame*.

EXAMPLE 4.6

The "A-frame" is subjected to the 150-N load as shown in Figure E4.6a. Find the pin reaction at A and the force exerted by the roller at C.

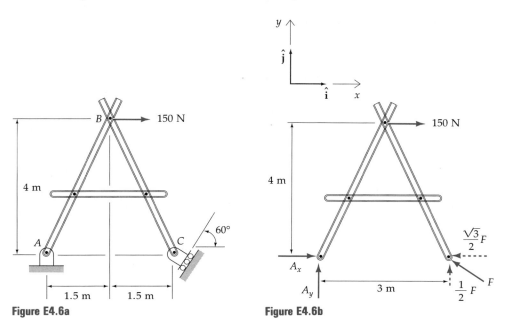

Figure E4.6a   Figure E4.6b

Solution

Referring to the free-body diagram in Figure E4.6b, we see that we may obtain the reaction at C by summing moments at A. Since the force system is coplanar we may easily use the "scalar" form of the moment equation. Thus, counting counterclockwise moments as positive,

$$\circlearrowleft_{+} \qquad \Sigma M_A = 0$$

$$3(F/2) - 4(150) = 0$$

$$F = 400 \text{ N}$$

The student should realize (or confirm) that this scalar equation is precisely what results when we write the vector equation $\Sigma \mathbf{M}_A = 0$ and then pick off coefficients of $\hat{\mathbf{k}}$. Were we instead to take *clockwise* moments as positive, then

$$\circlearrowright_{+} \qquad \Sigma M_A = 0$$

yields

$$-3(F/2) + 4(150) = 0$$

which is nothing more than the result of writing $\Sigma \mathbf{M}_A = 0$ and picking off coefficients of $(-\hat{\mathbf{k}})$.

The component forms of $\Sigma \mathbf{F} = 0$ are

$$\Sigma F_x = 0$$

and

$$\Sigma F_y = 0$$

From the first, we find

$$A_x + 150 - \frac{\sqrt{3}}{2}(400) = 0$$

so that

$$A_x = 196 \text{ N}$$

From the second,

$$A_y + F/2 = 0$$

so that

$$A_y = -200 \text{ N}$$

These are the same equations we would obtain by writing $\Sigma \mathbf{F} = 0$ and then picking off coefficients of $\hat{\mathbf{i}}$ and $\hat{\mathbf{j}}$, respectively.

In conclusion, the magnitude of the pin reaction at A is $\sqrt{A_x^2 + A_y^2}$ $= \sqrt{(196)^2 + (200)^2} = 280$ N, and vectorially the force exerted by the pin on the frame at A is $196\hat{\mathbf{i}} - 200\hat{\mathbf{j}}$ N. The force exerted on the frame at C is

$$400\left(-\frac{\sqrt{3}}{2}\hat{\mathbf{i}} + \frac{1}{2}\hat{\mathbf{j}}\right) = -346\hat{\mathbf{i}} + 200\hat{\mathbf{j}} \text{ N}$$

Before moving on, we mention the concept of *stability*. Let the sphere of Example 2.6 (see Figure 4.9(a)) be nonuniform so that the mass center C does not coincide with the geometric center O. Taking the surfaces to be smooth, let us find all possible equilibrium configurations of the sphere.

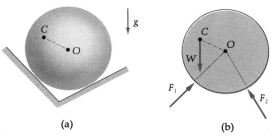

(a)   (b)

Figure 4.9

From the free-body diagram (Figure 4.9(b)) we see, as before, that the lines of action of the two reactions, F_1 and F_2, pass through the geometric center (O) of the sphere. Thus in order for $\Sigma M_O = 0$ to be satisfied, the mass center C must lie either directly above or below O, so that the always-vertical weight can also pass through O.

The reader may recognize that this problem is similar to that of finding the equilibrium positions of a body supported in the manner of a pendulum, as shown (see Figure 4.10) in the next illustration. The equations of equilibrium tell us that there are two such positions: one where the mass center is directly below the support and one where the mass center is directly above. Our experience tells us that the body will not remain (without additional restraint) in this second position. This configuration is said to be an *unstable* equilibrium configuration and the first is said to be a *stable* equilibrium configuration. An important point to recognize is that the equilibrium equations do not themselves distinguish these two types of equilibrium states.

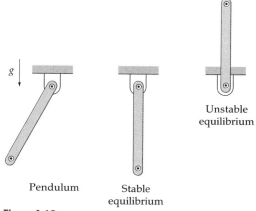

Figure 4.10

Special Results for Pulleys and Two-Force Members

We now wish to derive two extremely useful results before presenting another series of examples. One of these is concerned with the tensions in a belt on either side of a pulley; the other deals with the situation in which

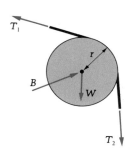

Figure 4.11

a body is held in equilibrium by the action of just two forces. These situations will arise many times in both examples and problems, and knowing the results will always facilitate the solutions.

Belt, Rope, Cord, or Cable Passing Over a Pulley on Frictionless Bearings

Referring to the free-body diagram (Figure 4.11), the equilibrium equation requiring that $\Sigma M_{\text{bearing axis}} = 0$ yields (provided the mass center of the pulley is on the bearing axis):

$$\circlearrowleft_{+} \qquad T_1 r - T_2 r = 0$$

(The symbol $\circlearrowleft_{+}$ means that the unit vector out of the page has been suppressed from each term in the equation.) Thus we obtain

$$T_1 = T_2$$

That is, the belt tension is the same on both sides of a pulley supported in equilibrium by frictionless bearings. We used this result in Example 2.8, deferring the proof to the current section.

Two-force Body (or Member)

Suppose that a light body is held in equilibrium by two (and only two) external forces. We let the plane of the page contain the points of application of these forces as shown below in Figure 4.12(a). In order that $\Sigma \mathbf{F} = 0$, we must have $\mathbf{F}_1 = -\mathbf{F}_2$ so that the free-body diagram might now appear as shown in Figure 4.12(b). But wait! These two forces, as drawn, constitute a couple and so $\Sigma \mathbf{M} \neq \mathbf{0}$ unless the lines of action of the two forces coincide. Thus the proper free-body diagram must be Figure 4.12(c) below:

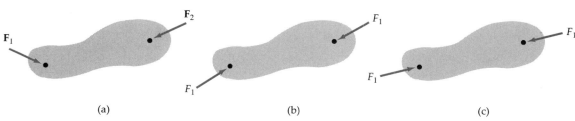

(a)   (b)   (c)

Figure 4.12

We now see that the two forces acting on the body must have a common line of action — *the line joining the two points of application* of the forces. Note in Figure 4.12(c) above that the forces are drawn acting *toward* each other, tending to compress the material between them. Alternately, they might in a given problem be directed *away from* each other, tending to stretch the material between them. These are the only two possibilities for the directions of the pair of forces acting on a two-force member.

In each of the next three examples a part of the body to be analyzed is either a pulley on bearings of negligible friction, or a two-force body.

EXAMPLE 4.7

The light pulley (radius 0.1 m) is supported by frictionless bearings at the end of the uniform 640-N beam AB, as shown in Figure E4.7a. A cable is wrapped around the pulley and connected to a ceiling at points C and D. Find the force in the cable and the reactions of the pin at A onto the beam.

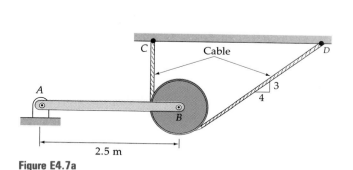

Figure E4.7a

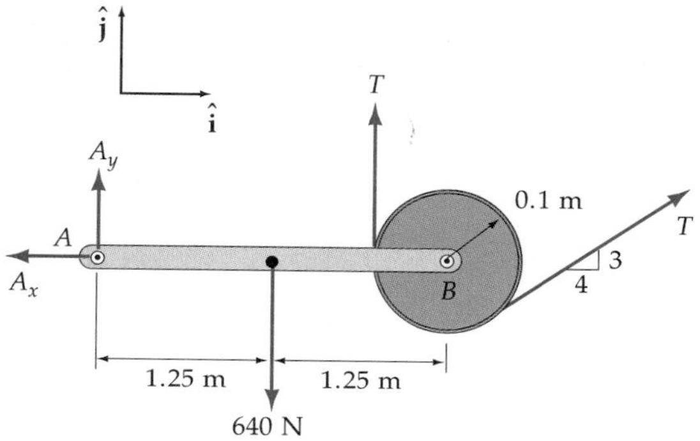

Figure E4.7b

Solution

In drawing the free-body diagram (see Figure E4.7b) we use the fact that the cable tension on either side of the pulley is the same, and we call it T. The pin reaction at A is represented by force components in the horizontal and vertical directions. We drew A_x acting to the left because in this case we are certain of its direction — it has to oppose the horizontal component of the right-most cable tension, which acts to the right. Observations like this are not essential, but they do cut down on the number of minus signs in the solution, and they encourage good thinking about such things as directions of forces.

If we now sum the moments about the pin A and equate the result to zero, the only unknown in the equation will be T, and we can thus find it with a single equation (the "moment equation"):

$$\circlearrowleft_+ \qquad \Sigma M_A = 0 = -(640 \text{ N})(1.25 \text{ m}) + T(2.4 \text{ m}) + 0.6T(2.5 \text{ m})$$
$$+ 0.8T(0) + T(0.1 \text{ m})$$

so that

$$T = \frac{800}{4.00} = 200 \text{ N}$$

In the above equation, we have for convenience replaced the rightmost tension by an equipollent system at B as shown in Figure E4.7c at the left. This is easier than resolving T into its components at E and then computing the moments of these components about A.* Recall from Section 3.6 that the equipollent system at B is the same force at the new point (B) accompanied by the moment ($0.1T$) that the force at its original point (E) exerts about B.

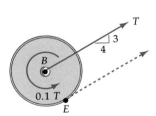

Figure E4.7c

---

* See problem 4.14.

To obtain the pin reaction at A, we use the "force equations":

$$\Sigma F_x = 0 = -A_x + 0.8\,\overset{200}{\cancel{T}}$$

so

$$A_x = 160 \text{ N}$$

$$+\uparrow \qquad \Sigma F_y = 0 = A_y - 640 + \overset{200}{\cancel{T}} + 0.6\,\overset{200}{\cancel{T}}$$

$$A_y = 640 - 200 - 120 = 320 \text{ N}$$

Thus the pin reaction at A onto the beam is the vector force $-160\hat{\mathbf{i}} + 320\hat{\mathbf{j}}$ N.

Let us carry the discussion about the equipollent system at B in the preceding example a bit further. If equipollent systems to *both* tensions were drawn at B, we would have (see Figure 4.13):

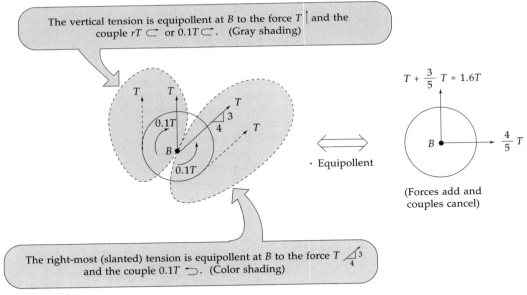

The vertical tension is equipollent at B to the force $T\,\uparrow$ and the couple $rT \curvearrowleft$ or $0.1T \curvearrowleft$.   (Gray shading)

$T + \dfrac{3}{5}\,T = 1.6T$

$\dfrac{4}{5}\,T$

⟺ · Equipollent

(Forces add and couples cancel)

The right-most (slanted) tension is equipollent at B to the force T and the couple $0.1T \curvearrowright$.   (Color shading)

Figure 4.13

This procedure would have made the solution for T even simpler; now we would have:

$$\curvearrowleft_{(+)} \qquad \Sigma M_A = 0 = (-640 \text{ N})(1.25 \text{ m}) + (1.6T)(2.5 \text{ m})$$

$$T = 200 \text{ N} \qquad \text{(as before)}$$

EXAMPLE 4.8

A dumping mechanism is shown in Figure E4.8a. The weight of the bed plus contents is 1200 lb. It is in equilibrium in the given position with mass center at C. Find the force in the strut AB, which contains a hydraulic cylinder for raising and lowering the bed.

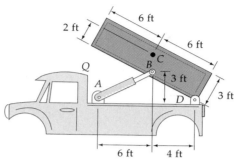

Figure E4.8a

Solution

Recognizing that the strut AB is a two-force member, we know that the force exerted on it at A by the pin lies along the line AB. Thus, we draw the free-body diagram as shown in Figure E4.8b. Note that the body we are analyzing consists of the bed, its contents, and the strut.

Summing moments about the pin at D will give the force in AB:

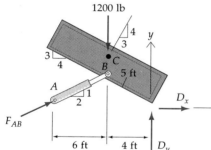

Figure E4.8b

$$\Sigma \mathbf{M}_D = 0 = \mathbf{r}_{DC} \times (-1200\hat{\mathbf{j}}) + \mathbf{r}_{DA} \times \mathbf{F}_{AB}$$

$$\left[6\left(-\frac{4}{5}\hat{\mathbf{i}} + \frac{3}{5}\hat{\mathbf{j}}\right) + 2\left(\frac{3}{5}\hat{\mathbf{i}} + \frac{4}{5}\hat{\mathbf{j}}\right) \right] \times (-1200\hat{\mathbf{j}})$$

$$+ (-10\hat{\mathbf{i}}) \times F_{AB}\left(\frac{2}{\sqrt{5}}\hat{\mathbf{i}} + \frac{1}{\sqrt{5}}\hat{\mathbf{j}}\right) = \mathbf{0}$$

$$(-3.60\hat{\mathbf{i}} + 5.20\hat{\mathbf{j}}) \times (-1200\hat{\mathbf{j}}) + F_{AB}(-4.47\hat{\mathbf{k}}) = 0$$

$$F_{AB} = 966 \text{ lb}$$

We say that the force "in" the two-force member AB is **compressive** since the strut is subjected to a pair of 966-pound forces that tend to cause it to shorten. Note that we defined (arbitrarily) the scalar F_{AB} in such a way that a positive value would indicate compression of the strut. Indeed it turned out that way as our intuition would suggest. Had F_{AB} turned out to be negative, we would have called the force **tensile** since the strut would then be subjected to a pair of forces tending to stretch it.

EXAMPLE 4.9

Find the reactions exerted on the frame at A and B by the pins. See Figure E4.9a.

Solution

The member BD is a two-force member because, even though it isn't a straight bar, it is loaded by forces at only two points. Therefore, we know the direction of the reaction at B to be along BD. The overall free-body diagram (Figure E4.9b) will help us determine this reaction, and those at A as well:

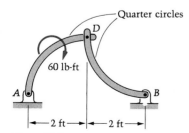

Figure E4.9a

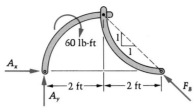

Figure E4.9b

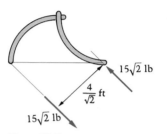

Figure E4.9c

$$\curvearrowleft_{+} \qquad \Sigma M_A = 0 = -60 \text{ lb-ft} + \left(\frac{F_B}{\sqrt{2}}\right)(4 \text{ ft})$$

$$F_B = 15\sqrt{2} \text{ lb}$$

$$\xrightarrow{+} \qquad \Sigma F_x = 0 = A_x - \frac{F_B}{\sqrt{2}} = A_x - 15$$

$$A_x = 15 \text{ lb}$$

and

$$+\uparrow \qquad \Sigma F_y = 0 = A_y + \frac{F_B}{\sqrt{2}} = A_y + 15$$

$$A_y = -15 \text{ lb}$$

Therefore, the reactions onto the frame are

$$\text{at } A, \quad 15\sqrt{2} \qquad \text{lb}$$

$$\text{at } B, \quad 15\sqrt{2} \qquad \text{lb}$$

Note from Figure E4.9c that these two reactions necessarily form a couple of 60 ↺ lb-ft.

Three-Dimensional Examples

We close the section now with a series of examples of equilibrium in three dimensions:

EXAMPLE 4.10

Find the force and couple reaction at the base of the advertising sign of Example 4.2, shown in Figure E4.10a.

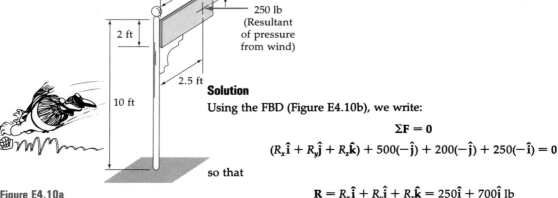

Figure E4.10a

Solution

Using the FBD (Figure E4.10b), we write:

$$\Sigma \mathbf{F} = 0$$

$$(R_x \hat{\mathbf{i}} + R_y \hat{\mathbf{j}} + R_z \hat{\mathbf{k}}) + 500(-\hat{\mathbf{j}}) + 200(-\hat{\mathbf{j}}) + 250(-\hat{\mathbf{i}}) = 0$$

so that

$$\mathbf{R} = R_x \hat{\mathbf{i}} + R_y \hat{\mathbf{j}} + R_z \hat{\mathbf{k}} = 250 \hat{\mathbf{i}} + 700 \hat{\mathbf{j}} \text{ lb}$$

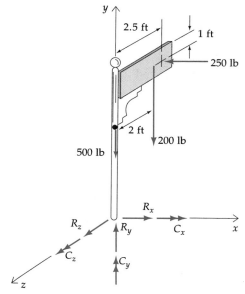

Figure E4.10b

is the resultant force exerted on the base of the post by the foundation. Thus

$$R_x = 250 \text{ lb} \qquad R_y = 700 \text{ lb} \qquad R_z = 0$$

Next, we sum the moments of all forces and couples about the base A of the sign:

$$\Sigma M_A = 0$$

so that

$$(C_x\hat{\mathbf{i}} + C_y\hat{\mathbf{j}} + C_z\hat{\mathbf{k}}) + (9\hat{\mathbf{j}} - 2\hat{\mathbf{k}}) \times [200(-\hat{\mathbf{j}})]$$
$$+ (9\hat{\mathbf{j}} - 2.5\hat{\mathbf{k}}) \times [250(-\hat{\mathbf{i}})] = 0$$

Therefore,

$$(C_x\hat{\mathbf{i}} + C_y\hat{\mathbf{j}} + C_z\hat{\mathbf{k}}) + 9(200)(0) + 2(200)(-\hat{\mathbf{i}})$$
$$+ 9(250)(\hat{\mathbf{k}}) + 2.5(250)\hat{\mathbf{j}} = 0$$

so that

$$\mathbf{C} = C_x\hat{\mathbf{i}} + C_y\hat{\mathbf{j}} + C_z\hat{\mathbf{k}} = 400\hat{\mathbf{i}} - 625\hat{\mathbf{j}} - 2250\hat{\mathbf{k}} \text{ lb-ft}$$

is the moment of the resultant couple on the base of the post. Therefore

$$C_x = 400 \text{ lb-ft} \qquad C_y = -625 \text{ lb-ft} \qquad C_z = -2250 \text{ lb-ft}$$

Question 4.6   Why doesn't the 500-lb force appear explicitly in the moment equilibrium equation, $\Sigma M_A = 0$?

Answer 4.6   Its line of action passes through A.

EXAMPLE 4.11

The boom, whose weight may be neglected, is supported by a ball-and-socket connection at A and two taut wires as shown in Figure E4.11a. Find the tensions in the wires and the reaction at A.

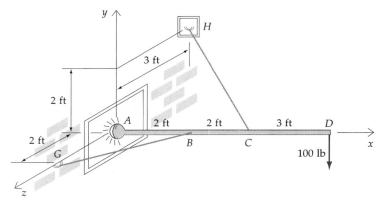

Figure E4.11a

Solution

From the free-body diagram (Figure E4.11b) we see that there are only five unknown scalars describing the reactions. For a three-dimensional problem such as this, we should then be concerned that there is not enough constraint provided against a rigid-body motion — that is, that we might not be able to satisfy the equations of equilibrium. However, in this case, one of the equations [the balance of moments about the axis of the boom (x)] is satisfied identically since each of the loads and reactions has a line of action intersecting that axis. The body would not be adequately restrained were there to be a loading that tended to turn the boom about its axis.

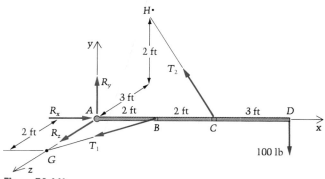

Figure E4.11b

The vector descriptions of the forces exerted by the wires on the boom are $T_1\hat{\mathbf{e}}_1$ and $T_2\hat{\mathbf{e}}_2$, where

$$\hat{e}_1 = \frac{\mathbf{r}_{BG}}{|\mathbf{r}_{BG}|} = \frac{-2\hat{i} + 2\hat{k}}{\sqrt{(2)^2 + (2)^2}} = -0.707\hat{i} + 0.707\hat{k}$$

and

$$\hat{e}_2 = \frac{\mathbf{r}_{CH}}{|\mathbf{r}_{CH}|} = \frac{-4\hat{i} + 2\hat{j} - 3\hat{k}}{\sqrt{(4)^2 + (2)^2 + (3)^2}} = -0.743\hat{i} + 0.371\hat{j} - 0.557\hat{k}$$

We first use the moment equation $\Sigma M_A = 0$, since it will not involve the unknowns R_x, R_y, and R_z:

$$\Sigma \mathbf{M}_A = 0$$

$$\mathbf{r}_{AB} \times T_1\hat{e}_1 + \mathbf{r}_{AC} \times T_2\hat{e}_2 + \mathbf{r}_{AD} \times (-100\hat{j}) = \mathbf{0}$$

$$2\hat{i} \times T_1(-0.707\hat{i} + 0.707\hat{k}) + 4\hat{i} \times T_2(-0.743\hat{i} + 0.371\hat{j} - 0.557\hat{k})$$
$$+ 7\hat{i} \times (-100\hat{j}) = \mathbf{0}$$

$$-2(0.707)T_1\hat{j} + 4(0.371)T_2\hat{k} + 4(0.557)T_2\hat{j} - 7(100)\hat{k} = \mathbf{0}$$

Note the absense of $\hat{i}$ terms! Thus we have

$$\hat{k}: \qquad 4(0.371)T_2 - 7(100) = 0$$

$$T_2 = 472 \text{ lb}$$

and

$$\hat{j}: \quad -2(0.707)T_1 + 4(0.577)T_2 = 0$$

$$T_1 = 744 \text{ lb}$$

Thus

$$T_1\hat{e}_1 = -526\hat{i} + 526\hat{k} \text{ lb}$$

and

$$T_2\hat{e}_2 = -351\hat{i} + 175\hat{j} - 263\hat{k} \text{ lb}$$

Next we equilibrate the forces:

$$\Sigma \mathbf{F} = \mathbf{0}$$

$$(R_x\hat{i} + R_y\hat{j} + R_z\hat{k}) + T_1\hat{e}_1 + T_2\hat{e}_2 + 100(-\hat{j}) = \mathbf{0}$$

$$(R_x\hat{i} + R_y\hat{j} + R_z\hat{k}) + (-526\hat{i} + 526\hat{k}) + (-351\hat{i} + 175\hat{j} - 263\hat{k}) - 100\hat{j} = \mathbf{0}$$

Thus, collecting like terms,

$$\hat{i}: \quad R_x - 526 - 351 = 0$$

$$R_x = 877 \text{ lb}$$

$$\hat{j}: \quad R_y + 175 - 100 = 0$$

$$R_y = -75 \text{ lb}$$

$$\hat{k}: \quad R_z + 526 - 263 = 0$$

$$R_z = -263 \text{ lb}$$

Therefore, the force exerted on the boom by the ball-and-socket connection is $877\hat{i} - 75\hat{j} - 263\hat{k}$ lb.

EXAMPLE 4.12

The uniform door weighing 1200 N is held in equilibrium in the horizontal position by the cable and by the smooth hinges at A and B (see Figure E4.12a). Find the force exerted on the door by the cable.

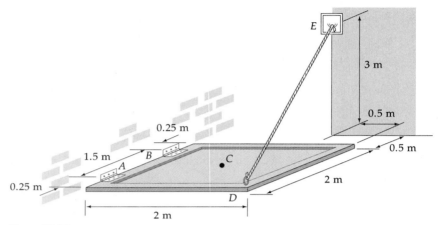

Figure E4.12a

Solution

As usual we express the unknown reactions in terms of the minimum number of unknown scalars (components). As indicated on the free-body diagram (Figure E4.12b), the hinge reactions are expressed as follows:

Force at A:

$$F_1\hat{\mathbf{i}} + F_2\hat{\mathbf{j}} + F_3\hat{\mathbf{k}}$$

Couple at A:

$$M_2\hat{\mathbf{j}} + M_3\hat{\mathbf{k}}$$

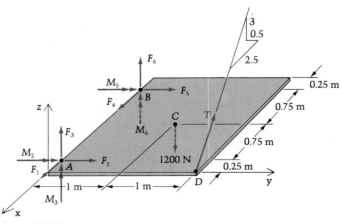

Figure E4.12b

Force at B:

$$F_4\hat{\mathbf{i}} + F_5\hat{\mathbf{j}} + F_6\hat{\mathbf{k}}$$

Couple at B:

$$M_5\hat{\mathbf{j}} + M_6\hat{\mathbf{k}}$$

The force exerted by the cable is

$$T\left(\frac{-2.5\hat{\mathbf{i}} - 0.5\hat{\mathbf{j}} + 3\hat{\mathbf{k}}}{\sqrt{(2.5)^2 + (0.5)^2 + (3)^2}}\right) = T(-0.635\hat{\mathbf{i}} - 0.127\hat{\mathbf{j}} + 0.762\hat{\mathbf{k}})$$

because we know that force to have the line joining points D and E as its line of action.

We see that ten unknown scalars are required to represent the hinge reactions. Including the unknown cable tension T, we have eleven unknown scalars, but we only have six independent equations available from the equations of equilibrium. Thus, the problem of finding all of the reactions is statically indeterminate. However, none of the hinge reactions produces a moment about the common axis of the hinges, but the cable force T does, and so it may be determined from the equilibrium equation requiring that the moments about the hinge axis sum to zero. This equation may be written as

$$\Sigma \mathbf{M}_A \cdot \hat{\mathbf{i}} = 0$$

Substituting the moments of the various forces and couples,

$$\hat{\mathbf{i}} \cdot [0 \times (F_1\hat{\mathbf{i}} + F_2\hat{\mathbf{j}} + F_3\hat{\mathbf{k}}) + M_2\hat{\mathbf{j}} + M_3\hat{\mathbf{k}}]$$
$$+ \hat{\mathbf{i}} \cdot [(-1.5\hat{\mathbf{i}}) \times (F_4\hat{\mathbf{i}} + F_5\hat{\mathbf{j}} + F_6\hat{\mathbf{k}}) + M_5\hat{\mathbf{j}} + M_6\hat{\mathbf{k}}]$$
$$+ \hat{\mathbf{i}} \cdot [(-0.75\hat{\mathbf{i}} + 1\hat{\mathbf{j}}) \times (-1200\hat{\mathbf{k}})]$$
$$+ \hat{\mathbf{i}} \cdot [(0.25\hat{\mathbf{i}} + 2\hat{\mathbf{j}}) \times T(-0.635\hat{\mathbf{i}} - 0.127\hat{\mathbf{j}} + 0.762\hat{\mathbf{k}})] = 0$$
$$0 + 0 - (1)(1200) + 2(0.762)T = 0$$
$$T = 787 \text{ N}$$

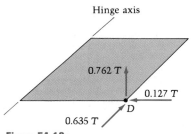

Hinge axis

0.762 T

0.127 T

D

0.635 T

Figure E4.12c

We could have obtained the same result with a little less formality. Suppose we first decompose the cable tension into three parts at D as shown in Figure E4.12c. We see that only the $0.762T$ component produces a moment about the hinge axis and that moment (using the "perpendicular distance" method) is $2(0.762T)\hat{\mathbf{i}}$. Similarly the moment of the 1200-N weight is seen to be $(1)(1200)(-\hat{\mathbf{i}})$. These two moments must sum to zero since we can see that the hinge reactions produce no moments about the hinge axis. Thus

$$2(0.762T)\hat{\mathbf{i}} + (1)(1200)(-\hat{\mathbf{i}}) = \mathbf{0}$$

or

$$2(0.762T) - (1)(1200) = 0$$
$$T = 787 \text{ N}$$

as before.

Let us now see what information the other equations of equilibrium yield.

$$\Sigma F = 0$$

$$(F_1\hat{i} + F_2\hat{j} + F_3\hat{k}) + (F_4\hat{i} + F_5\hat{j} + F_6\hat{k}) + 1200(-\hat{k})$$
$$+ 787(-0.635\hat{i} - 0.127\hat{j} + 0.762\hat{k}) = 0$$
$$(F_1\hat{i} + F_2\hat{j} + F_3\hat{k}) + (F_4\hat{i} + F_5\hat{j} + F_6\hat{k}) = 1200\hat{k} + 500\hat{i} + 100\hat{j} - 600\hat{k}$$

or

$$(F_1\hat{i} + F_2\hat{j} + F_3\hat{k}) + (F_4\hat{i} + F_5\hat{j} + F_6\hat{k}) = 500\hat{i} + 100\hat{j} + 600\hat{k} \ N$$

Notice that the left-hand side is the sum of the force parts of the hinge reactions at A and B; in other words, we have found the force part of the *resultant* of the hinge reactions.

Turning to the moment equation

$$\Sigma M_A = 0$$

$$(M_2\hat{j} + M_3\hat{k}) + (-1.5\hat{i}) \times (F_4\hat{i} + F_5\hat{j} + F_6\hat{k}) + (M_5\hat{j} + M_6\hat{k})$$
$$+ (-0.75\hat{i} + 1\hat{j}) \times (-1200\hat{k})$$
$$+ (0.25\hat{i} + 2\hat{j}) \times (-500\hat{i} - 100\hat{j} + 600\hat{k}) = 0$$

We can recognize the first three terms (all of the unknowns) as the couple part of the resultant at A of the hinge reactions.

We see then that the reactions at the individual hinges cannot be found here, but their resultant can be. Unfortunately this resultant doesn't tell us much about conditions at the hinges in a problem such as this one where it is reasonable to expect the reactions at the two hinges to be substantially different. In the next example we explore a situation in which the two-hinge resultant might be expected to provide valuable information.

Question 4.7   If in the preceding example the hinge at B broke, could the body remain in equilibrium? If so, would the problem of finding the hinge reaction at A then be statically determinate?

Answer 4.7   Yes, there is still enough constraint to prohibit rigid-body motion. Yes, the six scalar equations from $\Sigma F = 0$ and $\Sigma M_A = 0$ can now be solved for the six unknowns F_1, F_2, F_3, M_2, M_3, and T.

EXAMPLE 4.13

The hinged door of Example 4.12 is supported by the (different) cable as shown in Figure E4.13a. Find the tension in the cable and the resultant (at O) of the hinge reactions.

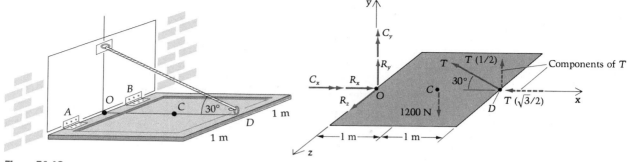

Figure E4.13a

Figure E4.13b

Solution

Example 4.12 illustrated the futility of attempting to find the individual hinge reactions; thus here we are seeking their resultant, drawn in the free-body diagram as a force and couple at O. Because in this example the hinge axis is the z axis, the component of the couple in the z direction is zero as suggested by the free-body diagram (Figure 4.13b). We use here the component forms of the equilibrium equations and encourage the reader to write out the vector equations $\Sigma \mathbf{F} = 0$ and $\Sigma \mathbf{M}_o = 0$ in terms of unit vectors so as to compare with the results below. First, we sum moments about the hinge axis:

$$+\;\circlearrowleft\quad (\Sigma M_O)_z = 0$$

so that

$$2(T/2) - (1)(1200) = 0$$
$$T = 1200 \text{ N}$$

Next, we sum moments about the x axis (through O):

$$+\;\circlearrowleft\quad (\Sigma M_O)_x = 0$$

This yields

$$C_x = 0$$

A third moment equation is written about the y axis:

$$+\;\circlearrowleft\quad (\Sigma M_O)_y = 0$$

It gives

$$C_y = 0$$

Next, we sum the forces; first, the x components:

$$\xrightarrow{+}\quad \Sigma F_x = 0$$

giving

$$R_x - T\left(\frac{\sqrt{3}}{2}\right) = 0$$

so that

$$R_x = \frac{\sqrt{3}}{2} T = \frac{\sqrt{3}}{2}(1200) = 1040 \text{ N}$$

Next, the y components:

$$+\uparrow \quad \Sigma F_y = 0$$

so that

$$R_y + T(\tfrac{1}{2}) - 1200 = 0$$
$$R_y = 1200 - 600 = 600 \text{ N}$$

Finally, the z components:

$$\nearrow\!\!\!\!/ \quad \Sigma F_z = 0$$

giving

$$R_z = 0$$

Thus the resultant of the hinge reactions is $1040\hat{\mathbf{i}} + 600\hat{\mathbf{j}}$ N, with a line of action through O since $C_x = C_y = C_z = 0$.

As we mentioned before, the equilibrium equations won't tell us the individual hinge reactions. However, it would seem reasonable to assume with identical hinges symmetrically placed about the xy plane (as is the case here) that the components of the hinge reactions parallel to this plane will be identical. These components would each equal $520\hat{\mathbf{i}} + 300\hat{\mathbf{j}}$ N. At the hinge A, the z-directional force and the x- and y-direction couples will each be equal in magnitude but opposite in direction to the corresponding force or couple at B. These reactions, however, result from the tendency of the fairly rigid door to deform and are typically small in a problem such as this one.

The fact that there is a plane (xy) of symmetry for loads and supports suggests that this problem could have been treated as two-dimensional as shown

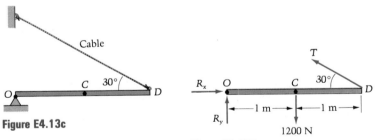

Figure E4.13c

Figure E4.13d

in Figure E4.13c and its free-body diagram (Figure E4.13d). Invoking the equilibrium equations in the form appropriate to a coplanar system of loads and reactions, we obtain

$$\curvearrowleft_{+} \quad (\Sigma M_O)_z = 0$$
$$2(T \sin 30°) - 1(1200) = 0$$
$$T = 1200 \text{ N} \quad \text{(as before)}$$
$$\xrightarrow{+} \quad \Sigma F_x = 0$$
$$R_x - T \cos 30° = 0$$
$$R_x = T(\sqrt{3}/2) = 600\sqrt{3} = 1040 \text{ N} \quad \text{(as before)}$$

and

$$+\uparrow \quad \Sigma F_y = 0$$

$$R_y - 1200 + T \sin 30° = 0$$

$$R_y = 1200 - 1200(\tfrac{1}{2}) = 600 \text{ N} \qquad \text{(as before)}$$

EXAMPLE 4.14

The uniform thin triangular plate of Figure E4.14a is supported by a slider-on-smooth-guide welded to the plate at A and a similar slider attached at B by a ball-and-socket connection. The plate weighs 10 lb per square foot of plan area and the mass center of the plate is at C. Find the reactions at A and B when the plate is subjected to the 400-lb force shown.

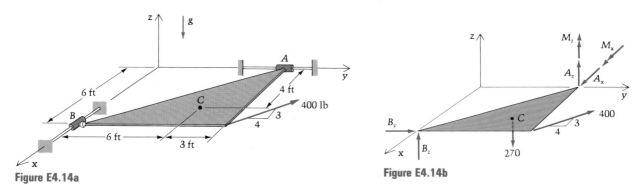

Figure E4.14a

Figure E4.14b

Solution

The weight of the plate is $10[(9)(6)/2] = 270$ lb and, of course, the line of action of this resultant of distributed gravitational attractions is downward through the mass center C as shown in the FBD of Figure E4.14b.

The ball-and-socket at B eliminates the possibility of a couple there, and the slider eliminates the possibility of an x component of force there. At A the slider eliminates the possibilities of a y component of force or of couple. Thus, we see from the free-body diagram that there are six unknown components of reactions. Since, in addition, the plate is adequately restrained against rigid motion, we should anticipate that this is a well-posed, statically determinate, three-dimensional problem; and thus that the six independent component-equations of equilibrium will provide a solution for the reactions.

Component equations guaranteeing that $\Sigma \mathbf{F} = \mathbf{0}$ are

$$\nearrow \quad \Sigma F_x = 0$$

$$A_x - \frac{3}{5}(400) = 0 \Rightarrow A_x = 240 \text{ lb} \tag{1}$$

$$\xrightarrow{+} \quad \Sigma F_y = 0$$

$$B_y + \frac{4}{5}(400) = 0 \Rightarrow B_y = -320 \text{ lb} \tag{2}$$

and

$$+\uparrow \quad \Sigma F_z = 0$$

$$A_z + B_z - 270 = 0 \Rightarrow A_z + B_z = 270 \tag{3}$$

Component equations corresponding to $\Sigma \mathbf{M}_A = \mathbf{0}$ are obtained by balancing moments about axes through A that are respectively parallel to x, y, and z:

$$+\nearrow \quad (\Sigma M_A)_x = 0$$

$$M_x + 3(270) - 9B_z = 0 \Rightarrow M_x - 9B_z = -810 \tag{4}$$

$$+\curvearrowleft \quad (\Sigma M_A)_y = 0$$

$$4(270) - 6B_z = 0 \Rightarrow B_z = 180 \text{ lb} \tag{5}$$

and

$$+\circlearrowleft \quad (\Sigma M_A)_z = 0$$

$$M_z + 6\left[\frac{4}{5}(400)\right] + 6B_y = 0 \tag{6}$$

or

$$M_z + 6(320) + 6(-320) = 0$$

and so

$$M_z = 0$$

Substituting the result $B_z = 180$ lb into Equations (3) and (4) we obtain

$$A_z + 180 = 270 \Rightarrow A_z = 90 \text{ lb}$$

and

$$M_x - 9(180) = -810 \Rightarrow M_x = 810 \text{ lb-ft}$$

We could have used other sets of component equations. For example, referring to the free-body diagram, if we had chosen to balance moments about the line AB, we would have been able to solve for M_x in one step. To show this, let $\hat{\mathbf{e}}_{AB}$ be a unit vector along AB:

$$\hat{\mathbf{e}}_{AB} = \frac{6\hat{\mathbf{i}} - 9\hat{\mathbf{j}}}{\sqrt{(6)^2 + (9)^2}} = \frac{2\hat{\mathbf{i}} - 3\hat{\mathbf{j}}}{\sqrt{13}}$$

The moment of the couple at A about line AB is

$$[(M_x\hat{\mathbf{i}} + M_z\hat{\mathbf{k}}) \cdot \hat{\mathbf{e}}_{AB}]\hat{\mathbf{e}}_{AB} = \frac{2}{\sqrt{13}} M_x \hat{\mathbf{e}}_{AB}$$

and the moment of the weight is, using $(\mathbf{M}_B \cdot \hat{\mathbf{e}}_{AB})\hat{\mathbf{e}}_{AB}$:

$$\{[(6\hat{\mathbf{j}} - 2\hat{\mathbf{i}}) \times (-270\hat{\mathbf{k}})] \cdot \hat{\mathbf{e}}_{AB}\}\hat{\mathbf{e}}_{AB} = (-1620\hat{\mathbf{i}} - 540\hat{\mathbf{j}}) \cdot \left(\frac{2\hat{\mathbf{i}} - 3\hat{\mathbf{j}}}{\sqrt{13}}\right)\hat{\mathbf{e}}_{AB}$$

$$= -\frac{1620}{\sqrt{13}}\hat{\mathbf{e}}_{AB}$$

Thus for $\Sigma \mathbf{M}_{AB} = \mathbf{0}$,

$$\frac{2}{\sqrt{13}} M_x - \frac{1620}{\sqrt{13}} = 0$$

or

$$M_x = 810 \text{ lb-ft} \qquad \text{(as before)}$$

The reader is encouraged to show that this equation expressing the balance of moments about AB is a linear combination of Equations (4) and (5). In this example we have chosen, out of convenience, to balance moments about axes that all intersect at the same point (A). Thus, the equations are all components of $\Sigma \mathbf{M}_A = 0$. However, because the sum of the moments about *any* line must vanish, we could have generated component equations by balancing moments about nonintersecting lines. The important thing is that we obtain three independent component equations that, together with $\Sigma \mathbf{F} = 0$, assure $\Sigma \mathbf{M}_P = 0$ for any point P, provided the three lines are nonparallel and nonplanar.

By this time the reader perhaps has developed some feel for the delicacy associated with providing adequate support for a body and at the same time having static determinacy. In the preceding example, if we were to fix the slider at B, a new component of reaction B_x would be introduced as shown on the free-body diagram in Figure 4.14. This problem is statically indeterminate, which we can clearly ascertain because we now have seven unknowns (B_x, B_y, B_z, A_x, A_z, M_x, and M_z) and only six independent scalar equations of equilibrium.

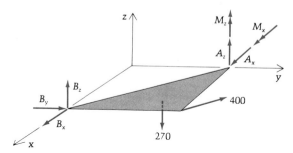

Figure 4.14

On the other hand, were we to provide only a "ball" support at B giving resistance only normal to the xz plane, so that $B_x = B_z = 0$, we would have only five unknowns. Then we would not be able to satisfy the equations of equilibrium unless, because of some special nature of the loading, one of the equations was identically satisfied. That would be the case, for example, were we here to neglect the weight (270 lb) of the plate. Without neglect of this force, we can see from the free-body diagram in Figure 4.15 that we cannot satisfy $\Sigma M_y = 0$! The inability to satisfy one (or more) equation(s) of equilibrium, except for special loadings, is characteristic of inadequate support (or constraint). When that is the case there usually will be one (or more) rigid motion(s) not prohibited by the support system. For the case at hand the supports provide no resistance to rotation of the plate about the y axis.

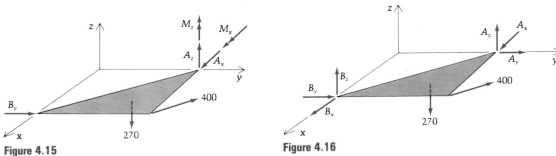

Figure 4.15

Figure 4.16

The preceding discussion might tempt us to conclude that if, in the three-dimensional situation, we have exactly six unknown reaction components, all is well. Unfortunately this is not the case. To illustrate this difficulty suppose that our plate is supported by ball-and-socket connections (no sliders) at both A and B as shown in Figure 4.16. The plate again is not adequately supported. We cannot satisfy the balance of moments about the line AB, and this correlates with the freedom to rotate about the line, which these (inadequate) supports do not curtail.

PROBLEMS ▶ Section 4.3

4.12 Is there a force $\mathbf{F}$ at P for which the system of three forces is in equilibrium? (See Figure P4.12.) If so, find it; if not, why not?

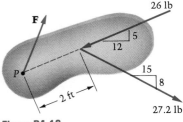

Figure P4.12

4.13 In Figure P4.13(a), the nail is on the verge of coming out of the board; that is, any larger force than 40 lb will pull it out. (a) Find the resultant of all the vertical differential friction forces acting on the nail, assuming the nail exerts no horizontal force on the claw of the hammer. (b) Suppose now that the nail turns out to be 3 in. long, and the claw hammer only succeeds in pulling it up 1 in. at first effort [see Figure P4.13(b)]. A board is then used as shown in Figure P4.13(c) to get enough leverage to finish extracting the nail. Assuming 150 lb of vertical friction

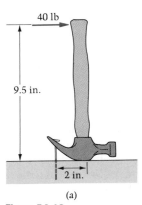

(a)

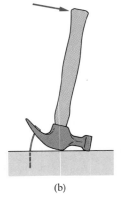

(b)

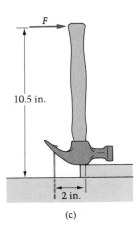

(c)

Figure P4.13

force must now be overcome, find the force **F** needed to complete the job. Again assume the nail exerts no horizontal force on the claw; that is, the nail has been straightened.

4.14 In Example 4.7, the moments about point A were summed and equated to zero in order to determine the tension T in the cable (see Figure P4.14). In that solution the right-most tension was replaced by an equipollent system of a force (itself) and couple $0.1T \curvearrowleft$ at point B. Without making this replacement, find the tension T (i.e., resolve T into its components at the point E where the cable leaves the pulley).

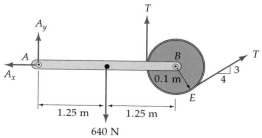

Figure P4.14

4.15 In Chapter 2, we found the wall and floor forces, and the internal force between the two cylinders (see Figures 2.9, 2.10), to be as shown in Figure P4.15. Show that when A and B are put back together as in Figure 2.10(a), the equilibrium equations $\Sigma F_x = 0$, $\Sigma F_y = 0$, and $\Sigma M_{\underset{\text{point}}{\text{any}}} = 0$ are satisfied, as they must be.

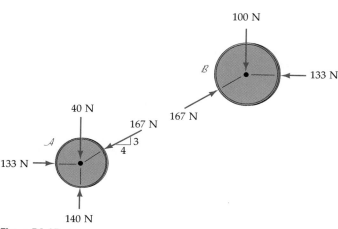

Figure P4.15

4.16 Find the pin reaction at A and the roller reaction at B onto the triangular block of weight 200 lb shown

in Figure P4.16. The center of gravity of the block is as shown.

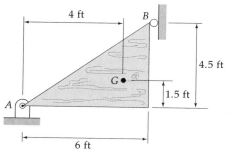

Figure P4.16

4.17 The uniform rod of mass m and length L is in equilibrium in the position shown in Figure P4.17. Find the tension in the string.

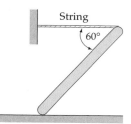

Figure P4.17

4.18 For the frame in Figure P4.18, find the reactions exerted by the ground onto member ABD at A.

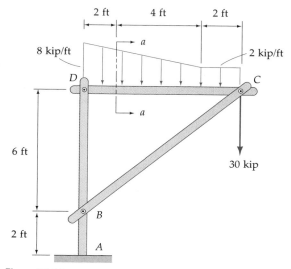

Figure P4.18

4.19 Find the force exerted by the slot on the pin at B (see Figure P4.19) which is fixed to the uniform bar ABCD. The weight of the bar is 100 N.

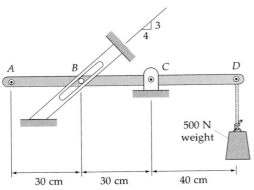

Figure P4.19

4.20 In Figure P4.20 joints A, B, and C are pinned, and the slender rods AC and BC are light in comparison to the applied forces. Determine the supporting force at B acting on member BC. Does this force put BC in tension or compression?

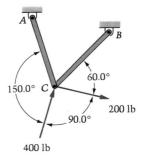

Figure P4.20

4.21 A straight uniform bar weighing 60 lb rests in a horizontal position against two frictionless slopes as shown in Figure P4.21. A concentrated vertical load of 200 lb acts at a distance x from the right end of the bar as shown. Find the distance x for equilibrium and determine the reactions at A and B.

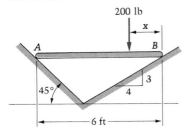

Figure P4.21

4.22 To problem 2.53, add the dotted cylinder 𝒟 of Figure P4.22, which though smaller is more dense and has the same mass. The contact between 𝒞 and 𝒟 is frictionless. Find the forces exerted by the planes on 𝒞 and 𝒟.

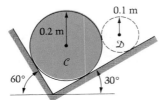

Figure P4.22

4.23 It is possible for the 20-N half-cylinder to be in equilibrium on the smooth plane for only one value of the angle ϕ. (See Figure P4.23.) For that angle, find the tension in the cord as a function of θ, and check your answer in the limiting cases $\theta = 0$ and $\theta \to \pi/2$.

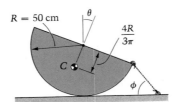

Figure P4.23

* **4.24** Show that the three scalar equations of equilibrium for a coplanar system of forces on a body ($\Sigma F_x = 0$, $\Sigma F_y = 0$, $\Sigma M_A = 0$) may be replaced by three moment equations $\Sigma M_A = 0$, $\Sigma M_B = 0$, $\Sigma M_C = 0$, provided the points A, B, and C (in the plane of the forces) are not collinear. *Hint:* Show that satisfaction of the three moment equations ensures a zero resultant. Use $\Sigma M_A = 0$ to establish the resultant as, at most, a force through A. Decompose this force into components parallel and perpendicular to the line joining A and B. Then apply the remaining moment equations. A sketch will help.

4.25 Find the conditions on the locations of points A and B so that the equations $\Sigma F_x = 0$, $\Sigma M_A = 0$, and $\Sigma M_B = 0$ are equivalent to $\Sigma F_x = 0$, $\Sigma F_y = 0$, $\Sigma M_P = 0$ as equilibrium equations of a body under a coplanar system of forces and couples.

4.26 What are the conditions on A and B in the previous exercise if the equivalent conditions include $\Sigma F_y = 0$ instead of $\Sigma F_x = 0$?

4.27 For equilibrium of the rectangular plate shown in Figure P4.27, what are the reactions at A and B?

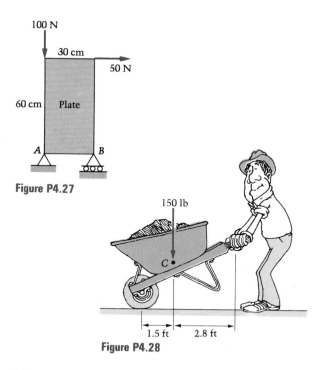

Figure P4.27

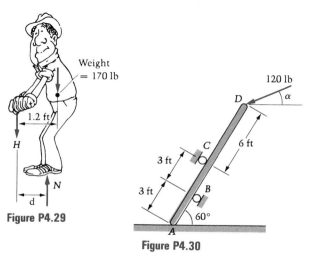

Figure P4.28

4.28 A wheelbarrow plus its load has the weight and center of mass location shown in Figure P4.28. Find the vertical force H exerted on the handle by the man and the force exerted on the tire by the ground. Assume the system is not moving.

4.29 In the preceding problem, determine the resultant N of the forces exerted on the man's shoes by the ground in the configuration shown in Figure P4.29. Also find the location of N; that is, find the distance d.

Figure P4.29

Figure P4.30

4.30 Rod $ABCD$ in Figure P4.30 rests against rollers at B and C, and against a smooth surface at A. (a) Find the angle α so the roller at B may be removed. (b) Find the resulting forces at A and C for this angle α.

4.31 The uniform bar BA in Figure P4.31 weighs 300 N and the block W weighs 500 N. The block rests on the bar but is secured by a taut wire wrapped around a smooth pulley and connected to the right end of the bar at A. Determine the tension in the wire and the reaction onto BA at B.

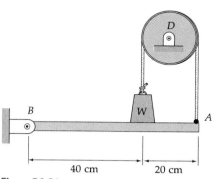

Figure P4.31

4.32 The bent bar $BAQRC$ of Figure P4.32 weighs 20 lb/ft. The cord is attached to the bar at A, and passes over pulleys P_1, P_2, P_3, and P_4, reattaching to the bar at Q. Find the tension in the cord using a single moment equation.

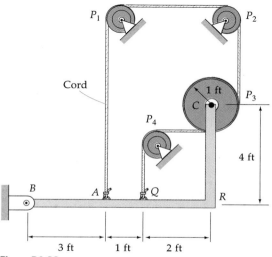

Figure P4.32

4.33 The pulley in Figure P4.33 is supported by friction-less bearings at the end of the uniform 750-N cantilever beam. The 1000-newton block is supported by the cable that passes over the pulley as shown. Neglect the weight of the pulley and find the reaction of the wall on the beam.

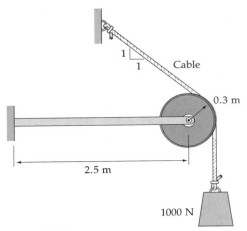

Figure P.4.33

4.34 The wrench in Figure P4.34 is applied to a hex-head bolt in an effort to loosen it. Determine the forces F_1 and F_2 on the bolt head if there is a slight clearance so that the contact is only at A and B.

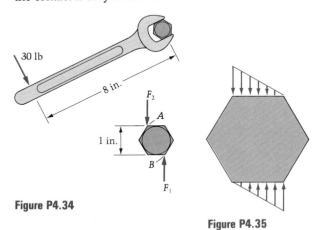

Figure P4.34

Figure P4.35

4.35 In the preceding problem, assume a tight fit between the bolt and the wrench and that the load is distributed linearly on the two faces as shown in Figure P4.35. Determine the force-alone equipollents of each of the two loadings. Explain the increases in these two values over the concentrated forces in the preceding problem.

4.36 Assume that the reaction of the wall onto the cantilever beam is the pair of linearly distributed forces shown in Figure P4.36. Find the intensities q_T and q_B in terms of F, L, and l.

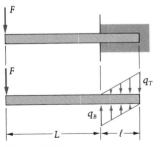

Figure P4.36

4.37 A slender uniform bar of length L and weight W is slung from the two cables shown in Figure P4.37. Find the tension in the cable on the right as a function of the ratio l/L. Show that equilibrium is impossible if $l < L/2$.

Figure P4.37

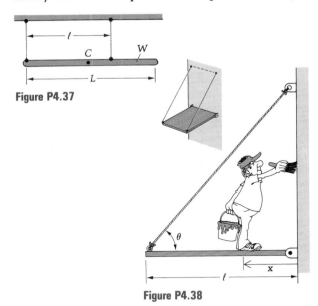

Figure P4.38

4.38 A man is painting a wall using a crude scaffold that consists of a pinned board and two cables. (See Figure P4.38.) (a) Find the tension T in each cable as a function of the man's weight and position (W and x), the length and weight of the board (l, w), and the angle θ. Investigate the following limiting cases: (b) the board is light and the man is at $x = 0$; (c) the board is light and the man is at $x = l$; (d) the man is light. Assume symmetry.

4.39 In the "one size fits all" wrench of Figure P4.39, the handle $\mathcal{H}$ and the member $\mathcal{B}$ are free to turn relative to each other about the pin P. Neglecting friction between the wrench and nut, find:

a. The force exerted on the nut by the handle if the nut is already tightened and nothing moves

b. The magnitude of the force carried by the pin P

c. The tightening moment (about the center line of the nut) that the wrench exerts on the nut.

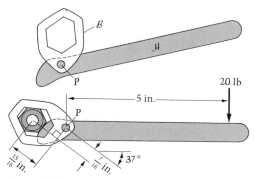

Figure P4.39

4.40 In Figure P4.40, find the reactions onto the beam at the pin (A) and roller (B).

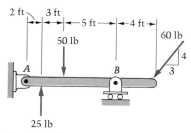

Figure P4.40

4.41 In Figure P4.41, find the reactions onto the bent bar at A, B, and C.

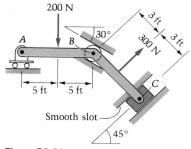

Figure P4.41

4.42 In Figure P4.42, find the reactions exerted by the wall on the beam.

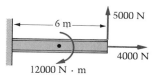

Figure P4.42

In Problems 4.43–4.54 find the reactions exerted on the members by the supports.

4.43

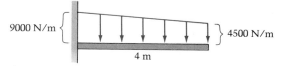

Figure P4.43

4.44

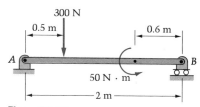

Figure P4.44

4.45

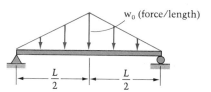

Figure P4.45

4.46

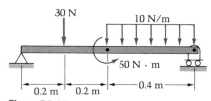

Figure P4.46

4.47

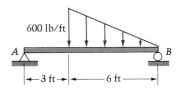

Figure P4.47

4.48

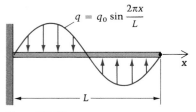

$q = q_0 \sin \dfrac{2\pi x}{L}$

Figure P4.48

4.49

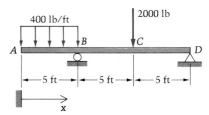

Figure P4.49

4.50

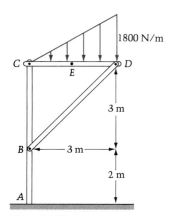

Figure P4.50

4.51

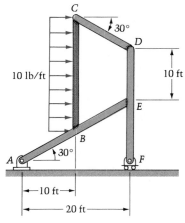

Figure P4.51

4.52

$w(x) = 2 \sin \dfrac{\pi x}{10}$ lb/ft

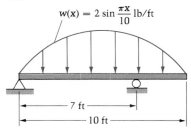

Figure P4.52

4.53

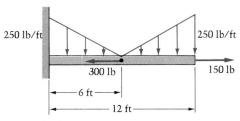

Figure P4.53

4.54

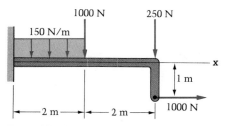

Figure P4.54

4.55 The crane in Figure P4.55 is pinned to the ground at O and to a screw-jack at A. The truss $\mathcal{J}$ and load $\mathcal{L}$ weigh 800 and 1200 pounds, respectively. In the given position, neglecting the distance between pin P and the nut, determine the tension in the screw and the pin reactions exerted onto $\mathcal{J}$ at O.

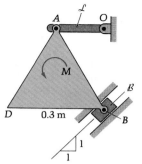

Figure P4.57

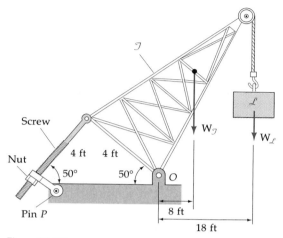

Figure P4.55

4.56 The crane of the preceding problem is lowered so that angle POA increases to $60°$. (See Figure P4.56.) Again find the tension in the screw-jack and the reactions at O.

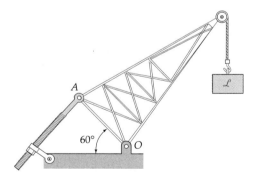

Figure P4.56

4.57 In Figure P4.57, the equilateral triangular plate has mass 80 kg and is supported by the light members $\mathcal{L}$ and $\mathcal{B}$, the latter of which is free to slide in a smooth slot. If the plate is in equilibrium, find the value of the couple M that is acting on it.

4.58 Repeat the preceding problem if M is replaced by a vertical force P at D whose magnitude is to be found.

4.59 The weight of $\mathcal{W}$ is 300 lb, and the weight of wheel $\mathcal{C}$ is 200 lb (see Figure P4.59). Find the height (y coordinate) of the center of C of $\mathcal{C}$. Assume sufficient friction to prevent slipping.

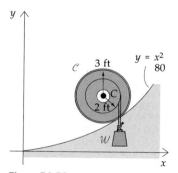

Figure P4.59

4.60 Repeat the preceding problem if the contact of the parabolic plane with the wheel is on its *inner* radius and if the weight is attached to a cord running over the *outer* radius. (See Figure P4.60)

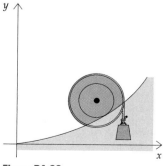

Figure P4.60

4.61 The 33-ft diameter antenna shown in Figure P4.61 (a) is supported at three pins. Pin *A* connects the "dish" to the jack *AB*, which extends to raise (and contracts to lower) the antenna in elevation. This rotation takes place about a horizontal line through two pins, shown as one at *D* in Figure P4.61(b). The entire structure suspended above *A* and *D* weighs 3500 lb, with mass center at *C*. Find the force in the elevation jack *AB*.

Figure P4.61(a)

4.62 To the preceding problem, we add the effect of the wind. Large antennas like the one shown in Figure P4.61(a) are designed to survive severe hurricanes and tornados of up to 125 mph wind velocities. At this wind speed, with the wind blowing horizontally into the dish (to the right), the wind forces on the dish have a resultant at the vertex *V* of: (1) Drag Force = 44,200 → lb; (2) "Lift" Force = 27,100 ↓ lb; and (3) Moment = 41,300 ↻ lb-ft. Find the force in the elevation jack and the combined reaction on the hub pin at *D*. [See Figure P4.61(b).]

* **4.63** Repeat the preceding problem, with the antenna pointing at 80° elevation. This time the 125 mph wind loads are: (1) Drag = 8790 → lb; (2) "Lift" = 19,100 ↓ lb; and (3) Moment = 106,000 ↻ lb-ft.

4.64 The fork-lift truck weighs 9500 lb. (a) If it carries a load of 5000 lb in the position shown in Figure P4.64, find the forces exerted by the ground on the tires. (b) How much greater load could be carried in the same position without the truck tipping forward?

4.65 The 50-kg uniform rod is pinned to the 80-kg wheel at *A* and *B*. (See Figure P4.65.) At *C*, the rod rests on a smooth floor, as does the wheel at *D*. Find the reactions at *C* and *D*.

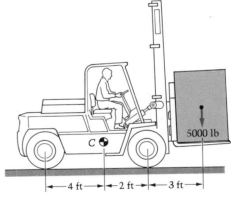

Figure P4.64

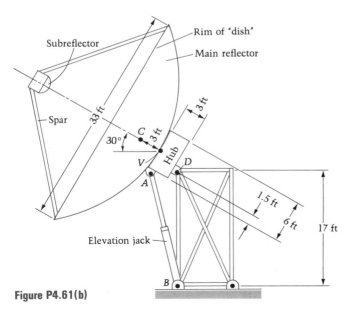

Figure P4.61(b)

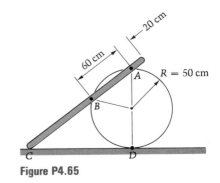

Figure P4.65

4.66 The rod weighs 64.4 lb and is held up by three cables as shown in Figure P4.66. Find the tension in each cable.

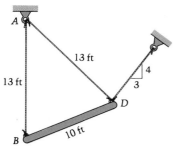

Figure P4.66

4.67 In Figure P4.67 the rod AB of length L and weight W is connected to smooth hinges at F and D by the light members AF and BD, each of length L. A cable CE completes the support of the rod. Find the force in the cable.

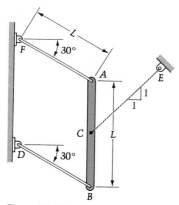

Figure P4.67

4.68 The man is holding the wheel of weight W in equilibrium on the rough inclined plane. (See Figure P4.68.) Find the tension in the rope, in terms of W. Assume sufficient friction to prevent slip.

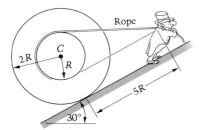

Figure P4.68

4.69 In Figure P4.69 disc $\mathcal{A}$ is pinned to the bent bar $\mathcal{B}$ at A; the bodies weigh 20 N ($\mathcal{A}$) and 30 N ($\mathcal{B}$). Body $\mathcal{B}$ has mass center at C and rests on the smooth plane. Find the force P required for equilibrium of the system of $\mathcal{A}$ and $\mathcal{B}$.

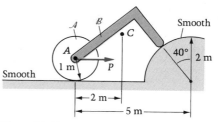

Figure P4.69

4.70 In Figure P4.70 the 150-kg cylinder is pinned at A to the 200-kg uniform bar. Find the force P required for equilibrium of the bodies.

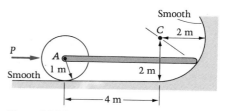

Figure P4.70

4.71 Repeat the preceding problem if a clockwise couple of 50 N · m acts upon the bar.

4.72 A beam with mass M and length l is supported by a smooth wall and floor, and a cable, as shown in Figure P4.72. Find the tension T in the cable as a function of θ (different cables for different θ's). Show from $T(\theta)$ that as $\theta \to 0$, $T \to 0$, and that T approaches ∞ as $\theta \to 90°$. Verify that these results make sense using free-body diagrams at $\theta = 0$ and $\theta =$ almost $90°$.

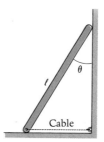

Figure P4.72

4.73 Show that if the cable in the preceding problem is instead attached to the top of the beam (see Figure P4.73), its tension is constant (i.e., this time it doesn't depend on θ).

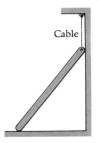

Cable

Figure P4.73

4.74 In Figure P4.74 the uniform circular disc of weight W and radius r has a uniform bar of length L and weight W welded to it at A such that the bar is perpendicular to AO. Find the angle θ for equilibrium of the combined body.

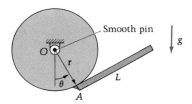

Smooth pin

g

O

r

θ

A

L

Figure P4.74

4.75 In Figure P4.75 the homogeneous, uniform bar weighs 10 lb/ft. Find the reaction the pin at A exerts on the bar.

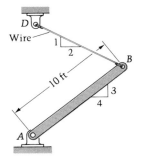

D

Wire

1

2

B

10 ft

3

4

A

Figure P4.75

4.76 The uniform bar weighs 1000 N and is in equilibrium in the horizontal position shown in Figure P4.76. Find the tension in the cable and the reactions exerted by the pin at A onto the bar.

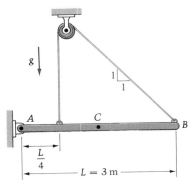

g

1

1

A C B

$\dfrac{L}{4}$

$L = 3$ m

Figure P4.76

4.77 The device in Figure P4.77 is called a crusher. With a hand-force H applied as shown, a large force P can be developed onto material within the enclosure resisting the block moving to the right. Find the force exerted by the pin at C onto the bar ABC.

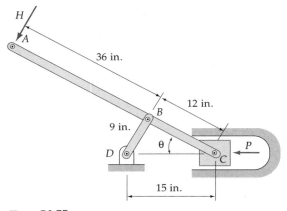

H

A

36 in.

12 in.

B

9 in.

D

θ

C

P

15 in.

Figure P4.77

4.78 What horizontal force F will pull the 100-kg lawn roller over the step? (See Figure P4.78.) What is the value of F if it is directed normal to AC as shown dotted?

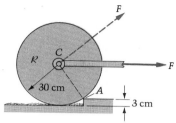

Figure P4.78

4.81 Repeat Problem 4.68 for the different man and rope arrangement shown in Figure P4.81.

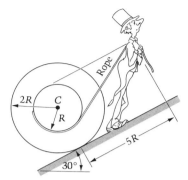

Figure P4.81

* **4.79** The 10-ton moving crane in Figure P4.79 has a mass center at C. It carries a maximum load of 18 tons. Find the smallest weight of counterweight C, and also the largest distance d, so that: (a) the crane doesn't tip clockwise when the maximum load is lifted; and (b) the crane doesn't tip counterclockwise when there is no load present. Also find, for $d = 2.5$ ft, the *range* of weights of C for which (a) and (b) can be satisfied.

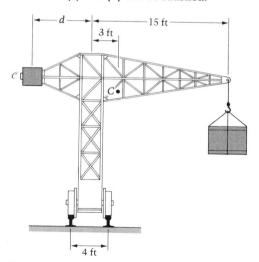

Figure P4.79

4.82 The car in Figure P4.82 has weight W. What is the resultant of the normal forces on the passenger-side tires if the car is parked on the indicated incline?

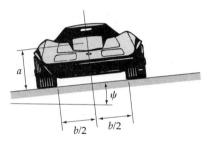

Figure P4.82

4.83 The 100-lb sign in Figure P4.83 is supported by a pin at A and a cable from B to C. If the cable breaks at 400-lb tension and the pin fails at 600 lb of force, find the safe values of θ.

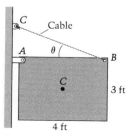

Figure P4.83

* **4.80** In the preceding problem, illustrate on a graph of W_C versus d the safe region of points (d, W_C) for which *both* conditions (a) and (b) will be satisfied. Consider only values of $d > 2$ ft for practical reasons.

4.84 In Figure P4.84 the member weighs 50 N and has its mass center at C. At B, a pin, fixed to the ground, bears against a slot. The spring carries a tensile load of 100 N. Find the vertical component of the reaction at A.

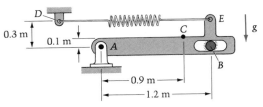

Figure P4.84

4.85 Zeke the moonshiner has built a "water" tower in which to store his liquid refreshment. The tank at the top is 5 ft high and 4.8 ft in diameter, and is mounted on four symmetrically located legs inclined to the horizontal. (See Figure P4.85.) The weight of the whole tower (legs plus tank) is 3800 lb. Wind force is to be computed on the basis of the pressure times the projected area of the tank on the plane perpendicular to the direction of the wind. (a) If this dynamic pressure is 62.5 lb/ft² (blowing left to right), find whether the tower will blow over or not. (Zeke sees the big storm coming and is running to get something with which to fill the tank.) (b) If Zeke begins filling the tank with moonshine having the density of water, how close to full must it get in order to prevent the tower from tipping over?

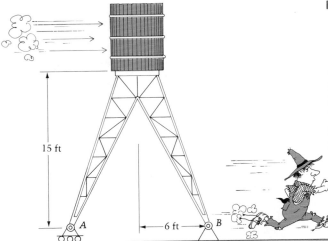

Figure P4.85

4.86 Bar AB in Figure P4.86 is supported by a roller at C, and by a smooth wall at A. The bar is uniform, has a mass of 20 kg and is 0.6 m long. What vertical force at B is necessary for equilibrium?

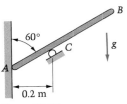

Figure P4.86

4.87 Find the reaction of the ground onto the pole in Problem 2.24, expressed as a force and couple at the point D, if the tension in the power line BE is 300 lb and the pole weighs 4000 lb.

4.88 In Figure P4.88 the beam AB and the pulley each weigh 30 lb. Find the tension in the rope, and the pin reaction at A.

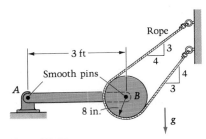

Figure P4.88

4.89 Five weights are suspended as shown in Figure P4.89 at equal distances along a uniform bar 40 cm long weighing 60 N. The first weight at the left end is 10 N, and each successive weight is 10 N heavier than the preceding one. At what distance x from the left end should the bar be suspended so as to remain horizontal?

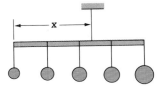

Figure P4.89

4.90 In Figure P4.90 the body *ADB* is a bell crank pinned to the floor at point *D*. Find the force *F* in the link $\mathcal{L}$, pinned to the bell crank at *B*, required for equilibrium.

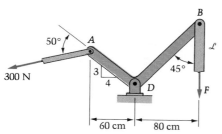

Figure P4.90

4.91 In Figure P4.91 the motor weighs 200 lb and is fixed to the 50-lb frame, which is pinned to the ground at *O*. The tension in the motor's belt prevents it from turning and falling. When the motor is turned off, the tensions T_a and T_b are equal if friction is neglected. In this case, find the tension in the belt.

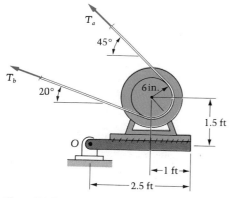

Figure P4.91

4.92 The frame in Figure P4.92 is supported by the pin at *A* and the cable. Find the pin reaction at *A* and the tension in the cable.

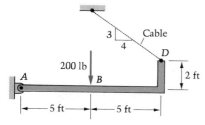

Figure P4.92

4.93 If the corner *Q* in Example 4.8 helps to support the 1200 lb weight with no force in *AB*, what is now the force in *AB* required to start the bed pivoting about the pin *D*? (See Figure P4.93.)

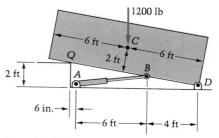

Figure P4.93

4.94 Bar *ABD* in Figure P4.94 is supported by a hinge at *B* and by a cable at *A*. Find (a) the tension in the cable, and (b) the components of the force at *B* on *ABD*.

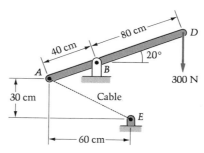

Figure P4.94

4.95 Find the pin reaction at *A* and the roller reaction at *B* for the bent bar situated and loaded as shown in Figure P4.95.

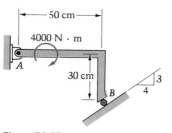

Figure P4.95

4.96 Find the force that the pin at A exerts on the rectangular plate shown in Figure P4.96.

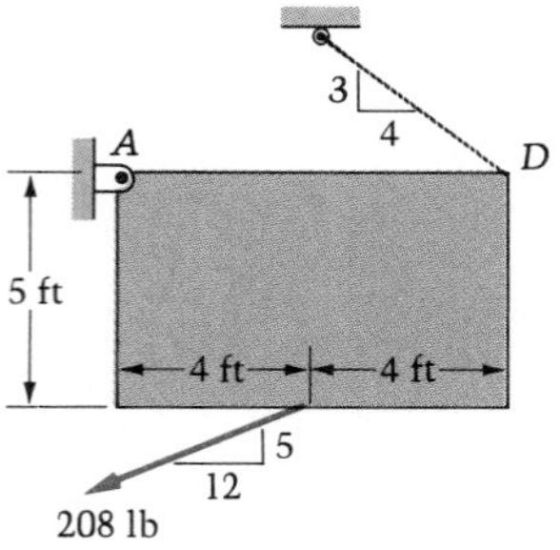

Figure P4.96

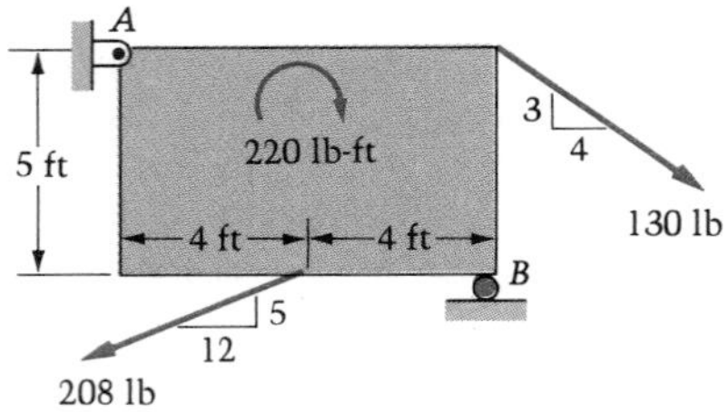

Figure P4.97

4.97 Repeat the preceding problem if a couple is applied as in Figure P4.97 and if the cable is replaced by an applied force of 130 lb. Also, the plate now rests on a roller at B as shown.

* **4.98** A davit is a crane on a ship for hoisting lifeboats, anchors, or cargo. A "gravity davit" uses gravity to help with the lowering. A boat is shown being held in the stow position by two gravity davits, one of which is shown in Figure P4.98(a). Assume that a removable pin is located at roller $\mathcal{A}$. The boat weighs 1500 lb, and each of the three 8-ft sections of the davit weighs 350 lb. Find the force in the cable (two per davit, i.e., four per boat) and also the reactions exerted on the smooth rollers $\mathcal{A}$ and $\mathcal{B}$. The rollers are pinned to the davit, as are the free pulleys $\mathcal{C}$, $\mathcal{D}$, and $\mathcal{E}$. (There are also pulleys on the other side of the davit at C, D, and E.) Neglect friction and the width of the davit, and note the photographs in Figures P4.98(b, c, d).

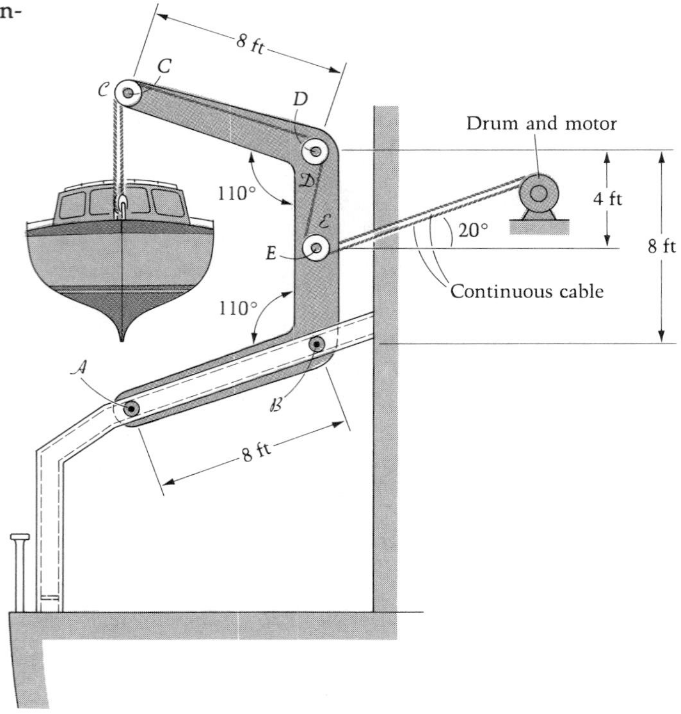

Figure P4.98(a)

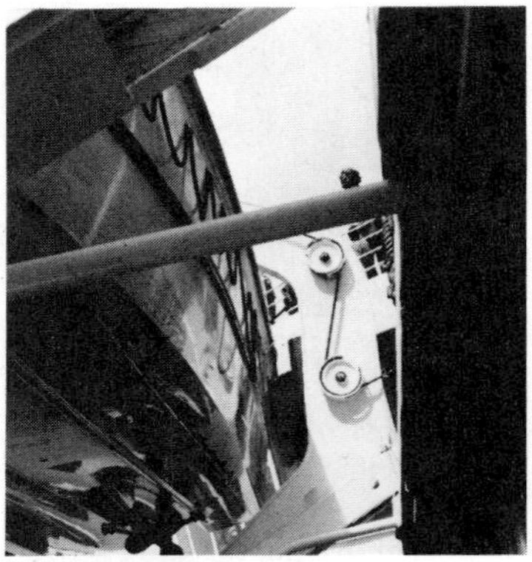

Figure P4.98(b)

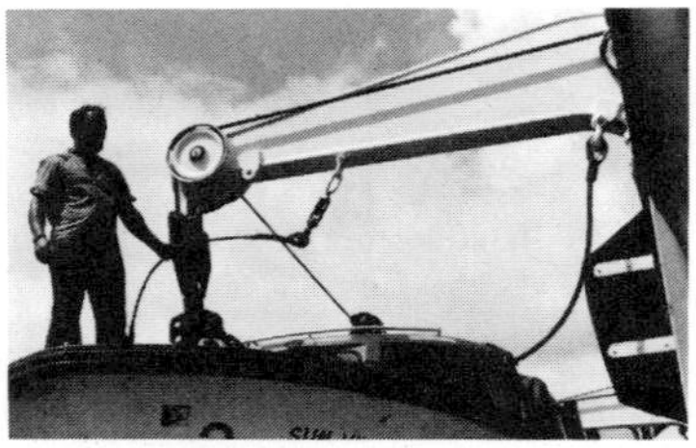

Figure P4.98(c)

Figure P4.98(d)

4.100 (a) If the child and swing together weigh 200 N, find the force in each of the two ropes. (See Figure P4.100(a).) (b) If the child's father uses a horizontal force to pull the swing back 30° from the vertical plane of the ropes (see Figure P4.100(b)), what is this force and what is now the force in each rope?

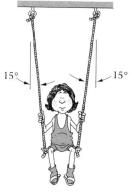

Figure P4.100(a)

* 4.99 When the cable is let out in the preceding problem, gravity causes the davit to move downward to the left in the slot. Eventually roller A reaches the bottom of the slot as shown in Figure P4.99, and the boat then lowers into the sea. After the cable is released from the boat, find the forces on the rollers A and B in the position shown.

Figure P4.100(b)

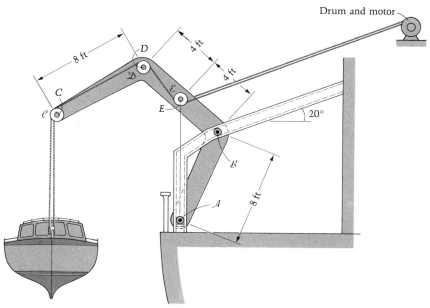

Figure P4.99

4.101 Find the reaction of the floor onto the roller in Figure P4.101. The block W weighs 200 N and the uniform bar ABD weighs 300 N.

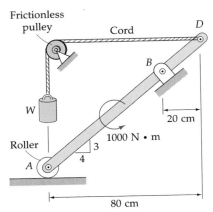

Figure P4.101

* 4.102 In Figure P4.102 determine the angle θ for which the ladder is in equilibrium if planes AB and BC are both smooth.

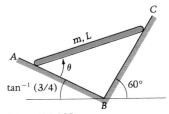

Figure P4.102

* 4.103 In Figure P4.103 the bar of mass m and length l is pinned at B to a smooth collar, which is free to slide on a fixed vertical rod. The other end rests on the smooth parabolic surface shown. Find the x coordinate of the contact point A.

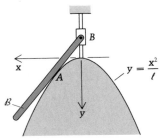

Figure P4.103

* 4.104 If the weight of the rod of length L is negligible, and if all surfaces are smooth, find the range of values of the angle β for which the rod will be in equilibrium. See Figure P4.104.

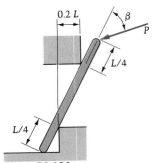

Figure P4.104

* 4.105 Find the angle θ assumed by the stirrer of length L, in equilibrium in a smooth hemispherical cup, for $L = 3R$. (See Figure P4.105.)

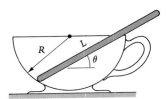

Figure P4.105

4.106 The boom in Figure P4.106, consisting of the identical struts S_1 and S_2 and the cable C, holds up a compressor weighing 1300 lb. Find the forces in the struts and in the cable.

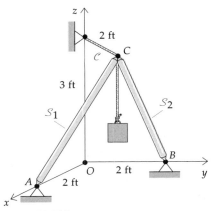

Figure P4.106

4.107 The plate in Figure P4.107 weighs 2 kN and is supported by the hinge at corner A and the cable BC. Find the tension in the cable, and the force and couple exerted by the hinge onto the plate.

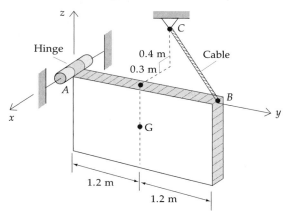

Figure P4.107

4.108 The 10-ft mast OD in Figure P4.108 can carry 10,000 lb without failing. The cables can each carry 4000 lb without breaking. Find the radius of the base circle, on which the cables are attached, for which both post and cables will reach their maximum loads simultaneously if the cables are gradually tightened.

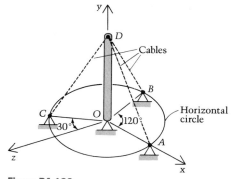

Figure P4.108

4.109 For the structure $ABCD$ loaded and supported as shown in Figure P4.109, find the reactions at A and the cable tensions.

4.110 The uniform block in Figure P4.110 weighs 500 N. It is supported by ball joints at O and A, and a cable from B to C. Find the tension in the cable.

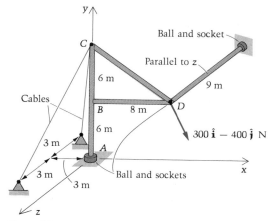

Figure P4.109

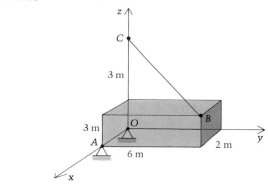

Figure P4.110

4.111 In Figure P4.111 determine the tensions in cables BC and BE. Neglect the weights of all members and assume that the support at A is a ball-and-socket joint. The 5200-N force has no x component.

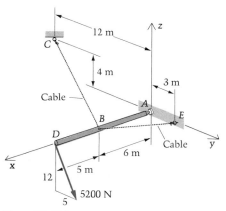

Figure P4.111

4.112 A 40-lb cellar door is propped open with a light stick, as shown in Figure P4.112. Find the force in the stick.

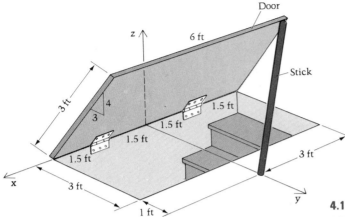

Figure P4.112

4.113 In Figure P4.113 the bar CD is welded to the center of another bar AB, with the end D resting against a smooth vertical wall in the yz plane. The bars aren't perpendicular, and AB is connected at A to a ball joint; at B it passes through a smooth eyebolt. The bars weigh 12 lb/ft. Find the force exerted on the bar at point D.

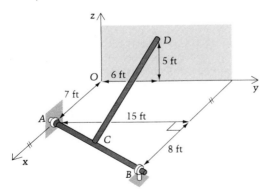

Figure P4.113

4.114 A rigid frame whose base is in the xy plane is shown in Figure P4.114. Calculate the forces in the cables, and the reactions on the frame at A, if the frame weighs 150 N/m. Neglect the cross-sectional dimensions of the beams constituting the frame.

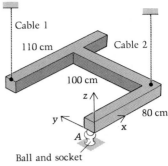

Figure P4.114

4.115 The triangular plate weighs 1000 N and is supported by the roller and hinge as shown in Figure P4.115. The plane of the plate is horizontal. Find the forces exerted by the roller, and the force- and couple-components exerted by the hinge (which cannot exert a couple in the y-direction).

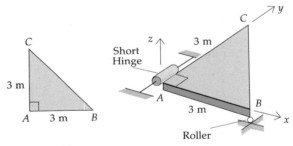

Figure P4.115

4.116 In Figure P4.116 the door weighs 120 lb. If there is no friction between the rope and the tree limb, what must the children collectively weigh to start the door swinging open about its hinges on the x axis?

4.117 The smooth collars (or sleeves) in Figure P4.117 are attached at C_1 and C_2 to the rod of mass m by ball and socket joints. In terms of mg, find the force P that, when applied parallel to the x-axis onto the lower collar, will result in equilibrium. Upon completing the solution, comment on why one of the six scalar component equations was redundant (i.e., yielded no new information).

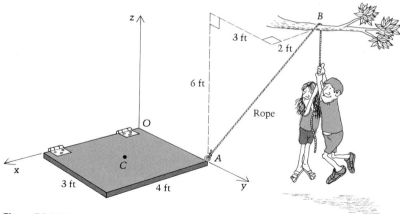

Figure P4.116

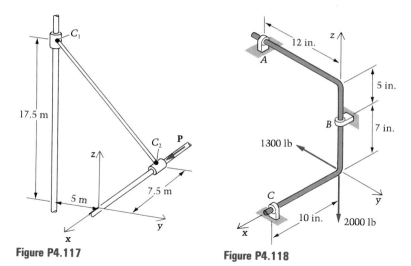

Figure P4.117

Figure P4.118

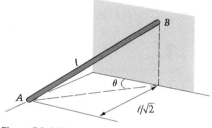

Figure P4.119

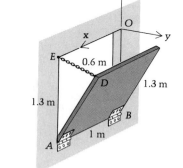

Figure P4.120

4.118 A shaft with two 90° bends is supported by three ball bearings as shown at A, B, and C in Figure P4.118. Find the reaction of bearing C on the shaft.

** **4.119** In Figure P4.119 a heavy uniform rod of length ℓ rests with one end, A, on the ground and the other, B, against a vertical wall. The vertical plane through the rod makes an angle θ with the wall. End A is $\ell/\sqrt{2}$ from the wall. Letting ρ and σ be the ratios of the tangential to normal reactions at the ground and wall, respectively, show that

$$\sqrt{\sigma^2 \sin^2 \theta - \cos^2 \theta} = \frac{1}{\rho} - 2\sqrt{2} \sin^2 \theta - 1$$

4.120 A 1 m × 1.3 m plate weighs 325 N and is supported by hinges at A and B. (See Figure P4.120.) It is held in the position shown by the 0.6-m chain ED. Find the tension in the chain.

4.121 Member *AD* in Figure P4.121 is supported by cables *CE*, *BG*, and *BF*, and by a ball-and-socket at *A*. Find the tensions in *CE* and *BF* if the tension in *BG* is 3000 N.

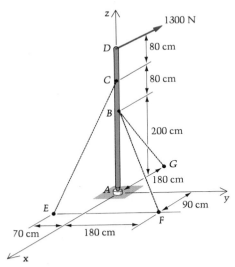

Figure P4.121

4.122 Find the tension *T* in the cable *DE*. This cable gives the same moment about line *AC* as does the weight of the uniform segment *BD*, which is 400 lb. (See Figure P4.122.)

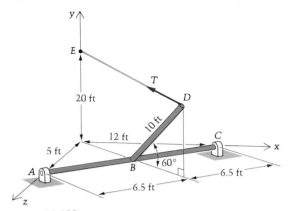

Figure P4.122

4.123 In Figure P4.123 find the force and couple exerted on the 200-lb plate by the hinge at *O*.

4.124 The bar in Figure P4.124 is supported by cables *BD* and *CE*, and by a ball-and-socket at *A*. Points *D* and *E* lie in the *xz* plane. Find the tension in each cable.

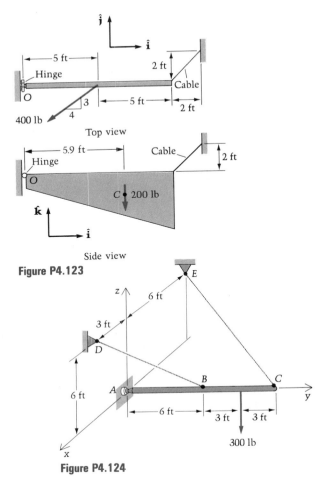

Top view

Side view

Figure P4.123

Figure P4.124

4.125 In Figure P4.125 the hinge at *A* has broken off. Find the tension in the cable. Then find the couples C_{Bz} and C_{By} that the remaining hinge must exert if the door remains in equilibrium.

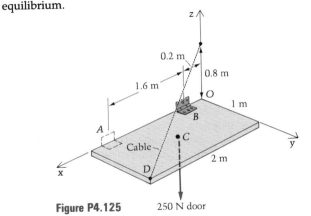

Figure P4.125   250 N door

4.126 Find the force F and the bearing reactions. (See Figure P4.126.) Assume that one of the bearings cannot exert force in the z direction and that neither can exert a couple.

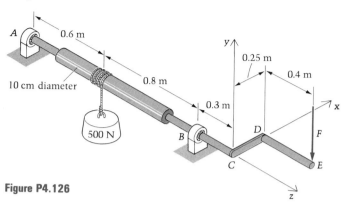

Figure P4.126

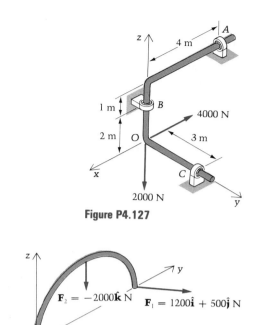

Figure P4.127

4.127 A shaft with two 90° bends is supported by three ball bearings at A, B, and C as shown in Figure P4.127. Find the reactions of the bearings on the shaft.

4.128 The semicircular bar of radius 30 cm is clamped at O. The 2000-N force acts downward through the bar's highest point, and the other force lies in the horizontal plane through its ends. Find the reactions $\mathbf{F}_O$ and $\mathbf{M}_O$ at the clamp. (See Figure P4.128.)

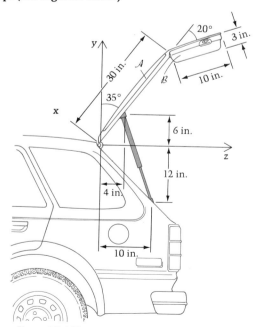

Figure P4.128

4.129 The rear door of the station wagon in Figure P4.129 is held up when open by the two gas-filled struts attached to the car by ball-joints. The door weighs 90 lb — 30 lb in part $\mathcal{A}$ and 60 lb in part $\mathcal{B}$. Find the forces in the two struts.

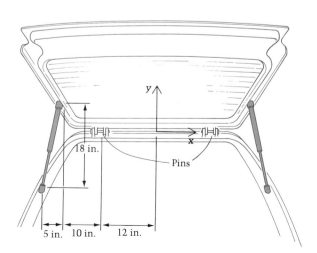

Figure P4.129

4.130 When the gas leaks out of one of the struts of Problem P4.130, it will no longer exert a force to hold up the door. If the right strut has become useless and the left pin breaks, find the resultant forces and couples exerted by the left strut and the right pin (hinge). (See Figure P4.130.)

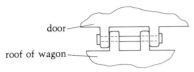

Figure P4.130

4.131 The horizontal homogeneous trap door in Figure P4.131 weighs 72 pounds. It is supported by the cable AB, a ball-and-socket at O, and a hinge at D that provides no support in the x direction. Find the force in the cable.

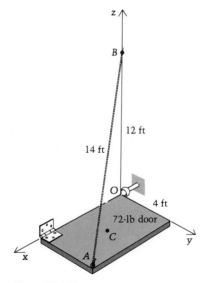

Figure P4.131

* 4.132 A ring is welded to a rod at a point A as shown in Figure P4.132. The cross-sectional area A and mass density of the rod are the same for the ring. The combined body is in equilibrium in the (precarious) position shown. Find the vertical reaction components exerted by the ball and socket onto the rod at B and by the smooth floor onto the ring at Q.

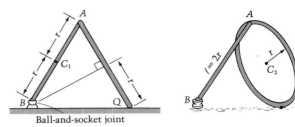

Ball-and-socket joint

Figure P4.132

4.133 Find the supporting force system at A on the bent bar. (See Figures P4.133.)

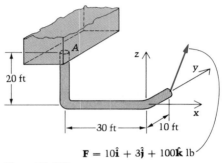

$$\mathbf{F} = 10\hat{\mathbf{i}} + 3\hat{\mathbf{j}} + 100\hat{\mathbf{k}} \text{ lb}$$

Figure P4.133

* 4.134 The plate weighs 100 lb and is supported as shown in Figure P4.134. Find the reactions at A, B, and D.

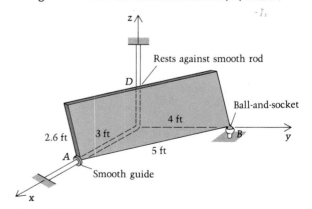

Rests against smooth rod

Ball-and-socket

Smooth guide

Figure P4.134

The spring is one of the most useful of all mechanical devices. It gives a resistive force proportional to its stretch if it is "linear," and the constant of proportionality is called the spring *modulus.* Problems 4.135–4.140 contain springs, and the last four are challenging.

4.135 The cart in Figure P4.135 weighs 500 N and is held in equilibrium on the inclined plane by a spring of modulus 5000 N/m. Find the force in the spring and its stretch.

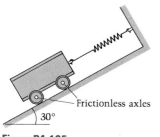

Frictionless axles

30°

Figure P4.135

4.136 In Figure P4.136 the slender homogeneous rod 12-ft long and weighing 5 lb is connected by a pin and a moment spring to the vertical wall. The moment (or rotational or torsional) spring exerts a moment when the angle is changed between the two bodies to which it is fixed; an example is the springs in most flexible, metal-link watch bands. In a moment spring the modulus has units such as lb-ft per radian. If the rod is in equilibrium in the given position, and the natural (zero moment) position of the spring is when the bar is vertical, find the modulus of the moment spring.

4.137 The light bar in Figure P4.137 is fixed to the ground at O by means of a smooth pin, and is subjected to the vertical force P at its other end. The spring of modulus k is constrained to remain horizontal, and is unstretched when $\theta = 0$. If $Pl < kh^2$, there is an equilibrium position for $0 < \theta < \pi/2$. Find this angle θ. Why must $Pl < kh^2$ hold for this configuration to exist?

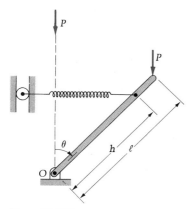

Figure P4.137

4.138 The spring in Figure P4.138 is unstretched with length l when the uniform bar of mass m is vertical. (a) Write the moment equilibrium equation for the bar at an angle θ. (b) Show that $\theta = 0$ satisfies the equation for any k, l, m, and g. (c) Let $kl = mg$ and show that there is another equilibrium angle θ between 0° and 180°. Find this angle numerically, using a calculator or computer.

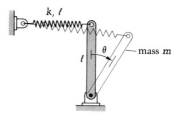

Figure P4.138

4.139 The bead of mass m can slide on the smooth semicircular hoop, and the spring connects the bead to the top A of the hoop (of radius R). (See Figure P4.139.) The spring has modulus $2mg/R$ and natural (unstretched) length $R/4$. Find the angle(s) θ at which the bead is in equilibrium. *Hint:* Prove and then use the fact that the three forces acting on the bead form a force triangle that is similar to *ABC*.

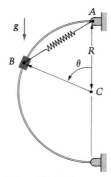

Figure P4.139

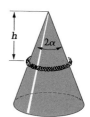

Figure P4.140

4.140 A linearly elastic endless spring with modulus k is placed around a smooth cone of vertex angle 2α. (See Figure P4.140.) The natural length of the spring is L, and it weighs W. Find the value of h for equilibrium.

25°

Figure P4.136

•4.141 Show that, if a body is held in equilibrium by three forces, the forces must have coplanar lines of action that are either parallel *or* concurrent. Note that we only need to consider the case in which none of the forces vanishes, for otherwise we have a two-force body that has already been discussed. *Hint:* We outline below a set of steps by which the result may be obtained; the student is encouraged also to think about alternative approaches.

1. Let F_1, F_2, and F_3 be the three forces. Consider the two possibilities: (a) two lines of action are parallel and (b) no two lines of action are parallel.

2. For case (a), apply $\Sigma F = 0$ to conclude that all three lines of action are parallel. Then consider the plane defined by the lines of action of F_1 and F_2 (if they are collinear then they, of course, must coincide with that of F_3). Requiring that the sum of moments about any line in that plane vanish, conclude that all three of the (parallel) lines of action lie in the same plane.

3. For case (b), conclude from $\Sigma F = 0$ that the three lines of action lie in parallel planes. Let a line, l, be parallel to the line of action of F_2 and intersect the line of action of F_1. From $\Sigma M_l = 0$ conclude that the line of action of F_3 intersects l, and hence lies in the same plane as do l and the line of action of F_1. Now apply the moment equation of equilibrium for moments about the point of intersection

of the lines of action of F_1 and F_3 to conclude that the three lines of action are concurrent and coplanar.

4.142 Illustrate the result of the preceding problem by showing that (a) the three forces at A, B, and G of Problem 4.16 have lines of action that all meet at one point, and (b) the three forces of Problem 4.23, for the correct angle ϕ, are parallel.

4.143 (a) If the light bar *ABCD* of Figure P4.143 is held in equilibrium by the two rollers at B and C and the smooth wall at A, show using the result of Problem 4.141 that none of the three support forces can vanish if $\theta = 30°$ as shown. (b) For what angle θ will the support force at B vanish?

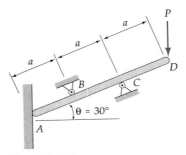

Figure P4.143

4.4 Applications of the Equations of Equilibrium to Interacting Bodies or Parts of a Structure

In Section 4.3, we studied the equilibrium of a body which was either a single physical object or else two or more objects which were left intact. These objects were connected in various ways, including pressing contact, pins, rollers, or cables. What was common about all those multiobject problems was that we could solve for the desired unknowns without separating the objects constituting the overall body. In this section, we shall see that sometimes we *must* separate the objects, or parts, of the body in order to obtain a solution.

As an example, suppose we wish to know the forces of interaction *between* the objects making up a body. These forces will never appear on free-body diagrams or in equations of the overall body because they are internal to it. But if we draw FBD's of the *separate* objects, treat each as a body in itself, and write equilibrium equations for each, these equations will now include the desired forces of interaction. This procedure makes use of the fact that *if a body is in equilibrium, each of its parts — being a body in its own right — is also in equilibrium.**

Let us re-examine Example 4.7 in the light of these ideas. Figure 4.17 (a) shows the problem, and (b) repeats the free-body diagram of the body consisting of the beam AB and the pulley supported on frictionless bearings at B:

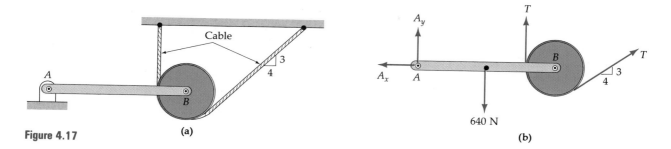

Figure 4.17   (a)   (b)

In the earlier example, we simply set the moments about A of the external forces acting on the intact body equal to zero, and thereby found the tension T in the cable from that single equation to be 200 N. We then used the force equations to find the components of the pin reaction at A: $A_x = 160$ N and $A_y = 320$ N.

But now suppose we desire to know the pin force exerted on the beam at B. That force can *only* be exposed, and subsequently solved for, by separating the beam and pulley, as shown below:

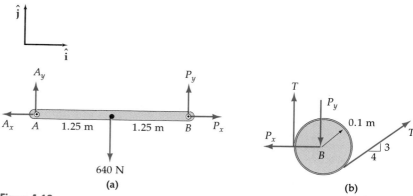

(a)

Figure 4.18

---

* More rigorously, when we say that a body is at rest (equilibrium) we mean that each of its material points (or particles) is at rest. Consequently, when a body is in equilibrium each of its parts, or subdivisions, is in equilibrium. Thus the equations of equilibrium apply to any part of a body as well as to the whole.

Now the desired pin force (expressed in terms of its x- and y-components P_x and P_y) appears as an *external force* on the separated bodies shown in Figure 4.18(a, b).

Note that the action-reaction principle (Newton's third law) has to be *rigorously enforced* when separating the objects or parts constituting a body. Thus P_x is shown acting to the left on the pulley but to the right on the bar, and P_y is depicted down on the pulley but up on the bar. Violation of the equal magnitude/opposite direction characteristic when dealing with interactive forces dooms a solution to failure.

On the FBD of the pulley, we have, using $T = 200$ N from Example 4.7:

$$\xrightarrow{+} \qquad \Sigma F_x = 0 = \tfrac{4}{5}(200) - P_x$$
$$P_x = 160 \text{ N}$$
$$+\uparrow \qquad \Sigma F_y = 0 = 200 + \tfrac{3}{5}(200) - P_y$$
$$P_y = 320 \text{ N}$$

Since the values of P_x and P_y came out positive, the directions of the pin forces onto the pulley are *as shown* in Figure 4.18(b), i.e., the force is $-160\hat{\mathbf{i}} - 320\hat{\mathbf{j}}$ N onto the pulley. Likewise, the equal magnitude/oppositely directed force exerted onto the bar is $160\,\hat{\mathbf{i}} + 320\hat{\mathbf{j}}$ N.

We have now reached a curious point: We still have an unused FBD [that of the beam, Figure 4.18(a)], but there are no more forces to compute! If we write the equilibrium equations from that free-body and substitute our previously obtained results, all we will find is a check on those answers:

$$\xrightarrow{+} \qquad \Sigma F_x = -A_x + P_x = -160 + 160 = 0 \quad \checkmark$$
$$+\uparrow \qquad \Sigma F_y = A_y + P_y - 640 = 320 + 320 - 640 = 0 \quad \checkmark$$
$$\overset{+}{\curvearrowright} \qquad \Sigma M_A = 2.5P_y - 640(1.25) = 800 - 800 = 0 \quad \checkmark$$

So these three equations are satisfied, but they give us no new information. This is because if we (a) divide a body into n parts; (b) consider separate FBD's of the n + 1 bodies consisting of the n parts plus the overall body; and (c) write a set of equilibrium equations for each of the n + 1 bodies, only n of the sets can be independent equations.

The proof of this statement is simple. If we picture all the parts and put them back together to form the overall original body, all of the interactions (like the two P_x's and the two P_y's in the foregoing example) between respective parts will disappear, leaving only the (external) loads and reactions on the overall body. In the same way, if we add all the separate bodies' $\Sigma F_x = 0$ equations,* all the x-forces of internal interaction will cancel, leaving only the x-components of loads and reactions on the overall body to add to zero. But that is precisely the $\Sigma F_x = 0$ equation for the overall body. Hence this equation is redundant. The same argu-

---

* Using the same sign convention for each!

ment is valid for the $\Sigma F_y = 0$ equation and for the moment equation (about a common point for all bodies). Therefore only n of the n + 1 possible sets of equations are independent.

Of course, instead of the equations of the n parts, we could use those of the overall body and n − 1 of the n parts. The point is that one set of equilibrium equations will always be redundant. In the foregoing example, that set was the equations of equilibrium for the bar, since we had already used the equations of the overall body and of the pulley.

Let us illustrate these general ideas for the problem of Figures 4.17,18 above. Note that if we start over and separately write the equations of equilibrium for the beam and the pulley using the two free-body diagrams above, we obtain:

Beam:

$$\Sigma F_x = 0 = P_x - A_x$$

$$\Sigma F_y = 0 = A_y + P_y - 640$$

$$\Sigma M_A = 0 = 2.5 P_y - 1.25(640)$$

Pulley:

$$\Sigma F_x = 0 = -P_x + \tfrac{4}{5}T$$

$$\Sigma F_y = 0 = T + \tfrac{3}{5}T - P_y$$

$$\Sigma M_A = 0 = -2.5 P_y + 2.4T$$
$$+ 2.5(\tfrac{3}{5})T + 0.1T$$

If we now add the respective pairs of equations, we have:

$$x\text{-equation:} \quad \Sigma F_x = 0 = (\tfrac{4}{5})T - A_x$$

$$y\text{-equation:} \quad \Sigma F_y = 0 = A_y + T + \tfrac{3}{5}T - 640$$

$$\text{moment-equation:} \quad \Sigma M_A = 0 = -640(1.25) + T(2.4)$$
$$+ \tfrac{3}{5}T(2.5) + 0.1T$$

These equations are identical to those that were written for the combined (intact) body in Example 4.7 consisting of beam plus pulley.

Thus when the separate FBDs are used to form the equilibrium equations, and those equations are then summed, the equations of equilibrium for the combined (or "overall") body will be reproduced. Therefore we can use the two parts independently, or the overall body and one of the parts, but not all three. The third set of equilibrium equations will be dependent on the other two. And if a body is separated into more than two parts, say seven parts, and we write equilibrium equations for the seven parts plus the overall (intact) body, only seven of the eight sets of equations will be independent.

The ideas expressed above are illustrated more rigorously for the general case of an arbitrary body $\mathcal{B}$ divided into parts $\mathcal{B}_1$ and $\mathcal{B}_2$ in Appendix E.

Sometimes, as previously mentioned, we find that the equilibrium equations for a combined body such as $\mathcal{B}$ contain more scalar unknowns than there are independent equations. This suggests that the problem might be statically indeterminate, but that is not always the case. The special nature of the connection between two parts ($\mathcal{B}_1$ and $\mathcal{B}_2$) of $\mathcal{B}$ may render one or more components of the interaction force (**R**) and couple (**C**) between the parts to vanish. If this happens, the equations of equilibrium for $\mathcal{B}$ together with those of $\mathcal{B}_1$ may yield an equal number of equations and unknowns, rendering the problem determinate. Some of the examples that follow illustrate such situations. We note, however, that when we know *nothing* in advance about **R** and **C** (for example, when $\mathcal{B}_1$ and $\mathcal{B}_2$ are "welded"), then the equilibrium equations for $\mathcal{B}_1$ will be of no assistance in solving for the external forces (reactions) on $\mathcal{B}$.

Question 4.8   Why won't they?

In most of the examples that follow, the body or device under consideration is an assembly of rigid (or near-rigid) parts.

When the assembly is intended to be a stationary structure for supporting loads, it is often called a **frame**, unless it is composed exclusively of straight two-force bodies or members. In that case it is called a **truss** (about which more will be said in Chapter 5). When the function of the device depends upon the freedom of the parts to move relative to one another, particularly for the purpose of doing mechanical work, it is often called a **machine**. The pliers of Example 4.15 will be seen to constitute such a device.

Answer 4.8   In this case, separating the body into two bodies will expose six new unknowns — three new force components and three new couple components. Thus the creation of six new equilibrium equations will gain us nothing in this case.

EXAMPLE 4.15

Find the forces transmitted through the pin in the pliers of Example 4.3 (see Figure E4.15a).

Solution

We first shall consider a free-body diagram of one of the parts of the pliers (Figure E4.15b) because the force we seek will be *external* to that body.

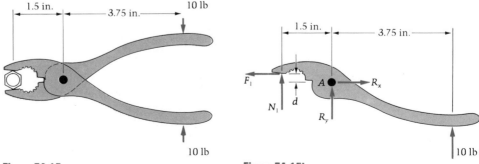

Figure E4.15a   **Figure E4.15b**

We can see immediately that there are more (four) unknown forces appearing than independent component-equations (three) of equilibrium for this two-dimensional problem. Writing these equations, we have

$$\xrightarrow{+} \quad \Sigma F_x = 0$$

$$R_x - F_1 = 0 \Rightarrow R_x = F_1 \tag{1}$$

and

$$+\uparrow \quad \Sigma F_y = 0$$

$$N_1 + R_y + 10 = 0 \tag{2}$$

and

$$\curvearrowleft{+} \quad \Sigma M_A = 0$$

$$-1.5N_1 + dF_1 + 3.75(10) = 0 \tag{3}$$

Turning to the free-body diagram (Figure E4.15c) of the whole pair of pliers, and summing moments at B, we obtain

$$\curvearrowleft{+} \quad \Sigma M_B = 0$$

$$F_1 b = 0 \Rightarrow F_1 = 0$$

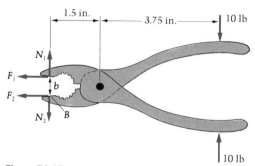

Figure E4.15c

The balance of forces yields

$$\xleftarrow{+} \quad \Sigma F_x = F_1 + F_2 = 0 \Rightarrow F_2 = -F_1 = 0$$

and

$$+\uparrow \quad \Sigma F_y = N_1 - N_2 + 10 - 10 = 0 \Rightarrow N_1 = N_2$$

Note that $F_1 = 0$ means from Equation (1) that R_x is also zero. Substituting $F_1 = 0$ into Equation (3),

$$1.5N_1 - 0 - 37.5 = 0$$

$$N_1 = 25 \text{ lb}$$

Then from Equation (2),

$$25 + R_y + 10 = 0$$

$$R_y = -35 \text{ lb}$$

The results displayed as free-body diagrams of the two parts are shown in Figure E4.15d.

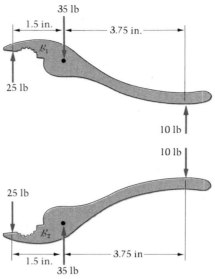

Figure E4.15d

In the preceding example, note that we have not actually specified which of the parts of the pliers (if either) would have the pin at A included in the free-body diagram. Which choice is made doesn't affect the analysis because the free-body diagram of the pin is as shown in Figure 4.19.

Figure 4.19

Note in the three figures that follow in Figure 4.20, the net result in each case is the pair of free-body diagrams of $\mathcal{B}_1$ and $\mathcal{B}_2$ just shown.

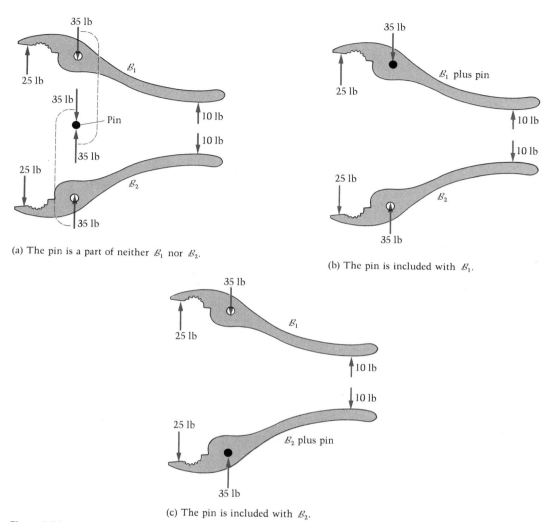

(a) The pin is a part of neither $\mathcal{B}_1$ nor $\mathcal{B}_2$.

(b) The pin is included with $\mathcal{B}_1$.

(c) The pin is included with $\mathcal{B}_2$.

Figure 4.20

We note also for completeness that the free-body diagram of the nut being gripped is Figure 4.21:

Figure 4.21

Question 4.9   If the hand force were different, would the gripping force still be 2.5 times it?

Answer 4.9   Yes.

EXAMPLE 4.16

For the frame of Example 4.6, shown in Figure E4.16a, find the force exerted by the pin at B onto member ADB. The 150 N force is applied to the pin at B.

Solution

In the earlier example, we found the reactions onto the frame at A to be $196\hat{i} - 200\hat{j}$ N and onto the frame at C to be $-346\hat{i} + 200\hat{j}$ N. Isolating member ADB and drawing its free-body diagram (see Figure E4.16b), we expose the components B_x and B_y of the force we seek. Note also that DE is a two-force member, so the force Q that it exerts at D lies along DE. Also, unlike Example 4.6, we now need to know the distance of DE above the baseline AC.

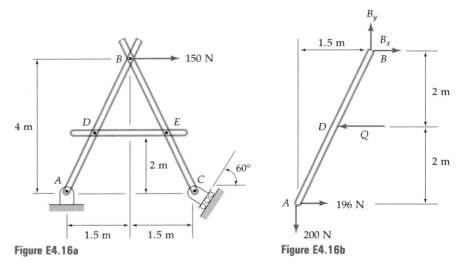

Figure E4.16a   **Figure E4.16b**

Question 4.10   Why wasn't the position of DE needed in the earlier example (4.6)?

We now write the equilibrium equations of the separated member ADB. Note that taking moments about B to first determine the force Q in the two-force member

Answer 4.10   In Example 4.6, force Q was internal, thus not exposed; hence its distance from A was not needed.

allows us to determine B_x and B_y in a series of "one equation in a single unknown" steps:

$$\curvearrowright \quad \Sigma M_B = 0$$

$$(200 \text{ N})(1.5 \text{ m}) + (196 \text{ N})(4 \text{ m}) - Q(2 \text{ m}) = 0$$

$$Q = 542 \text{ N}$$

$$\xrightarrow{+} \quad \Sigma F_x = 0$$

$$B_x - Q + 196 = 0$$

$$B_x = 542 - 196 = 346 \text{ N}$$

$$+\uparrow \quad \Sigma F_y = 0$$

$$B_y - 200 = 0$$

$$B_y = 200 \text{ N}$$

Thus the force exerted by the pin at B onto ADB is

$$B_x\hat{\mathbf{i}} + B_y\hat{\mathbf{j}} = 346\hat{\mathbf{i}} + 200\hat{\mathbf{j}} \text{ N}.$$

It is interesting to examine a FBD of the pin at B from the preceding example (see Figure 4.22):

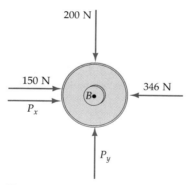

200 N

150 N

346 N

P_x

B•

P_y

Figure 4.22

Acting on this pin are the external force ($150 \rightarrow$ N), the (equal magnitude, opposite direction) reaction from ADB, and the reaction from bar CEB, expressed on the FBD as $P_x\hat{\mathbf{i}} + P_y\hat{\mathbf{j}}$. Summing forces gives:

$$\xrightarrow{+} \quad \Sigma F_x = 0 \qquad\qquad +\uparrow \quad \Sigma F_y = 0$$

$$150 + P_x - 346 = 0 \qquad\qquad P_y - 200 = 0$$

$$P_x = 196 \text{ N} \qquad\qquad P_y = 200 \text{ N}$$

Let us check these results by seeing if the (now completely known) force system on CEB places it in equilibrium. Note in the FBD below (Figure 4.23) how we must reverse the forces P_x, P_y, and Q so as not to violate the principle of action and reaction (Newton's Third Law):

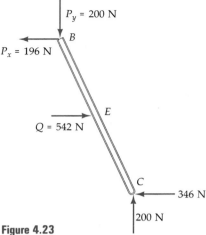

Figure 4.23

Checking,

$$\xrightarrow{+} \quad \Sigma F_x = 542 - 196 - 346 = 0 \quad \checkmark$$

$$+\uparrow \quad \Sigma F_y = 200 - 200 = 0 \quad \checkmark$$

$$\curvearrowleft{+} \quad \Sigma M_C = 196(4) + 200(1.5) - 542(2) = 0 \quad \checkmark$$

EXAMPLE 4.17

In Figure E4.17a, find the reactions at the supports A, B, and C, and the force between the beams transmitted by roller D.

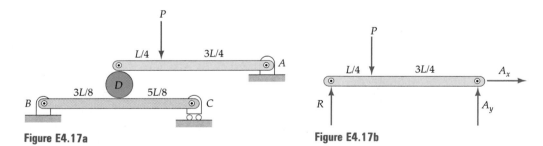

Figure E4.17a   **Figure E4.17b**

Solution

The free-body diagrams of the separate beams are shown in Figures E4.17b and E4.17c. We see from these FBD's that A_x and B_x must respectively vanish in order that $\Sigma F_x = 0$ for each beam. Using Figure E4.17b,

$$\curvearrowleft{+} \quad \Sigma M_A = 0 = (3L/4)P - LR$$

$$R = (3/4)P$$

$$+\uparrow \quad \Sigma F_y = 0 = R + A_y - P$$

$$A_y = P - (3/4)P = P/4$$

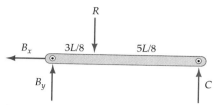

Figure E4.17c

And using Figure E4.17c,

$$\curvearrowleft_{+} \qquad \Sigma M_B = \; 0 = LC - (3/8)LR$$

$$C = (3/8)(3/4)P = (9/32)P$$

$$+\uparrow \qquad \Sigma F_y = \; 0 = B_y + C - R$$

$$B_y = (3/4)P - (9/32)P = (15/32)P$$

Putting the beams and roller back together in Figure E4.17e after displaying our results in Figure E4.17d, we see clearly how the roller forces, being internal, now cancel:

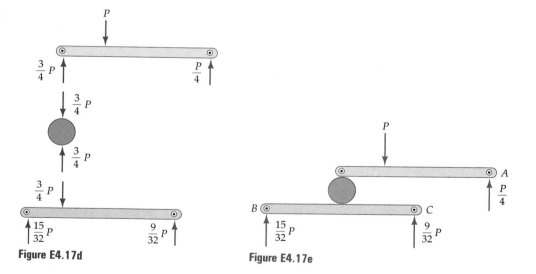

Figure E4.17d   **Figure E4.17e**

Checking the equilibrium of the overall body with the help of Figure E4.17e, we find three verifications of our solution:

$$\xrightarrow{+} \qquad \Sigma F_x \; = 0 \quad \checkmark$$

$$+\uparrow \qquad \Sigma F_y \; = (15/32 + 9/32 - 32/32 + 8/32)P = 0 \quad \checkmark$$

$$\curvearrowleft_{+} \qquad \Sigma M_B = -P(3/8 + 1/4)L + (9/32)PL + (P/4)(3/8 + 1)L$$

$$= PL[-20/32 + 9/32 + 11/32] = 0 \quad \checkmark$$

It is important that the reader note that we cannot determine all the desired forces in the preceding example without disassembling the structure. First of all, the overall FBD, shown below in Figure 4.24, does not

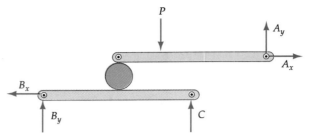

Figure 4.24

expose the roller force R which we seek. Secondly, all we know about the horizontal forces from this FBD is that $A_x = B_x$. Finally, even if we knew (from, say, mental FBD's of the separated beams) that $A_x = B_x = 0$, we would still be unable to find the three reactions A_y, B_y, and C from the two remaining equations $\Sigma F_y = 0$ and $\Sigma M_{\substack{any \\ point}} = 0$.

EXAMPLE 4.18

In Figure E4.18a, the pin P, which bears against the slot in bar BA, is fixed to rod RT. The weight of W is 125 lb. Find the force the pin exerts onto BA, and the reaction onto BA at B exerted by the ground.

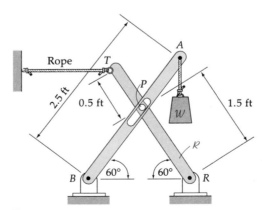

Figure E4.18a

Solution

The free-body diagrams of the structure and its two separate parts are shown in Figure E4.18b,c,d, although the only one we'll need in this problem is Figure E4.18c. Note that the line of action of the pin force P is known:

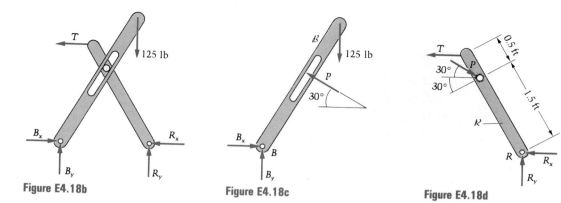

Figure E4.18b   **Figure E4.18c**   **Figure E4.18d**

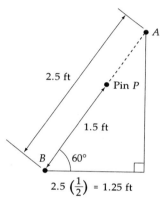

Figure E4.18e

Using Figures E4.18c and E4.18e, we may find the pin force immediately by summing moments about B:

$$\curvearrowright \quad \Sigma M_B = 0 = (1.5 \text{ ft})(P) - (125 \text{ lb})(1.25 \text{ ft})$$

$$P = 104 \text{ lb}$$

Summing forces in the x direction on body BA then yields

$$\xrightarrow{+} \quad \Sigma F_x = 0 = B_x - P \cos 30°$$

from which

$$B_x = (104)\sqrt{3}/2) = 90.1 \text{ lb}$$

Also, in the y direction,

$$+\uparrow \quad \Sigma F_y = 0 = B_y - 125 + P \sin 30°$$

so that

$$B_y = 125 - 104(0.5) = 73 \text{ lb}$$

In vector form, with the convention $\hat{\mathbf{j}}\overset{\uparrow}{\underset{\longrightarrow}{}}\hat{\mathbf{i}}$, the answers are: $\mathbf{P} = 104(-0.866\hat{\mathbf{i}} + 0.500\hat{\mathbf{j}})$ lb and $\mathbf{B} = 90.1\hat{\mathbf{i}} + 73.0\hat{\mathbf{j}}$ lb.

In contrast to Example 4.6, the preceding problem is an example of a **nonrigid frame.** Separation of the frame from its external supports by removal of the rope and removal of the pins at B and R leaves the two bars free to move relative to one another without deformation of the individual elements. Since T, R_x and R_y may be found with the help of the FBD of Figure E4.18d, the problem is statically determinate *in spite of the fact that the external reactions cannot be determined solely from equilibrium equations written for the assembled structure.*

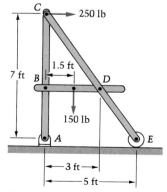

Figure E4.19a

EXAMPLE 4.19

The 250 lb force in Figure E4.19a is applied to the pin at C. Find the x and y components of the force exerted on frame member CDE at D by the member BD.

Solution

In frame problems, it is wise to develop the habit of drawing free-body diagrams of the overall frame and its constituent parts.* See the five FBD's of Figure E4.19b.

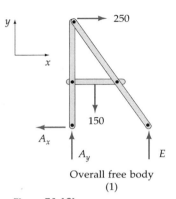

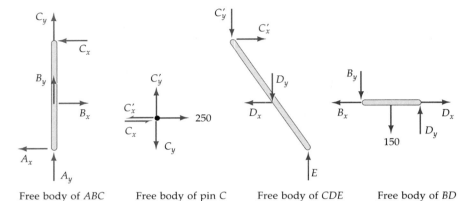

| Overall free body | Free body of *ABC* | Free body of pin *C* | Free body of *CDE* | Free body of *BD* |
| (1) | (2) | (3) | (4) | (5) |

Figure E4.19b

What we are seeking in this problem are the forces D_x and D_y (see Figure E4.19b(4). Here are the steps we will employ to obtain them:

1. Sum the moments about A in FBD (1) to obtain the roller force E;
2. Sum the moments about point B in FBD (5) to obtain D_y;
3. Sum the moments about point C in FBD (4) to get an equation in E, D_x, and D_y. Solve for D_x, using D_y from Step 2 and E from Step 1.

Step (1): On FBD (1),

$$\circlearrowleft_+ \quad \Sigma M_A = 0 = (5 \text{ ft})E - (150 \text{ lb})(1.5 \text{ ft}) - (250 \text{ lb})(7 \text{ ft})$$

$$E = \frac{1975}{5} = 395 \text{ lb}$$

Step (2): On FBD (5),

$$\circlearrowleft_+ \quad \Sigma M_B = 0 = (3 \text{ ft})D_y - (1.5 \text{ ft})(150 \text{ lb})$$

$$D_y = 75 \text{ lb}$$

Step (3): On FBD (4),

$$\circlearrowleft_+ \quad \Sigma M_C = 0 = (5 \text{ ft})\overset{395 \text{ lb}}{\cancel{E}} - (3 \text{ ft})\overset{75 \text{ lb}}{\cancel{D}_y} - (4.2 \text{ ft})D_x$$

$$D_x = 417 \text{ lb}$$

* As we have seen in general at the start of this section, the complete set of equilibrium equations that we could write using each of the free-body diagrams in Figure E4.19b won't all be independent. This is because, when combined (or "put back together"), the bodies of Figures E4.19b(2–5) form or constitute the "overall" body of Figure E4.19b(1). Thus if we draw a "redundant" free-body diagram [any *one* of Figures 1–5 is redundant], we must be aware that some of the possible equilibrium equations (three, here) will be redundant also, and will serve us only as checks on our solutions.

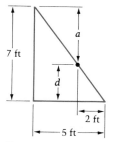

Figure E4.19c

where the dimension 4.2 ft for the moment arm of D_x was found as follows, using the similar triangles in Figure E4.19c:

$$\frac{2}{d} = \frac{5}{7} \Rightarrow d = \frac{14}{5} = 2.8 \text{ ft}$$

Thus

$$a = 7 - d = 4.2 \text{ ft}$$

Question 4.11 Draw a free-body diagram of the roller at E in the preceding example, and from it argue that the force exerted by the pin on CDE is the same as the normal force exerted by the ground onto the roller.

Answer 4.11

$$\left. \begin{array}{l} \Sigma M_E = 0 \Rightarrow F = 0 \\ \Sigma F_x = 0 \Rightarrow E_x = 0 \\ \Sigma F_y = 0 \Rightarrow N = E_y \end{array} \right\} \text{From the FBD:}$$

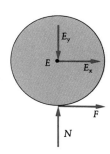

EXAMPLE 4.20

In Figure E4.20a find the forces exerted by the pin at B on (a) the pulley, (b) the bar ABE, and (c) the bar BC.

Solution

We note by inspection that BC is a two-force member. The importance of such an observation is in realizing the reduction in unknowns from four (two pin reactions at C and at B) down to just one — the tensile or compressive force in the member. Thus we need no free-body diagram of that member.

The overall free-body diagram, which will prove helpful in analyzing the frame in this problem, is sketched in Figure E4.20b. Equilibrium requires

$$\stackrel{\curvearrowleft}{+} \quad \Sigma M_A = 0 = C_y(3 \text{ m}) - (500 \text{ N})(2.5 \text{ m}) - (500 \text{ N})(4.5 \text{ m})$$

$$C_y = \frac{3500}{3}$$

$$= 1167 \text{ N}$$

where we start with four digits in this example. Also,

$$+\uparrow \quad \Sigma F_y = 0 = A_y + C_y - 500$$

$$A_y = 500 - 1167$$

$$= -667 \text{ N}$$

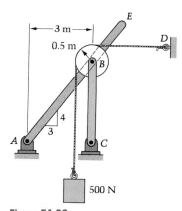

Figure E4.20a

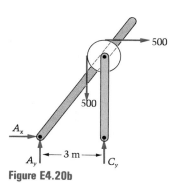

Figure E4.20b

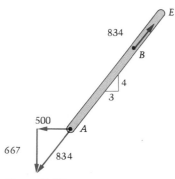

Figure E4.20c

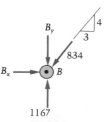

Figure E4.20d

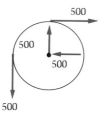

Figure E4.20e

Furthermore, we obtain:

$$\xrightarrow{+} \quad \Sigma F_x = 0 = A_x + 500$$

or

$$A_x = -500 \text{ N}$$

We note that the vector sum of $A_x\hat{\mathbf{i}}$ and $A_y\hat{\mathbf{j}}$ lies along BA because ABE, loaded by forces at just two points, is also a two-force member. Its free-body diagram is shown in Figure E4.20c. We see from this simple free-body diagram that the pin at B exerts the 834-N force shown on the bar.

The free-body diagram of the pin at B is shown in Figure E4.20d, with B_x and B_y representing the components of force extended on the pin by the pulley. The 1167-N and 834-N forces are the forces from the pair of two-force members. Equilibrium of the pin requires

$$\xrightarrow{+} \quad \Sigma F_x = 0 = B_x - 834 \left(\frac{3}{5}\right)$$

$$B_x = 500 \text{ N}$$

$$+\uparrow \quad \Sigma F_y = 0 = 1167 - B_y - 834 \left(\frac{4}{5}\right)$$

$$B_y = 500 \text{ N}$$

Putting the reverses of these two forces onto the pulley [see its free-body diagram (Figure E4.20e)] provides an immediate check on our solution by inspection.

In conclusion, the force exerted by the pin at B onto:

a. the pulley is $-500\hat{\mathbf{i}} + 500\hat{\mathbf{j}}$ N
b. the bar ABE is $500\hat{\mathbf{i}} + 667\hat{\mathbf{j}}$ N
c. the bar BC is $-1167\hat{\mathbf{j}}$ N

Note that in this problem, the forces exerted onto ABE and BC at B are not equal and opposite because of the presence of a third body (the pulley) there. What *is* true is that all three of these resultants add to zero because their negatives (Figure E4.20d) form the totality of external forces acting on the pin at B.

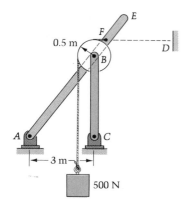

Figure E4.21a

EXAMPLE 4.21

Rework Example 4.20 if the horizontal portion of the cable is tied to ABE at F as shown in Figure E4.21a, instead of extending past it to the wall at D.

Solution

Recognizing again that BC is a two-force member, we have the overall free-body diagram shown in Figure E4.21b. Note that in order to isolate the frame, in this example we have to cut the rope only once. Thus $A_x = 0$ because

$$\xrightarrow{+} \quad \Sigma F_x = 0 = A_x$$

Note that ABE is not a two-force member in this example.

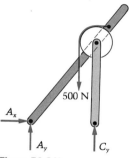

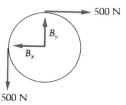

Figure E4.21b

Figure E4.21c

Figure E4.21d

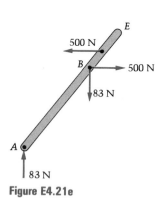

Figure E4.21e

Question 4.12   Why not?

Continuing,

$$\curvearrowleft_{+} \qquad \Sigma M_A = 0 = C_y(3 \text{ m}) - (500 \text{ N})(2.5 \text{ m})$$

$$C_y = \frac{1250}{3} = 417 \text{ N}$$

from which

$$+\uparrow \qquad \Sigma F_y = 0 = A_y + C_y - 500$$

$$A_y = 83 \text{ N}$$

The free-body diagram of the pulley shown in Figure E4.21c is the same as in Example 4.20, and we obtain

$$\left. \begin{array}{l} \Sigma F_x = 0 \Rightarrow B_x = 500 \text{ N} \\ \Sigma F_y = 0 \Rightarrow B_y = 500 \text{ N} \end{array} \right\}$$ Both as shown. These are the forces exerted on the pulley by its pin, at B.

The free-body diagram of the pin, with (B'_x, B'_y) representing the forces exerted on it by bar ABE, is shown in Figure E4.21d. Note that the force exerted on it by the bar BC is 417 $\uparrow$ N. It in turn exerts a force of 417 $\downarrow$ N on BC at B. Thus

$$\xrightarrow{+} \qquad \Sigma F_x = 0 = 500 - B'_x$$

$$B'_x = 500 \text{ N}$$

and

$$+\uparrow \qquad \Sigma F_y = 0 = 417 - 500 - B'_y$$

$$B'_y = -83 \text{ N or } 83 \uparrow \text{N} \qquad \text{(on the pin)}$$

Therefore the free-body diagram of member ABE is as shown in Figure E4.21e. It is obvious that $\Sigma F_x = 0 = \Sigma F_y$. Checking for moment equilibrium,

$$\curvearrowleft_{+} \qquad \Sigma M_A = -(83 \text{ N})(3 \text{ m}) - (500 \text{ N})(4 \text{ m}) + (500 \text{ N})(4.5 \text{ m})$$

$$= 1 \text{ N} \cdot \text{m} \qquad \text{(differing from zero due to roundoff error)}$$

We next take up an example of equilibrium of a body in three dimensions. Recall that for such a body, there are six scalar equations, these being the components of $\Sigma \mathbf{F} = \mathbf{0}$ and $\Sigma \mathbf{M}_P = \mathbf{0}$, where P is any one point. Consequently, 3D-problems are generally more difficult than their planar counterparts, especially when the body must be taken apart.

Answer 4.12   It is loaded at more than two points.

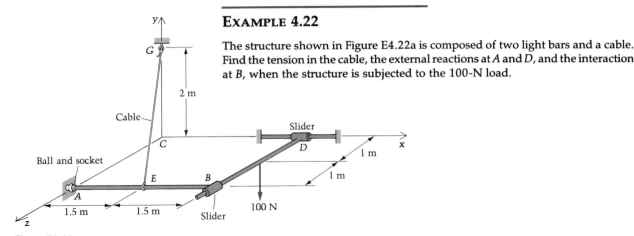

Figure E4.22a

EXAMPLE 4.22

The structure shown in Figure E4.22a is composed of two light bars and a cable. Find the tension in the cable, the external reactions at A and D, and the interaction at B, when the structure is subjected to the 100-N load.

Solution

First we shall consider the free-body diagram of the two bars taken together as shown in Figure E4.22b. We observe that, in this free-body diagram, eight unknown components of reaction appear. Thus we cannot determine all of the external reactions from the six corresponding equations of equilibrium. However, these equations taken together with those appropriate to *one* of the bars will turn out to be sufficient. Writing the equations of equilibrium for the two-bar system,

$$\Sigma \mathbf{F} = \mathbf{0}$$

$$(A_x\hat{\mathbf{i}} + A_y\hat{\mathbf{j}} + A_z\hat{\mathbf{k}}) + T\hat{\mathbf{e}}_{EG} + (D_y\hat{\mathbf{j}} + D_z\hat{\mathbf{k}}) + 100(-\hat{\mathbf{j}}) = \mathbf{0}$$

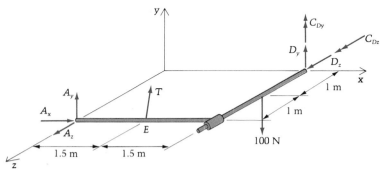

Figure E4.22b

Now $\hat{\mathbf{e}}_{EG}$, the unit vector along the line EG, is

$$\hat{\mathbf{e}}_{EG} = \frac{-1.5\hat{\mathbf{i}} + 2\hat{\mathbf{j}} - 2\hat{\mathbf{k}}}{\sqrt{(1.5)^2 + (2)^2 + (2)^2}}$$

$$= -0.469\hat{\mathbf{i}} + 0.625\hat{\mathbf{j}} - 0.625\hat{\mathbf{k}}$$

Thus from the coefficients of $\hat{\mathbf{i}}$, $\hat{\mathbf{j}}$, and $\hat{\mathbf{k}}$, respectively, in $\Sigma \mathbf{F} = \mathbf{0}$,

$$A_x - 0.469T = 0 \tag{1}$$

$$A_y + 0.625T + D_y - 100 = 0 \tag{2}$$

$$A_z - 0.625T - D_z = 0 \tag{3}$$

Summing moments at A,

$$\Sigma \mathbf{M}_A = \mathbf{0}$$

$$1.5\hat{\mathbf{i}} \times T(-0.469\hat{\mathbf{i}} + 0.625\hat{\mathbf{j}} - 0.625\hat{\mathbf{k}})$$
$$+ (3\hat{\mathbf{i}} - \hat{\mathbf{k}}) \times (-100\hat{\mathbf{j}}) + (3\hat{\mathbf{i}} - 2\hat{\mathbf{k}}) \times (D_y\hat{\mathbf{j}} + D_z\hat{\mathbf{k}})$$
$$+ (C_{Dy}\hat{\mathbf{j}} + C_{Dz}\hat{\mathbf{k}}) = \mathbf{0}$$

or

$$(1.5)(0.625T)(\hat{\mathbf{k}}) + (1.5)(-0.625T)(-\hat{\mathbf{j}}) + 3(-100)(\hat{\mathbf{k}})$$
$$+ (-1)(-100)(-\hat{\mathbf{i}}) + 3D_y(\hat{\mathbf{k}}) + 3D_z(-\hat{\mathbf{j}})$$
$$+ (-2D_y)(-\hat{\mathbf{i}}) + C_{Dy}\hat{\mathbf{j}} + C_{Dz}\hat{\mathbf{k}} = \mathbf{0}$$

From requiring that the coefficients of $\hat{\mathbf{i}}$, $\hat{\mathbf{j}}$, and $\hat{\mathbf{k}}$ vanish, respectively, we obtain:

$$-100 + 2D_y = 0 \Rightarrow D_y = 50 \text{ N} \tag{4}$$

$$+0.938T - 3D_z + C_{Dy} = 0 \tag{5}$$

$$0.938T - 300 + 3D_y + C_{Dz} = 0 \tag{6}$$

Holding Equations (1)–(6) in reserve, we now turn to the free-body diagram (Figure E4.22c) of bar BD. Note that our six equations of equilibrium will involve only four new unknowns, because of the special nature of the connection at B. Thus after writing six more equilibrium equations, we shall have $6 + 6 = 12$ equations in $8 + 4 = 12$ unknowns.

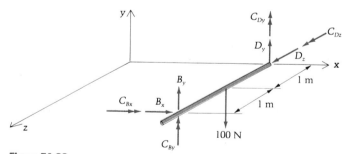

Figure E4.22c

The geometry here is so simple that it is relatively easy to write out the equations of equilibrium directly in component form:

$$\xrightarrow{+} \quad \Sigma F_x = 0 \Rightarrow B_x = 0 \tag{7}$$

$$+\uparrow \quad \Sigma F_y = 0 \Rightarrow B_y + D_y - 100 = 0$$

which, together with (4), yields

$$B_y + 50 - 100 = 0 \quad \text{or} \quad B_y = 50 \text{ N} \tag{8}$$

$$\cancel{+} \quad \Sigma F_z = 0 \Rightarrow D_z = 0 \tag{9}$$

From $\Sigma M_B = 0$, we obtain our final three equations:

$$\text{(i)} \qquad (\Sigma M_B)_x = 0$$

$$C_{Bx} + 2D_y - (1)(100) = 0$$

or, using (4)

$$C_{Bx} + 2(50) - 100 = 0 \Rightarrow C_{Bx} = 0 \qquad (10)$$

$$\text{(ii)} \qquad (\Sigma M_B)_y = 0 \Rightarrow C_{By} + C_{Dy} = 0 \qquad (11)$$

$$\text{(iii)} \qquad (\Sigma M_B)_z = 0 \Rightarrow C_{Dz} = 0 \qquad (12)$$

Substituting (4) and (12) into (6), we obtain the cable tension:

$$0.938T - 300 + 3(50) + 0 = 0$$

$$T = \frac{150}{0.938} = 160 \text{ N}$$

This, together with (9), yields for (5) an equation we can solve for C_{Dy}:

$$0.938 \left(\frac{150}{0.938} \right) - 3(0) + C_{Dy} = 0$$

$$C_{Dy} = -150 \text{ N} \cdot \text{m}$$

and hence from (11),

$$C_{By} = 150 \text{ N} \cdot \text{m}$$

Note that now we have determined the cable tension and the forces and moments associated with the connections at B and D. We now may return to Equations (1)–(3) to obtain the components of reaction at the ball-and-socket joint:

$$A_x - 0.469(160) = 0$$

$$A_x = 75.0 \text{ N}$$

$$A_y + 0.625(160) + 50 - 100 = 0$$

$$A_y = -50 \text{ N}$$

$$A_z - 0.625(160) + 0 = 0$$

$$A_z = 100 \text{ N}$$

Our final results, displayed on free-body diagrams of the individual bars, are shown in Figure E4.22d:

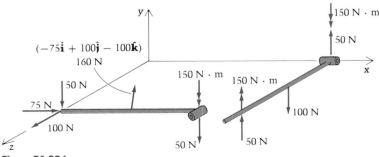

Figure E4.22d

The reader may wish to show that bar *AB*, which was not analyzed alone, is indeed in equilibrium under the forces acting on it in the final figure of the preceding example.

In the last two examples of this section, we shall have a preview of what is to come in Chapter 5. In the examples thus far, we have only removed supports, pins, ropes, and so on, leaving the various constituent members of the body intact (or whole). We have also made use of the special features of two-force bodies and pulleys. But there is nothing to prevent us from actually slicing through a member in order to expose, on a free-body diagram of only part of it, the forces and couples it transmits to the *other* part. This will be a distinguishing feature of Chapter 5. The following two examples will introduce the idea, although it is important to realize that there is nothing really new here in concept.* We again are using the fact that a body in equilibrium has each of its parts in equilibrium. If we wish to determine a certain force (and/or couple), we must choose a free-body diagram that exhibits the force as *external* so that it will appear in the corresponding equations of equilibrium.

Example 4.23

For the beam of Example 4.4 (shown in Figure E4.23a), find the force-couple resultant transmitted at a cross section 3 feet from the left end.

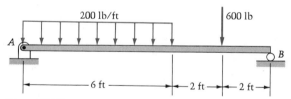

Figure E4.23a

Solution

We first separate (cut) the beam at the cross section of interest (point *C* of Figure E4.23b) and then sketch the free-body diagram of the material either to the left or to the right of the cut. Here we choose the material on the left. The free-body diagram is then shown with the dashed 600-lb force representing the resultant of the distributed load over the 3 feet. The 960-lb force at *A* is the reaction already determined in Example 4.4. Only a portion (3 feet) of the 200-lb/ft distributed load is external to the body we have chosen to analyze. The resultant of that distributed loading is 600 pounds as shown on the free-body diagram. For this two-dimensional problem the arrow code indicates that we have chosen to represent the force part of the resultant exerted on the material to the left of the cut by the material to the right of the cut as

---

* These distinctions—a single physical object, two or more objects left intact, physical objects separated, and a physical object imagined to be sliced into two parts—all obey the same simple laws of equilibrium. We have separated these topics because it is our experience that students master them more easily in graduated steps.

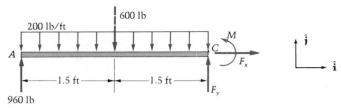

Figure E4.23b

by the material to the right of the cut as

$$F_x \hat{\mathbf{i}} + F_y \hat{\mathbf{j}}$$

The couple part of the resultant is expressed (vectorially) as $M\hat{\mathbf{k}}$.
Applying the equations of equilibrium in component form,

$$\xrightarrow{+} \quad \Sigma F_x = 0$$

so that

$$F_x = 0$$

and

$$+\uparrow \quad \Sigma F_y = 0$$

or

$$F_y + 960 - 600 = 0$$

so that

$$F_y = -360 \text{ lb}$$

The minus sign means that the vertical component of force acting on the cross section at C in Figure E4.23b is 360↓ lb. Finally, we sum moments to determine M:

$$\curvearrowleft_{+} \quad \Sigma M_c = 0$$

or

$$M + 1.5(600) - 3(960) = 0$$

so that

$$M = 1980 \text{ lb-ft}$$

It is instructive to see what would have happened in the preceding example had we chosen instead to apply the equations of equilibrium to the material to the *right* of the cut. The appropriate free-body diagram is shown in Figure 4.25. Note the 840-lb right-end reaction previously found in Example 4.4. Note further that the arrow code and letters (F_x, F_y, M) represent automatic satisfaction of the action-reaction principle because the forces and couple they represent are equal in magnitude but opposite in direction to those of Figure E4.23b.

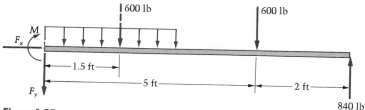

Figure 4.25

The equilibrium equations yield

$$\xrightarrow{+} \quad \Sigma F_x = 0 = -F_x$$

$$F_x = 0 \quad \text{(as before)}$$

and

$$+\uparrow \quad \Sigma F_y = 0$$

$$-F_y - 600 - 600 + 840 = 0$$

$$F_y = -360 \text{ lb} \quad \text{(as before)}$$

and

$$\curvearrowleft{+} \quad \Sigma M_C = 0$$

$$-M - 1.5(600) - 5(600) + 7(840) = 0$$

$$M = 1980 \text{ lb-ft} \quad \text{(as before)}$$

Thus the answers are independent of which part of the cut body we use to obtain them.

Question 4.13   In the preceding example, why is there no force component F_z, nor couple components M_x or M_y, acting on the cut section in addition to the forces F_x and F_y and the couple M parallel to $\hat{k}$?

Answer 4.13   The equations $\Sigma F_z = 0$, $\Sigma M_{Cx} = 0$ and $\Sigma M_{Cy} = 0$ would have respectively given $F_z = 0$, $M_x = 0$ and $M_y = 0$ if these components had been included on the FBD.

EXAMPLE 4.24

For the boom of Example 4.11, shown in Figure E4.24a, find the force-couple resultant transmitted at a cross section 4 feet from the right end of the boom.

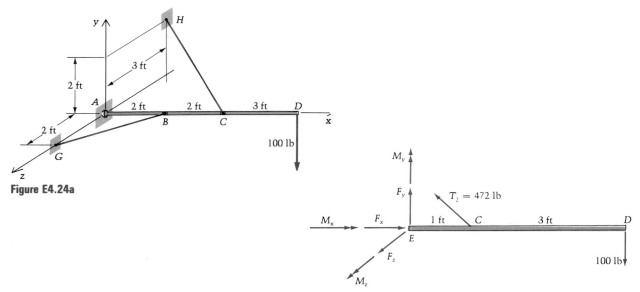

Figure E4.24a

Figure E4.24b

Solution

Isolating the part of the boom to the right of the cross section in question, we obtain the free-body diagram shown in Figure E4.24b. We found in Example 4.11 that the force exerted by the wire is

$$T_2\hat{e}_2 = T_2 \frac{\mathbf{r}_{CH}}{|\mathbf{r}_{CH}|}$$

$$= 472(-0.743\hat{i} + 0.371\hat{j} - 0.577\hat{k})$$

$$= -351\hat{i} + 175\hat{j} - 263\hat{k} \text{ lb}$$

Applying the equations of equilibrium,

$$\Sigma\mathbf{F} = \mathbf{0}$$

$$(F_x\hat{i} + F_y\hat{j} + F_z\hat{k}) + (-351\hat{i} + 175\hat{j} - 263\hat{k}) + 100(-\hat{j}) = \mathbf{0}$$

Thus

$$\hat{i}: \qquad F_x - 351 = 0$$

$$F_x = 351 \text{ lb}$$

$$\hat{j}: \quad F_y + 175 - 100 = 0$$

$$F_y = -75 \text{ lb}$$

$$\hat{k}: \qquad F_z - 263 = 0$$

$$F_z = 263 \text{ lb}$$

Taking moments about point E will yield the couples there:

$$\Sigma \mathbf{M}_E = 0$$

$$(M_x\hat{\mathbf{i}} + M_y\hat{\mathbf{j}} + M_z\hat{\mathbf{k}}) + \mathbf{r}_{EC} \times T_2\hat{\mathbf{e}}_2 + \mathbf{r}_{ED} \times (-100\hat{\mathbf{j}}) = 0$$

$$M_x\hat{\mathbf{i}} + M_y\hat{\mathbf{j}} + M_z\hat{\mathbf{k}} + 1\hat{\mathbf{i}} \times (-351\hat{\mathbf{i}} + 175\hat{\mathbf{j}} - 263\hat{\mathbf{k}})$$
$$+ 4\hat{\mathbf{i}} \times (-100\hat{\mathbf{j}}) = 0$$

$$M_x\hat{\mathbf{i}} + M_y\hat{\mathbf{j}} + M_z\hat{\mathbf{k}} + (175\hat{\mathbf{k}} + 263\hat{\mathbf{j}}) - 400\hat{\mathbf{k}} = 0$$

Thus

$$\hat{\mathbf{i}}: \qquad M_x = 0$$

$$\hat{\mathbf{j}}: \qquad M_y + 263 = 0$$

$$M_y = -263 \text{ lb-ft}$$

$$\hat{\mathbf{k}}: \quad M_z + 175 - 400 = 0$$

$$M_z = 225 \text{ lb-ft}$$

Question 4.14   Why did the force exerted by the wire BG onto the boom appear neither on the free-body diagram nor in the equilibrium equations?

Answer 4.14   That force is not acting on the material that has been isolated here (free-body diagram) for analysis.

PROBLEMS ▶ Section 4.4

4.144 The uniform bar in Figure P4.144 weighs 100 lb and the man 140 lb. Find the tension in the cable for equilibrium. (Can the man exert this much force?)

1.5 ft   1.5 ft   3 ft

Figure P4.144

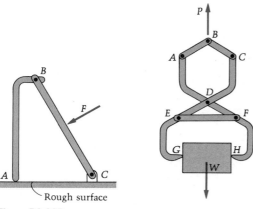

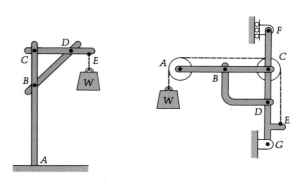

Figure P4.145

4.145 Name the two-force members (not counting cables) in the four structures shown in Figure P4.145 (three have one and one has three).

4.146 If b/a is 5, show that the compound lever system in Figure P4.146 will hold up a weight W that is 125 times the magnitude of the force F.

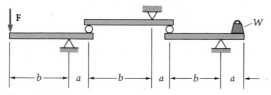

Figure P4.146

4.147 The uniform 1-ton (2000 lb) beam BD rests on a roller pinned to the truss structure at point B in Figure P4.147. Find the reactions onto the structure at A and C.

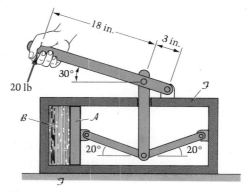

Figure P4.148

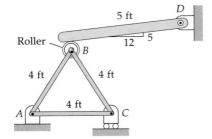

Figure P4.147

4.148 The object $\mathscr{B}$ is compressed by the toggle device. If a person pulls with 20 lb as shown in Figure P4.148, what is the compressive force exerted on $\mathscr{B}$ by block $\mathscr{A}$? Neglect friction between $\mathscr{A}$ and frame $\mathscr{I}$.

4.149 If the 3000-lb spherical boulder is in equilibrium, and if friction is negligible except at the fulcrum F, find the reaction of the wall on the boulder, and the angle θ. (See Figure P4.149.)

4.150 In Figure P4.150 the cord passes over a pulley at C and supports the block (mass 15 kg), in contact with the uniform bar AB (mass 10 kg). Find the horizontal and vertical components of the force at A on AB, and the force exerted on the bar by the block.

4.151 If the archer's left hand is pushing against the bow handle with a force $P = 50$ lb, what is the tension in the string (as a function of α)? What are the vertical forces exerted on the ends of the bow by the string? What is the horizontal component of the resultant force exerted on the archer's feet by the ground? (See Figure P4.151.)

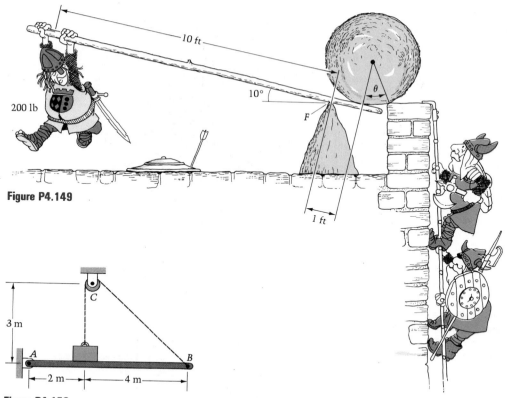

Figure P4.149

Figure P4.150

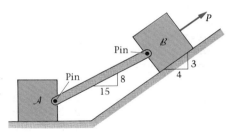

Figure P4.151

4.152 In Figure P4.152, the blocks A and B respectively weigh 223 N and 133 N, and the planes are smooth. The connecting rod is light. (a) Find the force P that will hold the system in equilibrium. (b) Repeat the problem with P applied instead horizontally to block A.

Figure P4.152

4.153 A block is lifted by the tongs shown in Figure P4.153. Find the force exerted on member $\mathcal{L}$, at C, by member $\mathcal{R}$.

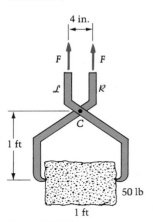

Figure P4.153

4.154 The identical wheels in Figure P4.154 each have mass 80 kg, and the rod has mass 40 kg and is pinned to the wheels as shown at A and B. The plane is smooth. Using free-body diagrams, show that the three bodies cannot be in equilibrium in the given position.

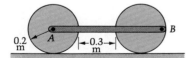

Figure P4.154

4.155 In the preceding problem, compute the moment of a couple that, when applied to the wheel on the right in the given position, results in equilibrium.

4.156 The ring gear $\mathcal{A}$ is fixed in a reference frame to which the centers of the sun gear $\mathcal{C}$ and spider arm $\mathcal{S}$ are pinned at C. (See Figure P4.156.) A clockwise couple M_o is applied to $\mathcal{S}$. Find the couple that must be applied to $\mathcal{C}$ in order that all bodies be in equilibrium.

4.157 In Figure P4.157 the pulley weighs 15 lb, the beam weighs 60 lb, the man weighs 160 lb, and the system is in equilibrium. Find the force in one of the two ropes at A, assuming the force is the same in each of the two ropes at an end.

4.158 Determine the gripping forces on (a) the nail in the most closed position of the pliers if $F = 20$ lb; (b) the pipe in the most open position of the pliers if $F = 20$ lb. (See Figure P4.158.)

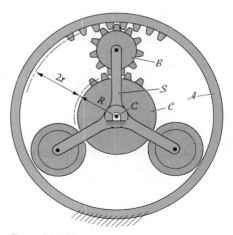

Figure P4.156

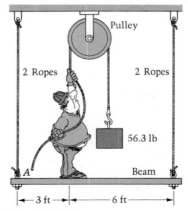

Figure P4.157

4.159 The rod AB in Figure P4.159 weighs 30 lb, and the block weighs 50 lb. Friction is negligible. What force P will hold the system of two bodies in equilibrium?

4.160 Find the forces in whichever of $\mathcal{B}_1$, $\mathcal{B}_2$, $\mathcal{B}_3$, and $\mathcal{B}_4$ are two-force members. (See Figure P4.160.)

• **4.161** (a) Two marbles, each of radius R and weight W, are placed inside a hollow tube of diameter D as shown in Figure P4.161. Note that $D < 4R$, so that only one marble touches the floor. Find the minimum weight of the tube such that it will not turn over. (b) Show that if the bottom of the tube is capped, then it will not turn over regardless of the weights or dimensions. All surfaces are smooth.

4.162 Find the force exerted on member $ABCD$ by the pin at B in Figure P4.162.

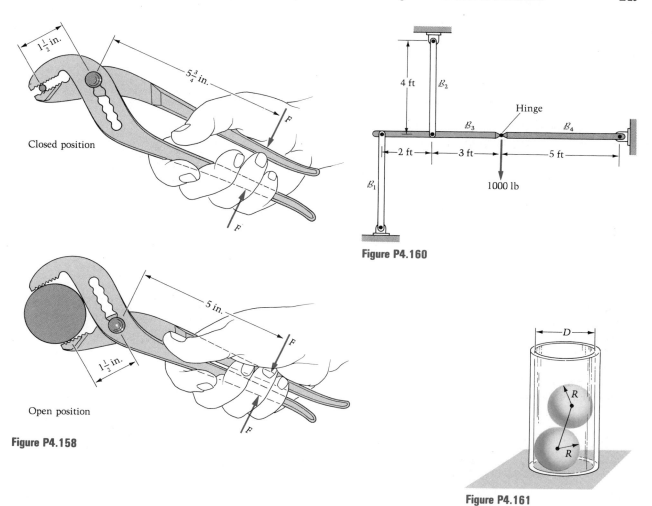

Closed position

Open position

Figure P4.158

Figure P4.160

Hinge

4 ft $\mathcal{B}_2$

$\mathcal{B}_3$ $\mathcal{B}_4$

2 ft   3 ft   5 ft

$\mathcal{B}_1$ 1000 lb

$5\frac{3}{4}$ in. $1\frac{1}{2}$ in.   5 in. $1\frac{1}{2}$ in.

F

Figure P4.161

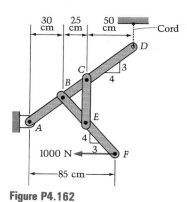

Figure P4.159   **Figure P4.162**

A   10 ft   4   3   B   12   5   P

30 cm   25 cm   50 cm   Cord

D   C   3   4   B   A   E   4   1000 N   3   F

85 cm

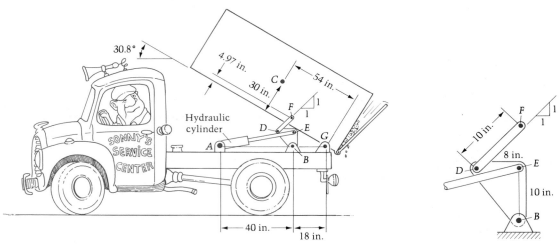

Figure P4.163

4.163 The dumptruck in Figure P4.163 is ready to release a load of gravel. The total weight being hoisted is 15,000 lb with mass center at C. Treating the cylinder rod AE as a two-force member, find the force in it at the indicated position.

4.164 Find the force that the bar CE exerts on the bar AF at C. (See Figure P4.164.)

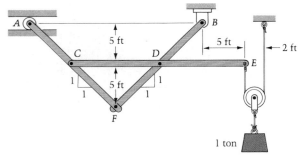

Figure P4.164

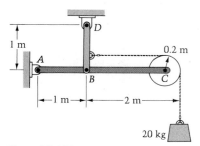

Figure P4.165

4.165 In the frame shown in Figure P4.165, the members are pin-connected and their weights can be neglected.

   a. Find the external reactions on the frame at A and D.

   b. Find the forces at B and C on member ABC.

4.166 In the frame shown in Figure P4.166, the members are pin-connected and their weights can be neglected. The 42 kN force is applied to the pin at C. Find:

   a. The reactions on the frame at A and E

   b. The components of the forces exerted by the pins at B and C on member ABC.

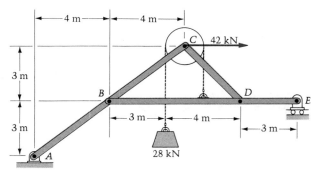

Figure P4.166

4.167 In Figure P4.167 find the reaction onto the frame at point F, and the force exerted on the pin at D by the member CF.

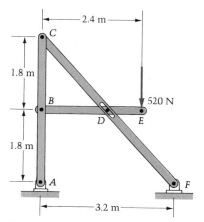

Figure P4.167

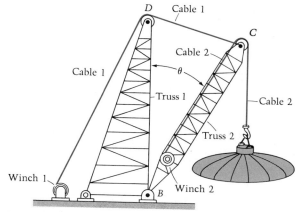

Figure P4.169

4.168 A slender, homogeneous, 20-ft rod weighing 64.4 lb is supported as shown in Figure P4.168. The bars AB and DE are of negligible mass. In terms of θ, find the force P that must be applied to the right end for equilibrium.

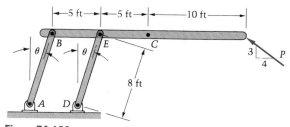

Figure P4.168

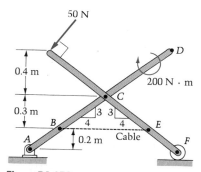

Figure P4.170

4.169 The crane in Figure P4.169 consists of two trusses, two cables, and two winches. Cable 1, let out by winch 1 to lower truss 2, passes over the small pulley at D and is attached to truss 2 at the center of the small pulley at C. Cable 2, let out by winch 2 attached to truss 2, lowers the 2000-lb antenna reflector after passing over the pulley at C. The lengths BD and BC are equal, and the weights of the trusses are to be neglected.

    a. Find the tension in cable DC as a function of θ.

    b. Find the force exerted by pin B on truss 2, and note that it is independent of θ.

4.170 In Figure P4.170 find the force exerted on $ABCD$ by the pin at C.

4.171 Find the magnitude of the force of interaction between the two bars of Figure P4.171.

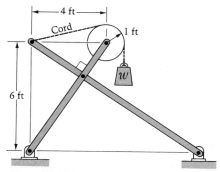

Figure P4.171

4.172 Find the pin reactions at B on member AB in Figure P4.172.

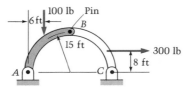

Figure P4.172

4.173 The blocks in Figure P4.173 each weigh 1250 N, with centers of mass at C_1 and C_2. The (shaded) platform on which they rest weighs 800 N with mass center at C_3, and is supported by two pairs of crossbars (one pair shown). Neglecting the weights of the crossbars, find the magnitude of the force transmitted by the pin that connects these two members at F. Assume that half the load is carried by each pair of crossbars.

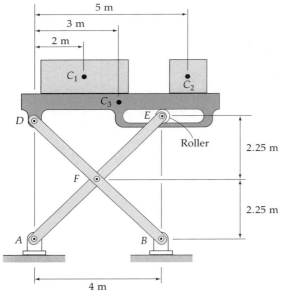

Figure P4.173

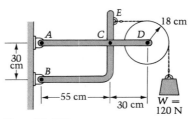

Figure P4.174

4.175 For the frame shown in Figure P4.175 find:

a. The reactions onto the frame members at A and E

b. The force exerted on BDF at D by the pin that joins the two members BDF and CDE to the pulley there.

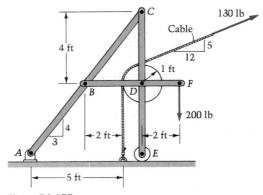

Figure P4.175

4.176 Find the reactions at A and B in Figure P4.176 when the horizontal force P is applied to the three-hinged arch. Neglect the weight of the arch.

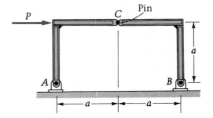

Figure P4.176

4.174 Draw the free-body diagrams of all members in Figure P4.174 and compute:

a. The force exerted by the pin at C on member ACD

b. The reactions at A and B.

4.177 In Figure P4.177, the pin at D is a part of member EB. Find:

a. The reactions at A and B on the frame members

b. The force exerted by CD on EB at D.

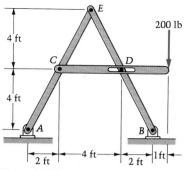

Figure P4.177

4.178 The length of the connecting rod R of the steam engine is 5 ft, and its crank C, to which it is pinned at B, has length 10 in. (See Figure P4.178.) The front and back pressures on either side of the piston are indicated. Find the force in member R, neglecting friction and assuming all bodies to be in equilibrium. *Hint:* Note that C carries a moment and is not a two-force member!

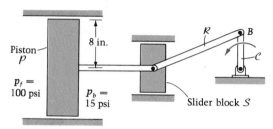

Figure P4.178

4.179 In Figure P4.179 determine the components of the pin force at C on member A.

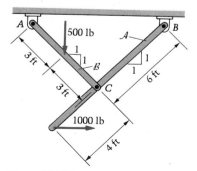

Figure P4.179

4.180 Find the reactions exerted on the bent bar AB at A and B. (See Figure P4.180.)

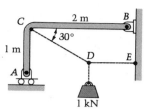

Figure P4.180

4.181 In Figure P4.181 what is the compressive force exerted by the nutcracker on the pecan? What is the force in the link AB?

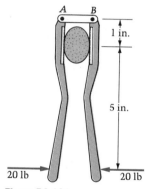

Figure P4.181

4.182 Find the force exerted by the pin at G onto member BEG. (See Figure P4.182.) Neglect the weights of the bars, but consider the weight of the 200-lb drum D.

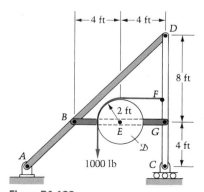

Figure P4.182

4.183 Find the force exerted by the pin at C onto member ABC. (See Figure P4.183.)

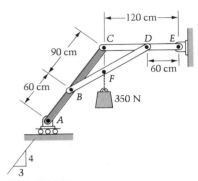

Figure P4.183

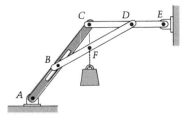

Figure P4.184

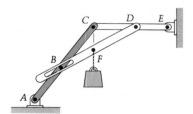

Figure P4.185

4.184 Repeat the preceding problem if the connections at A and B are changed to those shown in Figure P4.184. Pin B is attached to BD and bears against a smooth slot cut in ABC.

4.185 Repeat Problem 4.183 if the connections at A and B are changed to those shown in Figure P4.185. Pin B is attached to ABC and bears against a smooth slot cut in BD.

4.186 A worker in a "cherry picker" is installing cable TV equipment. (See Figure P4.186.) If the man plus bucket weigh 400 lb and the extendable member CE weighs 800 lb, find the force in the hydraulic cylinder BD and the pin reactions at C onto the extendable member CE. Neglect the weight of BD.

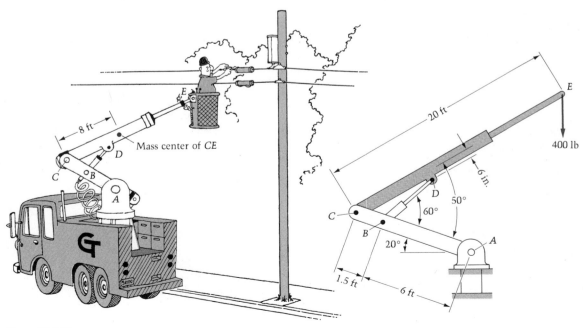

Figure P4.186

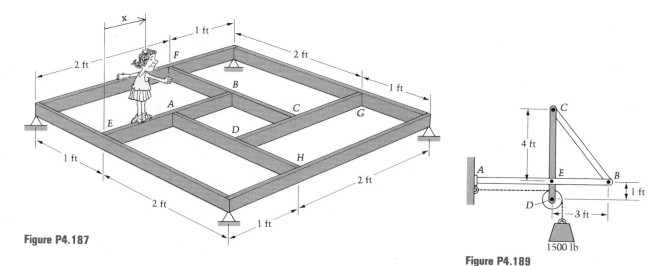

Figure P4.187

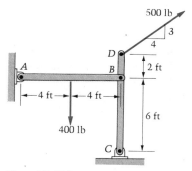

Figure P4.189

* **4.187** A carpenter has built a form for making concrete patio slabs. A little girl of weight W walks on the inside boards of the form as shown in Figure P4.187. Assuming no moments are exerted at the eight connections A, B, ... , H, find the largest force ever exerted by any connection, and note that it is larger than the weight of the child. *Hint:* You need only consider all positions (use x) on board EAB, by symmetry.

4.188 A frame is loaded as shown in Figure P4.188. Find the forces exerted on member DG at D and at B.

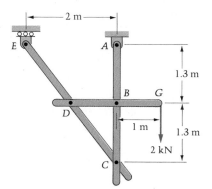

Figure P4.188

4.189 In Figure P4.189 find the force exerted on member AB at E.

4.190 In the preceding problem, find the reactions exerted on member AB at A if the distance AE is 6 ft.

4.191 Find: (a) the force that AB exerts on CD at B and (b) the force that the wall exerts on AB at A. (See Figure P4.191.)

Figure P4.191

4.192 In Figure P4.192 find the vertical component of the reaction at C. Can the horizontal component be found by separating the members of the frame? If so, find it; if not, why not?

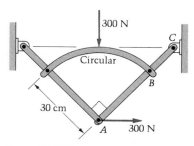

Figure P4.192

4.193 For the frame in Figure P4.193, find the forces exerted onto the member *ACE* at (a) *A*; (b) *C*; (c) *E*.

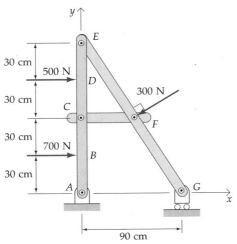

Figure P4.193

4.194 The three members *ABC*, *CDE*, and *BDF* are connected to form the frame in Figure P4.194. The cable over the pulley is fastened to the pin at *C*. The pin at *B* is fastened to member *ABC* and is free to slide in the horizontal slot in member *BDF*. Determine the force exerted on member *ABC* at *C*.

4.195 Find the reactions exerted on the structure shown in Figure P4.195 at (a) *A*; (b) *C*; (c) *D*. Also, (d) find the force exerted by the pin at *E* (which is fixed to member *CEF*) onto member *BD*.

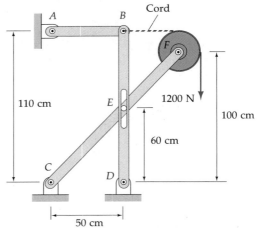

Figure P4.195

4.196 A woman weighing 120 lbs has climbed 60% of the way up a lightweight folding ladder as shown in Figure P4.196. The ladder rests on a frictionless surface. Assume that the woman's weight acts through her body's center of gravity as shown in the figure. Find the force in the two identical symmetrical cross-braces (one on each side).

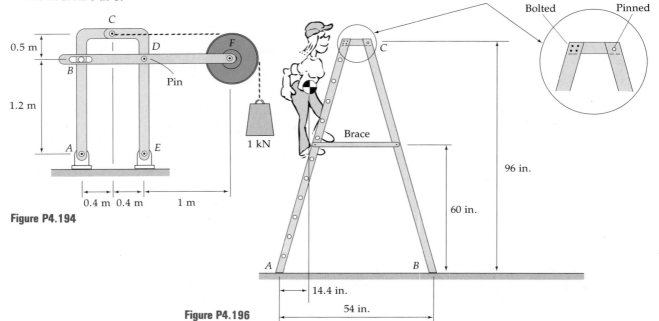

Figure P4.194

Figure P4.196

4.197 Two quarter-rings, each of mass m, are pinned smoothly together at Q and held in place by the two forces of magnitude P shown in Figure P4.197. The plane is smooth. Find the value of P for equilibrium.

Figure P4.197

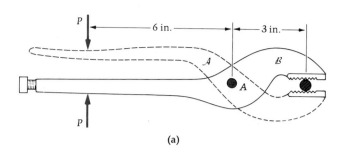

(a)

4.198 Two cylinders $\mathcal{A}$ and $\mathcal{B}$ are joined as shown by a stiff, light rod $\mathcal{R}$ and rest in equilibrium on two smooth planes (Figure P4.198). What is the angle between $\mathcal{R}$ and the horizontal?

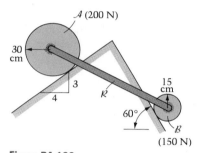

Figure P4.198

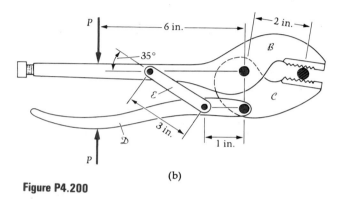

(b)

Figure P4.200

4.199 Repeat the preceding problem if the cylinders are turned around ($\mathcal{A}$ on the right, $\mathcal{B}$ on the left).

4.200 Show that if the dashed member $\mathcal{A}$ is pinned to $\mathcal{B}$ at A to form a simple pair of pliers [Figure P4.200(a)], the clamping force is $2P$. Then for the actual locking pliers comprising members $\mathcal{B}$, $\mathcal{C}$, $\mathcal{D}$, and the link $\mathcal{E}$ [Figure P4.200(b)], find and compare the new clamping force to the simple-pliers answer.

4.201 In the preceding problem, let the length of $\mathcal{E}$ be variable, with its upper end allowed to pin to $\mathcal{B}$ anywhere from directly above its fixed-position lower end to the point of application of P. find the largest possible clamping force.

4.202 Find the compressive clamping force on the object at C in Figure P4.202. (Note that the pin at B joins $\mathcal{B}$ and $\mathcal{D}$, but not C!) Then rework the (simpler) problem if member $\mathcal{A}$ is pinned directly to $\mathcal{B}$ as shown in the lower figure. Compare the mechanical advantage of these simple snips with the compound snips.

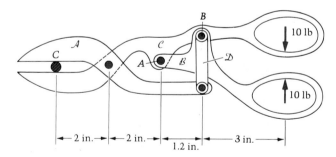

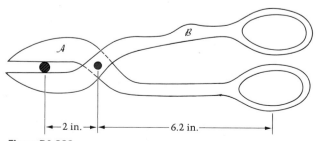

Figure P4.202

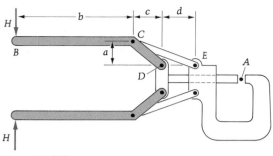

Figure P4.203

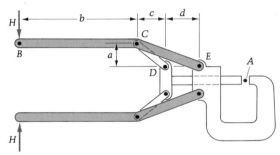

Figure P4.204

4.203 For the rivet squeezer shown in Figure P4.203, find the ratio of the clamping force at the point A to the gripping force H.

4.204 Repeat the preceding problem if the handle is member BCE instead of BCD. (Refer to Figure P4.204.) For the same dimensions, does this give a greater or lesser clamping force?

4.205 The toggle device in Figure P4.205 is being used to crush rocks. If the pressure in the chamber, p, is 70 psi and the radius of the piston is 6 in., find the force that the rock crusher exerts upon the rock. Members AB, BC, and BD are pinned at their ends.

4.206 For the frame shown in Figure P4.206, find the magnitude of the force exerted on each of the connecting pins at B, C, and D.

4.207 The uniform slender bars in Figure P4.207 are identical and each weighs 20 N. Find the angles α and β for equilibrium.

4.208 In the frame shown in Figure P4.208, (a) find the reactions at A and E and (b) find the components of the forces at B and C on member ABC.

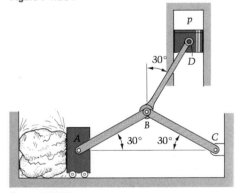

Figure P4.205

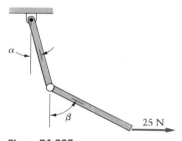

Figure P4.207

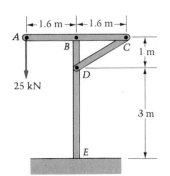

Figure P4.206

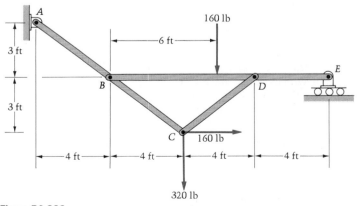

Figure P4.208

4.209 Find the reactions beneath each wheel, assuming symmetry, and the reaction on the ball (attached to the truck) of the ball-and-socket joint. (See Figure P4.209.)

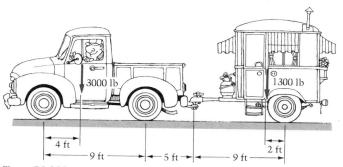

Figure P4.209

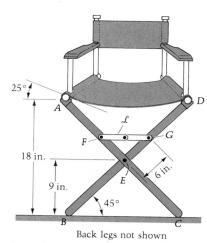

Figure P4.211

* **4.210** In the preceding problem the answers for the forces under the front tires are 753 lb (each of two), and under the back tires, 891 lb (each of two). A "load-leveler," or "equalizer," hitch can be used to more evenly distribute these forces. Each of two angle bars $\mathscr{A}$ fits and "bottoms" into a socket on the trailer side (see Figure P4.210), and its chain is pretensioned to 400 lb. Assume the ball and socket to be in the same position relative to truck and trailer as in the preceding problem. Find the distances a and b in the figure (which add to 2 ft) for which the reactions of the road on the four truck tires will be equal. For these values, show that the force between ball and socket (which was previously 289 lb) is greatly increased whereas the reactions of the road on the trailer tires (which were 506 lb each) are slightly increased.

4.211 Shown in Figure P4.211 is a sketch of a director's chair. If the director is well-fed at 260 lb and if he sits with each of 4 legs supporting 20% of his weight, find:

a. The force exerted by the floor onto a leg, neglecting friction there

b. The force in the link member $\mathscr{L}$

c. The force exerted by the pin at E onto member BED.

4.212 In Figure P4.212 the sleeve is pinned to bar $\mathscr{B}$ and can slide smoothly on the rod $\mathscr{R}$. Find the force in the cord if the system is in equilibrium. Then repeat the problem if the sleeve is pinned to $\mathscr{R}$ and free to slide on $\mathscr{B}$.

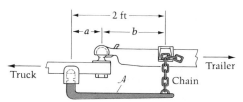

Figure P4.210

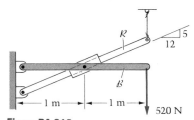

Figure P4.212

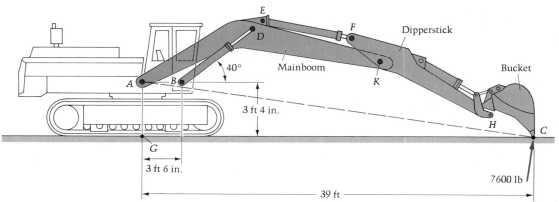

Figure P4.213

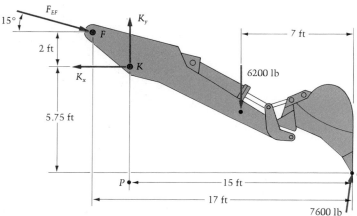

Figure P4.214

4.213 In Figure P4.213 the excavator is beginning to re-move a section of earth. If the force from the ground is 7600 lb, approximately normal to the line AC as shown, find the force in the hydraulic cylinder strut BD, and the pin reactions onto the mainboom at A. The weight of the mainboom, dipperstick, bucket, and lift cylinders is 15,000 lb; assume the horizontal mass center location of the weight to be halfway between G and C.

4.214 In the preceding problem, find the force F_{EF} in the hydraulic cylinder between the mainboom (ADK) and the dipperstick (FH). *Hint:* Consider Figure P4.214, which is a free-body diagram of dipperstick plus bucket, which to-gether weigh 6200 lb.

4.215 In Figure P4.215 compute the force in the cross-member $\mathscr{L}$ of the lifting tongs. Also find the horizontal component of the force at A acting on the 200-N block.

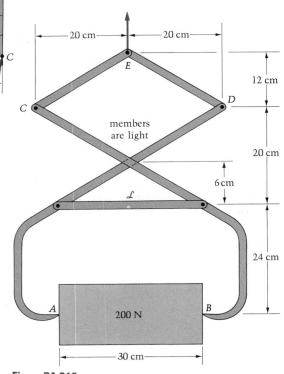

Figure P4.215

4.216 In the preceding exercise, let the link $\mathcal{L}$ be removed and let the members DA and CB be pinned where they cross. Find (a) the force exerted by this pin on DA, and (b) the horizontal component of the force at A acting on the block.

4.217 The utility loader (Figure P4.217) has raised 1600 lb of earth and rocks to its highest possible position. Exclusive of the bucket, each of two loader arms weighs 150 lb (including the dumping strut JH) with mass center at C, while the bucket weighs 280 lb with the mass center of it plus its contents at E. Determine the force in one of the hydraulic lifting struts AB, and the pin reactions at D, in the given position.

4.218 In the preceding problem, find the force in the hydraulic dumping strut JH and the reactions at pin P onto the bucket. (Refer to Figure P4.218.) Recall that the bucket plus contents weighs 1880 lb and that there are two of each of the struts in the figure.

4.219 A counterclockwise couple of 1000 lb-in. is applied to disk G, which is free to turn about a pin at O', as shown in Figure P4.219. Pin P, attached to G, bears against the slot in body $\mathcal{A}$. If the mass center of $\mathcal{A}$ is at C and the system is in equilibrium, find the weight of $\mathcal{A}$.

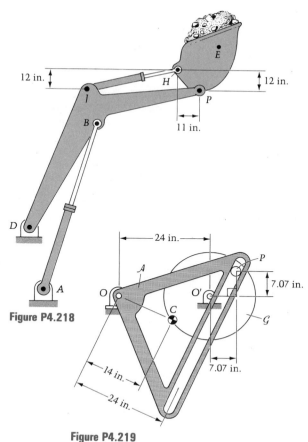

Figure P4.218

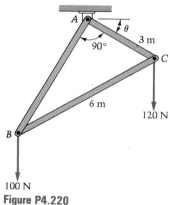

Figure P4.219

4.220 In Figure P4.220, AB, AC, and BC are light, slender bars, joined at their ends and supported by a hinge at A. Find (a) the angle θ for equilibrium and (b) the force in each bar.

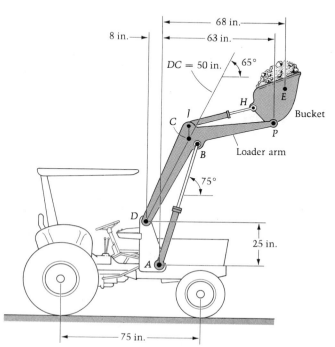

Figure P4.217

Figure P4.220

4.221 In Figure P4.221 find the reactions of (a) roller A onto bar $\mathscr{B}_1$ and (b) roller B onto bar $\mathscr{B}_2$.

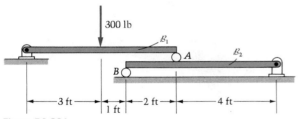

Figure P4.221

4.222 Find the compressive force in the spreader bar shown in Figure P4.222.

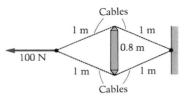

Figure P4.222

4.223 The shaded structure in Figure P4.223 holds up body $\mathscr{B}$, which weighs $2w$ with half the weight supported by each roller. Each half of the structure weighs W; the centers of gravity are at G_1 and G_2. Find the force exerted by the ground at E, and the magnitude of the roller reaction at C, in terms of b, L, e, d, w and W.

4.224 When running, the clothes dryer drum $\mathscr{D}$ is turned by means of a belt that passes around a motor pulley $\mathscr{M}$ as indicated in Figure P4.224. The belt also passes under an idler pulley $\mathscr{I}$ that is pinned to the bracket $\mathscr{B}$. The bracket is supported by the floor $\mathscr{I}$ of the dryer, which bears against

extensions of $\mathscr{B}$ that fit through slots in $\mathscr{I}$. If the force at A is vertical, and if that at B has both x and y components, find these reaction forces when the dryer is turned off, if the belt tension then is 1 lb. Neglect the weights of $\mathscr{B}$ and $\mathscr{I}$, and note that this means you are actually finding the *differences* between the reactions with and without the belt.

4.225 The suspension shown in Figure P4.225 supports one-half of the front of a car. Find the force in the spring and the force exerted on the frame by the members AB and CD. The wheel and its brakes and support weigh 100 lb with mass center at G.

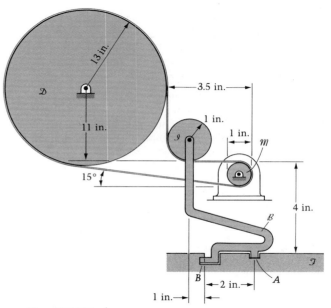

Figure P4.224

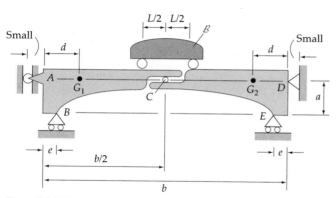

Figure P4.223

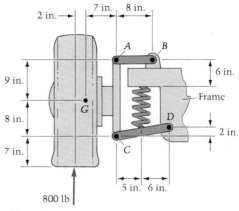

Figure P4.225

4.226 In Figure P4.226 winch 1 is used to raise and lower the boom of the derrick. After the desired angle θ is reached, winch 2 is then used to raise and lower the load. If $\theta = 50°$, find the forces in cables 1 and 2, and the compressive force in the boom.

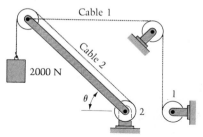

Figure P4.226

4.227 In Figure P4.227 find the force P that will hold the system of two 200-N cylinders and two light bars in equilibrium. (The other bar is behind the one shown.)

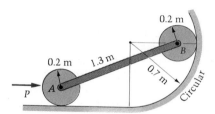

Figure P4.227

4.228 Determine the components of the pin forces onto the bars at A, B, and C. (See Figure P4.228.)

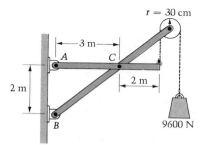

Figure P4.228

4.229 In Figure P4.229 find the force F needed for equilibrium of the system (called "Roberval's Balance"), and show that the value of F is independent of its position (i.e., doesn't depend on x). Neglect the weights of the members.

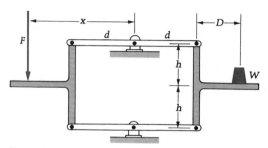

Figure P4.229

4.230 The linear spring exerts a force at each of its ends that is proportional to the amount of stretch it undergoes. In Figure P4.230, the spring modulus (proportionality constant) is 2 N/cm and its natural (unstretched) length is 1.5 m. Find the normal and friction forces (a) between cylinders A and B and (b) between B and the ground, if the weight of A is 500 N and that of each of B and C is 200 N.

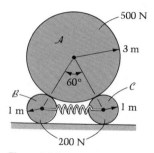

Figure P4.230

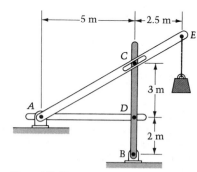

Figure P4.231

4.231 The weight (500 N) is suspended from point E by a cord. Find the force exerted at point D onto the shaded member BDC by the pin. (See Figure P4.231.)

4.232 Repeat the preceding problem if the slot is cut in BDC and the pin that slides in this slot is fixed to ACE.

4.233 Find the torque (twisting moment) carried by the sections S_1, S_2, and S_3 of the stepped shaft shown in Figure P4.233.

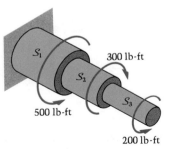

Figure P4.233

4.234 A cylinder weighing 2000 N is symmetrically lodged between two pairs of cross pieces of negligible weight. (See Figure P4.234.) Find the tension in the rope AB. (AD and BC are each continuous bars.)

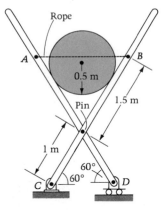

Figure P4.234

4.235 The 100-lb bar in Figure P4.235 rests in equilibrium against the 200-lb cubical block. The contact is smooth (frictionless) between the two bodies. Find the reaction of the plane onto the block.

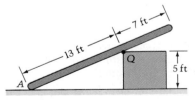

Figure P4.235

4.236 The two rods B_1 and B_2 are pinned as shown, and B_2 is fit (with friction) through a sleeve in body B_3, which is pinned smoothly to the ground. The 12 N · cm couple is applied to B_1 as shown in Figure P4.236. Find the resultant of the force system exerted by B_3 on B_2, expressed at Q as a normal force, friction force, and a couple.

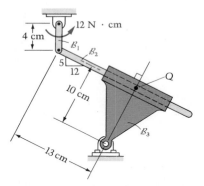

Figure P4.236

4.237 In Figure P4.237, the positioner supports a large paraboloidal antenna that is not shown. The antenna exerts the forces and couple (caused by wind and weight) shown at Q onto the positioner. [For information's sake, the antenna is "positioned" by (a) turning about the vertical around the azimuth bearing and (b) turning around a horizontal (elevation) axis normal to the page through O

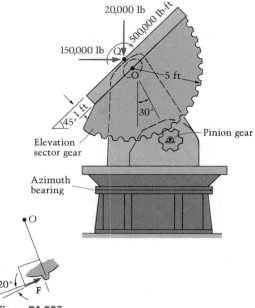

Figure P4.237

about *two* elevation bearings. This is called an elevation over azimuth positioner. Sometimes there is yet another azimuth rotation (for polarization) at the top.] If the tooth force **F** from the pinion onto one of *two* elevation sector gears is as shown, find the magnitude of **F**. Neglect the weight of the elevation assembly, which is in equilibrium, and assume just one tooth on each side is in contact.

4.238 The slender rod *AC* in Figure P4.238 is pinned to the small block at *C*. Friction prevents the block from sliding within the slotted body *B*. Find the reactions onto the bar at *A* and onto *B* at *B*.

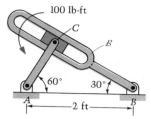

Figure P4.238

4.239 Repeat the preceding problem if the 60° angle is changed to 90°.

4.240 Figure P4.240 illustrates a jib crane. Its beam weighs 600 lb and is 10 ft in length. The weight of the suspended object is 400 lb. Plot the tension in the upper cable as a function of distance *d*, and find the pin reaction at *B* when $d = 6$ ft.

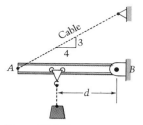

Figure P4.240

* **4.241** Four identical light rods are pinned together to form a square as shown in Figure P4.241, and supported by the four smooth platforms at the corners *A, B, C,* and *D*. A smooth sphere of radius *R* ($2R > a$) is then placed on the square. Show that the horizontal reaction between two adjacent rods has magnitude $Wa/(8\sqrt{2}\,h)$, where *h* is the height of the center of the sphere above the plane of the square, and *W* is the weight of the sphere.

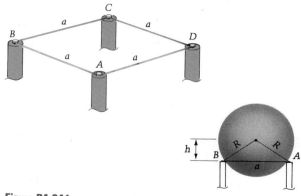

Figure P4.241

4.242 The two identical sticks shown in Figure P4.242 are pinned together at *A* and placed as shown onto the smooth block, the width of which is $\ell/2$. What is the angle ϕ for equilibrium?

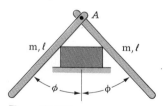

Figure P4.242

* **4.243** What is the maximum overhang for each identical slab shown in Figure P4.243 so that they are in equilibrium? There can be any number of slabs, each of length *b*. *Hint*: Start at the top instead of the bottom.

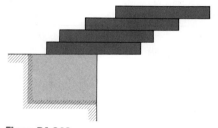

Figure P4.243

4.244 On a ten-speed bicycle, measure the following lengths (shown in Figure P4.244):

R_1, R_2 = large gear radii (measure to the middle of the teeth);

r_1, r_2, r_3, r_4, r_5 = small gear radii;

R = wheel radius;

R_f = radius to the middle of the pedal.

Let the bike and rider be traveling at constant speed, so that the force f is balanced by air resistance and we may consider the problem to be one of statics. Let F_f be a foot force, taken here to be constant for all gear combinations, with the pedal in the same position (shown) for each. Show, using appropriate free-body diagrams, that:

   1. $TR_i = F_f R_f$ (where R_i is R_1 or R_2, depending on the gear being used);

and

   2. $Tr_j = fR$ (where r_j is r_1, r_2, r_3, r_4 or r_5, depending on the gear being used);

so that

   3. $\dfrac{F_f R_f}{R_i} = \dfrac{fR}{r_j}$

or

$$f = \text{friction force that} \atop \text{moves the bike} = \frac{F_f R_f r_j}{R R_i}$$

Thus the driving force f is largest when r_j / R_i is largest, i.e., the "easiest" gear ratio is r_5 / R_1. Make a table ordering the ratios from the easiest to the "hardest," r_1 / R_2. Then re-do the chart by using ratios of numbers of teeth and compare. Ideally, the answers should be the same. Why should they?

4.245 In Figure P4.18, write one moment equation and find the force in the two-force member BC, without using the result of Problem 4.18.

4.246 Find the force exerted by the smooth pin at B onto bar ABC in Figure P4.246. Then find the force exerted by the smooth roller at E onto bar DBE.

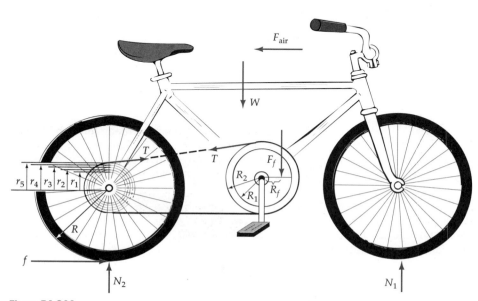

Figure P4.244

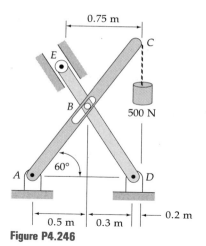

0.75 m

500 N

60°

0.5 m   0.3 m   0.2 m

Figure P4.246

4.247 Find the force in the 2-force member of the frame in Figure P4.247.

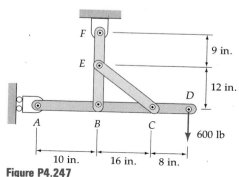

9 in.

12 in.

600 lb

10 in.   16 in.   8 in.

Figure P4.247

4.248 In Example 4.18, find the tension in the rope and the reactions exerted by the ground onto the structure at R. Then enter values for B_x, B_y, R_x, R_y, and T onto the overall FBD (Figure E4.18b) in the example, and verify its equilibrium by summing forces in the x and y directions, and moments about any point you choose.

4.249 Show that if in Examples 4.6 and 4.16 the roller at C is removed and the end C of bar BEC is instead pinned to the ground, the reactions onto ADB at B cannot be found. Show this by separating the members of the frame, writing their equilibrium equations, and attempting to solve them. The reason no solution is forthcoming from the equations of statics alone is that the structure contains more members than are needed for it to be stable. Note that DE *was* needed when there was a roller at C, but not now.

4.250 In the preceding problem, remove member DE and solve for the reactions B_x and B_y exerted by the pin at B onto member ADB. Do this by making use of the fact that AB and CB are now two-force members. Note the large differences in the x-components of the reactions when compared to the Examples 4.6 and 4.16.

4.251 In Example 4.19, find the forces exerted on members ABC and EDC at point C, assuming again that the 250 lb force is applied to the *pin* at C. Then draw the FBD of the pin at C to explain why the x-components of the forces found above are necessarily *not* equal and opposite.

4.252 The three light bars AB, CD, and EF of Figure P4.252 are connected together and to the ground by ball-and-socket joints. The 10-kN and 20-kN loads are applied parallel to the y- and x-axes, respectively. Find the force exerted onto member AB at C by member CD.

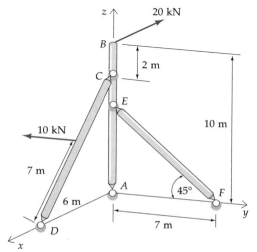

20 kN

2 m

10 kN

10 m

7 m

45°

6 m

7 m

Figure P4.252

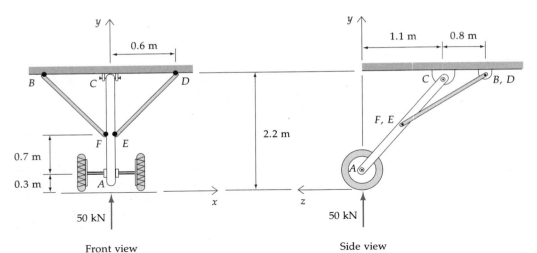

0.6 m

1.1 m   0.8 m

2.2 m

0.7 m

0.3 m

50 kN

50 kN

Front view

Side view

Figure P4.253

4.253 The struts *BF* and *ED* in Figure P4.253 are connected to the continuous member *AC* by ball-and-socket joints. The resultant force on the two wheels is 50↑ kN. Find the force in either strut (their magnitudes being equal by symmetry).

4.254 Bars *AB* and *DE* in Figure P4.254 are connected to *BDC* by ball-and-socket joints at *B* and *D*, respectively.

Find the forces in *DE* and *AB*, the latter by mental inspection of a free-body diagram of bar *BDC*.

4.255 The three rods *AB*, *DB*, and *CB* (Figure P4.255) are joined together at *B*, and are also joined by the three horizontal bracing members connected to them at *F*, *G*, and *E*. Find the forces in the bracing members if a weight of 1200 lb is hung from *B*. Treat the joints as ball-and-socket connections. The floor is smooth.

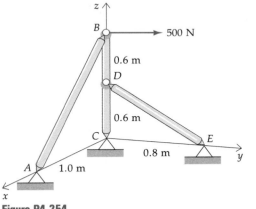

500 N

0.6 m

D

0.6 m

C

E

0.8 m

A   1.0 m

Figure P4.254

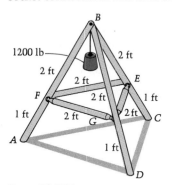

1200 lb   2 ft

2 ft

2 ft

F   2 ft   *E*

1 ft

1 ft   2 ft   2 ft   *C*

G

A

1 ft

D

Figure P4.255

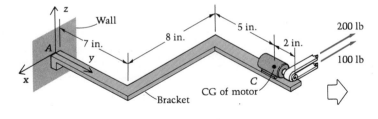

Wall

7 in.

A

8 in.

5 in.

2 in.

200 lb

100 lb

CG of motor

Bracket

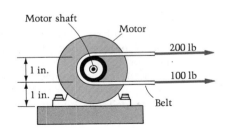

Motor shaft

Motor

1 in.

1 in.

200 lb

100 lb

Belt

Figure P4.256

4.256 A steel bracket is bolted to a wall at A, and supports a 10-lb motor at B. The motor is delivering torque, with the forces in the belt shown in Figure P4.256 on the preceding page. The bracket weighs 5 lb. Find:

a. The torque exerted by the two belt forces about the centerline of the motor shaft
b. The force exerted on the bracket by the wall at A
c. The moment exerted on the bracket by the wall at A.

4.257 So as not to interfere with other bodies, an antenna was designed and built with an offset axis as shown in Figure P4.257. The antenna is composed of a 12-ft, 1200-lb parabolic reflector $\mathcal{B}_1$, a counterweight $\mathcal{B}_2$, a reflector support structure $\mathcal{B}_3$ and a positioner. The positioner consists of (1) a pedestal $\mathcal{B}_4$ that is fixed to the ground; (2) an azimuth bearing at O and ring gear by means of which the housing $\mathcal{B}_5$ is made to rotate about the vertical; and (3) an elevation torque motor at E that rotates the support structure $\mathcal{B}_3$ with respect to $\mathcal{B}_5$. The purpose of the counterweight is to place the center of gravity of the combined body $\mathcal{B}_6(\mathcal{B}_1 + \mathcal{B}_2 + \mathcal{B}_3)$ on its elevation axis (x). If the reflector is modeled as indicated by a simple disk, determine the reactions onto $\mathcal{B}_6$ at E, and onto $\mathcal{B}_5$ at O, if the system is in equilibrium in the given position. Neglect the weights of $\mathcal{B}_3$ and $\mathcal{B}_5$.

* **4.258** In Figure P4.258 a quarter-ring is formed by two sections (AB and BC) of a circular bar being connected by a ball joint at B. The other ends are fixed to the reference frame, also by ball joints, at A and C. The three cables then hold the bar in the xz plane as shown. The radius R = 2 m, and the ring weighs 10 N/m. Find the cable tensions.

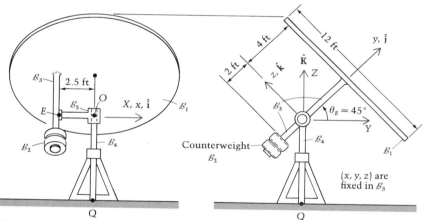

Figure P4.257

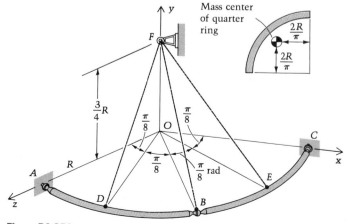

Figure P4.258

COMPUTER PROBLEMS ▶ Chapter 4

4.259 There is a range of values of L/R for which the uniform stick in Figure P4.259 can rest in equilibrium in the smooth hemispherical bowl. Show, using the equations of equilibrium, that this range is $\frac{2}{3}\sqrt{6} < \frac{L}{R} < 4$.

Then write a computer program that will divide this range into 101 equally spaced values of L/R and, for each such value, to calculate and print the value of θ for equilibrium.

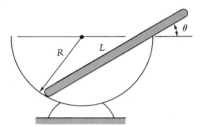

Figure P4.259

4.260 The uniform stick of mass m and length L rests in equilibrium between two smooth walls as shown in Figure P4.260. Write a computer program to help you construct a plot of the equilibrium angle θ of the rod versus the angle α of the right-hand plane.

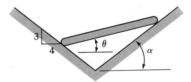

Figure P4.260

4.261 The distributed load shown in Figure P4.261 has a resultant of 300 ↓ lb acting 4 ft from the left end. The roller reaction is 200 ↑ lb. Note that two concentrated loads increasing proportionately and adding to 300 ↓ lb result in a roller reaction of 250 ↑ lb; three loads in 233 ↑ lb; four loads in 225 ↑ lb; etc. Write a computer program that will compute the value of this reaction for *any number* of concentrated loads. Use the program to compute the smallest number of concentrated loads needed to make the roller reaction within 1% of the continuously distributed limiting case.

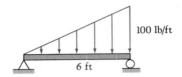

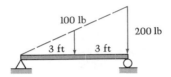

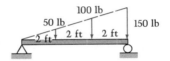

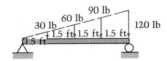

Figure P4.261

4.262 Clearly, when the boom angle θ in Figure P4.262(a) is either O or π, no force in the winch cable is needed for equilibrium. Somewhere between these values, the winch cable tension T is a maximum. We wish to find the

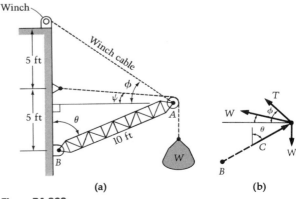

Figure P4.262

value of θ at which this occurs. First, use the free-body diagram (Figure 4.262(b)) of the pin at A, sum moments about B, and obtain an equilibrium equation, free of the boom compression C. Solve the equation for the ratio T/W as a function only of the angles θ, ϕ, and ψ in the figure. Prove that $\phi = \theta/2$ and that

$$\sin \psi = \frac{5(1 - 2\cos\theta)}{\sqrt{125 - 100\cos\theta}} \quad \text{and}$$

$$\cos \psi = \frac{10\sin\theta}{\sqrt{125 - 100\cos\theta}}$$

Use the computer to generate data for a plot of T/W versus θ for the range $0 < \theta < \pi$.

4.263 In Problem 4.77, DBC was a right triangle with the angle θ (see Figure P4.77) equalling $\tan^{-1}(3/4)$. Consider now other values of θ obtained by moving pin D to different points on the horizontal line through C as suggested by Figure P4.263.

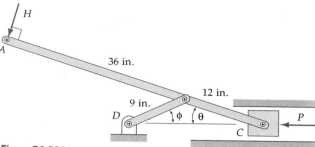

Figure P4.263

Show that: (a) $\phi = \sin^{-1}(\frac{4}{3}\sin\theta)$; (b) the force B in the 2-force member is $4H/\sin(\theta + \phi)$, compression; (c) the crushing force $P = B\cos\phi - H\sin\theta$. (d) Write a computer program and generate data for a plot of P/H versus θ, for θ in the range from $2°$ to $47°$. (e) Show that the largest angle θ for which crushing is possible is, to five significant digits, $47.929°$.

SUMMARY ▶ **Chapter 4**

This Chapter has introduced the student to a large number of realistic equilibrium problems, each one solvable through the use of only two equations: the sum of the external forces acting on a body is zero ($\Sigma\mathbf{F} = \mathbf{0}$), and the sum of the moments of the external forces about any common point is zero ($\Sigma\mathbf{M_P} = \mathbf{0}$). This process is facilitated immensely by the use of the free-body diagram (FBD), covered in minute detail in Section 4.2.

We had used $\Sigma\mathbf{F} = \mathbf{0}$ alone to solve some particle equilibrium problems in Chapter 2. In Section 4.3, we expanded our knowledge to the solution of equilibrium problems of finite-sized bodies requiring the moment equation as well as the force equation. In Section 4.3, however, we restricted ourselves to problems involving only a single body (which includes the idea of two or more physical objects left intact and thus considered as a single body).

In Section 4.4, we took yet another step and learned to separate bodies joined by simple connections by removing the pins, rollers, cables, etc., and drawing FBD's of the separate bodies. With the help of these FBD's, we then wrote equilibrium equations of the various separated bodies as well as of the overall (unseparated) "combined body." We noted that this process contains both good news and bad news: It is bad news that if the equilibrium equations are written for the combined body and all its separated bodies, not all these equations will be independent. But it is good news that the separation process often renders a problem solvable which was indeterminate on the basis of the combined body's equations alone.

In Chapter 5, we will take a final step in the "particle / single finite-sized body / finite-sized body separated into two or more parts" sequence. It will be to imagine slices through bodies which expose, on the cut-through cross sections of the separated parts, the internal forces within the body. These forces are of paramount importance in determining the stresses existing in bodies under load.

REVIEW QUESTIONS ▶ Chapter 4

True or False?

1. Free-body diagrams help us considerably in writing correct equations of equilibrium in statics.

2. If a free-body diagram of body $\mathcal{B}$ is to be useful, then $\mathcal{B}$ must be in equilibrium.

3. On a two-force member, the two forces are equal in magnitude, opposite in direction, and each acts along the line joining their two points of application.

4. If a body is in equilibrium under the action of three forces, and two of these intersect at a point P, then the line of action of the third *also* passes through P.

5. If a body is in equilibrium under the action of three forces, and two of these are parallel, then the third need not be parallel to the first two.

6. If a body is in equilibrium under four forces — two of which form a couple — then the other two also form a couple.

7. If a body is in equilibrium under the action of three forces, the forces need not be coplanar.

8. A body acted on only by a single couple cannot be in equilibrium.

9. Let the external forces on a body be such that $\Sigma\mathbf{F} = 0$ and $\Sigma\mathbf{M} = 0$; then the body must be at rest.

10. On a three-force member in equilibrium, the forces are either (a) coplanar and concurrent or (b) coplanar and parallel.

11. The tensions in a cable passing over a pulley in equilibrium are always equal.

12. It is possible for a body $\mathcal{B}$ to be in equilibrium with two separate parts comprising $\mathcal{B}$ not being in equilibrium separately.

13. One of the most important things to keep in mind when drawing free-body diagrams of various parts of a body is the action-reaction principle.

Answers: 1. T 2. F 3. T 4. T 5. F 6. T 7. F 8. T 9. F 10. T 11. F 12. F 13. T

Model-Based Problems in Engineering Mechanics

▶
▶
▶

Statics

The study of classical mechanics is a profound experience. The deeper one delves into it, the more he appreciates the contributions of the great masters. — Y. C. FUNG

INTRODUCTION

COMPREHENDING MECHANICS GOES beyond reading the textbook and working problems. Knowing statics means that you understand the physics embodied in the laws of mechanics, recognize their limitations and assumptions, and can correctly apply them to situations you encounter in practicing engineering. Knowing mechanics requires that you also develop a reasonable sense of the physical consequences of the fundamental principles along with the mathematical consequences.

GREAT MASTERS OF MECHANICS such as Galileo, Leonardo da Vinci, Hooke, Kepler, and Newton formulated the laws of statics and dynamics from the results of numerous observations and experiments. They devised simple experiments to test and clarify their ideas. Using empirical findings, they developed theories for predicting the behavior of mechanical systems. The principles they discovered and the mathematical expressions that describe these principles are the cornerstones of engineering mechanics.

STUDENTS (AND TEACHERS) of statics often overemphasize analysis and pay too little attention to the relationship between theory and the actual physical behavior of mechanical systems. Understanding both aspects of mechanics is essential. Engineers cannot successfully model the behavior of a mechanical system if they are unsure of the physics of the system. And the ability to predict successfully the behavior of physical systems is fundamental to the process of engineering design.

▲ ▲

OVERVIEW

THIS SECTION PRESENTS experiments not unlike those used in early empirical studies of mechanics. These experiments demonstrate actual behaviors of simple mechanical systems and are intended to strengthen your understanding of the basic laws of statics. These exercises emphasize physical reality to help you develop qualitative intuitive skills that are essential in the practice of engineering. In addition, the demonstrations provide a way to check the soundness of certain mathematical models that are used to describe real-world problems.

▶

THESE EXPERIMENTS PROVIDE only a starting point for your explorations and will raise additional questions as you conduct them. Do not leave those questions unanswered. To answer them you may need to modify a demonstration, design new experiments, or simply concentrate on interpreting a mathematical model. The important point is that you should pursue the answers. Along the way you will develop new insights into statics, you will become more proficient, and your ability to explain and predict the physical world will improve.

THE EXERCISES IN statics are keyed to specific sections and problems in the text. Be sure to review the text material before attempting these exercises. Each demonstration requires that you compare your observations with behavior predicted from a mathematical model. In most cases, experimental and theoretical results should be reasonably close. Remember, however, that models are only approximate and that experiments are never perfect. So, if your results disagree, find out why. To do so, verify your measurements and, if necessary, repeat or redesign the demonstration. Review the assumptions and limitations of the theory and check your analysis or computer program for mistakes. If you still find a disparity between the results, you may be applying the wrong principles or using incorrect equations.

▲ ▲ ▲ ▲ ▲ ▲ ▲ ▲ ▲ ▲ ▲ ▲ ▲ ▲ ▲ ▲ ▲ ▲ ▲

MATERIALS

THE EXPERIMENTAL setups are simple and easy to construct. To conduct them, however, you will need some materials, all of which are readily available and can be obtained at little or no cost. These materials can be found at hardware stores, hobby shops, toy stores, and in the engineering shop at your college. We encourage the use of scrap materials and creative scrounging!

FOR MECHANICAL PARTS you should collect an assortment of cylinders, tubes, spheres, wheels, and rectangular blocks. The only requirement is that the parts be homogeneous and reasonably uniform. For example, if you need a cylindrical tube, select one that is straight and has a constant diameter and thickness. Manufactured tubes such as the following are excellent:

> Wood dowel
> PVC pipe, copper pipe, steel pipe
> Conduit tubing
> Aluminum rod, steel rod
> Empty coffee can with ends removed, tennis ball can
> Cardboard tube from roll of paper towels or toilet tissue
> Cardboard mailing tube
> Hockey puck
> Thread spool, metal adhesive tape spool

Be creative and resourceful when selecting materials.

IN ADDITION, you will need some laboratory supplies. They include string, duct tape, protractor, graph paper, stopwatch, tape measure or ruler, scissors, inexpensive calipers, and a scale or access to a scale for weighing parts.

▲ ▲ ▲ ▲ ▲ ▲ ▲ ▲ ▲ ▲ ▲ ▲ ▲ ▲ ▲ ▲ ▲ ▲ ▲

STATIC EQUILIBRIUM
OF A SYSTEM OF BODIES

The external forces acting on a body at rest satisfy the equilibrium equations $\Sigma F = 0$ and $\Sigma M = 0$. This demonstration examines static equilibrium for a system of bodies in contact. Use a cylindrical cardboard tube from a roll of paper towels or toilet tissue and two identical balls with a diameter of 60% to 80% of the tube diameter. Hold the hollow tube vertically on a flat surface and drop the two balls into the tube as shown in the figure. Slowly release the tube and observe its behavior. If the tube remains upright, reduce its length by 10% and repeat the process. At some point the system will not be able to remain in static equilibrium and the tube will fall over when you release it. Can you predict when this will occur? Show that the relative diameters and weights of the balls and tube are the parameters that determine whether the system remains at rest. Cap the bottom end of the tube with stiff paper and tape and repeat the experiment. Explain why the tube does not tip over when the bottom is capped.

You can perform this demonstration with different tubes and balls of various weights and sizes. In this case the balls need not be identical; however, the sum of their diameters must be greater than the inside diameter of the tube. Use the equations of static equilibrium to verify or predict your observations and to study the relationships among the system parameters. Be sure to include the wall thickness of the tube and the order of placement of the balls if their sizes or weights are different.

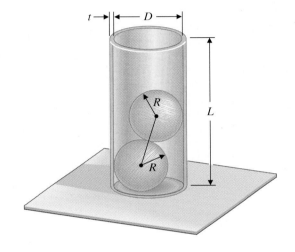

REFERENCE: *Statics* Sections 4.2-4.4

MEASURING COEFFICIENTS
OF STATIC FRICTION

Recall from statics that the coefficient of static friction, μ_s, between two materials is the tangent of the angle θ_s at which a sample of one material will start to slide down a plane made of the other. (See the figure.) Use this relationship to determine μ_s for different combinations of materials. Compare your values with some of those given in the text. Explain how the following affect μ_s: the contact area between samples, the roughness of their surfaces, and the weight and geometry of the sample that slides. Devise a method to measure μ_s if one of the samples is a length of pipe.

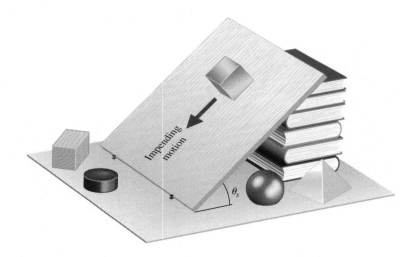

REFERENCE: *Statics* Section 6.2

STATIC EQUILIBRIUM OF AN
UNBALANCED CYLINDER

Build an unbalanced but round cylinder. One approach shown in the figure is to tape or glue a rod or small cylinder to the inside of a cylindrical tube. (Be sure that the two axes are parallel.) Equal lengths of 3 1/2" diameter PVC pipe and 1" diameter steel conduit tubing work well for this demonstration. Measure and calculate the eccentricity e of the mass center. Place the cylinder on a slightly inclined plane, remove the support provided by your finger as shown below, and observe the cylinder's motion. If the cylinder continues to roll, reduce the slope of the plane until the cylinder remains at rest. Slowly increase the slope and find the maximum angle of inclination β_{max} for which the cylinder will not roll down the incline. Measure β_{max} and the corresponding

resting angle ϕ which is defined in the bottom figure. Using the equations of static equilibrium, derive an expression that relates the angles ϕ and β in terms of the ratio e/R. Calculate the maximum inclination angle β_{max} and the corresponding resting angle from that equation. Note that other values of ϕ and β are possible if β is less than β_{max}. Plot these predicted values of ϕ vs. β for several values of e/R between 0 and 1. Be sure to include a curve corresponding to the value of e/R for the unbalanced cylinder. Measure ϕ vs. β for the cylinder and plot those values over your predictions. Do the results agree? Explain.

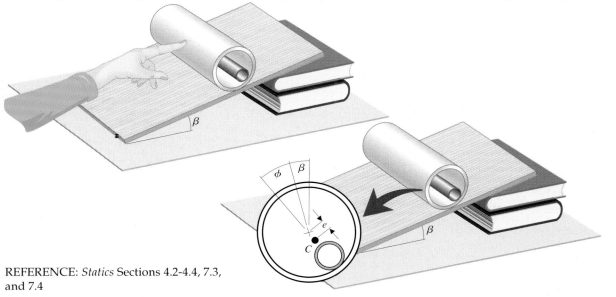

REFERENCE: *Statics* Sections 4.2-4.4, 7.3, and 7.4

HOLDING A SPOOL
ON AN INCLINE

Assemble a spool using two identical disks or cylinders and a single axle as shown in the figure. A metal adhesive tape spool works well for this demonstration. Tape a string to the center of the axle. Bring the string over the top of the axle and hold it parallel to the surface of an inclined board as in the top figure. Make sure the string pulls in the midplane of the axle. Notice that for small values of slope θ the spool remains in static equilibrium. Confirm this observation and predict the angle θ_{max} at which the spool will begin to move down the incline. You will need to know the coefficient of friction between the spool and the board. Measure θ_{max} and compare it to your prediction. Repeat the above procedures when the string is pulled from the bottom of the axle. (See the bottom figure.) Explain why these two cases are markedly different. Plot the theoretical relationships between tan θ_{max} and the axle/spool radius ratio r/R for the two cases. Can you anticipate the initial motion of the spool when θ exceeds θ_{max}? Could you use this experimental procedure to determine μ_s?

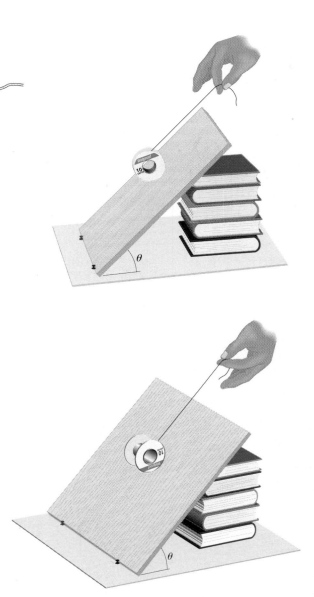

REFERENCE: *Statics* Section 4.3 and 6.2

BALANCING A CYLINDER
ON AN INCLINE

This demonstration explores the static equilibrium of a cylinder resting against a step on an inclined plane. Select a cylinder of radius R and install a step of height $H \cong 0.1R$ on a flat board. Place the cylinder against the step as shown in the figure below. Slowly increase the inclination of the board and measure the angle θ_{max} at which the cylinder just rolls over the step. Repeat the experiment for several values of step heights $H \leq R$. One easy way to vary the height is to build the steps from layers of stiff cardboard cut from the backing of a note pad. Use the equations of static equilibrium to obtain an expression for $\tan \theta_{max}$ in terms of the ratio H/R. Graph the measured and calculated values of $\tan \theta_{max}$ vs. H/R and compare the results. Show that your results are independent of the coefficient of friction and weight of the cylinder.

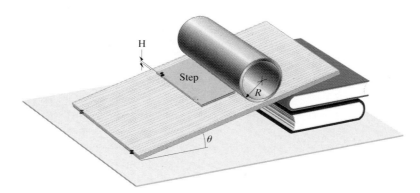

REFERENCE: *Statics* Sections 4.2-4.4, and 6.2

BALANCING A RECTANGULAR BLOCK ON AN INCLINE

Install a small lip or bump on an inclined flat board and place a rectangular block against the lip, as shown in the figure below. Slowly increase the inclination of the board and measure the angle θ_{max} at which the block just tips over. Use the equations of static equilibrium to calculate a theoretical value for θ_{max} in terms of the length-to-height ratio L/H. Compare the two values and explain any difference. Why should the bump be very small? Why do you not need to know the coefficient of friction? Repeat the demonstration for rectangular blocks of various dimensions. Graph the measured and predicted values of $\tan \theta_{max}$ vs. L/H and compare the results.

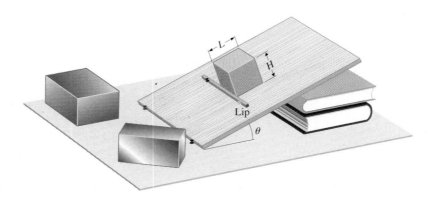

REFERENCE: *Statics* Section 4.3 and 6.2

5

STRUCTURAL APPLICATIONS

5.1   Introduction

In building a structure, we must always ensure that its members are designed with sufficient strength to carry the loads intended (and then some, to account for a factor of safety). This "strength" of the materials, covered in detail in later courses in the mechanics of deformable bodies, is measured by a quantity called stress, which has units of force per unit of area.

To be properly prepared for such later studies, a student must exit statics knowing how to find the resultants of forces *within* elements of structures such as trusses, beams, and frames. In other words, if a beam is imagined sliced into two parts, we must be able to determine the resultant forces (and/or couples) exerted on each "half" by the other. These forces are called internal forces for obvious reasons.

We will again use the result that if a body is in equilibrium, then every part of it is also in equilibrium. Thus for the beam in the preceding paragraph, we will learn to write the equilibrium equations for one of the parts, and to then solve them for the forces and moments exerted on the "face" that was exposed by the cut. These forces, while *external* to the part of the structure that is being analyzed, are *internal* to the complete (uncut) structure, and are resultants of the stresses on the face.

This chapter differs from Chapter 4 only in the level of sophistication in separating a body into parts: the bodies here will be cut completely through to expose internal forces, whereas in Chapter 4 we were simply separating bodies by removing simple connection devices such as pins, rollers and cables. We emphasize, however, that the *equations are the same* ($\Sigma \mathbf{F} = \mathbf{0}$ and $\Sigma \mathbf{M}_P = \mathbf{0}$ for each part analyzed).

The chapter is divided into three parts. In Part I, we deal with the truss. Each member of a truss is a straight bar in which the internal force resultant is a tensile or compressive force parallel to the member's axis. In Sections 5.2–5.6, we will examine the various means of determining these forces in truss members, or elements. Then in Section 5.7, we present a brief introduction to the mechanics of deformable bodies, treating the uniaxial stress and strain found in truss members.

The second part of Chapter 5 deals with multiforce members. Frames are structures comprising a number of members (or elements) connected together for some structural purpose, at least one member of which is not a straight, two-force, truss-like element. Such members, which in general can carry shear forces and bending and twisting moments in addition to axial forces, are called multiforce members. How to find these more complicated internal forces in frames is the subject of Section 5.8, and in Sections 5.9 and 5.10 this study is continued for another very common structural element known as a beam.

Part III deals with cables, in Sections 5.11 and 5.12. The first of these two sections looks at cables under distributed loads; in the second we examine cables acted on by a finite number of concentrated forces.

I / TRUSSES

5.2 Definition of a Truss; Examples of Trusses

We define a **truss** to be an *idealized* structure consisting of straight and slender bars, each of which is pinned to the rest of the structure and/or to the ground at its two endpoints by frictionless pins (or, in three-dimensional trusses, by ball-and-socket joints). In addition, the structure is loaded only by forces at such pins. Thus trusses are composed entirely of two-force members. See the examples in Figure 5.1.

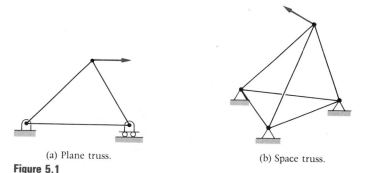

(a) Plane truss.   (b) Space truss.

Figure 5.1

We hasten to mention that the above idealization is a "mathematical truss," and the connections are, in reality, rarely smooth pins or ball joints. However, this idealization of a physical truss gives good results if the centerlines of the members at each connection all intersect at a single point; more will be said about this later. For now, the reader is invited to note, in the set of photographs on the next two pages, just a few of the many uses of trusses. These photographs indicate that there are many practical applications of the truss. Furthermore, even within a given application, there are often a large number of different types of trusses. For example, Figure 5.2 (see page 278) shows a number of common roof truss configurations. It should be mentioned that the names of the trusses may vary from one manufacturer to another. Each type has its own special use and span capability.

If the members of the truss all lie in a plane [as in Figure 5.1(a)], then we have a "plane truss"; if not [Figure 5.1(b)], the truss is called a "space truss." We shall introduce the methods of truss analysis with plane trusses, then consider the more complex space trusses later in the chapter.

Water towers.

Signs.

Lighting.

Backing and mounting
structures for antennas.

Crane supports in steel mills.

Roof structures.

Highway signs.

Temporary support for new highways.

Conveyors.

Electric power transmission towers.

Construction cranes.

Radio and TV towers.

Supports for amusement park rides.
(*Courtesy of Six Flags Over Georgia*)

Derricks.

Bridges.

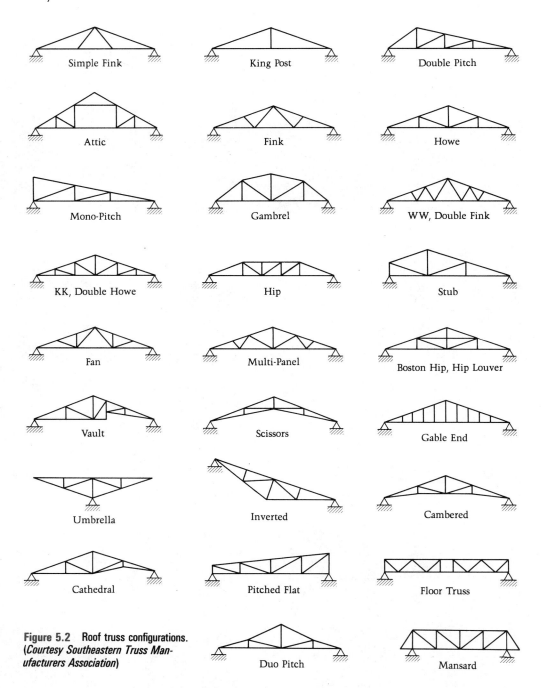

Figure 5.2 Roof truss configurations. (*Courtesy Southeastern Truss Manufacturers Association*)

Forces in Truss Members

By its definition, a truss is made up only of two-force members, and the force distribution across any cross section of a member has a very simple resultant. To determine it, we note, recalling the discussion of two-force members in Chapter 4, that a truss member carries only a pair of equal magnitude, oppositely directed, forces along its length (see Figure 5.3).

Therefore, if we cut a section through the respective members shown in Figure 5.3, we obtain the free-body diagrams shown in Figure 5.4.

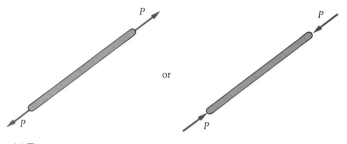

(a) Tension member of a truss.   (b) Compression member of a truss.

Figure 5.3

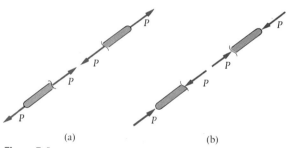

(a)   (b)

Figure 5.4

In both parts of Figure 5.4, the resultant is simply an axial force. This is because, if we were to assume the more general distribution of internal forces across the lower section of Figure 5.4(a) to be as shown in Figure 5.5, equilibrium would immediately require that the shear force (V) and the bending moment (M) vanish and that the axial force (A) be equal to P.*

A truss member in the condition of Figure 5.4(a) is being stretched and is said to be in tension; the member of Figure 5.4(b), however, is being compressed and is said to be in compression. (If $P = 0$, of course, the member is not loaded.) These are the only possibilities for truss member forces, and thus the answers are easy to present. For example, if we say that the force "in" a member extending from A to B (member AB) is 647 N Ⓣ, we mean that a free-body diagram of AB looks like Figure 5.3(a) with $P = 647$ N. Similarly, if we say the force in a member

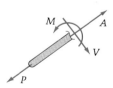

Figure 5.5

---

* Figure 5.5 refers to a member of a plane truss. If the member is from a space truss, then equilibrium would require that *two* shear forces vanish, that *two* bending moments *and* a twisting couple also vanish, and that, again, $A = P$.

DE is 212 lb ©, we mean that a free-body diagram of *DE* looks like Figure 5.3(b) with *P* = 212 lb.

For example, the truss shown in Figure 5.6 has seven members that, if the truss is solved,* carry the forces indicated in Figure 5.7 (all in kips, or "kilo-pounds"; 1 kip = 1 k = 1000 lb). The forces exerted on the various pins are also shown.

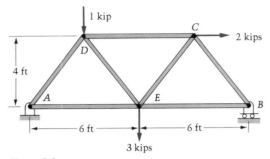

Figure 5.6

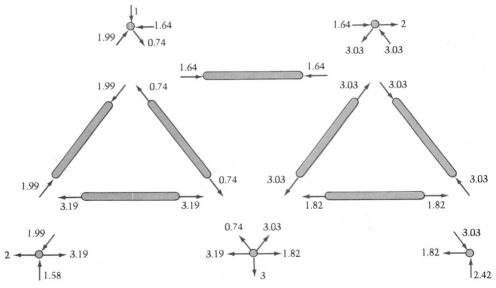

Figure 5.7   (All forces in kips)

We shall return to that truss in Example 5.1 and actually compute the indicated member forces after more preliminary discussion.

Though truss members are always assumed to be pinned at their ends, in reality this is seldom the case. The members of roof trusses and bridges, for example, are normally connected by means of a plate to

---

* Meaning all the forces in its members have been found.

which the members are joined by nails, rivets, pins, welds, or bolts at a number of points, as suggested in Figure 5.8.

Although it may seem like a bad assumption to replace a plate and a large number of bolts with a single pin, this is not the case. If the center-lines of the members intersect approximately at a point as shown in Figure 5.8, then the structure turns out to behave very much as though it were an ideal truss; that is, the transverse (shear) forces and the bending moments in the members will be small. Such structures are thus usually analyzed as ideal trusses.

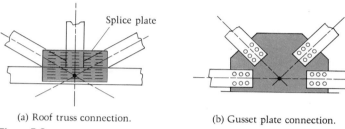

(a) Roof truss connection.   (b) Gusset plate connection.

Figure 5.8

There are two main methods commonly used in truss analysis: the method of joints and the method of sections. We shall begin our study with the method of joints (or pins).

5.3   The Method of Joints (or Pins)

In this section, we will be studying the use of the method of joints in truss analysis. This method is simply to isolate one pin at a time (with a free-body diagram) and to write the equilibrium equations for it. Often-times, we can at the outset find at least one pin on which only two unknown member forces act; when this is the case,* both these forces may be found from $\Sigma F_x = 0$ and $\Sigma F_y = 0$. After doing so, we repeat the procedure at another joint and thus work our way into and through the truss.

Before moving into some examples, we wish to first discuss the directions of unknown member forces acting on pins. Consider the free-body diagram on the next page of the pin at point B [Figures 5.9(a,b)], where the 2.42-kip force is an already-determined roller reaction. Now, when we sketch a force such as F_{CB}† pushing on the pin of a joint, we are assuming the member which is doing the pushing (CB in this case) to be in compression (ⓒ). This is because if the member pushes on the pin, then by action and reaction, the pin pushes *back*, compressing the member. Thus the four forces associated with member CB (the two it exerts, and

---

* This always happens with "simple plane trusses," as we shall see later.
† We attach no meaning to the order of the subscripts on the force; thus $F_{CB} = F_{BC}$.

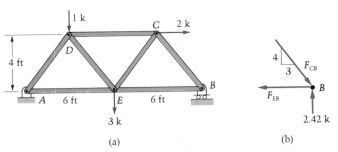

(a)

(b)

Figure 5.9

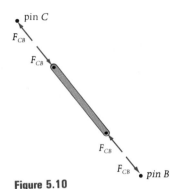

Figure 5.10

the two exerted on it) are in the directions shown in Figure 5.10 if member *CB* turns out to be in compression. [In this particular case, we are actually *sure* that *CB* is in compression because in the free-body diagram of joint *B*, only member *CB* can have a (vertical) component to balance the upward roller reaction of 2.42 kips, and the vertical component of F_{CB} onto the pin at *B* can be downward only if *CB* is in compression.]

On the other hand, if we draw a force such as F_{EB} *pulling* on a pin as in Figure 5.9(b), then we are assuming the member (*EB* here) to be in tension (ⓣ). Again the reason is action and reaction. If the member pulls on the pin, the pin likewise pulls back on the member. (See Figure 5.11.) We are also sure here that *EB* is in tension, once we have seen that *CB* pushes down and to the right onto pin *B*. In the free-body diagram of pin *B* [Figure 5.9(b)], only F_{EB} can balance the horizontal component of F_{CB}, which is to the right.

Figure 5.11

Sometimes, however, we are not certain of the direction of a force. (For example, does it push, or pull, on a pin?) In such a case, we simply guess one or the other, and if the solution for the force *F* turns out in the algebraic solution to be negative, this means the bar is in tension if we "assumed" compression, and vice-versa.* For example, suppose we assumed that a member *PQ* was in compression and later found F_{PQ} $= -300$ lb. It is common to then communicate that the force in *PQ* is

---

* Usually in an engineering analysis, if we make an assumption and then find it to be false, we must revise our assumption and repeat the analysis. That is not the case here because the "assumption" isn't really an assumption in the usual sense. Rather, it is nothing more than a statement of the physical significance of a positive value of the scalar used (together with a unit vector) to represent a certain force vector. There is actually no prejudgment about the sign of the scalar since the laws of mechanics and mathematics will dictate its sign. Because the practice is widespread, however, and because it provides such a concise means of communication, we use this weak form of the word "assume."

tensile by writing $F_{PQ} = 300$ lb ⓣ. This *second* use of the symbol F_{PQ} is not an algebraic statement, but merely a shorthand means of reporting that the force in PQ has been found to be tensile with magnitude 300 lb.

We are now ready for our first example, in which we shall solve (for all the forces in the members of) the truss of Figure 5.9(a).

EXAMPLE 5.1

Find the forces in each member of the truss in Figure E5.1a by the method of joints.

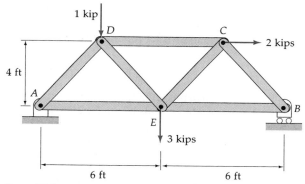

Figure E5.1a

Solution

We begin by finding the reactions at A and B, which are the forces exerted on the truss by its supports. Using the free-body diagram of the overall structure, we obtain (with F_B up at B, F_{Ay} up at A, and F_{Ax} to the left at A):

$$\curvearrowleft_+ \quad \Sigma M_A = 0 = (12 \text{ ft})F_B - (6 \text{ ft})(3 \text{ k}) - (3 \text{ ft})(1 \text{ k}) - (4 \text{ ft})(2 \text{ k})$$

Thus the reaction at the roller is

$$F_B = \frac{29}{12} = 2.42 \text{ kips}$$

Then,

$$+\uparrow \quad \Sigma F_y = 0 = F_{Ay} + \overset{2.42 \text{ k}}{\cancel{F_B}} - 1 \text{ k} - 3 \text{ k}$$

so that the vertical component of the pin reaction is

$$F_{Ay} = 1.58 \text{ k}$$

(The reader is encouraged to check this value of F_{A_y} by using $\Sigma M_B = 0$.) Continuing,

$$\xrightarrow{+} \quad \Sigma F_x = 0 = 2 \text{ k} - F_{Ax} \Rightarrow F_{Ax} = 2 \text{ k}$$

We are now ready to use the method of joints. We use the free-body diagram [Figure 5.9(b)] discussed earlier and repeated in Figure E5.1b. We obtain the forces F_{CB} and F_{EB} by satisfying the equilibrium equations of the pin:

Figure E5.1b

$$+\uparrow \quad \Sigma F_y = 0 = 2.42 - \frac{4}{5} F_{CB}$$

$$F_{CB} = \frac{5}{4}(2.42) = 3.03 \text{ k} \quad (\text{or } F_{CB} = 3.03 \text{ k } ©)$$

Because F_{CB} came out positive, the bar CB is in compression as assumed. Then,

$$\xrightarrow{+} \quad \Sigma F_x = 0 = \frac{3}{5} F_{CB} - F_{EB}$$

$$F_{EB} = 0.6(3.03) = 1.82 \text{ k} \quad (\text{or } F_{EB} = 1.82 \text{ k } ⓣ).$$

Again the answer came out positive; this time we had assumed FB to be in tension, so it actually is.

As we have mentioned, letters © and ⓣ beside the force in a truss respectively indicate to us whether the member is in compression or tension. In the case of a truss member, these letters tell much more than a direction arrow or even than a vector representation. For example, if we were to say

$$\mathbf{F}_{CB} = 3.03 \,^4\!\!\diagdown_3 \text{ kips} \quad \text{or} \quad 3.03 \left(\frac{3\hat{\mathbf{i}} - 4\hat{\mathbf{j}}}{5} \right) \text{ kips},$$

then this is OK if what is meant is the force exerted by CB onto the pin at B, *and* if such is stated clearly. But if this force vector was used as the force exerted *by* the pin onto CB at B, *or* as the force exerted onto the pin at C by CB, then the result would be 180° away from the correct direction. Therefore, a © beside the answer "$F_{CB} = 3.03$ kips" removes all this uncertainty.

Next we analyze the pin at C. Its free-body diagram is shown in Figure E5.1c. Enforcing the equilibrium of joint C,

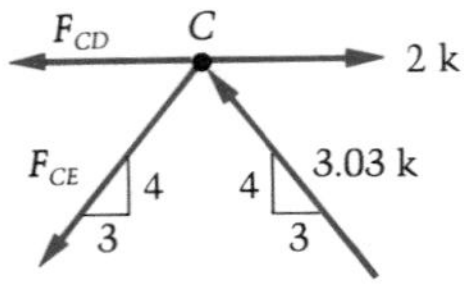

$$+\uparrow \quad \Sigma F_y = 0 = -F_{CE} \left(\frac{4}{5} \right) + 3.03 \left(\frac{4}{5} \right)$$

$$F_{CE} = 3.03 \text{ k} \quad (\text{or } F_{CE} = 3.03 \text{ k } ⓣ)$$

and

$$\xrightarrow{+} \quad \Sigma F_x = 0 = 2 - F_{CD} - 3.03 \left(\frac{3}{5} \right) - F_{CE} \left(\frac{3}{5} \right)$$

$$F_{CD} = -1.64 \text{ k} \quad (\text{or } F_{CD} = 1.64 \text{ k } ©)$$

Figure E5.1c

Question 5.1   Why do we use pin C prior to E at this stage of the solution?
Question 5.2   Why is the 3.03 kip force acting upward and to the left on pin C in Figure E5.1c?
Question 5.3   Why is the direction associated with F_{CE} in the diagram bound to be correct here, and not just a guess?

Answer 5.1   Pin C will at this stage have two unknown forces acting on it; pin E has four.
Answer 5.2   Because we found it to be compressive when we examined pin B. If we were to now draw it tensile on pin C, we would be in violation of the action-reaction principle!
Answer 5.3   Because unless F_{CE} is tensile, ΣF_y cannot vanish for pin C.

This time, we have encountered an incorrect guess for the first time. By the way we drew the force F_{CD} in Figure 5.1c, we had "assumed" member CD to be in tension. We see now that it is in compression, as evidenced by the solution $F_{CD} = -1.64$ k. Thus, the force in CD is expressed as 1.64 k ©.

We next examine pin D. Assuming members DA and DE to be in compression yields the free-body diagram in Figure E5.1d.

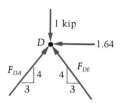

Figure E5.1d

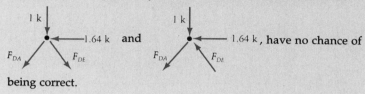

Question 5.4   Explain why these two sets of assumed directions:

and   , have no chance of

being correct.

Writing the equilibrium equations,

$$\xrightarrow{+} \quad \Sigma F_x = 0 = F_{DA}\left(\frac{3}{5}\right) - F_{DE}\left(\frac{3}{5}\right) - 1.64$$

$$+\uparrow \quad \Sigma F_y = 0 = F_{DA}\left(\frac{4}{5}\right) + F_{DE}\left(\frac{4}{5}\right) - 1$$

or

$$F_{DA} - F_{DE} = +\frac{5}{3}(1.64) = +2.73 \text{ k}$$

$$F_{DA} + F_{DE} = \frac{5}{4}(1) = 1.25 \text{ k}$$

Adding, we eliminate F_{DE} and find F_{DA}:

$$2F_{DA} = +3.98 \Rightarrow F_{DA} = 1.99 \text{ k}$$

Subtracting,

$$2F_{DE} = -1.48 \Rightarrow F_{DE} = -0.74 \text{ k}$$

Thus we have guessed the wrong direction for the force in DE. Because we assumed compression, it is actually in tension:

$$F_{DE} = 0.74 \text{ k } \textcircled{T} \qquad F_{DA} = 1.99 \text{ k } \textcircled{C}$$

At the pin at A in Figure E5.1e there is now but one unknown, which is the force in AE. Checking the equilibrium in the vertical direction, we see that

$$+\uparrow \quad \Sigma F_y = 1.58 - 1.99\left(\frac{4}{5}\right) = -0.01 \approx 0$$

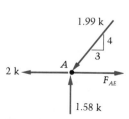

Figure E5.1e

the difference being due to numerical roundoff. In the x direction,

$$\xrightarrow{+} \quad \Sigma F_x = 0 = F_{AE} - 2 - 1.99\left(\frac{3}{5}\right)$$

$$F_{AE} = 3.19 \text{ kips} \qquad (\text{or } F_{AE} = 3.19 \text{ kips } \textcircled{T})$$

Answer 5.4   If both F_{DA} and F_{DE} were to turn out positive as indicated, then: (a) the first sketch leaves $\Sigma F_y \neq 0$; (b) the second sketch leaves $\Sigma F_x \neq 0$.

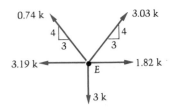

Figure E5.1f

Finally, we may use pin E as a check on the results. We come into it with calculated forces from five different directions (see Figure E5.1f).

The four bar forces in the free-body diagram were all computed to be in tension; thus, they each pull on the pin at E. Checking the equilibrium of the pin, we see that

$$\xrightarrow{+} \quad \Sigma F_x = 1.82 + 3.03 \left(\frac{3}{5}\right) - 0.74 \left(\frac{3}{5}\right) - 3.19 = 0.00$$

and

$$+\uparrow \quad \Sigma F_y = 3.03 \left(\frac{4}{5}\right) + 0.74 \left(\frac{4}{5}\right) - 3 = 0.02 \approx 0$$

Thus we have successfully solved the truss.

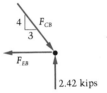

Figure 5.12

In the preceding example, the first free-body diagram was of the pin at B. (See Figure 5.12.) It could equally well have been drawn as shown in Figure 5.13, which is a free-body diagram of the connection, plus short, cut lengths of members CB and EB. With this approach, one sees more clearly the tension or compression in the bars.

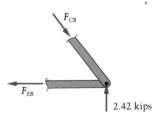

Figure 5.13

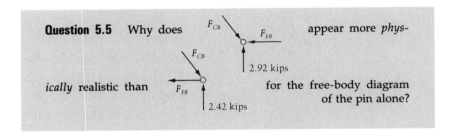

Question 5.5 Why does [F_{CB} F_{EB} 2.92 kips] appear more *phys-ically* realistic than [F_{CB} F_{EB} 2.42 kips] for the free-body diagram of the pin alone?

In Example 5.1 the reader may have noticed that no member weights were considered. If a structure is to be analyzed as a truss, then, by definition, its members can be loaded only by forces at their pinned ends. Weights of truss members will be neglected in the truss analyses in this book.

In our second example, we shall not explain each step in quite as much detail.

Answer 5.5 The force exerted by the inside of the hole (in member EB) onto the pin comes from the right side of the hole; that is, it pushes to the left instead of pulling to the left. Nonetheless, because forces may be translated along their lines of action, the free-body diagram of the pin is often drawn as in the second sketch.

EXAMPLE 5.2

Find the forces in the members of the truss in Figure E5.2a.

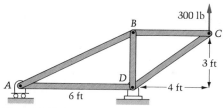

Figure E5.2a

Solution

We calculate the reactions, but only for use at the end in checking our solution. We use the overall free-body diagram; Figure E5.2b, and obtain:

$$\curvearrowleft_{+} \qquad \Sigma M_D = 0 = (300\text{ lb})(4\text{ ft}) - F_A\,(6\text{ ft})$$

$$F_A = 200\text{ lb} \qquad \text{(the roller reaction)}$$

$$\xrightarrow{+} \qquad \Sigma F_x = 0 = F_{Dx}$$

$$+\uparrow \qquad \Sigma F_y = 0 = F_{Dy} + F_A + 300$$

$$F_{Dy} = -300 - 200 \quad = -500\text{ lb}$$

Thus the pin reaction is 500 lb ↓. We now use the method of joints to determine the forces in the members. From a free-body diagram of pin C (see Figure E5.2c), we have

$$+\uparrow \qquad \Sigma F_y = 0 = 300 - F_{CD}\left(\frac{3}{5}\right) \Rightarrow F_{CD} = 500\text{ lb} \qquad \text{(or 500 lb ⓣ)}$$

$$\xrightarrow{+} \qquad \Sigma F_x = 0 = F_{BC} - F_{CD}\left(\frac{4}{5}\right) = F_{BC} - (500)\left(\frac{4}{5}\right)$$

$$F_{BC} = 400\text{ lb} \qquad \text{(or } F_{BC} = 400\text{ lb ⓒ)}$$

(Note that in finding F_{CD} or F_{BC}, we did not need the reactions.) Next, from a free-body diagram of pin B (Figure E5.2d), we obtain:

$$\xrightarrow{+} \qquad \Sigma F_x = 0 = F_{AB}\,\frac{2}{\sqrt{5}} - 400 \Rightarrow F_{AB} = 200\sqrt{5}\text{ lb (or } 200\sqrt{5}\text{ lb ⓒ)}$$

$$+\uparrow \qquad \Sigma F_y = 0 = F_{AB}\left(\frac{1}{\sqrt{5}}\right) - F_{BD} = 200 - F_{BD}$$

or

$$F_{BD} = 200\text{ lb} \qquad \text{(or 200 lb ⓣ)}$$

And from a free-body diagram of A (Figure E5.2e), we may write

$$\xrightarrow{+} \qquad \Sigma F_x = 0 = -200\sqrt{5}\,\frac{2}{\sqrt{5}} + F_{AD} \Rightarrow F_{AD} = 400\text{ lb} \qquad \text{(or } F_{AD} = 400\text{ lb ⓣ)}$$

and for a check, using the precalculated 200-lb roller reaction,

$$+\uparrow \qquad \Sigma F_y = 200 - 200\sqrt{5}\,\frac{1}{\sqrt{5}} = 0 \quad ✓$$

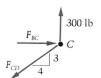

Figure E5.2b

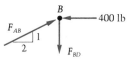

Figure E5.2c

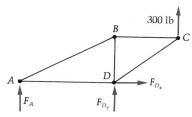

Figure E5.2d

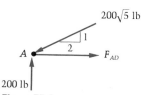

Figure E5.2e

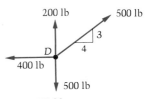

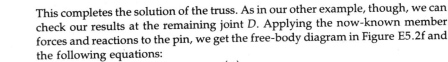

This completes the solution of the truss. As in our other example, though, we can check our results at the remaining joint D. Applying the now-known member forces and reactions to the pin, we get the free-body diagram in Figure E5.2f and the following equations:

$$\xrightarrow{+} \quad \Sigma F_x = 500 \left(\frac{4}{5}\right) - 400 = 0 \quad \checkmark$$

$$+\uparrow \quad \Sigma F_y = 200 + 500 \left(\frac{3}{5}\right) - 500 = 0 \quad \checkmark$$

Figure E5.2f

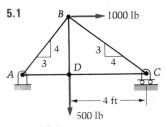

Because of the way its members are loaded, a truss is an extremely efficient, lightweight structure. To emphasize this idea, we encourage the reader to think about the difference between the weights W_1 and W_2 that could be safely supported by a yardstick in the two manners shown in Figure 5.14. In the diagram at the left, the member is loaded as it would be in a truss; at the right, the member is a beam and is not a two-force member because the reaction at the wall includes a couple and a force component normal to the member's axis.

We shall work through one more plane truss example with the method of joints in the next section, after first discussing some shortcuts that sometimes make truss analysis easier.

Figure 5.14

PROBLEMS ▶ Section 5.3

Find the forces in each member of the trusses in Problems 5.1–5.11.

Figure P5.1

5.1

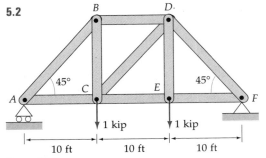

5.2

Figure P5.2

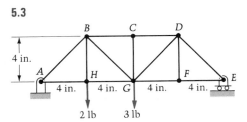

5.3

Figure P5.3

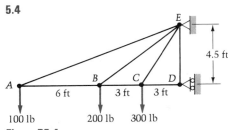

5.4

Figure P5.4

5.5

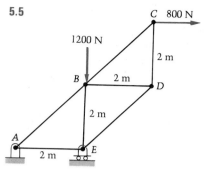

Figure P5.5

5.6

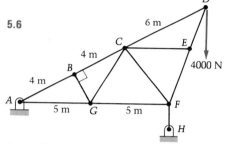

Figure P5.6

5.7

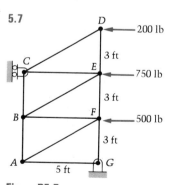

Figure P5.7

5.8

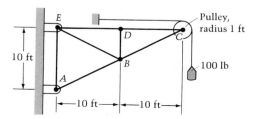

Figure P5.8

5.9

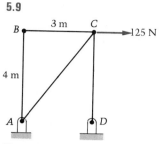

Figure P5.9

5.10

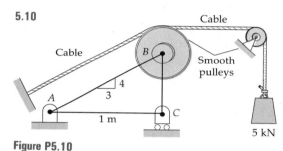

Figure P5.10

5.11

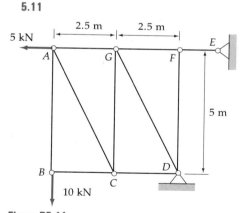

Figure P5.11

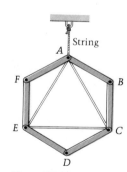

Figure P5.12

* **5.12** In Figure P5.12 the hexagon *ABCDEF* of six uniform pinned rods *AB, BC, CD, DE, EF,* and *FA* of equal lengths and weights *W,* is stiffened by light rods *AC, CE,* and *EA* and suspended by the string at *A.* Find the tension in *AC,* and then the upward force that must be applied at *D* to reduce the force in *AC* to zero.

---

* Asterisks identify the more difficult problems.

5.13 Members *AE* and *EQ* are pinned together as shown in Figure P5.13 to form a billboard. It is subjected to the given distributed wind load. Find the forces in each of the two-force members.

5.15 Find the forces in truss members *AB*, *BH*, and *OH* shown in Figure P5.15. Note that the external reactions are indeterminate and cannot be found by statics alone.

5.16 The truss in Figure P5.16 is pin-connected and is supported by the pin at *A*, and by the cable attached at *B* and *C*. The cable passes over a pulley that is connected to the reference frame by a smooth pin. Neglecting the weights of the truss members, find the force in *BD*.

⁕5.17 In Figure P5.17 find the forces in all the members of the truss, in terms of *W*.

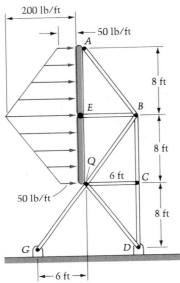

Figure P5.13

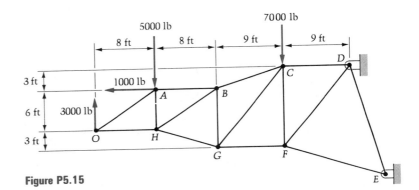

Figure P5.15

5.14 Find the force in member *BD* of the truss shown in Figure P5.14. Note that the external reactions are indeterminate and cannot be found by statics alone.

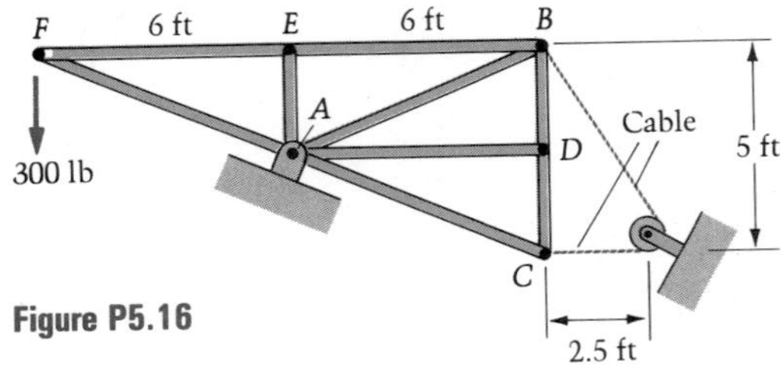

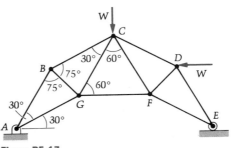

Figure P5.16

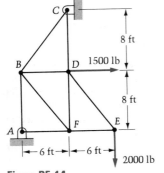

Figure P5.14

Figure P5.17

5.18 Determine the force in each member of the truss shown in Figure P5.18 if $a = 4$ ft, $b = 8$ ft, and $c = d = 3$ ft.

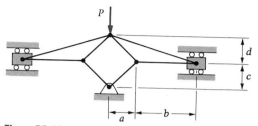

Figure P5.18

5.19 Find the forces in all members of the truss shown in Figure P5.19.

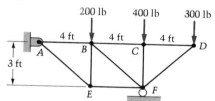

Figure P5.19

5.20 Find the force in member CG of the truss in Figure P5.20.

5.21 Find the forces in members AB, AF, AG, and CD of the truss shown in Figure P5.21.

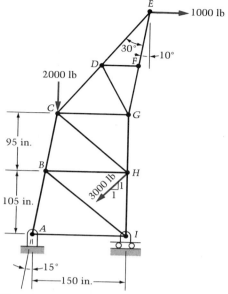

Figure P5.20

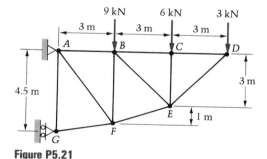

Figure P5.21

5.4 Shortcuts and Rigidity/Determinacy Results

Zero-force Members

There are several common shortcuts to watch for in analyzing a truss. The first involves what are called "zero-force members." Sometimes in a truss we will find at a joint J that only one member could carry a component of force in a certain direction and that there are no external loads at J in that direction. For example, at joint A in Figure 5.15, only member AE can have a force component normal to line CAD. If we call this normal direction u, then from the free-body diagram of pin A (see Figure 5.16 on the next page), it is seen that F_{AE} must be zero:

$$\searrow^+ \quad \Sigma F_u = 0 = F_{AE} \cos \theta$$

and so

$$F_{AE} = 0 \quad \text{(note } \theta \neq 90°)$$

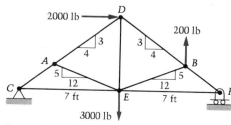

Figure 5.15

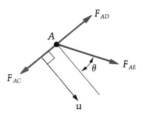

Figure 5.16

Thus *AE* is a "zero-force member" of the truss. In general, this only happens when there is no external loading at the pin with a component along the direction of the bar (such as *AE*) that is being examined for a possible zero value. For example, at *B* (Figure 5.17) we have only *BE* able

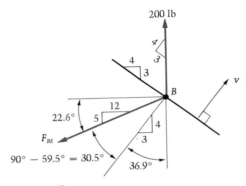

Figure 5.17

to carry a force normal to *DBF*. This time, however, this normal component balances the component of the 200-lb external force in that direction:

$$\nwarrow \quad \Sigma F_v = 0 = 200 \left(\frac{4}{5} \right) - F_{BE} \cos 30.5°$$

$$F_{BE} = 186 \text{ lb}$$

Even though $F_{BE} \neq 0$ in this case, its value is nonetheless determined from a single equation, which is still a help.

> **Question 5.6**   Give reasons why a zero-force member like *AE* in Figure 5.16 is still an important part of the truss even if it isn't carrying any load.

Answer 5.6   For *other* external loads, the member may carry a force. Also, the member really does carry load due to the (neglected) weights of the members. It also adds stability to the structure.

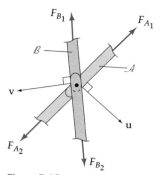

Figure 5.18

Other Shortcuts

A second shortcut in truss analysis arises in a situation where four members that are lined up in pairs as indicated in Figure 5.18 meet at a pin. If there are no other members or external forces at that joint, then we see from Figure 5.18 that the forces F_{A_1} and F_{A_2} are equal, from the equation $\Sigma F_v = 0$ applied to the free body shown. In the same way, $\Sigma F_u = 0$ gives $F_{B_1} = F_{B_2}$.

A third shortcut arises when two non-collinear members are joined at a pin where no other bar or external force appears (see Figure 5.19(a)). At pin B the equilibrium equations, written with the help of the free-body diagram, show that both F_{BA} and F_{BC} are zero (see Figure 5.19(b)):

$$\Sigma F_u = 0 = -F_{BC} \cos \theta$$
$$F_{BC} = 0$$
$$\Sigma F_v = 0 = -F_{BA} \cos \beta$$
$$F_{BA} = 0$$

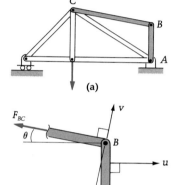

Figure 5.19

EXAMPLE 5.3

Find the forces in the members of the truss in Figure E5.3a.

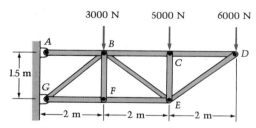

Figure E5.3a

Solution

We shall retain four digits in this example for numerical checking purposes. Two of the answers are known by inspection:

1. $F_{BF} = 0$, by considering $\Sigma F_y = 0$ at joint F.
2. $F_{CE} = 5000$ N Ⓒ, by considering $\Sigma F_y = 0$ at joint C.

In a simple truss such as this one, as we shall see later in general, we can always find a joint where only two unknown bar forces act. Here, it is joint D shown in Figure E5.3b:

$$+\uparrow \quad \Sigma F_y = 0 = F_{ED}\left(\frac{3}{5}\right) - 6000 \Rightarrow F_{ED} = 10{,}000 \text{ N Ⓒ}$$

$$\xrightarrow{+} \quad \Sigma F_x = 0 = F_{ED}\left(\frac{4}{5}\right) - F_{CD} = 10{,}000\left(\frac{4}{5}\right) - F_{CD}$$

or

$$F_{CD} = 8000 \text{ N} \qquad (\text{or } F_{CD} = 8000 \text{ N Ⓣ})$$

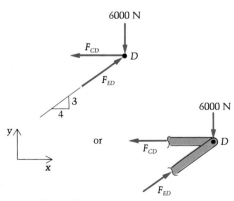

Figure E5.3b

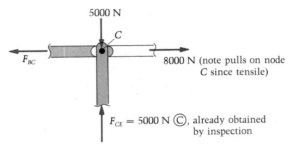

Figure E5.3c

Next, we isolate joint C (see Figure E5.3c), and obtain:

$$\xrightarrow{+} \quad \Sigma F_x = 0 = 8000 - F_{BC}$$

$$F_{BC} = 8000 \text{ N} \quad \text{(or 8000 N } \textcircled{T}\text{)}$$

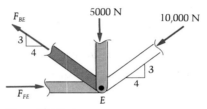

Figure E5.3d

We next use joint E (see Figure E5.3d), because at present B has three unknown forces acting on it.

Question 5.7   In this case, we know ahead of time that F_{BE} is $\textcircled{T}$ and F_{FE} is $\textcircled{C}$. Why?

The equilibrium equations are:

$$+\uparrow \quad \Sigma F_y = 0 = F_{BE}\left(\frac{3}{5}\right) - 10{,}000\left(\frac{3}{5}\right) - 5000$$

$$F_{BE} = 18{,}330 \text{ N} \quad \text{(or } F_{BE} = 18{,}330 \text{ N } \textcircled{T}\text{)}$$

$$\xrightarrow{+} \quad \Sigma F_x = 0 = F_{FE} - F_{BE}\left(\frac{4}{5}\right) - 10{,}000\left(\frac{4}{5}\right)$$

$$F_{FE} = 18{,}330(0.8) + 8000$$

$$F_{FE} = 22{,}660 \text{ N} \quad \text{(or } F_{FE} = 22{,}660 \text{ N } \textcircled{C}\text{)}$$

Answer 5.7   First, F_{BE} has to be tensile so that ΣF_y can be zero. Then F_{FE} has to be compressive in order that ΣF_x can vanish.

By inspection of joint F, the force in GF is the same as FE. This is because no other horizontal forces act on the pin at F, so that GF and FE have to equilibrate each other there. Thus,

$$F_{GF} = 22{,}660 \text{ N} \qquad \text{(or } F_{GF} = 22{,}660 \text{ N ©)}$$

At B, the pin has the forces acting as shown in Figure E5.3e. We obtain:

$$+\uparrow \qquad \Sigma F_y = 0 = F_{GB}\left(\frac{3}{5}\right) - 18{,}330\left(\frac{3}{5}\right) - 3000$$

$$F_{GB} = 23{,}330 \text{ N} \qquad \text{(or } F_{GB} = 23{,}330 \text{ N ©)}$$

$$\xrightarrow{+} \qquad \Sigma F_x = 0 = F_{GB}\left(\frac{4}{5}\right) + 18{,}330\left(\frac{4}{5}\right) + 8000 - F_{AB}$$

$$F_{AB} = 41{,}330 \text{ N} \qquad \text{(or } F_{AB} = 41{,}330 \text{ N Ⓣ)}$$

In practice, truss members such as AB that carry much more load than others for typical expected loadings will be made larger in cross section.

Finally, free-body diagrams of the pins at A and G (see Figures E5.3f and E5.3g) allow us to compute the reactions there (onto the pins from the clevis attached to the wall).

$$+\uparrow \qquad \Sigma F_y = 0 \Rightarrow A_y = 0$$

$$\xrightarrow{+} \qquad \Sigma F_x = 0 \Rightarrow A_x = 41{,}330 \text{ N}$$

$$\xrightarrow{+} \qquad \Sigma F_x = 0 = G_x - 22{,}660 - 23{,}330\left(\frac{4}{5}\right)$$

$$G_x = 41{,}320 \text{ N}$$

$$+\uparrow \qquad \Sigma F_y = 0 = G_y - 23{,}330\left(\frac{3}{5}\right)$$

$$G_y = 14{,}000 \text{ N}$$

Note that the pin at G actually *feels* the vector sum of G_x and G_y in shear; that is, the clevis exerts on it the force

$$\mathbf{G} = 41{,}320\hat{\mathbf{i}} + 14{,}000\hat{\mathbf{j}} \text{ N} \qquad \text{(or } 43{,}630 \quad \angle 18.72° \text{ N)}$$

Back-checking with the overall free-body diagram shown in Figure E5.3h, we see that our calculations look correct:

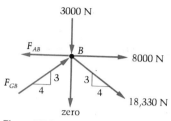

3000 N

F_{AB}

B

8000 N

F_{GB} 3   3

4   4

18,330 N

zero

Figure E5.3e

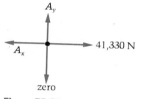

A_y

A_x 41,330 N

zero

Figure E5.3f

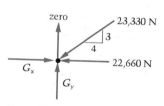

zero   23,330 N

3

4

G_x 22,660 N

G_y

Figure E5.3g

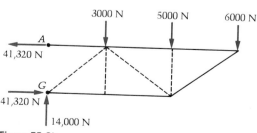

3000 N   5000 N   6000 N

A

41,320 N

G

41,320 N

14,000 N

Figure E5.3h

It is obvious by inspection that ΣF_x and ΣF_y each vanish. Checking moments,

$$\curvearrowleft_+ \quad \Sigma M_G = (41{,}320 \text{ N})(1.5 \text{ m})$$
$$- (3000 \text{ N})(2 \text{ m}) - (5000 \text{ N})(4 \text{ m}) - (6000 \text{ N})(6 \text{ m})$$
$$= -20 \text{ N} \cdot \text{m} \quad \text{(differing from zero due to rounding to four digits)}$$

Note from the final figure that when considering the entire "overall" truss as a free-body diagram, one need not take time to draw in all the dashed internal members; the outside profile is sufficient unless there are forces applied at "internal" joints. We must, of course, remember that all the elements of the truss are actually being included.

Numbers of Members and Pins in a Truss

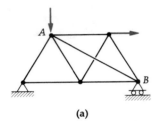

(a)

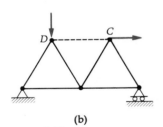

(b)

Figure 5.20

In Example 5.1 we recall that there were three equations used as checks (one at A and two at E) because they involved no new unknowns. These three redundant equations resulted because there are obviously 2 × (number of pins) = $2p$ independent equilibrium equations, and we used up three independent equations in finding the external reactions (using the overall free-body diagram). If a plane truss with three statically determinate reactions has more than $m = 2p - 3$ members, then we cannot solve for the forces in them all from the equilibrium equations alone, and the truss is then appropriately deemed statically indeterminate.* Though we may find the reactions, we cannot solve the truss if $m > 2p - 3$. On the other hand, if there are *fewer* than $2p - 3$ members with three statically determinate reactions, we then do not have enough member forces to satisfy all the equilibrium equations at the pins. In this case, the truss is not rigid. Figures 5.20(a) and 5.20(b) illustrate these two ideas. In Figure 5.20(a), adding member AB to the truss of Figure 5.6 makes the truss statically indeterminate; p still = 5, but m now = 8 and $2p - 3 = 7$, which is now <m. In Figure 5.20(b), on the other hand, removing member DC gives $m = 6$ while p still = 5, so that $2p - 3 = 7$, which is now >m. This leaves a non-rigid structure, which is unstable and will collapse as the reader may visualize.

Of course, the word *rigid* does not mean that a truss will not deform at all under loading. It will undergo very small deformations, very nearly retaining its original shape, as we shall see in Section 5.7.

We have seen that in a two-dimensional (or plane) truss, the pin connections are called joints (or nodes). They are the points where the members, or bars, that constitute the truss are joined together. If, for example, we begin with a single triangle (Figure 5.21(a)), it has three pins (p) (or joints) and three members (m). Thus $p = 3$ and $m = 3$. If we add,

---

* In Section 5.7, we will see how an indeterminate structure *can* be solved if *deformations* are included in the analysis.

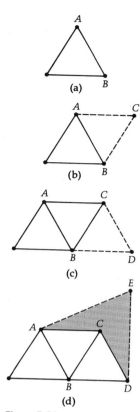

Figure 5.21

from two distinct joints A and B of this triangle, two new members and pin them at a new, common joint C not on line AB, then we have (see Figure 5.21(b)):

$$p = 3 + 1$$
$$m = 3 + 2$$

Repeating this procedure once more, we obtain a truss similar to that of Figure 5.6 (see Figure 5.21(c)):

$$p = 3 + 1 + 1 = 5$$
$$m = 3 + 2 + 2 = 7$$

For a plane truss constructed in this way (and they often are), the number of pins (joints) and the number of members are related by

$$m = 3 + 2(p - 3) = 2p - 3$$

a relation we have seen before. For the last truss, $m = 2(5) - 3 = 7$. A plane truss constructed in the above manner is called a **simple truss.** If a simple truss is supported with a pin and roller (or their equivalent) so as to satisfy overall equilibrium for any loading, it will be a rigid, stable, determinate structure. Note in Figure 5.21(d) that the simple plane truss need not be made up of a series of connected triangles (though it often is). Note also that in a simple plane truss, we can always find at least one joint where there are only two unknown member forces.

Question 5.8   How?

As we have already seen, this is a natural starting place for solving such a truss by the method of joints.

It is important to note that the condition $m = 2p - 3$ is not generally either sufficient or necessary for a *non*-simple plane truss to be rigid and statically determinate. For example, the (silly) truss shown in Figure 5.22 has $m = 2p - 3$ but is not rigid. The truss shown in Figure 5.23, however, has $m \neq 2p - 3$ but is, while not rigid, both stable *and* statically determinate. (Note the *two* support pins!)

There are also well-known and commonly used plane trusses that with three determinate support reactions are both rigid and statically determinate but not simple (buildable from a triangle by successively adding two new members and one new pin at a time). One such truss is shown in Figure 5.24. Note again that the equation $m = 2p - 3$ does not guarantee a truss to be simple.

The truss of Figure 5.24 is called a compound truss, which is a truss comprising two or more simple trusses connected together so as to leave a rigid, determinate truss as the result. If the connection is made as

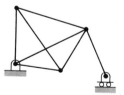

Figure 5.22 $p = 5$; $m = 7$
$= 2p - 3$

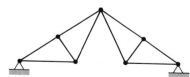

Figure 5.23 $p = 7$; $m = 10 \neq 2p - 3$

Answer 5.8   The last vertex drawn has but two members terminating at it.

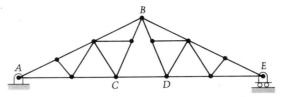

Figure 5.24   A Fink truss. $p = 15$; $m = 27 = 2p - 3$

in Figure 5.24—that is, by joining the simple trusses ABC and BED at B and adding one member (CD)—then, in general, if there are p_{ABC} pins and m_{ABC} members in ABC, and p_{BDE} pins and m_{BDE} members in BDE, we have

$$2p_{ABC} - 3 = m_{ABC} \qquad \text{and} \qquad 2p_{BDE} - 3 = m_{BDE}$$

and we see that for the combined truss,

$$\begin{aligned}
2p - 3 &= 2(p_{ABC} + p_{BDE} - 1) - 3 \\
&= (3 + m_{ABC}) + (3 + m_{BDE}) - 2 - 3 \\
&= m_{ABC} + m_{BDE} + 1
\end{aligned}$$

or

$$2p - 3 = m$$

Thus $m = 2p - 3$ even though (as the reader may wish to show) the compound truss is itself not simple. With proper support reactions such as a pin and a roller, a compound truss, like the simple trusses of which it is made, will be rigid (non-collapsible upon release from its supports) and statically determinate.

> **Question 5.9**   Three of the entries in Figure 5.2, if made only of pinned-together two-force members, would not be able to maintain their shapes if detached from their supports. Which ones are they?

Answer 5.9   Attic, gable end, floor truss.

PROBLEMS ▶ Section 5.4

5.22   Find the forces in members GB and DF of the truss shown in Figure P5.22.

5.23   Find the forces in all members of the truss shown in Figure P5.23.

5.24   By inspection, identify six zero-force members in Figure P5.24 and explain why each vanishes.

5.25   Find the forces in members 1, 2, 3, 4, and 5 of the truss shown in Figure P5.25.

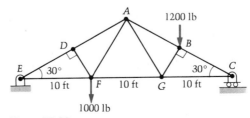

Figure P5.22

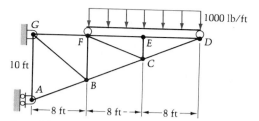

Figure P5.23

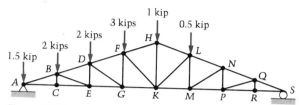

Figure P5.24

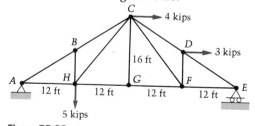

Figure P5.25

5.26 Find the forces in members *AB*, *BH*, *BC*, and *DF* of the truss shown in Figure P5.26.

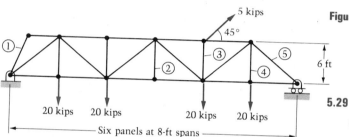

Figure P5.26

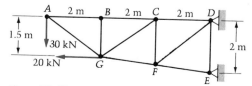

Figure P5.27

5.27 Find the forces in members *AG*, *BG*, and *CG* of the truss shown in Figure P5.27.

5.28 Find the force in each of the members of the truss in Figure P5.28.

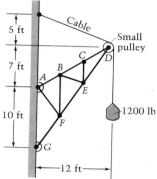

Figure P5.28

5.29

a. How many zero-force members can be found by inspection for the truss shown in Figure P5.29?

b. Determine the force in member *HF*.

c. The roller at *G* is replaced by a pin, and the pin at *A* is replaced by a roller. With no calculations, argue by means of a free-body diagram of pin *A* that the force in *AJ* will have changed.

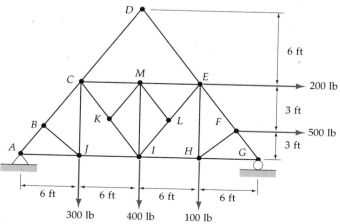

Figure P5.29

5.30 For the truss in Figure P5.30.

   a. Show that half the members are zero-force members for the given loading;

   b. find the reaction at G;

   c. find the force in member DF.

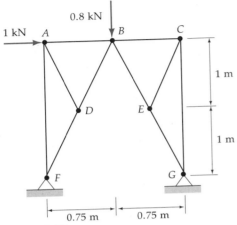

Figure P5.30

5.31 Show with a series of sketches that the K-truss shown in Figure P5.31 is a simple truss.

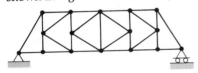

Figure P5.31   K-truss. $p = 16$; $m = 29 = 2p - 3$

5.32 Show that the Baltimore truss in Figure P5.32 is not a simple truss, even though $m = 2p - 3$.

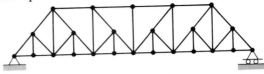

Figure P5.32   Baltimore truss. $p = 24$; $m = 45 = 2p - 3$

5.33 Show that if two simple trusses are joined to form a compound truss by simply bringing two joints together as shown in Figure P5.33 and omitting bar $B'C'$, the resulting truss will have $m = 2p - 3$.

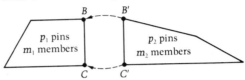

Figure P5.33

5.34 Two simple trusses are connected by the dashed bars in Figure P5.34 to form a compound truss. Prove that $m = 2p - 3$ for this compound truss.

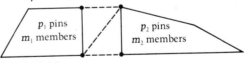

Figure P5.34

5.5 The Method of Sections

Another method commonly used to determine the forces in the members of a truss is called the method of sections. This name comes from the fact that in using this method, the truss is divided into two sections, which are both in equilibrium. (*If a body is in equilibrium, any and all parts of it are.*)

The advantage in using the method of sections is that member forces of interest may be found very quickly without solving the entire truss. For example, suppose we wished to know only the force in member GH of the (symmetrical) truss shown in Figure 5.25. By the method of joints, we would probably arrive at F_{GH} by:

1. Finding the roller reaction at F by $\Sigma M_A = 0$ on the overall truss.

2. Finding F_{EF} and F_{GF} by enforcing equilibrium of pin F.

3. Finding F_{GE} by enforcing equilibrium of pin E.

4. Finding F_{GH} by enforcing equilibrium of pin G.

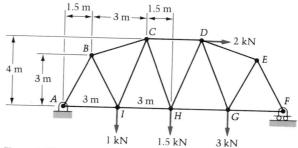

Figure 5.25

Cutting the Truss into Two Distinct Sections

A much quicker way to determine F_{GH} is to use the method of sections, which will be described in the example to follow. In this method, we cut the truss into two separate sections. One of the cut members is the one whose force we seek. This force thus appears as an *external* force on each of the two "halves" of the cut truss. If only three members have been cut and the external reactions are known, then the three equilibrium equations for either "half" will yield the desired member force. Sometimes we can find it by just summing moments about the point of intersection (if there is one) of the other unknown forces; this is the case in the following example.

EXAMPLE 5.4

Determine the force in member GH of the truss shown in Figure E5.4a.

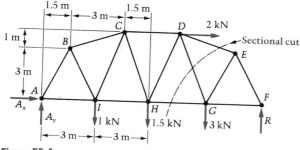

Figure E5.4a

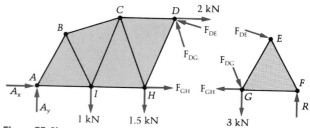

Figure E5.4b

Solution

In Figure E5.4b, we have divided the truss into two sections. In each section one of the unknown forces exposed by the cut is the desired F_{GH}.

As we mentioned (following Example 5.1), we can think of the forces of the cut members in either of the ways shown in Figure E5.4c at the top of the next page.

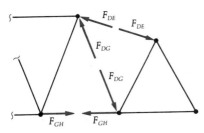

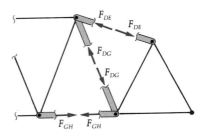

Figure E5.4c

The three bars DE, DG, and GH have been removed and replaced by the forces they exert on their respective pins.

The bars are still there, but they are each cut in two. Thus their internal (axial) forces become exposed and are *external* to the separate sections, each of which is in equilibrium alone.

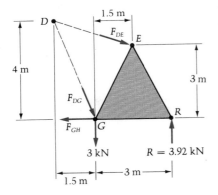

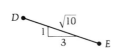

Figure E5.4d

Summing moments about A, we find the roller reaction on the overall (uncut!) free-body diagram (Figure E5.4a):

$$\circlearrowleft^+ \quad \Sigma M_A = 0 = R(12\ \text{m}) - (1\ \text{kN})(3\ \text{m}) - (1.5\ \text{kN})(6\ \text{m})$$
$$- (3\ \text{kN})(9\ \text{m}) - (2\ \text{kN})(4\ \text{m})$$

$$R = \frac{47}{12} = 3.92\ \text{kN}$$

We are ready to use one of the cut sections of Figure E5.4b to find F_{GH}. We see that if we use the section in Figure E5.4d and simply equate the sum of the moments about D to zero, we shall find the desired force in a single step. This is because the other two unknown forces F_{DG} and F_{DE} both pass through point D, and thus the only unknown appearing in the moment equation will be the desired F_{GH}:

$$\circlearrowleft^+ \quad \Sigma M_D = 0 = -F_{GH}(4\ \text{m}) - (3\ \text{kN})(1.5\ \text{m}) + (3.92\ \text{kN})(4.5\ \text{m})$$

$$F_{GH} = 3.29\ \text{kN}$$

(The force is tensile because it was drawn that way and the scalar F_{GH} turned out to be positive in the solution.)

Note that if we were looking for F_{DE}, the same cut section could be used, and the summation of moments about G would give that force, again in one step (i.e., one equation in the single desired unknown)*.

$$\circlearrowleft^+ \quad \Sigma M_G = 0 = (3.92\ \text{kN})(3\text{m}) - \left(F_{DE}\frac{3}{\sqrt{10}}\right)(3\ \text{m}) - \left(F_{DE}\frac{1}{\sqrt{10}}\right)(1.5\ \text{m})$$

$$F_{DE} = 3.54\ \text{kN}$$

(The force is compressive because it was drawn that way and the scalar F_{DE} turned out positive in the solution.)

---

* We could, once F_{GH} has been found, solve for F_{DE} and F_{DG} by $\Sigma F_x = 0$ together with $\Sigma F_y = 0$. But $\Sigma M_G = 0$ gives F_{DE} in one equation, even if F_{GH} has not been previously determined. Thus errors we may have made in F_{GH} will not propagate into our solution for F_{DE}.

Figure E5.4e

In the $\Sigma M_G = 0$ equation, note that the force in DE was resolved into its two components at E. The horizontal component, $F_{DE}(3/\sqrt{10})$, has a moment arm (or perpendicular distance) from G of 3 m. The vertical component, $F_{DE}(1/\sqrt{10})$, has a moment arm of 1.5 m. The reader should note that if the point Q in Figure E5.4e is used, then the vertical component of F_{DE} has no moment about G and the computation is shortened to

$$\curvearrowleft+ \quad \Sigma M_G = (3.92 \text{ kN})(3 \text{ m}) - \left(F_{DE}\frac{3}{\sqrt{10}}\right)(3.5 \text{ m})$$

$$F_{DE} = 3.54 \text{ kN} \qquad (\text{or } 3.54 \text{ kN } \copyright), \text{ again}$$

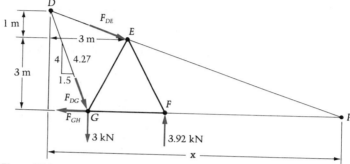

Figure E5.4f

If we wanted F_{DG}, we could now solve for it from $\Sigma F_x = 0$ or $\Sigma F_y = 0$. But, to further emphasize the idea of summing moments about the point of intersection P of other unknown forces, we find the intersection of F_{DE} and F_{GH} as shown in Figure E5.4f above. By similar triangles,

$$\frac{1}{3} = \frac{4}{x} \qquad \text{or} \qquad x = 12 \text{ m}$$

Therefore,

$$\curvearrowleft+ \quad \Sigma M_P = 0 = (3 \text{ kN})(10.5 \text{ m}) - (3.92 \text{ kN})(7.5 \text{ m}) + F_{DG}\left(\frac{4}{4.27}\right)(10.5 \text{ m})$$

$$F_{DG} = -0.214 \text{ kN} \qquad \text{or} \qquad 0.214 \text{ kN } \textcircled{T}$$

Let us now check our three results by using the force equilibrium equations on the cut section:

$$\xrightarrow{+} \quad \Sigma F_x = F_{DE}\frac{3}{\sqrt{10}} + F_{DG}\frac{1.5}{4.27} - F_{GH}$$

$$= 3.54(0.949) + (-0.214)(0.351) - 3.29$$

$$= -0.006 \approx 0 \qquad (\text{roundoff!})$$

and

$$+\uparrow \quad \Sigma F_y = 3.92 - 3 - F_{DE}\frac{1}{\sqrt{10}} - F_{DG}\frac{4}{4.27}$$

$$= 0.92 - 3.54(0.316) + 0.214(0.937)$$

$$= 0.002 \approx 0$$

The reader is encouraged to find the three forces F_{DE}, F_{DG}, and F_{GH} using the *left* section of Figure E5.4b, after determining A_x and A_y from the equilibrium equations for the overall free-body diagram. The answers for the three forces should, of course, agree with those found above.

Question 5.10   Why would a sectional cut that only went partly through a truss (i.e., didn't divide it into two parts) be useless?

EXAMPLE 5.5

Find the force in member *HI* of the truss shown in Figure E5.5a.

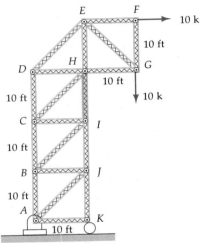

Figure E5.5a

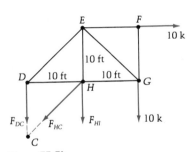

Figure E5.5b

Solution

We make a horizontal cut through the truss that exposes the force in the member *HI*, as suggested by the free-body diagram of Figure E5.5b. In this problem, we use the "upper half" of the truss because by so doing we avoid having to find the reactions. We see that summing moments at point *C* will eliminate the unknown forces F_{DC} and F_{HC} and allow an immediate solution for F_{HI}:

$$\circlearrowleft_+ \quad \Sigma M_C = 0 = -(10 \text{ k})(20 \text{ ft}) - (10 \text{ k})(20 \text{ ft}) - F_{HI}(10 \text{ ft})$$

$$F_{HI} = -40 \text{ k}$$

Because the force in member *HI* was drawn as if the member were in tension, and F_{HI} turned out negative, the force in *HI* is 40 kips Ⓒ.

Answer 5.10   Depicted as external forces on any legitimate free-body diagram thus produced would be a cancelling pair of forces associated with each cut member. Only the whole truss can be used as a free-body unless a part of it is completely cut away.

EXAMPLE 5.6

Find the forces in members *EF*, *RF*, *RN*, and *ON* of the K-truss shown in Figure E5.6a.

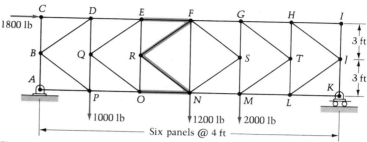

Figure E5.6a

Solution

In this truss we will not be able to cut sections and find all the desired forces one by one. For instance, a mental vertical cut through the four bars of interest shows that on each resulting part we will have three equations and four unknowns. Furthermore, there is no point where three of the four bars intersect. However, we shall use another section to find two of the four member forces, then employ force equilibrium on a second section to complete the solution.

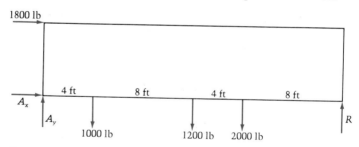

Figure E5.6b

First, though, we find the reactions; we use the overall *FBD* in Figure E5.6b:

$$\curvearrowleft_+ \quad \Sigma M_A = 0 = R(24 \text{ ft}) - (1000 \text{ lb})(4 \text{ ft}) - (1200 \text{ lb})(12 \text{ ft})$$
$$- (2000 \text{ lb})(16 \text{ ft}) - (1800 \text{ lb})(6 \text{ ft})$$
$$R = 2550 \text{ lb}$$

$$+\uparrow \quad \Sigma F_y = 0 = A_y + 2550 - 1000 - 1200 - 2000$$
$$A_y = 1650 \text{ lb}$$

$$\xrightarrow{+} \quad \Sigma F_x = 0 = A_x + 1800$$
$$A_x = -1800 \text{ lb}$$

Next, we consider the sectioning of the truss shown in Figure E5.6c on the following page.

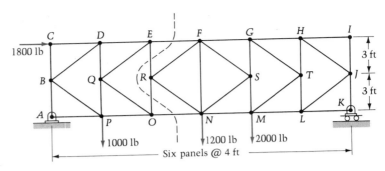

Figure E5.6c

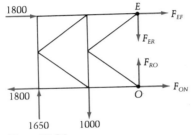

Figure E5.6d

Using the left section (see Figure E5.6d), we can determine two of the desired member forces: $\Sigma M_E = 0$ will give us F_{ON}; $\Sigma F_x = 0$ will then give us F_{EF}. Thus,

$$\circlearrowleft \quad \Sigma M_E = 0 = F_{ON}(6 \text{ ft}) + (1000 \text{ lb})(4 \text{ ft}) - (1650 \text{ lb})(8 \text{ ft}) - (1800 \text{ lb})(6 \text{ ft})$$

$$F_{ON} = 3330 \text{ lb} \qquad (\text{or } F_{ON} = 3330 \text{ lb } \textcircled{T})$$

and

$$\xrightarrow{+} \quad \Sigma F_x = 0 = F_{EF} + F_{ON} + 1800 - 1800$$

$$F_{EF} = -F_{ON} = -3330 \text{ lb} \qquad (\text{or } F_{EF} = 3330 \text{ lb } \textcircled{C})$$

Now we examine the previously mentioned vertical cut shown in Figure E5.5e. By inspection, the forces F_{RF} and F_{NR} have equal magnitudes. This is because all the other external forces with x components are in balance; thus the horizontal components of F_{RF} and F_{NR} must cancel. Because these bars make the same angle $[\cos^{-1}(\frac{4}{5})]$ with the horizontal through R, then F_{RF} must be equal to F_{NR}. As to whether they are

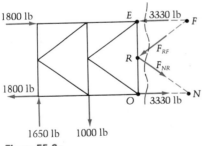

Figure E5.6e

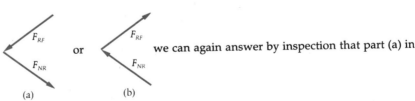

(a)   or   (b)

we can again answer by inspection that part (a) in

the figure is the only possibility. This is because the vertical components of these two forces must add to $1650 - 1000 = 650$ lb downward. Therefore,

$$+\uparrow \quad \Sigma F_y = 0 = 1650 - 1000 - F_{RF}\left(\frac{3}{5}\right)2$$

or

$$F_{RF} = 542 \text{ lb} \qquad (\text{or } 542 \text{ lb } \textcircled{C})$$

so that

$$F_{NR} = 542 \text{ lb} \qquad (\text{or } 542 \text{ lb } \textcircled{T})$$

It is unnecessary (as we have seen a number of times) to try to determine in advance whether an unknown load is $\textcircled{T}$ or $\textcircled{C}$ as we are drawing the free-body diagram. However, it is very instructive and improves the student's feel for the equilibrium analysis.

The other section can always be used as a check on our solutions for the bar forces; using Figure E5.6f, we see:

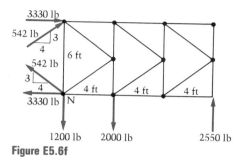

3330 lb

542 lb

542 lb

3330 lb   N

6 ft

4 ft   4 ft   4 ft

1200 lb   2000 lb   2550 lb

Figure E5.6f

$$\xrightarrow{+} \quad \Sigma F_x = 0 \quad \text{(by inspection)}$$

$$+\uparrow \quad \Sigma F_y = 542 \left(\frac{3}{5}\right) 2 - 1200 - 2000 + 2550$$

$$= 0.40 \text{ lb} \approx 0$$

$$\curvearrowright_+ \quad \Sigma M_N = -(3330 \text{ lb})(6 \text{ ft}) - (542 \text{ lb})\left(\frac{4}{5}\right)(6 \text{ ft})$$

$$-(2000 \text{ lb})(4 \text{ ft}) + (2550 \text{ lb})(12 \text{ ft})$$

$$= 18.4 \text{ lb-ft}$$

These values differ from zero due to roundoff error. (If four digits are retained instead of three, then ΣM_N becomes less than 0.5 lb-ft.)

EXAMPLE 5.7

Find the force in member EL, the only member of the truss in Figure E5.7a that has a length other than 1 m.

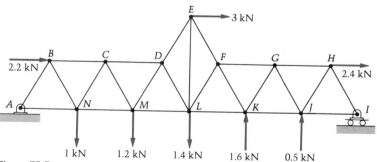

Figure E5.7a

Solution

Except for isolating joint E where there are three unknown bar forces acting, any complete section through EL will have to cut at least three other members that are not all concurrent. Thus the solution cannot be obtained with just one moment equation using the method of sections; nor could we obtain it by a combination of equilibrium equations from one sectioning. We shall have to use a combination of the methods of sections and joints before arriving at F_{EL}.

First, we find the roller reaction from equilibrium of the overall truss shown in Figure E5.7b on the following page.

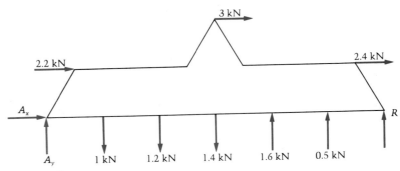

Figure E5.7b

$$\text{\Large$\curvearrowleft$}_{+} \qquad \Sigma M_A = 0 = R(6\text{ m}) + (0.5\text{ kN})(5\text{ m}) + (1.6\text{ kN})(4\text{ m})$$
$$- (1.4\text{ kN})(3\text{ m}) - (1.2\text{ kN})(2\text{ m}) - (1\text{ kN})(1\text{ m})$$
$$- (2.4 + 2.2)\text{ kN}\ (1\sin 60°\text{ m})$$
$$- (3\text{ kN})(2\sin 60°\text{ m})$$
$$R = 1.31\text{ kN}$$

Next we cut a section that will allow us to find F_{DE} (see Figure E5.7c). Then we can isolate pin E and find F_{EL}.

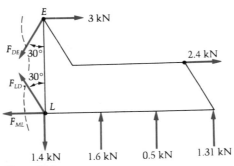

Figure E5.7c

$$\text{\Large$\curvearrowleft$}_{+} \qquad \Sigma M_L = 0 = (1.6\text{ kN})(1\text{ m}) + (0.5\text{ kN})(2\text{ m}) + (1.31\text{ kN})(3\text{ m})$$
$$- (2.4\text{ kN})(1\sin 60°\text{ m})$$
$$- (3\text{ kN})(2\sin 60°\text{ m}) + (F_{DE}\sin 30°)(2\sin 60°\text{ m})$$
$$F_{DE} = 0.860\text{ kN} \qquad (\text{or } 0.860\text{ kN }\text{\textcircled{T}})$$

Now we shift our attention to the pin at E shown in Figure E5.7d, using the method of joints to find F_{EL}.

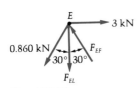

Figure E5.7d

$$\xrightarrow{+} \qquad \Sigma F_x = 0 = 3 - 0.860\sin 30° - F_{EF}\sin 30°$$
$$F_{EF} = 5.14\text{ kN} \qquad (\text{or } 5.14\text{ kN }\text{\textcircled{C}})$$

$$+\uparrow \qquad \Sigma F_y = 0 = -0.860\cos 30° + \overset{5.14}{\cancel{F_{EF}}}\cos 30° - F_{EL}$$
$$F_{EL} = 3.71\text{ kN} \qquad (\text{or } 3.71\text{ kN }\text{\textcircled{T}})$$

As a check, consider the free-body diagram* shown below in Figure E5.7e:

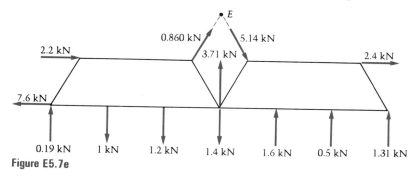

Figure E5.7e

$$\xrightarrow{+} \quad \Sigma F_x = (2.2 + 2.4 - 7.6 + 5.14 \cos 60° + 0.860 \cos 60°) \text{ kN}$$
$$= 0 \ \checkmark$$

$$+\uparrow \quad \Sigma F_y = (0.19 - 1 - 1.2 - 1.4 + 1.6 + 0.5 + 1.31 + 3.71$$
$$+ \ 0.860 \sin 60° - 5.14 \sin 60°) \text{ kN}$$
$$= 0.003 \text{ kN} \approx 0 \ \checkmark$$

$$\curvearrowleft_+ \quad \Sigma M_E = (2.2 + 2.4) \text{ kN } (1 \sin 60° \text{ m}) + (1.31 \text{ kN})(3 \text{ m})$$
$$- \ (0.19 \text{ kN})(3 \text{ m}) - (7.6 \text{ kN})(2 \sin 60° \text{ m})$$
$$+ \ (1.6 + 1.2) \text{ kN } (1 \text{ m}) + (0.5 + 1) \text{ kN } (2 \text{ m})$$
$$= -0.02 \text{ kN} \cdot \text{m} \approx 0 \ \checkmark$$

We note that these checks will *usually* indicate an error. However, they are only necessary conditions for equilibrium and are *no guarantee* that an error has not been made. For instance, if the 0.860 kN ⓉT force F_{DE} had been incorrectly drawn as compressive on Figure E5.7d, erroneous results would have resulted for F_{EF} and F_{EL}. Applying these (with $F_{DE} = 0.860$ kN ©C) onto the free-body diagram of Figure E5.7e would interestingly *and erroneously* have given the same three "checks" in this case!

Question 5.11   Return to the place in the solution where the cut was made that resulted in the solution for F_{DE}. Describe a different sectioning that would have allowed us to find a *different* force at E by means of a simpler free-body diagram (with one less contribution to ΣM_L).

Answer 5.11   Use a vertical cut through *EF*, *FL*, and *LK*. $\Sigma M_L = 0$ now gives F_{EF}, and the 3-kN force is not used.

PROBLEMS ▶ Section 5.5

5.35 Each truss member in Figure P5.35 has length 2 m. What is the force in *CD*?

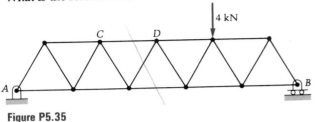

Figure P5.35

5.36 Find the force in members *BD* and *CD* of the truss shown in Figure P5.36.

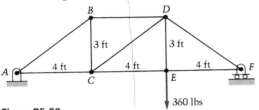

Figure P5.36

5.37 Determine the force in member *GH* of the truss shown in Figure P5.37.

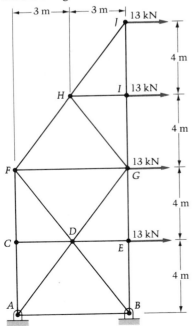

Figure P5.37

5.38 Find the forces in members *BC*, *BH*, and *AH* of the truss, loaded as shown in Figure P5.38.

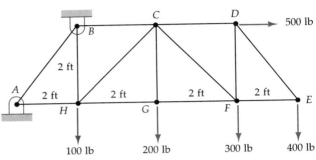

Figure P5.38

5.39 Find the forces in the darkened members of the trusses in Problems 5.39–5.43.

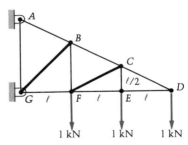

Figure P5.39

* 5.40

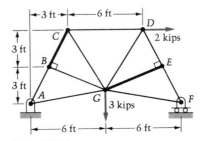

Figure P5.40

5.41

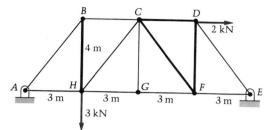

Figure P5.41

5.42

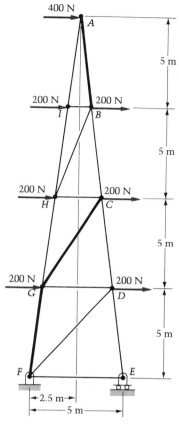

Figure P5.42

5.43

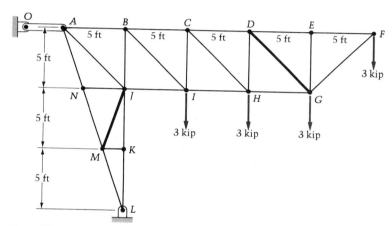

Figure P5.43

5.44 Find the force in members *FH*, *IH*, and *JK* of the truss shown in Figure P5.44.

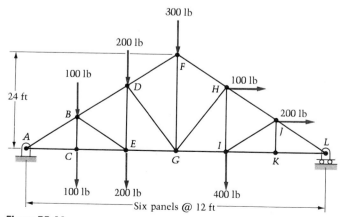

Figure P5.44

5.45 For the pin connected truss shown in Figure P5.45:

a. Using the method of joints, find the force in member *BC*.

b. Using the method of sections, find the force in member *CE*.

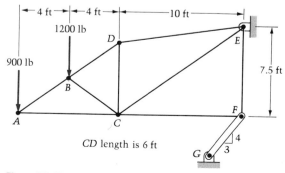

Figure P5.45

5.46 Find the forces in members *IE, JD, KJ*, and *CJ* of the pin-connected truss shown in Figure P5.46.

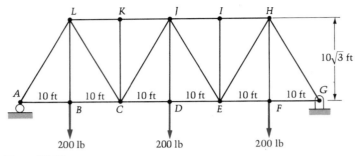

Figure P5.46

5.47 Find the forces in members *CF* and *AF* shown in Figure P5.47.

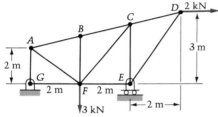

Figure P5.47

5.48 In Figure P5.48 find the forces in members *AD, ED,* and *AB*. Can the forces in all members of this truss be found from statics equations alone?

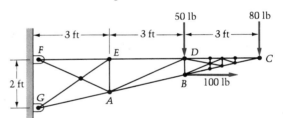

Figure P5.48

5.49 Find the forces in members *IE, JC, KC,* and *DE* of the truss of Problem 5.46, shown again in Figure P5.49 with a new loading.

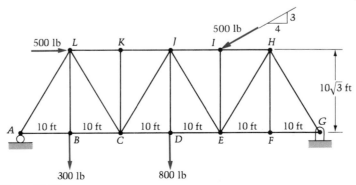

Figure P5.49

5.50 An antenna on a testing range, together with its supporting structure, weighs 300 lb with mass center at C. Assume that $\frac{1}{4}$ of the weight is supported by each of the pins at A and B that connect the structure to the truss tower. See Figure P5.50. (The other half is carried by the pins on an identical truss behind the one in the figure.)

  a. Compute the horizontal components of the pin reactions at A and B on the truss.

  b. Find the forces in members DF, DG, AH, and AB. Which of these four forces is dependent on the assumption about "$\frac{1}{4}$ of the weight"?

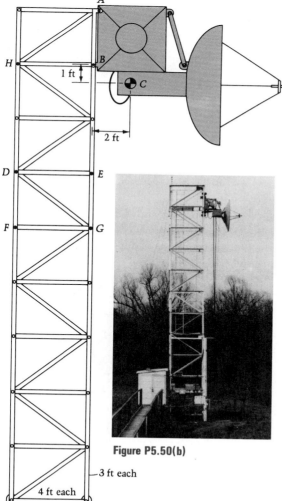

Figure P5.50(b)

Figure P5.50(a)

5.51 For the truss in Figure P5.51 calculate the forces in members BF and BC.

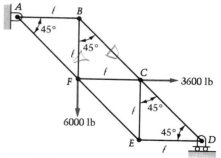

Figure P5.51

5.52 Given: $GK = KE = EB = BF = FL = LH = 10$ ft. Find the force in member AB for the plane truss shown in Figure P5.52.

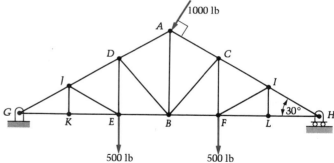

Figure P5.52

5.53 Find the forces in members GH, CH, and BC of the roof truss shown in Figure P5.53.

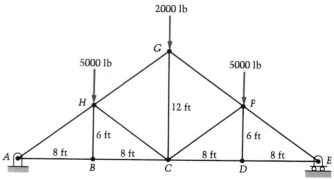

Figure P5.53

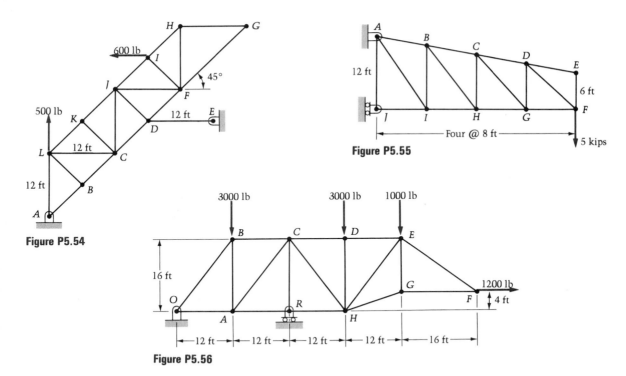

Figure P5.54

Figure P5.55

Figure P5.56

5.54 Find the forces in members *CJ*, *BL*, *FI*, and *HG* shown in Figure P5.54. All members have length 12 ft or 12/√2 ft, and all angles between members are 0°, 45°, 90°, 135° or 180°.

5.55 Find the force in member *BH* of the truss in Figure P5.55. Identify three zero-force members.

5.56 Find the forces in members *DE*, *CD*, and *CH* of the truss in Figure P5.56.

5.57 Find the force in member *EF* of the truss shown in Figure P5.57.

5.58 Demonstrate with a series of sketches that the truss in Figure P5.58 is simple. Find the force in *BE*.

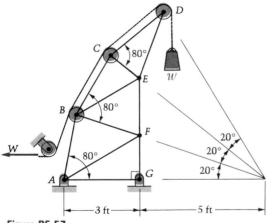

Figure P5.57

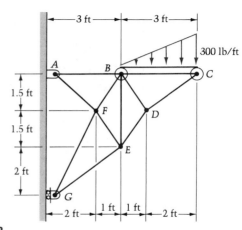

Figure P5.58

5.59 Find the forces in members *DC*, *DG*, and *DF* of the truss in Figure P5.59.

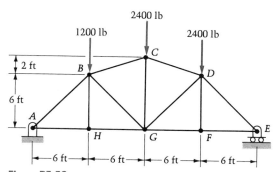

Figure P5.59

5.60 Find the force in member *BD* of the truss in Figure P5.60 if the weight of *W* is 1000 newtons.

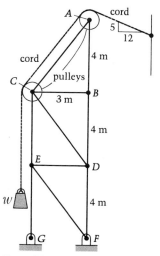

Figure P5.60

5.61 Find the forces in members *CD*, *DH*, and *CH* of the truss in Figure P5.61.

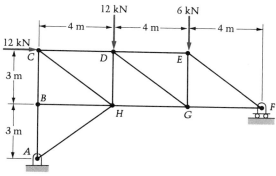

Figure P5.61

5.62 Find the forces in members *AB* and *BC* of the truss in Figure P5.62.

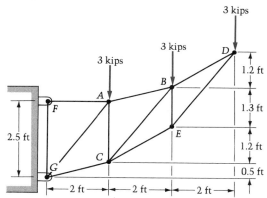

Figure P5.62

5.63 Find the forces in members *DC* and *DE* of the truss in Figure P5.63.

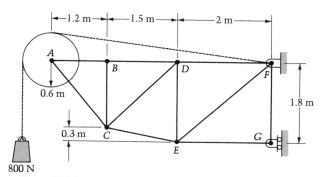

Figure P5.63

5.64   Find the force in member *CF* of the truss in Figure P5.64.

5.65   Find the forces in members *CD*, *KJ*, and *LJ* of the truss in Figure P5.65.

5.66   Find the forces in *FE*, *BE*, and *BC* of the truss shown in Figure P5.66.

5.67   Find the forces in members *CD* and *DJ* of the truss shown in Figure P5.67.

5.68   Find the forces in members *DE*, *QE*, and *OP* of the truss of Example 5.6 shown in Figure P5.68. Use the reactions from that example.

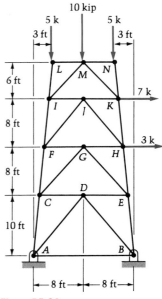

Figure P5.64

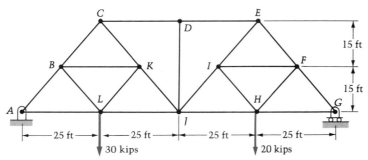

Figure P5.65

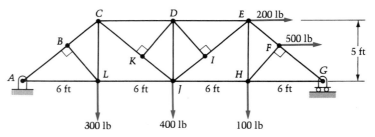

Figure P5.67

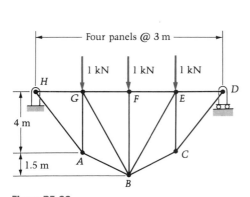

Figure P5.66

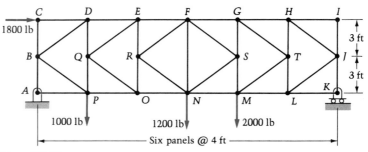

Figure P5.68

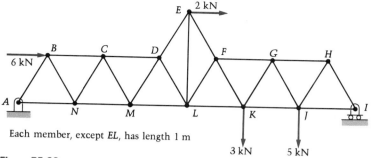

Each member, except *EL*, has length 1 m

Figure P5.69

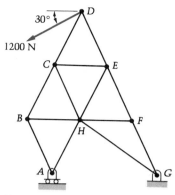

Figure P5.70

5.69 Find the forces in members *EL* and *GH* of the truss of Example 5.7 under the new loading shown in Figure P5.69. (Do not use the reactions of that example!)

5.70 Determine the force in member *HE* for the truss in Figure P5.70. All members are of length 0.8 m except *GH*.

* **5.71** In Figure P5.71(a), the two pulleys at *A* are independently mounted.

a. Show that three members of the truss are zero-force members.

b. Find the forces in members *FG*, *FD*, *FE*, and *CD*.

Hint:

$$\tan \theta = \frac{1 + 0.9 \cos \theta}{6 - 0.9 \sin \theta} \qquad \text{[See Figure P5.71(b).]}$$

which can be solved by trial-and-error for θ.

5.72 The cable in Figure P5.72 extends from a winch at *A* around two pulleys P_1 and P_2 to the weight *W*. Find the forces in truss members *CJ* and *IJ*. The pulleys each have a radius of 0.2 m.

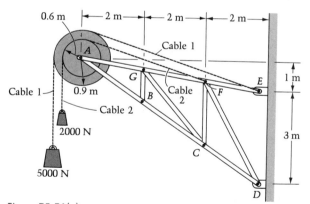

Figure P5.71(a)

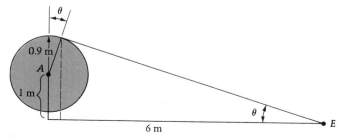

Figure P5.71(b)

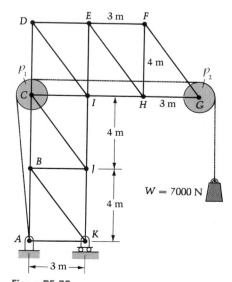

Figure P5.72

5.6   Space Trusses

A nonplanar truss is called a "space truss." Its analysis is complicated by the fact that at each joint there are not two, but three nontrivial equations of equilibrium.

The basic space truss is a tetrahedron, formed by six members pinned at their ends as shown in Figure 5.26. By adding three members at a time and connecting them to create a new joint, a gradually larger structure is formed, called (as in the plane case) a **simple space truss**. This time, instead of $2p - 3$ members as in the plane case, the space truss has $3p - 6$ members, as the reader may wish to show. Instead of the three constraint reactions required in the plane case, in three dimensions we need six to completely constrain the body.

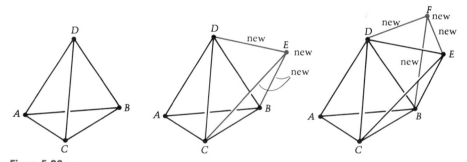

Figure 5.26

The methods of joints and sections apply as well to space trusses as they did to plane trusses. The problem with the space truss, however, is the much larger number of unknown member forces. In a simple space truss, a joint (the last one added in the above procedure) with only three member forces acting on it can always be found. This joint is a good place to start if the truss is to be completely solved. On the other hand, the method of sections, as in the plane case, can sometimes be used to great advantage in finding an isolated member force somewhere in the middle of a space truss. This time, if we are to find this force with a single equation, we have to look for a *line* (instead of a point) through which all the undesired unknown forces pass. If we can find one, then equating the moments about the line to zero will yield the desired member force. We now examine some examples of space trusses.

EXAMPLE 5.8

Find the forces in the three members AB, AC, and AD of the space truss shown in Figure E5.8a.

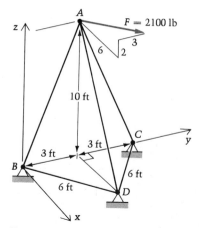

Figure E5.8a

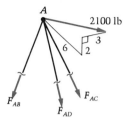

Figure E5.8b

Solution

Just as in a plane truss, the bars of a space truss are all two-force members. Therefore, isolating pin A, we have the free-body diagram shown in Figure E5.8b. The three bar forces are expressible as:

$$\mathbf{F}_{AB} = F_{AB}\left(\frac{-3\hat{\mathbf{j}} - 10\hat{\mathbf{k}}}{\sqrt{109}}\right) = F_{AB}(-0.287\hat{\mathbf{j}} - 0.958\hat{\mathbf{k}})$$

$$\mathbf{F}_{AC} = F_{AC}\left(\frac{3\hat{\mathbf{j}} - 10\hat{\mathbf{k}}}{\sqrt{109}}\right) = F_{AC}(0.287\hat{\mathbf{j}} - 0.958\hat{\mathbf{k}})$$

$$\mathbf{F}_{AD} = F_{AD}\left(\frac{3\sqrt{3}\hat{\mathbf{i}} - 10\hat{\mathbf{k}}}{\sqrt{127}}\right) = F_{AD}(0.461\hat{\mathbf{i}} - 0.887\hat{\mathbf{k}})$$

The applied force can be written as

$$\mathbf{F} = 2100\left(\frac{6\hat{\mathbf{i}} + 3\hat{\mathbf{j}} + 2\hat{\mathbf{k}}}{7}\right) = 1800\hat{\mathbf{i}} + 900\hat{\mathbf{j}} + 600\hat{\mathbf{k}}$$

Equilibrium of joint A requires that

$$\Sigma\mathbf{F} = \mathbf{F}_{AB} + \mathbf{F}_{AC} + \mathbf{F}_{AD} + \mathbf{F} = 0$$

Therefore, equating the coefficients of $\hat{\mathbf{i}}$, $\hat{\mathbf{j}}$, and $\hat{\mathbf{k}}$ to zero,

$$\hat{\mathbf{i}}\text{-coefficients} \Rightarrow 0 + 0 + 0.461\,F_{AD} + 1800 = 0$$

from which

$$F_{AD} = -3900 \text{ lb}$$

Therefore, member AD is in compression (because we assumed tension and obtained a negative answer), and it carries 3900 lb of compressive force.

$$\hat{\mathbf{j}}\text{-coefficients} \Rightarrow -0.287F_{AB} + 0.287F_{AC} + 900 = 0$$

$$\hat{\mathbf{k}}\text{-coefficients} \Rightarrow -0.958F_{AB} - 0.958F_{AC} - 0.887(-3900) + 600 = 0$$

The solution to these two equations is

$$F_{AB} = 3690 \text{ lb}$$

$$F_{AC} = 554 \text{ lb}$$

Both these members are in tension, because we assumed they were and positive answers were then obtained for F_{AB} and F_{AC}.

EXAMPLE 5.9

The space truss shown in Figure E5.9a supports a parabolic antenna. The antenna is positioned in azimuth* by altering the lengths of AQ and BQ, and in elevation by the screwjack member OR. The wind and gravity loads for a certain orientation of the antenna are equipollent to the forces at P, Q, and R given in the

---

* "Azimuth" is the rotation around the z axis (local vertical).

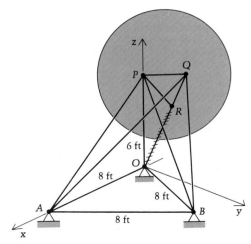

Figure E5.9a

following equations. The locations of Q and R are also indicated. Find the force in the screwjack.

$$\text{Coordinates of } Q: \quad (-1.5, 2, 6) \text{ ft}$$

$$\text{Coordinates of } R: \quad (-2.75, -0.25, 3) \text{ ft}$$

$$\mathbf{F}_P = 2000\hat{\mathbf{i}} + 500\hat{\mathbf{j}} - 800\hat{\mathbf{k}} \text{ lb}$$

$$\mathbf{F}_Q = -1400\hat{\mathbf{i}} - 300\hat{\mathbf{j}} - 800\hat{\mathbf{k}} \text{ lb}$$

$$\mathbf{F}_R = 700\hat{\mathbf{i}} + 500\hat{\mathbf{j}} - 400\hat{\mathbf{k}} \text{ lb}$$

Solution

The free-body diagram of the antenna dish is shown in Figure E5.9b.

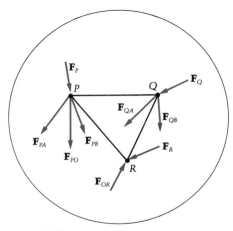

Figure E5.9b

A moment summation about the line PQ will yield the screwjack force F_{OR}:

$$\Sigma M_P \cdot u_{PQ} = \Sigma M_{\text{line }PQ} = 0$$

The only forces contributing to this equation are F_{OR} and F_R. Therefore,

$$r_{PR} \times (F_R + F_{OR}) \cdot \hat{u}_{PQ} = 0$$

$$[(-2.75 - 0)\hat{i} + (-0.25 - 0)\hat{j} + (3 - 6)\hat{k}]$$

$$\times \left[(700\hat{i} + 500\hat{j} - 400\hat{k}) + F_{OR}\left(\frac{-2.75\hat{i} - 0.25\hat{j} + 3\hat{k}}{4.08} \right) \right]$$

$$\cdot \left[\frac{(-1.5 - 0)\hat{i} + (2 - 0)\hat{j} + (6 - 6)\hat{k}}{2.50} \right] = 0$$

The scalar triple product can be expressed as a determinant:

$$
\begin{array}{l}
\hat{u}_{PQ} \text{ components} \rightarrow \\
r_{PR} \text{ components} \rightarrow \\
F_R + F_{RO} \text{ components} \rightarrow
\end{array}
\begin{vmatrix}
-0.6 & 0.8 & 0 \\
-2.75 & -0.25 & -3 \\
(700 - 0.674F_{OP}) & (500 - 0.0613F_{OP}) & (-400 + 0.735F_{OR})
\end{vmatrix} = 0
$$

Adding $\frac{4}{3}$ times the first column to the second column will simplify the evaluation of the determinant:

$$
\begin{vmatrix}
-0.6 & 0 & 0 \\
-2.75 & -3.92 & -3 \\
(700 - 0.674F_{OR}) & (1430 - 0.960F_{OR}) & (-400 + 0.735F_{OR})
\end{vmatrix} = 0
$$

Thus,

$$-0.6[-3.92(-400 + 0.735F_{OR}) + 3(1430 - 0.960F_{OR})] = 0$$

$$1570 - 2.88F_{OR} + 4290 - 2.88F_{OR} = 0$$

$$F_{OR} \frac{5860}{5.76} = 1020 \text{ lb} \qquad \text{(compressive as assumed)}$$

Question 5.12   At the outset of the preceding example, is there a line about which the moments, summed and equated to zero, will yield by this single equation the force in BP? OP? AP? AQ? BQ?

Answer 5.12   No to all five questions.

EXAMPLE 5.10

The space truss in Figure E5.10a (see on the next page) was used to support a large "cross log periodic antenna" (which is shown in Figure P7.155(a) of Chapter 7). A certain loading under heavy wind, self-weight, and ice weight is transmitted to the truss as indicated in the figure. Find the forces in the truss members.*

---

* The actual truss had a seventh member from point B to D. Because this makes the truss statically indeterminate, we omit it in this example.

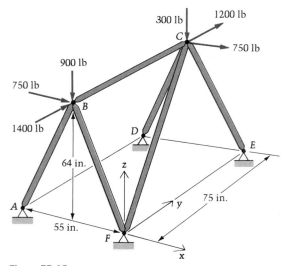

Figure E5.10a

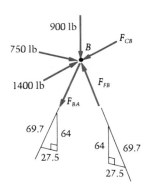

Figure E5.10b

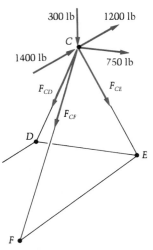

Figure E5.10c

Solution

We consider first a free-body diagram of joint B (see Figure E5.10b). Writing the equilibrium equations,

$$\Sigma F_y = 0 = 1400 - F_{CB} \Rightarrow F_{CB} = 1400 \text{ lb } \textcircled{C}$$

$$\Sigma F_x = 0 = 750 - F_{BA}\left(\frac{27.5}{69.7}\right) - F_{FB}\left(\frac{27.5}{69.7}\right) \qquad (1)$$

$$\Sigma F_z = 0 = -900 - F_{BA}\left(\frac{64}{69.7}\right) + F_{FB}\left(\frac{64}{69.7}\right) \qquad (2)$$

Solving Equations (1) and (2),

$$F_{FB} = 1440 \text{ lb } \textcircled{C} \qquad \text{and} \qquad F_{BA} = 460 \text{ lb } \textcircled{T}$$

We note that F_{FB} could have been found without the need to solve two equations simultaneously:

$$+\curvearrowleft \qquad \Sigma M_{AD} = 0 = -\left(F_{BP}\frac{64}{69.7}\right)(55 \text{ in.}) + (750 \text{ lb})(64 \text{ in.})$$

$$+ (900 \text{ lb})(27.5 \text{ in.})$$

$$F_{BF} = +1440 \text{ lb } \textcircled{C}$$

Then F_{BA} would follow from either force equation.

Moving next to a free-body diagram of joint C (see Figure E5.10c), we see that we can avoid three equations in three unknowns (which we would get from $\Sigma \mathbf{F} = 0$) by summing moments about line FE:

$$\curvearrowright+ \qquad \Sigma M_{FE} = 0 = (750 \text{ lb})(64 \text{ in.}) - (300 \text{ lb})(27.5 \text{ in.})$$

$$- F_{CD}\left(\frac{64}{69.7}\right)(55 \text{ in.})$$

$$F_{CD} = 787 \text{ lb } \textcircled{T}$$

Similarly, moments about line DE will yield F_{CF}:

$$+\circlearrowleft \quad \Sigma M_{DE} = [(1400 + 1200)\text{lb}](64 \text{ in.})$$

$$+ F_{CF}\left(\frac{75}{\sqrt{64^2 + 27.5^2 + 75^2}}\right)(64 \text{ in.}) = 0 \qquad (3)$$

$$F_{CF} = 3550 \text{ lb } \text{(T)}$$

Note that the force F_{CF} was broken up into components at joint C. If we had broken it up at F, the last term in Equation (3) would have been

$$F_{CF}\left(\frac{64}{\sqrt{64^2 + 27.5^2 + 75^2}}\right)(75 \text{ in.})$$

which is the same result but with a different lever arm. Finally,

$$+\nwarrow \quad \Sigma F_x = 0 = F_{CE}\left(\frac{27.5}{69.7}\right) - 787\left(\frac{27.5}{69.7}\right) + 3550\left(\frac{27.5}{102}\right) + 750$$

$$F_{CE} = -3540 \text{ lb} \qquad \text{(or 3540 lb } \text{(C))}$$

As checks,

$$\swarrow \quad \Sigma F_y = 1200 + 1400 - 3550\left(\frac{75}{102}\right)$$

$$= 2600 - 2610 = -10 \text{ lb } \checkmark$$

$$+\uparrow \quad \Sigma F_z = -300 + 3540\left(\frac{64}{69.7}\right) - 3550\left(\frac{64}{102}\right) - 787\left(\frac{64}{69.7}\right)$$

$$= -300 + 3250 - 2230 - 720^* = 0 \checkmark$$

The -10-lb result for ΣF_y differs from zero due to rounding to three digits. Had we kept more digits, ΣF_y would have been smaller, as the reader may wish to show, but we must always remember that an answer is no more accurate than the least accurate number in the data.

---

* To three digits, 787(64 / 69.7) is 723; however, the units digit in the second and third terms (3250 and -2230) is insignificant.

PROBLEMS ▶ Section 5.6

5.73 A set of three forces in coordinate directions is applied at point C in Figure P5.73. Find the forces in the three bars, which are pinned together at C and rest on ball joints at A, O, and B.

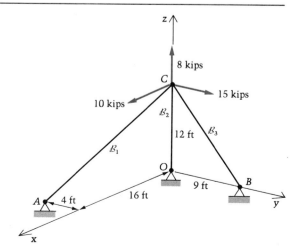

Figure P5.73

5.74 Two uniform light bars BD and DC are pinned together at D, where a cable is also attached as shown in Figure P5.74. The bars are connected to the reference frame at B and C by smooth ball joints. The system supports the 300-lb force pulling at D in the negative x direction. Find the force in the cable, and the reactions at B and C. The bars are in the yz plane.

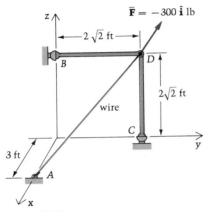

Figure P5.74

5.75 The portable jack stand in Figure P5.75 carries a downward load of 600 lb. How much compressive force is there in each of the four symmetrical legs OA, OB, OC, and OD? Assume that all connections are ball joints, and that the only reactions from the ground are equal vertical forces (using symmetry) at A, B, C, and D. Also, neglect the dimensions of the top plate.

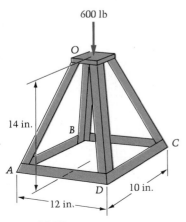

Figure P5.75

5.76 Consider all connections of the weightless struts in Figure P5.76 as being ball-and-socket. The points A, C, and D all lie in the xz plane. A vertical load of 960 lb is applied downward at B. Determine the force in member AB.

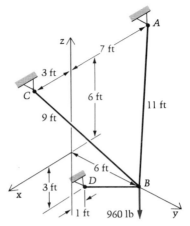

Figure P5.76

5.77 The light tripod in Figure P5.77 supports a weight of 144 lb. If the legs A, C, and D are pinned at B and if the ground is rough, find the force in leg C.

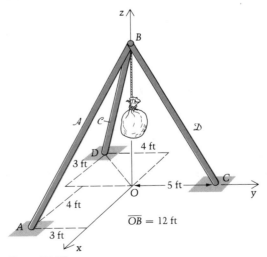

Figure P5.77

5.78 By inspection (it takes thinking, but not writing!), find the force in member AB of the truss in Figure P5.78. Then find the force in DB.

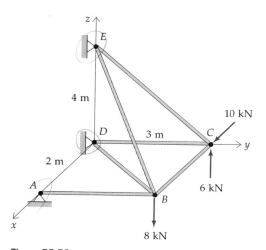

Figure P5.78

5.79 For the space truss shown in Figure P5.79:

   a. Find the reactions at A, B, and C.
   b. Using the method of sections, with a section cut by a plane parallel to xz, find the force in members AD and CD.
   c. Find the forces in members AB, BC, and BD using the method of joints.
   d. Find the force in AC by inspection.

(F_{A_x}, F_{A_y}, and F_{B_z} represent the only nonzero reaction components at A and B.)

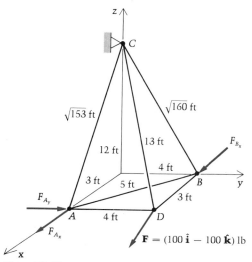

Figure P5.79

* 5.80 In Example 5.9 find the forces in the adjustable legs AQ and BQ.

* 5.81 In Figure P5.81 the nine members all have the same length ℓ. Find the forces in the members, noting that, by symmetry, (a) the forces in AB, AC, AD, EB, EC, and ED are equal; and (b) the forces in BC, CD, and DB are equal.

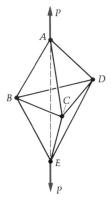

Figure P5.81

5.82 The 2000-N force is in the negative x direction; the 2400-N force is in the yz plane, and the 2100-N force has direction cosines $(-\frac{3}{7}, \frac{2}{7}, -\frac{6}{7})$. Find all the members' forces that are not indeterminate in the space truss shown in Figure P5.82.

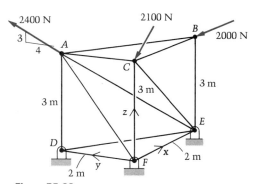

Figure P5.82

Problems 5.83–5.87 are based upon Figure P5.83 and the following text: Bars $\mathcal{B}_1$, $\mathcal{B}_2$, and $\mathcal{B}_3$ are pinned at A to form a space truss. They are likewise pinned (ball-jointed) to the ground at O_1, O_2, and O_3, respectively. Find the forces in the three members if the force $\mathbf{F}$ applied at A is as given.

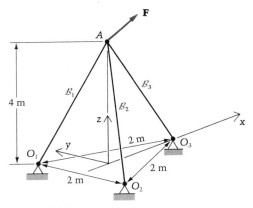

Figure P5.83

5.83 $\mathbf{F} = -1000\hat{\mathbf{k}}$ N

5.84 $\mathbf{F} = 1000\hat{\mathbf{i}}$ N

5.85 $\mathbf{F} = 1000\hat{\mathbf{j}}$ N

5.86 $\mathbf{F} = 707\hat{\mathbf{i}} + 707\hat{\mathbf{j}}$ N

5.87 $\mathbf{F} = 577\hat{\mathbf{i}} + 577\hat{\mathbf{j}} + 577\hat{\mathbf{k}}$ N

5.88 The space truss in Figure P5.88 carries the loads

$$\mathbf{F}_1 = -1000\hat{\mathbf{i}} + 1200\hat{\mathbf{j}} + 2000\hat{\mathbf{k}} \text{ lb}$$

and

$$\mathbf{F}_2 = 500\hat{\mathbf{i}} - 1000\hat{\mathbf{j}} + 1500\hat{\mathbf{k}} \text{ lb}$$

a. By inspection, find a zero-force member.

b. Find the force in member AD.

5.89 In Figure P5.89 line AB of the space truss is vertical, and the 2000-lb force is parallel to the x axis.

a. Show that the structure is a simple space truss.

b. Find the force in member EF.

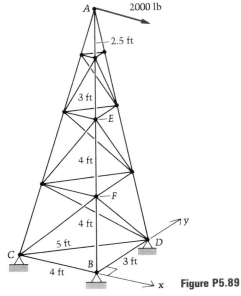

Figure P5.89

5.90 For the space truss in Figure P5.90, find:

a. The reactions at A, B, C, and D

b. The force in member BE using the method of sections with the section shown and a single equation

c. The force in member BE by the method of joints.

Note: The support ⟋⟍ transmits force only normal to the plane defined by the two perpendicular lines l_1 and l_2.

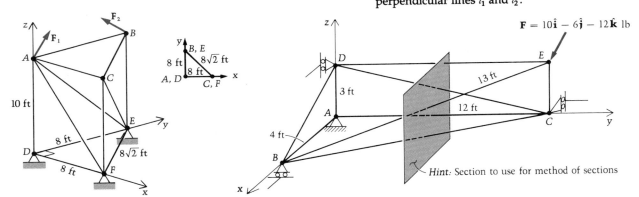

Figure P5.88

Figure P5.90

5.7   A Brief Introduction to the Mechanics of Deformable Bodies

We will now use the truss to give the reader a brief introduction to the mechanics of deformable bodies, also known as strength of materials, or mechanics of materials.

In this section, we shall introduce — in a one dimensional analysis appropriate to the truss — the concepts of stress, strain, and Hooke's law. We will then demonstrate how a consideration of deformations allows the solution of certain types of problems previously deemed statically indeterminate. Lastly, we will look at the effect of temperature.

Normal Stress

The major reason why we have learned in Sections 5.2–5.6 to find the forces in a truss is so that we can later determine the *stresses* within the members. Stresses, which have units of force per unit of area, are important because materials fail when stresses exceed certain limits; thus in design we are interested in knowing the maximum stress within each of the members of a structure. It turns out that this is very easy to do in a truss, for two reasons:

1. In each member of a truss, the resultant force F, acting on every cross section (of area A) normal to the member's axis, is the same.

2. The normal stress at every point in the above cross section, in the direction of the member's axis, is simply F/A, as we shall shortly see.

Suppose now that the member AB in Figure 5.27(a) is a truss member carrying the 10 kN tensile force as shown.

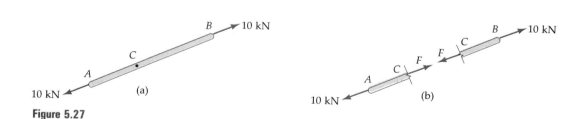

Figure 5.27

We know from previous sections of the chapter that if we were to mentally slice through the member at point C as in Figure 5.27(b), then using either of the two free-body diagrams in that figure, we would find F to be a 10-kN tensile force. But let us now look more carefully at the cross-sectional cut on the FBD of CB.

We see from Figure 5.28(a), on the next page, that F is merely the *resultant* of infinitely many small forces ΔF, each acting on a small area ΔA. At a point P within one of these ΔA's (see Figure 5.28(b)), we define the **normal stress** σ to be the limit of the resultant force ΔF acting on ΔA, divided by ΔA, as ΔA shrinks to zero while always including point P:

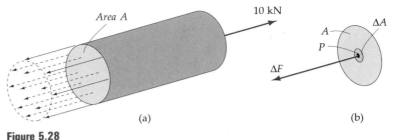

Figure 5.28

$$\sigma = \lim_{\Delta A \to 0} \frac{\Delta F}{\Delta A}$$

A small distance away from the ends of the truss member (where the load typically is applied through a pin), the force distribution across the cross section becomes uniform, so that

$$\sigma = \frac{F}{A} \tag{5.1}$$

and thus the normal stress at every point of a truss member in the direction of its axis is the force in the member divided by its cross-sectional area. Furthermore, because σ is the same for all points within the member, we may simply speak of the stress in the truss member as F/A. Thus if the member in Figure 5.27 had a cross-sectional area of 0.01 m², the stress in the member would be

$$\sigma = \frac{10,000 \text{ N}}{0.01 \text{ m}^2} = 1 \times 10^6 \text{ N/m}^2$$

This is a tensile stress, because (see Figure 5.28) it "pulls" on the area on which it acts. If the two 10 kN forces were compressing the bar AB instead of stretching it, the force F on the section at C would be "pushing" on the area, and the stress in that case would be called a compressive stress. All normal stresses σ are either tensile or compressive, and the sign convention is that they are positive if tensile, and thus negative if compressive.

In the SI system of units, one N/m² is called a Pascal (Pa); thus the above stress is $\sigma = 1 \times 10^6$ Pa or 1MPa (megapascal). In the U.S. system of units, stress is usually written in pounds per square inch, or psi. In an inflated tire, for example, 32 psi is a typical recommended air pressure.

Before moving to an example of stress calculation in a truss, we mention that in structures other than trusses, stress analysis is in general *much* more complicated. For one thing, as forces vary from point to point, so does the resulting stress. Secondly, we may have normal stress in two and even three dimensions to be concerned about. And thirdly, there is yet another kind of stress, known as shear stress and which we shall not treat in this brief section, which complicates the picture still further. While normal stress is force per unit area perpendicular to an area, shear stress is force per unit area *in the plane* of an area. Thus, for example, the intensity of friction force is shear stress because it resists the tendency of one body to slide past another in their plane of contact.

EXAMPLE 5.11

In the truss of Figure 5.6, shown in Figure E5.11a, the members are bars with cross-sectional areas 3 in.² (*AE* and *EB*), 2 in.² (*AD*, *DE*, *CE*, and *CB*), and 1.5 in.² (*DC*). Find which member is being subjected to the largest stress for the given loading.

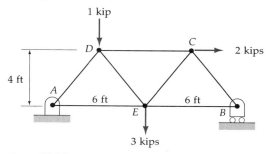

Figure E5.11a

Solution

The forces in the members were given in kips near the end of Section 5.2, and they are shown pictorially in Figure E5.11b:

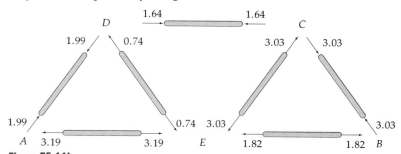

Figure E5.11b

Thus the stresses carried by the members are:

$$\sigma_{AE} = P_{AE} / A_{AE} = \frac{3190}{3} = 1063 \text{ psi, tensile}$$

$$\sigma_{EB} = \frac{1820}{3} = 607 \text{ psi, tensile}$$

$$\sigma_{AD} = \frac{1990}{2} = 995 \text{ psi, compressive}$$

$$\sigma_{DE} = \frac{740}{2} = 370 \text{ psi, tensile}$$

$$\sigma_{CE} = \frac{3030}{2} = 1515 \text{ psi, tensile}$$

$$\sigma_{CB} = \frac{3030}{2} = 1515 \text{ psi, compressive}$$

$$\sigma_{DC} = \frac{1640}{1.5} = 1093 \text{ psi, compressive}$$

Members CE and CB are seen to carry the largest stress, and it is worth noting that these two members do not carry the largest force, nor do they have the smallest area. It is the *combination* of force and area that counts in computing the maximum stress.

Extensional Strain

Normal stress, which we have discussed above in conjunction with trusses, causes line elements in a body to extend or contract, depending on whether σ is positive (tensile) or negative (compressive). These extensions and contractions strain a truss element (for example, the one depicted previously in Figure 5.27), and in fact we may give a formal definition for **extensional strain** ϵ in the direction of the member axis at a point such as C (see Figure 5.29(a)) located at the coordinate x along the axis:

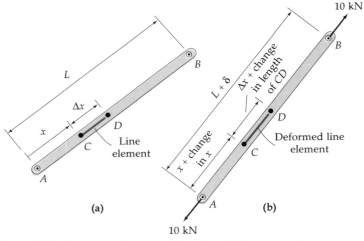

Figure 5.29   (Line element length and deformations highly exaggerated)

The extensional strain ϵ is the limit of the change in length of a line element at C divided by the original length of that line element, as the length of the element approaches zero. In Figure 5.29(a), the short line element CD at point C has a length Δx before the truss (of which AB is a member) is loaded. Therefore,

$$\begin{array}{c}\text{Strain}\\\text{at } C\end{array} = \epsilon = \lim_{\Delta x \to 0} \left(\frac{\text{Change in Length of } CD}{\Delta x} \right)$$

After the truss is loaded, member AB carries its 10 kN load, and member AB has moved slightly in space and has elongated. (It would have shortened had the 10 kN force been compressive.) As will be seen in the next section, stress and strain are proportional for linearly elastic materials. Thus the strain in a truss element such as AB is also constant, meaning that the change in length per original length is the same for every line element from one end of the member to the other. If the *total* elongation of member AB is δ, then we may write for a truss element:

$$\epsilon = \frac{\delta}{L} \tag{5.2}$$

i.e., the constant strain in the member is its change in length divided by its original length. If member AB elongates 0.0020 m under the 10 kN load, then if its original length was 1.5 m, the strain in the member is

$$\epsilon = \frac{0.0020}{1.5} = 0.0013 \text{ m/m}$$

Note that strain is a dimensionless quantity, although if we expressed the above answer as 1.3 mm/m, the units would need to be displayed. Also, sometimes strain is written as "0.13% strain," which simply means $\epsilon = 0.0013$ (or 0.0013 m/m, or 0.0013 in./in., etc.)

We will do more with Equation (5.2) after we study the relationship between stress and strain in the following sub-section.

Hooke's Law

For elastic materials, stress and strain are proportional — up to a point. The linear relationship, discovered by Robert Hooke in the 17th century and known as **Hooke's Law,** is written for uniaxial stress and strain, such as we have in a truss element, as

$$\sigma = E\epsilon \tag{5.3}$$

where the constant of proportionality, E, is called Young's modulus of elasticity. As strain is dimensionless, E has the same units as stress, namely pascals (N/m²) or psi (lb/in.²). Figure 5.30 shows the relationship pictorially, in a drawing called a stress-strain diagram. A figure like this will be produced if a mild steel specimen is mounted in grips in a testing machine (see Figure 5.31 on the next page) and subjected to a tensile test. The machine stretches the specimen with a gradually increasing load and measures its increasing elongation until it fails. From the force and elongation data, σ and ϵ are calculated and plotted. With respect to the stress-strain diagram, we note the following:

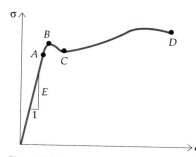

Figure 5.30

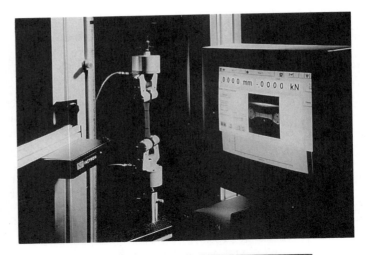

Figure 5.31   Tensile Tests (Courtesy of Instron Corporation.)

1. The slope of the "straight-line portion" is $\sigma / \epsilon = E$ at any point.

2. Beyond a point A called the proportional limit, stress and strain are no longer proportional.

3. Beyond a point B, called the elastic limit, the curve will not retrace its path if the load is then reduced back to zero.

4. Beyond a point C, called the yield point, the material begins to give, or yield, with large increase in strain for very little increase in stress.

5. At a point D, the material finally breaks (or fractures, or fails) at a value of stress called the ultimate strength.

Some materials such as aluminum behave elastically up to an elastic limit but do not map out a straight line portion like that of Figure 5.30 if stress and strain are plotted from a tensile test. The curve for such a material, which might look more like Figure 5.32, may be used to obtain a yield stress by the offset method. It turns out that if a line is drawn from the point A of 0.002 strain on the ϵ-axis in the figure, with the same slope as the curve has at the origin, this dotted line will intersect the stress-strain curve at a value of σ very close to the yield stress σ_y of the material. Furthermore, the slope of this line, i.e., the slope of the curve at the origin, serves well for Young's modulus E in calculations.

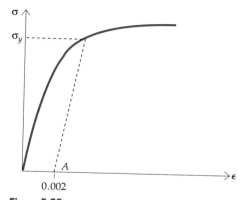

Figure 5.32

Table 5.1 below gives values of Young's modulus of elasticity for several commonly used elastic materials:

Table 5.1

Material	Young's Modulus of Elasticity E		Yield Stress σ_y	
	ksi	MPa	ksi	MPa
Aluminum Alloy 6061-T6	10,000	69,000	40	280
Structural Steel ASTM-A36	30,000	210,000	36	250
Titanium Alloy	16,000	110,000	120	830
Brass	15,000	104,000	70	480
Bronze	15,000	104,000	65	450
Hard-drawn Copper	17,000	117,000	48	330
Medium-Strength Concrete in Compression	3,600	25,000	—	—

Elongation of a Truss Element

If we now assume that our truss members are made of elastic material, we may use Hooke's law to link the results (Equations 5.1,2,3) of the previous three subsections, as follows:

$$\underbrace{\frac{\delta}{L}}_{\substack{\text{strain-deformation} \\ \text{relation, Eq. (5.2)}}} = \epsilon = \overbrace{\frac{\sigma}{E}}^{\substack{\text{stress-strain relation} \\ \text{(Hooke's law), Eq. (5.3)}}} \underbrace{\frac{F}{AE}}_{\substack{\text{Eq. (5.1) for stress in} \\ \text{terms of force and area}}} \tag{5.4}$$

Therefore, we obtain the following relation between the elongation δ and the force F in the truss element:

$$\delta = \frac{FL}{AE} \tag{5.5}$$

We note that all of the equations (5.1–5.5) of this section are equally applicable to any straight two-force member that we may encounter within a frame. Such a member, like a truss member, is also in a state of uniaxial stress and strain. Other members of frames, however, may carry forces in lateral directions as well as bending and twisting moments; their analysis is beyond the scope of this brief "one-dimensional" treatment and must await a course in the mechanics of deformable bodies.

As an example of the use of Equation (5.5), the steel truss members DE and CB of Example 5.11 respectively extend and contract the amounts:

$$\delta_{DE} = \frac{P_{DE}\,L_{DE}}{A_{DE}\,E_{DE}} = \frac{740\,(5 \times 12)}{2\,(30 \times 10^6)} = 0.000740 \text{ in.}$$

$$\delta_{CB} = \frac{P_{CB}\,L_{CB}}{A_{CB}\,E_{CB}} = \frac{3030\,(5 \times 12)}{2\,(30 \times 10^6)} = 0.00303 \text{ in.}$$

which are tiny in comparison to the 60-inch original lengths of DE and CB, but are by no means unimportant.

In modern times, structures large and small may be analyzed by means of computer programs that utilize advanced techniques known as finite elements to quickly solve for stresses in the members, and deflections of the joints, of a truss. Just the same, for the insight it will afford us, let us use Equation (5.5) to compute the deflection of a joint in a simple truss.

EXAMPLE 5.12

Find the location of joint A of the truss after the 300 N load is applied as shown in Figure E5.12a. The members are each made of steel with $E = 207,000$ MPa and have cross-sectional area $= 0.004$ m^2.

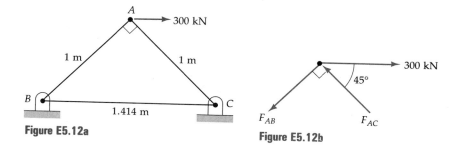

Figure E5.12a

Figure E5.12b

Solution

At joint A, we have the FBD shown in Figure E5.12b. If we sum forces along the CA-direction (to get rid of one of the forces), we obtain:

$$\nwarrow \qquad \Sigma F_{CA} = 0 = F_{AC} - 300{,}000\,\frac{\sqrt{2}}{2}$$

so that

$$F_{AC} = 212{,}000 \text{ N} \quad \copyright$$

Then,

$$+\uparrow \qquad \Sigma F_y = 0 = 212{,}000\left(\frac{\sqrt{2}}{2}\right) - F_{AB}\left(\frac{\sqrt{2}}{2}\right)$$

so

$$F_{AB} = 212{,}000 \text{ N} \quad \text{\textcircled{T}}$$

From these results, we find

$$\delta_{AB} = \frac{212{,}000(1)}{(0.004)207 \times 10^9} \qquad \text{and} \qquad \delta_{AC} = \frac{-212{,}000(1)}{(0.004)207 \times 10^9}$$

$$= 0.256 \times 10^{-3} \text{ m (extension)} \qquad \qquad = -0.256 \times 10^{-3} \text{ m (contraction)}$$

Separating the members for clarity in Figure E5.12c, we see that the extended end A of AC, and the contracted end A of AC, must lie on the circles shown in the figure in order that they have the respective proper final lengths

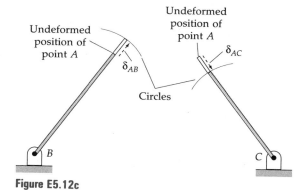

Figure E5.12c

$1 + 0.256 \times 10^{-3} = 1.0003$ m and $1 - 0.256 \times 10^{-3} = 0.9997$ m. But these deflections are so small with respect to the original lengths of AB and AC that we may treat these curves as straight lines; when we then put the members back together (Figure E5.12d), we find that the final position of point A has to lie on the intersection of the two straight lines shown. Thus point A experiences a deflection of $.000256\sqrt{2} = 0.000362$ m→. The fact that the deflection is in the exact direction as the applied force is a peculiarity here of the symmetry and the equal "AE's"; the reader should note, for example, that if the "AE" of bar AB were different from that of AC, the δ's would differ and the resulting displacement of A would contain a vertical component (see Problem 5.118).

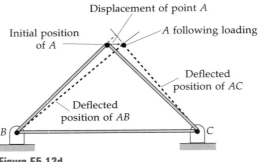

Figure E5.12d

Note in the preceding example that the stresses in the members AB and AC of the truss are:

$$\sigma_{AB} = \frac{F}{A} = \frac{212,000}{0.004} = 53 \times 10^6 \text{ Pa} = 53 \text{ MPa} \quad \text{(tensile)}$$

and

$$\sigma_{AC} = \frac{F}{A} = \frac{-212,000}{0.004} = -53 \text{ MPa, or } 53 \text{ MPa} \quad \text{(compressive)}$$

These values are less than one-fourth the yield stress for steel.

Statically Indeterminate Problems

In Chapters 2 (just below Figure 2.8) and 4 (surrounding Figure 4.7) we learned that not all equilibrium problems can be solved by means of the equations of statics alone. Those that cannot, because there are more unknowns than equations, were called statically indeterminate, for obvious reasons. For example, for the light bar in Figure 5.33,

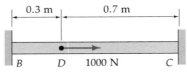

Figure 5.33

the free-body diagram is as shown in Figure 5.34:

Figure 5.34

Clearly, we know that

$$\xrightarrow{+} \quad \Sigma F_x = 0 = 1000 - F_L - F_R$$

$$F_L + F_R = 1000 \tag{5.6}$$

but that is *all* we know from statics, because any moment equation will contain the sum of F_L and F_R and thus F_L and F_R cannot be separately found.

However, our brief introduction to mechanics of materials will allow us to determine F_L and F_R by considering the *deformation* of the bar. Note from the sectionings in Figure 5.35 that:

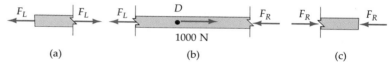

Figure 5.35

a. on any cross-section to the left of D (see Figure 5.35(a or b)) the force is F_L (tension); and

b. on any cross-section to the right of D (see Figure 5.35(b or c)), the force is F_R (compression).

Therefore, using Equation (5.5), the elongations of the two portions of the bar are:

$$\delta_{BD} = \frac{F_L \, 0.3}{AE} \quad \text{and} \quad \delta_{DC} = \frac{-F_R \, 0.7}{AE}$$

where F_R is positive because it is already drawn as compressive.

Now we are ready for the big step: The total elongation of BC has to be zero, because it is built into the (assumed-to-be) rigid walls. Thus:

$$\delta_{BC} = 0 = \delta_{BD} + \delta_{DC} \tag{5.7}$$

or

$$\frac{F_L \, 0.3}{AE} - \frac{F_R \, 0.7}{AE} = 0$$

Therefore,

$$F_L = (7/3) \, F_R \tag{5.8}$$

and we have found a second equation in the two unknown forces.

Substituting Equation (5.8) into (5.6), we arrive at:

$$(7/3)F_R + F_R = 1000$$
$$F_R = 300 \text{ N}$$

and from (3),

$$F_L = 700 \text{ N}$$

The deformations of a typical elastic structural member are nearly always very, very, small. Yet we have used them above to solve a problem which was hopeless using the equations of Statics alone. Now let us look at a second example, this time with bars of different materials and areas constituting a composite member between the walls:

EXAMPLE 5.13

Find the forces exerted by the rigid walls on the ends of the composite member BDC in Figure E5.13a.

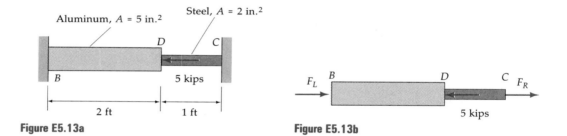

Figure E5.13a   Figure E5.13b

Solution

From the FBD of the entire composite bar in Figure E5.13b, we have

$$\xrightarrow{+} \quad \Sigma F_x = 0 = F_L + F_R - 5$$

or

$$F_L + F_R = 5 \tag{1}$$

And from the FBD's in Figure E5.13c, we see that everywhere to the right of D, the force on a cross-section is F_R (tensile); and on all cross-sections to the left of D, the force is F_L (compressive).

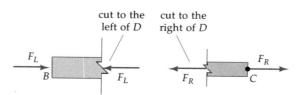

Figure E5.13c

Enforcing now the fact that the total elongation δ between the walls must again vanish, we obtain (using Table 5.1 for the elastic moduli):

$$\delta_{BC} = 0 = \delta_{BD} + \delta_{DC}$$

$$\frac{-F_L\,24}{5(10 \times 10^6)} + \frac{F_R\,12}{2(30 \times 10^6)} = 0$$

from which $F_L = (5/12)F_R$, and substituting this result into Equation (1) yields

$$\left(\frac{5}{12} + 1\right) F_R = 5$$

or

$$F_R = \frac{60}{17} = 3.53 \text{ kips}$$

and thus

$$F_L = \frac{5}{12}(3.53) = 1.47 \text{ kips}$$

In the preceding example, the steel carries more load because even though it has only 40% the area of the aluminum, it has three times the modulus and only half the length; these last two factors combine to make the steel "spring" (see Figure 5.36) $3 \times 2 = 6$ times as stiff, while the area makes it 40% as stiff. All things considered, then, the steel is $6 \times 0.4 = 2.4$ times stiffer and indeed the force in it, 3.53 kips, is 2.4 times that (1.47 kips) in the aluminum. A straight, two-force structure such as a truss element is perfectly analogous to a spring with modulus $k = AE/L$.

Spring modulus

$= k_{alum} = \dfrac{AE}{L}$

$= \dfrac{5(10 \times 10^6)}{24}$

$= 2.08 \times 10^6 \text{ lb/in.}$

Spring modulus

$= k_{steel} = \dfrac{AE}{L}$

$= \dfrac{2(30 \times 10^6)}{12}$

$= 5 \times 10^6 \text{ lb/in.}$

Figure 5.36

EXAMPLE 5.14

The three bars in Figure E5.14a support the 100,000-lb weight. The two outer bars, symmetrically placed, are steel, with areas 6 in.^{2} each; the central bar is copper with area 8 in.^{2} Find the forces in the bars.

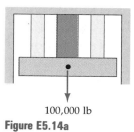

100,000 lb

Figure E5.14a

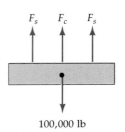

F_s F_c F_s

100,000 lb

Figure E5.14b

Solution

The vertical walls force the weight to deflect straight down (which if all is symmetrical it would do anyway), so that the deflections δ_{steel} and δ_{copper} will be equal. We have from Figure E5.14b:

$$+\uparrow \qquad \Sigma F_y = 0 = 2F_s + F_c - 100,000$$

$$2F_s + F_c = 100,000 \tag{1}$$

The fact that the δ's are the same here yields (note the lengths are the same):

$$\delta_s = \delta_c$$

$$\frac{F_s L}{A_s E_s} = \frac{F_c L}{A_c E_c}$$

$$F_s = \frac{A_s E_s F_c}{A_c E_c} = \frac{(6)30 \times 10^6}{(8)17 \times 10^6} F_c = 1.32 F_c \tag{2}$$

> **Question 5.13**   Why did we use the area of just one steel bar in the preceding equation?

Substituting Equation (2) into (1),

$$2(1.32 F_c) + F_c = 100,000$$

$$F_c = 27,500 \text{ lb}$$

$$F_s = 1.32 F_c = 36,300$$

As a check, $2F_s + F_c = 100,100$ lb, off by a tenth of a percent due to roundoff error.

Now that we have seen that previously indeterminate problems can be solved by considering deformations in addition to equilibrium equations; we also see the importance of realizing that these equations are not restricted to rigid bodies.

Thermal Effects

The strain due to a temperature change ΔT in a truss member free to expand is given by

$$\epsilon = \alpha \Delta T \tag{5.9}$$

where α is the coefficient of thermal expansion, in units of $1/°F$ (in U.S. units) or $1/°C$ (in SI units). Table 5.2 gives some typical values for the coefficient of thermal expansion. It is perhaps helpful to think of the units of α as, for example, "inches/inch per °F," because then one

Table 5.2

Material	$\alpha \times 10^6$	
	1/°F	1/°C
Structural Steel	6.5	11.7
Aluminum	13	23.4
Copper	9.5	17.1
Brass	11	19.8
Bronze	10.5	18.9
Titanium Alloy	5	9
Concrete	6	10.8

Answer 5.13   While *both* steel bar forces add to the copper force to support the 100 kip weight, the elongation of *each* steel bar equals that of the copper bar.

realizes that once α is multiplied by the ΔT, what remains is in. / in., i.e., the change in length per original length, or dimensionless strain.

The change in length of a free-to-expand member is, since the strain due to temperature is uniform, ϵL or $\alpha L(\Delta T)$. If, however, the member is in any way constrained from expanding (or contracting if ΔT is negative), then there is also strain present from the tensile or compressive load carried by the member. Consider the following example:

EXAMPLE 5.15

The brass bar in Figure E5.15a has area 2 in.² If subjected to a temperature rise of 30°F, what are the forces exerted by the rigid walls that squeeze the ends together, preventing the expansion?

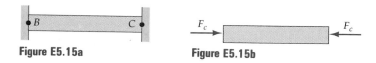

Figure E5.15a   Figure E5.15b

Solution

It is obvious from Figure E5.15b that this problem is statically indeterminate, since $\Sigma F_x = 0$ yields $0 = 0$. Thus we must again consider the deformation of the member. We mentally free the right end C, allow the expansion to occur, then apply the precise wall force which will bring end C back to its given distance from B, as suggested by Figure E5.15c:

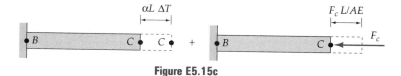

Figure E5.15c

As in the preceding sub-section, we have $\delta_{BC} = 0$:

$$\delta_{BC} = 0 = \delta_{BC_{\text{due to temp.}}} + \delta_{BC_{\text{due to } F_C}} = \alpha L(\Delta T) - F_C L / (AE)$$

or

$$\alpha L(\Delta T) = F_C L / (AE)$$

Therefore

$$F_C = AE \, \alpha \, \Delta T$$
$$= 2(15 \times 10^6)(11 \times 10^{-6})(30)$$
$$= 9900 \text{ lb,}$$

a surprisingly large force.

EXAMPLE 5.16

The truss (see Figure E5.16a) of Example 5.12, unloaded, experiences a temperature drop of 20°C. Find the deflection of joint A.

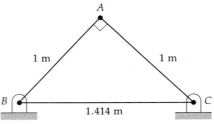

Figure E5.16a

Solution

The members AB and AC, if the pin at A were removed, would each contract the amount

$$\delta = \alpha L \, \Delta T$$

$$= 11.7 \times 10^{-6}(1)20 = 234 \times 10^{-6} \text{ m} \quad \text{(or 234 } \mu\text{m*)}$$

The final position of A lies on the intersection of the two lines drawn perpendicular to the bars after their shortenings, as seen in Figure E5.16b. Therefore, the displacement of A is $(234 \times 10^{-6}) \dfrac{\sqrt{2}}{2}$ (2), or 331 μm$\downarrow$.

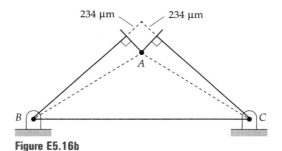

Figure E5.16b

In the preceding example, the bar BC, constrained by the pins at its ends, cannot relieve its thermal stress by straining, as did AB and AC. Instead it is left with the tensile stress (see Example 5.15):

$$\sigma_{BC} = \frac{F_{BC}}{A_{BC}} = \frac{\cancel{A}E \, \alpha \, \Delta T}{\cancel{A}} = 207(10^9)(11.7 \times 10^{-6})(20) \text{ Pa}$$

$$= 48.4 \text{ MPa}$$

---

* One μm ("micro-meter") is 10^{-6} m.

PROBLEMS ▶ Section 5.7

Find the stresses in each member of the trusses in Problems 5.91–93.

5.91

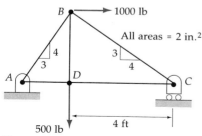

500 lb

4 ft

Figure P5.91

5.92

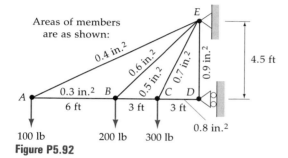

100 lb   200 lb   300 lb

Figure P5.92

5.93

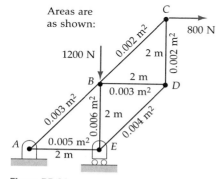

Figure P5.93

5.94 A wooden box-beam has the cross-section shown in Figure P5.94(a). Its modulus of elasticity is 1800 ksi, and it is centrally loaded through a plate as shown in Figure P5.94(b). Find the stress in the member.

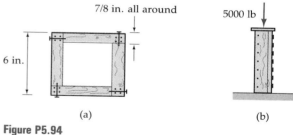

(a)   **(b)**

Figure P5.94

5.95 Find the stresses in members *BG* and *CF* of the truss in Figure P5.95 if their respective areas are 0.002 m² and 0.0025 m².

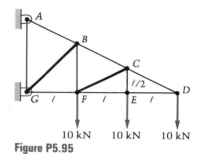

10 kN   10 kN   10 kN

Figure P5.95

5.96 Find the stress in member *DF* of the truss in Figure P5.96 if the area of each member is 3 in.² and it is made of a material having $E = 24 \times 10^6$ psi.

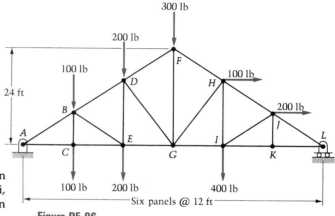

Figure P5.96

5.97 The 500-N force is applied to a bar made of two connected parts as shown in Figure P5.97. Find the normal stresses in the sections LC and CR.

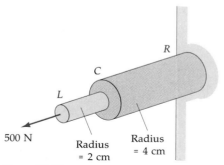

Figure P5.97

5.98 In Figure P5.98, the nylon cords each have area 0.1 in.^{2} If the breaking stress of the cord is 8 ksi, determine how much weight W can be suspended.

Figure P5.98

5.99 Members DE, DG, and HG of the truss in Figure P5.99 have respective areas 0.0038 m^2, 0.0030 m^2, and 0.0060 m^2. Which of the three members is being subjected to the largest stress?

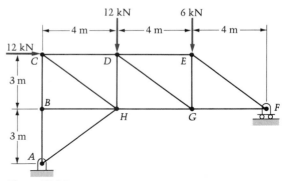

Figure P5.99

5.100 Find the stress in the two-force member AB of the frame in Figure P5.100, if AB is a 1-inch diameter steel rod.

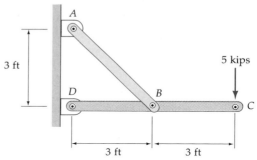

Figure P5.100

5.101 Find the stress in the two-force member BD of the frame in Figure P5.101, if BD is a 2-cm diameter aluminum rod.

5.102 Find the stress in member AB of the truss of Figure P5.102. The area of AB is 3 in.^{2}.

5.103 The toggle device in Figure P5.103 is being used to crush rocks. If the pressure in the chamber, p, is 70 psi and the radius of the piston is 6 in., find the stresses in the three two-force members AB, BC, and BD. The piston rod BD has diameter 1 in., and AB and BC each have diameter 1.5 in.

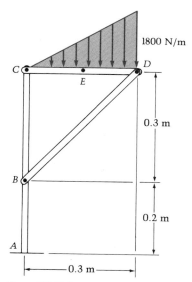

Figure P5.101

1800 N/m

0.3 m

0.2 m

0.3 m

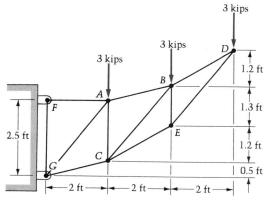

Figure P5.102

3 kips

3 kips

3 kips

1.2 ft

1.3 ft

1.2 ft

0.5 ft

2.5 ft

2 ft   2 ft   2 ft

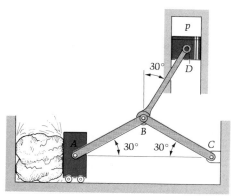

Figure P5.103

5.104 Find the maximum stress in the hollow cylinder in Figure P5.104 due to its own weight.

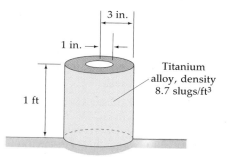

3 in.

1 in.

Titanium alloy, density 8.7 slugs/ft³

1 ft

Figure P5.104

5.105 The solid steel bar (density 7850 kg/m³) in Figure P5.105 is suspended from a ceiling. Find the maximum stress in the bar caused by its own weight.

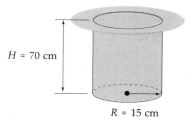

$H = 70$ cm

$R = 15$ cm

Figure P5.105

* 5.106 Where on line xx in Figure P5.106 should the pin D be placed (with B staying beneath it) so that the stress in DC is minimized? The volume of DC, i.e. $A_{DC}L_{DC}$, is to remain equal to that of DB which is $A_{DB}L_{DB} = (4 \text{ in.}^2)(48 \text{ in.}) = 192 \text{ in.}^3$.

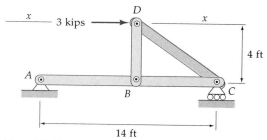

x 3 kips D x

4 ft

A B C

14 ft

Figure P5.106

5.107 In Problem 5.100, find the elongation of the two-force member AB.

5.108 In Problem 5.101, find the elongation of member BD.

5.109 In Problem 5.102, find the elongation of member *AB* if the truss is constructed of aluminum.

5.110 Find the changes in length of the steel members *AB*, *BC*, and *BD* in Problem 5.103 if they each have length 2 ft.

5.111 In Problem 5.95, find the elongations of the two members *BG* and *CF* caused by the applied loads. The members are made of structural steel.

5.112 In Problem 5.99, if the members are all made of aluminum, which of the three members *DE*, *DG*, and *HG* has the largest elongation or contraction under the given loading? (See Figure P5.99.) Which has the smallest? Would the order be the same if the members were made of steel?

5.113 If the box-beam in Problem 5.94 is 6 feet long, how much does it shorten under the applied 5-kip loading?

5.114 In Problem 5.97 (see Figure P5.114), find the total elongation of the bar (made of two parts) if:

    a. section *LC* is made of steel and *CR* is of aluminum;

    b. section *LC* is made of aluminum and *CR* is of steel.

5.115 The truss member of Figure P5.115 increases in length by 1 cm after the 10 kN load is applied. From this information, find the modulus of elasticity of the member.

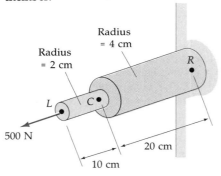

Figure P5.114

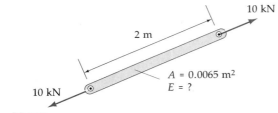

Figure P5.115

5.116 The bar *AB* in Figure P5.116 is a brass rod of diameter 1.5 in. Find the total change in length of *AB* after the four forces are applied as shown.

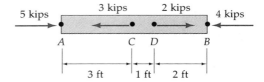

Figure P5.116

5.117 The beam *BD* in Figure P5.117 is supported by the wires *AB* and *CD*, which have equal areas and elastic moduli. Find the length of *CD*, in terms of *L*, for which *BD* stays horizontal after the force *P* is applied as shown.

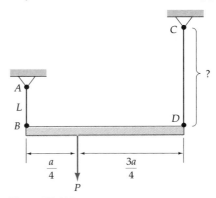

Figure P5.117

5.118 Find the displacement of joint *A* in Example 5.12 if the areas are changed to 0.003 m² for member *AB* and 0.005 m² for *AC*.

* 5.119 Find the displacement of joint *C* following application of the 2000 lb force shown in Figure P5.119. Data on the members is:

    BC: aluminum, *L* = 3 ft, *A* = 3 in.²

    AC: steel, *L* = 4 ft, *A* = 2 in.²

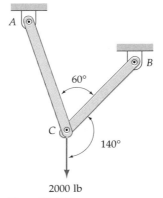

Figure P5.119

5.120 Find the final position of joint *C* in Figure P5.120 if the rods are constituted as follows:

    a. *AC*: bronze, Area 0.5 in.²

    b. *BC*: copper, Area 0.4 in.²

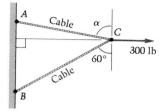

Figure P5.120

5.121 Show that the vertical component δ_y of the deflection of joint *Q* caused by the force *P* acting on the simple plane truss in Figure P5.121 is given by the equation

$$\delta_y = \frac{PL}{AE}\left(\frac{1}{\sin^2\theta\cos\theta} + \frac{1}{\tan^2\theta}\right)$$

where *P*, *L*, and θ are shown in the figure, *A* is the area of members *BQ* and *CQ*, and *E* is a constant.

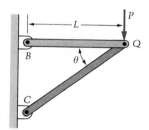

Figure P5.121

5.122 Find the displacements of joints *B* and *C* of the truss of Problem 5.10 (see Figure P5.122). The bars are made of steel, each with an area 0.004 m².

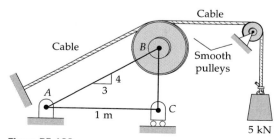

Figure P5.122

5.123 Find the displacement of joint *A* in Example 5.12 if there is a roller at *C* as shown in Figure P5.123, instead of the joint being pinned to the ground.

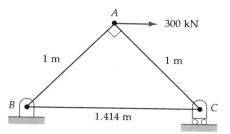

Figure P5.123

5.124 Repeat Example 5.13 if the 5 kN force is applied at the mid-point of *BD* (in the same direction), instead of at the juncture of the two bars.

5.125 Find the downward displacement of the 100-kip rigid block *W* in Figure P5.125. Also find the force in each bar.

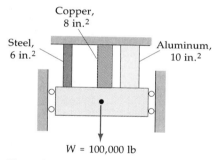

Figure P5.125

5.126 The rigid bar *AB* is supported by the pin at *A* and the two members *CE* and *DF* in Figure P5.126. Find the small angular clockwise rotation of *AB* about *A* when the 5-kN force is applied as shown.

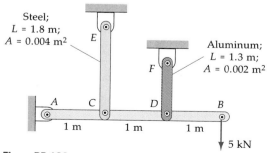

Figure P5.126

5.127 Find the temperature increase in °C that will expand the bar so as to barely close the gap in Figure P5.127.

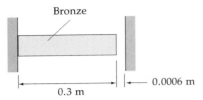

Bronze

0.0006 m

0.3 m

Figure P5.127

5.128 In the preceding problem, if the bar has area 0.005 m² and if the temperature change from the position shown in the figure is 110°C, what are the reaction forces from the walls onto the bar?

* **5.129** In Problem 5.125, find the forces in the bars if an 80°F temperature rise is imposed on the system.

* **5.130** In Problem 5.126, determine the angular rotation change if the structure is cooled by the amount $\Delta T = -20°C$ with the 5kN force acting as shown.

II / SYSTEMS CONTAINING MULTIFORCE MEMBERS

5.8   Axial and Shear Forces and Bending Moments

The Multiforce Member

In this section we extend the work we have done on trusses to structures containing "multiforce members." Such a member is not a straight two-force bar, as were all the members of a truss. Consequently, on cross-sectional cuts, a member will be subjected to more than just an axial force.

Consider Figure 5.37. The half-ring is pulled at the ends by forces P as shown. Though the half-ring is a two-force member, it is not a *straight* two-force member. Hence its various cross sections are subjected to more than just the now-familiar axial forces of a truss. Consider the free-body diagrams shown in Figure 5.38. For equilibrium, we need equal magnitude, oppositely directed forces P as shown, together with the indicated moments M. Considering the shaded section, equilibrium requires

P P

Figure 5.37

$$\curvearrowleft_{+} \quad \Sigma M_A = 0 = M - P(R \sin \theta)$$

or

$$M = PR \sin \theta$$

This moment (called a bending moment because it tends to bend the ring) is seen to vanish only at $\theta = 0$ and 180°.

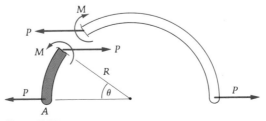

Figure 5.38

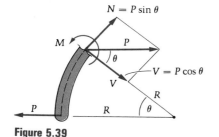

Figure 5.39

$N = P \sin \theta$

M

P

θ

$V = P \cos \theta$

V

R

θ

P

R

> **Question 5.14**   Looking at Figure 5.38, why is it clear that M must vanish at $\theta = 0$ and $180°$?

Our member above thus differs from a truss member in that it carries a bending moment. But also, the force acting on the cross sections has become complicated.

We see in Figure 5.39 that if we resolve the force at the cut into its two components (a) tangent to the center line of the curved ring, and (b) normal to the center line, we obtain, respectively, in addition to the familiar axial force (N) studied in trusses, a shear force (V). Note from the figure that N and V, as was the case with M, vary with θ. This is another distinct difference from a truss, in which the (axial) member forces were constant from end to end.

The body need not be curved for V and M to be present. Consider a second example, in which we seek N, V, and M at isolated points.

Answer 5.14   Because the distance between the pair of forces P goes to zero at those two points.

EXAMPLE 5.17

Find the axial force, shear force, and bending moment at points P and Q of the frame in Figure E5.17a.

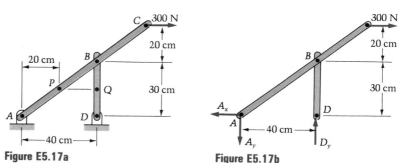

Figure E5.17a   **Figure E5.17b**

Solution
The free-body diagram of the complete frame is shown in Figure E5.17b. Note that BD is a two-force member, which will allow us to determine A_x and A_y, and also the force D_y in BD, from this single free-body diagram.

$$\curvearrowleft{+} \quad \Sigma M_A = 0 = D_y(40 \text{ cm}) - (300 \text{ N})(50 \text{ cm})$$

$$D_y = 375 \text{ N}$$

Thus

$$+\uparrow \quad \Sigma F_y = 0 = D_y - A_y = 375 - A_y$$

$$A_y = 375 \text{ N}$$

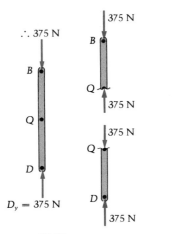

∴ 375 N

$D_y = 375$ N

Figure E5.17c

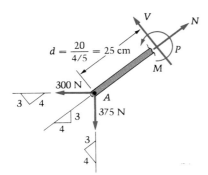

$d = \dfrac{20}{4/5} = 25$ cm

Figure E5.17d

Also,

$$\xrightarrow{+} \qquad \Sigma F_x = 0 = 300 - A_x \Rightarrow A_x = 300 \text{ N}$$

The force at Q is therefore 375 N ©, as shown on both the upper and lower cut sections of BD (Figure E5.17c).

Sectioning member ABC at P will expose the forces N, V, and M that we seek (see Figure E5.17d, the free-body diagram of AP):

$$\Sigma F_x = 0 = N - 0.8(300) - 0.6(375)$$

$$N = 465 \text{ N}$$

$$\Sigma F_y = 0 = V + 0.6(300) - 0.8(375)$$

$$V = 120 \text{ N}$$

$$(+) \qquad \Sigma M_P = 0 = M + 0.8(375)25 - 0.6(300)25$$

$$M = -3000 \text{ N} \cdot \text{m}$$

Therefore the forces are in the directions sketched in Figure E5.17d, while the bending moment is in the opposite (or clockwise) direction to that shown on the free-body diagram.

The next example is similar to the last one, but it is a bit longer and requires more free-body diagrams.

EXAMPLE 5.18

In the frame of Example 4.19, find the axial force, shear force, and bending moment at the points P, Q, and R in Figure E5.18a.

Solution

We had found the results in Example 4.19 by dismantling the frame, as shown in the three figures E5.18b, E5.18c and E5.18d below. Making a cut at P in member

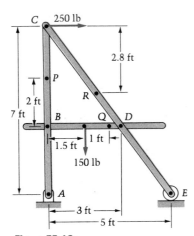

Figure E5.18a

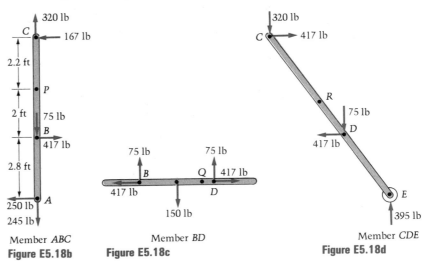

Member ABC
Figure E5.18b

Member BD
Figure E5.18c

Member CDE
Figure E5.18d

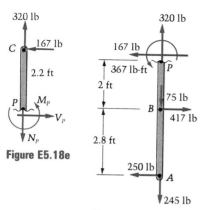

Figure E5.18e

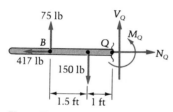

Figure E5.18f

ABC of Figure E5.18b, we have, with the help of the resulting free-body diagram in Figure E5.18e:

$$\xrightarrow{+} \qquad \Sigma F_x = 0 = V_P - 167 \Rightarrow V_P = 167 \text{ lb}$$

$$+\uparrow \qquad \Sigma F_y = 0 = 320 - N_P \Rightarrow N_P = 320 \text{ lb}$$

$$\curvearrowleft_{+} \qquad \Sigma M_P = 0 = M_P + (167 \text{ lb})(2.2 \text{ ft}) \Rightarrow M_P = -367 \text{ lb-ft}$$

Note that if we had chosen to use the lower section of *ABC*, shown in Figure E5.18f, we would have had more forces to deal with but we would have obtained consistent results — the opposites* of those on the upper section, in agreement with action and reaction. Either set of answers (shown on the proper section) is correct.

For the internal force system at *Q*, we cut the bar *BD* of Figure E5.18c there. We can use either side of the cut:

Using the left section, Figure E5.18g below,

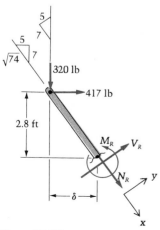

Figure E5.18g

$$\xrightarrow{+} \qquad \Sigma F_x = 0 = N_Q - 417 \Rightarrow N_Q = 417 \text{ lb}$$

$$+\uparrow \qquad \Sigma F_y = 0 = 75 - 150 + V_Q \Rightarrow V_Q = 75 \text{ lb}$$

$$\curvearrowleft_{+} \qquad \Sigma M_Q = 0 = M_Q + (150 \text{ lb})(1 \text{ ft}) - (75 \text{ lb})(2.5 \text{ ft})$$

$$M_Q = 37.5 \text{ lb-ft}$$

Alternatively, using the right section, Figure E5.18h below,

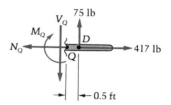

Figure E5.18h

$$\xrightarrow{+} \qquad \Sigma F_x = 0 = 417 - N_Q \Rightarrow N_Q = 417 \text{ lb}$$

$$+\uparrow \qquad \Sigma F_y = 0 = 75 - V_Q \Rightarrow V_Q = 75 \text{ lb}$$

$$\curvearrowleft_{+} \qquad \Sigma M_Q = 0 = -M_Q + (75 \text{ lb})(0.5 \text{ ft})$$

$$M_Q = 37.5 \text{ lb-ft}$$

The results agree, as they must, for each of the axial force (N_Q), shear force (V_Q), and bending moment (M_Q). Note that we took care of "Newton's Third Law" when we assigned the directions on the free-body diagrams.

For the internal force system at *R*, we use a cut section of *CDE* from Figure E5.18d, as Figure E5.18i. From similar triangles, $\dfrac{2.8}{\delta} = \dfrac{7}{5} \Rightarrow \delta = 2$ ft. We can avoid having to solve two equations in V_R and N_R by summing forces in the slanted coordinate directions *x* and *y* in the figure.

$$\searrow^{+} \qquad \Sigma F_x = 0 = N_R + 320 \left(\frac{7}{\sqrt{74}} \right) + 417 \left(\frac{5}{\sqrt{74}} \right)$$

$$N_R = -503 \text{ lb}$$

Figure E5.18i

---

* The moment is off by 1 lb-ft due to rounding to three digits in the original example.

$$\measuredangle \qquad \Sigma F_y = 0 = V_R - 320\left(\frac{5}{\sqrt{74}}\right) + 417\left(\frac{7}{\sqrt{74}}\right)$$

$$V_R = -153 \text{ lb}$$

$$\curvearrowleft{(+)} \qquad \Sigma M_R = 0 = M_R + (320 \text{ lb})(2 \text{ ft}) - (417 \text{ lb})(2.8 \text{ ft})$$

$$M_R = 528 \text{ lb-ft}$$

(Note that for the moment calculation we have used the original horizontal and vertical force and moment arm components.)

The reader is encouraged to check the above results using the lower cut section of member *CDE*.

In the last example in this section, we take a detailed look at the variation in the internal force system elements (*N*, *V*, and *M*) along the axial direction of a member. This example is in preparation for our study of beams, to follow in the next section.

EXAMPLE 5.19

Determine the distribution of internal forces in the bent member *ABC* shown in Figure E5.19a.

Solution

In contrast to the first two examples, we are looking for *N*, *V*, and *M* at *all* points instead of just two or three. In other words, we wish to know how *N*, *V*, and *M* vary along the member.

The three reactions are found first. Using the overall free-body diagram in Figure E5.19b,

$$\curvearrowleft{(+)} \qquad \Sigma M_A = 0 = D(3 \text{ ft}) - (18 \text{ lb})(1.5 \text{ ft})$$

$$D = 9 \text{ lb}$$

$$\xrightarrow{+} \qquad \Sigma F_x = 0 = 18 \text{ lb} - A_x \Rightarrow A_x = 18 \text{ lb}$$

$$+\uparrow \qquad \Sigma F_y = 0 = A_y + D = A_y + 9$$

or

$$A_y = -9 \text{ lb}$$

Next we consider a free-body diagram of a portion of the horizontal section *AB* of the bar, extending from the left end *A* to a cut section with *x* < 3 ft (see Figure E5.19c).

For equilibrium, we see that not only do we need the normal force N_{H1} (as with trusses), but also a shear force V_{H1} and bending moment M_{H1}:

$$\xrightarrow{+} \qquad \Sigma F_x = 0 = N_{H1} - 18 \text{ lb} \Rightarrow N_{H1} = 18 \text{ lb}$$

$$+\uparrow \qquad \Sigma F_y = 0 = -9 \text{ lb} + V_{H1} \Rightarrow V_{H1} = 9 \text{ lb}$$

$$\curvearrowleft{(+)} \qquad \Sigma M_Q = 0 = M_{H1} + (9 \text{ lb})(x \text{ ft}) \Rightarrow M_{H1} = -9x \text{ lb-ft}$$

After *x* becomes greater than 3 ft, the free-body diagram changes. At a cut section such that 3 ft < *x* < 5 ft, we now have (see Figure E5.19d):

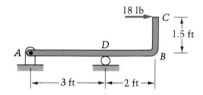

Figure E5.19a

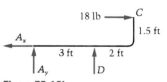

Figure E5.19b

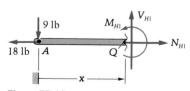

Figure E5.19c

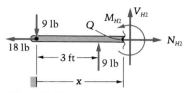

Figure E5.19d

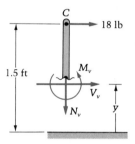

Figure E5.19e

$$\xrightarrow{+} \quad \Sigma F_x = 0 = N_{H2} - 18 \text{ lb} \Rightarrow N_{H2} = 18 \text{ lb} \quad \text{(still the same)}$$

but

$$+\uparrow \quad \Sigma F_y = 0 = 9 \text{ lb} - 9 \text{ lb} + V_{H2} \Rightarrow V_{H2} = 0$$

and

$$\overset{\curvearrowleft}{+} \quad \Sigma M_Q = 0 = M_{H2} + (9 \text{ lb})(x \text{ ft}) - (9 \text{ lb})(x - 3) \text{ ft}$$

Note that this equals the constant 27 lb-ft, the strength of the couple formed by the 9-lb forces!

$$M_{H2} = -27 \text{ lb-ft}$$

Note that N_{H2} and M_{H2} are continuous across the cut at $x = 3$ ft, but that there has occurred a jump (or discontinuity) in the shear force V_{H2} caused by the concentrated reaction at the roller.

The above expressions for N_{H2}, V_{H2}, and M_{H2} are valid until $x = 5$ ft when the bend occurs. On a cut past the bend at P, we may use the upper section from P to C to most easily determine what is happening in the vertical part (BC) of the bar. (See Figure E5.19e.) Enforcing equilibrium gives

$$\xrightarrow{+} \quad \Sigma F_x = 0 = 18 \text{ lb} + V_v \Rightarrow V_v = -18 \text{ lb}$$

$$+\uparrow \quad \Sigma F_y = 0 = -N_v \Rightarrow N_v = 0$$

$$\overset{\curvearrowleft}{+} \quad \Sigma M_P = 0 = M_v - (18 \text{ lb})(1.5 - y) \text{ ft}$$

$$M_v = (27 - 18y) \text{ lb-ft}$$

If we now look at a free-body diagram of the bend alone, we see that it, too, is of course in equilibrium. (See Figure E5.19f.) We note that the discontinuity of V and N in "rounding the bend" is somewhat artificial. It is due to their exchange of roles caused by the discontinuity at the corner in the orientation of the center line.

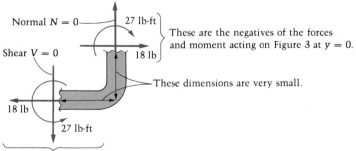

These are the negatives of the forces and moment acting on Figure 2 at $x = 5$ ft.

Figure E5.19f

We mention again that when we wish to know the internal force resultant (N, V, and M) at a point, we may use either of the two free-body diagrams formed by the cut section. For instance, if the section to the right of the first cut is used (as in Figure E5.19g) instead of the section in Figure E5.19c, then equilibrium requires:

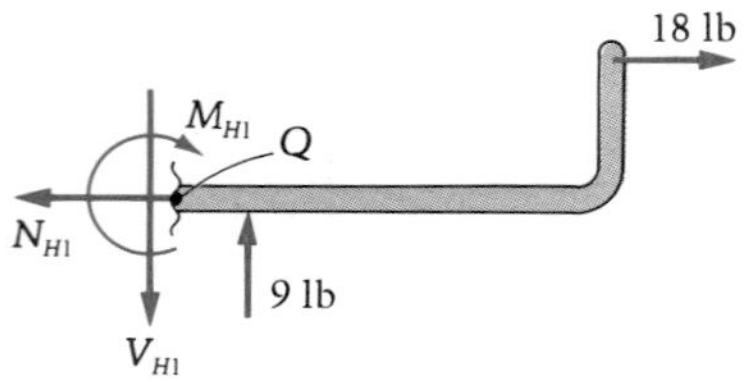

18 lb

9 lb

N_{H1}

M_{H1}

Q

V_{H1}

Figure E5.19g

$$\xrightarrow{+} \quad \Sigma F_x = 0 = 18 - N_{H1} \Rightarrow N_{H1} = 18 \text{ lb} \quad \text{(as before)}$$

$$+\uparrow \quad \Sigma F_y = 0 = -V_{H1} + 9 \Rightarrow V_{H1} = 9 \text{ lb} \quad \text{(as before)}$$

$$\curvearrowleft_{+} \quad \Sigma M_Q = 0 = -M_{H1} - (18 \text{ lb})(1.5 \text{ ft}) + (9 \text{ lb})(3 - x) \text{ ft}$$

$$M_{H1} = -9x \text{ lb-ft} \quad \text{(as before)}$$

Thus all three results are in agreement with those we obtained using the material on the other side of the cut.

PROBLEMS ▶ Section 5.8

5.131 Find the internal forces at section a-a on bar B shown in Figure P5.131.

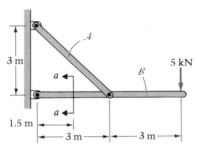

Figure P5.131

5.132 Find the internal forces at section a-a on bar BCD shown in Figure P5.132.

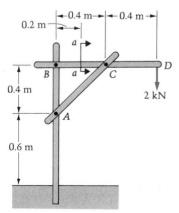

Figure P5.132

5.133 In Figure P5.133 a weight of 3000 N is suspended from beam B at point C. The beam is connected to a vertical wall by a frictionless pin and a cable as shown.

If the beam (of negligible weight) has length 3 m and the length AC is 1 m, find the internal forces and moment transmitted at section E-E.

5.134 In Figure P5.134:

a. Find the force exerted on the shaded member in Figure P5.134 by pin A.

b. Find the internal forces (axial, shear, and bending moment) at the midpoint M of the horizontal part of the shaded member.

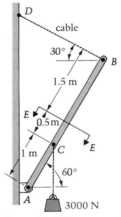

Figure P5.133

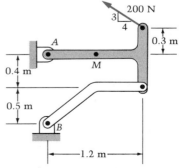

Figure P5.134

5.135 The two boards in Figure P5.135 have been glued together and are being held as they dry by a C-clamp. If the compressive force holding the boards together is 20 lb, find the internal force system at sections A-A, B-B, and C-C.

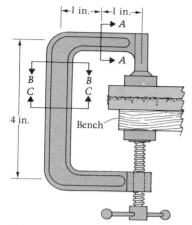

Figure P5.135

5.136 In Figure P5.136 a beam of length L is fastened into a wall at angle α with respect to the horizontal. An oil drum of weight W is slung under the beam using a cable as shown fastened at $L/4$ and $3L/4$ to the beam. If $\alpha = 30°$ and $\beta = 30°$, find the axial force, shearing force, and bending moment at the midpoint of the beam. Assume that the beam is weightless.

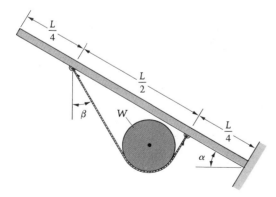

Figure P5.136

5.137 Find the internal force resultants (axial, shear, and bending moment) at sections A-A, B-B, and C-C in Figure P5.137.

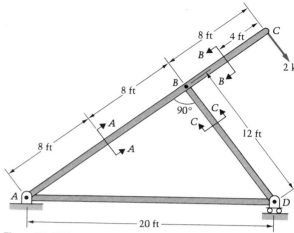

Figure P5.137

5.138 In Figure P5.138 a weight of 1000 lb is suspended from the beam at point C. The beam is connected to a vertical wall by a frictionless pin at A and by a cable between B and D. Neglecting the weight of the beam, determine the internal forces and moment transmitted across section E-E.

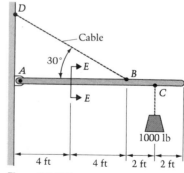

Figure P5.138

In Problems 5.139–5.144 find the shear force and bending moment in the beam at point P. Show the result on a free-body diagram of the part of the beam between P and B.

5.139

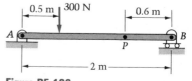

Figure P5.139

(For Problems 5.140–144, see the instructions preceding problem 5.139.)

5.140

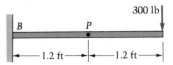

Figure P5.140

5.141

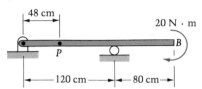

Figure P5.141

5.142

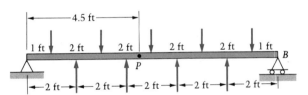

Each of the nine forces has magnitude 25 lb.

Figure P5.142

5.143

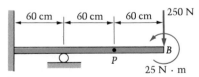

Figure P5.143

5.144

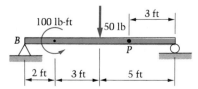

Figure P5.144

5.145 Find the shear force and bending moment on section A-A of the beam in Figure P5.145.

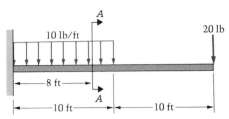

Figure P5.145

5.146 In Figure P5.146 find the values of the shear and axial forces and the bending moment at point E midway between C and D. Show your results on a sketch of the cut section at E.

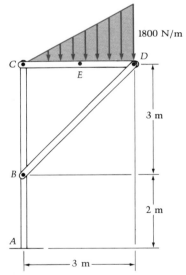

Figure P5.146

5.147 For the frame in Figure P5.147, compute the internal forces and bending moment acting within the member DC on section a-a.

5.148 Find the internal forces and moments at points E and F of the light frame shown in Figure P5.148.

5.149 Calculate the axial force, shear force, and bending moment at the midpoint D of the portion AB of the bent bar ABC in Figure P5.149. Show the results on a complete free-body diagram of AD.

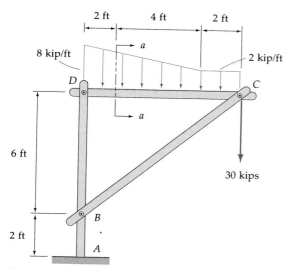

Figure P5.147

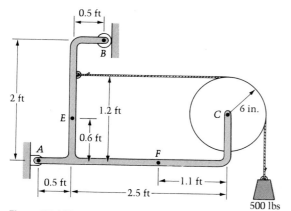

Figure P5.148

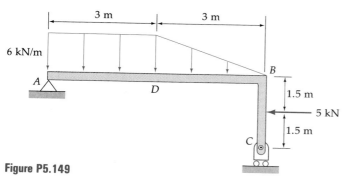

Figure P5.149

5.150 In Problem 4.195, find the internal forces and moments acting on a cross-section of member *BED* 30 cm above point *D* (see Figure P5.150).

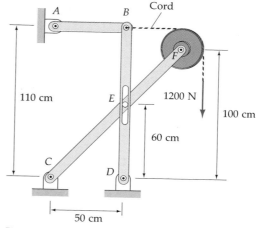

Figure P5.150

5.151 In Problem 4.195, find the axial force, shear force, and bending moment acting on a cross-section of member *CEF* 40 cm from point *C* (see Figure P5.150).

5.152 In Examples 4.11 and 4.24, draw a free-body diagram of the part of the boom between 1 ft and 3 ft from the left end. Use results from the two examples to determine all forces acting on the ends of the free-body, then verify that the equilibrium equations are satisfied.

5.153 In Problem 4.32, find the internal forces (V, N, M) within the bar at a point 5 ft to the right of pin *B*. Show the results on a sketch which includes the part of the bar to the left of the cut section (i.e., which includes points *B*, *A*, and *Q*). See Figure P5.153.

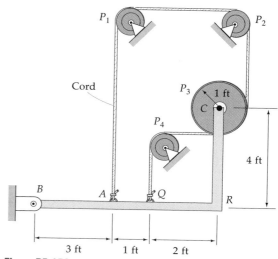

Figure P5.153

5.154 The semicircular arch in Figure P5.154 is loaded as shown by the two forces P. Find the internal forces acting on the cross-section at D, and show them on a FBD of the half DB of the arch.

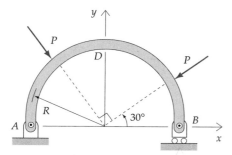

Figure P5.154

5.9   Beams/Shear and Moment Distributions

In the preceding section, we studied the internal force system (N, V, M) acting within a cross section of a planar structure. More precisely, N, V, and M represent a force and couple system equipollent to the infinitely many differential forces exerted on the material on one side of the cut by the material on the other side.

To describe the resultant internal force system under a more general loading than the planar case, we need not just three, but *six* components at a selected point in the section: three of force and three of moment. To see this, let us slice a plane section through a body $\mathcal{B}$, producing two parts $\mathcal{B}_1$ and $\mathcal{B}_2$ as shown in Figure 5.40.

The forces exerted upon $\mathcal{B}_1$ by $\mathcal{B}_2$ are equipollent to a force and a couple at any selected point (such as A). In general, these vectors each have three components, as shown in Figure 5.41 at the point A in the plane of the cut.

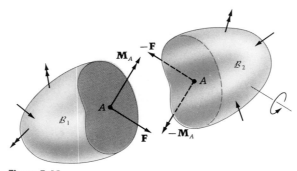

Figure 5.40

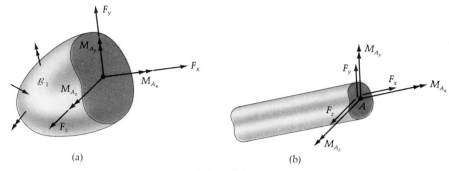

Figure 5.41

The two force components lying *in* the plane of the cut, F_y and F_z in this case (see Figure 5.41), are called shear forces. The force perpendicular to the plane of the cut (here, F_x) is known as a normal force. When $\mathcal{B}_1$ is a slender member, as shown in Figure 5.41(b), the two moment components whose vector directions (thumb of right hand) lie *in* the plane of the cut are called bending moments (M_{Ay}, M_{Az} here), while the other one, M_{Ax}, with vector *normal* to the plane of the cut and turning effect in the plane, is called a twisting moment.* The same six components acting at A on $\mathcal{B}_2$ would each be respectively opposite in direction, by the principle of action and reaction.

When $\mathcal{B}$ is a slender member that carries shear force(s), bending moment(s), and / or a twisting moment, in addition to the axial force of a truss member, it is called a **beam.**

A beam can be loaded in two planes such that it carries a perpendicular pair of shearing force and bending moment components. It may also be loaded so that the twisting moment component is present. In this elementary look at beams, we shall, however, restrict our attention to the case in which the beam is loaded in just one (xy) plane. That is, all the forces will lie in a plane and all the couples will be normal to that plane (in the $\pm z$ direction).

Thus, only one shear force (F_y, which we shall call V) and one bending moment (M_{Az}, which we shall call M) will be produced, in addition to the axial load (F_x, called N). The twisting moment M_{Ax} will necessarily be absent, as will the other component of shear force (F_z) and of bending moment (M_{Ay}). In Figure 5.42, on the following page, we illustrate these ideas. With all the forces in the plane of the paper, and with the couples normal to this plane, the only resultant of all the internal forces at a cut section at x is expressible as the shear force V, axial force N, and bending moment M.

---

* The student can readily appreciate this by successively applying moment components to a slender body such as a yardstick.

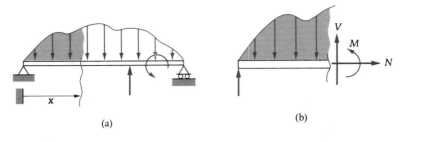

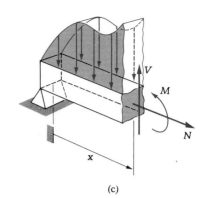

(a)

(b)

(c)

Figure 5.42

Sign Conventions for Internal Forces in Beams

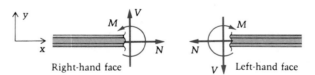

Outward normals

Right-hand face   Left-hand face

Figure 5.43

We next set down our sign conventions for N, V, and M.* Suppose a beam has been cut, or "sectioned," into two parts. If we are considering the section on the left, then the exposed face is called a "right-hand face" (with outward normal to the right). If we are considering the section on the right, then the exposed face is on the left (outward normal to the left), and is called a "left-hand face" (see Figure 5.43).

We are now in a position to define our sign conventions for N, V, and M (see Figure 5.44).

Figure 5.44   Sign convention for axial force, shear force, and bending moment in beams.

As seen in Figure 5.44, the normal (or axial) force N is defined as positive on either type of face if it is in the direction of the outward normal — that is, to the right for a right-hand face and to the left for a left-hand face. In other words, in both cases N is positive if it tends to produce tension in (or to stretch) the axial fibers aligned with x. Saying that "N is in the direction of the outward normal" is shorthand for saying that the scalar N multiplies a unit vector in the direction of the outward normal to form the axial force.

As for the shear force V, it is defined to be positive if directed upward on a right-hand face and positive if downward on a left-hand face. The bending moment is positive if counterclockwise on a right-hand face and if clockwise on a left-hand face. Thus M is positive in both instances if it

---

* Unfortunately, these conventions vary from book to book, especially for the shear force V, and one must be aware of this when referring to other texts.

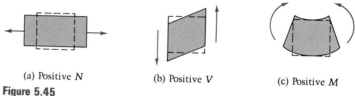

(a) Positive N (b) Positive V (c) Positive M

Figure 5.45

bends the section upward (i.e., toward a concave upward configuration). Perhaps the summary sketches shown in Figure 5.45 will be helpful.

Consider next the typical beam shown in Figure 5.46. Such a beam, supported on one end by a pin and on the other by a roller, is said to be "simply supported."

In presenting beam problems, the pin is usually drawn as either

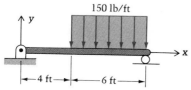

Figure 5.46

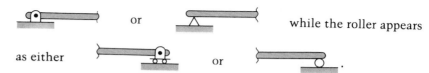

or   while the roller appears

as either   or   .

As we have already seen in Chapter 4, the pin is able to exert two components of force (for example, horizontally and vertically) on the beam, while the roller exerts only a component normal to its contact plane with the beam. (This is a vertical force for the roller in Figure 5.46.) The purpose of the roller is to allow the beam to slightly move so as to prevent or relieve large axial stretching stresses when it is loaded, and / or the potentially large axial stretching or compressing stresses when it is cooled or heated. Thus for a simply supported beam without applied axial (x direction) loads, the horizontal component of the pin reaction will be zero. Let us proceed to find the distribution of shear force and bending moment in the beam of Figure 5.46.

EXAMPLE 5.20

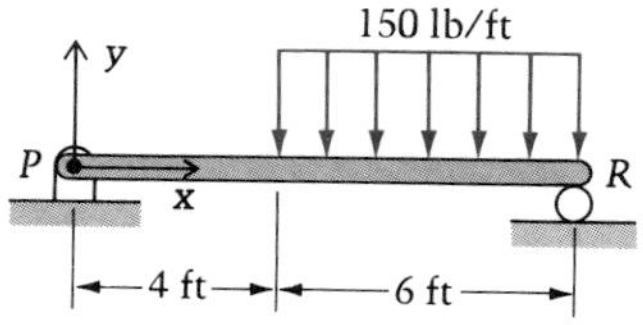

Find the distribution (as functions of x) of the normal (N) and shear (V) forces and bending moment (M) for the beam shown in Figure E5.20a.

Solution

For finding the reactions, we replace the uniformly distributed load by the equipollent single force shown in the free-body diagram (Figure E5.20b). Note that $P_x = 0$ because it is the only external force in the x direction. Summing moments about the pin gives the roller reaction:

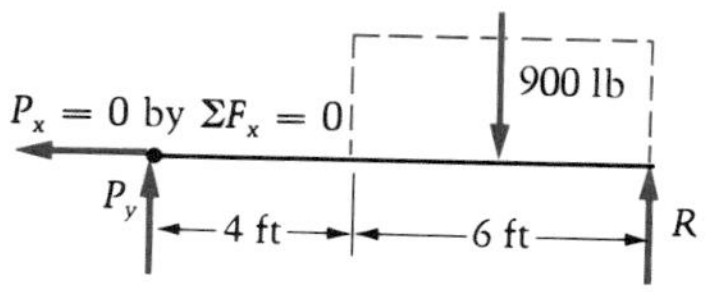

$$\circlearrowleft{+} \quad \Sigma M_P = 0 = R(10 \text{ ft}) - (900 \text{ lb})(7 \text{ ft})$$

$$R = 630 \text{ lb}$$

Figure E5.20a

Figure E5.20b

Then

$$+\uparrow \qquad \Sigma F_y = 0 = P_y + R - 900 \Rightarrow P_y = 270 \text{ lb}$$

As a check,

$$\overset{\curvearrowright}{(+)} \qquad \Sigma M_R = 0 = (900 \text{ lb})(3 \text{ ft}) - (270 \text{ lb})(10 \text{ ft}) = 0$$

Now we proceed to the determination of $V(x)$ and $M(x)$; note that $N(x) \equiv 0$ in this beam.

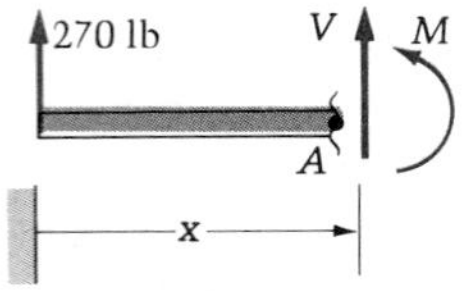

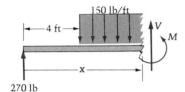

Figure E5.20c

Question 5.15   Why?

If we make a sectional cut between $x = 0$ and $x = 4$ ft, we can obtain expressions for V and M valid throughout that interval (see Figure E5.20c).

We always place V and M (and N, in beams where it is present) on the cut sections in positive directions according to our sign conventions. Hence in Figure E5.20c, V is drawn upward and M counterclockwise on the exposed right-hand face. Equilibrium then requires

$$+\uparrow \qquad \Sigma F_y = 0 = 270 + V \Rightarrow V = -270 \text{ lb} \qquad (1)$$

$$\overset{\curvearrowright}{(+)} \qquad \Sigma M_A = 0 = M - 270x \Rightarrow M = 270x \text{ lb-ft} \qquad (2)$$

The reader should note that at $x = 0$, these expressions give the correct V and M — namely, a negative shear force of 270 lb (up on a left-hand face is negative V) and a zero moment.

From $x = 4$ ft to $x = 10$ ft, Equations (1) and (2) will not be valid because the distributed load "starts" at $x = 4$ ft and was not included in the free-body diagram of Figure E5.20c. Thus we must make a new cut whenever something changes, such as a distributed load starting or ending, or a concentrated load or couple appearing. The new free-body diagram (Figure E5.20d) will be valid for the rest of the beam (4 ft $< x <$ 10 ft).

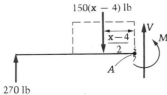

Figure E5.20d

For purposes of finding V and M from the equilibrium equations, we may replace the (part of the) distributed load by its equipollent "force-alone resultant," shown in Figure E5.20e. Then we obtain, from the equilibrium equations for the material to the left of the cut section,

$$+\uparrow \qquad \Sigma F_y = 0 = 270 - 150(x - 4) + V$$

$$V = 150(x - 4) - 270 \text{ lb}$$

Figure E5.20e

and

$$\overset{\curvearrowright}{(+)} \qquad \Sigma M_A = 0 = -270 \text{ lb}(x \text{ ft}) + [150(x - 4) \text{ lb}]\left[\left(\frac{x - 4}{2}\right) \text{ ft}\right] + M$$

$$M = 270x - 75(x - 4)^2 \text{ lb-ft}$$

Note as a check that at $x = 10$ ft,

$$V|_{x=10\text{ft}} = 150(10 - 4) - 270 = 630 \text{ lb}$$

which is the shear (value and sign) produced by the roller reaction R on the right-hand face at $x = 10$ ft. Also,

$$M|_{x=10\text{ft}} = 270(10) - 75(10 - 4)^2 = 0$$

Answer 5.15   Because on any free-body diagram such as Figure E5.20c, N would be the only force in the x direction so that $\Sigma F_x = 0 = N$.

This result also agrees with what is actually going on at the right end of the beam, for there is no moment there. These "self-checks" should always be made. They are not foolproof, but the chance of making errors in each of $V(x)$ and $M(x)$ that still give the correct values at the right end of the beam is highly unlikely.

The reader is asked to note for future reference that, in both segments of the beam in Example 5.20, we have $dM/dx = -V$. Also, if we define $q(x)$ to be the distributed load intensity, positive if upward, then $dV/dx = -q$:

$0 < x < 4$ ft	4 ft $< x < 10$ ft
$\dfrac{dV}{dx} = \dfrac{d(-270)}{dx}$	$\dfrac{dV}{dx} = \dfrac{d[150(x-4)-270]}{dx}$
$= 0$	$= 150$ lb/ft
This is $-q(x)$ because $q(x) = 0$ here.	This is $-q(x)$ because $q(x) = -150$ lb/ft here.
$\dfrac{dM}{dx} = \dfrac{d(270x)}{dx}$	$\dfrac{dM}{dx} = \dfrac{d[270x - 75(x-4)^2]}{dx}$
$= 270$ lb	$= 270 - 150(x-4)$ lb
This is $-V(x)$ because $V(x) = -270$ lb here.	This is $-V(x)$ because $V(x) = -270 + 150(x-4)$ lb here.

These relationships turn out to be true in general; we shall derive and study them in the next section of this chapter.

EXAMPLE 5.21

Obtain expressions for the shear force $V(x)$ and bending moment $M(x)$ for the cantilever beam loaded as shown in Figure E5.21a.

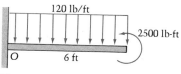

Figure E5.21a

Solution

A cantilever beam is one which is fixed at one end into a structure (such as a wall) and juts out with its other end unsupported, or "free." The free-body diagram in Figure E5.21b will help us compute the reactions at the wall. [Note that $N(x)$ is again identically zero here.]

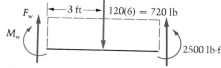

Figure E5.21b

$$+\uparrow \qquad \Sigma F_y = 0 = -720 + F_w \Rightarrow F_w = 720 \text{ lb} \qquad (1)$$

$$\curvearrowright+ \qquad \Sigma M_O = 0 = 2500 \text{ lb-ft} - (720 \text{ lb})(3 \text{ ft}) - M_w \qquad (2)$$

$$M_w = 340 \text{ lb-ft}$$

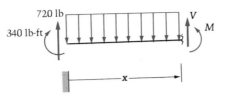

Figure E5.21c

In this problem we are lucky because a single cut section (see Figure E5.21c) will yield $V(x)$ and $M(x)$ expressions valid for the entire beam.

To find V and M, we replace the distributed load by the force $120x$ lb located at $x/2$ as shown in Figure E5.21d below:

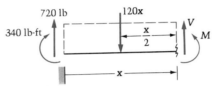

Figure E5.21d

Equilibrium of the cut section requires

$$+\uparrow \quad \Sigma F_y = 0 = V + 720 - 120x$$

$$V = 120x - 720 \text{ lb} \tag{3}$$

$$\curvearrowleft_+ \quad \Sigma M_A = 0 = M + (120x \text{ lb})\left(\frac{x}{2} \text{ ft}\right) - 340 \text{ lb-ft} - (720 \text{ lb})(x \text{ ft})$$

$$M = -60x^2 + 720x + 340 \text{ lb-ft} \tag{4}$$

Note that:

a. At $x = 0$, $V = -720$ lb, which checks.

b. At $x = 0$, $M = +340$ lb-ft, which checks.

c. At $x = 6$ ft, $V = 120(6) - 720 = 0$, which checks.

d. At $x = 6$ ft, $M = -60(6)^2 + 720(6) + 340 = 2500$ lb-ft, which checks.

It is our experience that students often have trouble, at least at first, in distinguishing between the meanings of:

a. The sign beside an equilibrium equation, such as the $+\uparrow$ in Equation (1) of the above example; and

b. The sign given to the scalar V (or to the scalar M) representing a shear force (or a bending moment).

The difference between these signs is explained as follows:

The sign $+\uparrow$ is a temporary convenience, where we are agreeing on which unit vector is factored when the forces (or moments) are summed and equated to zero. In the preceding example, Equation (1) could be written more formally as the y component of $\Sigma \mathbf{F} = \mathbf{0}$:

$$720(-\hat{\mathbf{j}}) + F_w(\hat{\mathbf{j}}) = 0$$

We see that if we suppress the "$\hat{\mathbf{j}}$," Equation (1) of the example remains. Thus the sign $+\uparrow$ in front of that equation is merely saying it is "plus $\hat{\mathbf{j}}$" that is omitted, and has nothing to do with a shear force sign convention. If we, however, want to know the shear force on the left end of that same beam, it comes from an upward force (of $F_w = 720$ lb) on a left-hand face, which is negative shear force by the sign convention for V; that is, $V = -720$ lb there at $x = 0$.

In our final example of this section, we shall consider a simply supported beam loaded in four commonly occurring ways: by concentrated axial and lateral forces, by a concentrated couple, and by a distributed loading.

EXAMPLE 5.22

Find the axial force, shear force, and bending moment distributions throughout the beam shown in Figure E5.22a.

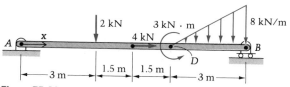

Figure E5.22a

Solution

For the reactions, we shall use the free-body diagram in Figure E5.22b, in which the distributed load (see Section 3.9) has been replaced, solely for purposes of using the equilibrium equations, by the area under the loading curve acting along the indicated line.

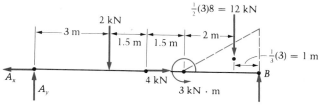

Figure E5.22b

Because the force-alone resultant of 12 kN ↓ acting 1 meter from the right end of the beam produces the same contribution to the equilibrium equations as does the actual distributed loading, it can be used to advantage in computing the three reactions A_x, A_y, and B:

$$\xrightarrow{+} \quad \Sigma F_x = 0 = 4 \text{ kN} - A_x \Rightarrow A_x = 4 \text{ kN} \qquad \text{(to the left as assumed)}$$

$$+\!\uparrow \quad \Sigma F_y = 0 = A_y - 2 \text{ kN} - 12 \text{ kN} + B$$

$$\curvearrowleft_{+} \quad \Sigma M_A = 0 = -2 \text{ kN}(3 \text{ m}) + 3 \text{ kN} \cdot \text{m} - 12 \text{ kN}(8 \text{ m}) + B(9 \text{ m})$$

$$B = \frac{(6 - 3 + 96) \text{ kN} \cdot \text{m}}{9 \text{ m}} = 11 \text{ kN}$$

Therefore, from the "y equation,"

$$A_y = (2 + 12 - 11) \text{ kN}$$

$$A_y = 3 \text{ kN}$$

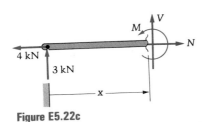

Figure E5.22c

and the reactions have been found. Now we proceed to find $N(x)$, $V(x)$, and $M(x)$ in the various segments of the beam. There will be four such segments: (0, 3), (3, 4.5), (4.5, 6), and (6, 9) because at $x = 0$, 3, 4.5, and 6 m, we have concentrated forces, couples, and / or the beginning of the distributed load.

We first cut the beam at a value of x between 0 and 3 meters, $0 < x < 3$ m. We consider the section to the left of the cut shown in Figure E5.22c.

Note that the sign convention is always in agreement with Figure 5.44 for the "right-hand face" above. To determine N, V, and M, we simply ensure that the section is in equilibrium:

$$\xrightarrow{+} \quad \Sigma F_x = 0 = -4 \text{ kN} + N \Rightarrow N = 4 \text{ kN}$$

$$+\!\uparrow \quad \Sigma F_y = 0 = 3 \text{ kN} + V \Rightarrow V = -3 \text{ kN}$$

$$\curvearrowleft_{+} \quad \Sigma M_A = 0 = M - (3 \text{ kN})(x \text{ m})$$

$$M = 3x \text{ kN} \cdot \text{m}$$

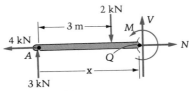

Figure E5.22d

Therefore, over the first 3 meters of the beam, the axial and shear forces are constant while the bending moment grows linearly with x. Note too that the derivative of M with respect to x equals the value of $(-V)$. As we have mentioned, this will later be seen to be true in general.

We next cut the beam between the 2-kN and 4-kN forces. From the equilibrium equations, we obtain (see Figure E5.22d):

$$\xrightarrow{+} \quad \Sigma F_x = 0 = N - 4 \text{ kN} \Rightarrow N = 4 \text{ kN}$$

$$+\!\uparrow \quad \Sigma F_y = 0 = (3 - 2) \text{ kN} + V$$

$$V = -1 \text{ kN}$$

$$\curvearrowleft_{+} \quad \Sigma M_Q = 0 = M + (2 \text{ kN})(x - 3) \text{ m} - (3 \text{ kN}) (x \text{ m})$$

$$M = (6 + x) \text{ kN} \cdot \text{m}$$

We note that the values of N are the same on either side of the point $x = 3$ m. This is also true for M; approaching $x = 3$ m from the left, $M \to 3(3) = 9$ kN $\cdot$ m, and from the right, $M \to 6 + 3 = 9$ kN $\cdot$ m. Thus N and M are continuous through the point, but not so for V.* A free body surrounding the point shows the situation, and why there is no single value of V at the point $x = 3$ (see Figure E5.22e).

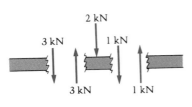

(showing only shear forces)

Figure E5.22e

---

* This is to be expected because we have no concentrated longitudinal force or couple applied at the point, but we do have a concentrated transverse external force there. Concentrated forces or moments will always and only give rise to discontinuities in the respective related quantity N, V, or M.

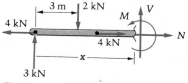

Figure E5.22f

Passing next over the concentrated axial force of 4 kN at $x = 4.5$ m, we see (Figure E5.22f) that to the right of this force, there is no axial force in the beam:

$$\xrightarrow{+} \quad \Sigma F_x = 0 = N - 4 \text{ kN} + 4 \text{ kN}$$

$$N = 0$$

V and M this time retain their earlier expressions, as the reader may note:

$$V = -1 \text{ kN}$$

$$M = (6 + x) \text{ kN} \cdot \text{m}$$

When we reach point D, two things happen:

1. There is a concentrated couple there that causes a jump in the expression for $M(x)$.

2. A distributed load begins that will affect both V and M.

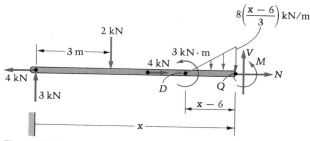

Figure E5.22g

Shown in Figure E5.22g is a free-body diagram of a section of the beam to the left of a cut at a point with $x > 6$ m. Equilibrium requires

$$\xrightarrow{+} \quad \Sigma F_x = 0 = -4 + 4 + N \Rightarrow N = 0$$

(Only the portion of the beam being stretched carries any axial force.)

$$+\uparrow \quad \Sigma F_y = 0 = 3 - 2 + V - \underbrace{\frac{1}{2}(x - 6)\left[8\left(\frac{x-6}{3}\right)\right]}_{\substack{\text{area of triangular loading} \\ \text{curve (see Example 3.34)}}}$$

$$V = -1 + \frac{4}{3}(x - 6)^2 \text{ kN}$$

(Note that as x grows from 6 m, the shear must become more and more positive.)

$$\curvearrowleft{+} \quad \Sigma M_Q = 0 = -3x + 2(x - 3) + 3 + M$$

$$+ \underbrace{\frac{4}{3}(x - 6)^2}_{\substack{\text{downward force} \\ \text{from distributed} \\ \text{load from} \\ \text{preceding equation}}} \quad \bullet \quad \underbrace{\left(\frac{x-6}{3}\right)}_{\substack{\text{location of} \\ \text{resultant of} \\ \text{distributed load} \\ \text{(see Example 3.34)}}}$$

$$M = 3 + x - \frac{4}{9}(x - 6)^3 \text{ kN} \cdot \text{m}$$

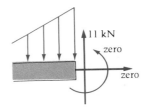

Figure E5.22h

As checks on the solution, note that at $x = 9$ m:

$$N = 0$$

$$V = -1 + \frac{4}{3}(3^2) = 11 \text{ kN}$$

$$M = 3 + 9 - \frac{4}{9}(3)^3 = 0$$

This check is at the right end of the beam shown in Figure E5.22h. We have:

N should be zero. ✓
V should be 11 kN. ✓
M should be zero. ✓

PROBLEMS ▶ Section 5.9

5.155 At what point(s) on the beam shown in Figure P5.155 do the following occur?

a. Zero shear
b. Maximum shear (give the value)
c. Zero moment
d. Maximum moment (give the value)

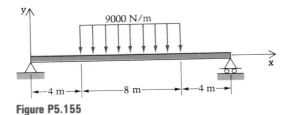

Figure P5.155

5.156 Repeat the preceding problem for the beam of Figure P5.156. Assume the reaction of the ground onto the beam is uniformly distributed.

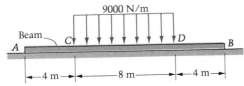

Figure P5.156

5.157 For the beam in Figure P5.157, what is the bending moment of largest magnitude and where does it occur?

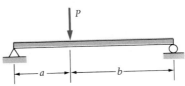

Figure P5.157

5.158 How do the shear force and bending moment vary in the central 3-ft segment of the beam in Figure P5.158?

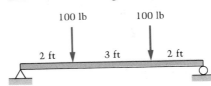

Figure P5.158

5.159 Find the maximum shear and bending moment in the beam shown in Figure P5.159.

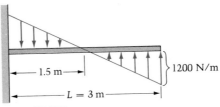

Figure P5.159

5.160 Given the beam shown in Figure P5.160:

a. Find the axial force (N), shear force (V), and bending moment (M) in the portion of the beam along x, as functions of x.

b. Sketch the graph of the functions in part (a) and label the values at $x = 0$, 2 m, and 4 m.

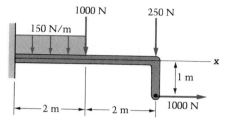

Figure P5.160

5.161 In Figure P5.161:

a. Find the reactions at the wall for the cantilever beam shown.

b. Find the shear force, bending moment, and axial force in the beam as functions of x.

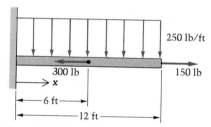

Figure P5.161

5.162 A signboard, BC, is to withstand a windload of $w = 80$ lb/ft. It is supported by the structure shown in Figure P5.162.

a. Determine the magnitude of the windload on the signboard, BC.

b. Identify each of the two-force members(s), if any.

c. Determine the force in each two-force member.

d. Determine the horizontal and vertical components of the reaction forces at A and F.

e. Show the axial and shear forces and the bending moment at point H of member DEF, and show the results on a free-body diagram that includes the cut section there.

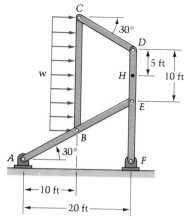

Figure P5.162

5.163 In Figure P5.163:

a. Find the reactions at the wall for the cantilever beam.

b. Find the shear force, bending moment, and axial force as functions of x.

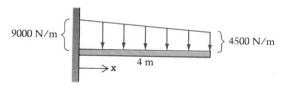

Figure P5.163

In terms of P and L, find the extreme values of the shear force V and bending moment M in the beams in Problems 5.164–5.169. Also give the locations along the beam where these extreme values occur. (It will help to graph $V(x)$ and $M(x)$.)

5.164

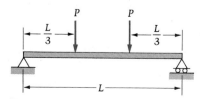

Figure P5.164

(For Problems 5.165–169, see the instructions preceding problem 5.164.)

5.165

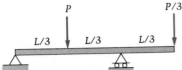

Figure P5.165

5.166

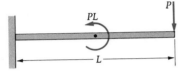

Figure P5.166

5.167

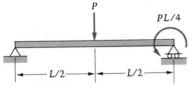

Figure P5.167

5.168

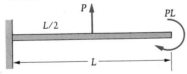

Figure P5.168

5.169

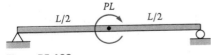

Figure P5.169

5.170 For the cantilever beam loaded as indicated in Figure P5.170, find the shear force V and bending moment M as functions of x. Then find the shear force and bending moment at the point where the bending moment in the beam is the largest in absolute value.

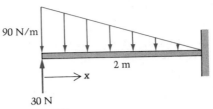

Figure P5.170

Write expressions for the shear force $V(x)$ and bending moment $M(x)$ in the beams shown in Problems 5.171–5.174.

5.171

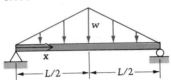

Figure P5.171

5.172

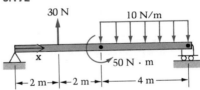

Figure P5.172

5.173

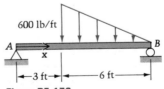

Figure P5.173

5.174

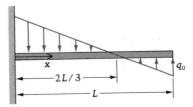

Figure P5.174

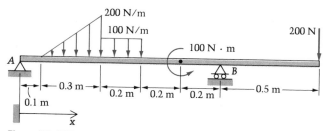

Figure P5.175

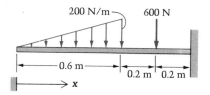

Figure P5.177

5.175 For the beam shown in Figure P5.175, write equations for the shear force V and bending moment M as functions of x, in the intervals $0 < x < 0.1$ m, $0.1 < x < 0.4$ m, $0.4 < x < 0.6$ m, $0.6 < x < 0.8$ m, $0.8 < x < 1.0$ m, and $1.0 < x < 1.5$ m.

5.176 For the beam shown in Figure P5.176, write equations for the shear force V and bending moment M as functions of x, in the intervals $0 < x < 5$ ft, $5 < x < 25$ ft, and $25 < x < 31$ ft.

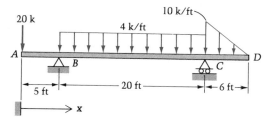

Figure P5.176

5.177 In Figure P5.177:

  a. Write algebraic expressions for the shear force and bending moment in the beam.

  b. Find the shear force of largest magnitude in the beam.

  c. Find the bending moment of largest magnitude in the beam.

5.178 Find expressions for the shear force and bending moment in the beam shown in Figure P5.178. What is the bending moment of largest magnitude and where does it occur?

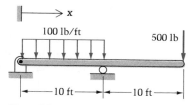

Figure P5.178

5.179 In Figure P5.179 find the shear force $V(x)$ in the segment of the beam $0 < x < 5$ ft and the bending moment $M(x)$ in the segment $5 < x < 10$ ft.

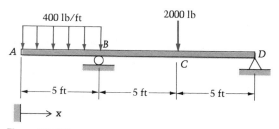

Figure P5.179

5.10 Differential Relationships Between $q(x)$, $V(x)$, and $M(x)$ in a Beam/Shear and Bending Moment Diagrams

We noted just after Example 5.20 that in both of the analyzed segments of the beam, the derived expressions for $q(x)$, $V(x)$, and $M(x)$ satisfied the equations

$$\frac{dV}{dx} = -q \quad \text{and} \quad \frac{dM}{dx} = -V$$

The same is true in Example 5.21, where

$$\frac{dV}{dx} = \frac{d(120x - 720)}{dx} = 120 = -q \qquad \text{(since } q = -120)$$

and

$$\frac{dM}{dx} = \frac{d(-60x^2 + 720x + 340)}{dx} = -120x + 720 = -V$$

$$\text{(since } V = 120x - 720)$$

The reader is encouraged to observe that the same two equations hold in Example 5.22 in each of the four analyzed segments of the beam.

Relationship Between Shear Force and Distributed Lateral Loading

The previous observations are *no coincidence.* There is this definite relationship between the shear force and bending moment in a beam. Let us assume we don't know the result and derive it "from scratch." We begin by isolating an elemental segment of a uniformly loaded beam as shown in Figure 5.47:

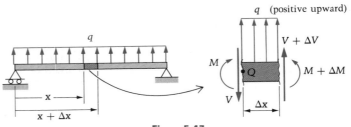

Figure 5.47

Note that, as we have seen, the values of V and M are generally functions of x, so that from the left-hand face (at x) to the right-hand face (at $x + \Delta x$), V and M respectively change to $V + \Delta V$ and $M + \Delta M$.

We now establish the equilibrium of the element:

$$\uparrow + \qquad \Sigma F_y = 0 = -V + (V + \Delta V) + q\Delta x \qquad (5.10)$$

Dividing Equation (5.10) by Δx and simplifying,

$$\frac{\Delta V}{\Delta x} = -q$$

Taking the limit of both sides,

$$\lim_{\Delta x \to 0} \frac{\Delta V}{\Delta x} = \frac{dV}{dx} = -q \qquad (5.11)$$

When q is $q(x)$, the value of the distributed load intensity at x, the result is the same.*

Thus the slope of the shear diagram (a plot of V versus x) at any point equals the negative of the distributed load value at the point. Let us see by two examples how we might use this result in obtaining the distribution of V throughout a beam.

EXAMPLE 5.23

Draw the shear diagram (V versus x) for the beam loaded as shown in Figure E5.23a.

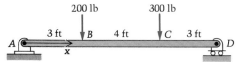

Figure E5.23a

Solution

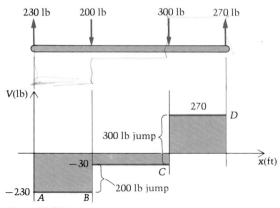

Figure E5.23b

In this problem, the reactions are 230 lb ↑ at the pin A and 270 lb ↑ at the roller B. At the left end of the beam, which is a left-hand face, an upward force corresponds to negative shear. Thus the shear diagram (Figure E5.23b) starts at the point $V = -230$ at $x = 0$. In this beam, there is no $q(x)$ anywhere, so the slope of the V diagram — that is, dV/dx or V' — is zero throughout. Hence we draw a horizontal line from point A of the diagram to the point where $x = 3$ ft, point B.

At B, there will be a discontinuity, or jump, in V because of the 200-lb downward load there. As soon as we pass the point B, the shear force on a

---

* Note that if the distributed load $q(x)$ is continuous *but not uniform* (i.e., not constant), the argument is slightly more complicated. By the mean-value theorem for integrals, there is a number $\bar{X}$ such that $x \le \bar{X} \le x + \Delta x$ and $\int_{\xi=x}^{x+\Delta x} q(\xi)\,d\xi = \Delta x\, q(\bar{X})$. Then, as the limit is taken, $\bar{X} \to x$ so that $q(\bar{X}) \to q(x)$, and the result is the same.
$\Delta x \to 0$

right-hand face will have to increase by 200 lb to maintain equilibrium (see Figure E5.23c).

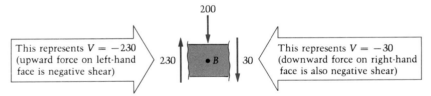

Figure E5.23c

Therefore the value of V becomes -30 lb (or $-230 + 200$) just past B, and stays at that value until we reach point C. There, we get a second jump, this time of 300 lb in V by the same reasoning as at B. This puts us at $V = -30 + 300 = 270$ lb. Again, because $q(x) = 0$ in this beam, we extend the line straight across to D, and end at $V = 270$ lb.

It is important to note that we have an automatic check on our solution because we know the answer for V at the right end of the beam, a right-hand face. In this problem, we have an upward reaction of 270 lb there. Thus (up on the right is positive) V has to be $+270$ lb at $x = 10$ ft; otherwise we have made a mistake. This automatic check should always be made because if you end at the correct value of V, the chances are very good that the diagram was drawn correctly.

EXAMPLE 5.24

Draw the shear diagram (V versus x) for the beam of Example 5.22.

Solution

We have found the reactions to be as indicated in Figure E.5.24a. We present the shear diagram and proceed in the series of steps below to explain how it was constructed.

1. The reaction at the left end gives the starting value of V. It is 3 kN upward on a left-hand face, which is negative shear. Thus we begin with a dot at $V = -3$ on the V axis (see Figure E5.24b).

2. There is no change in V between $x = 0$ and $x = 3$ m. This is because in that interval, $q = 0$ (no distributed load), so that $dV/dx = 0$. Thus the slope of V vs. x is constant at zero, and we draw a horizontal line across from the starting point to B.

3. As we cross B, there is a sudden jump in V of 2 kN because of the downward 2 kN load there.

4. Again, there is no $q(x)$ between $x = 3$ m and 6 m, so $dV/dx = 0$ again. We draw another horizontal line from C to D on the shear diagram.

5. At D, the couple has no effect on V. But starting there, the value of $q(x)$ begins to get more and more negative, linearly. Because $dV/dx = -q$, the *slope* of

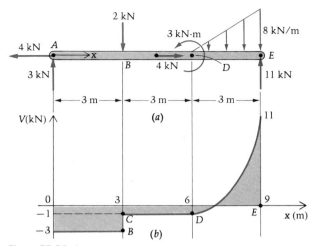

Figure E5.24a,b

the V diagram must get more and more *positive,* linearly. The *change* in V between two points is seen to be the negative of the area beneath the q diagram between those points:

$$\frac{dV}{dx} = -q \Rightarrow \int_{V_1}^{V_2} dV = -\int_{x_1}^{x_2} q \, dx$$

$$V_2 - V_1 = -\int_{x_1}^{x_2} q \, dx$$

From $x = 6$ m to $x = 9$ m, the area beneath "q versus x" is $\frac{1}{2}(3 \text{ m})(-8 \text{ kN}/\text{m})$ $= -12$ kN. Thus V climbs by 12 kN from -1 kN, ending at 11 kN.

6. We have an automatic check on our diagram because the reaction at the right end is applying an upward force of 11 kN on a right-hand face, which corresponds to $V = 11$ kN, a built-in check.

Relationship Between Bending Moment and Shear Force

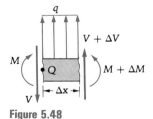

Figure 5.48

Now that we have the shear diagram in hand, we are ready to study its companion, the (bending) moment diagram. We begin by writing the moment equilibrium equation for the segment of Figure 5.47, reproduced at the left as Figure 5.48.

$$\Sigma M_Q = 0 = -M + (M + \Delta M) + (V + \Delta V) \, \Delta x + [q \, \Delta x]\left(\frac{\Delta x}{2}\right)$$

Therefore,

$$\frac{\Delta M}{\Delta x} = -V - \Delta V - \frac{q(\Delta x)}{2}$$

As the limit is taken ($\Delta x \to 0$), the last two terms in this equation vanish,*
leaving

$$\frac{dM}{dx} = -V \qquad (5.12)$$

Thus at each point where M' is defined, the slope of the bending
moment diagram is equal to the negative of the value of the shear force at
that point.

Also, we have

$$\int_{x_1}^{x_2} \frac{dM}{dx} \, dx = -\int_{x_1}^{x_2} V \, dx$$

or

$$M_2 - M_1 = -\int_{x_1}^{x_2} V \, dx$$

so that the *change* in the bending moment between two points x_1 and x_2
equals the negative of the area beneath the shear diagram between those
two points. Let us use these results to extend the last two examples to
include moment diagrams.

EXAMPLE 5.25

Draw the moment diagram for the beam of Example 5.23, which carried the
loading shown in Figure E5.25a and had the shear diagram in Figure E5.25b.

Solution

It is helpful to draw the various diagrams beneath one another because we can
use the relationships such as $V' = -q$ and $M' = -V$ more easily. Also, we can for
our convenience extend lines downward at places of importance, such as points
where loads appear.

The moment diagram starts at the origin (see Figure E5.25c) because there is
no moment applied to the beam by the pin and no external moment is being
exerted at the end $x = 0$. The slope of M versus x, for $0 < x < 3$ ft, is $-V$, or
$+230$ lb. Over 3 ft, this slope will result in a buildup of M from zero to 690 lb-ft.
Note that the *change* in M from its value of zero at $x = 0$ is the negative of
the area beneath the V diagram, or $-(3 \text{ ft})(-230 \text{ lb})$, which is, again, 690 lb-ft.

The jump in V at $x = 3$ ft means to us that there is a discontinuity in the *slope*
of the moment diagram. The M versus x curve will come out of the point B with a
different slope [namely $-(-30)$ or 30] than it had when it went in. Over the next
four feet, the change in M, then, will be the negative of the area under the shear
diagram between B and C, or $-(4 \text{ ft})(-30 \text{ lb}) = 120$ lb-ft. Hence the new (and
the maximum) value of M is the old plus the change, or $690 + 120 = 810$ lb-ft.

---

* If $q(x)$ is continuous but not constant, then again (as in the footnote on page 373)
using the mean-value theorem, there is a number X such that $x \leq X \leq x + \Delta x$ and
$\int_{\xi=x}^{x+\Delta x} (\xi - x)q(\xi) \, d\xi = \Delta x(X - x)q(X)$. Then, as the limit is taken, $X \to x$ and the
$\Delta x \to 0$
moment of the distributed load $q(x)$ about Q again vanishes.

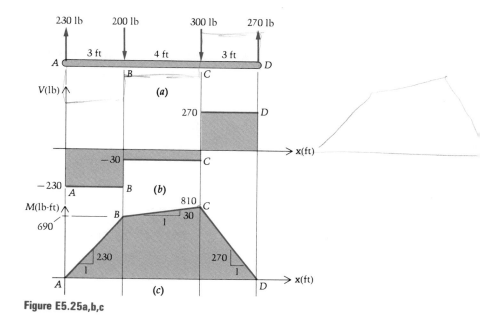

Figure E5.25a,b,c

Another change in slope occurs in the M diagram, this time at point C. To the right of this point, the slope is $-(270) = -270$ lb. The change in M between C and D is $-(270)(3 \text{ ft}) = -810$ lb-ft. Thus the value of M at the right end of the beam is $810 + (-810) = 0$. This result serves as a check on our diagram (just as we had with the shear diagram examples) because the value of M at $x = 10$ ft is indeed zero at a roller with no external couple applied on the end of the beam.

EXAMPLE 5.26

Draw the moment diagram (M versus x) for the beam of Example 5.24 (see Figure E5.26a), which had the shear diagram depicted in Figure E5.26b.

Solution

We now detail the steps we underwent in sketching the moment diagram (M versus x) in Figure E5.26c.

1. There is no moment applied to the left end of the beam, and the pin cannot transmit a moment to the beam there. Thus the moment diagram starts at zero.

2. The shear force is -3 kN between $x = 0$ and $x = 3$ m. Thus the slope of the M diagram equals a constant $+3$ kN/m (it is minus the value of V) for $3 > x \geq 0$. Thus at $x = 3$, $M = 3(3) = 9$ kN $\cdot$ m.

3. At $x = 3$, the concentrated 2-kN force does not produce a discontinuity in the M diagram. However, as soon as we pass over $x = 3$ m, the negative of the value of V (and hence the slope of M vs. x) changes to 1 kN. Over three more meters, this positive slope causes a further rise in M of $1(3) = 3$ kN $\cdot$ m, which

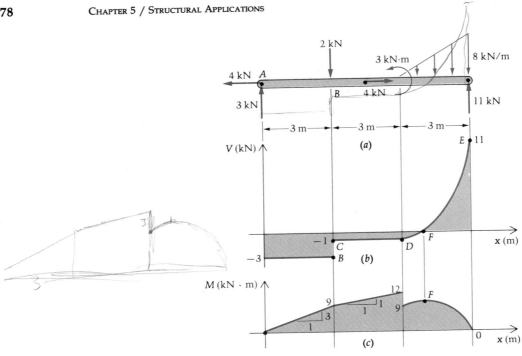

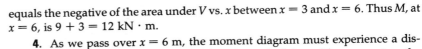

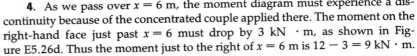

Figure E5.26a,b,c

equals the negative of the area under V vs. x between $x = 3$ and $x = 6$. Thus M, at $x = 6$, is $9 + 3 = 12$ kN · m.

4. As we pass over $x = 6$ m, the moment diagram must experience a discontinuity because of the concentrated couple applied there. The moment on the right-hand face just past $x = 6$ must drop by 3 kN · m, as shown in Figure E5.26d. Thus the moment just to the right of $x = 6$ m is $12 - 3 = 9$ kN · m.

5. Between D and F on the shear diagram of Figure E5.26b the values of the shear V get less and less negative, ending at zero. Thus the slope of M must get less and less positive ending at zero because $dM / dx = -V$; the M diagram is thus sketched with a horizontal tangent at F.

6. Between F and E, the shear gets more and more positive, ending at 11; thus dM / dx, the slope of the moment diagram, gets more and more negative, ending with a slope value of -11 at the right end of the beam.

Just as with the shear diagram, there is an automatic check available for the moment diagram. We simply see if the computed value of M at the right end of the beam is the actual (known) value there. In this case, we know that M must be zero at $x = 9$ m. To see if it calculates out to be zero, we need the area beneath V vs. x in the interval $6 < x < 9$. In that interval,

$$q = -\frac{8}{3}(x - 6)$$

so that, integrating,

$$V = -\int q\, dx = \frac{4(x - 6)^2}{3} + C_1 = \frac{4(x - 6)^2}{3} - 1$$

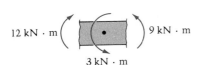

Figure E5.26d

where C_1 is found by the condition $V = -1$ at $x = 6$. Integrating a second time,

$$M = -\int V \, dx = -\frac{4(x-6)^3}{9} + x + C_2$$

$$= -\frac{4(x-6)^3}{9} + x + 3,$$

because $M = 9$ at $x = 6^+$ (just to the right of $x = 6$). Thus the change in M that we need is

$$\left[-\frac{4(x-6)^3}{9} + x + 3 \right]_6^9 = -12 + 9 + 3 + 0 - 6 - 3 = -9$$

This result gives a change in M of -9 from its $+9$ value, and it ends at zero at $x = 9$ m correctly. Alternatively, we could simply substitute into $M(x)$ to get $M = 0$ at $x = 9$ m:

$$M = -\frac{4(9-6)^3}{9} + 9 + 3 = 0$$

Or finally, if we are just interested in the change in M between $x = 6$ and 9 m, we could integrate the function $-V$ with definite integral limits:

$$\Delta M = -\int_{x=6}^9 V \, dx = -\int_{x=6}^9 \left[\frac{4(x-6)^2}{3} - 1 \right] dx$$

$$= -\left[\frac{4(x-6)^3}{9} - x \right]_6^9 = -[12 - 9 - (0 - 6)] = -9$$

as before.

It is also very important to know the value of M at a point such as F where M is a local maximum. Maxima of moment will often correspond to locations of highest stress in sections of the beam. At F, the shear is zero:

$$V = \frac{4(x-6)^2}{3} - 1 = 0$$

so that

$$(x_F - 6)^2 = 0.75$$

$$x_F = 6.87 \text{ m}$$

Now that F has been located, we can find the bending moment there:

$$M_F = \frac{-4(6.87-6)^3}{9} + 6.87 + 3 = 9.58 \text{ N} \cdot \text{m}$$

which is less than the 12 N $\cdot$ m at $x = 6$ m, but which is still a local maximum.

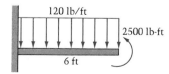

Figure E5.27a

EXAMPLE 5.27

For the beam of Example 5.21 (see Figure E5.27a), draw the shear and moment diagrams.

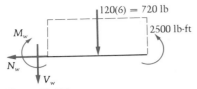

Figure E5.27b

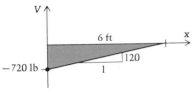

Figure E5.27c

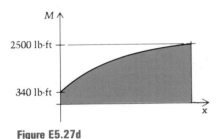

Figure E5.27d

Solution

We could, of course, plot the curves $V = 120x - 720$ and $M = -60x^2 + 720x + 340$ from Example 5.21. However, we shall work this problem as though we did not know these relationships, and then use them as a check.

The free-body diagram will help us find the reactions at the wall. Note that for "statics purposes" (i.e., use of equilibrium equations for the whole beam) we may replace the distributed load by the "force-alone resultant" of 720 lb as indicated in Figure E5.27b. The equations are:

$$\xrightarrow{+} \quad \Sigma F_x = 0 = N_w$$

$$+\uparrow \quad \Sigma F_y = 0 = -V_w - 720 \Rightarrow V_w = -720 \text{ lb}$$

$$\overset{+}{\curvearrowright} \quad \Sigma M_A = 0 = 2500 - 720(3) - M_w \Rightarrow M_w = 340 \text{ lb-ft}$$

Thus the shear diagram (Figure E5.27c) starts at a value of -720 lb when $x = 0$. The distributed load $q(x)$ is a constant (-120 lb/ft), hence the slope of V will be constant too $(V' = -q = 120)$. Thus the value reached by the curve at the free end is $V = -720 + 120(6) = 0$, which checks because there is no shear force on the cantilever beam at $x = 6$ ft. Note also that the change in V from $x = 0$ to 6 ft is the negative of the area beneath the $q(x)$ diagram — that is, $-(-120)6 = +720$.

For the moment diagram, we begin with the wall moment of $+340$ lb-ft (see Figure E5.27d). As x increases, the values of V become less and less negative. Thus the slopes of M versus x become less and less positive. The *value* of M changes over the beam by the negative of the area beneath the V diagram, which is $-[\frac{1}{2}(6)(-720)] = 2160$ lb-ft. Therefore $M|_{x=6\text{ft}} = 340 + 2160 = 2500$ lb-ft, which checks the applied moment at the free end.

Note that the V vs. x curve has the equation $V = 120x - 720$, which checks the result in Example 5.21. The M vs. x curve is a parabola (it is second order in x because it is the integral of a straight line equation) with vertex at $x = 6$. Thus $M - 2500 = -k(x - 6)^2$, and the constant is evaluated using $M = 340$ when $x = 0$:

$$340 - 2500 = -k(-6)^2 \Rightarrow k = 60 \text{ lb/ft}$$

Thus

$$M = 2500 - 60(x - 6)^2$$

$$= -60x^2 + 720x + 340 \text{ lb-ft}$$

This result also agrees with Example 5.21.

EXAMPLE 5.28

Draw the shear and moment diagrams for the beam loaded as shown in Figure E5.28a.

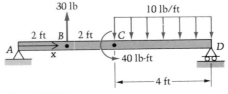

Figure E5.28a

Solution

First we find the reactions. The uniformly distributed load is replaced by its force-alone resultant, which has:

a. Magnitude = area beneath loading curve
$$= (10 \text{ lb}/\text{ft})(4 \text{ ft}) = 40 \text{ lb}$$

b. Location 2 ft to the left of D.

Thus we have the following free-body diagram (Figure E5.28b) and equilibrium equations:

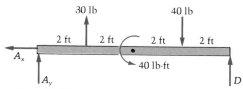

Figure E5.28b

$$\xleftarrow{+} \quad \Sigma F_x \; = 0 \Rightarrow A_x = 0$$
$$+\uparrow \quad \Sigma F_y \; = 0 = A_y + 30 + D - 40$$
$$\overset{+}{\curvearrowleft} \quad \Sigma M_A = 0 = (30 \text{ lb})(2 \text{ ft}) + 40 \text{ lb-ft} - (40 \text{ lb})(6 \text{ ft}) + (8 \text{ ft})D$$
$$D = 17.5 \text{ lb} \tag{1}$$

Thus from Equation (1), $A_y = -7.5$ lb.

> **Question 5.16** How might the reactions be checked before proceeding?

Here are the steps we have followed in constructing the V and M diagrams on the next page:

1. Redraw the original loading diagram, being sure to return to the distributed load — i.e., do *not* use the force-alone resultant (see Figure E5.28c).

> **Question 5.17** Why not stick with the force-alone resultant?

2. Draw the axes beneath one another.

> **Question 5.18** What is the advantage of this?

Answer 5.16 By $\Sigma M_D = 0$, we get the same answer for A_y.

Answer 5.17 The force-alone resultant makes the same contribution to the equilibrium equations, hence giving the correct external reactions. But it does not create the same internal force distributions within the beam!

Answer 5.18 Whenever $q = 0$, V will have a zero slope; whenever $V = 0$, M will have a zero slope. Thus if the curves are drawn beneath one another, these locations can be projected down to advantage.

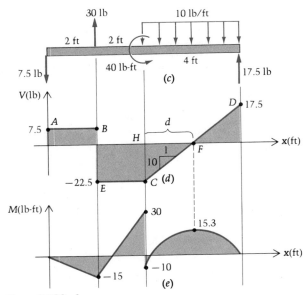

Figure E5.28c,d,e

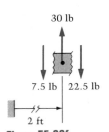

Figure E5.28f

3. Start (Figure E5.28d) at $V = +7.5$. (Recall that a downward force on a left-hand face is positive!)

4. Since $q = 0$ between $x = 0$ and 2 ft, draw the line AB with $dV/dx = 0$ as shown.

5. The shear must decrease by 30 as we cross $x = 2$ ft to equilibrate the 30-lb concentrated force there. Thus put a dot at E, which has a V coordinate of $7.5 - 30 = -22.5$ lb; see Figure E5.28f at the left which shows a small element at $x = 2$ ft.

6. Again draw a horizontal line, EC this time, because $dV/dx = 0$ in the interval $2 < x < 4$ ft.

7. Note that the couple has no effect on V at C (zero resultant force!).

8. Because $dV/dx = -q$, and because $q(x) = -10$ for $4 < x < 8$ ft, the slope of the V diagram between $x = 4$ and 8 ft is $+10$ lb/ft. Thus it ends at $V = -22.5 + 10(4) = +17.5$ lb.

9. The check is that V really is $+17.5$ lb at $x = 8$ ft because the reaction is 17.5 lb ↑, and an upward force on a right-hand face is positive shear.

10. Preparatory to discussing the moment diagram, let us find the location of the point F where V is zero. This will correspond to a horizontal tangent in the M diagram. By the similar triangles CDG and CFH, we have

$$\frac{22.5}{d} = \frac{10}{1}$$

so that

$$d = 2.25 \text{ ft}$$

11. The moment M (see Figure E5.28e) is zero at $x = 0$ (the pin cannot transmit a moment, and none is externally applied directly to the beam).

12. $dM/dx = -V = -7.5$ lb between $x = 0$ and 2 ft, so that M reaches a value of -15 lb-ft at $x = 2$ ft.

13. Between $x = 2$ and 4 ft, the slope of M changes to the negative of the new value of V, or $M' = +22.5$. The change in M from $x = 2$ ft to $x = 4$ ft is the negative of the area beneath V vs. x in that interval, or $(22.5$ lb$)(2$ ft$) = 45$ lb-ft. Thus M at $x = 4$ ft is $-15 + 45 = 30$ lb-ft.

14. At $x = 4$ ft, the moment diagram drops by the amount of the 40 lb-ft couple; see Figure E5.28g below of a small element at $x = 4$ ft:

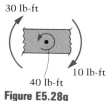

30 lb-ft

10 lb-ft

40 lb-ft

Figure E5.28g

Therefore, at $x = 4^+$ ft, M is $30 - 40 = -10$ lb-ft.

15. As we move from H to F(see the V diagram), the values of V are negative numbers heading toward zero. The slopes of M vs. x, being equal to $-V$, are thus positive numbers heading toward zero. The change in M is

$$-\left(\begin{array}{c}\text{Area of } V \text{ diagram}\\ \text{between } H \text{ and } F\end{array}\right) = \frac{1}{2}(2.25 \text{ ft})(22.5 \text{ lb})$$

$$= 25.3 \text{ lb-ft}$$

Thus M at F is $-10 + 25.3 = 15.3$ lb-ft.

16. From F to D, V is now getting more and more positive, ending at $+17.5$ lb. The slope of M thus gets more and more negative, and the change in M is, this time,

$$-\left(\begin{array}{c}\text{Area of } V \text{ diagram}\\ \text{between } F \text{ and } D\end{array}\right) = -\frac{1}{2}(4 - 2.25) \text{ ft } (17.5 \text{ lb})$$

$$= -15.3 \text{ lb-ft}$$

Thus M at $x = 8$ ft is $15.3 - 15.3 = 0$, as it should be because no moment is applied to the right end of the beam externally or by the roller.

We note that the maximum magnitude of moment M is 30 lb-ft just to the left of $x = 4$ ft; the maximum magnitude of shear force V is 22.5 lb throughout the interval $2 < x < 4$ ft.

The Other Sign Convention for Shear Force

In closing, we note that some prefer to define the shear force as positive if downward on a right-hand face. This convention, although not in agreement with definitions used in the theory of elasticity, has the advantage that the Equations (5.11) and (5.12) become $V' = +q$ and $M' = +V$. This makes it easier to construct loading, shear, and moment diagrams using the other curves, because one needn't keep up with the minus signs. The only difference between the diagrams drawn with V positive according to ↑☐↓ and our convention ↓☐↑ is that the shear diagram is "flipped," as shown in Figure 5.49:

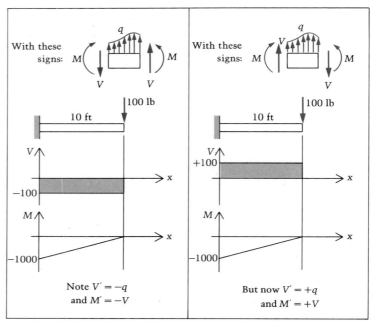

Figure 5.49

Note that the moment diagram is unaffected by the change in the sign of V.

PROBLEMS ▶ Section 5.10

5.180 A distributed load in a beam varies linearly from zero on each end to a maximum value of 8000 N / m in the center, as shown in Figure P5.180. Draw the shear and moment diagrams for this beam.

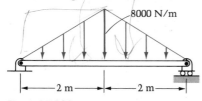

Figure P5.180

In Problems 5.181–5.185 draw the shear and bending moment diagrams for the beams loaded as shown.

5.181

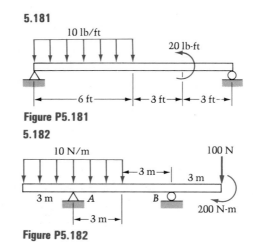

Figure P5.181

5.182

Figure P5.182

5.183

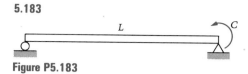

Figure P5.183

5.184

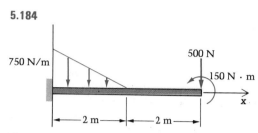

Figure P5.184

5.185

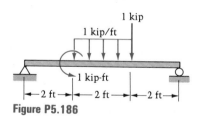

Figure P5.185

5.186 Draw the shear and moment diagram for the beam shown in Figure P5.186.

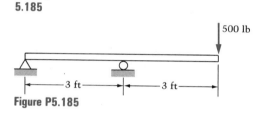

Figure P5.186

5.187 Draw the shear and bending moment diagrams for the beam shown in Figure P5.187. Select an appropriate free-body and write the equations for shear and bending moment for 2 < x < 6 m.

* **5.188** In Figure P5.188:

   a. Draw shear and moment diagrams for the beam shown when a = 6 ft. Indicate the values of all pertinent ordinates.

   b. Where should the roller be located (a = ?) in order that the greatest magnitude of bending moment in the beam be as small as possible?

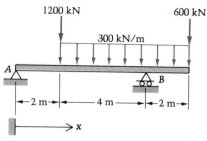

Figure P5.187

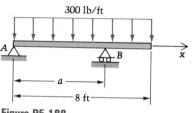

Figure P5.188

5.189 Draw the shear and bending moment diagrams for the beam shown in Figure P5.189.

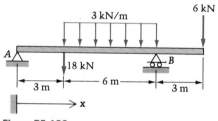

Figure P5.189

5.190 Assuming the upward reaction of the ground to be uniformly distributed, draw the shear and moment diagrams for the beam shown in Figure P5.190.

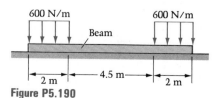

Figure P5.190

5.191 Draw the shear and moment diagrams for the beam in Figure P5.191.

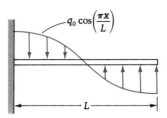

Figure P5.191

Draw shear and moment diagrams for the beams shown in Problems 5.192–5.195.

5.192

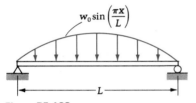

Figure P5.192

5.193

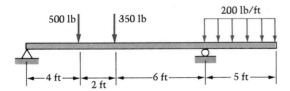

Figure P5.193

5.194

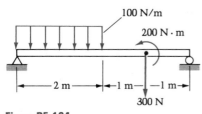

Figure P5.194

5.195

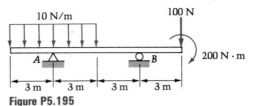

Figure P5.195

* **5.196** The simply supported beam of length L in Figure P5.196 is loaded as indicated with a uniformly distributed moment of intensity μ per unit length. Draw the shear and bending moment diagrams. *Hint: dM/dx = −V* does not apply since distributed moments were not included in its derivation.

Figure P5.196

5.197 *ABE* in Figure P5.197 is a single continuous member. Draw the shear and moment diagrams for the portion *AB* of the member.

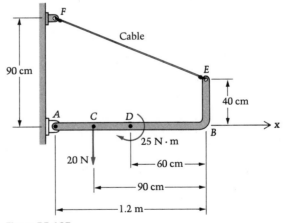

Figure P5.197

* **5.198** In Figure P5.198 the uniform rod has weight W and rests on two props as shown, with $b > a$. Draw the shear and bending moment diagrams and show that the largest bending moment is at B if $a\sqrt{2} > b$.

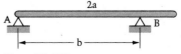

Figure P5.198

III / CABLES

5.11 Parabolic and Catenary Cables

Vertical Loadings Depending upon the Horizontal Coordinate

The flexible cable is yet another common method of supporting loads. For example, the suspension bridge has been used for many centuries and is perhaps the best example of the engineering use of cables. We shall first consider a cable suspended in a plane between the two points A and B, as indicated in Figure 5.50. Let us investigate the shape of the curve in which the cable will hang, under the action of a loading $q(x)$ that is everywhere vertical. In our analysis of cables, we shall take $q(x)$ to be positive *downward*.

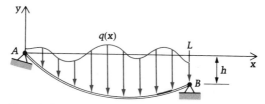

Figure 5.50

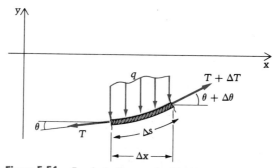

Figure 5.51   Free-body diagram of cable element.

We consider the equilibrium of a small element of such a cable, as shown in Figure 5.51. Summing forces,

$$\xrightarrow{+} \qquad \Sigma F_x = 0 = (T + \Delta T) \cos (\theta + \Delta \theta) - T \cos \theta = 0$$

Dividing by Δx and taking the limit as $\Delta x \to 0$, we see that

$$\frac{d(T \cos \theta)}{dx} = 0$$

which implies that

$$T \cos \theta = \text{constant} = T_H \tag{5.13}$$

Thus the horizontal component of cable tension does not change from one end to the other. Continuing, if $q(x)$ is constant,*

$$+\uparrow \qquad \Sigma F_y = 0 = (T + \Delta T) \sin(\theta + \Delta\theta) - T \sin\theta - q\,\Delta x$$

Dividing again by Δx and letting Δx approach zero, we get this time

$$\frac{d(T \sin\theta)}{dx} = q \qquad\qquad (5.14)$$

and substituting $T = T_H / \cos\theta$ from Equation (5.13),

$$T_H \frac{d(\tan\theta)}{dx} = q \qquad\qquad (5.15)$$

or because $\tan\theta = dy/dx = y'$,

$$y'' = \frac{q}{T_H} \qquad\qquad (5.16)$$

If we know the loading as a function of x, this equation may then be integrated twice to yield the cable deflection, as shown in the following example.

EXAMPLE 5.29

In the foregoing theory, let $q(x) = \text{constant} = q_0$ and let h (Figure 5.50) be zero; that is, let the suspension points lie on the same horizontal level. Compute the deflection of the cable and the tension in it as functions of x. Also find the length of the cable for a given sag H and span L. (See Figure E5.29b.)

Solution

This problem is similar to that of a suspension bridge held up by many vertical cables connected to the main cable as shown in Figure E5.29a. The more vertical

Figure E5.29a

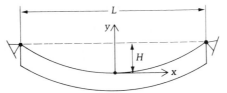

Figure E5.29b

---

* If $q(x)$ is not constant, the resulting Equations (5.14)–(5.16) are the same. The proof uses the mean-value theorem as described in the footnote on page 373.

cables there are, the closer we are to the distributed loading $q(x)$ that was acting on the main cable in the preceding text.

The change to new coordinates shown in Figure E5.29b will allow us to take advantage of symmetry when we evaluate our integration constants. Integrating Equation (5.16),

$$y' = \frac{q_0 x}{T_H} + C_1 \overset{\text{0, because } y' = 0 \text{ at } x = 0}{\nearrow} \tag{1}$$

And integrating again,

$$y = \frac{q_0 x^2}{2T_H} + C_2 \overset{\text{0, because } y = 0 \text{ at } x = 0}{\nearrow} \tag{2}$$

Thus this cable loading results in a parabolic deflection shape.

To find the tension, Equations (5.13) and (5.14) give [with $q = q_0$ in Equation (5.14)]

$$T \cos\theta = T_H = \text{constant} \tag{3}$$

$$T \sin\theta = q_0 x + C_3 \overset{\text{0, because } \theta = 0 \text{ at } x = 0}{\nearrow} \tag{4}$$

Squaring and adding,

$$T = \sqrt{T_H^2 + q_0^2 x^2} \tag{5}$$

The tension is seen to be minimum at the lowest point, where $x = 0$:

$$T_{\text{MIN}} = T_H \tag{6}$$

and maximum at the support points A and B, where $x = L/2$:

$$T_{\text{MAX}} = \sqrt{T_H^2 + q_0^2 L^2/4}^* \tag{7}$$

If we wish, we can express T_H in terms of the sag H, at the lowest point of the cable, by Equation (2):

$$H = \frac{q_0 L^2}{8T_H} \Rightarrow T_H = \frac{q_0 L^2}{8H} \tag{8}$$

> **Question 5.19** Does Equation (8) make sense in the limiting case of large values of q_0, L, and T_H? How about small values?

Thus, alternatively,

$$T_{\text{MIN}} = \frac{q_0 L^2}{8H} \quad \text{and} \quad T_{\text{MAX}} = \frac{q_0 L}{2}\sqrt{1 + \frac{L^2}{16H^2}} \tag{9}$$

The required length of the cable for given endpoints and sag is also of interest. It is (see Figure E5.29c):

$$\ell = 2\int_{s=0}^{s_B} ds = 2\int_{s=0}^{s_B} \sqrt{dx^2 + dy^2} = 2\int_{x=0}^{L/2} \sqrt{1 + y'^2}\, dx$$

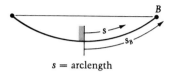

$s = $ arclength

Figure E5.29c

---

* Note that $q_0 L$ is the total load supported by the cable.

Answer 5.19 Yes. The larger the values of loading q_0 and length L, the larger the deflection. And the larger the tension, the smaller the deflection. The reverses also make sense, with each producing the opposite effect on deflection.

Using Equations (1) and (8),

$$\ell = 2 \int_0^{L/2} \sqrt{1 + \left(\frac{8Hx}{L^2}\right)^2} \, dx \tag{10}$$

Using integral tables,*

$$\ell = \left[\frac{L^2}{8H} \sinh^{-1} \frac{8Hx}{L^2} + x\sqrt{1 + \left(\frac{8Hx}{L^2}\right)^2}\right]\Bigg|_{x=0}^{L/2}$$

$$\ell = \frac{L^2}{8H} \sinh^{-1} \frac{4H}{L} + \frac{L}{2}\sqrt{1 + \left(\frac{4H}{L}\right)^2} \tag{11}$$

Defining $\mathcal{H}$ to be the ratio H/L (sag-to-span),

$$\ell = \frac{L}{2}\left[\frac{1}{4\mathcal{H}} \sinh^{-1}(4\mathcal{H}) + \sqrt{1 + (4\mathcal{H})^2}\right] \tag{12}$$

For example, if $\mathcal{H} = H/L = 0.2$, then $\sinh^{-1} 0.8 \approx 0.733$ and $\ell = 1.10L$. Thus the cable is about 10% longer than the span in this case.

Vertical Loadings Depending upon Arclength Along the Cable

Even if the cable in the preceding example is not supported with A and B (Figure 5.50) at the same level, the solution is still a parabolic curve as long as the distributed load, viewed as a function of x, is uniform. As just mentioned, this is the case for a suspension bridge.† But what if the cable is an electric power transmission line, not supporting anything but its own weight and perhaps that of adhered ice in a winter storm? In this instance, the important difference is that q, while still a vertical loading, is now force per unit of *arclength s* along the cable, and not of its horizontal projection x. The cable now hangs in a different shape than a parabola, though the two curves are close together if the maximum of y is small in relation to the span L. Let us, then, determine the shape of the cable when $q = q(s)$.

If $q = q(s)$, then appyling $\Sigma F_x = 0$ to the element of Figure 5.51 still yields

$$T \cos \theta = \text{constant} = T_H \tag{5.17}$$

The y-equilibrium equation, however, changes:

$$+\!\uparrow \quad \Sigma F_y = 0 = (T + \Delta T)\sin(\theta + \Delta\theta) - T \sin\theta - q(s)\,\Delta s$$

so that

$$\frac{(T \sin\theta)|_{x+\Delta x} - T \sin\theta|_x}{\Delta x} = q(s)\frac{\Delta s}{\Delta x} \tag{5.18}$$

---

* An approximate answer for ℓ may be obtained by expanding the radical using the binomial theorem, integrating each term, and truncating the series (which converges for $|8Hx/L^2| < 1$, that is, for $|y'| < 1$).

† Provided the suspended load far outweighs the cable itself.

Taking the limit,*

$$\frac{d(T \sin \theta)}{dx} = q(s) \frac{ds}{dx} \tag{5.19}$$

$$= q(s) \sqrt{1 + y^2} \frac{dx}{dx}$$

$$= q(s) \sqrt{1 + y'^2} \tag{5.20}$$

Substituting for T from Equation (5.17) as before,

$$T_H \frac{d(\tan \theta)}{dx} = q(s) \sqrt{1 + y'^2} \tag{5.21}$$

or

$$T_H y'' = q(s) \sqrt{1 + y'^2} \tag{5.22}$$

Therefore, we find that we must solve a different differential equation this time, which is

$$y'' = \frac{q}{T_H} \sqrt{1 + y'^2} \tag{5.23}$$

The solution to (5.23) is usually accomplished via a change of variable. We let y' be renamed

$$y' = p \tag{5.24}$$

so that

$$y'' = p' \tag{5.25}$$

Then Equation (5.23) becomes

$$\frac{p'}{\sqrt{1 + p^2}} = \frac{q}{T_H} \tag{5.26}$$

or

$$\frac{dp}{\sqrt{1 + p^2}} = \frac{q}{T_H} dx \tag{5.27}$$

Integrating (with integral tables), for the case when $q = q_0 = $ constant,†

$$\ln(p + \sqrt{1 + p^2}) = \frac{q_0}{T_H} x + K_1 \tag{5.28}$$

from which

$$\sqrt{1 + p^2} = e^{\left(\frac{q_0}{T_H} x + K_1\right)} - p \tag{5.29}$$

---

* See the footnote on page 373.
† This would be the case if the cable were both uniform prior to suspension and also practically inextensible.

Squaring,

$$1 + p^2 = e^{\left(\frac{2q_0x}{T_H} + 2K_1\right)} - 2pe^{\left(\frac{q_0x}{T_H} + K_1\right)} + p^2 \tag{5.30}$$

Replacing p by y' and solving for this quantity,

$$y' = \frac{e^{\left(\frac{2q_0x}{T_H} + 2K_1\right)} - 1}{2e^{\left(\frac{q_0x}{T_H} + K_1\right)}}$$

$$y' = \frac{e^{\left(\frac{q_0x}{T_H} + K_1\right)} - e^{-\left(\frac{q_0x}{T_H} + K_1\right)}}{2} \tag{5.31}$$

the right-hand side of which can be recognized as the hyperbolic sine. Therefore,

$$y' = \sinh\left(\frac{q_0x}{T_H} + K_1\right) \tag{5.32}$$

Integrating one last time,

$$y = \frac{T_H}{q_0} \cosh\left(\frac{q_0x}{T_H} + K_1\right) + K_2 \tag{5.33}$$

If we now again choose the origin to be the lowest point of the curve (see Figure 5.52), then $y = 0$ and $y' = 0$ at $x = 0$, so that

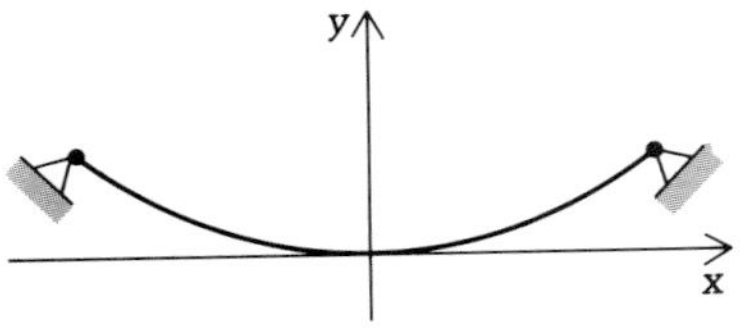

$$(1) \quad 0 = \frac{T_H}{q_0} \cosh K_1 + K_2 \tag{5.34}$$

and

$$(2) \quad 0 = \sinh K_1 \tag{5.35}$$

Equation (5.35) gives $K_1 = 0$, after which it follows from Equation (5.34) that $K_2 = -T_H/q_0$. Therefore,

$$y = \frac{T_H}{q_0}\left(\cosh\frac{q_0x}{T_H} - 1\right)^* \tag{5.36}$$

Nondimensionalizing the deflection, we obtain

$$\frac{q_0y}{T_H} = \cosh\frac{q_0x}{T_H} - 1 \tag{5.37}$$

Note that because

$$\cosh\left(\frac{q_0x}{T_H}\right) = 1 + \left(\frac{q_0x}{T_H}\right)^2 \Big/ 2! + \left(\frac{q_0x}{T_H}\right)^4 \Big/ 4! + \cdots \tag{5.38}$$

Figure 5.52

---

* The shape represented by this answer is often called the "catenary curve."

we have, if $q_0 L / T_H \ll 1$ (where L is the span),

$$\frac{q_0 y}{T_H} \approx \left(1 + \frac{q_0^2 x^2}{2T_H^2}\right) - 1$$

or

$$y = \frac{q_0 x^2}{2T_H} \tag{5.39}$$

which was the solution to Example 5.29!

EXAMPLE 5.30

The cable in Figure E5.30 holds the balloon in equilibrium with a tension at the ground attachment of 100 lb. If the cable weighs 0.25 lb/ft and is 150 ft long, what is the height H of the balloon?

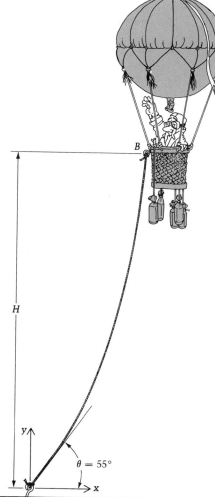

Figure E5.30

Solution

This is a "catenary cable" because it is loaded by its weight — that is, $q = q(s) = q_0$. Therefore we begin with the general solution for this class of cables, which is Equation (5.33):

$$y = \frac{T_H}{q_0} \cosh\left(\frac{q_0 x}{T_H} + K_1\right) + K_2 \tag{1}$$

Question 5.20 Why not start with Equation (5.36)?

Using the coordinates shown in the figure, we have $y = 0$ at $x = 0$, so that

$$0 = \frac{T_H}{q_0} \cosh K_1 + K_2 \tag{2}$$

Because $q_0 = 0.25$ lb/ft, and $T_H = $ constant $= 100 \cos 55° = 57.4$ lb, Equation (2) becomes

$$K_2 = -230 \cosh K_1 \tag{3}$$

Also, $y' = \tan \theta = \tan 55°$ at $x = 0$, so that

$$\tan 55° = \left[\sinh\left(\frac{q_0 x}{T_H} + K_1\right)\right]\Bigg|_{x=0} \tag{4}$$

or

$$1.43 = \sinh K_1$$

Therefore, after evaluating the inverse hyperbolic sine,

$$K_1 = 1.16$$

Answer 5.20 Equation (5.36) is based on the condition $y = y' = 0$ at $x = 0$.

Thus from Equation (3),

$$K_2 = -403$$

And so we get

$$y = 230 \cosh\left(\frac{x}{230} + 1.16\right) - 403$$

The length l of the cable is given by

$$l = 150 = \int_{s=0}^{150} ds = \int_{s=0}^{150} \sqrt{dx^2 + dy^2} = \int_{x=0}^{x_B} \sqrt{1 + y'^2}\, dx$$

where x_B is the x coordinate of the attachment (at B) to the balloon. Because

$$y' = \sinh\left(\frac{x}{230} + 1.16\right)$$

and because $1 + \sinh^2\theta = \cosh^2\theta$, the integral is

$$150 = \int_{x=0}^{x_B} \cosh\left(\frac{x}{230} + 1.16\right) dx = 230 \sinh\left(\frac{x}{230} + 1.16\right)\Big|_0^{x_B}$$

or

$$0.652 = \sinh\left(\frac{x_B}{230} + 116\right) - 1.44$$

or

$$\sinh^{-1} 2.09 = \frac{x_B}{230} + 1.16$$

Thus

$$1.48 = \frac{x_B}{230} + 1.16$$

or

$$x_B = 73.6 \text{ ft}$$

Therefore, the height of the balloon is

$$H = y\Big|_{x=73.6} = 230 \cosh\left(\frac{73.6}{230} + 1.16\right) - 403$$

$$H = 531 - 403 = 128 \text{ ft}$$

Question 5.21   Can we summarize the cable study by saying that we get the parabolic shape for small deflections and the catenary for large deflections?

Question 5.22   Would the analyses of this section apply to suspended chains?

Answer 5.21   No! The parabolic curve comes from a load uniformly distributed with respect to x; the catenary from a load uniformly distributed with respect to arclength s. In fact, transmission cables have *small* deflections but hang in the catenary shape, while suspension bridge cables have *large* deflections with $q(x) = q_0$.

Answer 5.22   Yes. The difference between cables and chains is that there is a small bending stiffness in a cable, and none in a chain; but because this stiffness was neglected, the results apply to chains. Of course, the lengths of (near-rigid) links must be very small compared to the span of the chain.

PROBLEMS ▶ Section 5.11

5.199 The cable in Figure P5.199 is to support a uniform load of 50 lb / ft between P and Q, with the lowest point B being 10 ft below the level of Q. Find the horizontal location of B and then the maximum cable tension.

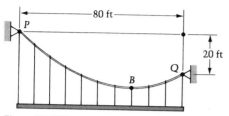

Figure P5.199

5.200 Two cables of the suspension bridge shown in Figure P5.200 symmetrically support a total load of 80×10^6 lb. The sag in the cables at mid-span is 500 ft. Find the mid-span tension and the angle θ the cables make with the tower. Neglect the cable weight.

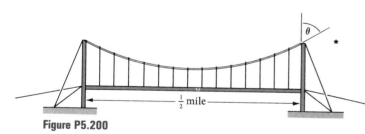

Figure P5.200

5.201 A clothesline is deflected in an ice storm as suggested by Figure P5.201. Assume that the ice load varies cosinusoidally with x, with intensity q_0 at $x = 0$. Neglect the weight of the clothesline and determine its deflection curve.

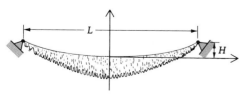

q_0 = maximum intensity

Figure P5.201

5.202 For the cable in Figure P5.202 carrying a uniform load $q(x) = q_0$, prove that at the supports, the tensions are

$$T_A = q_0 L \sqrt{1 + \frac{L^2}{4H_L^2}}$$

and

$$T_B = q_0 R \sqrt{1 + \frac{R^2}{4H_R^2}}$$

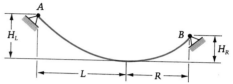

Figure P5.202

5.203 Obtain the result $T = \sqrt{T_H^2 + q_0^2 x^2}$ (from Example 5.29) by using the free-body diagram shown in Figure P5.203.

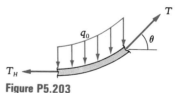

Figure P5.203

5.204 Show that summing moments about the left end of the element of Figure 5.51 adds nothing to the results of the analysis.

5.205 Note that if the ends of the cable are not at the same level, as shown in Figure P5.205, then we don't know the location of the lowest point ahead of time, and cannot

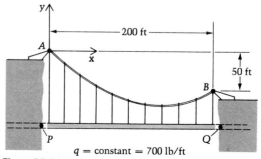

q = constant = 700 lb/ft

Figure P5.205

take advantage of symmetry. For the parabolic cable, begin with

$$y = \frac{q_0 x^2}{2T_H} + C_1 x + C_2$$

and solve the following problem: The pipeline weighs 700 lb/ft. It is to be assumed that over the gorge it is supported entirely by cables (i.e., neglect any forces from the rest of the pipe entering the earth at P and Q). Find the required length of cable if the lowest point of the parabola has an x coordinate of 150 ft. Also find the tensions at the cable suspension points A and B.

5.206 In the preceding problem, find the length of cable from the lowest point to the support at B.

• 5.207 A light cable shown in Figure P5.207 carries a load that grows with x according to $q = kx^n$, where k is a constant and n is a positive integer. If the cable crosses the x axis at $x = L/2$, find the deflection $y(x)$.

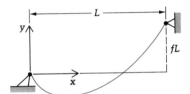

Figure P5.207

5.208 Expand the integrand of Equation (10) of Example 5.29 into the following power series:

$$\sqrt{1 + \left(\frac{8Hx}{L^2}\right)^2} = 1 + \frac{1}{2}\left(\frac{8Hx}{L^2}\right)^2 - \frac{1}{8}\left(\frac{8Hx}{L^2}\right)^4 \pm \cdots$$

Integrate term by term. For $H/L = 0.2$ compare ℓ/L from the example with this series result for: (a) one term; (b) two terms; (c) three terms.

5.209 Given a cable with maximum allowable tension T_0, length ℓ, and weight per unit length q_0, find the maximum horizontal span that can exist for this cable without exceeding T_0.

5.210 In the preceding problem, find the maximum span if $T_0 = q_0\ell$.

• 5.211 In Figure P5.211 determine the vertical distance from B (the center of the pulley) to A.

5.212 If a rope 22 ft long is to be suspended from a tree limb to make a swing, and the attachment points are 3 ft apart, what is the sag distance H? Assume that 1 ft is used on each side for the knotting. Compare the answer with the straight line result of 9.89 ft shown in Figure P5.212.

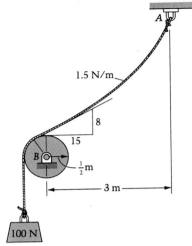

Figure P5.211

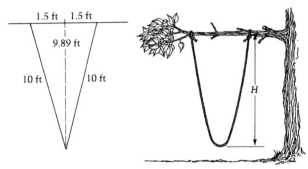

Figure P5.212

• 5.213 Find the error incurred in using the parabolic solution [Equation (5.39)] to get the center sag of a symmetrical catenary cable if the sag ÷ span is: (a) $\frac{1}{10}$; (b) $\frac{1}{4}$; (c) 1.

5.214 A cable of length 80 m hangs between the tops of two identical towers that are 66 m apart. Find the maximum sag in the cable.

5.215 A cable of length 80 m hangs between the tops of two towers. If the maximum sag is 5 m and the maximum tension is 480 N, what is the weight per unit length of the cable?

5.216 Repeat Problem 5.199 if there is no suspended load, but rather the cable *itself* weighs 50 lb/ft.

5.217 In Example 5.30, if the diameter of the base (where the cable is attached) is 8 ft, and the wind is horizontal, find:

   a. The difference between the vertical force of the air and the weight

b. The resultant horizontal force caused by the wind and the height of its line of action above B.

* 5.218 In Figure P5.218 the balloon of Example 4.24 has come loose from the ground and has begun very slowly moving horizontally across an open field. The balloonist lets out some air, and the balloon lowers to a height of 30 ft and begins to drag its cable and its spike. If the net (horizontal) wind force is 50 lb, find the length of cable on the ground and the maximum cable tension.

* 5.219 The cable in Figure P5.219 has a weight per unit length of q_0. A length L hangs over the pulley as shown. Find the maximum tension in the cable, the total length of the cable, and the center sag.

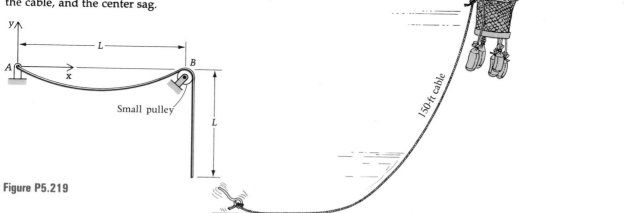

Figure P5.219

Figure P5.218

5.12   Cables Under Concentrated Loads

Sometimes we are faced with the necessity of analyzing a cable of negligible weight from which a number of forces ($W_1, W_2, \ldots, W_n$) are suspended at a series of load points $P_1, P_2 \ldots, P_n$. (See Figure 5.53.) Suppose we are given the lengths of the segments ($L_1, L_2 \ldots, L_{n+1}$ in Figure 5.53) and coordinates (D, H) of the right end with respect to the left. Then if we wish to know the shape in which the cable will hang (given by $\theta_1, \theta_2, \ldots, \theta_{n+1}$) and the tensions in the segments ($T_1, T_2, \ldots, T_{n+1}$), we have an extremely difficult set of equations to solve, consisting of $2n$ equilibrium equations, two for each load point, like (see Figure 5.54)

$$\xrightarrow{+} \qquad \Sigma F_x = 0 = -T_1 \cos \theta_1 + T_2 \cos \theta_2 \qquad (5.40)$$

$$+\uparrow \qquad \Sigma F_y = 0 = T_1 \sin \theta_1 - T_2 \sin \theta_2 - W_1 \qquad (5.41)$$

plus two geometric equations relating the L_i's and θ_i's to D and H:

$$D = \sum_{i=1}^{n+1} L_i \cos \theta_i \qquad H = -\sum_{i=1}^{n+1} L_i \sin \theta_i \qquad (5.42)$$

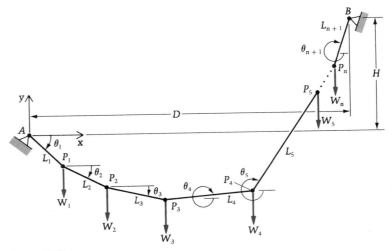

Figure 5.53

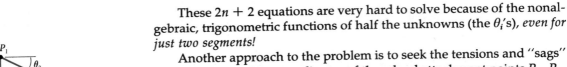

These $2n + 2$ equations are very hard to solve because of the nonalgebraic, trigonometric functions of half the unknowns (the θ_i's), *even for just two segments!*

Another approach to the problem is to seek the tensions and "sags" when we know the x coordinates of the n load attachment points $P_1, P_2, \ldots, P_n$, as suggested by Figure 5.55.

First we note that the horizontal components of all the tensions in the $n + 1$ segments are the same. From Equation (5.40) we have

$$T_1 \cos \theta_1 = T_2 \cos \theta_2$$

Because this can be repeated for each load point, we have

$$T_1 \cos \theta_1 = T_2 \cos \theta_2 = \cdots = T_{n+1} \cos \theta_{n+1} = T_H \qquad (5.43)$$

Figure 5.54

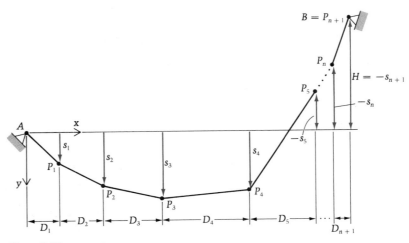

Figure 5.55

which is the same result we obtained in Section 5.11 for continuously loaded cables. We can use this "common horizontal component of tension" result to write the vertical components of each tension in terms of the unknown sags s_1, s_2, ..., s_{n+1}. Consider segment 2 in Figure 5.56 (from P_1 to P_2) and the force that its tension exerts on P_1:

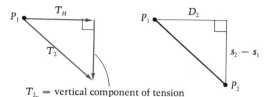

T_{2_v} = vertical component of tension

Figure 5.56

These triangles are similar, so that

$$\frac{T_{2v}}{T_H} = \frac{s_2 - s_1}{D_2} \Rightarrow T_{2v} = T_H \left(\frac{s_2 - s_1}{D_2} \right)$$

and in general*

$$T_{iv} = T_H \left(\frac{s_i - s_{i-1}}{D_i} \right) \tag{5.44}$$

Thus if we know T_H and the n sags s_1, s_2, ..., s_n, then the problem is solved because the tensions may be found as follows:

$$T_i = \sqrt{T_H^2 + T_{iv}^2} = T_H \sqrt{1 + \left(\frac{s_i - s_{i-1}}{D_i} \right)^2} \tag{5.45}$$

It is clear that the largest tension will occur in either the first or last segment.

Note now that in using Equation (5.43) we automatically satisfy $\Sigma F_x = 0$ at all n load points. If we now set the summation of forces in the y direction at each load point equal to zero, n equations will be generated. But all except the first and last will involve three of the sags. For example, at P_2, we have (see Figure 5.57):

$$+\uparrow \qquad \Sigma F_y = 0 = T_{2v} - T_{3v} - W_2$$

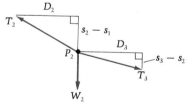

Figure 5.57

Using Equation (5.44),

$$T_H \left(\frac{s_2 - s_1}{D_2} \right) - T_H \left(\frac{s_3 - s_2}{D_3} \right) = W_2$$

or

$$s_2(D_2 + D_3) - s_1 D_3 - s_3 D_2 = \frac{W_2 D_2 D_3}{T_H}$$

---

* Note that $s_i - s_{i-1} < 0$ for the segments with negative slopes.

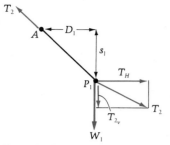

Figure 5.58

Figure 5.59

Figure 5.60

which involves s_1, s_2, and s_3 as well as T_H. If instead of summing forces we sum moments on the following series of free-body diagrams,* only two sags at most are involved in each equation. For one full segment (see Figure 5.58 at the left),

$$\circlearrowright \quad \Sigma M_A = 0 = T_H s_1 - T_{2v} D_1 - W_1 D_1$$

or

$$T_H s_1 - T_H \left(\frac{s_2 - s_1}{D_2}\right) D_1 = W_1 D_1$$

or

$$s_1(D_2 + D_1) - s_2 D_1 = \frac{W_1}{T_H} D_1 D_2$$

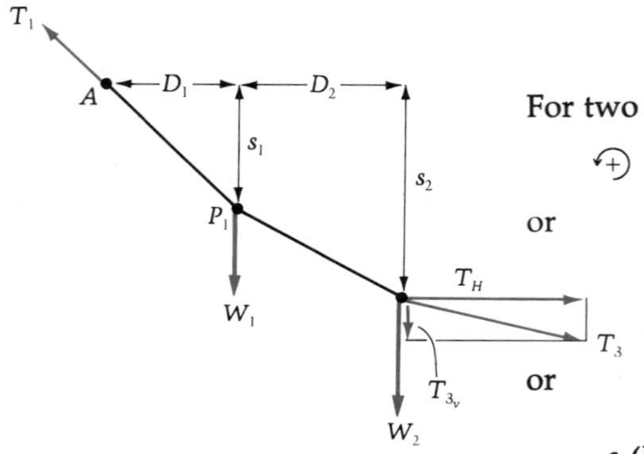

For two full segments (see Figure 5.59),

$$\circlearrowright \quad \Sigma M_A = 0 = T_H s_2 - T_{3v}(D_1 + D_2) - W_1 D_1 - W_2(D_1 + D_2)$$

or

$$T_H s_2 - T_H \left(\frac{s_3 - s_2}{D_3}\right)(D_1 + D_2) = W_1 D_1 + W_2(D_1 + D_2)$$

or

$$s_2(D_3 + D_2 + D_1) - s_3(D_1 + D_2) = \frac{[W_1 D_1 + W_2(D_1 + D_2)]}{T_H} D_3$$

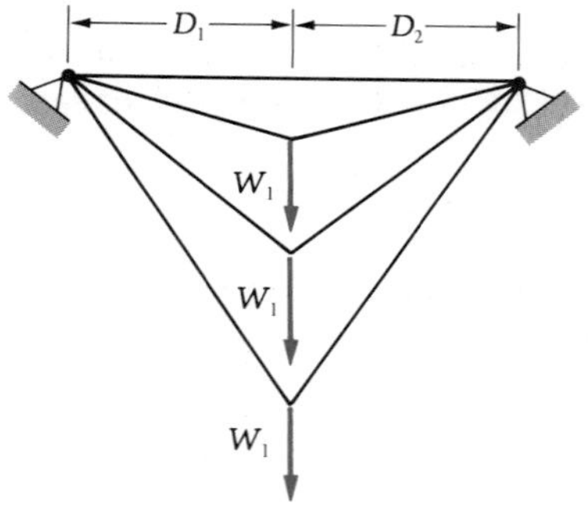

This procedure will give us n equations, as we sum moments on a series of free-body diagrams that involve successive cuts of segments $2, 3, \ldots, n + 1$. Our problem, however, is that we have $n + 1$ unknowns ($s_1, s_2, \ldots, s_n$, and T_H), and until we specify one more condition on the geometry, there is no unique solution; that is, there are lots of cables that could satisfy the conditions. For example, each of the cables in Figure 5.60 has the same set values of D_1, D_2, and W_1, yet clearly the sought-after sags and tensions are different for each. If for our $(n + 1)$st equation we specify the overall length of the cable,

$$L = \sqrt{D_1^2 + s_1^2} + \sqrt{D_2^2 + (s_2 - s_1)^2} + \cdots + \sqrt{D_{n+1}^2 + (H - s_n)^2}$$

then we are back to formidable analytical difficulties. A better condition is to specify one of the sags. Then we can use the moment equations above and solve for the other sags in terms of T_H, one at a time, never having over one unknown sag per equation. Another more realistic possibility for the $(n + 1)$st equation is to be given a maximum allowable tension. This tension will occur in the first or last segment, as we have noted. We now give examples of these ideas.

---

* Note that the reactions at A and B necessarily equal the tensions T_1 and T_{n+1} and act along the lines of the first and last segments.

EXAMPLE 5.31

The endpoints of a cable are tied to the points A and B in Figure E5.31a. The cable carries the indicated loads at $x = 2$ ft, 4 ft, and 6 ft. The maximum tension in the cable is 1000 lb. Find the tensions in all four segments of the cable, and the sags of the load points P_1, P_2, and P_3.

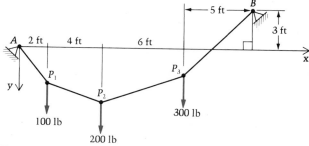

Figure E5.31a

Solution

We shall work to four significant digits in this example. We write the moment equations described in the foregoing discussion (see Figure E5.31b):

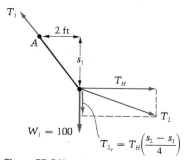

Figure E5.31b Free-body diagram of segment 1 and part of 2.

$$\curvearrowleft^{+} \qquad \Sigma M_A = T_H s_1 - \left[T_H \left(\frac{s_2 - s_1}{4} \right) \right] 2 - 100(2) = 0$$

$$s_1(4 + 2) - s_2 2 = \frac{200(4)}{T_H}$$

$$6s_1 - 2s_2 = \frac{800}{T_H} \qquad\qquad (1)$$

and using Figure E5.31c,

$$\curvearrowleft^{+} \qquad \Sigma M_A = T_H s_2 - \left[T_H \left(\frac{s_3 - s_2}{6} \right) \right] 6 - 100(2) - 200(6) = 0$$

$$s_2(6 + 6) - s_3(6) = \frac{1400}{T_H} \quad (6)$$

$$12s_2 - 6s_3 = \frac{8400}{T_H} \qquad\qquad (2)$$

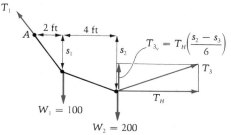

Figure E5.31c Free-body diagram of segments 1, 2, and part of 3.

Next, using Figure E5.31d,

$$\overset{\curvearrowleft}{(+)} \quad \Sigma M_A = T_H s_3 - \left[T_H \left(\frac{-3 - s_3}{5} \right) \right] 12 - 200 - 1200 - 3600 = 0$$

$$17 s_3 + 12\,(3) = \frac{5000}{T_H}\,(5) \tag{3}$$

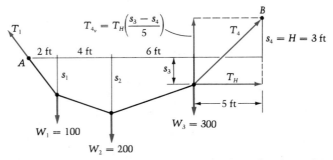

Figure E5.31d   Free-body diagram of segments 1, 2, and 3, and part of 4.

We now have four equations [(1), (2), (3), and (5.45)] in four unknowns. First we find the sags in terms of T_H:

From (3),

$$s_3 = \frac{1471}{T_H} - 2.118 \tag{4}$$

From (2),

$$s_2 = \frac{700.0}{T_H} + \frac{1}{2} \overbrace{\left(\frac{1471}{T_H} - 2.118 \right)}^{s_3}$$

or

$$s_2 = \frac{1436}{T_H} - 1.059 \tag{5}$$

From (1),

$$s_1 = \frac{133.3}{T_H} + \frac{1}{3} \overbrace{\left(\frac{1436}{T_H} - 1.059 \right)}^{s_2}$$

$$s_1 = \frac{612.0}{T_H} - 0.3530 \tag{6}$$

We now assume (and must later check) that the maximum tension is in segment AP_1. (The other possibility is $P_3 B$.) Then $T_1 = 1000$ lb, and Equation (5.45) gives

$$1000 = T_H \sqrt{1 + \left(\frac{s_1}{2} \right)^2} \tag{7}$$

or

$$\frac{10^6}{T_H^2} = 1 + \frac{s_1^2}{4} \Rightarrow s_1 = \sqrt{\frac{4 \times 10^6}{T_H^2} - 4} \qquad (8)$$

Equating the two expressions (6) and (8) for s_1 and squaring,

$$\frac{4 \times 10^6}{T_H^2} - 4 = \frac{374,500}{T_H^2} - \frac{432.1}{T_H} + 0.1246$$

Rearranging, $4.125T_H^2 - 432.1\,T_H - 3,626,000$. Solving with the quadratic formula gives

$$T_H = 991.4 \text{ lb}$$

Back-substituting,

$$s_1 = 0.2643 \text{ ft}$$

$$s_2 = 0.3895 \text{ ft}$$

$$s_3 = -0.6342 \text{ ft}$$

(meaning the third load point is *above* the x axis through A)

The other tensions are given by Equation (5.45):

$$T_2 = T_H \sqrt{1 + \left(\frac{s_2 - s_1}{D_2}\right)^2}$$

$$= 991.9 \text{ lb}$$

$$T_3 = T_H \sqrt{1 + \left(\frac{s_3 - s_2}{D_3}\right)^2}$$

$$= 1006 \text{ lb}$$

$$T_4 = T_H \sqrt{1 + \left(\frac{s_4 - s_3}{D_4}\right)^2} \qquad (s_4 = -H = -3 \text{ ft})$$

$$= 1097 \text{ lb}$$

We therefore see that we have guessed wrong and that the maximum tension is instead in the *right-most* cable. Rewriting Equation (7) (the three moment equations are still valid):

$$1000 = T_H \sqrt{1 + \left(\frac{s_4 - s_3}{5}\right)^2} \qquad (s_4 = -3 \text{ ft})$$

$$\frac{10^6}{T_H^2} = 1 + \left(\frac{-3 - s_3}{5}\right)^2$$

$$(3 + s_3)^2 = 25\frac{10^6}{T_H^2} - 25$$

$$s_3 = 5\sqrt{\frac{10^6}{T_H^2} - 1} - 3$$

Equating expressions for s_3 gives

$$5\sqrt{\frac{10^6}{T_H^2} - 1} = \frac{1471}{T_H} + 0.8820$$

Squaring,

$$25\left(\frac{10^6}{T_H^2} - 1\right) = \frac{2,164,000}{T_H^2} + \frac{2595}{T_H} + 0.7779$$

$$1.031T_H^2 + 103.8T_H - 913,400 = 0$$

$$T_H = 892.2 \text{ lb}$$

so that, using Equations (6), (5), and (4) in order,

$$s_1 = 0.3329 \text{ ft}$$

$$s_2 = 0.5505 \text{ ft}$$

$$s_3 = -0.4693 \text{ ft}$$

and

$$T_1 = T_H \sqrt{1 + \left(\frac{s_1}{D_1}\right)^2} = 904.5 \text{ lb}$$

$$T_2 = T_H \sqrt{1 + \left(\frac{s_2 - s_1}{D_2}\right)^2} = 893.5 \text{ lb}$$

$$T_3 = T_H \sqrt{1 + \left(\frac{s_3 - s_2}{D_3}\right)^2} = 905.0 \text{ lb}$$

This time all the tensions are ≤ 1000 lb, and we have also solved for the sags.

To illustrate the procedure when a sag is specified instead of a maximum tension, we shall specify that $s_2 = 0.5505$ ft, which came out of the preceding solution, and solve for all the tensions in the following example.

EXAMPLE 5.32

If the sag s_2 of load point P_2 is 0.5505 ft, find the tensions in the four cable segments and the sags of P_1 and P_3 (see Figure E5.32a).

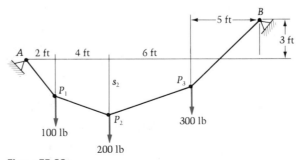

Figure E5.32a

Solution

For the overall free-body diagram shown in Figure E5.32b, we have:

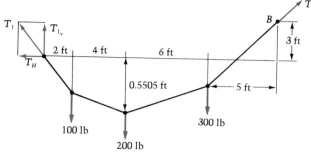

Figure E5.32b

$$\Sigma M_B = 0 = (300 \text{ lb})(5 \text{ ft}) + (200 \text{ lb})(11 \text{ ft})$$
$$+ (100 \text{ lb})(15 \text{ ft}) - T_H(3 \text{ ft}) - T_{1v}(17 \text{ ft})$$

or

$$3T_H + 17T_{1v} = 5200$$

And using the free-body diagram of AP_1P_2 in Figure E5.32c,

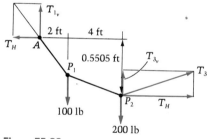

Figure E5.32c

$$\overset{+}{\curvearrowleft} \quad \Sigma M_{P_2} = 0 = (100 \text{ lb})(4 \text{ ft}) + T_H(0.5505 \text{ ft}) - T_{1v}(6 \text{ ft})$$

or

$$0.5505T_H - 6T_{1v} = -400$$

Solving,

$$\left. \begin{array}{l} T_H = 891.9 \text{ lb} \\ T_{1v} = 148.5 \text{ lb} \end{array} \right\} T_1 = 904.2 \text{ lb}$$

From the same free-body, remembering that $T_{3x} = T_H = 891.9$ lb,

$$\overset{+}{\curvearrowleft} \quad \Sigma M_A = 0 = (891.9 \text{ lb})(0.5505 \text{ ft}) + T_{3v}(6 \text{ ft}) - (100 \text{ lb})(2 \text{ ft})$$
$$- (200 \text{ lb})(6 \text{ ft})$$

$$T_{3v} = 151.5 \text{ lb}$$

$$T_3 = \sqrt{891.9^2 + 151.5^2} = 904.7 \text{ lb}$$

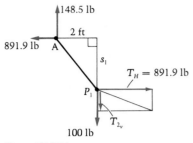

Figure E5.32d

Then we use other free-body diagrams to get the unknown sags; from Figure E5.32d,

$$\curvearrowleft+ \quad \Sigma M_{P_1} = 0 = (891.9 \text{ lb})(s_1) - (148.5 \text{ lb})(2 \text{ ft})$$

$$s_1 = 0.3330 \text{ ft}$$

For T_2, note that its horizontal component is $T_H = 891.9$ lb and

$$+\uparrow \quad \Sigma F_y = 0 = 148.5 - 100 - T_{2v}$$

$$T_{2v} = 48.5 \text{ lb}$$

Thus

$$T_2 = \sqrt{891.9^2 + 48.5^2} = 893 \text{ lb}$$

Then, using the free-body diagram in Figure E5.32e,

$$\curvearrowleft+ \quad \Sigma M_{P_3} = 0 = (100 \text{ lb})(10 \text{ ft}) + (200 \text{ lb})(6 \text{ ft}) + (891.9 \text{ lb})(s_3)$$

$$- (148.5 \text{ lb})(12 \text{ ft})$$

$$s_3 = -0.4687 \text{ ft} \qquad \text{(as before)}$$

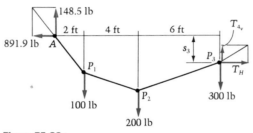

Figure E5.32e

Also, T_{4v} can be computed from

$$+\uparrow \quad \Sigma F_y = 0 = 148.5 - 100 - 200 - 300 + T_{4v}$$

$$T_{4v} = 451.5 \text{ lb}$$

so that

$$T_4 = \sqrt{891.9^2 + 451.5^2} = 999.7 \text{ lb}$$

differing from 1000 lb due to rounding to four digits.

Partial checks of the two solutions obtained in Example 5.31 are afforded by writing the overall vertical equations of equilibrium (see Figure 5.61):

$$+\uparrow \quad \Sigma F_y = -600 + 1000 \left(\frac{0.2643}{\sqrt{0.2643^2 + 2^2}} \right)$$

$$+ 1097 \left(\frac{2.366}{\sqrt{2.366^2 + 5^2}} \right)$$

$$= -600 + 131.0 + 469.2 = 0.200 \text{ lb} \quad \checkmark$$

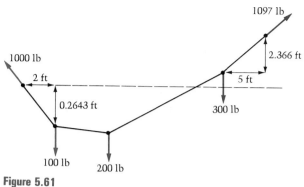

Figure 5.61

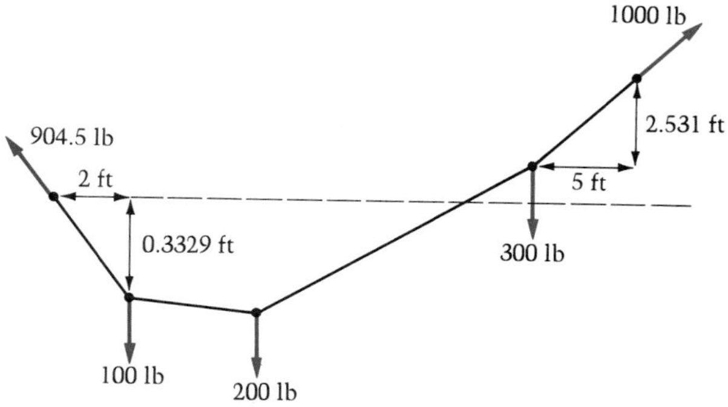

Figure 5.62

and (see Figure 5.62):

$$+\uparrow \quad \Sigma F_y = -600 + 904.5 \left(\frac{0.3329}{\sqrt{0.3329^2 + 2^2}} \right)$$

$$+ 1000 \left(\frac{2.531}{\sqrt{2.531^2 + 5^2}} \right)$$

$$= -600 + 148.5 + 451.6 = 0.100 \text{ lb} \quad \checkmark$$

PROBLEMS ▶ Section 5.12

5.220 The cable supports two vertical loads at P_1 and P_2 as shown in Figure P5.220. Find the elevation of the load point P_1 with respect to the horizontal line through A, and the tensions in the three segments (AP_1, P_1P_2, and P_2B) of the cable.

5.221 In Figure P5.221 find the sags of P_1 and P_2, and the tensions in the three cable segments, if the maximum tension is known to be 1400 lb.

5.222 In Figure P5.222:

  a. If the horizontal force P is holding the cable in equilibrium, what is the value of P?

  b. Find the distances d_1 and d_2.

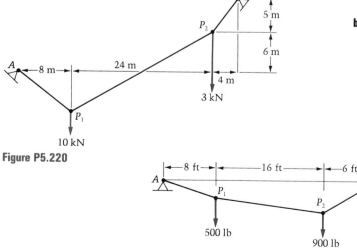

Figure P5.220

Figure P5.221

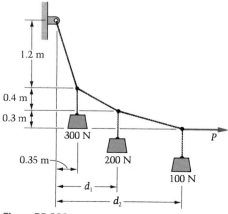

Figure P5.222

5.223 The beam weighs 200 lb and is to be supported in a horizontal position by two cables as shown in Figure P5.223. Find:

  a. The angles ϕ_1 and ϕ_2
  b. The tensions in segments AP_1 and P_2B
  c. The elevation H of B.

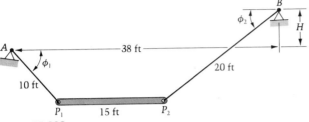

Figure P5.223

5.224 The light cable is loaded by the two downward forces as indicated in Figure P5.224. Find:

  a. The elevation of the load application point P_1 with respect to A.
  b. The tensions in the three cable segments.

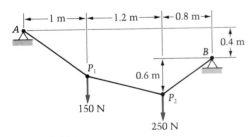

Figure P5.224

5.225 Repeat the preceding problem if the 0.6-m dimension is deleted and replaced by the condition that the maximum cable tension is to be 600 N.

5.226 Find the angles θ and ϕ and the tensions in the two segments of the cable loaded by the 200-lb force shown in Figure P5.226(a). Check your answers for θ and ϕ by making the construction suggested in Figure P5.226(b) and measuring the angles with a protractor.

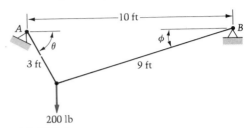

Figure P5.226(a)

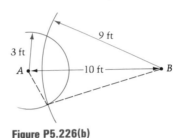

Figure P5.226(b)

COMPUTER PROBLEMS ► Chapter 5

5.227 Use a computer to find the angle θ for which δ_y is a minimum in Problem 5.121. You may either list values of the parenthesized function of θ and watch for the mini-mum, or else use calculus and then solve a resulting cubic polynomial for the cosine of θ.

5.228 Solve Problem 5.213 using a computer.

SUMMARY ► **Chapter 5**

Chapter 5 has been concerned with finding the resultants of internal force distributions within structural elements. In Part I (Sections 5.2–5.7) we studied the truss, and learned in Section 5.2 that all its elements are by definition straight, two-force members. We found in Sections 5.3–5.5 that there are two basic means of determining the forces in a truss: (1) the method of joints (or pins), in which we isolate one pin at a time, writing and solving equilibrium equations until we have found the member force(s) of interest to us; and (2) the method of sections, in which we imagine the truss cut completely in two, with the slice exposing the forces in the cut members. Equilibrium equations on one or the other "half" of the sliced-through truss are then written and solved to determine this force.

Space (3D) trusses are examined in Section 5.6, and in Section 5.7. We introduced the reader to the mechanics of deformable bodies by studying the uniaxial stress and strain found in a truss element.

Part II of the chapter dealt with the internal forces and moments in frames (Section 5.8) and beams (Sections 5.9, 5.10). Here we again slice through members, this time at locations where we wish to know the axial force, shear force and bending moments. (Unlike trusses, where there is but one constant axial force in each member from end to end, the internal forces and moments in frames and beams generally vary from point to point.) For the beam, we also studied in Section 5.10 the differential relationships between the bending moment, the transverse shearing force, and the externally applied lateral distributed loading. This led to the concepts of shear and bending moment diagrams. Once completed, these diagrams tell us at a glance how these two stress-related quantities vary over the beam's length.

We examined the cable in Part III of the chapter. We found in Section 5.11 that when the distributed load on a cable is a uniform function of the x-coordinate, the cable hangs in a parabolic arc, while when the load varies uniformly with arclength along the cable, it hangs in a catenary (hyperbolic cosine) curve. In Section 5.12 we studied the light cable under a set of concentrated parallel forces, and found that, surprisingly, such problems are quite complicated. The reason is that the equations are non-algebraic with variables occurring as arguments of sines and cosines.

REVIEW QUESTIONS ▶ Chapter 5

True or False?

1. Any truss can be solved (i.e., the forces in all its members can be found) by enough applications of the method of joints.
2. In a simple plane truss, a joint can always be found with only two unknown member forces acting on it.
3. For the method of sections to be helpful, the truss has to be sectioned into two distinct, unconnected parts.
4. It is always possible to cut a section that allows the force in any truss member to be found by a single moment equation.
5. No cross section of a two-force member ever has to resist a bending moment or a shear force.

6. No member of a truss ever has to resist a bending moment or a shear force.

7. If a truss member pulls on a pin, then it is in tension; if it pushes on a pin, it is in compression.

8. If a truss member pulls on a pin at one of its ends, then it must push on the pin at its other end.

9. The normal stress in a truss member is the force in the member divided by its cross-sectional area.

10. In a tensile test, if stress versus strain is plotted, we will get a straight line for any material.

11. Statically indeterminate problems may be solved by including consideration of the deformations of the structure.

12. The rate of change of bending moment in a beam with respect to x equals the axial force.

13. The area beneath the loading curve between two points P_1 and P_2 of a beam equals the negative of the change in the shear force between P_1 and P_2.

14. If shear and moment diagrams are sketched and correct values at the right end of the beam are obtained, then the diagrams have probably been drawn correctly.

15. If a cable load is uniformly distributed with respect to x, then the shape of the cable is parabolic. If the load is uniformly distributed with respect to arclength s, the shape of the cable is a catenary.

16. For a cable under any vertical loading, the horizontal component of tension is constant across the cable.

Answers: 1. F   2. T   3. T   4. F   5. F   6. T   7. T   8. F   9. T   10. F   11. T   12. F   13. T   14. T   15. T   16. T

6

▶
▶
▶

FRICTION

6.1 Introduction

This chapter is concerned with one and only one topic: Coulomb friction. Friction is the force resisting one body's tendency to slide past another. If the two bodies are dry, then the friction is often known as Coulomb friction.

We have actually already drawn friction forces on free body diagrams in Chapter 4. Each was the "tangent-plane-of-contact component" of the interactive force of one body onto another. But there, we didn't know those forces as friction forces, nor did we study the principle of Coulomb friction (postponing that until now), which is that a Coulomb friction force f is limited to a maximum value f_{MAX} which is proportional to the normal (pressing) force N between the bodies, according to:

$$0 \leq f \leq f_{MAX} = \mu N$$

In this inequality, μ is the "coefficient of static friction," which depends on the types of materials in contact and on the roughnesses of their surfaces. A low value of μ means not much of a friction force can be developed; the limiting value of $\mu = 0$ corresponds to "smooth" surfaces on which friction is absent. A high value of μ (say about 0.5 and above), conversely, indicates that a relatively large friction force may be developed between the two surfaces.

It is only when a body is on the verge of slipping past another (or is actually doing so) that $f = \mu N$. Usually this is not the case, and the magnitude of the friction force f lies somewhere between zero and the maximum possible value: $f_{MAX} = \mu N$. Beneath a wheel, for example, the friction force is rarely ever equal to μN.

In actual *slipping* situations, the friction force is written as $f = \mu_k N$ where μ_k is the coefficient of kinetic friction, generally less than the static coefficient μ; the latter is written μ_s when a distinction between it and μ_k needs to be made.

Section 6.2 is concerned with basic applications of Coulomb friction. The examples and problems of this section are like those of Chapter 4, *except that now there is the possibility of slipping (sliding)*. In the other section, 6.3, we introduce the reader to a few special engineering applications of Coulomb friction: wedges, flexible flat belts and V-belts, square-threaded screws, and disk friction.

The reader should already appreciate the fact that friction is of enormous importance in engineering.

6.2 Laws, Coefficients, and Basic Applications of Coulomb Friction

The world would continue to go round and round without friction, but the lives of just about everything living on it would change drastically if friction were suddenly to disappear. Friction is therefore a very important topic, worthy of much study. It serves both good and bad purposes.

Without friction, automobiles, trains, and bicycles would not work at all because their wheels need friction to produce acceleration of the vehicles to which they are attached. And without friction on the soles of your shoes, you could not walk out of the room in which you are sitting or lying. We wish to maximize the friction force in situations where it is being used to our advantage, such as in the use of tires, brakes, wedges, screws, belts, clutches, and in simple but important applications such as at the foot of a ladder. Conversely, friction has many deleterious effects: it causes mechanical parts to wear out, and much energy is expended in overcoming frictional resistance to motions of machines and vehicles.

We shall now examine the law of dry friction between a pair of surfaces; this law was first set forth by Coulomb in the 18th century and is called **Coulomb's Law of Dry Friction.**

Normal and Friction Forces

For two bodies $\mathcal{B}_1$ and $\mathcal{B}_2$ in contact, the **friction force** on, say, $\mathcal{B}_1$, is the component of the resultant force (exerted on $\mathcal{B}_1$ by $\mathcal{B}_2$) that lies in the tangent plane of contact (see Figure 6.1). The other component, perpendicular to this tangent plane, is called the **normal force** exerted on $\mathcal{B}_1$ by $\mathcal{B}_2$.

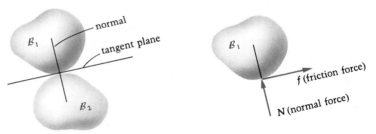

Figure 6.1

Let us now imagine a crate resting on a horizontal surface. The free-body diagram is shown in Figure 6.2(a). The resultant of all the "pressures times areas" at each of the points of contact beneath the box is depicted as a single resultant force N, which equals the weight mg of the crate and which acts upward through the mass center C. In this state, there is no friction at all. Friction resists the tendency to slide and if there is no such tendency, it will not be present.

Now suppose a child exerts a slowly and linearly increasing (with time t) horizontal force P in an effort to move the crate to the right as shown in Figure 6.2(b) (see the following page). Two things of importance to our study then happen at once. They are:

1. A friction force f develops simultaneously that opposes the tendency of the crate to slide to the right.

2. The normal force N is still equal to mg because there is no vertical acceleration and the y forces are still in balance. But the *position* of

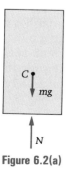

Figure 6.2(a)

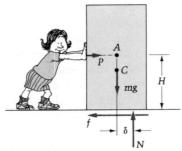

Figure 6.2(b)

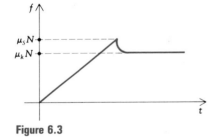

Figure 6.3

N — that is, of the resultant of the many pressing forces under the crate — shifts to the right. The *way* this happens is by the pressure increasing on the right and decreasing on the left. The *reason* it happens is that N must act so as to overcome the tendency of the crate to tip forward. Said another way, the moments of f and N about A^* must add to zero to prevent rotation. Thus N moves to the right so that the moment $fH\circlearrowright$ can be balanced by the moment $N\delta\circlearrowleft$, where δ is (see the figure) the shift, to the right, of the force N from the centerline of the crate.

As the applied force of the child grows, so do f and δ. If the crate is wide enough and/or short enough, it will not tip before sliding. Let us assume this to be the case for now. The friction force f will continue to balance the force P for as long as it is able.

Maximum Value of Friction; Coefficient of Friction

At some point in the above process, a maximum value is reached for f, as suggested by Figure 6.3, and the crate will start to slide. This value was observed by Coulomb to be proportional to the normal force:

$$f_{\text{MAX}} = \mu_s N \tag{6.1}$$

in which the proportionality constant μ_s is the **coefficient of static friction,** which depends upon the types of materials in contact and upon the roughnesses of their surfaces. The value of μ_s varies from near zero for very smooth surfaces to 1 or even more than 1 for very rough surfaces.

Now let us return to the child and the crate. As most of us have experienced, it is harder to "get something moving than to keep it moving." When P reaches a value equal to the static maximum of f, motion is said to be impending. Any further increase in P will cause the crate to slide, and the friction force normally drops off to a second, usually slightly lower value.

---

* Note that P and mg produce no moment about A.

> **Question 6.1** Sketch a possible plot of P vs. t for the case when the block accelerates over Δt to a constant velocity.

This reduced value of f is assumed to be the normal force N times a different coefficient of friction, this time called the coefficient of kinetic friction, μ_k:

$$f_k = \mu_k N \qquad (6.2)$$

where f_k is the value of friction to be used if sliding has started and is actually ongoing. The force f_k is thus used in dynamics. There, if no subscript is placed on the μ, it is to be assumed that $\mu_s \approx \mu_k$. But actually, μ_k is normally dependent upon the speed of sliding, and, as the above discussion suggests, it is less than μ_s. A simple model that is often used is to take μ_k to be a constant (independent of sliding speed) less than μ_s.

In statics, however, we are usually only concerned with μ_s, so that we may drop the subscript on μ in most cases. The student should always remember that

$$0 \le f \le f_{\text{MAX}} = \mu N \qquad (6.3)$$

and that in the absence of relative motion, $f = \mu N$ *if and only if* the bodies at rest and in contact are at the point (on the verge) of slipping relative to each other. This is called "impending slipping." When surfaces are called "smooth," it means μ is very small, approaching zero. In this case,

$$0 \le f \le \mu N \approx 0 \qquad (6.4)$$

so that no friction is available and f is to be assumed as zero.

The static coefficient of friction may be determined for two materials by constructing a plane of one of the materials and a block of the other (see Figure 6.4):

As the angle θ of the plane is increased from zero, the friction force beneath $\mathcal{B}_1$ will increase (see the free-body diagram of $\mathcal{B}_1$) because it has to equilibrate an increasing component of the weight (mg) down the plane. Friction always acts on each body in the direction that opposes sliding of the two contacting bodies past each other.

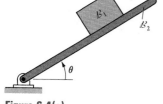

Figure 6.4(a)

Answer 6.1 A plot of P versus time will be identical to f vs. t *except* during the short interval Δt during which the block accelerates up to constant speed. (See the figure.) During that time P is larger than f, and their difference, $P - f$, produces the acceleration of the block.

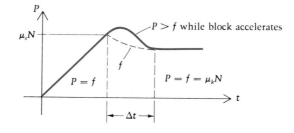

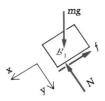

Figure 6.4(b)

Using the indicated directions for x and y, the equilibrium equations of $\mathcal{B}_1$ are seen from Figure 6.4(b) to be

$$\Sigma F_x = 0 = mg \sin \theta - f \Rightarrow f = mg \sin \theta$$

$$\Sigma F_y = 0 = mg \cos \theta - N \Rightarrow N = mg \cos \theta$$

(Note that the moment equation would give us the location of the normal force N.)

If we slowly increase θ until the block slides, then at this critical angle (call it θ_s), we have $f = \mu N$. Therefore, at $\theta = \theta_s$,

$$f = mg \sin \theta_s = \mu N = \mu\, mg \cos \theta_s$$

or

$$\mu = \tan \theta_s \qquad (6.5)$$

so that the tangent of the angle at which slipping begins is the coefficient of friction for the two materials of interest. The angle θ_s is called the **angle of friction,** or angle of repose.

In the following table, some typical values of μ_s are presented for various selected pairs of materials.

Table 6.1

Approximate* Values of the Coefficient of Static Friction for Clean, Dry Surfaces	
Surface Materials	**Coefficient of Friction**
Steel on steel	0.75
Rubber on concrete	0.8
Rubber on asphalt	0.85
Rubber on ice	0.1
Oak on oak	0.6
Wood on metal	0.4
Cast iron on cast iron	1.1
Cast iron on copper	1.05
Teflon on teflon or steel	0.04
Copper on steel	0.5
Aluminum on aluminum	1.1
Steel on lead	0.95
Aluminum on steel	0.6
Brass on brass	0.9
Hemp rope on wood	0.7
Metal on stone	0.5
Metal on leather	0.45
Metal on ice	0.04
Oak on leather	0.6
Bonded carbide on iron	0.8
Glass on glass	0.9
Copper on copper	1.2

---

* Values of μ may vary widely for a pair of materials, depending on such things as whether the steel is mild or hard, whether the friction on the wood is with or across the grain, and so forth. The tabulated values are intended just to give the reader a general idea of values of μ.

We now proceed to some examples in which we have friction acting between various flat surfaces. In the first one, we shall make use of Equation 6.5 at the outset.

EXAMPLE 6.1

In Figure E6.1a find the smallest force P that will prevent the crate from sliding* down the plane. The crate weighs 200 lb and the friction coefficient between it and the plane is 0.4.

Solution

We note first from Equation 6.5 that the crate will not remain in equilibrium without force P because tan $30° = 0.577 > \mu = 0.4$. For the smallest force P, the friction force will be at its maximum, acting up the plane as shown in the free-body diagram of Figure E6.1b. Equilibrium requires:

$$\nearrow \quad \Sigma F_x = 0 = 200 \sin 30° - 0.4N - P \tag{1}$$

and

$$\nwarrow \quad \Sigma F_y = 0 = 200 \cos 30° - N \tag{2}$$

From Equation (2), $N = 173$ lb. Substituting this result into Equation (1) yields

$$P = 31 \text{ lb}$$

Question 6.2   What change in Equation (1) would result in the *maximum* value of P for equilibrium of the crate?

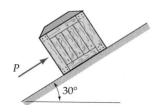

P

$30°$

Figure E6.1a

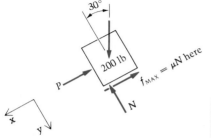

$30°$

200 lb

$f_{MAX} = \mu N$ here

P

N

x

y

Figure E6.1b

In the preceding example, we did not concern ourselves with the possibility of the crate tipping over. In the next example, we examine both cases: slipping *and* tipping.

---

* Slipping and sliding mean the same thing here, and the two verbs are used interchangeably throughout the chapter.

Answer 6.2   If $0 = 200 \sin 30° \oplus 0.4N - P$, then the crate would be on the verge of slipping up the plane. This would result in the maximum P, for then P would be opposing friction as well as gravity, instead of being aided by friction as in the example.

EXAMPLE 6.2

Find the condition for which the uniform crate of Figure E6.2a pushed to the right by a slowly increasing force P, will slide before it tips over.

Solution

The equations of equilibrium are seen to be (note the free-body diagram, Figure E6.2b):

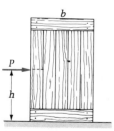

b

P

h

Figure E6.2a

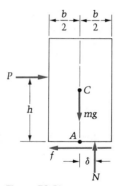

Figure E6.2b

$$\xrightarrow{\;+\;} \quad \Sigma F_x \;= 0 = P - f \Rightarrow P = f$$

$$+\uparrow \quad\;\; \Sigma F_y \;= 0 = N - mg \Rightarrow N = mg$$

$$\overset{+}{\curvearrowleft} \quad\;\; \Sigma M_A = 0 = N\delta - Ph \Rightarrow N\delta = Ph$$

If we assume that the block slips before tipping, then at the point of slipping (or sliding), $f = \mu N$. The equations then give

$$N\delta = Ph = fh = \mu Nh$$

or

$$\delta = \mu h \qquad \text{and} \qquad P = \mu mg$$

To check whether our assumption of slipping was correct, we must determine whether or not N acts between the centerline and the lower right corner of the block; that is, δ must lie between zero and $b/2$ for slipping to occur first.

$$\delta = \mu h < \frac{b}{2}$$

or

$$b > 2\mu h$$

Thus if the base dimension b is greater than $2\mu h$, the block will slip. Equality of b and $2\mu h$ means slipping and tipping occur simultaneously, and $2\mu h > b$ implies that tipping happens first.

Let us check the results of the preceding example by starting over with a different assumption: that tipping occurs first. If so, both N and f act at the lower right-hand corner of the block as shown in Figure 6.5. We obtain, from the moment equation,

$$\overset{+}{\curvearrowleft} \quad \Sigma M_B = 0 = N\frac{b}{2} - fh$$

or

$$f = \frac{Nb}{2h}$$

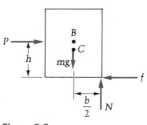

Figure 6.5

To check our assumption and see whether slipping has already occurred, we must find out if $f < \mu N$. If so, our tipping assumption was correct:

$$f = \frac{Nb}{2h} < \mu N$$

or $b < 2\mu h$ for tipping, as we had previously obtained. If $b = 2\mu h$, the crate slips and tips together, and if $b > 2\mu h$, the assumption was incorrect and it slides first, again as previously determined.

Sometimes we have to decide between more than one sliding possibility, as in the next example:

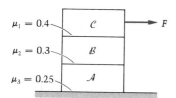

Figure E6.3a

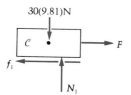

Figure E6.3b

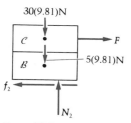

Figure E6.3c

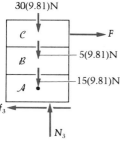

Figure E6.3d

EXAMPLE 6.3

The masses of blocks C, B, and A of Figure E6.3a are 30 kg, 5 kg, and 15 kg, respectively. Find the largest value of the force F for which no sliding will take place between any pair of surfaces. Assume the blocks to be wide enough so that tipping will not occur.

Solution

There are three cases to consider:

1. Impending sliding of C on B, with A and B stationary (see Figure E6.3b):

$$+\!\uparrow \quad \Sigma F_y = 0 = N_1 - 30(9.81) \Rightarrow N_1 = 294 \text{ N}$$

$$f_1 = f_{1\,MAX} = \mu_1 N_1 = 0.4(294) = 118 \text{ N}$$

$$\xrightarrow{+} \quad \Sigma F_x = 0 = F - f_1 \Rightarrow F = 118 \text{ N}$$

2. Impending motion of C and B together as one body, with impending sliding of B on a stationary block A (see Figure E6.3c):

$$+\!\uparrow \quad \Sigma F_y = 0 = N_2 - 35(9.81) \Rightarrow N_2 = 343 \text{ N}$$

$$f_2 = f_{2\,MAX} = \mu_2 N_2 = 0.3(343) = 103 \text{ N}$$

$$\xrightarrow{+} \quad \Sigma F_x = 0 = F - f_2 \Rightarrow F = 103 \text{ N}$$

which is less than the F value to move C alone.

> **Question 6.3**   The results of cases 1 and 2 prove that C will not be on the verge of slipping on B with B simultaneously on the verge of slipping on A. Why?

3. All three blocks are about to move together, with impending slipping of A on the plane (see Figure E6.3d):

$$+\!\uparrow \quad \Sigma F_y = 0 = N_3 - 50(9.81) \Rightarrow N_3 = 491 \text{ N}$$

$$f_3 = f_{3\,MAX} = \mu_3 N_3 = 0.25(491) = 123 \text{ N}$$

$$\xrightarrow{+} \quad \Sigma F_x = 0 = F - f_3 \Rightarrow F = 123 \text{ N},$$

the highest of the three values of F. Therefore the maximum force F for which equilibrium of the stack of three blocks can exist is 103 newtons.

In the preceding example, intuition tells us that the case of A and C sliding to the right with B remaining stationary is impossible. It is interesting to study the reason why.

The free-body diagrams in Figure 6.6 show the directions of the various friction forces f_1, f_2, and f_3 upon application of F. Note that if B remains stationary, then f_1 balances f_2 in the figure. But for A to slide to

---

Answer 6.3   Because when F reaches 103 N, we have impending slippage of B on A; however, 118 N is required to simultaneously cause impending slippage of C on B.

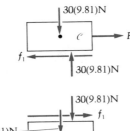

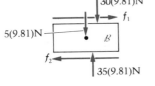

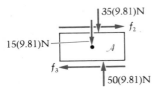

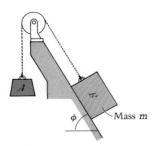

Figure 6.6

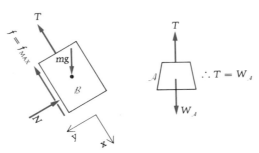

Figure E6.4a

the right relative to B would require f_2 to be *opposite* to the directions indicated on both A and B opposing the relative sliding. Two contradictions would then result: (a) A would no longer have an applied force to the right to tend to make it move that way; (b) on B, f_1 and f_2 would add and thus not balance; B could not then remain still. Thus B cannot remain stationary unless A does also.

The next example illustrates the fact that often a *range* of values of a quantity (here, of weights of body A) will be possible for equilibrium:

EXAMPLE 6.4

From the discussion surrounding Equation 6.5, we know that without the rope and body A present in Figure E6.4a block B will slide down the incline if $\tan \phi > \mu$. Let $\phi = 60°$ and $\mu = 0.6$, so that

$$\tan \phi = \sqrt{3} > \mu = 0.6$$

Determine the range of values of the weight of A that will result in equilibrium of the system.

Solution

There is a minimum value of A that will prevent B from sliding down the plane; similarly, there is also a *maximum* weight of A above which A will pull B *up* the plane. Between these two weights, the bodies will be in equilibrium with the friction force f beneath B lying between its maxima up and down the plane, respectively. In the first case, the motion of B impends downward (see the FBDs in Figure E6.4b). Equilibrium requires

$$\searrow^+ \quad \Sigma F_x = 0 = mg \sin 60° - T - f_{\text{MAX}} \tag{1}$$

$$\nearrow^+ \quad \Sigma F_y = 0 = mg \cos 60° - N \Rightarrow N = \frac{mg}{2} \tag{2}$$

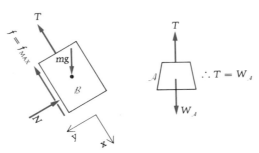

Figure E6.4b

But $f_{\text{MAX}} = \mu N = 0.6mg / 2 = 0.3mg$, so that, from Equation (1),

$$T = mg \frac{\sqrt{3}}{2} - f_{\text{MAX}} = mg\left(\frac{\sqrt{3}}{2} - 0.3\right) = 0.566mg$$

and

$$W_{A\text{MIN}} = T = 0.566mg$$

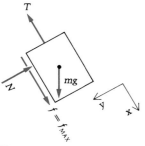

Figure E6.4c

In the second case, the motion of B impends upward, with friction resisting this tendency, acting downward (see Figure E6.4c and note the change in direction of f). The equations become:

$$\searrow^+ \quad \Sigma F_x = 0 = mg \sin 60° - T + f_{MAX} \tag{3}$$

$$\nearrow^+ \quad \Sigma F_y = 0 = mg \cos 60° - N \Rightarrow N = \frac{mg}{2} \quad \text{as before.} \tag{4}$$

Again, $f_{MAX} = \mu N = 0.3mg$. This time, the first equation (3) gives

$$T = mg\frac{\sqrt{3}}{2} + f_{MAX} = mg\left(\frac{\sqrt{3}}{2} + 0.3\right) = 1.17mg$$

and

$$W_{A\,MAX} = T = 1.17mg$$

Any weight of A between $0.566mg$ and $1.17mg$ will result in equilibrium of the two bodies. Note that one of the values of the friction force, for W_A in this range, is zero. This happens when $T = mg \sin 60°$, or for $W_A = mg\sqrt{3}/2 = 0.866mg$.

Our first four examples have been concerned with impending sliding of flat surfaces in contact. Friction also develops in rolling situations, as we shall see in the next few examples. In the first one, we have to determine whether the two lower cylinders are about to roll on the ground (while slipping on A), or about to roll on A (while slipping on the ground):

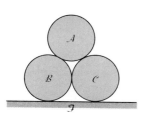

Figure E6.5a

EXAMPLE 6.5

Three identical cylinders are stacked as shown in Figure E6.5a. Assuming that the friction coefficient μ is the same for all pairs of contacting surfaces, determine the minimum value of μ for which the cylinders will remain stacked in equilibrium.

Solution

From the free-body diagram of the lower-left cylinder B in Figure E6.5b, we can sum moments about B and immediately find a relation between the forces f and N:

$$\overset{+}{\curvearrowleft} \quad \Sigma M_B = 0 = N\frac{r}{2} - fr(1 + \sqrt{3}/2)$$

from which

$$f = 0.268N \tag{1}$$

Since f is limited by μN, the above result means that the critical (smallest) value of μ that will prevent B and C from rolling outward on the ground (while each slips on A) is 0.268.

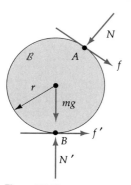

Figure E6.5b

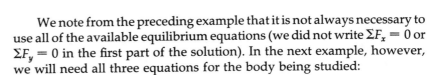

Question 6.4   In the free-body diagram of cylinder $\mathcal{B}$, why is there no force exerted by $\mathcal{C}$?

We must also investigate the possibility of $\mathcal{B}$ and $\mathcal{C}$ *skidding* outward on the plane while maintaining rolling (i.e., no slip) contact with the upper cylinder $\mathcal{A}$. To this end, we sum moments about A on the same free-body (Figure E6.5b):

$$\Sigma M_A = 0 = (mg - N')\frac{r}{2} + f'r(1 + \sqrt{3}/2) \qquad (2)$$

If we knew the relation between mg and N', we could obtain the relation between f' and N' that we need. From the "overall" free-body in Figure E6.5c, we see that

$$\Sigma F_y = 0 = 2N' - 3mg$$

so that

$$mg = \frac{2}{3}N'$$

Substituting $\frac{2}{3}N'$ for mg in Equation (2) gives:

$$f' = \frac{0.268}{3}N' = 0.089N' \qquad (3)$$

Therefore, to prevent "skidding" of $\mathcal{B}$ and $\mathcal{C}$ on the plane, the friction coefficient μ needs to be only one-third that of the other case. Therefore the value needed to prevent *both* motions is the larger value, 0.268. Note that Equations (1) and (3) tell us the manner in which equilibrium can be lost.

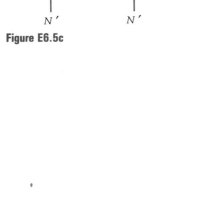

Figure E6.5c

We note from the preceding example that it is not always necessary to use all of the available equilibrium equations (we did not write $\Sigma F_x = 0$ or $\Sigma F_y = 0$ in the first part of the solution). In the next example, however, we will need all three equations for the body being studied:

Answer 6.4   For impending motion of $\mathcal{B}$ to the left and $\mathcal{C}$ to the right, the force of $\mathcal{C}$ on $\mathcal{B}$ has gone to zero.

EXAMPLE 6.6

The truck in Figure E6.6a is preparing to push a drum up an 8° incline. The coefficient of friction between the drum and the incline is 0.3. Find the range of positive values of the friction coefficient μ between the drum and truck bumper so that the drum rolls (does not slip) on the ground.

Solution

Let us assume that the drum is in equilibrium — that is, that its motion has not quite yet begun. Making use of the FBD in Figure E6.6b, the equilibrium equations are:

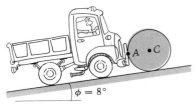

Figure E6.6a

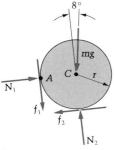

Figure E6.6b

$$\nearrow \quad \Sigma F_x = 0 = N_1 - f_2 - mg \sin 8° \tag{1}$$

$$\nwarrow \quad \Sigma F_y = 0 = N_2 - f_1 - mg \cos 8° \tag{2}$$

$$\curvearrowleft{+} \quad \Sigma M_C = 0 = f_1 r - f_2 r \Rightarrow f_1 = f_2 \tag{3}$$

Thus we have three equations in the four unknowns N_1, N_2, f_1, and f_2. Assuming that slipping impends at A (which it will if the drum is about to roll on the ground) means

$$f_1 = \mu N_1 \tag{4}$$

Thus (1) and (3) give

$$N_1 (1 - \mu) = mg \sin 8°$$

$$N_1 = \frac{mg \sin 8°}{1 - \mu} \tag{5}$$

Equation (2) gives

$$N_2 = \mu N_1 + mg \cos 8°$$

$$= \frac{\mu}{1 - \mu} mg \sin 8° + mg \cos 8°$$

$$N_2 = mg \left(\cos 8° + \frac{\mu}{1 - \mu} \sin 8° \right) \tag{6}$$

And from Equations (3), (4), and (5) we have

$$f_2 = f_1 = \frac{\mu mg \sin 8°}{1 - \mu} \tag{7}$$

If the drum is not to slip on the incline, then

$$f_2 \le 0.3 N_2 \tag{8}$$

Substituting for f_2 from Equation (7) and N_2 from (6), and canceling mg,

$$\frac{\mu}{1 - \mu} \sin 8° \le 0.3 \left[\cos 8° + \frac{\mu}{1 - \mu} \sin 8° \right]$$

$$0.7 \frac{\mu}{1 - \mu} \sin 8° \le 0.3 \cos 8°$$

$$\mu 0.0974 \le (1 - \mu)(0.297)$$

$$\mu \le \frac{0.297}{0.394} = 0.754$$

In the preceding example, note that if the "$f \le \mu N$" inequality is carried forward (as was done by substituting into (8)), then the final result is an entire *range* of values of μ ("anything less than or equal to 0.754") instead of just a number. The inequality thus gives us a more complete result than if we were to set $f = \mu N$ and solve for one number for μ, for then we would have to think about whether the desired range consisted of all values above this μ, or all values below it.

In the next example, the cylinder is prevented from rolling on the plane; it slips there instead, for a large enough θ, because of the constraint imposed by the cord.

EXAMPLE 6.7

The angle θ in Figure E6.7a is slowly increased from zero. Find the value of θ at which the slotted cylinder of mass m will begin to slip.

Solution

The cylinder cannot roll on the plane because it is restrained by the cord, which is assumed to be wrapped tightly and not slipping in the slot. At a certain value of θ, the cylinder will slip downward on the plane while pivoting about the point A shown on the free-body diagram. It will do this when the value of f needed for equilibrium exceeds the maximum value possible for f, which is μN. Referring to the FBD, Figure E6.7b, we write the equilibrium equations:

$$\nearrow \quad \Sigma F_x = 0 = mg \sin \theta - T - f \tag{1}$$

$$\nwarrow \quad \Sigma F_y = 0 = mg \cos \theta - N \Rightarrow N = mg \cos \theta \tag{2}$$

$$\circlearrowleft_+ \quad \Sigma M_A = 0 = -rmg \sin \theta + f(R + r) \Rightarrow f = \frac{rmg \sin \theta}{R + r} \tag{3}$$

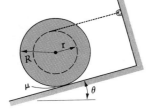

Figure E6.7a

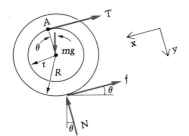

Figure E6.7b

Enforcing the no-slipping condition,

$$f \le \mu N$$

$$\frac{rmg \sin \theta}{R + r} \le \mu mg \cos \theta$$

$$\tan \theta \le \frac{\mu(R + r)}{r}$$

$$\theta_{\text{slip}} = \tan^{-1}\left(\frac{\mu(R + r)}{r}\right)$$

This angle could be substituted if desired into Equations (3) and (1) to yield the values of the friction force f and the tension T at the slipping angle θ_{slip}.

In the next example, because a moment equation is involved, we must locate the center of mass before we can proceed to the friction part of the problem.

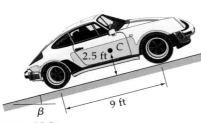

Figure E6.8a

EXAMPLE 6.8

The front wheels of a certain car (see Figure E6.8a) support 55% of its weight W on level ground. The coefficient of friction between the tires and an inclined road surface is 0.3. What is the largest incline angle β for which the car will not slip when parked? (Only the rear wheels are locked.)

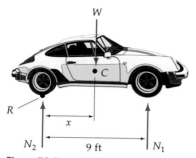

Figure E6.8b

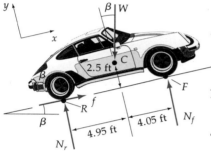

Figure E6.8c

Solution

In equilibrium on level ground, by summing moments about R we find that the mass center C of the car is 4.95 ft forward of the rear wheels (see the free-body diagram in Figure E6.8b):

$$\curvearrowleft{+} \qquad \Sigma M_R = 0 = 9N_1 - W$$

$$= 9(0.55\ W) - xW$$

$$x = 4.95\ \text{ft}$$

Putting the car on the incline, we sum moments about the bottom (F) of the front wheel and determine N_r (see the free-body diagram in Figure E6.8c):

$$\curvearrowleft{+} \quad \Sigma M_F = 0 = W\cos\beta(4.05\ \text{ft}) + W\sin\beta(2.5\ \text{ft}) - N_r(9\ \text{ft}) \qquad (1)$$

$$N_r = W[0.45\cos\beta + 0.278\sin\beta] \qquad (2)$$

Also,

$$\Sigma F_x = 0 = f - W\sin\beta$$

$$f = W\sin\beta \qquad (3)$$

But for the car not to slip,

$$f \le \mu N_r \qquad (4)$$

Thus, substituting N_r from Equation (2) and f from (3) into (4),

$$W\sin\beta \le 0.3W[0.45\cos\beta + 0.278\sin\beta]$$

or

$$0.917\sin\beta \le 0.3(0.45)\cos\beta$$

$$\tan\beta \le 0.147$$

$$\beta \le 8.36°$$

In the preceding example we note that, all other parameters remaining equal, raising the mass center (increasing the 2.5-ft dimension) increases the "$W\sin\beta$" coefficient in Equation (1). This then has the effect of raising both N_r and the angle β that comes out of the inequality above. Physically, a larger N_r with the same μ gives more maximum frictional resistance to slipping.

In the next example, the beam and the attached shoe act as a brake on the cylinder:

EXAMPLE 6.9

The uniform beam $\mathcal{B}$ in Figure E6.9a weighs 1000 N, and is attached to the light shoe $\mathcal{S}$, which rests on the cylinder $\mathcal{C}$. How much weight can be suspended in equilibrium by a cord wrapped tightly around the hub of $\mathcal{C}$ as shown?

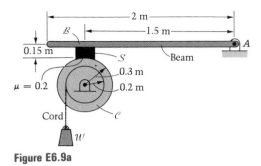

Figure E6.9a

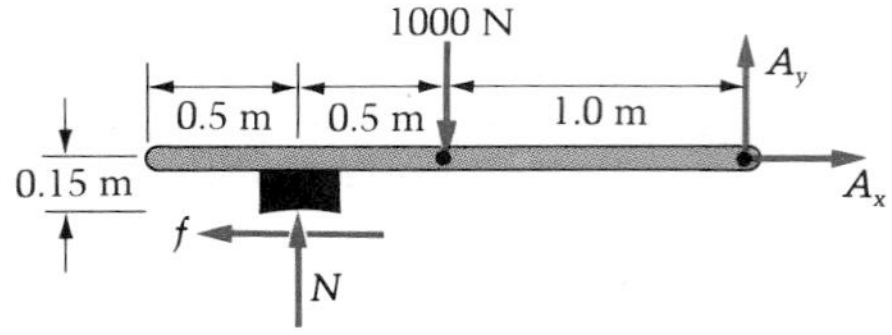

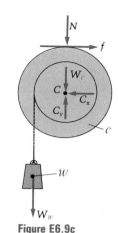

Figure E6.9c

1000 N

| 0.5 m | 0.5 m | 1.0 m |

A_y

0.15 m

f

N

A_x

Figure E6.9b

Solution

Free-body diagrams of (a) beam plus shoe, and of (b) cylinder plus weight are shown in Figure E6.9b and E6.9c.

If we imagine the weight W to be slowly increased, it will eventually reach a critical value at which cylinder $\mathcal{C}$ will begin to turn counterclockwise. This has to happen because the free-body diagram of $(\mathcal{C} + W)$ shows that the only moment about C available to equilibrate that of W_w is the moment of the friction force f. That moment is bounded by μN times the radius 0.3 m. Thus we first need N. From the free-body diagram of $(\mathcal{B} + \mathcal{S})$, Figure E6.9b, we see that:

1. Writing $\Sigma F_x = 0$ and $\Sigma F_y = 0$ will not be of help because they will include A_x and A_y, which are undesired unknowns.
2. Moments about A will allow us to determine N without finding A_x and A_y:

$$\curvearrowleft_{+} \qquad \Sigma M_A = 0 = (1000 \text{ newtons})(1 \text{ m}) - N(1.5 \text{ m}) - f(0.15 \text{ m}) \quad (1)$$

At the critical value of W, which we seek, f becomes f_{MAX}, or μN, so that

$$N = \frac{1000}{1.53}$$

$$= 654 \text{ newtons}$$

We return to the free-body diagram of $(\mathcal{C} + W)$, Figure E6.9c. The same remark (1) above again holds, this time with the pin reactions C_x and C_y unknown and unwanted. Thus again we proceed directly to the moment equation that eliminates these reactions from consideration:*

$$\curvearrowleft_{+} \qquad \Sigma M_C = 0 = W (0.2 \text{ m}) - (0.2N)(0.3 \text{ m}) \qquad (2)$$

so that

$$W = \frac{(0.2)(654)(0.3)}{0.2} = 196 \text{ newtons}$$

---

* If we were seeking G_x and G_y in addition to W, we would *still* write $\Sigma M_C = 0$ first to get W. Then, writing $\Sigma F_x = 0$ and $\Sigma F_y = 0$ would yield the pin reactions.

It is instructive to consider what changes might occur in the solution if the weight comes off the *right* side of the hub of the cylinder instead of the left side.

Question 6.5   Will the answer then be the same?

Self-locking Mechanisms

In the next example, we consider the phenomenon called **self-locking.** This refers to a mechanism in which there will be no slipping regardless of increases in the size of the applied force. The reason for self-locking, as you will see, is that in such situations the normal force and friction force increase proportionately so that f can never exceed μN if μ is large enough to begin with.

Answer 6.5   No, it won't. The friction will reverse direction on both C and S, and Equation (1) will then give a different (larger) N. Thus f will be larger and so will W.

EXAMPLE 6.10

A cam consisting of a cylinder mounted off-center (see Figure E6.10a) is to be used as a self-locking mechanism. Whenever the force P acts downward on the plate, the cam tends to rotate clockwise and grip the plate, holding it fast. Neglecting the weights of cam and plate in comparison with the large force P, and assuming the wall that bears against the plate is smooth, find the smallest value of the coefficient of friction between cam and plate for which the self-locking will occur.

Solution

Free-body diagrams of the cam and plate are shown in Figures E6.10b and E6.10c:

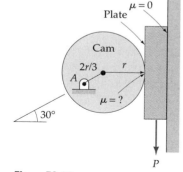

Figure E6.10a

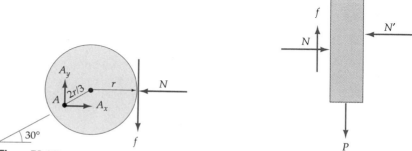

Figure E6.10b

Figure E6.10c

We see by inspection of Figure E6.10c that $f = P$ for equilibrium of the plate.*

---
* Also, $N' = N$ and, for a balance of moments, f and P form a clockwise couple which is the negative of another formed by N and N'.

Thus the magnitude of f cannot be limited if equilibrium is to exist regardless of the size of P. We move next to the FBD of the cam, Figure E6.10b.

As the pin reactions A_x and A_y are not of interest to us here, we sum moments about A in Figure E6.10b and thereby eliminate them:

$$\circlearrowleft^{+} \quad \Sigma M_A = 0 = -f\left(r + \frac{2r}{3}\frac{\sqrt{3}}{2}\right) + N\frac{2r}{3}\frac{1}{2}$$

or

$$f = 0.211N \tag{1}$$

But

$$f \le f_{MAX} = \mu N$$

or, substituting for f using Equation (1),

$$0.211N \le \mu N$$

so that if $\mu > 0.211$, the plate will never slip downward past the cam regardless of the size of P. As P gets larger, (a) so does f since $f = P$, and (b) so does N by Equation (1).

Question 6.6   In the preceding example, could the self-locking occur if the cam were pinned at the center? to the left of the center?

The last example in the section is longer than the others. It involves a body (tool) that must be taken apart to complete the solution. The reader should note that working such a problem in terms of symbols gives a designer much more flexibility when it comes time to analyze the effects of different dimensions on the answers.

Answer 6.6   No, it would develop no friction. No, it would lose contact.

EXAMPLE 6.11

The pipe wrench (or Stillson wrench) in Figure E6.11a is gripping a pipe $\mathcal{P}$ that is about to be unscrewed from a fitting by the force P as shown in the figure. The member $\mathcal{B}$ is pinned to $\mathcal{H}$ and is loose-fitting on $\mathcal{J}$. The friction coefficients μ between $\mathcal{J}$ and $\mathcal{P}$ and between $\mathcal{H}$ and $\mathcal{P}$ are given to be equal. Find the relationship between μ, D, h, b, and L for which the pipe wrench will be "self-locking," meaning that it will not slip on the pipe *regardless of the size of* P. Neglect all weights, and for the purposes of this example, make the assumption that the teeth of the wrench do not significantly dig into the surface of the pipe.

Solution

Separate free-body diagrams of the wrench as well as two of the parts of the wrench are shown in Figures E6.11b–d. We see immediately from Figure E6.11b that

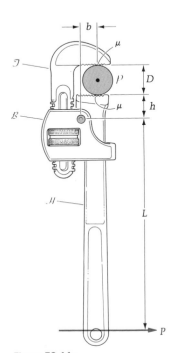

Figure E6.11a

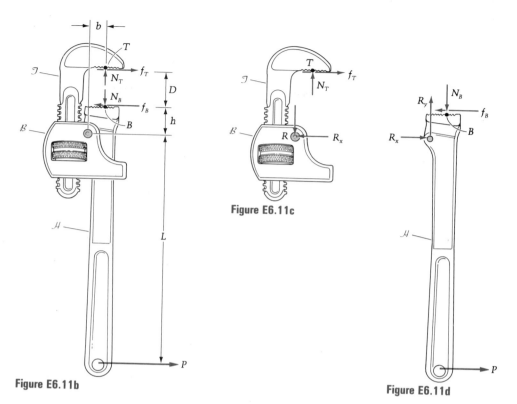

Figure E6.11b

Figure E6.11c

Figure E6.11d

$$\xrightarrow{+} \qquad \Sigma F_x = 0 = f_T - f_B + P \Rightarrow P = f_B - f_T \tag{1}$$

$$+\uparrow \qquad \Sigma F_y = 0 = N_T - N_B \Rightarrow N_T = N_B \tag{2}$$

$$\curvearrowleft+ \qquad \Sigma M_T = 0 = (D + h + L)P - f_B D \tag{3}$$

Notice that we are not yet stating that either of the friction forces f is equal to μN. We shall first get the equilibrium equations written, then investigate the slipping possibilities.

Using the free-body diagram (Figure E6.11c) of the combined body ($\mathcal{B} + \mathcal{J}$), we get

$$\xrightarrow{+} \qquad \Sigma F_x = 0 = f_T - R_x \Rightarrow f_T = R_x \tag{4}$$

$$+\uparrow \qquad \Sigma F_y = 0 = N_T - R_y \Rightarrow N_T = R_y \tag{5}$$

$$\curvearrowright+ \qquad \Sigma M_R = 0 = N_T b - f_T(D + h) \tag{6}$$

Using Equation (6) we can immediately obtain the minimum value of the coefficient of friction μ needed to prevent slipping at the top of the pipe (point T):

$$f_T = N_T \frac{b}{D + h} \leq f_{TMAX} = \mu N_T \tag{7}$$

or

$$\mu \geq \frac{b}{D + h} \tag{8}$$

We proceed to obtain a similar condition on μ for which slipping will not occur at the bottom point (B) of the pipe. Equations (1) and (3) give, eliminating P,

$$f_B - f_T = \frac{f_B D}{D + h + L} \tag{9}$$

or, using Equation (7),

$$f_B\left(\frac{h + L}{D + h + L}\right) = f_T = N_T \frac{b}{D + h} \tag{10}$$

But $N_T = N_B$ by Equation (2); thus

$$f_B\left(\frac{h + L}{D + h + L}\right) = \frac{b}{D + h} N_B \tag{11}$$

or

$$f_B = \frac{b(D + h + L)}{(h + L)(D + h)} N_B \leq f_{B_{MAX}} = \mu N_B \tag{12}$$

so that, this time,

$$\mu \geq \frac{b(D + h + L)}{(h + L)(D + h)} \tag{13}$$

This coefficient is slightly higher than the one given in inequality (8), so there will be no slipping on either contact point (T or B) with the pipe if $\mu \geq b(D + h + L)/[(h + L)(D + h)]$.

For the larger wrench in the photo (see Figure E6.11e), for example, we see that $D, h, b,$ and L are, respectively, 1 in., 1.25 in., 0.75 in., and 10.3 in. Therefore, the inequalities (8) and (13) give

$$\mu \geq \frac{0.75}{1 + 1.25} = 0.33 \quad \text{needed to prevent slip at } T$$

and

$$\mu \geq \frac{0.75(1 + 1.25 + 10.3)}{(1.25 + 10.3)(1 + 1.25)} = 0.36 \quad \text{needed to prevent slip at } B.$$

With the teeth that are actually built into the jaws of a pipe wrench, 0.36 is easily obtained.

Note from the photos that the jaws are not exactly parallel; resistance to slip is further increased because of this angle. Note also that this result, $\mu = 0.36$, is for a 1-inch pipe ($D = 1$ in.); for other diameters, different answers for μ will be obtained.

Question 6.7 In designing a pipe wrench, which should be used in determining the minimum coefficient of friction needed: the largest or the smallest pipe expected to be turned by the wrench?

The reader may wish to show that for the smaller wrench in the figure, less friction is needed for self-locking against a 1-inch pipe. Another follow-up exer-

---

Answer 6.7 The smallest D, using Equation (13), will give us the minimum μ needed.

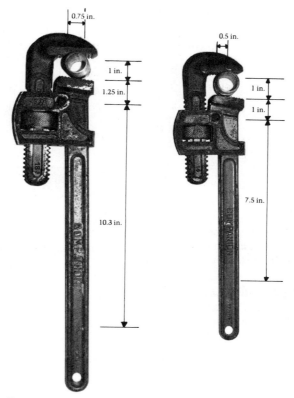

Figure E6.11e

cise is to solve Equations (1)-(6) for f_T, N_T, f_B, N_B, R_x, and R_y in terms of P, and then to show that the equilibrium equations for body $\mathcal{H}$ (Figure E6.11d) are satisfied by these results.

PROBLEMS ▶ **Section 6.2**

6.1 The electronics cabinet in Figure P6.1 is 7 ft high, weighs 200 lb and has its mass center at C. The coefficient of friction between the cabinet and the floor is 0.3. Without disturbing equilibrium, how large a force P can be applied

    a. to the right?

    b. to the left?

6.2 The friction coefficients for the bodies in Figure P6.2 are: 0.4 between $\mathcal{B}_1$ and $\mathcal{B}_2$ and 0.2 between the floor and $\mathcal{B}_2$. The mass of $\mathcal{B}_1$ is 10 kg. Find the minimum force P needed to disturb equilibrium if (a) mass of $\mathcal{B}_2 = 8$ kg; (b) mass of $\mathcal{B}_2 = 12$ kg.

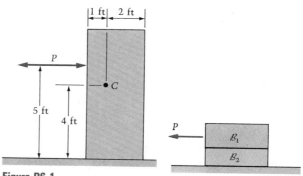

Figure P6.1

Figure P6.2

6.3   If each of W_1 and W_2 weighs 100 lb, and if the coefficient of friction for all surfaces of contact is 0.2, find the angle α for which W_1 begins to slide downward. (See Figure P6.3.)

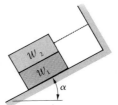

Figure P6.3

6.4   Return to Example 6.1 and find the *largest* force P for which the block will not slide *up* the plane. What do you conclude will happen for any value of P between 31 lb and your answer?

6.5   The block of weight $W = 100$ newtons is in equilibrium on the inclined plane (see Figure P6.5). The force T is applied as shown, and slowly increases from zero. At what value of T will equilibrium cease to exist? How is the equilibrium lost at this value of T?

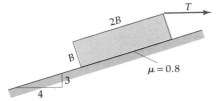

Figure P6.5

6.6   In Figure P6.6 blocks A and B weigh 50 lb and 100 lb, respectively. The coefficients of friction are: $\mu_1 = 0.2$ between A and B, and $\mu_2 = 0.35$ between B and the horizontal plane. Find the smallest force P that will cause A to move to the left.

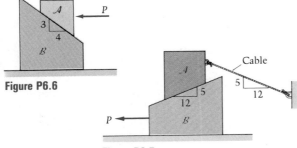

Figure P6.6

Figure P6.7

6.7   Block A, weighing 50 lb, rests on block B, which weighs 100 lb. (see Figure P6.7.) The coefficient of friction between A and B is 0.4. Neglecting friction between B and the floor, find the largest force P for which B does not move.

6.8   In Figure P6.8 find the smallest force P for which the blocks will slide if angle ϕ is $20°$.

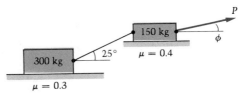

Figure P6.8

6.9   Find the maximum force P for which no slipping will occur anywhere for the system of blocks in Figure P6.9.

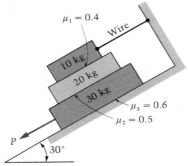

Figure P6.9

6.10   Repeat the preceding problem, if the values of μ_1 and μ_3 are reversed.

6.11   Two identical triangular blocks, each of mass m, are put together to form a rectangular solid, and they are being held in equilibrium by the force P (see Figure P6.11). If P is then increased until motion impends, at what value of P will this occur? Consider *all* possibilities.

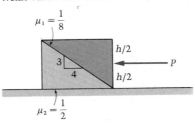

Figure P6.11

6.12 In Figure P6.12 a small box rests on the parabolic incline at the point $(1, \frac{1}{2})$. If the box is on the verge of slipping, what is the coefficient of friction between it and the incline? Also, find the force P that must be exerted on the box, tangent to the incline, to start it moving *upward*.

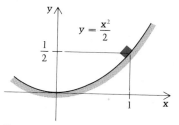

$$y = \frac{x^2}{2}$$

Figure P6.12

6.13 In Figure P6.13 a bug falls into a bowl that has the shape of an inverted spherical cap of radius R. The coefficient of friction between the bowl and the bug's feet is $\mu = \frac{1}{4}$. The bug is barely able to crawl out without slipping. Find the depth D of the bowl.

6.14 The uniform rod of mass m and length l rests in the position shown in Figure P6.14. Find the minimum coefficient of friction μ between the rod and the wall for which the equilibrium can occur, as a function of angle α.

6.15 In a strength test of fiber optic cable, the nut in Figure P6.15(a) was attached to the jacket of the cable, which was then slipped through a slot in the channel (Figure P6.15(b)). The coefficient of friction between nut and channel was 0.1. The load T in Figure P6.15(c) was then slowly increased, and at 200 lb it was observed that a bend induced in the channel caused the nut to slip off the channel. At what angle θ did this occur?

6.16 The 140-N triangular block in Figure P6.16 is pulled at the top by a gradually increasing horizontal force P. The coefficient of friction between the block and the floor is $\mu = 0.4$. Find the smallest base dimension B for which the uniform block will slide instead of tipping.

6.17 The cylinder in Figure P6.17 weighs 200 lb and is in equilibrium on the inclined plane. What is the friction force at its point of contact? If it is on the verge of slipping on the plane, what is the coefficient of friction?

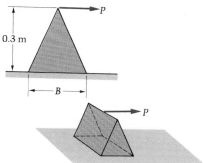

Figure P6.16

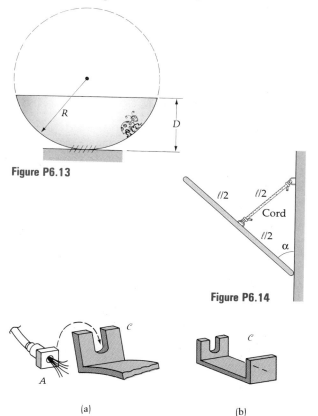

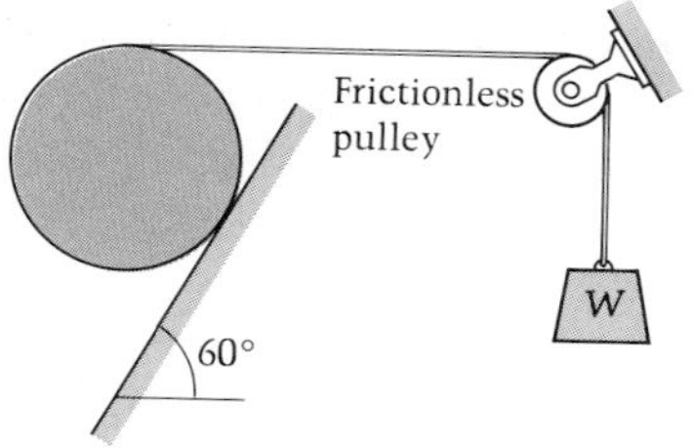

Figure P6.13

Figure P6.14

Figure P6.17

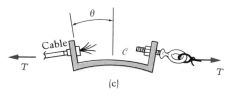

(a)

(b)

Cable

(c)

Figure P6.15

6.18 In Figure P6.18 block $\mathcal{B}$ weighs 50 lb. Determine the range of values of the weight of $\mathcal{A}$ that will maintain equilibrium of the system. (Be sure to check all cases!)

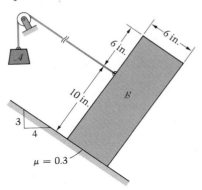

Figure P6.18

6.19 In Figure P6.19 body $\mathcal{A}$ weighs 270 N and body $\mathcal{B}$ weighs 180 N. The coefficient of static friction between body $\mathcal{A}$ and the plane is 0.30. Determine the range of values of the force P for which body $\mathcal{A}$ will be in equilibrium.

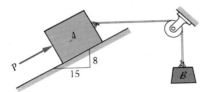

Figure P6.19

6.20 In Figure P6.20 what is the largest angle ϕ for which the slotted cylinder will remain in equilibrium when released from rest on the inclined plane? The coefficient of friction between the cylinder and the plane is $\mu = \frac{1}{6}$.

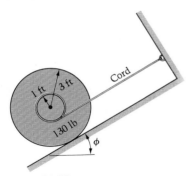

Figure P6.20

6.21 The slotted cylinder is in equilibrium on the inclined plane in Figure P6.21. If the spring is stretched 4 in., what is its modulus? What is the minimum friction coefficient between the cylinder and the plane for which the equilibrium can exist?

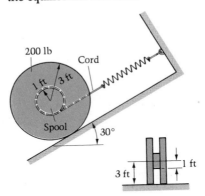

Figure P6.21

6.22 What force P will cause motion of the 100-lb cylinder in Figure P6.22 to impend?

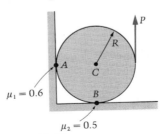

Figure P6.22

6.23 The cylinder in Figure P6.23 has a mass of 30 kg and a radius of 0.6 meter. The coefficients of static friction are 0.2 between the cylinder and the inclined plane, and 0.3 between the cylinder and the floor. Find the value of the counterclockwise couple M_0 that will cause the cylinder to begin to rotate.

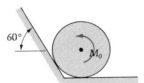

Figure P6.23

6.24 In Figure P6.24 a force P is applied to the rope by a man who erroneously thinks that if he pulls to the left, the 200-lb wheel should roll to the left. If the friction coefficient between the wheel and both the wall and floor is 0.35, determine the force P that will cause the wheel to move. *How* will it move?

6.26 In Figure P6.26 find the largest force that can be developed in the tow cable if the coefficient of friction between the back tires of the truck and the ground is 0.25. The truck weighs 5000 lb and is rear-wheel driven.

6.27 In Figure P6.27 find the minimum coefficient of friction between the bar and the floor for which the bar can be in equilibrium in the given position.

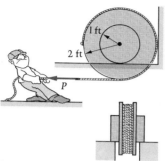

Figure P6.24

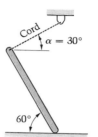

Figure P6.27

* 6.25 The spool is in equilibrium as shown in Figure P6.25. Find the minimum value of the coefficient of friction for which this is possible. *Hint:* Angle θ can be found from the dimensions given. Also, assume that the rope is tightly wrapped.

6.28 Repeat the preceding problem if the upper angle α is 60°.

6.29 The stick of mass m was originally in equilibrium with the string force S acting vertically ($\theta = 90°$). The force S is now slowly turned clockwise as indicated in Figure P6.29, with the stick remaining in equilibrium in the same position.

   a. Find the force S as a function of θ (and mg and ϕ), and show that it must increase in magnitude.

   b. Let $\phi = 30°$ and $mg = 5$ lb. If the stick is observed to slip at $\theta = 30°$, find the coefficient of friction between the stick and the floor.

Figure P6.25

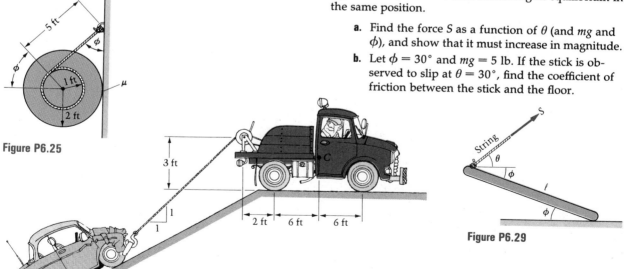

Figure P6.26

Figure P6.29

6.30 Find the weight of the block that will prevent slipping of the 300-N ladder if the coefficients of friction are: 0.3 at A, 0.1 at B, zero at C, and 0.4 between the block and the floor. (See Figure P6.30).

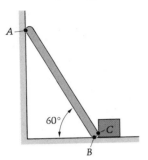

Figure P6.30

6.31 Given a homogeneous block on an inclined plane as shown in Figure P6.31, find the angle θ for which the block is on the verge of tipping. Then find the coefficient of friction μ for which the block is also on the verge of sliding. Note that C_1 and C_2 are the centers of gravity of the triangular and rectangular parts of the block, respectively.

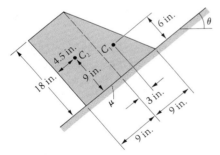

Figure P6.31

6.32 The girl in Figure P6.32 is learning to use a wheelchair. What force must she exert tangent to the handwheel (radius r) to make the wheelchair begin to roll up the incline (angle ϕ)? The radius of the large wheels is R; the masses of patient and wheelchair are M and m, respectively.

6.33 In Figure P6.33 bodies $\mathcal{A}$ and $\mathcal{B}$ respectively weigh 2000 N and 1500 N. The various coefficients of friction are: between $\mathcal{A}$ and the plane, 0.2; between $\mathcal{A}$ and $\mathcal{B}$, 0.3; between $\mathcal{B}$ and the plane, 0.4. Find the magnitude of couple $\mathcal{C}$ that will cause body $\mathcal{B}$ to have impending motion.

Figure P6.32

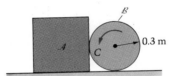

Figure P6.33

6.34 A 100-kg man climbs a 30-kg ladder that is 11 m in length and makes a 30° angle with the vertical. (See Figure P6.34.) The man has placed a solid block at the foot of the ladder to prevent slipping.

  a. Is the block necessary?

  b. If so, what is the minimum block mass to keep the man and ladder from slipping?

Figure P6.34

6.35 In Figure P6.35 a solid circular disk of weight 65 lb is connected to a 130-lb block by a light bar AB. Find the smallest coefficient of friction between the block and the plane for which the system will be in equilibrium. Assume that the block is wide enough so that it will not tip.

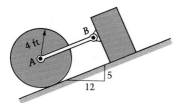

Figure P6.35

6.36 In Figure P6.36, the light rod is connected to the drum and the block at pins. The block weighs 26 lb and the coefficient of static friction for all surfaces is 0.6. If the system is to be in equilibrium, what is the maximum allowable value for the weight of the drum?

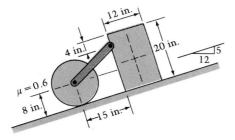

Figure P6.36

6.37 The coefficient of friction between the homogeneous beam and the corner at A is $\mu = 0.3$. (See Figure P6.37.) Find the coordinates of the mass center of the beam if it is in equilibrium but on the verge of slipping at A.

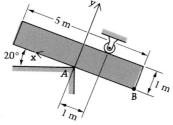

Figure P6.37

6.38 Using Figure P6.37, suppose that the distance from A to B is 1.6 m and that the beam is now uniform. Find the minimum coefficient of friction between the beam and the corner support that will prevent slipping of the beam.

6.39 The cylinder C and ring R in Figure P6.39 are of the same radius r and W, and are connected by the light rod AB.

   a. Find the couple M, which when applied to C results in equilibrium. Assume no slipping.

   b. Argue from new free-body diagrams that if M is applied instead to R, the same answer is obtained for the required couple.

6.40 In Figure P6.40 the coefficient of friction μ is the same between A and B as it is between B and the plane. Find the minimum value of μ for which the system will remain in equilibrium, and the tension in the cord for this value of μ.

6.41 Repeat the preceding problem with the altered pulley arrangement shown in Figure P6.41.

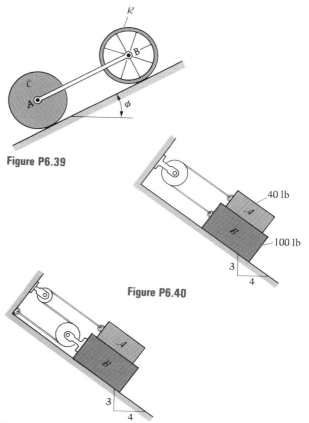

Figure P6.39

Figure P6.40

Figure P6.41

Figure P6.43

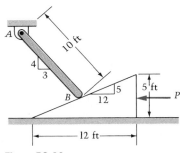

Figure P6.44

6.42 In Problem 6.40, suppose the common friction co-efficient is $\mu = 0.2$, and the angle of the plane is now variable. Find the largest angle of the plane for which the blocks will remain in equilibrium.

6.43 In Figure P6.43 Mr. A weighs 160 lb and Mr. B weighs 200 lb. Each man pulls on the rope with a force of 20 lb.

   a. What force does the fish scale register?
   b. What are the magnitudes of the friction forces on Mr. A and Mr. B?
   c. If Messrs. A and B are wearing the same kind of shoes and if there is impending slipping, which man is about to slip? What is the coefficient of static friction μ_s?

6.44 In Figure P6.44, the 50-lb rod is pinned smoothly to the ceiling at A, and it rests on the 100-lb wedge at B. The coefficient of friction between the two bodies is 0.2, and that between the wedge and the floor is 0.1. Find the force P for which motion of the wedge to the left will be impending.

6.45 In the preceding problem, is any force P needed to prevent the wedge sliding to the right? If so, find it; if not, find the force P *to the right* that will start motion of the wedge to the right.

6.46 In Problems 4.238 and 4.239, find the minimum coefficient of friction between the block and the slotted body for which the equilibrium can exist.

6.47 The uniform blocks A and B weigh 90 lb and 60 lb, respectively. The coefficients of friction at the three contacting surfaces are shown in Figure P6.47. Find the maximum value of P for which equilibrium is possible. Is the answer the same if P is pushing to the left on B instead of pulling to the right?

6.48 Repeat the preceding problem if the coefficient of friction between A and B is changed from 0.3 to 0.1.

6.49 In Figure P6.49 a heavy man of mass m has placed a light ladder of length L at a dangerous angle, and has

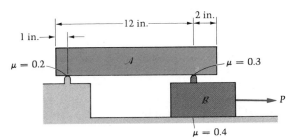

Figure P6.47

begun to climb it. If the coefficients of friction are 0.25 between both the ladder and ground, and the ladder and wall, find how far up the ladder (fL in the figure) the man can climb before the ladder slips. Show further that prior to the slipping position, the ground and wall reactions cannot be found (i.e., that the problem is statically indeterminate).

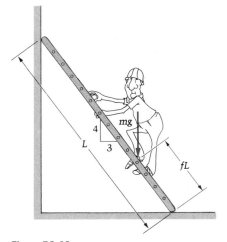

Figure P6.49

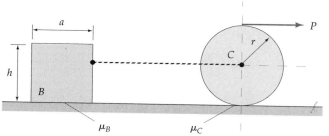

Figure P6.50

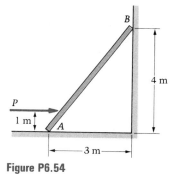

Figure P6.54

6.50 Block B and the homogeneous cylinder C in Figure P6.50 each weigh W. A light cable connects the center of the cylinder to the block B. The coefficients of static friction are μ_B and μ_C, as indicated.

a. If $\mu B = 0.5$ and $\mu_C = 0.2$, what is the largest force P (in terms of W) that can be applied without disturbing equilibrium? (Assume that dimension "a" is large enough to prevent the block from tipping.) Describe in a sentence the impending motion at this value of P.

b. Repeat the problem if the friction coefficients are reversed, i.e., $\mu_B = 0.2$ and $\mu_C = 0.5$ (again assume no tipping).

6.51 What ratio of the coefficients of friction, μ_B / μ_C, would be necessary for *simultaneous* impending motion of B and C in the preceding problem?

6.52 In Problem 6.50, let $\mu_C = 0.5$ and $\mu_B = 0.2$. In terms of r, find the smallest value of the width "a" of the block so that it will not tip over for any value of P less than $0.1W$.

6.53 In Figure P6.53 the block and cylinder each weigh 300 N.

a. Show that the block will be on the verge of slipping when P is increased to the value $120/\cos\theta$ N.

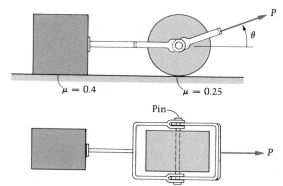

Figure P6.53

b. At this value of P, what is the cylinder on the verge of doing?

6.54 A uniform ladder, mass 12 kg, rests against a wall and against a horizontal surface, as shown in Figure P6.54. If the coefficient of friction at A and B is 0.3, find the smallest force P that will prevent motion of the ladder to the left at A.

6.55 The bodies $\mathcal{A}$, $\mathcal{B}$ and $\mathcal{C}$ in Figure P6.55 each weigh 20 lb and are in equilibrium. $\mathcal{A}$ and $\mathcal{C}$ are pinned to $\mathcal{B}$ at its ends. Find the force P (magnitude and sense) and the forces exerted by the slot on $\mathcal{A}$ and by the circular track on $\mathcal{C}$.

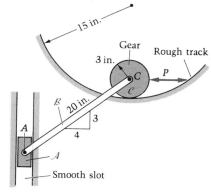

Figure P6.55

6.56 In the preceding problem, let P be replaced by a couple M applied to $\mathcal{C}$. Find M. Also find the minimum coefficient of friction between $\mathcal{C}$ and the rough track for equilibrium.

6.57 The uniform 10-lb rod and 100-lb cylinder are pinned smoothly at C. If the 2-lb force is applied as shown in Figure P6.57 and the system remains in equilibrium, then the coefficient of friction between rod and floor has to be larger than a certain value. What is this value?

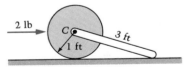

Figure P6.57

6.58 In Figure P6.58 the collar C is very light compared with the force P. It is fitted over the rough vertical shaft S with a slight amount of clearance. Show that if the distance d is greater than $H/(2\mu) - r$, then C will not slip down the shaft regardless of the size of P. *Hint:* There will be contact at two points of S.

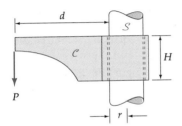

Figure P6.58

6.59 In Example 6.8 suppose the car is to be parked on a 15° incline. (See Figure P6.59.) What must be the coefficient of friction between tires and street in order to prevent slipping of the car if only the rear wheels are locked?

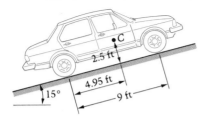

Figure P6.59

6.60 Repeat the preceding problem if all four wheels are locked.

6.61 The homogeneous cylinder in Figure P6.61 weighs 425 lb and the block weighs 300 lb. The coefficient of friction for all surfaces of contact is 0.25. Moment T_0 is

slowly increased from zero. Determine the value of T_0 that will cause the cylinder to have impending motion, assuming that the block does not tip.

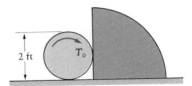

Figure P6.61

6.62 In Figure P6.62 if $\mu_1 = \mu_2$, then as P increases from its lowest possible equilibrium value, will the cylinder C first roll, or slide, on the plane for (a) $\theta = 30°$. (b) $\theta = 45°$, (c) $\theta = 60°$? *Hint:* Show that for rolling,

$$P_1 = \frac{W \sin \theta}{1 - \mu_1}$$

and for sliding,

$$P_2 = W\left[\sin \theta + \frac{\mu_2 \cos \theta}{1 - \mu_2}\right]$$

and compare the numerical values.

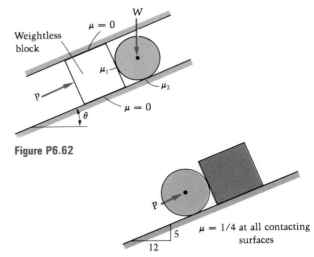

Figure P6.62

Figure P6.63

6.63 The cylinder and the block each weigh 26 N. (See Figure P6.63.) The friction coefficient is $\frac{1}{4}$ between block and cylinder, $\frac{1}{4}$ between cylinder and plane, and $\frac{1}{4}$ between block and plane. Find the smallest force P for which motion will begin up the plane if it is known that the block won't tip.

6.64 In the preceding problem, find the value of the common friction coefficient below which the cylinder will roll on the plane and above which it will slip on the plane, provided $\mu \neq 0$.

6.65 Without block $\mathcal{B}$ present, the center of the wheel $\mathcal{C}$ will move to the right on the plane. If $\mathcal{A}$, $\mathcal{B}$, and $\mathcal{C}$ each weigh W, and if the friction coefficients at the three contacting surfaces are all equal (μ), find the minimum value of μ for which equilibrium will exist. (See Figure P6.65.) Assume that $\mathcal{B}$ is wide enough so that it will not tip.

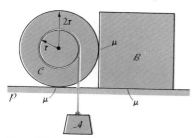

Figure P6.65

6.66 Two identical cylinders are placed together on an incline as shown in Figure P6.66. Show that the cylinders cannot remain in equilibrium. *Hint:* Write the three equilibrium equations for each cylinder. Show that they require the normal force between them to be zero. Thus conclude that: (a) the friction between them is also zero; (b) there can be no friction beneath the cylinders if there is equilibrium; and finally (c) the component of the weights down the plane is unbalanced, which is a contradiction.

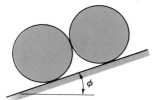

Figure P6.66

6.67 In the preceding problem, suppose that a normal force between the cylinders is created by a pair of springs (see Figure P6.67), one at each end of the cylinders. If $m = 20$ kg, $\mu = 0.6$ at all surfaces, and $\phi = 20°$, find the spring tension required for equilibrium of the two cylinders.

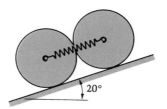

Figure P6.67

6.68 A vehicle of mass m in Figure P6.68 is driven by a gear and pinion as follows: A motor exerts a constant moment $M_0 \circlearrowright$ onto the pinion $\mathcal{P}$. It then drives the gear $\mathcal{G}$ that is attached to a rear wheel, thereby turning the wheel counterclockwise and moving the vehicle up the plane. What value of M_0 is needed to barely move the vehicle if:

   a. Axle friction is negligible?

   b. Axle friction gives a constant resistive moment of M_f opposing the rolling?

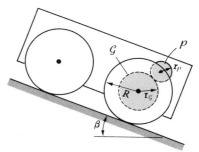

Figure P6.68

Figure P6.69

6.69 In Example 6.6 let the two friction coefficients be μ_P between drum and plane and μ_T between drum and truck. Find the relationship between μ_P and μ_T for which the drum will roll up the plane (i.e., will not slip on the ground). (See Figure P6.69.) The answer should be of the form $\mu_T \leq f(\mu_P)$.

6.70 In Example 6.6 let both friction coefficients be 0.3. Find the maximum incline angle ϕ for which the drum will roll (i.e., not slip on the inclined plane).

6.71 In Figure P6.71 the cylinder is in equilibrium. Determine:

   a. The stretch in the spring.

   b. The minimum coefficient of friction μ for which the equilibrium is possible.

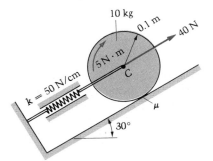

Figure P6.71

6.72 The 130-lb homogeneous plank of length 6 ft is placed on the two homogeneous cylindrical rollers, each of weight 30 lb, and held there by the force P. (See Figure P6.72.)

   a. Find P for equilibrium of the system, assuming no slipping occurs anywhere.

   b. If the coefficient of friction μ at all surfaces of contact is the same, find the smallest μ needed for equilibrium.

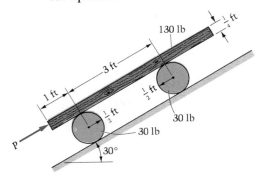

Figure P6.72

In Problems 6.73–6.76 the left support is 1.2 m from the left end of the beam, and the force P is holding the beam in equilibrium. Find the force P required to start the beam moving to the left. The thin beam in each figure weighs 2600 N.

6.73

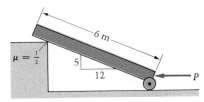

Figure P6.73

6.74

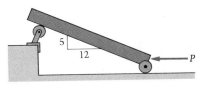

Figure P6.74

6.75

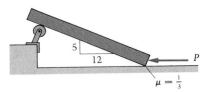

Figure P6.75

6.76

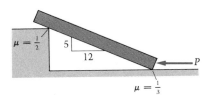

Figure P6.76

6.77–6.80 Reverse the direction of P in Problems 6.73–6.76, respectively, and find the value of P required to start the beam moving to the right.

6.81 The two identical sticks are pinned together at A and are in equilibrium as shown in Figure P6.81 (on the following page). What is the lowest coefficient of friction (as a function of α) for which this can take place?

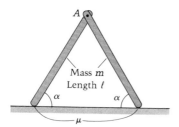

Figure P6.81

6.82 The cylinder has mass M, including an attached hub around which a cord is wrapped. The cord, after passing over a pulley, is attached to a mass m. (See Figure P6.82.)

a. Show that if the plane is smooth, the cylinder cannot be in equilibrium, regardless of the values of the parameters.

b. Find the minimum coefficient of friction μ between the cylinder and the plane for which the system is in equilibrium. Find the relationship between M, m, ϕ, R, and r for equilibrium.

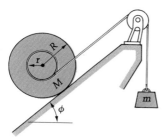

Figure P6.82

6.83 In Figure P6.83 the cord is wrapped around the light hub attached to the 130-lb cylinder. Find the minimum coefficient of friction between the cylinder and the plane for which the cylinder will be in equilibrium.

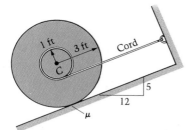

Figure P6.83

6.84 Find the braking force B if motion impends between the brake and drum at A, with friction coefficient 0.4. Neglect the weight of the brake and its thickness. (See Figure P6.84.) The couple T has magnitude 80 N · m.

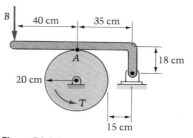

Figure P6.84

6.85 Repeat the preceding problem if the torque T is applied clockwise.

6.86 In Problem 6.84, if $T = 100$ N · m and if the slender brake arm weighs 3 N/cm, find the force B if motion is impending.

6.87 The cable exerts 320 lb on the 400-lb weight as shown in Figure P6.87. The truck then moves slowly to the right, while rotating the arm $\mathcal{A}$ counterclockwise so as to keep the tension approximately constant. At what angle ϕ will the weight begin to move? Will it slide, or will it tip?

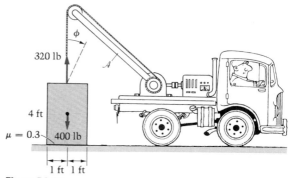

Figure P6.87

6.88 The coefficient of limiting static friction between the 100-lb block and the plane is 0.5. The force P is applied horizontally as shown in Figure P6.88 (see the following page). Determine:

a. The magnitude of the force P for which the friction force vanishes.

b. The force P required to start the block moving up the plane.

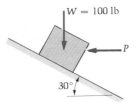

Figure P6.88

6.89 A 200-lb cylinder is in equilibrium in the position shown in Figure P6.89. Find:

 a. The friction force at A.

 b. The coefficient of friction if slipping impends at A in the position shown.

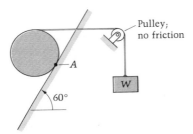

Figure P6.89

6.90 Determine the force P that will move the two identical 10-kg cylindrical rollers up the inclined plane shown in Figure P6.90. The friction coefficient is 0.3 at A, B, and C.

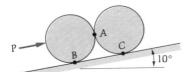

Figure P6.90

6.91 The hydraulic cylinder in Figure P6.91 exerts a force to the right on the brake arm at point B. This brings the shoe into contact with a turning wheel, in order to stop it by friction. While the wheel is not in equilibrium (indeed, it is being decelerated), the arm is. Find the ratio of the force exerted by the hydraulic cylinder to the normal force between the shoe and the wheel if the wheel rotates (a) clockwise; (b) counterclockwise. Neglect the weights of arm and shoe.

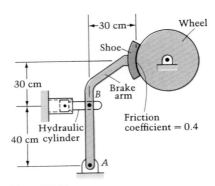

Figure P6.91

6.92 In Figure P6.92 find the smallest couple M that will cause the cylinder of weight W and radius R to have impending motion. The coefficient of friction for all surfaces is μ.

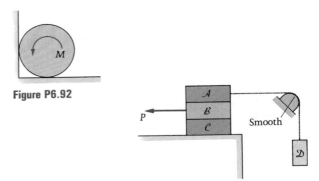

Figure P6.92

Figure P6.93

6.93 The blocks $\mathcal{A}$, $\mathcal{B}$, $\mathcal{C}$, and $\mathcal{D}$ in Figure P6.93 weigh 100 lb, 50 lb, 75 lb, and 40 lb, respectively. The friction coefficients are 0.5 between $\mathcal{A}$ and $\mathcal{B}$, 0.3 between $\mathcal{B}$ and $\mathcal{C}$, and 0.25 between $\mathcal{C}$ and the plane. Because the peg is smooth, the tension in the cord is the same on both sides. Force P is slowly increased from zero. At what value of P will the equilibrium be disturbed?

6.94 A slender uniform rod of length 17 in. and weight 50 lb is smoothly hinged to a fixed support at A and rests on a half-cylinder block at B, where the contact is smooth. (See Figure P6.94.)

 a. Find the smallest coefficient of friction μ between the 70-lb block and the plane for which the system can be in equilibrium as shown.

 b. For this value of μ, find the location of the normal force resultant between the block and the plane.

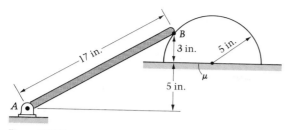

Figure P6.94

6.95 The homogenous rod AB in Figure P6.95 weighs 50 pounds and is supported by the half-cylinder C and by the pin at A. If $D = 6$ ft, $R = 3$ ft, and $L = 8$ ft and if the contact between C and the floor is smooth, find

  a. the normal and friction forces acting on C at E;

  b. the minimum coefficient of friction between the rod and half-cylinder for which the equilibrium can occur.

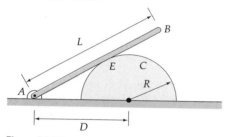

Figure P6.95

6.96 The rod in Figure P6.96 has a mass of 3 slugs. It rests on a rough floor at A and a smooth, fixed half-cylinder at B. Find:

  a. The reactions exerted on the rod at A and B.

  b. The minimum coefficient of friction for which the rod will not slip at A.

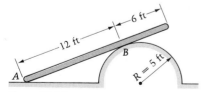

Figure P6.96

6.97 The bent bar in Figure P6.97 is pinned to a disc at one end (A) and rests on a semicircular fixed support at the other end (B). The bar weighs 20 N and its center of

gravity is indicated in the figure. There is also a force P acting on the pin at the center of the disc. The coefficient of friction μ between the bar and the support is 0.35, and the pin is smooth.

  a. Find the normal force exerted by the semicircular support on the bar as a function of the applied force P.

  b. Likewise, find the frictional force exerted by the semicircular support on the bar in terms of P.

  c. Using the results of (a) and (b), find the least value of P for which the bar-disc assembly will remain in equilibrium.

  d. Using the results of (a) and (b), find the greatest value of P for which the bar-disc assembly will remain in equilibrium.

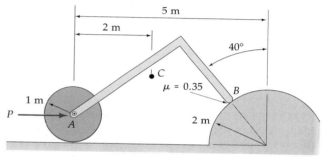

Figure P6.97

6.98 The weightless members in Figure P6.98 are supported by pins C and B and by a rough surface at A. Determine:

  a. The minimum coefficient of friction at A to assure equilibrium of the structure.

  b. The forces exerted on member BC by the pins at B and C.

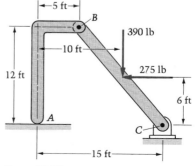

Figure P6.98

6.99 A person plugs in an electric appliance, pushing the plug into the receptacle with a horizontal force F. (See Figure P6.99.) The prongs initially touch the contacts as shown, and when the force F reaches 3 lb, the prongs slide past them, completing the contact. When the sliding starts, what is the component of force on each contact perpendicular to the prong? Assume a friction coefficient of 0.25.

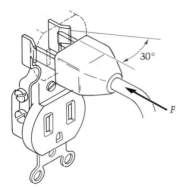

Figure P6.99

6.100 The coefficient of friction between the rod and the triangular block is μ_0, and the plane is smooth. Show that the only condition for the block to remain in equilibrium is that $\tan \theta \le \mu_0$. (See Figure P6.100.)

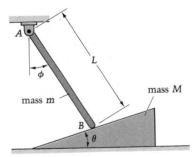

Figure P6.100

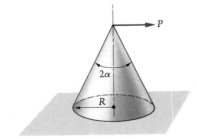

Figure P6.101

6.101 The force P is applied to the cone of mass m as shown in Figure P6.101, and gradually increased in magnitude from zero. If the coefficient of friction between cone and the plane is μ_0, find the value of P at which motion of the cone impends. Is it tipping, or slipping?

6.102 The edge AB of block ABC in Figure P6.102 is a parabola whose vertex is at A. Its equation is $x^2 = \frac{64}{3}y$. If the coefficient of friction between the shaft and the block ABC is μ_0, find the largest value of x for which the system can be in equilibrium. Neglect friction beneath the rollers and let m be the mass of the shaft plus the plate attached above it.

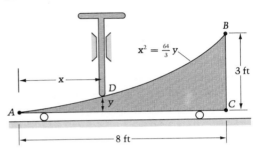

Figure P6.102

6.103 In the preceding problem, let a force P be applied to the left on face BC of block ABC. Neglecting friction in the vertical slot in which the shaft slides, find the value of P, in terms of x, for which motion of the shaft will be impending to the left.

6.104 Show that the bar $\mathcal{B}$ in Figure P6.104 will not slip to the right, regardless of the size of F, if the friction coefficient μ is greater than or equal to $\cot \theta$. The weight of $\mathcal{R}$ is W.

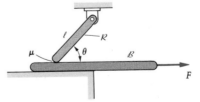

Figure P6.104

6.105 The uniform rod rests atop a uniform circular cylinder of the same weight W. The coefficients of friction at the contact points A and B are each equal to μ. Calculate the smallest value of μ for which the cylinder can stay in equilibrium with the thin rod maintaining its horizontal position. (See Figure P6.105 on the following page.)

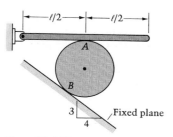

Figure P6.105

6.106 In Figure P6.106 find the minimum coefficient of friction between the small block and the wall for which equilibrium of the system is possible.

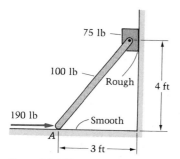

Figure P6.106

6.107 The 200-lb block rests between the smooth vertical guides as shown in Figure P6.107. It is supported by the 50-lb rod. The coefficients of static friction between the rod and the block at A, and the rod and the ground at B, are each 0.5. Determine the maximum angle θ for equilibrium.

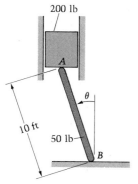

Figure P6.107

6.108 The uniform bar of mass 50 kg is in equilibrium. The wall is smooth and the floor is rough. (See Figure P6.108.) Find the reactions at A and B, and the smallest coefficient of friction between the bar and the floor for which the equilibrium is possible.

Figure P6.108

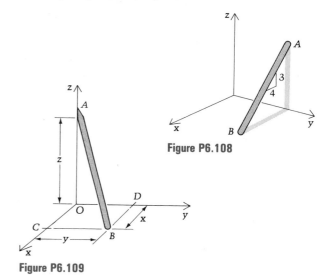

Figure P6.109

6.109 In the preceding problem, suppose the stick is sharpened at A and that this upper end is in the corner (touching the z axis). (See Figure P6.109).

a. Show that the reaction from the corner is in the plane containing the triangle AOB.

b. Find, in terms of x, y, and z, the minimum coefficient of friction between the bar and the floor for which equilibrium is possible.

6.110 The block weighs 200 lb and rests on the 300-lb wheel with a friction coefficient between them of 0.4. (See Figure P6.110.) If the friction coefficient between the wheel and the plane is 0.3, find the lowest value of T for which the wheel will move. *How* will it start to move?

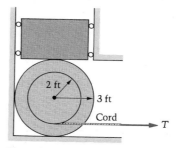

Figure P6.110

6.111 Repeat the preceding problem if the coefficient of friction between the wheel and the plane is (a) 0.4; (b) 0.9.

6.112 Repeat Problem 6.110 if the cord comes off the *top* of the inner radius instead of the bottom.

6.113 The frame in Figure P6.113 consists of the bar *AD*, the curved bar *DC* (quarter circle), and the pulley. Determine whether the frame is in equilibrium without the box. If not, find the minimum weight of the box for equilibrium. Assume that all the forces exerted on *DC*, by the floor and by the box, act at the point *C*. Also find the force exerted on *AD* by the pin at *D*, for your equilibrium case.

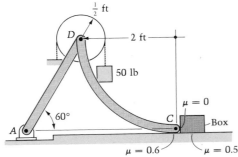

Figure P6.113

6.114 The ladder *AB* in Figure P6.114 is 30 ft long and weighs 60 lb. If the coefficient of friction is 0.5 between the ladder and both the floor and the wall, find the force *P* required to start the ladder moving up at *A*.

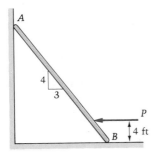

Figure P6.114

6.115 In the preceding problem, suppose the force *P* is applied at height *y* instead of 4 ft. Find the smallest value of *y* for which the ladder will not slide upward at *A* no matter how large the force *P*. *Hint:* Write the three equilibrium equations, eliminate the normal forces, and examine carefully the remaining equation.

6.116 The coefficient of friction between the drum and brake shown in Figure P6.116 is 0.5. (a) Show that there is adequate friction for equilibrium. (b) Determine the horizontal and vertical components of the pin reaction at *O* on the brake. The drum weighs 100 lb and the weight of the brake is 10 lb/ft. The brake is a uniform bar.

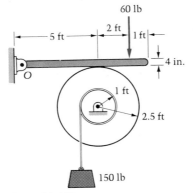

Figure P6.116

6.117 The man in Figure P6.117 is trying to ride a one-speed bicycle up a steep incline. He is tiring, and the bike is moving *very* slowly. Find the normal forces beneath each tire, and the friction force exerted up the plane by the pavement on the rear tire. Neglect the friction beneath the non-driven front wheel, and take the total weight of bike and man to be 190 lb with mass center at *C*. Also find the minimum coefficient of friction between tire and incline for which the wheel will not slip.

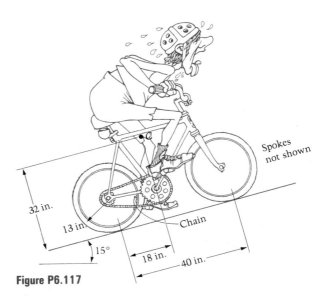

Figure P6.117

6.118 Extend the preceding problem. Knowing the friction force, use the free-body diagram of the rear wheel to determine the force T in the upper segment of the chain. Then use the free-body diagram of the pedal sprocket to find the force F that the man's left foot is exerting at the instant shown in Figure P6.118.

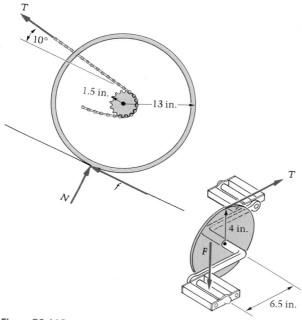

Figure P6.118

6.119 As P increases from an equilibrium value, where does the cylinder slip first, and at what value of P does this occur? (See Figure P6.119.)

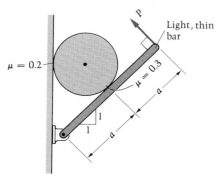

Figure P6.119

6.120 Repeat the preceding problem with the friction coefficients reversed.

6.121 The 240-N · m couple is applied to the bar DB. Bar AC rests on DB in equilibrium as shown in Figure P6.121. Find the minimum coefficient of friction between the two bars for which the equilibrium can exist.

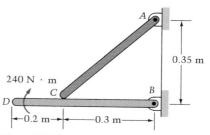

Figure P6.121

6.122 Repeat the preceding problem if the 240-N · m couple is replaced by an upward force of 480 N at D. Tell why the answer for μ_{min} is the same in the two problems, but not the reaction at B.

6.123 In Problem 4.235, suppose the bar and block are made of the same material and their coefficients of friction μ with the plane are the same. What is the minimum value of μ for which the equilibrium can exist?

6.124 The rope is held securely by the self-clamping device in Figure P6.124.

  a. Explain how both the actual and maximum friction forces increase with P.

  b. Find the minimum coefficient of friction between device and rope to prevent slipping.

  c. Find the resultant force exerted by the pin at A onto the support if $P = 500$ lb.

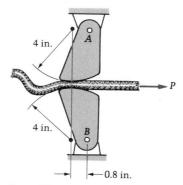

Figure P6.124

6.125 The light rod *AB* in Figure P6.125 is pinned smoothly to the ground at *A* and to the cylinder at *B*. The cylinder rests against the block, which in turn rests against the wall. The cylinder and block each weigh *W*. Find the minimum coefficient of friction between the block and the wall so that the block will not slip downward regardless of the value of *W*.

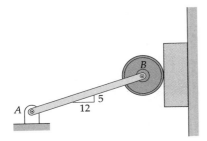

Figure P6.125

6.126 Repeat the preceding problem if the cylinder weighs W_C and the block weighs W_B.

6.127 Repeat the preceding problem if in addition the rod weighs W_R and has length ℓ.

6.128 Show that the identical, light cylindrical rollers in Figure P6.128 will hold the heavy bar of weight *W* in equilibrium in a self-locking manner (i.e., independent of the value of *W*) if the friction coefficients μ_1 and μ_2 between roller and bar, and roller and wall, respectively, are greater than a certain number. Find this common value of the two μ's, which depends on the angle α. Note that the smaller the angle α, the better the self-locking, because then the smaller μ needs to be to prevent slip regardless of the size of *W*.

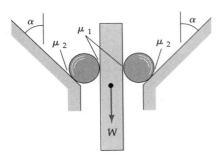

Figure P6.128

6.129 The light scaffolding hook in Figure P6.129 rests against a rough ceiling at *A* and a smooth wall at *B*. (a) What must be the friction coefficient between the hook and the ceiling to prevent the hook from slipping at *A* regardless of the size of *P*? (b) Change the 0.4m dimension (i.e., alter the shape of the hook) so as to make the answer become $\mu = 0.2$, for a safer result.

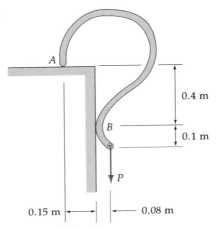

Figure P6.129

* 6.130 The plank rests on the floor at *A* and on the smooth, fixed block at *C*. (See Figure P6.130.) If the coefficient of friction between the plank and the floor at *A* is 0.3, find the range of heights *H* of the block for which the plank won't slip on the floor.

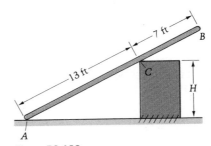

Figure P6.130

* 6.131 In Problem 6.130, let the coefficient of friction be 0.3 at *both* surfaces of contact. Find the lowest height of the block for which the plank will slip.

6.132 Solve Example 6.9 if the cord to which $\mathcal{W}$ is attached comes off the *right* side of the hub of $\mathcal{C}$.

---

* Asterisks identify the more difficult problems.

6.133 The tongs in Figure P6.133 are holding the light cylinder in equilibrium. What is the minimum coefficient of friction between the cylinder and the tongs such that the cylinder can't slip upward regardless of the value of *P*?

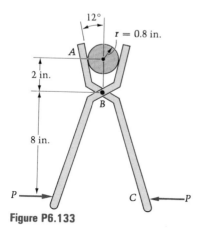

Figure P6.133

6.134 In the preceding problem, find the force exerted by the pin at *B* onto member *ABC* when *P* = 15 lb. Assume a friction coefficient large enough to prevent slip.

6.135 The 20-N half-cylinder is in equilibrium on the plane, with a cord between the center of the right edge and the plane as shown in Figure P6.135. If the cylinder is on the verge of slipping with $\theta = 30°$ and $\phi = 20°$, find the tension in the cord and the coefficient of friction between the cylinder and the plane.

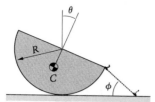

Figure P6.135

6.136 Find the angle ϕ (in terms of *T*, *W*, and *R*), for which the cylinder of weight *W* can be in equilibrium in the cylindrical cavity under the action of the couple of moment *T* (see Figure P6.136). Also find the required minimum coefficient of friction μ for which this equilibrium can exist. It is given that $WR > T$.

* **6.137** In Figure P6.137 find the smallest coefficient of friction for which the thin rod *ABC* will not slip on the cylindrical surface.

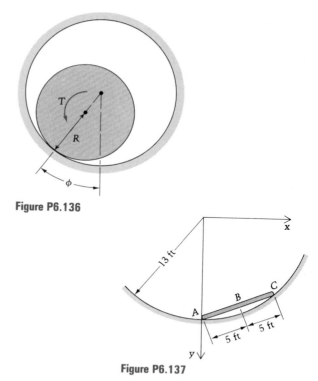

Figure P6.136

Figure P6.137

** **6.138** The friction coefficient between the stick and the wall is μ. The stick (mass *m*), connected to the floor at *A* with a ball-joint, can clearly rest on the wall at *P* in equilibrium. Find the largest angle between the stick and line *AP* for which the stick can be in equilibrium. (See Figure P6.138.) Assume that the friction force at *Q* acts in a direction so as to oppose movement of that end of the stick.

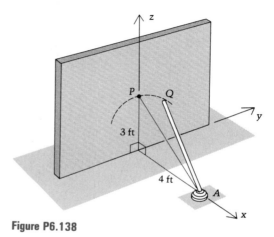

Figure P6.138

6.139 The block of mass m is at rest on an inclined plane as shown in Figure P6.139. The coefficient of friction between the block and the plane is μ, which is therefore $\geq \tan \phi$. If a horizontal force H is applied parallel to the incline, find the largest value of H for which the block will not slip.

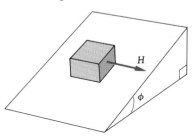

Figure P6.139

* **6.140** Repeat the preceding problem if H is applied at a 45° angle downward from its previous horizontal position (but still parallel to the incline). (See Figure P6.140.)

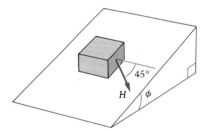

Figure P6.140

Figure P6.141

* **6.141** In Figure P6.141 the coefficient of friction μ between any two of the four identical spheres is the same as between any sphere and the ground. The three lower spheres are as shown in the lower figure, and the upper sphere rests on the top. What is the minimum value of μ for which the spheres can remain in equilibrium? *Hint:* Use symmetry.

* **6.142** Two small balls of equal masses m are in equilibrium on a cylinder as shown in Figure P6.142, connected by a light string of length $1.5R$. The balls are on the verge of slipping clockwise; their friction coefficient with the cylinder is 0.3 for each. Find the angles θ_1 and θ_2, the (constant) tension in the string, and the normal forces N_1 and N_2, respectively, exerted on the left and right masses.

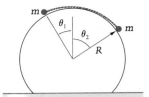

Figure P6.142

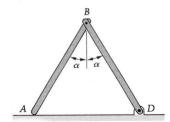

Figure P6.143

6.143 Two identical uniform heavy rods AB and BD, each of weight W, are freely pivoted at B and stand in a vertical plane. (See Figure P6.143). End A rests on a rough horizontal ground, to which D is pinned. In terms of α, find the minimum friction coefficient between AB and the ground for which equilibrium can exist.

* **6.144** In the preceding problem, let $\alpha = 30°$ and $\mu = 0.4$. Let a force P be applied downward at B and turned downward slowly toward BD (B ). Find the value of P if slipping occurs at A when $\theta = 30°$.

* **6.145** The pinned, identical bars in Figure P6.145 rest symmetrically on the cylinder. What is the range of angles between the bars for equilibrium if $L = 2R$ and the coefficient of friction is $\mu = 0.2$?

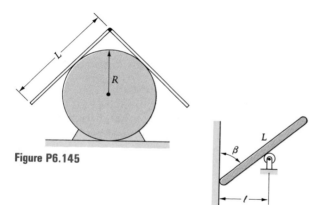

Figure P6.145

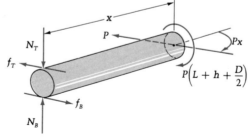

Figure P6.146

6.147 In the preceding problem, find the *largest* angle β for which the bar can be in equilibrium.

* **6.148** In Figure P6.148:

    a. Show that if the system in the figure is in equilibrium, then $m_1 \geq m_2 / 2$.

    b. If the friction coefficient between the bars and the plane is μ, find the minimum value of μ for equilibrium of the system, as a function of θ, m_1, and m_2.

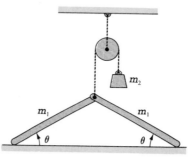

Figure P6.148

* **6.146** The friction coefficient between the bar and the wall is 0.5. The uniform bar has length $L = 2\ell$, where ℓ is the distance of the small, fixed pulley from the wall as shown in Figure P6.146. determine the smallest angle β for which the bar can be in equilibrium.

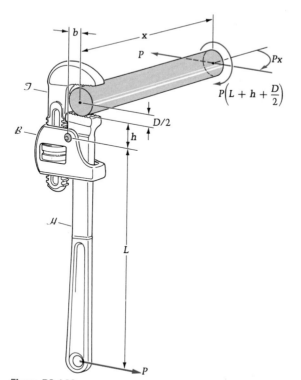

Figure P6.149

6.149 This is not a "friction problem" but an equilibrium exercise based on example 6.11. The free-body diagram in Figure P6.149 of the pipe wrench plus pipe shows the necessary resultant of the reactions onto the pipe from its fitting. Note the free-body diagram of the pipe alone, which includes f_T, N_T, f_B, and N_B in addition to the force and couples in the figure. From Equations (2), (6), and (11) of Example 6.11, write f_T, f_B, and N_B in terms of N_T. Then from equilibrium of forces on the free-body diagram of the pipe, show that $N_T = R(D + h)(h + L)/(bD)$. Finally, check this result by summing moments about any point of the pipe and obtaining zero.

6.3 Special Applications of Coulomb Friction

In this section, we are going to see how Coulomb friction is used to great advantage in four special ways: wedges, belts, screws, and disks.

Wedges

Wedges are simple machines in which large normal and friction forces are developed between two bodies by means of one ($\mathcal{B}_1$) being wedged between the other ($\mathcal{B}_2$) and a fixed surface $\mathcal{I}$. This is done in order to effect a (usually) small change in the position of $\mathcal{B}_2$, as in the following two examples.

___

EXAMPLE 6.12

Find the force P for which the 200-lb block $\mathcal{B}_2$ in Figure E6.12a will be on the verge of sliding upward. The wedge $\mathcal{B}_1$ weighs 50 lb.

Solution

In order for block $\mathcal{B}_2$ to be at the point of impending slipping, block $\mathcal{B}_1$ must also be at the point of slipping at both of its surfaces of contact with $\mathcal{B}_2$ and $\mathcal{I}$. Sometimes, as was seen in previous examples of Section 6.2, two or more cases of slipping are possible in a problem. Normally in wedge problems, slip occurs at all surfaces if it occurs at all.

The free-body diagrams and resulting equilibrium equations, with each friction force being maximum, are as follows:

For $\mathcal{B}_2$ (see Figure E6.12b):

$$\xrightarrow{+} \quad \Sigma F_x = N_1 - \overset{0.3\,N_2}{\cancel{f_2}}\cos 20° - N_2 \sin 20° = 0$$

$$+\uparrow \quad \Sigma F_y = -200 - \overset{0.2\,N_1}{\cancel{f_1}} - \overset{0.3\,N_2}{\cancel{f_2}}\sin 20° + N_2 \cos 20° = 0$$

For $\mathcal{B}_1$ (see Figure E6.12c):

$$\xrightarrow{+} \quad \Sigma F_x = N_2 \sin 20° + \overset{0.3\,N_2}{\cancel{f_2}}\cos 20° + \overset{0.4\,N_3}{\cancel{f_3}} - P = 0$$

$$+\uparrow \quad \Sigma F_y = N_3 - N_2 \cos 20° + \overset{0.3\,N_2}{\cancel{f_2}}\sin 20° - 50 = 0$$

(Note that here we have not used any moment equations, as they would merely serve to locate the lines of action of the resultants of the normal forces; we assume the blocks to be large enough that tipping will not occur.)

The solution to the above four equations in the unknowns N_1, N_2, N_3, and P is*

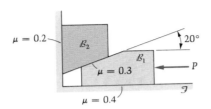

$\mu = 0.2$

$\mathcal{B}_2$

$\mathcal{B}_1$

$\mu = 0.3$ $\longleftarrow P$

$20°$

$\mathcal{I}$

$\mu = 0.4$

Figure E6.12a

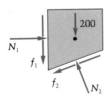

200

N_1

f_1

f_2 N_2

Figure E6.12b

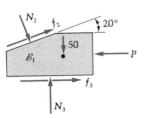

N_2 f_2 $20°$

50

$\mathcal{B}_1$ $\longleftarrow P$

f_3

N_3

Figure E6.12c

___

* Note that if P is given and motion is not impending or occurring, the problem of finding the three f's and the three N's is statically indeterminate.

$$N_1 = 175 \text{ lb}$$
$$N_2 = 281 \text{ lb}$$
$$N_3 = 285 \text{ lb}$$
$$P = 289 \text{ lb}$$

Note that all three normal forces have to be positive in a problem such as this, which serves as a partial check on the algebra. Also, on the system of $\mathcal{B}_1$ *plus* $\mathcal{B}_2$, note that (see the free-body diagram below, Figure E6.12d):

$$\xrightarrow{+} \quad \Sigma F_x = N_1 + f_3 - P$$
$$= 175 + 0.4(285) - 289$$
$$= 0 \checkmark$$

$$+\uparrow \quad \Sigma F_y = 0 = N_3 - f_1 - 200 - 50$$
$$= 285 - 0.2(175) - 250$$
$$= 0 \checkmark$$

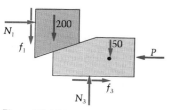

Figure E6.12d

which are further checks. In the above two equations, note that f_2 and N_2 are internal forces within the system that do not therefore appear in the "overall" equilibrium equations. Only when $\mathcal{B}_1$ and $\mathcal{B}_2$ are separated do f_2 and N_2 become *external* forces, and thus appear on the free-body diagrams.

EXAMPLE 6.13

A wedge is being forced into a log by a maul (see Figure E6.13a) in an effort to split the log into firewood. What is the minimum coefficient of friction between the log and the wedge for which the wedge will not pop out of the log when it is not in contact with the maul?

Solution

Note that very large normal forces N are developed on the faces of the wedge. Without friction, when $P = 0$ the vertical components of the N's could pop the wedge upward because these components would ordinarily be considerably greater than the weight (w) of the wedge. The friction opposes the entry of the wedge during the striking of it by the maul. (See Figure E6.13b.) But then, when the force P is absent, the forces f change direction as shown and try to keep the wedge within the log. (Friction is a peacemaker; it always opposes the relative motion, or tendency for relative motion, of the contacting points of the bodies.)

Figure E6.13a

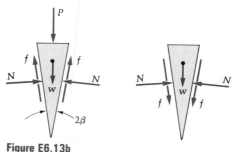

Figure E6.13b

Only the vertical equilibrium equation is of value in this problem. It is:

$$+\uparrow \quad \Sigma F_y = 0 = 2N \sin \beta - 2f \cos \beta - w \tag{1}$$

If the wedge is on the verge of popping out, then $f = \mu N$:

$$N(2 \sin \beta - 2\mu \cos \beta) - w = 0$$

If, now, the weight is neglected (it is typically much smaller than the wedging forces in this application), we obtain

$$\mu = \tan \beta \tag{2}$$

Thus any friction coefficient $\geq \tan \beta$ will insure that the wedge is "self-locking"—that is, that it will not move in the absence of the driving force P.*

Another slightly different approach to this problem, which makes use of the friction inequality, is to use Equation (1) when w is small to obtain

$$f = N \tan \beta$$

and then from $0 \leq f \leq f_{MAX} = \mu N$, we get

$$N \tan \beta \leq \mu N \Rightarrow \mu \geq \tan \beta$$

for equilibrium.

> **Question 6.8** Using equation (1), and noting that the weight of the wedge was neglected, determine whether or not $\mu = \tan \beta$ is a conservative answer for the minimum friction coefficient for self-locking.

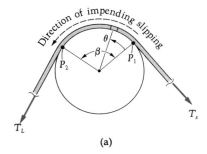

(a)

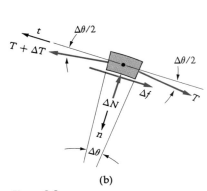

(b)

Figure 6.6

Flexible Flat Belts and V-Belts

Belts are used to turn pulleys, sheaves, and drums in countless applications. For example, a fan belt turns the alternator shaft beneath the hood of an automobile. A string on pulleys once allowed us to tune to different stations on an old radio. A flat belt surrounds the rotating parts of a clothes dryer and makes it turn when a motor shaft drives the belt.

In this section we shall derive the relationship between the tensions in the two sides of a belt wrapped around a circular drum or restrained pulley when slipping is impending.

As shown in Figure 6.6(a), the angle between the entering and leaving points P_1 and P_2 is traditionally called β. We isolate an element of the belt at the angle θ, somewhere between 0 and β, and show the forces acting on it in Figure 6.6(b). The belt is assumed to be perfectly flexible, which means it has no bending stiffness. Thus no moments (and in turn no shear forces) appear on the free-body diagram. For the element, the

---

* One of the authors has split hundreds of logs and has never seen a wedge pop out of one, though he has heard of it happening. Wedges have half angles (β) of about 6°, so that μ between the wood and wedge need only be about 0.1 for the wedge to stay in place.
Answer 6.8 If w is larger than zero, then f must decrease for the equation to remain valid. Therefore, the maximum f need not be as great, and so $\tan \beta$ was a conservative answer for μ.

equilibrium equations in the normal (n) and tangential (t) directions are

$$+\swarrow \quad \Sigma F_n = T \sin \frac{\Delta\theta}{2} + (T + \Delta T) \sin \frac{\Delta\theta}{2} - \Delta N = 0 \qquad (6.6)$$

$$+\nwarrow \quad \Sigma F_t = (T + \Delta T) \cos \frac{\Delta\theta}{2} - T \cos \frac{\Delta\theta}{2} - \Delta f = 0 \qquad (6.7)$$

Dividing Equations (6.6) and (6.7) by $\Delta\theta$ yields

$$T \frac{\sin \dfrac{\Delta\theta}{2}}{\left(\dfrac{\Delta\theta}{2} \right)} + \frac{\Delta T}{2} \frac{\sin \dfrac{\Delta\theta}{2}}{\left(\dfrac{\Delta\theta}{2} \right)} - \frac{\Delta N}{\Delta\theta} = 0 \qquad (6.8)$$

$$\frac{\Delta T}{\Delta\theta} \cos \frac{\Delta\theta}{2} - \frac{\Delta f}{\Delta\theta} = 0 \qquad (6.9)$$

Taking the limits of these two equations as $\Delta\theta \to 0$ gives

$$T = \frac{dN}{d\theta} \qquad (6.10)$$

and

$$\frac{dT}{d\theta} = \frac{df}{d\theta} \qquad (6.11)$$

where we have used:

$$\lim_{\Delta\theta \to 0} \left[\frac{\sin \left(\dfrac{\Delta\theta}{2} \right)}{\left(\dfrac{\Delta\theta}{2} \right)} \right] = 1; \quad \lim_{\Delta\theta \to 0} \left[\cos \left(\frac{\Delta\theta}{2} \right) \right] = 1; \quad \lim_{\Delta\theta \to 0} (\Delta T) = 0$$

We are interested in the case when the belt is at the point of slipping, for which $f = \mu N$. Thus from Equation (6.11):

$$\frac{1}{\mu} \frac{dT}{d\theta} = \frac{dN}{d\theta} \qquad (6.12)$$

and Equations (6.10) and (6.12) give

$$\frac{1}{T} \frac{dT}{d\theta} = \mu \qquad (6.13)$$

from which, by integrating, we find

$$\ln T = \mu\theta + C_1 \qquad (6.14)$$

The constant can be evaluated to be $\ln T_s$ because the tension is T_s when $\theta = 0$. Therefore,

$$\ln \frac{T}{T_s} = \mu\theta \qquad (6.15)$$

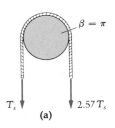

$\beta = \pi$

T_s $2.57\,T_s$

(a)

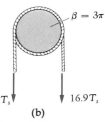

$\beta = 3\pi$

T_s $16.9\,T_s$

(b)

Figure 6.7

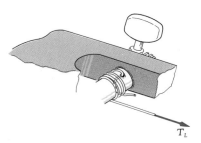

Figure 6.8

And the tension at any angle θ (necessarily in radians!) is given by

$$T = T_s e^{\mu\theta} \qquad (6.16)$$

The maximum T (called T_L) is seen to be, when $\theta = \beta$,

$$T_L = T_s e^{\mu\beta} \qquad (6.17)$$

The ratio of large to small tensions (leaving to entering the contact surface) is therefore an exponential function of the product of the coefficient of friction and the angle of wrap β. This can build up dramatically. For example, if a rope passes over a tree limb (Figure 6.7(a)) and $\mu = 0.3$ between the two materials, then for $\beta = \pi$,

$$T_L = T_s e^{0.3\pi} = 2.57 T_s$$

But with a wrap and a half (Figure 6.7(b)),

$$T_L = T_s e^{0.3(3\pi)} = 16.9 T_s$$

an increase of about 5.6 times!

As another example, just a few wraps of a guitar or banjo string around its peg enables it to be tightened to large tensions without the string being securely tied or clamped at all. (See Figure 6.8.) The friction buildup, even on such a relatively smooth shaft, is sufficiently great to prevent slippage.

> **Question 6.9**   What stiffness of the belt was neglected in the development of Equation (6.17)?

We now examine two formal examples of the use of Equation (6.17).

Answer 6.9   Its bending stiffness, which has a slight effect on the angle of wrap. If the belt is moving, its inertia also becomes a factor.

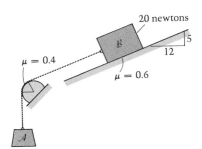

20 newtons

$\mathcal{B}$

5

12

$\mu = 0.4$

$\mu = 0.6$

Figure E6.14a

β

Figure E6.14b

EXAMPLE 6.14

Determine the minimum weight of block $\mathcal{A}$ for which block $\mathcal{B}$ will have impending motion down the plane in Figure E6.14a.

Solution

We note that $\tan\phi = \frac{5}{12} = 0.417 < 0.6 = \mu$, so that $\mathcal{B}$ will not begin to slide on its own. The angle of wrap of the rope is (see Figure E6.14b):

$$\beta = 90° - \tan^{-1}(5/12) = 67.4° = 1.18 \text{ rad}$$

Thus the tension T_s in the portion of rope between the fixed drum and $\mathcal{B}$ is obtained from Equation 6.17, with T_L equalling the weight of $\mathcal{A}$:

$$W_{\mathcal{A}} = T_s e^{0.4(1.18)} = 1.60 T_s$$

or

$$T_s = W_{\mathcal{A}}/1.60$$

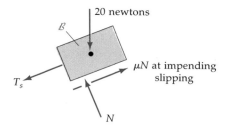

20 newtons

μN at impending slipping

T_s

N

Figure E6.14c

The free-body diagram of $\mathscr{B}$ is shown in Figure E6.14c. The equilibrium equations give:

$$+\uparrow \quad \Sigma F_y = 0 = N - 20\left(\frac{12}{13}\right) \Rightarrow N = 18.5 \text{ newtons}$$

$$\overset{+}{\swarrow} \quad \Sigma F_x = 0 = T_s + 20\left(\frac{5}{13}\right) - \underbrace{0.6(18.5)}_{\mu N}$$

from which

$$T_s = 3.41 \text{ newtons} = \frac{W_{\mathscr{A}}}{1.60}$$

$$W_{\mathscr{A}} = 5.46 \text{ newtons}$$

This is the weight of $\mathscr{A}$ for which slipping of $\mathscr{B}$ down the plane will be impending.

Answer 6.10   Because if $\mathscr{B}$ is to slide downward, then the rope must be at the point of slipping counterclockwise on the drum.

EXAMPLE 6.15

The oil-filter wrench in Figure E6.15a consists of a handle and a thin, flexible band that fits over the cylindrical filter. To remove the filter, force P is applied as shown in the figure. The tensions in the band produce friction forces around the filter whose resultant is a force together with a counterclockwise couple which

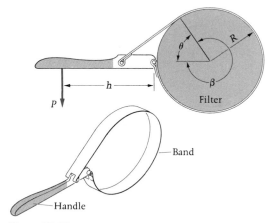

Figure E6.15a

loosens it. For a given filter radius R, friction coefficient μ, and wrap angle $\beta\,(2\pi - \theta)$, there is a minimum handle length h above which the band will not slip on the filter. Find $h = h(R, \mu, \beta)$. Then for $R = 2$ in. and $\mu = 0.3$, find h_{min} for $\theta = 60°, 45°$, and $30°$, and note that practical wrench handles are all longer than this.

Solution

Taking moments about the point E of the handle (using its free-body diagram, Figure E6.15b), we eliminate P and therefore can relate T_L and T_s immediately:

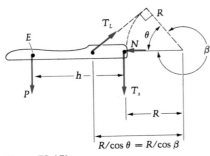

$$\curvearrowright_{(+)} \quad \Sigma M_E = 0 = T_L \cos\theta \left[h - \left(\frac{R}{\cos\theta} - R \right) \right] - T_s h \tag{1}$$

$$T_L = \frac{T_s h}{(h + R) \cos\theta - R}$$

This gives the relationship between the large and small tensions at the ends of the band. Now we know that if T_L never reaches $T_s e^{\mu\beta}$, the band will never slip on the filter. Therefore,

$$T_L = \frac{T_s h}{(h + R) \cos\theta - R} < T_s e^{\mu\beta} \tag{2}$$

or

$$h < e^{\mu\beta}[(h + R)\cos\theta - R] \tag{3}$$

Figure E6.15b

The denominator in Equation (2) is positive for a practical wrench so that the

θ (deg)	β (rad)	h (in.)	Sketch of Wrench
60°	5.24	3.42	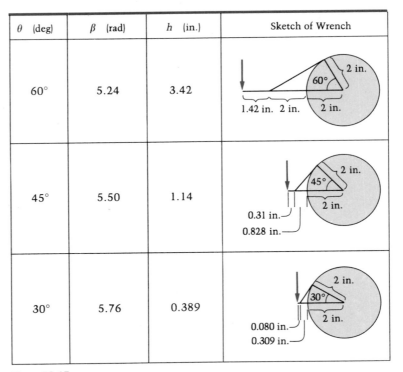
45°	5.50	1.14	
30°	5.76	0.389	

Figure E6.15c

sense in inequality (3) is unchanged. Also, we note that the cosines of θ and β are identical because $\beta = 2\pi - \theta$ so that, rearranging inequality (3),

$$e^{\mu\beta}(R - R\cos\beta) < h(e^{\mu\beta}\cos\beta - 1)$$

If $e^{\mu\beta}\cos\beta > 1$,* we obtain

$$\frac{h}{R} > \frac{e^{\mu\beta}(1 - \cos\beta)}{e^{\mu\beta}\cos\beta - 1}$$

as our final equation. For $\mu = 0.3$ and $R = 2$ in., we then compute h for the three required values of θ. The results are shown in Figure E6.15c on previous page.

We see that even at $\theta = 60°$, the average handle is longer than the minimum h calculated for no slip. This is to accommodate the hand width of an average person; note that h is the dimension to the *resultant* of the force of the hand.

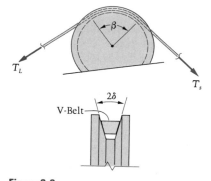

V-Belt

Figure 6.9

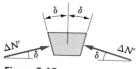

Figure 6.10

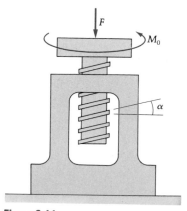

Figure 6.11

If the belt in our derivation is a V-belt instead of a flat belt, then the relationship between T_L and T_s changes and much larger tensions are possible. Suppose the angle of the "V" is 2δ as shown in Figure 6.9. It is seen that in contrast to our previous derivation, there is now no normal force *beneath* the belt at all. Instead, all the normal forces are exerted on the slanted faces on the *sides* of the belt as seen in Figure 6.10.

The outward component of the total increment ΔN of normal force acting on the element of Figure 6.6(b) is seen now in Figure 6.10 to be replaced, for a V-belt, by ($2\,\Delta N'\sin\delta$). Furthermore, it follows that the friction increment Δf will be $2\mu\,\Delta N'$, which is the total friction developed on *both* faces of the element of the V-belt. Substituting these new values of ΔN and Δf into Equations (6.6) and (6.7) leads to the following revised relation between the large and small tensions for a V-belt:

$$T_L = T_s e^{[\mu\beta/\sin\delta]} \tag{6.18}$$

Typical groove angle (2δ) values range from $30°$ to $38°$. For a leather belt on a fixed iron pulley, $\mu \approx 0.26$. With an angle of wrap of $180°$ and with $2\delta = 30°$, the value of T_L/T_s is

$$e^{(0.26)(\pi)/\sin 15°} = 23.5$$

as compared to a value for the flat belt that is less than one-tenth this amount:

$$e^{(0.26)\pi} = 2.26$$

Screws

We shall consider here only the square-threaded screw, which is used with more efficiency than the V-thread in transmitting power or in causing motion.

Suppose that a load F is to be raised by applying a moment M to a square-threaded jackscrew as shown in Figure 6.11.

---

* Which is usually the case. For example, if $\theta < 45°$ ($\beta > 315°$), it is true for any $\mu > 0.063$.

We consider the forces acting on the bottom surfaces of an elemental segment S of the screw thread in Figure 6.12.

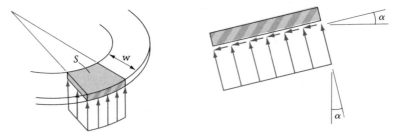

Figure 6.12

We emphasize that Figure 6.12 is not a free-body diagram of the segment S because it only shows the forces on one "face" of S. However, the accumulation of these forces, and the load F and moment M_0, are the only external forces acting on the overall screw. Thus they are the only forces that are important in expressing its equilibrium equations.

In what follows we shall assume that the pressure p and the shearing (frictional) stress s are constants. These stresses are exerted by the jack threads onto the screw threads, over whatever number of threads are in contact. On the elemental segment, the resultant normal and friction forces respectively caused by the stresses p and s are simply the stresses times the area beneath s (see Figure 6.13):

$$dN = pdA \qquad df = sdA \qquad (6.19)$$

If we exert the minimum moment that will turn the screw, then we have impending slipping and

$$df = \mu \, dN$$

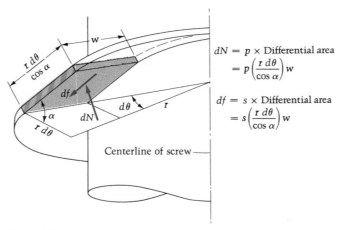

$$dN = p \times \text{Differential area}$$
$$= p\left(\frac{r \, d\theta}{\cos \alpha}\right) w$$

$$df = s \times \text{Differential area}$$
$$= s\left(\frac{r \, d\theta}{\cos \alpha}\right) w$$

Centerline of screw

Figure 6.13

Summing the vertical components of the external forces on the *entire screw* then gives

$$+\uparrow \quad \Sigma F_y = 0 = -F + \int\limits_{\substack{\text{contacting} \\ \text{threads}}} \cos \alpha \; dN - \int\limits_{\substack{\text{contacting} \\ \text{threads}}} \sin \alpha \; df$$

or

$$F = \int\limits_{\substack{\text{contacting} \\ \text{threads}}} (\cos \alpha - \mu \sin \alpha) \; dN$$

$$= \int_0^{\theta_c} (\cos \alpha - \mu \sin \alpha) \frac{prw}{\cos \alpha} \; d\theta$$

$$F = rw(1 - \mu \tan \alpha) \int_0^{\theta_c} p \; d\theta \tag{6.20}$$

where θ_c is the total angle of contact between the threads of the screw and jack. For example, if $5\frac{1}{2}$ threads are in contact, then $\theta = 11\pi$ rad.

Equation (6.20) then gives the average pressure p in terms of the load:

$$p_{ave} = \frac{F}{rw\theta_c(1 - \mu \tan \alpha)} \tag{6.21}$$

We also have to ensure that the total of all the differential moments caused by df and dN equilibrate the applied moment M_0. Summing moments about the axis, z, of the screw,

$$\substack{\circlearrowleft \\ +|} \quad (\Sigma M)_z = 0 = M_0 - \int\limits_{\substack{\text{contacting} \\ \text{threads}}} r \; dN \sin \alpha - \int\limits_{\substack{\text{contacting} \\ \text{threads}}} r \; df \cos \alpha$$

or

$$M_0 = \int_0^{\theta_c} r \sin \alpha \left(\frac{prw \; d\theta}{\cos \alpha} \right) + \int_0^{\theta_c} r \cos \alpha \left(\frac{\mu prw \; d\theta}{\cos \alpha} \right)$$

$$M_0 = r^2 w(\tan \alpha + \mu) \int_0^{\theta_c} p \; d\theta \tag{6.22}$$

Using Equation (6.20) to eliminate the pressure,

$$M_0 = \frac{Fr(\mu + \tan \alpha)}{1 - \mu \tan \alpha} \tag{6.23}$$

This is the moment required to begin to move the load upward.

If M_0 is now removed, let us determine the condition under which the screw will unwind itself. In this case, the motion is impending *downward*, and in Figure 6.13 the friction direction would have to be reversed. This reversal can be accomplished in Equation (6.20) by changing the sign of the "friction force term" (the term with μ). Therefore, for impending motion downward without M_0, we get

$$F = rw(1 + \mu \tan \alpha) \int_0^{\theta_c} p \, d\theta \tag{6.24}$$

and

$$\tan \alpha = \mu$$

Note that this time, the friction *helps* the normal force to support the load. Therefore, if

$$\tan \alpha > \mu$$

then df will exceed $\mu \, dN$ and the screw will "unwind itself," moving downward without any applied moment M_0. Such a screw is called an "overhauling" screw.

If we have $\tan \alpha < \mu$—that is, $\alpha < \tan^{-1}\mu$—then, regardless of the size of the load F, the screw will not move downward. Such a screw is said to be "self-locking." In this case, a moment opposite in direction to that of Figure 6.11 is needed to lower the load and the screw. Equation (6.24) applies, together with the moment equation of equilibrium:

$$-M_0 - r(\sin \alpha - \mu \cos \alpha) \left[\left(\int_0^{\theta_c} p \, d\theta \right) rw / \cos \alpha \right] = 0 \tag{6.25}$$

Thus if we use Equation (6.24) to eliminate p, we obtain

$$M_0 = Fr \left(\frac{\mu - \tan \alpha}{1 + \mu \tan \alpha} \right) \tag{6.26}$$

which has both a smaller numerator and a larger denominator than Equation (6.23), indicating that not nearly as much moment is needed to lower the load as to raise it.

We note that the path of a screw is a helix. We can think of it as an inclined plane spiraling around a cylinder. A property of the helix is that the *lead* L (the amount the screw will advance or retract in one turn) is a constant. The pitch P of the screw is the axial distance between similar points on successive threads of the screw. Thus, if we imagine the inclined plane to be "unwound," we have (see Figure 6.14) $\tan \alpha = L/(2\pi r)$. For a single-threaded screw, we have $L = P$, but for a double* or triple or n-threaded screw, we would have $L = nP$. Equations (6.23) and (6.26) may be rewritten in terms of the pitch P:

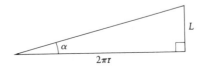

Figure 6.14

$$M_0 \text{ (advancing against the load)} = Fr \left(\frac{2\pi r\mu + nP}{2\pi r - nP\mu} \right) \tag{6.27}$$

$$M_0 \text{ (retracting away from the load)} = Fr \left(\frac{2\pi r\mu - nP}{2\pi r + nP\mu} \right) \tag{6.28}$$

In these formulas, we would use the static coefficient of friction μ_s to compute the moments required to begin the turning. For very slow rota-

---

* A double-threaded screw is one that has two separate threads intertwined.

tions so that inertial effects are negligible, the kinetic coefficient μ_k must be used, but the equilibrium equations still apply.

A final remark prior to two examples is that screws, like wedges, are statically indeterminate when motion is not impending. We can see this from our analysis, because without the relationship $df = \mu\, dN$, we would not be able to relate M_0 and F.

EXAMPLE 6.16

The scissors jack is supporting an automobile while a flat tire is being changed. The screw of the jack has a single square thread with 5-mm radius and 2-mm pitch, with a friction coefficient of 0.3. The owner of the car notices that he has jacked it up high enough to remove the flat tire, but not high enough to put on the inflated spare. What force B will begin to raise the car higher? (See Figure E6.16a.)

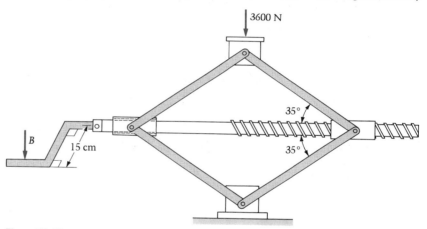

Figure E6.16a

Solution

The two free-body diagrams (see Figures E6.16b and E6.16c) and the accompanying equilibrium equations show that the force F developed in the jackscrew is 5140 N:

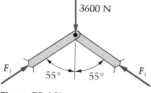

Figure E6.16b

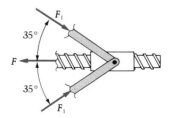

Figure E6.16c

$$+\!\!\uparrow \quad \Sigma F_y = 0 = 2F_1 \cos 55° - 3600 \qquad \xrightarrow{+} \quad \Sigma F_x = 0 = 2F_1 \cos 35° - F$$

$$F_1 = 3140 \text{ N} \qquad\qquad\qquad F = 5140 \text{ N}$$

In other words, *more* than the 3600 N supported weight of part of the car must be exerted by the jackscrew because of the angles involved.

Next, the moment that turns the jack is seen from the original figure to be $0.15B$ N · m, with B in newtons. To raise the car higher, Equation (5.27) gives

$$0.15B = 5140(0.005)\left(\frac{2\pi(0.005)(0.3) + 1(0.002)}{2\pi(0.005) - 1(0.002)(0.3)}\right)$$

$$B = 63.5 \text{ N}$$

When the tire has been changed and the owner is ready to lower the car, the force B required then is obtained from equation (5.28):

$$0.15B = 5140(0.005)\left(\frac{2\pi(0.005)(0.3) - 1(0.002)}{2\pi(0.005) + 1(0.002)(0.3)}\right)$$

$$B = 39.7 \text{ N}$$

about 63% of the load that was required to lift the car.

EXAMPLE 6.17

A single-threaded (or single-pitch) bolt has square threads* with 0.5-inch mean radius and $\frac{1}{16}$-inch pitch. The bolt is being used to securely fasten two plates which have kinetic coefficients of friction with the bolt of $\mu_k = 0.3$. (a) Find the axial force in the bolt if it is slowly turning under a 12 lb-ft torque. (b) If the rotation stops, what torque is required to start the bolt turning again if the static coefficient of friction is $\mu_s = 0.4$?

Solution

(a) Equation (6.27) gives

$$F = \frac{M_0}{r}\left(\frac{2\pi r - nP\mu}{2\pi r\mu + nP}\right) = \frac{12(12)}{0.5}\left(\frac{2\pi(0.5) - 1(\frac{1}{16})0.3}{2\pi(0.5)0.3 + 1(\frac{1}{16})}\right)$$

$$F = 895 \text{ lb}$$

(b) To begin again, we would use μ_s instead of μ_k. The same equation, but this time solved for M_0, gives

$$M_0 = Fr\left(\frac{2\pi r\mu + nP}{2\pi r - nP\mu}\right) = (895)\left(\frac{0.5}{12}\right)\left(\frac{2\pi(0.5)0.4 + 1(\frac{1}{16})}{2\pi(0.5) - 1(\frac{1}{16})0.4}\right)$$

$$= 15.8 \text{ lb-ft, almost a 32% increase!}$$

Disk Friction

Disk friction is developed when two flat circular or annular surfaces come into contact, with each either turning or tending to turn relative to the other, about a central axis normal to their contact plane. Common examples are disk brakes and clutch plates.

---

* Note that most fastenings are made with V-threaded screws, which provide slightly more friction. See, for example, *Statics* by Goodman and Warner, Wadsworth, Belmont, CA, 1963, p. 325.

Suppose we wish to know how much moment resistance to the relative motion (or tendency toward it) is developed over the area of contact by the friction. We shall assume that:

a. the contact is over an annular area having inner and outer radii R_i and R_0;

b. the friction coefficient and pressure p are constant over the area;

c. there is no lubrication between the surfaces.

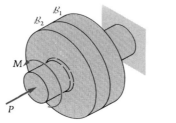
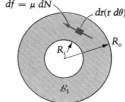

Figure 6.15

With these assumptions, the normal force on a differential area $r\,dr\,d\theta$ of one of the surfaces, say S, of body $\mathcal{B}_1$, is (see Figure 6.15)

$$dN = p\,dA$$

so that the moment developed on $\mathcal{B}_1$ by all the elemental friction forces at the point of slipping is

$$M = \int \underbrace{(\mu\,dN)}_{df}r = \int \mu p r\,dA = \int_{\theta=0}^{2\pi}\int_{r=R_i}^{R_0} \mu p r^2\,dr\,d\theta \qquad (6.29)$$

Therefore, after integrating we find

$$M = \frac{\mu p(R_0^3 - R_i^3)2\pi}{3} \qquad (6.30)$$

Or, in terms of the compressing force P between the bodies, which is

$$P = pA = p\pi(R_0^2 - R_i^2), \qquad (6.31)$$

we may eliminate the pressure p from Equation (6.30) and get

$$M = \frac{2\mu P}{3}\frac{R_0^3 - R_i^3}{R_0^2 - R_i^2} = \frac{2\mu P}{3}\frac{(R_0^2 + R_0 R_i + R_i^2)}{R_0 + R_i} \qquad (6.32)$$

where it is seen that if $R_i = 0$ (solid circular area of contact),

$$M = \frac{2\mu P R_0}{3} \qquad (6.33)$$

which is the moment that can be transmitted by friction across two circular disks. It decreases somewhat as the surfaces wear, as one might expect. Initially the material at larger r will wear faster because it has greater speed. This will alter the pressure distribution. If instead of constant pressure we reach a condition of constant rate of wear, it can be shown that the moment is reduced by 25% to $\frac{1}{2}\mu PR_0$.

Here is an example of the use of Equation (6.32):

EXAMPLE 6.18

A clutch is being tested. Its contacting surfaces are annular disks of inner and outer radii two and three inches, respectively. If with an axial force of 500 lb the clutch slips when the transmitted moment reaches 400 lb-in., what is the coefficient of friction?

Solution

We use Equation (6.32), which when solved for μ is

$$\mu = \frac{3M}{2P} \frac{R_0 + R_i}{R_0^2 + R_0 R_i + R_i^2}$$

Therefore,

$$\mu = \frac{3(400)}{2(500)} \frac{3 + 2}{3^2 + 3(2) + 2^2} = 0.316$$

PROBLEMS ▶ Section 6.3

6.150 The bodies $\mathcal{A}$ and $\mathcal{B}$ in Figure P6.150 have masses 50 kg and 75 kg, respectively. The various coefficients of friction between pairs of surfaces are shown in the figure. Find the force F that will cause $\mathcal{A}$ to begin to move up the wall.

6.151 In the preceding problem, let the force F be instead applied downward on the top of $\mathcal{A}$. Show that regardless of the value of F, block $\mathcal{B}$ will not move to the right.

6.152 In Figure P6.152 block $\mathcal{A}$, mass 20 kg, rests on block $\mathcal{B}$, mass 30 kg. If the coefficient of friction is 0.25 for all surfaces, find the force P necessary to raise block $\mathcal{A}$.

6.153 In Figure P6.153 find the smallest force F that will raise the body $\mathcal{W}$, which weighs 1000 lb. The friction coefficients are 0.2 between $\mathcal{A}$ and $\mathcal{B}$, and 0.15 between $\mathcal{A}$ and $\mathcal{W}$. Neglect the weight of $\mathcal{A}$.

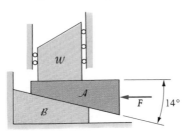

Figure P6.153

6.154 In the preceding problem, will the system remain in equilibrium if force F is removed? If not, how much force is needed?

6.155 In Figure P6.155 find the smallest force F that will cause the center C of the cylinder to move up the inclined plane. The cylinder has a mass of 300 kg, and the wedge is light.

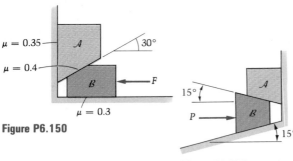

Figure P6.150

Figure P6.152

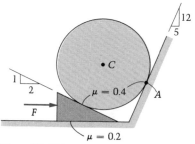

Figure P6.155

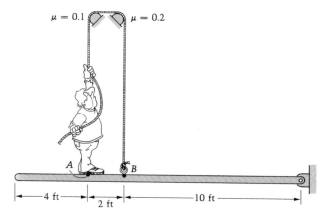

Figure P6.159

6.156 In the preceding problem, determine whether equilibrium is possible if force F is removed. If not, how much force is needed?

6.157 Find the range of values of the weight of $\mathcal{A}$ that will hold $\mathcal{B}$ in equilibrium. The friction coefficient is 0.45 between the cable and the fixed drum $\mathcal{D}$. (See Figure P6.157.)

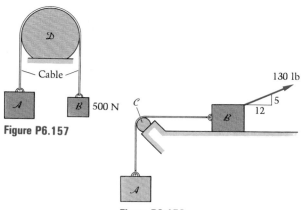

Figure P6.157

Figure P6.158

6.160 In the preceding problem, find the *largest* force the man can exert on the cable if the bar is to remain horizontal.

6.161 In Figure P6.161 a rope is wrapped around the fixed drum $1\frac{1}{4}$ times. If the girl can pull with 40 lb, what maximum weight W can be held in equilibrium if $\mu = 0.3$ between rope and drum?

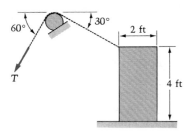

Figure P6.161

6.158 Homogeneous blocks $\mathcal{A}$ and $\mathcal{B}$ are connected by a belt that passes over the fixed drum $\mathcal{C}$. (See Figure P6.158.) The force of 130 lb acts upon block $\mathcal{B}$ as shown. The coefficients of friction are 0.4 between drum and belt, and 0.2 between $\mathcal{B}$ and the plane. Block $\mathcal{B}$ weighs 150 lb and is wide enough so that it will not tip. Find the range of weights that $\mathcal{A}$ can have for equilibrium of the system.

6.159 A 150-lb man stands on a 60-lb bar, to which a cable is fixed at B. (See Figure P6.159.) The cable passes over two fixed pegs with coefficients of friction as indicated. Assume that the normal force exerted by the man on the bar acts downward through A and find the smallest force the man can exert on the cable if the bar is to remain horizontal.

6.162 In Figure P6.162 the block weighs 40 lb. The friction coefficients are 0.5 between the rope and drum, and 0.35 between the block and the plane. Find the lowest force T that will cause the block to move.

Figure P6.162

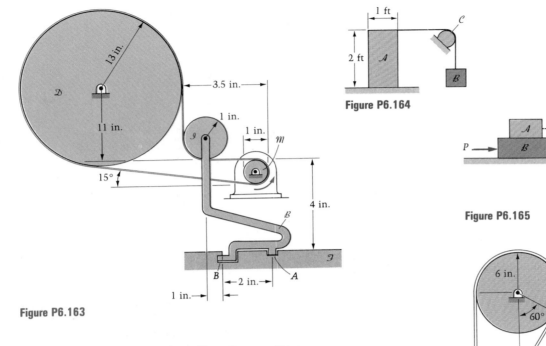

Figure P6.163

Figure P6.164

Figure P6.165

Figure P6.166

6.163 In the clothes dryer, the pulley of motor $\mathcal{M}$ turns counterclockwise as shown in Figure P6.163. If the dryer drum $\mathcal{D}$ is turning at constant angular velocity, then the moments of the large (lower) and the small (upper) tensions are in balance with the moments due to friction at the support wheels (not shown). At the point of impending slipping of the belt on the drum, find the ratio of large to small tensions with and without the idler pulley $\mathcal{I}$ present. Let $\mu = 0.3$.

6.164 The homogeneous block $\mathcal{A}$ in Figure P6.164 is connected by a belt to another homogeneous block, $\mathcal{B}$. Block $\mathcal{A}$ weighs 350 lb, and the coefficients of friction are 0.3 between $\mathcal{A}$ and the horizontal plane and 0.2 between the belt and the fixed drum $\mathcal{C}$. Find the maximum weight $\mathcal{B}$ can have without disturbing the equilibrium of the two bodies.

6.165 In Figure P6.165 body $\mathcal{B}$ weighs 1200 lb and body $\mathcal{C}$ weighs 680 lb. The coefficients of friction μ are $2/\pi$ between rope and drum, 0.4 between $\mathcal{A}$ and $\mathcal{B}$, and 0.3 between $\mathcal{B}$ and the plane. The force P is 350 lb. Determine the minimum weight of $\mathcal{A}$ that will prevent downward motion of $\mathcal{C}$. Neglect the possibility of tipping.

6.166 The coefficient of friction between the drum and the band is 0.4. (See Figure P6.166.) Find the friction moment on the drum developed by the band brake, in terms of force P, for impending slipping. The drum is held fixed by a couple that is not shown.

6.167 Can the 100-lb painter in Figure P6.167 stand on the left end of the scaffold and drink his lemonade without the rope slipping and dumping him to the ground below?

6.168 In Figure P6.168, given that $\mathcal{A}$ weighs 1000 lb, $\mathcal{B}$ weighs 750 lb, and the coefficients of friction μ are 0.3 between $\mathcal{A}$ and $\mathcal{B}$, 0.3 between $\mathcal{B}$ and the plane, and $1/\pi$ between rope and drum, find the minimum force P that will cause $\mathcal{A}$ to move.

6.169 Repeat Problem 6.93 if the peg is not smooth but has a friction coefficient with the rope of 0.4. (See Figure P6.169.) The blocks $\mathcal{A}$, $\mathcal{B}$, $\mathcal{C}$, and $\mathcal{D}$ weigh 100 lb, 50 lb, 75 lb, and 40 lb, respectively. Force P is slowly increased from zero. At what value of P will the equilibrium be disturbed?

6.170 The flat belt in Figure P6.170 is transmitting power from the small pulley P_1 to the large pulley P_2. The applied torque T_a turns P_1 clockwise, and the belt then turns P_2 in the same direction, with the load (being rotated by P_2 and

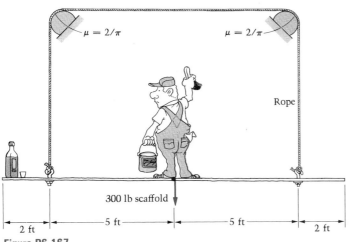

Figure P6.167

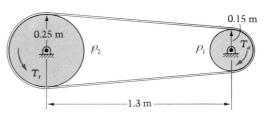

Figure P6.170

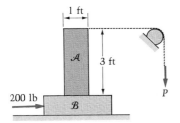

Figure P6.168

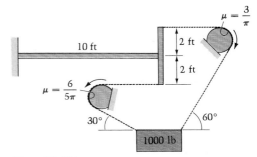

Figure P6.171

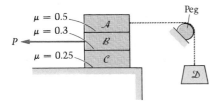

Figure P6.169

the belt can be neglected, which really means that the pulley speeds are not too large.

6.171 The two cables are on the verge of slipping in the indicated directions in Figure P6.171. Find the reactions exerted by the wall onto the left end of the flexible beam, whose weight is to be neglected.

6.172 The sandbag in Figure P6.172 weighs 50 lb. The coefficient of friction between the rope and the tree branch is 0.4. What range of force can the man exert and maintain equilibrium if he weighs 150 lb and if the coefficient of friction is 0.3 between his shoes and the ground?

not shown) resisting the rotation with torque T_R. The friction coefficient between the belt and each pulley is 0.25. If the largest safe belt tension is 200 N, find:

   a. The maximum torque T_a that can be applied without slipping the belt on P_1.

   b. The maximum load torque T_R to which the pulley P_2 should be connected.

Note: Strictly speaking, this is a dynamics problem. However, for the pulleys turning at constant rates, the equilibrium equations hold for each pulley. Conditions in the belt are as developed in this section provided the inertia of

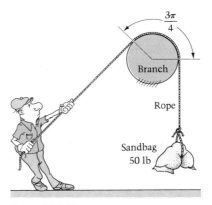

Figure P6.172

6.173 The ferry boat in Figure P6.173 is being docked, and exerts 1000 lb on each of the ropes. How many wraps around the posts (mooring bitts) must each dock worker make in order to hold the boat in position with less than 40 lb? The friction coefficient is $\mu = 0.3$.

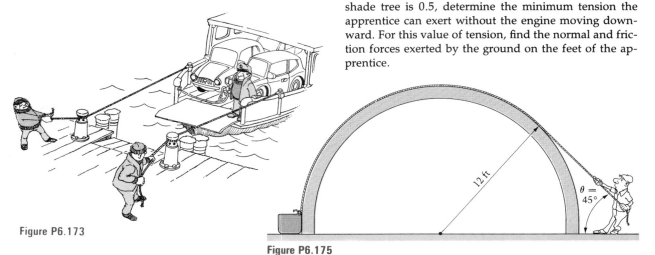

Figure P6.173

Figure P6.174

Figure P6.175

6.174 The band brake in Figure P6.174 is designed to stop the drum $\mathcal{D}$ from rotating. A linkage provides the force F_0 as shown. The friction coefficient between the band and the drum is $\mu = 0.3$. Neglecting inertial effects, what is the friction moment that can be developed by the brake, in terms of F_0 and R?

6.175 The man in Figure P6.175 is trying to pull the block over the round hut. He is able to exert 50 lb of force on the rope; in the position shown he is unable to lift the block. He backs up until the angle θ becomes 35°, and finds he is then barely able, with his 50 lb, to raise the block. The friction coefficient between rope and hut is 0.1. Neglecting friction between the block and the hut, find the weight of the block.

6.176 The shade tree mechanic in Figure P6.176 is removing some concrete blocks from beneath an engine, preparatory to reinstalling it in his car. The engine weighs 700 lb, and is being held in equilibrium by a 150-lb apprentice. If the coefficient of friction between rope and shade tree is 0.5, determine the minimum tension the apprentice can exert without the engine moving downward. For this value of tension, find the normal and friction forces exerted by the ground on the feet of the apprentice.

6.177 An old bench vise has a square-threaded screw $\mathcal{S}$ with a mean diameter of $\frac{1}{2}$ inch and a pitch of $\frac{1}{8}$ inch. (See Figure P6.177.) The screw turns within the guide $\mathcal{G}$, which is fixed to one jaw ($\mathcal{J}_1$) of the vise. The other jaw, $\mathcal{J}_2$, is part of body $\mathcal{A}$, which is clamped to the bench as shown. Body $\mathcal{A}$ contains the threaded sleeve $\mathcal{C}$ that engages the screw with coefficients of friction 0.25 (kinetic) and 0.3 (static).

When the vise is clamping something between its jaws, the guide $\mathcal{G}$ contacts $\mathcal{A}$ at B and at D, and the head of the screw bears against $\mathcal{G}$ at A. If a board is being held in the vise with a clamping force of 200 lb:

a. What are the forces exerted at B and D onto guide $\mathcal{G}$, neglecting friction at B and D and the weight of the vise?

b. What was the force at H on handle $\mathcal{H}$ just before the tightening was completed?

6.178 In the preceding problem, after the handle was released (clamping force = 200 lb), it was decided to increase the clamping action.

a. What force must be applied at H to initiate further compression of the board?

b. What force must be applied to initiate a reduction (from 200 lb) of the clamping force?

6.179 The power input while raising the load of Figure 6.11 at a constant angular speed ω_0 rad/sec of the

Figure P6.176

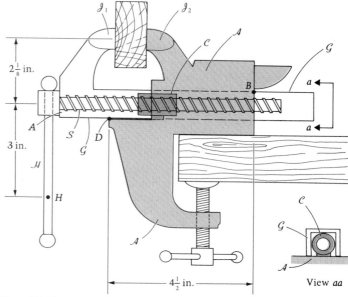

Figure P6.177

screw is $M\omega_0$. The "power output" is the product of F and the (vertical) velocity of the load, or $F[(r\omega_0)\tan\alpha]$. If the efficiency of the jackscrew is the ratio of the power output to the power input, find the efficiency in terms of α and μ.

6.180 In the preceding problem, if $\mu = 0.3$, find the pitch angle giving maximum efficiency to the jackscrew. What is this maximum efficiency?

6.181 In Example 6.17 find the torque required to retract the bolt when it has stopped with the 895-lb tensile force in it.

6.182 The single-threaded turnbuckle in Figure P6.182 is being used to move the heavy crates slightly closer. When the tension in the turnbuckle is at the point of causing impending slipping of the crates, what moment M_0 applied to the turnbuckle will start them slipping? The turn-

buckle data are: pitches = 0.06 in., radii = 0.25 in., and static friction coefficients = 0.35. Assume symmetry.

6.183 In the preceding problem, let the respective weights of A and B be 1200 lb and 700 lb, with friction coefficients of 0.2 between A and the floor and 0.3 between B and the floor. Find the couple being applied to the turnbuckle at the point when one of the crates moves.

6.184 Show that for the square-threaded jackscrew, the moment M to turn the screw against the load F may be expressed in terms of the angle of friction ϕ as $M = Fr\tan(\alpha + \phi)$.

6.185 Prove the statement, made in the text just prior to Example 6.18, that the moment transferred once the condition of constant rate of wear is reached is reduced by 25% to $\mu PR_0/2$. *Hint:* Constant rate of wear means that the rate of work done per unit area by friction is constant,

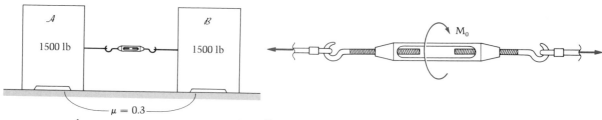

Figure P6.182

so that

$$\mu p v = \text{constant} = k_1$$

where v = velocity of the point in question. This velocity magnitude is the radius times a constant angular speed. Therefore, $pr = \text{constant} = k_2$, and the constant k_2 can be found by $P = \int_A p \, dA$. Then the moment M is balanced by

$$M = \int_A r(p \, dA)\mu = \int_{\theta=0}^{2\pi} \int_{r=0}^{R} r \frac{k_2}{r} \mu \, r \, dr \, d\theta$$

Integrate and find the required moment.

6.186 If a person pushes a disk sander against a surface with a force of 10 lb, and if the friction coefficient is 0.4 between the sandpaper and the surface, find the moment the motor must exert to overcome the developed friction. (See Figure P6.186.) Assume the pressure to be constant over the surface.

8 in.

Figure P6.186

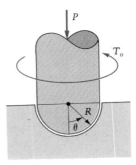

P

T_o

R

θ

Figure P6.188

6.187 In the preceding problem, assume the pressure drops off linearly with radial distance from the center to a value at the rim that is 60% of that at the center. Find the new value of the moment, and compare.

* **6.188** Find the largest torque that may be applied to the shaft in Figure P6.188 without it slipping in the hemispherical seat. Assume the pressure varies as $k\left(\dfrac{\pi - 2\theta}{\pi}\right)$ and the coefficient of friction is μ.

6.189 Repeat the preceding problem if the pressure distribution is cosinusoidal: $p = k \cos \theta$.

6.190 Find the maximum torque T_0 that may be applied to the shaft in Figure P6.190 without it slipping in the conical seat. Assume a uniform pressure distribution. Check your answer against Equation (6.30) for the case when $d = 0$ and $\beta = 90°$.

6.191 The collar bearing in Figure P6.191 supports the 200-lb thrust in the shaft. What is the value of the torque T at which the shaft will begin to turn?

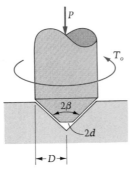

P

T_o

2β

$2d$

$\leftarrow D \rightarrow$

Figure P6.190

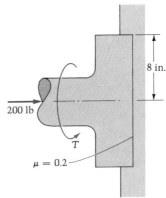

8 in.

200 lb

T

$\mu = 0.2$

Figure P6.191

6.192 Repeat the preceding problem for the new support conditions on the flange shown in Figure P6.192.

6.193 In Figure P6.193 the shaft A and disk B are welded together to form one body that is free to rotate around its horizontal axis in bearings not shown in the figure. Disk C, larger than B, is free to slide on A but is normally held apart from B. When the separating forces are removed, the tension in four equally spaced springs (two are shown) cause C to contact B. The friction then stops the rotation. The lugs attached to C slide in fixed slots and prevent rotation of C during contact. If the coefficient of friction between B and C is 0.3, find the force in each spring required to produce a disk friction moment of $5 \text{ N} \cdot \text{m}$.

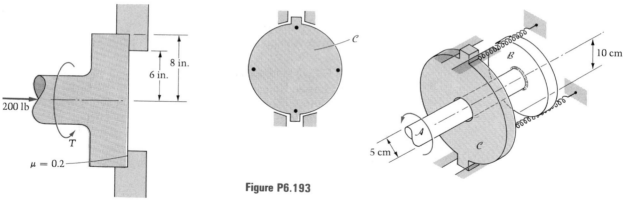

Figure P6.192

Figure P6.193

COMPUTER PROBLEMS ▶ Chapter 6

6.194 Work Problem 6.142 with the help of a computer. *Hint*: There will be two equilibrium equations for each mass. The fifth equation in the five unknowns is simply that $1.5R = R(\theta_1 + \theta_2)$. After eliminating the tension, the two normal forces, and the angle θ_2, you should be left with the following equation in the remaining unknown θ_1:

$$0.3 \cos \theta_1 + 0.3 \cos(1.5 - \theta_1) \\ + \sin(\theta_1 - 1.5) + \sin \theta_1 = 0.$$

Use the computer to solve for θ_1, then backtrack through your equations to obtain θ_2, T, N_1, and N_2.

6.195 Read Problem 4.119. Note that if the bar is on the verge of slipping at both surfaces of contact, then the

friction coefficients at these surfaces are ρ and σ. Assume that these coefficients are equal, and plot the value of ρ required for equilibrium of the stick as a function of θ. Note from the equation that for values of θ smaller than 45°, *no* amount of friction can result in equilibrium.

6.196 Solve Problem 6.145 with the help of a computer. *Hint*: In each case (impending slipping up and down), after eliminating the two normal forces from your equilibrium equations for one of the sticks, you will obtain a cubic equation in $\sin \alpha$ to solve, where α is half the angle between the sticks for equilibrium.

6.197 Solve Problem 6.146 with the help of a computer.

SUMMARY ▶ Chapter 6

In this chapter we learned that resistance to the tendency of two bodies to slide past each other is called friction, and if the surfaces are dry, it is often called Coulomb friction. This resistance is in the form of a friction force f that obeys the inequality

$$0 \le f \le f_{\text{MAX}} = \mu N$$

In other words, with Coulomb friction the friction force is limited to a maximum value proportional to the normal, or pressing force between the two bodies. The proportionality constant μ is called the coefficient of friction, and it is dependent on the types of materials in contact and the roughnesses of their surfaces. The table on page 416 gives approximate values of μ for a number of different pairs of materials. We learned that if a block of one material is placed on a plane made of the other material and the angle of the plane is slowly increased, the angle at which the block will slip on the plane is $\tan^{-1}\mu$; thus this is one way of determining μ.

In Section 6.2 we studied a number of examples of equilibrium problems involving impending slipping. As we did so, we saw that the study of friction often entails decisions—whether a body is about to slip or not, whether a body slips or tips first, whether it slips on one surface or another (or yet another, etc.), whether a wheel rolls or slips against a surface, and so on. It is for the most part these various cases which must be considered that make this chapter's problems different from those of Chapter 4.

Some special applications of Coulomb friction were taken up in Section 6.3. These were wedges, flexible flat belts and V-belts, square-threaded screws, and disk friction. Wedges are used to create large normal and friction forces that effect a small change in the position of one body when another is wedged between it and a second, usually fixed, surface.

When a flat belt is wrapped around a fixed circular surface and then placed under tension, the large tension T_L was seen to be related to the small tension T_s by the formula

$$T_L = T_s \, e^{\mu\beta}$$

where e is the base of natural logarithms and β is the angle (in radians) of wrap. For a V-belt with angle 2δ, the corresponding formula is

$$T_L = T_s \, e^{\mu\beta/\sin\delta}$$

The above two formulae show the great effect of the buildup of friction around wrapped belts.

We developed a number of useful formula for the square-threaded screw, including the following, which gives the ratio of M_o to (Fr), where F is a weight to be raised, M_o is the moment applied to the screw to raise the weight, and r is the radius of the screw:

$$\frac{M_o}{Fr} = \frac{\mu + \tan\alpha}{1 - \mu\tan\alpha}$$

where μ is the coefficient of friction and α is the angle of the thread.

Disk friction occurs when two flat circular or annular surfaces (such as disk brakes or clutch plates) come into contact, with each turning or tending to turn relative to the other about a central axis normal to their contact plane. If μ is the coefficient of friction as usual, R_i and R_o are the inner and outer radii of contact, and P is the pressing force between the bodies, then the resisting moment that can be developed by the friction between the bodies at the point of impending slipping is

$$M = \frac{2\,\mu\,P}{3} \frac{R_o^2 + R_o\,R_i + R_i^2}{R_o + R_i}$$

which if $R_i = 0$ (solid circular area of contact) reduces to

$$M = \frac{2 \mu P R_o}{3}$$

REVIEW QUESTIONS ▶ **Chapter 6**

True or False?

1. The Coulomb law is appropriate for surfaces lubricated by grease.
2. The coefficient of friction for two surfaces is independent of the materials comprising the surfaces.
3. The friction force always has magnitude μN, unless it is zero.
4. Friction is never helpful.
5. Friction is a force that resists the tendency of two bodies to slide past each other.
6. If the normal force between two bodies is zero at the point of contact, there can be no friction force developed there according to the Coulomb law.
7. It is not possible for an object to start to slide and tip at the same time.
8. If body A is rolling on body B, the friction force between A and B has to be zero.
9. In working friction problems, we always know in advance whether or not there is slipping between two surfaces.
10. Suppose we assume a certain state of impending slipping exists between two bodies, and we then draw the force as μN on the free-body diagram of one of the bodies. It is OK if we get the direction of the force wrong because the sign of the answer will indicate it is really the other way.
11. The formula $T_L = T_s e^{\mu \beta}$ relates the small and large end tensions in a flat belt wrapped on a drum, at the point of slipping of the belt.
12. The formula in the preceding question applies only for angles β between 0 and 2π radians.
13. For the same coefficient of friction and angle of wrap, the quotient T_L / T_s of the large and small tensions, at the point of slipping, is smaller for a V-belt than for a flat belt.
14. It is harder (takes more moment) to advance a double-threaded screw against a load than a single-threaded screw, but it is easier (less moment needed) to retract the double-threaded screw away from the load than the single-threaded screw.
15. The moment that can be transmitted by disk friction is independent of the radii of the contacting surfaces.

Answers: 1. F   2. F   3. F   4. F   5. T   6. T   7. F   8. F   9. F   10. F   11. T   12. F   13. F
14. T   15. F

7 ▶▶▶ CENTROIDS AND MASS CENTERS

7.1 Introduction

In this chapter we will learn about the center of arclength, center of area, center of volume, and center of mass. The first three of these, involved with spatial quantities, are commonly called centroids (of arclength, area, and volume).

All four of the above quantities may be defined by the same formula, in which the quantity (be it arclength, area, volume, or mass) appears each time in an integral which we therefore call the recurring integral. Section 7.2 is concerned with finding centroids using the recurring integral. Also in Section 7.2 we shall revisit some results from Section 3.9 about distributed forces because we are able now to understand their relationship to centroids.

It happens that when an entity is made up of several known parts (e.g., an area made up of several smaller, known areas), and the centroid (or center of mass, if the entity is mass) of each of the parts is known, we can find the centroid of the "overall" entity without integration. This procedure, called the Method of Composite Parts, is presented and studied in detail in Section 7.3. This interesting method not only works in combining parts to form the total entity whose centroid is of interest, but also in *subtracting* one part from another to yield the entity of interest.

The center of mass, or mass center, covered in Section 7.4, is of monumental importance in engineering mechanics. In Statics, for uniform gravity fields it becomes indistinguishable from the center of gravity — which of course is the point of application* of the equipollent single force representing a body's weight. In a later course in dynamics, the mass center is the point of a body whose acceleration equals the external force exerted on the body divided by its mass. Clearly, then, we could get nowhere without the mass center concept.

When the density of a body is constant (or "uniform"), its center of mass will be seen in Section 7.4 to be the very same point as the centroid of volume, but if the density varies with x, y, or z, this will ordinarily not be the case. The reader will recall that density is "mass per unit volume," so that if a density function varies over the body's volume, the integrand of the recurring integral will generally be more complicated and, as a result, mass center computation will be more difficult.

A pair of very old theorems, known as The Theorems of Pappus, are at once so elegant and useful that we have included them in their own section, 7.5. These theorems allow us to find surface areas and volumes which are respectively formed by rotating plane curves and plane areas around an axis; alternately, if we know the areas and volumes of revolution, we may use the theorems to find the centroids of the generating curves and areas.

---

* Meaning, for a body B of mass m attracted by a particle P of mass M, the point Q on the line of action of the weight force $\mathbf{F}$ such that $r_{PQ} = \left(\dfrac{GmM}{|\mathbf{F}|} \right)^{1/2}$.

7.2   Centroids of Lines, Areas, and Volumes

The Recurring Integral

In this section we shall learn the meaning of the word centroid, or, more generally, the "center of Q," where the letter Q denotes any scalar quantity associated with a region of space. The most commonly occurring quantities Q are arclength, area, volume, and mass; we shall examine the first three of these in this section, using s, A, and V as the respective symbols to represent them. The center of mass will be examined in Section 7.3, even though it occurs naturally in the development of dynamics.

The "center of Q" is a point C whose location (see Figure 7.1) is defined by the equation

$$Q\mathbf{r}_{OC} = \int_{R} \mathbf{r} \, dQ \tag{7.1}$$

where Q itself is given by

$$Q = \int_{R} dQ \tag{7.2}$$

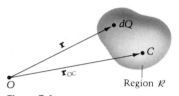

Figure 7.1

and R denotes the region in space over which the integration is carried out; it may be a one-, two-, or three-dimensional region. If, for example, we are seeking a "center of volume," the region of integration is three-dimensional and Equation (7.1) becomes

$$V\mathbf{r}_{OC} = \int_{R} \mathbf{r} \, dV$$

where R is the spatial region whose volume is of interest. Similarly, when Q denotes an area the region of integration is two-dimensional, and when Q denotes length of a line (either curved or straight), the region of integration is one-dimensional. When, as in each of these cases, the quantity Q is geometric — that is, arclength, area, or volume — then the "center of Q" is more commonly known as the **centroid** (of arclength, area, or volume). An important fact is that the centroid is unique; that is, point C located by Equation (7.1) is independent of the choice of origin O. The proof of this is left to the student as Problem 7.1.

To locate the centroid, we use the scalar orthogonal components of the vectors in Equation (7.1). If we set up a rectangular coordinate system with origin at O, then $\mathbf{r} = x\hat{\mathbf{i}} + y\hat{\mathbf{j}} + z\hat{\mathbf{k}}$ and $\mathbf{r}_{OC} = \bar{x}\hat{\mathbf{i}} + \bar{y}\hat{\mathbf{j}} + \bar{z}\hat{\mathbf{k}}$ where $\bar{x}$, $\bar{y}$, and $\bar{z}$ are the coordinates of the centroid C in the rectangular coordinate system. Thus

$$Q\bar{x} = \int x \, dQ$$

$$Q\bar{y} = \int y \, dQ \tag{7.3}$$

$$Q\bar{z} = \int z \, dQ$$

Scalar integrals* of the type in Equation (7.3) (and occasionally the vector form $\int \mathbf{r} \, dQ$) naturally arise again and again in different areas of mechanics as well as in other branches of mathematical physics and engineering. The recurrence of these integrals is the principal reason for identifying and studying centroids. Suppose, for example, that we encounter and must evaluate the integral $\int y \, dA$, which naturally arises in the analysis of stresses in solids. If we know the location of the centroid of the area in question, we need only multiply the total area by the y coordinate of the centroid to obtain the value of the integral.

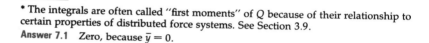

Question 7.1   What then is $\int y \, dA$ if the origin of the coordinate system is *at* the centroid C?

We now proceed to a series of examples in each of which we shall locate the centroid of a line, an area, or a volume. It will be seen that sometimes one or more of $(\bar{x}, \bar{y}, \bar{z})$ will be zero by symmetry, and we will not have to perform as many integrations in order to find the centroid.

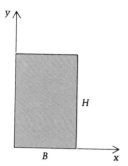

Figure E7.1a

---

* The integrals are often called "first moments" of Q because of their relationship to certain properties of distributed force systems. See Section 3.9.
Answer 7.1   Zero, because $\bar{y} = 0$.

EXAMPLE 7.1

Locate the centroid of the area of a rectangle of base B and height H (see Figure E7.1a).

Solution

Our use of the term "center" suggests that the centroid will be at $x = B/2$, $y = H/2$. We are simply going to confirm that this is so.

To find $\bar{x}$, the x-coordinate of the centroid we seek, we use the first of Equations (7.3), with Q replaced by area (A):

$$A\bar{x} = \int x \, dA$$

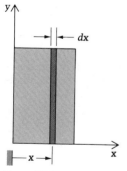

Figure E7.1b

Using the vertical strip of width dx and height H for our differential area dA (see Figure E7.1b), and substituting $A = BH$,

$$\bar{x} = \frac{\displaystyle\int_0^B xH \, dx}{BH} = \frac{H\,\dfrac{x^2}{2}\Big|_0^B}{BH} = \frac{B}{2}$$

We see that the centroid is, as expected, on a vertical line halfway across the rectangle.

To obtain $\bar{y}$, we shall use the horizontal strip shown in Figure E7.1c so that $dA = B \, dy$ this time:

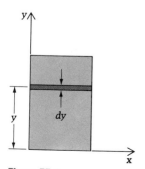

Figure E7.1c

$$A\bar{y} = \int y\, dA$$

$$BH\bar{y} = \int_0^H yB\, dy$$

$$\bar{y} = \frac{B\left.\dfrac{y^2}{2}\right|_0^H}{BH} = \frac{H}{2}$$

and so the centroid is indeed in the middle of the rectangle.

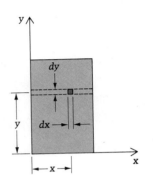

Figure E7.1d

Question 7.2   What could be the problem in using the vertical strip as the dA in calculating $\bar{y}$?

For readers familiar with double integrals at this point of their study, then with $dA = dx\, dy$ we have (see Figure E7.1d):

$$A\bar{y} = (BH)\bar{y} = \int_0^H \int_0^B y\, dx\, dy$$

$$= \int_0^H y\left(\left.x\right|_0^B\right) dy$$

$$= \int yB\, dy = B\left.\frac{y^2}{2}\right|_0^H$$

Thus

$$(BH)\bar{y} = BH^2/2$$

so that $\bar{y} = H/2$ as expected. Note that the first integration (on x) generates coverage of the horizontal strip, B by dy, that was used above in the single integration solution.

Answer 7.2   The coordinate y would not be the same for all parts of the element.

EXAMPLE 7.2

Find the centroid of the semicircle shown in Figure E7.2a.

Solution

This time the quantity Q is arclength s, and we want to find the point C for which

$$s\mathbf{r}_{OC} = \int \mathbf{r}\, ds$$

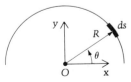

Figure E7.2a

or, for which

$$s\bar{x} = \int x\, ds \qquad s\bar{y} = \int y\, ds \qquad s\bar{z} = \int z\, ds$$

where the total arclength s is πR.

In this problem, by symmetry, $\bar{x} = 0$. This is because for every element ds to the right of the y axis with x coordinate "x," there is an element to the *left* of the y axis with x coordinate "$-x$" (and vice-versa). Thus the entire integral vanishes. We never *have* to use symmetry, however. Letting $ds = R\, d\theta$, and noting that $x = R \cos \theta$,

$$\pi R \bar{x} = \int_{\theta=0}^{\pi} (R \cos \theta) R\, d\theta$$

$$= R^2 \sin \theta \Big|_0^{\pi}$$

$$= 0$$

or

$$\bar{x} = 0$$

The z coordinate of C is also zero, this time because the line is in the xy plane, so that the z coordinate of each "ds" is zero. Thus $\int z\, ds = \int (0)\, ds = 0$.

Thus the only coordinate not immediately obvious is $\bar{y}$. Noting that the y coordinate of ds is $R \sin \theta$, we find, from $s\bar{y} = \pi R \bar{y} = \int y\, ds$,

$$\bar{y} = \frac{\int_{\theta=0}^{\theta=\pi} yR\, d\theta}{\pi R} = \frac{\int_0^{\pi} R^2 \sin \theta\, d\theta}{\pi R} = \frac{R^2(-\cos \theta)\Big|_0^{\pi}}{\pi R} = \frac{2R}{\pi}$$

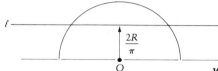

Figure E7.2b

which is slightly less than two-thirds of the way from O to the top of the semicircle, as shown in Figure E7.2b.

Question 7.3   Does the arclength above the line ℓ equal that below?

Answer 7.3   No. The *integrals* of arclength *times* y-distance are what are balanced; i.e., the "first moments" are balanced above and below the line ℓ.

EXAMPLE 7.3

Find the centroid of the triangular area shown in Figure E7.3a.

Solution

The differential area dA may be taken to be $Y\, dx$, as suggested by the shaded strip in the figure.

$$A\bar{x} = \int x\, dA$$

$$\frac{BH}{2}\bar{x} = \int_0^B x(Y\, dx)$$

But Y is a function of x, given by $Y = \dfrac{H}{B}x$; therefore,

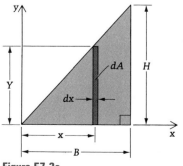

Figure E7.3a

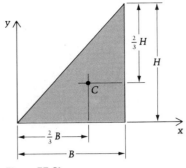

Figure E7.3b

$$\left(\frac{BH}{2}\right)\bar{x} = \int_0^B x\left(\frac{H}{B}x\right)dx$$

$$= \frac{HB^2}{3}$$

or

$$\bar{x} = \frac{2}{3}B$$

Therefore the centroid of a right triangle is $\frac{2}{3}$ of the distance from a vertex on either end of the hypotenuse to the opposite side, as shown in Figure E7.3b:

Question 7.4   How do we know the y coordinate of C without integrating?

We shall later see in Example 7.12 that the "$\frac{2}{3}$ rule" stated above is in fact true for *any* triangle.

Answer 7.4   Relabeling dimensions, $\bar{x} = 2H/3$ in the figure.

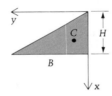

EXAMPLE 7.4

Find the centroid of the shaded area beneath the parabola shown in Figure E7.4a.

Solution

Using the vertical strip as shown in Figure E7.4b, we have:

$$A\bar{x} = \int x\,dA$$

$$= \int_0^B x(Y\,dx)$$

$$= \int_0^B x\left(\frac{H}{B^2}x^2\right)dx$$

or

$$A\bar{x} = \frac{HB^2}{4}$$

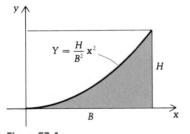

Figure E7.4a

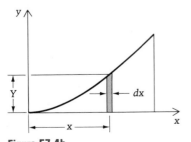

Figure E7.4b

But

$$A = \int_0^B Y\, dx$$

$$= \int_0^B \frac{H}{B^2} x^2\, dx$$

$$A = \frac{BH}{3}$$

so that $\bar{x} = \frac{3}{4} B$.

For the y coordinate, we may still use the above area element $Y\, dx$, provided we recognize that *its own centroid* is at $Y/2$ (see Example 7.1). We shall "add up" the areas times the y coordinates of the centroids of the vertical strips:

$$A\bar{y} = \int \frac{Y}{2}\, dA \qquad (\text{again, } Y/2 \text{ is the centroid distance for "}dA\text{"})$$

Thus

$$\bar{y} = \frac{\displaystyle\int \frac{Y}{2} Y\, dx}{BH/3}$$

$$= \frac{3}{2BH} \int_{x=0}^B \frac{H^2}{B^4} x^4\, dx$$

So

$$\bar{y} = \frac{3}{2BH} \frac{H^2}{B^4} \frac{B^5}{5} = \frac{3}{10} H$$

The student is encouraged to confirm this result either by double integration or by single integration using a horizontal strip dA.

EXAMPLE 7.5

Find the location of the centroid of the area enclosed by the semicircle and the (diametral) x axis as shown in Figure E7.5a.

Solution

For the same reasons as in Example 7.2, $\bar{x} = \bar{z} = 0$. To find $\bar{y}$, we again use Equation (7.1) where this time the quantity Q is area; and we shall also use the horizontal strip, shown in Figure E7.5b, for our "dA." We get, then, with $dA = 2\sqrt{R^2 - y^2}\, dy$,

$$A\bar{y} = \int y\, dA$$

$$\frac{\pi R^2}{2} \bar{y} = \int_0^R y(2\sqrt{R^2 - y^2}\, dy)$$

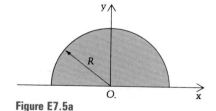

Figure E7.5a

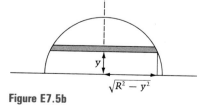

Figure E7.5b

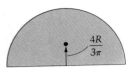

Figure E7.5c

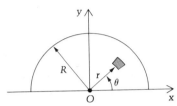

Figure E7.5d

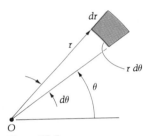

Figure E7.5e

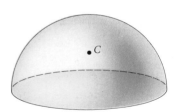

Figure E7.6a

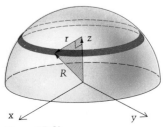

Figure E7.6b

Thus

$$\frac{\pi R^2}{2}\,\bar{y} = \int_0^R 2y\,\sqrt{R^2 - y^2}\,dy = \left.\frac{(R^2 - y^2)^{3/2}}{-\frac{3}{2}}\right|_0^R$$

$$= \frac{2}{3}\,R^3$$

And therefore

$$\bar{y} = \frac{4R}{3\pi}$$

This centroid (see Figure E7.5c) is lower than the centroid of the semicircle (see Example 7.2).

Question 7.5   Explain this, from the standpoint of where the quantities Q (arclength, area) are located in the two problems.

For the reader familiar with double integration, polar coordinates (see Figure E7.5d) offer a nice alternative to the solution above. This time, the differential area is seen in Figure E7.5e to be $dA = r\,dr\,d\theta$. We obtain

$$A\bar{y} = \int y\,dA$$

$$\left(\frac{\pi R^2}{2}\right)\bar{y} = \int_0^\pi \int_0^R (r \sin\theta)r\,dr\,d\theta$$

$$= \int_0^\pi \left.\frac{r^3}{3}\right|_0^R \sin\theta\,d\theta$$

$$= \frac{R^3}{3}\,(-\cos\theta)\Big|_0^\pi$$

$$= \frac{2R^3}{3}$$

$$\bar{y} = \frac{4R}{3\pi} \qquad \text{(as before)}$$

Answer 7.5   This time it is "y-distance times *area*," which is balanced above and below the horizontal line through the centroid. The semicircle is now "filled-in" and there is much more quantity Q (area) lower than there was when Q was arclength. This time, the "first moments of *area*" are balanced.

EXAMPLE 7.6

Find the centroid of the volume of a hemisphere of radius R (see Figure E7.6a).

Solution

By symmetry it is clear that $\bar{x} = \bar{y} = 0$. To find $\bar{z}$, we shall use the volume element shown in Figure E7.6b, which is a disk having the differential volume dV

$= \pi r^2\, dz$. But we see from the shaded triangle that

$$R^2 = r^2 + z^2$$

so that as a function of z, the radius is expressible as

$$r = \sqrt{R^2 - z^2}$$

Therefore, because all points of the disk have equal z coordinates,

$$V\bar{z} = \int z\, dV = \int z\pi r^2\, dz$$

$$= \pi \int_0^R z(R^2 - z^2)\, dz = \pi \int_0^R (R^2 z - z^3)\, dz$$

$$= \pi \left(\frac{R^2 z^2}{2} - \frac{z^4}{4} \right) \Big|_0^R$$

$$= \frac{\pi R^4}{4}$$

Because the volume V is $\frac{1}{2}(\frac{4}{3}\pi R^3)$, we obtain

$$\bar{z} = \frac{\pi R^4 / 4}{2\pi R^3 / 3} = \frac{3}{8} R$$

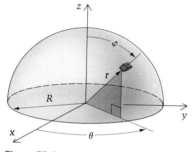

Figure E7.6c

Question 7.6 Why should this answer of $\dfrac{3}{8} R$ be less than the $4R/3\pi$ for the semicircular area?

The centroid of the hemisphere may also be computed using the spherical coordinates (r, φ, θ) (see Figure E7.6c). For those familiar with multiple integration, this alternative approach is presented next:

$$V\bar{z} = \int z\, dV = \int_0^{2\pi} \int_0^{\pi/2} \int_0^R \overbrace{(r \cos \varphi)}^{z}\, \overbrace{(r^2 \sin \varphi\, dr\, d\varphi\, d\theta)}^{dV}$$

$$= \frac{R^4}{4} \int_0^{2\pi} \int_0^{\pi/2} \sin \varphi \cos \varphi\, d\varphi\, d\theta$$

$$= \frac{R^4}{4} \int_0^{2\pi} \left[\frac{\sin^2 \varphi}{2} \right]_0^{\pi/2} d\theta$$

$$= \frac{R^4}{8} \int_0^{2\pi} d\theta$$

$$= \frac{\pi R^4}{4}$$

Answer 7.6 This time it is the first moments of *volume* that are balanced above and below the plane $z = \frac{3}{8}R$. There is proportionately much more volume near the plane $z = 0$ than there was area near the line $y = 0$ in Example 7.5.

so that

$$\bar{z} = \frac{\pi R^4/4}{2\pi R^3/3} = \frac{3}{8}R \qquad \text{(as before)}$$

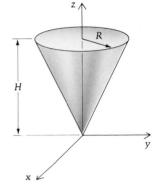

Figure E7.7a

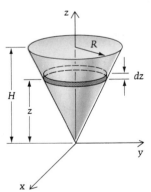

Figure E7.7b

EXAMPLE 7.7

Find the centroid of the conical volume shown in Figure E7.7a.

Solution

As in the previous example we shall use a disk as the element of volume for integration. The radius of the disk is $(R/H)z$ so that the element of volume is $\pi[(R/H)z]^2\, dz$ (see Figure E7.7b). We take this opportunity to derive the cone's volume:

$$V = \int_0^H \pi\left(\frac{R}{H}z\right)^2 dz$$

$$= \frac{\pi R^2}{H^2}\int_0^H z^2\, dz$$

$$= \frac{\pi R^2 H}{3}$$

Noting next that $\bar{x} = \bar{y} = 0$ by symmetry, we proceed to compute $\bar{z}$:

$$V\bar{z} = \int z\, dV$$

$$= \int_0^H z\left(\frac{\pi R^2}{H^2}z^2\right) dz$$

$$= \frac{\pi R^2 H^2}{4}$$

Thus

$$\bar{z} = \frac{\dfrac{\pi R^2 H^2}{4}}{\dfrac{\pi R^2 H}{3}} = \frac{3}{4}H$$

Therefore the centroid of a conical volume is on its axis, three-fourths of the distance from the vertex to the base.

The reader familiar with multiple integration is encouraged to reproduce this result using cylindrical coordinates.

Use of Centroids in Expressing Resultants of Distributed Forces

We close this section by reviewing some results from Section 3.9 about distributed forces. In that section, we saw that for a distributed line loading of parallel forces, the resultant force F_r equals the "area" beneath

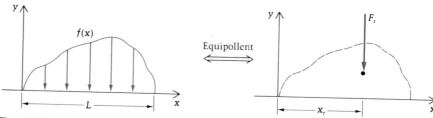

Figure 7.2

the loading diagram (see Figure 7.2), with a line of action given by Equation (3.25):

$$x_r = \frac{\int_0^L xf(x)\,dx}{\int_0^L f(x)\,dx} \qquad (7.4)$$

By comparing Equations (7.4) and the first of (7.3), we now see clearly what we were only able to mention in Chapter 3 — namely, that the force-alone resultant acts through the centroid of the "area" beneath the loading curve.

For pressure acting on a plane surface we found, in Section 3.9, a force-alone resultant as indicated in Figure 7.3:

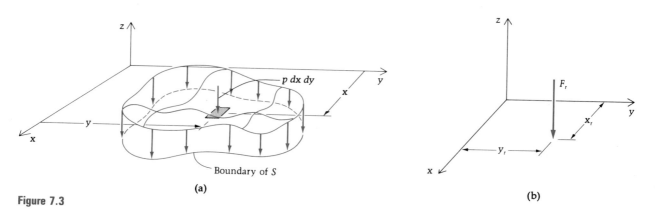

Figure 7.3

which was

$$F_r = \int p(x,\,y)\,dA$$

with a line of action given by $(x_r,\,y_r)$:

$$F_r x_r = \int xp(x,\,y)\,dA \qquad \text{and} \qquad F_r y_r = \int yp(x,\,y)\,dA$$

Thus we see that the resultant force is, this time, the "volume" of the region for which the loaded plane surface is the base and the altitude is the varying pressure $p(x, y)$. Furthermore, the resultant has a line of action through the centroid of this volume.

A special case of the above with great practical importance arises when the pressure is uniform. Then, we obtain

$$F_r = p \int dA = pA$$

and

$$Ax_r = \int x \, dA$$

$$Ay_r = \int y \, dA$$

so that the resultant force has a line of action through the centroid of the plane area on which the pressure acts.

EXAMPLE 7.8

Find the force-alone resultant of the distributed line loading shown in Figure E7.8a, and its line of action.

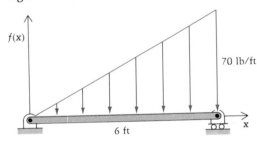

Figure E7.8a

Solution

The resultant is:

$$F_r = \underbrace{\frac{\text{Area beneath}}{\text{loading curve}}}_{} \quad \left(\text{or} \quad \int_0^6 f(x) \, dx \right)$$

$$\underbrace{}_{\frac{70}{6}x}$$

$$= \frac{1}{2} (6)70 \qquad \left(\text{or} \quad \frac{70}{6} \frac{x^2}{2} \Big|_0^6 \right)$$

$$= 210 \text{ lb}$$

This force acts downward (parallel to the given loading) at the centroid of the area beneath the loading curve.

$$x_r = \frac{2}{3}(6) \qquad \left(\text{or } \frac{\int_0^6 \overbrace{xf(x)}^{\frac{70}{6}x} \, dx}{210} = \frac{\frac{70}{6} \frac{x^3}{3} \Big|_0^6}{210} \right)$$

$$= 4 \text{ ft}$$

Therefore, we have found the results shown in Figure E7.8b with the aid of centroids:

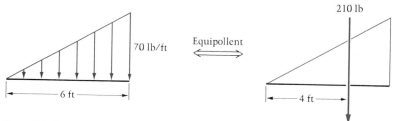

Figure E7.8b

It is interesting to contrast this problem with Example 3.30, which consisted of the six forces shown in Figure E7.8c, and which was shown to be equipollent to the single force shown in Figure E7.8d. We see that even though the six forces sum to the same 210 pounds, and even though the forces' magnitudes are proportional to their distances from O, the single force resultants of the two systems are *not the same because the lines of action differ.*

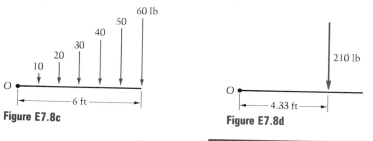

Figure E7.8c

Figure E7.8d

PROBLEMS ▶ Section 7.2

7.1   Show that the centroid C of the area in Figure P7.1 does not depend upon the choice of origin in the reference frame. *Hint:* Consider the two centroids resulting from Equation (7.1) with two separate origins O_1 and O_2

$$\mathbf{r}_{O_1 C_1} = \frac{1}{A} \int \mathbf{r}_1 \, dA, \qquad \mathbf{r}_{O_2 C_2} = \frac{1}{A} \int \mathbf{r}_2 \, dA$$

and relate $\mathbf{r}_1$ to $\mathbf{r}_2$. Show then that $\mathbf{r}_{O_1 C_1} = \mathbf{r}_{O_1 C_2}$, which means that C_1 and C_2 are the same point.

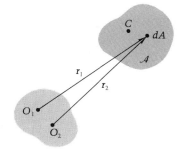

Figure P7.1

7.2   Find the centroid of the area bounded by the parabola and the line $x = a$ in Figure P7.2.

7.3   Find the centroid of the circular arc shown in Figure P7.3.

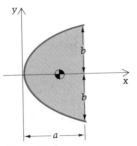

Figure P7.2

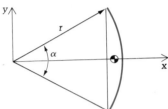

Figure P7.3

7.4   Which of the following expressions may be used to find the centroid location $\bar{y}$ of the area under the curve $y = Y(x)$? (See Figure P7.4.)

(a) $\dfrac{\displaystyle\int_0^{x_1} xY\,dx}{\displaystyle\int_0^{x_1} Y\,dx}$

(b) $\dfrac{\frac{1}{2}\displaystyle\int_0^{x_1} xY\,dx}{\displaystyle\int_0^{x_1} Y\,dx}$

(c) $\dfrac{\displaystyle\int_0^{x_1} xY\,dx}{\displaystyle\int_0^{x_1} Y^2\,dx}$

(d) $\dfrac{\frac{1}{2}\displaystyle\int_0^{x_1} Y^2\,dx}{\displaystyle\int_0^{x_1} Y\,dx}$

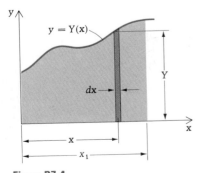

Figure P7.4

7.5   In Problem 7.4, find $\bar{y}$ if $Y(x) = e^x$ and $x_1 = 2$.

In Problems 7.6–7.9 find the centroids of the shaded areas.

7.6

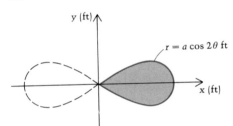

Figure P7.6

* 7.7

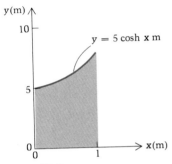

Figure P7.7

7.8

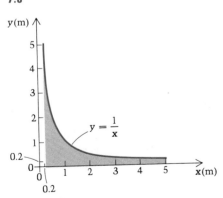

Figure P7.8

---

* Asterisks identify the more difficult problems.

7.9

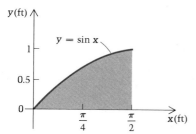

Figure P7.9

7.10 Find the y coordinate of the centroid of the shaded area shown in Figure P7.10.

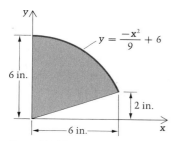

Figure P7.10

7.11 In Problem 7.10, find the x coordinate of the centroid.

In Problems 7.12 and 7.13 find the x coordinates of the centroids of the shaded areas.

7.12

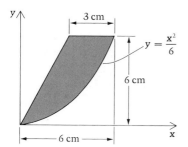

Figure P7.12

7.13

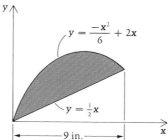

Figure P7.13

7.14 Find the y coordinate of the centroid of the shaded area in Problem 7.12.

7.15 Find the y coordinate of the centroid of the shaded area in Problem 7.13.

In Problems 7.16–7.20 find the x coordinates of the centroids of the *lines* by integration. *Hint* for 7.17–7.19: $ds = \sqrt{dx^2 + dy^2} = \sqrt{1 + (y')^2}\, dx$.

7.16

Figure P7.16

7.17

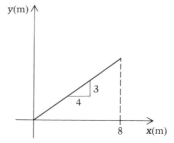

Figure P7.17

7.18

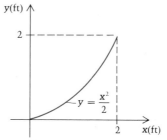

Figure P7.18

For Problems 7.19 and 7.20, see instructions above Problem 7.16.

7.19

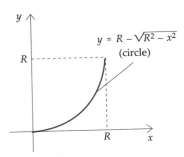

Figure P7.19

7.20

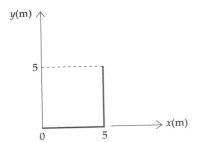

Figure P7.20

7.21 Find by integration the y-coordinate of the centroid of the straight line in Problem 7.17.

7.22 Find by integration the y-coordinate of the centroid of the bent line in Problem 7.20.

7.23 Find the centroid of the shaded area in Figure P7.23.

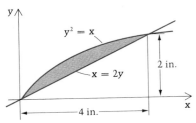

Figure P7.23

7.24 Find the centroid of the shaded area shown in Figure P7.24.

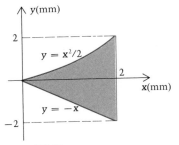

Figure P7.24

7.25 Determine the x coordinate of the centroid of the shaded area in Figure P7.25. The equation of the curve is $y^2 = x$.

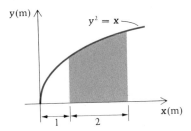

Figure P7.25

7.26 Find the centroid of the solid of revolution shown in Figure P7.26.

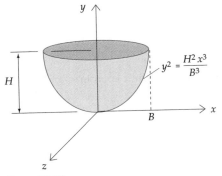

Figure P7.26

7.27 Find the centroid of the volume of the paraboloid of revolution shown in Figure P7.27.

* 7.28 In Problem 7.27 find the centroid of the outer surface area of the paraboloid. *Hint:* You'll need integral tables!

---

* Asterisks identify the more difficult problems.

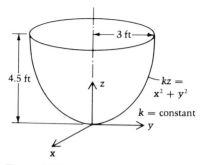

Figure P7.27

7.29 Show that the centroid of the area under the "nth degree parabola" in Figure P7.29 is at

$$(\bar{x}, \bar{y}) = \left(\frac{n+1}{2n+1} b, \frac{n+1}{2(n+2)} h \right),$$

and that the area is

$$\frac{n}{n+1} bh.$$

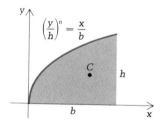

Figure P7.29

7.30 Find the centroid of the volume of the pyramid shown in Figure P7.30.

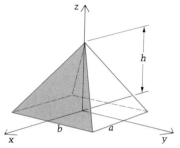

Figure P7.30

7.31 Find the centroid of the volume of the tetrahedron shown in Figure P7.31.

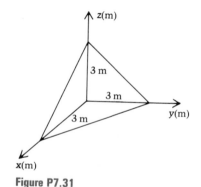

Figure P7.31

7.3 The Method of Composite Parts

When two (or more) familiar areas* A_1 and A_2 are to be considered together as a single area, we do not have to integrate over this combined area in order to find the centroid. Fortunately, an approach called the method of composite parts allows us to use the known areas and centroid locations of the separate parts to establish the centroid of the combined area ($A_1 + A_2 = A$).

The theory goes as follows, with areas A_1 and A_2 having respective centroidal x coordinates $\bar{x}_1$ and $\bar{x}_2$:

$$A\bar{x} = \int_R x\, dA$$

---

* Or volumes or lines, as the case may be.

But the integral over region $\mathcal{R}$ may be taken separately over subregions $\mathcal{R}_1$ and $\mathcal{R}_2$ for which the areas are, respectively, A_1 and A_2:

$$(A_1 + A_2)\bar{x} = \int_{\mathcal{R}_1} x \, dA + \int_{\mathcal{R}_2} x \, dA$$

Now separately invoking the basic centroid equation (7.3) over each of A_1 and A_2, we obtain

$$(A_1 + A_2)\bar{x} = A_1\bar{x}_1 + A_2\bar{x}_2 \qquad (7.5)$$

and we see that all need for integration has been circumvented if we know A_1, A_2, $\bar{x}_1$, and $\bar{x}_2$. Any number of areas may be combined in this way.

In the more general form (where the quantity whose centroid we seek is any scalar Q), the steps to the similar vector expression are

$$Q\mathbf{r}_{OC} = \int \mathbf{r} \, dQ$$

$$= \int_{\mathcal{R}_1} \mathbf{r} \, dQ + \int_{\mathcal{R}_2} \mathbf{r} \, dQ$$

$$= Q_1\mathbf{r}_{OC_1} + Q_2\mathbf{r}_{OC_2}$$

where

$$Q = Q_1 + Q_2$$

We now consider a number of examples of the use of "composite parts."

EXAMPLE 7.9

Find the centroid of the three line segments connected as shown in Figure E7.9a.

Solution

Labeling the segments as shown, we have, using composite parts,

$$(h + b + h)\bar{y} = L_1\bar{y}_1 + L_2\bar{y}_2 + L_3\bar{y}_3$$

$$= h\left(\frac{h}{2}\right) + b(0) + h\left(\frac{h}{2}\right)$$

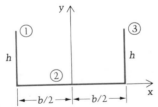

Figure E7.9a

where the middle term vanishes because the y coordinate of the centroid of element ② is zero. Therefore,

$$(2h + b)\bar{y} = \frac{h^2}{2} + \frac{h^2}{2} = h^2$$

so that

$$\bar{y} = \frac{h^2}{2h + b}$$

To illustrate the vanishing of "first moments" when measurements are made from the centroid, consider the special case in this example for which $b = h$;

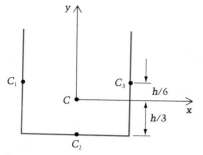

Figure E7.9b

then $\bar{y} = h/3$. Now let us set up *new* axes with origin at the centroid as shown in Figure E7.9b.

Relative to these new axes, we see that the "first moment" is

$$\int y\,ds = h(h/6) + h(-h/3) + h(h/6)$$

$$= 0$$

as, of course, must be the case since the y coordinate of the centroid in this system is zero.

EXAMPLE 7.10

The circles in Figure E7.10 represent cross sections of bolts, with the bolts holding down a gear-box cover whose shape is the curve shown. It is important to locate the centroid C of the bolt pattern because, for example, the distance from C to each bolt is used both to compute the stiffness of the overall pattern against twisting and also to calculate the shear stress in each of the bolts. Letting each bolt's cross-sectional area be A, find the centroid of the pattern of ten identical bolts.

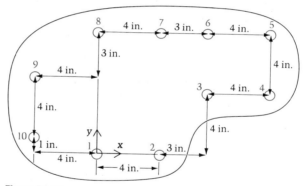

Figure E7.10

Solution
Using composite areas,

$$\left(\sum_{i=1}^{10} A_i\right)\bar{x} = \sum_{i=1}^{10} A_i\bar{x}_i$$

$$10A\bar{x} = A\Sigma\bar{x}_i$$

$$\bar{x} = \frac{\Sigma\bar{x}_i}{10}$$

and similarly,

$$\bar{y} = \frac{\Sigma\bar{y}_i}{10}$$

while $\bar{z}$ is zero because all of the areas lie in the x-y plane.

Substituting (using the bolt-numbering system shown in the figure), and moving counterclockwise from bolt ① at the origin,

$$\bar{x} = (0 + 4 + 7 + 11 + 11 + 7 + 4 + 0 - 4 - 4)/10$$

$$\bar{x} = 3.60 \text{ in.}$$

Question 7.7   Why do the contributions of bolts 2, 7, 9, and 10 cancel in the $\bar{x}$ computation?

Continuing,

$$\bar{y} = (0 + 0 + 4 + 4 + 8 + 8 + 8 + 8 + 5 + 1)/10$$

$$\bar{y} = 4.60 \text{ in.}$$

Therefore $(\bar{x}, \bar{y}, \bar{z}) = (3.60, 4.60, 0)$ in.

Answer 7.7   Because, with the four areas equal, two $+4$ coordinates cancel two -4 coordinates.

EXAMPLE 7.11

Find the centroid of the cross-sectional area (of a channel beam) shown in Figure E7.11a.

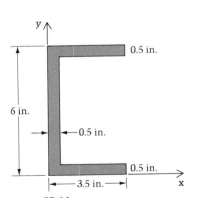

Figure E7.11a

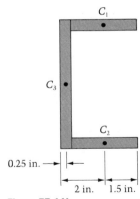

Figure E7.11b

Solution

The axis of symmetry ($y = 3$ in.) allows us to conclude by inspection that $\bar{y} = 3$ in. We shall determine the x coordinate of C in two ways. First we note that the area in question is the composite of three rectangles as shown in Figure E7.11b, and we obtain:

$$A\bar{x} = A_1\bar{x}_1 + A_2\bar{x}_2 + A_3\bar{x}_3$$

$$\bar{x} = \frac{[3(0.5)](2) + [3(0.5)](2) + [6(0.5)](0.25)}{3(0.5) + 3(0.5) + 6(0.5)}$$

$$= \frac{6.75}{6} = 1.12 \text{ in.}$$

An alternative approach is to note that the area of interest is obtained by removing a 3 in. × 5 in. rectangle from a 3.5 in. × 6 in. rectangle as shown in Figure E7.11c below:

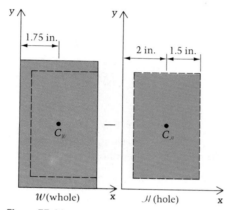

Figure E7.11c

Applying the method of composite parts to the whole (3.5 in. × 6 in.) rectangle, we have

$$A_w\bar{x}_w = A\bar{x} + A_{_H}\bar{x}_{_H}$$

where A and $\bar{x}$ refer to the (net) area of interest. Rearranging,

$$A\bar{x} = A_w\bar{x}_w - A_{_H}\bar{x}_{_H}\text{*}$$

$$[(6)(3.5) - 5(3)]\bar{x} = [6(3.5)](1.75) - [5(3)](2)$$

or

$$\bar{x} = (3.5)(1.75) - 5$$

$$= 6.12 - 5$$

$$= 1.12 \text{ in.} \qquad \text{(as before)}$$

---

* Some like to see this as a direct application of the method of composite parts by thinking of the "hole" as a negative area; that is,

$$A\bar{x} = A_w\bar{x}_w + (-A_{_H})\bar{x}_{_H}$$

EXAMPLE 7.12

Show that the centroid of *any* triangle is on the line that is two-thirds the distance from any vertex V to the opposite side and parallel to that side.

Solution

This example provides an opportunity to obtain a useful, general result without using integration. Consider the actual shaded, scalene triangle of Figure E7.12a to be made up of the right triangle OVD less the smaller right triangle EVD. Then, using what we have learned about the centroids of right triangles,

$$A_{OVD}\bar{y}_{OVD} = A\bar{y} + A_{EVD}\bar{y}_{EVD}$$

or

$$A\bar{y} = A_{OVD}\bar{y}_{OVD} - A_{EVD}\bar{y}_{EVD}$$

$$\left(\frac{1}{2}BH\right)\bar{y} = \left[\frac{1}{2}(B+Q)H\right]\frac{H}{3} - \left(\frac{1}{2}QH\right)\frac{H}{3}$$

$$\bar{y} = \frac{BH^2/6}{BH/2} = \frac{H}{3} \qquad \text{(again)}$$

This means that all three of the lines ℓ_1, ℓ_2, ℓ_3 below will intersect at the centroid C of the triangle, as shown in Figure E7.12b.

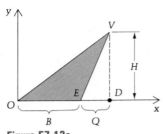

Figure E7.12a

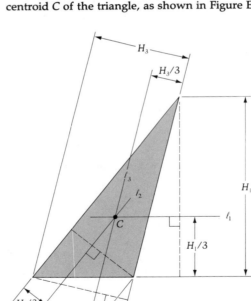

Figure E7.12b

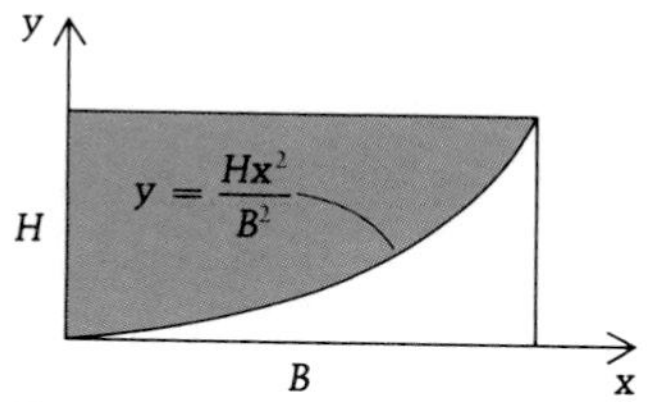

EXAMPLE 7.13

Use the method of composite areas and the result of Example 7.4 to find the centroid of the shaded area in Figure E7.13a.

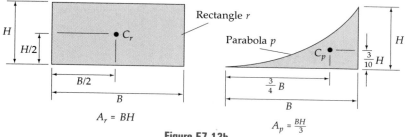

$$y = \frac{Hx^2}{B^2}$$

Figure E7.13a

Figure E7.13b

Solution

The shaded area is obtained by subtracting the area beneath the parabola (p) in Figure E7.13b from the rectangle (r). Their areas and centroid locations are shown in the figures. Therefore,

$$A = A_r - A_p$$

$$= BH - \frac{BH}{3} = \frac{2}{3}BH$$

and

$$A_r \bar{x}_r = A\bar{x} + A_p \bar{x}_p$$

$$(BH)\left(\frac{B}{2}\right) = \left(\frac{2}{3}BH\right)\bar{x} + \left(\frac{BH}{3}\right)\left(\frac{3}{4}B\right)$$

$$\bar{x} = \frac{3}{8}B$$

The same procedure will yield the y coordinate of the centroid:

$$A_r \bar{y}_r = A\bar{y} + A_p \bar{y}_p$$

$$(BH)\left(\frac{H}{2}\right) = \left(\frac{2}{3}BH\right)\bar{y} + \left(\frac{BH}{3}\right)\left(\frac{3}{10}H\right)$$

$$\bar{y} = \frac{3}{5}H$$

EXAMPLE 7.14

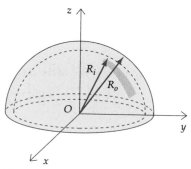

Figure E7.14

Locate the centroid of the volume of a hemispherical shell of inner radius R_i and outer radius R_o (see Figure E7.14).

Solution

From symmetry we recognize that $\bar{x} = \bar{y} = 0$. The volume in question may be viewed as a hemisphere of radius R_o out of which a hemisphere of radius R_i has been removed. From Example 7.6, we know that the z coordinates of centroids for these are $\bar{z}_o = \frac{3}{8}R_o$ and $\bar{z}_i = \frac{3}{8}R_i$. Using the method of composite parts,

$$V_o\bar{z}_o = V\bar{z} + V_i\bar{z}_i$$

or

$$V\bar{z} = V_o\bar{z}_o - V_i\bar{z}_i$$

$$\left(\frac{2}{3}\pi R_o^3 - \frac{2}{3}\pi R_i^3\right)\bar{z} = \left(\frac{2}{3}\pi R_o^3\right)\left(\frac{3}{8}R_o\right) - \left(\frac{2}{3}\pi R_i^3\right)\left(\frac{3}{8}R_i\right)$$

$$\bar{z} = \frac{3(R_o^4 - R_i^4)}{8(R_o^3 - R_i^3)} = \frac{3(R_o^2 + R_i^2)(R_o + R_i)(R_o - R_i)}{8(R_o^2 + R_oR_i + R_i^2)(R_o - R_i)} \qquad (1)$$

Note that if we set $R_i = 0$, we recover $\bar{z} = \frac{3}{8}R_o$.

A limiting case sometimes of interest is that for which $R_i \approx R_o$, where the volume of interest is a thin shell. In that circumstance we obtain from Equation (1)

$$\bar{z} \approx \frac{3(2)R_o^2(2R_o)}{8(3R_o^2)} = \frac{R_o}{2}$$

which is precisely the z coordinate of the centroid of the (curved) surface *area* of the hemisphere. (The student is encouraged to verify this by direct calculation.)

Question 7.8   Is this a coincidence?

Answer 7.8   No. As the thickness gets smaller and smaller (let $R_i \to R_o$), the volume elements get closer and closer to the surface of the outer hemisphere.

PROBLEMS ▶ Section 7.3

7.32 Find the centroid of the shaded area in Figure P7.32.

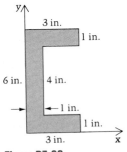

Figure P7.32

7.33 Find the y coordinate of the centroid of the T-shaped area shown in Figure P7.33.

7.34 In Figure P7.34 locate the centroid of the shaded area.

7.35 In Figure P7.35 show that the centroid of the trapezoidal area has a y coordinate of $\dfrac{H}{3}\left(\dfrac{a + 2b}{a + b}\right)$, independent of the angles α_1 and α_2.

7.36 Find the centroid of the area shown in Figure P7.36.

7.37 Find the centroid of the area shown in Figure P7.37.

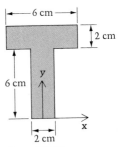

Figure P7.33

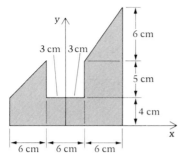

Figure P7.34

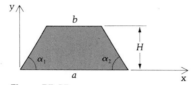

Figure P7.35

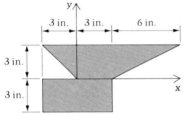

Figure P7.36

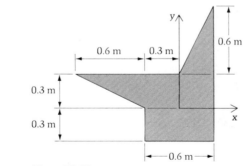

Figure P7.37

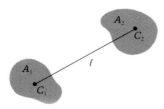

Figure P7.38

7.38 In Figure P7.38, A_1 and A_2 are two areas with centroids C_1 and C_2 separated by the distance l as shown. Show that the centroid of the composite area (A_1 plus A_2) lies on line C_1C_2, and is a distance $\dfrac{A_2}{A_1 + A_2}\, l$ from C_1.

7.39 Find the centroid of the three *lines* in Figure P7.39, and compare its location with the centroid of the enclosed *area*.

7.40 Find the centroid of the tee rail cross section shown in Figure P7.40. (We are neglecting the fact that the web sides and the top of the head are actually segments of large circles for this particular rail.)

7.41 Locate the centroid of the shaded area in Figure P7.41.

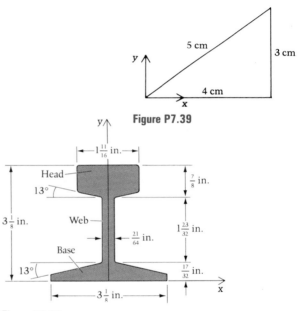

Figure P7.39

Figure P7.40

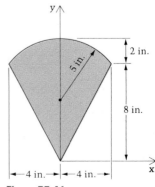

Figure P7.41

7.42 In Figure P7.42 locate the centroid of the homogeneous square plate with a triangular cut-out.

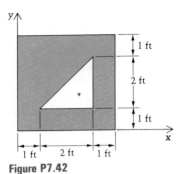

Figure P7.42

7.43 Find the centroid of the shaded area in Figure P7.43.

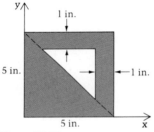

Figure P7.43

7.44 Find the centroid of the shaded area in Figure P7.44.

7.45 Find the centroid of the shaded area in Figure P7.45.

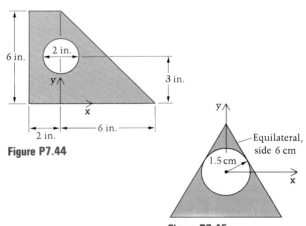

Figure P7.44

Figure P7.45

7.46 Find the centroid of the shaded area in Figure P7.46.

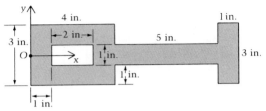

Figure P7.46

7.47 Find the centroid of the shaded area in Figure P7.47.

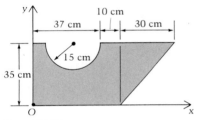

Figure P7.47

7.48 Find the x and y coordinates of the centroid of the shaded area in Figure P7.48.

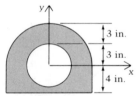

Figure P7.48

7.49 The circular cutout shown in Figure P7.49 has an area of 12 in.^{2} Find the centroid of the shaded area.

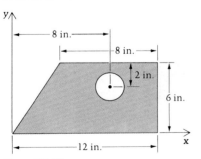

Figure P7.49

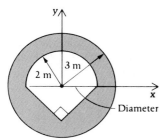

Figure P7.50

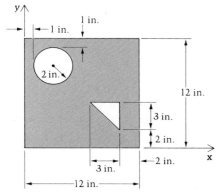

Figure P7.51

7.50 Find the centroid of the shaded area in Figure P7.50.

7.51 In Figure P7.51 find the centroid of the shaded square, which has a circular and triangular cutout as shown.

7.52 Find the centroid of the shaded area in Figure P7.52.

7.53 Find the centroid of the shaded area in Figure P7.53.

7.54 Find the y coordinate of the centroid of the shaded area in Figure P7.54.

7.55 In Figure P7.55 find the centroid of the shaded triangular area with the semicircular cutout shown.

7.56 In Figure P7.56 the holes in the trapezoidal area have 1-in. radii. Find the x coordinate of the centroid of the area.

7.57 Find the centroid of the shaded area in Figure P7.57.

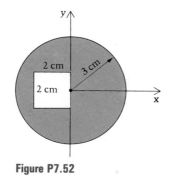

Figure P7.52

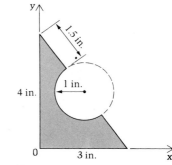

Figure P7.55

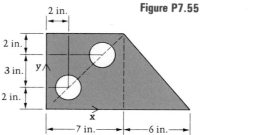

Figure P7.56

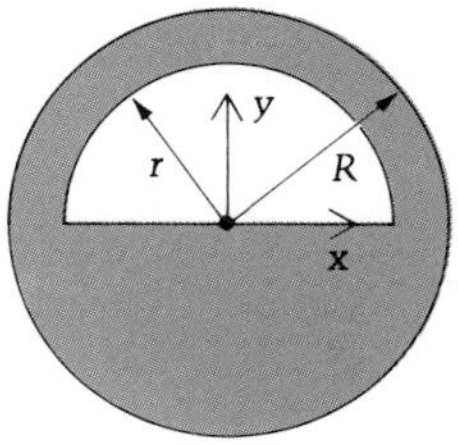

Figure P7.53

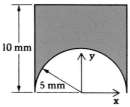

Figure P7.54

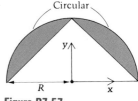

Figure P7.57

7.58 For the angle section shown in Figure P7.58, find the coordinates of the centroid.

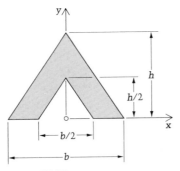

Figure P7.58

7.59 Find the centroid of the composite area shown in Figure P7.59.

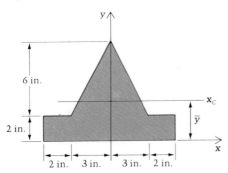

Figure P7.59

7.60 In Figure P7.60 find the centroid of the five circular areas. Each area is 0.75 in.^{2}

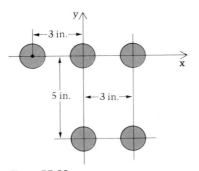

Figure P7.60

7.61 Find the centroid of the cross-sectional area of the structural steel angle beam shown in Figure P7.61. Assume a constant thickness of $\frac{3}{8}$ in., and compare your answers with the actual (considering rounded corners and fillets) values of $(\bar{x}, \bar{y}) = (1.03, 2.03)$ in.

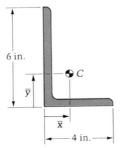

Figure P7.61

7.62 In Figure P7.62 determine the height h of the "T" (formed by two perpendicular lines as shown) in order that the centroid be 2 cm below the top.

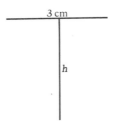

Figure P7.62

7.63 Find the centroid of the center line of the bar consisting of three straight sections parallel to the (x, y, z) axes as shown in Figure P7.63.

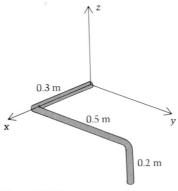

Figure P7.63

7.64 Rework Problem 7.20 using the method of composite parts.

7.65 Find the centroid of the "bent line" *ABCD* in Figure P7.65.

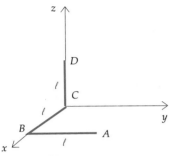

Figure P7.65

7.66 Find the centroid of the set of five lines shown in Figure P7.66.

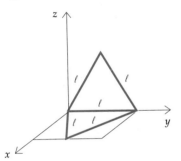

Figure P7.66

7.67 Find the centroid of the set of nine lines shown in Figure P7.67. (Which one of $\bar{x}, \bar{y}, \bar{z}$ can actually be seen by inspection?)

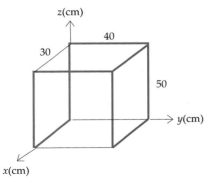

Figure P7.67

7.68 Find the centroid of the bent line *ABCDA* in Figure P7.68.

7.69 Find the centroid of the set of three areas shown in Figure P7.69. The triangle lies in the *yz*-plane, the rectangle in the *xy*-plane, and the semicircle in the *xz*-plane.

7.70 Locate the centroid of the enclosed area *ABCD* shown in Figure P7.70.

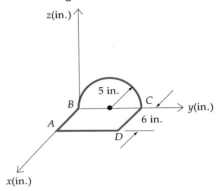

Figure P7.68

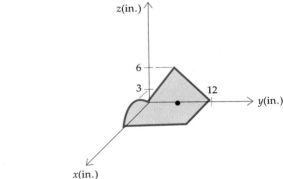

Figure P7.69

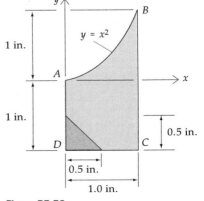

Figure P7.70

7.71 If the shaded triangle is removed from the area in the preceding problem, find the centroid of the remaining area.

7.72 Locate the centroid of the enclosed area shown in Figure P7.72.

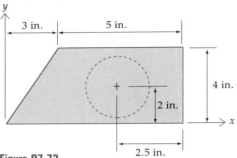

Figure P7.72

7.73 In the preceding problem, a circular hole of radius 1.5 inches is cut from the rectangular part of the plate at the location indicated by the dashed circle. Determine the new location of the centroid.

7.74 In Figure P7.74 determine the height h of the cutout rectangle that will place the centroid C of the *shaded* area at the position shown.

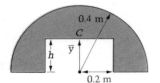

Figure P7.74

7.75 In Problem 7.33, to what value should the 6-cm vertical dimension be changed in order that the new centroid be on the line separating the two rectangular areas?

7.76 Locate the centroid of the shaded area shown in Figure P7.76.

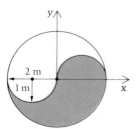

Figure P7.76

7.77 In Figure P7.77 find the centroid of the shaded area.

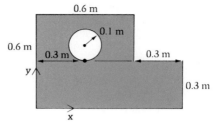

Figure P7.77

7.78 In Figure P7.78 find $\bar{x}$ for the shaded area, cut from a circle.

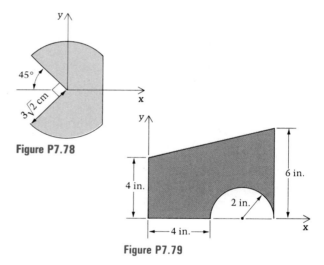

Figure P7.78

Figure P7.79

7.79 In Figure P7.79 find the centroid of the shaded area.

7.80 Find the area and the centroid of the *dotted* area in Figure P7.80.

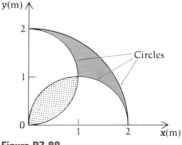

Figure P7.80

7.81 Find the centroid of the *shaded* area in Figure P7.80.

7.82 A circular plate with radius r_1 and unit thickness has five holes drilled in it as shown in Figure P7.82. The centers of the holes lie on an arc of radius r_2. The spacing between the centers is $60°$. Locate the centroid of the volume of the plate if $r_1 = 0.5$ m, $r_2 = 0.3$ m, and $r_3 = 0.1$ m.

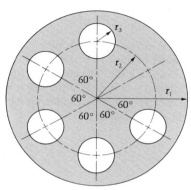

Figure P7.82

7.83 Show by (a) redrawing Figure P7.29 with the y-axis horizontal and x vertical; then (b) changing the names of x, y, b, and h; and finally (c) letting n in the answer be 2, that the centroid of the shaded area in Figure P7.83 is as shown, and that its area is $(2/3)bh$. Use the results of Problem 7.29, and note that the solution agrees with that of Example 7.13.

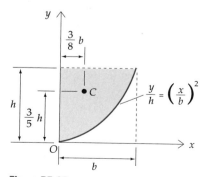

Figure P7.83

7.84 Use the results of the preceding problem together with the method of composite parts to show that the centroid of the dotted area beneath the parabolic segment in Figure P7.84 is $(\bar{x}, \bar{y}) = (3b/4, 3h/10)$ and that its area is $bh/3$, all in agreement with Example 7.4.

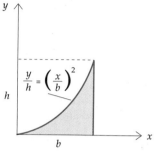

Figure P7.84

7.85 Using the results of the preceding problem and the method of composite parts, find the x-coordinate of the centroid of the shaded area in Figure P7.85.

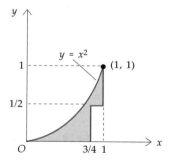

Figure P7.85

7.86 Using the results of Problem 7.84 and the method of composite parts, find the y-coordinate of the centroid of the same shaded area as in the preceding problem.

7.87 Find the ratio of R_i to R_o for which the centroid of the semicircular annulus in Figure P7.87 is at the highest point of the inner bounding circle.

Figure P7.87

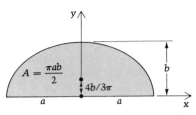

$$A = \frac{\pi ab}{2}$$

$4b/3\pi$

Figure P7.88a

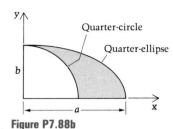

$$A = \frac{\pi R^2}{2}$$

$\frac{4R}{3\pi}$

R

Figure P7.88b

Quarter-circle

Quarter-ellipse

b

a

7.88 Given the two results in Figure P7.88a for the y coordinates of the centroids, use the method of composite areas to find the y coordinate of the centroid of the shaded area in Figure P7.88b. Do you really need the areas in this problem?

7.89 Find the x coordinate of the centroid in Problem 7.88. Are the areas needed now?

7.90 In Figure P7.90 show by integration that the centroid of the circular sector is given by

$$\bar{y} = \frac{2r \sin \alpha}{3\alpha}$$

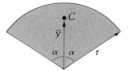

C

$\bar{y}$

α α r

Figure P7.90

C

$\bar{y}$

r

2α

Figure P7.92

7.91 In Problem 7.90 use L'Hospital's rule to show that as the sector becomes a sliver ($\alpha \to 0$), the result for $\bar{y}$ approaches that for a triangle, as it should.

7.92 Using the result of Problem 7.90, show by composite areas (sector minus triangle) that the centroid of the shaded circular segment shown in Figure P7.92 is given by

$$\bar{y} = \frac{2r \sin^3 \alpha}{3(\alpha - \sin \alpha \cos \alpha)}$$

* Show that your answer approaches r as $\alpha \to 0$, as it must.

* **7.93** Find the centroid of the shaded area in Figure P7.93.

7.94 Find the centroid of the shaded area in Figure P7.94.

7.95 The flat circular cam in Figure P7.95 rotates about an axis $\frac{3}{8}$ in. from the center of the 6-in. radius circle. To balance the cam — that is, place the centroid of its cross section at the axis of rotation — a hole of 2-in. radius is machined through the cam. Determine the distance b to the hole center that will accomplish this.

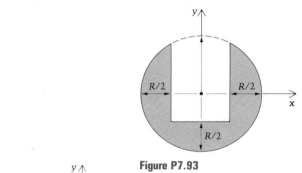

$R/2$ $R/2$

$R/2$

Figure P7.93

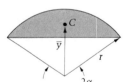

40 cm

40 cm

50 cm

30 cm

Figure P7.94

$\frac{3}{8}$ in.

b

6 in.

2 in.

Figure P7.95

* **7.96** Approximate the *y* coordinate of the centroid of the (shaded) cross-sectional area of the cam, which is symmetrical about the *y* axis. (See Figure P7.96.)

7.97 Find the centroid of the volume of the very thin piece of sheet metal in Figure P7.97.

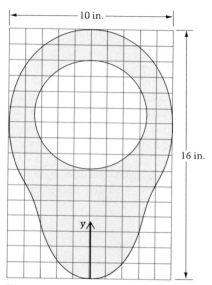

Figure P7.96

Top view

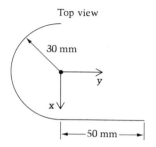

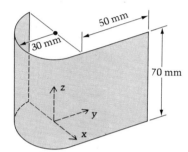

Figure P7.97

7.4 Center of Mass

The center of mass, or mass center, of a body $\mathcal{B}$ is the point C defined by Equation (7.1) when the quantity Q becomes mass:

$$m\mathbf{r}_{OC} = \int \mathbf{r}\, dm \qquad (7.6)$$

where $m = \int_{\mathcal{B}} dm$ is the mass of the body and the differential mass dm is related to its volume dV through the mass density ρ according to $dm = \rho\, dV$ (see Figure 7.4). Thus

$$m\mathbf{r}_{OC} = \int \mathbf{r}\rho\, dV \qquad (7.7)$$

If the density ρ is constant, then $m = \rho V$, where V is the volume of body $\mathcal{B}$ and

$$\rho V \mathbf{r}_{OC} = \rho \int \mathbf{r}\, dV$$

or

$$V\mathbf{r}_{OC} = \int \mathbf{r}\, dV \qquad (7.8)$$

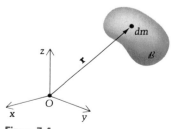

Figure 7.4

which is seen to be the equation defining the location of the centroid of the volume of $\mathcal{B}$. *Thus the mass center always coincides with the centroid of volume for a body of constant density.*

> **Question 7.9**   Can you think of a case in which the density is *not* constant and yet the two points coincide?

The mass center of a body plays a central role in dynamics, and the student will encounter it naturally there. In statics its significance is that, assuming a uniform field of gravity, then regardless of the orientation of a body $\mathcal{B}$ on the earth's surface, the resultant of the distributed gravity forces (weight) exerted on $\mathcal{B}$ by the earth will have a line of action intersecting the center of mass of $\mathcal{B}$.

We now consider some examples of locating mass centers. The first example contains a body with variable density, and the second uses composite parts.

Answer 7.9   A sphere whose density depends only upon the radius.

EXAMPLE 7.15

The mass density of a uniform, circular cylindrical rod of constant cross-sectional area A varies linearly with x as indicated in Figure E7.15. Find the center of mass.

Solution

The equation for the spatial variation of the density is

$$\rho = \frac{\rho_2 - \rho_1}{L} x + \rho_1$$

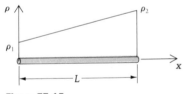

Figure E7.15

Therefore, with $\bar{y} = \bar{z} = 0$ by symmetry,

$$\underbrace{(\int \rho \, dV)}_{m} \bar{x} = \int \rho x \, dV$$

$$\left\{ \int \left[\left(\frac{\rho_2 - \rho_1}{L} \right) x + \rho_1 \right] A \, dx \right\} \bar{x} = \int \left[\left(\frac{\rho_2 - \rho_1}{L} \right) x + \rho_1 \right] x A \, dx$$

$$\left[\frac{\rho_2 - \rho_1}{L} \frac{x^2}{2} \Big|_0^L + \rho_1 x \Big|_0^L \right] \bar{x} = \left(\frac{\rho_2 - \rho_1}{L} \right) \frac{x^3}{3} \Big|_0^L + \rho_1 \frac{x^2}{2} \Big|_0^L$$

$$\left[(\rho_2 - \rho_1) \frac{L}{2} + \rho_1 L \right] \bar{x} = \left(\frac{\rho_2 - \rho_1}{L} \right) \frac{L^3}{3} + \rho_1 \frac{L^2}{2}$$

Therefore,

$$\bar{x} = \left(\frac{\rho_1 + 2\rho_2}{3(\rho_1 + \rho_2)} \right) L$$

As a check, if $\rho_1 = \rho_2$, then $\bar{x} = L/2$, as it should be for a uniform rod.

Answer 7.10   Yes; the answers are $\frac{2}{3}L$ and $\frac{1}{3}L$. These solutions are the same as for the thin plates △ and ◥ of uniform density. The solutions should match, for with these plates we have a linear variation of height instead of density.

EXAMPLE 7.16

Find the center of mass of the body shown in Figure E7.16a, composed of two uniform slender bars and a uniform sphere.

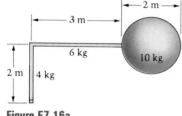

Figure E7.16a

Solution

We shall use the method of composite parts. The masses and mass-center coordinates (see Figure E7.16b) are:

Body	Mass	Mass-Center Coordinates (x, y)
1	10 kg	$(0, 0)$ m
2	6 kg	$(-2.5, 0)$ m
3	4 kg	$(-4, -1)$ m

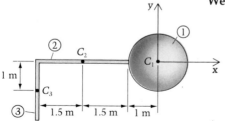

Figure E7.16b

We have, for the x coordinate of the mass center,

$$m\bar{x} = m_1\bar{x}_1 + m_2\bar{x}_2 + m_3\bar{x}_3$$
$$(10 + 6 + 4)\bar{x} = 10(0) + 6(-2.5) + 4(-4)$$
$$20\bar{x} = -31$$
$$\bar{x} = -1.55 \text{ m}$$

And for the y coordinate,

$$m\bar{y} = m_1\bar{y}_1 + m_2\bar{y}_2 = m_3\bar{y}_3$$
$$20\bar{y} = 10(0) + 6(0) + 4(-1)$$
$$\bar{y} = -0.200 \text{ m}$$

And, of course, $\bar{z} = 0$ by symmetry.

> **Question 7.11**   In the preceding example, it turns out that the mass center lies inside the sphere. Does the mass center of a body have to be a material point of the body? If not, give an example.
>
> **Question 7.12**   Give an example to illustrate the fact that the mass center need not always have the same location relative to material points of the body. *Hint:* Clearly for this to be true, the body has to be deformable!

Answer 7.11   No, it does not. A length of straight pipe has a mass center on its axis, in the space inside it.

Answer 7.12   Bend a pipe cleaner. The mass center before bending is on the axis of the pipe cleaner, while afterwards it is not.

PROBLEMS ▶ Section 7.4

7.98   Find the mass center of the body in Figure P7.98, which is a hemisphere glued to a solid cylinder of the same density, if $L = 2R$.

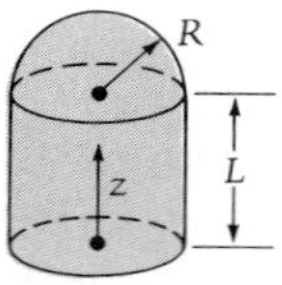

Figure P7.98

7.99   In the preceding problem, for what ratio of L to R is the mass center in the interface between the sphere and the cylinder?

7.100 In Figure P7.100 find the distance d such that the center of mass of the uniform thin wire is at the center Q of the semicircle.

7.101 Repeat Problem 7.100 if instead of a wire, the body is a uniform thin plate with the same periphery.

7.102 Determine the center of mass of the composite body shown in Figure P7.102. The density of $\mathcal{A}$ is 2000 kg/m³, the density of $\mathcal{B}$ is 3000 kg/m³, and the density of $\mathcal{C}$ is 4000 kg/m³.

7.103 In Figure P7.103 find the center of mass of the rigid frame, comprising seven identical rigid rods, each of length l.

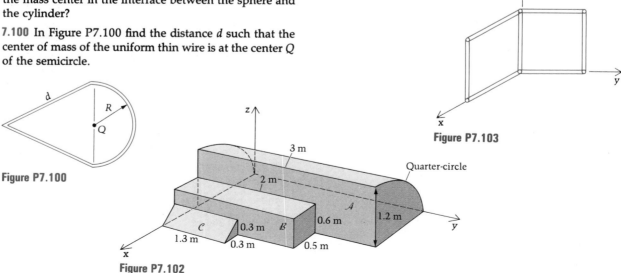

Figure P7.100

Figure P7.102

Figure P7.103

7.104 Find the center of mass of the bent bar, each leg of which is parallel to a coordinate axis and has uniform density and mass m. (See Figure P7.104).

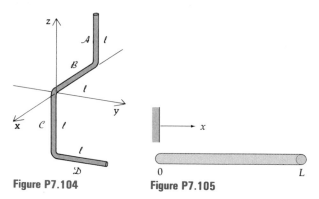

Figure P7.104   Figure P7.105

7.105 Find the mass center of the wire in Figure P7.105 if its density varies according to $\rho = \rho_o x / L$, where ρ_o is a constant, and so is the cross-sectional area A.

7.106 The density of the thin plate of thickness t in Figure P7.106 is given by $\rho = \rho_0 xy$, where ρ_0 is a constant. Find the mass center of the plate. Note the differences between your $(\bar{x}, \bar{y})$ and the answer $(\frac{2}{3}, \frac{1}{3})$ for a uniform density.

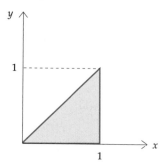

Figure P7.106

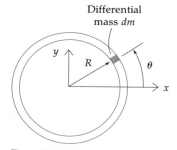

Figure P7.107

7.107 Find the mass center of the thin circular wire in Figure P7.107 if its area A is constant and its density varies according to $\rho = \rho_o \, \theta / (2\pi)$. In the figure, θ measures to a differential element of mass $dm = \rho A R \, d\theta$.

7.108 Repeat Problem 7.104 if the four legs have uniform, but different, densities, so that the masses of $\mathcal{A}, \mathcal{B}, \mathcal{C},$ and $\mathcal{D}$ are respectively, $m, 2m, 3m,$ and $4m$.

7.109 See Figure P7.109.

    a. How far over the edge can the can extend without falling over? It is open on the left and closed on the right, and the thickness is the same throughout.

    b. How far can it extend if the closed end goes first?

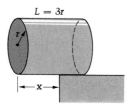

Figure P7.109

7.110 Use spherical coordinates, as suggested in Figure P7.110, to show that the center of mass of the uniform body formed by the intersection of the sphere of radius R, and the cone having vertex angle 2α, is at

$$\bar{z} = \frac{3R}{8} (1 + \cos \alpha)$$

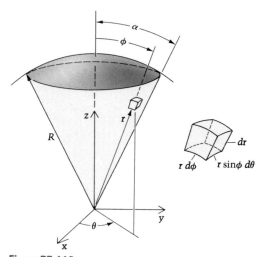

Figure P7.110

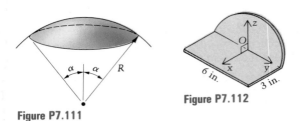

Figure P7.111

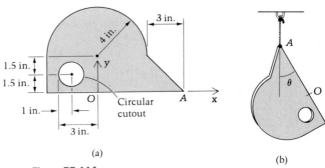

Figure P7.112

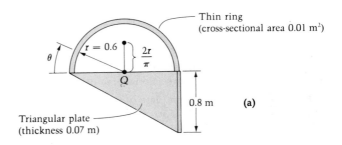

(a)

Figure P7.114

(b)

7.111 Use the results of Problem 7.110, together with the knowledge that the centroid of a cone is three-fourths the distance from vertex to base, to find the mass center of a uniform spherical cap without further integration. (See Figure P7.111.)

7.112 The body in Figure P7.112 is made of a single piece of sheet metal, cut into a rectangle and semicircle and bent at 90° as shown. Find the mass center.

7.113 A trailer carries a load of specific weight γ lb/ft³, which varies with y (but not with z) as shown in Figure P7.113.

   a. Find the weight and the position of the mass center of the load.

   b. Find the reactions F_A, F_B, and F_C due to the load only (do not include the trailer weight in this exercise).

7.114 Find the angle between line OA and the vertical if the thin plate in Figure P7.114(a) is suspended at A by a string, as shown in the sketch (b).

7.115 The object in Figure P7.115(a) is constructed of a thin but rigid uniform semicircular ring and a right triangular plate. The ring and the plate are made of the same material. Locate, with the angle θ, the point on the ring that would contact a smooth horizontal cylinder if the object was hung on the cylinder as shown in Figure P7.115(b). The mass center C of the semicircular ring is at a distance of $2r/\pi$ from point Q.

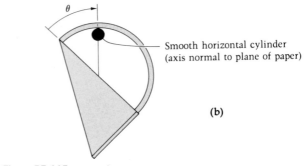

Figure P7.115

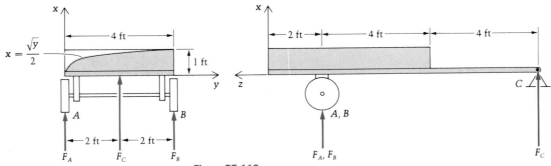

Figure P7.113

* **7.116** Shown in Figure P7.116 is a scale model of a thin parabolic shell structure. The shell's outer surface is a curve formed by revolving the indicated parabola about the y axis. Assume a constant mass density ρ and that the constant thickness t of the model is very small. Find the center of mass of the shell model. Note that because of the two assumptions you are actually seeking the centroid of the outer (curved) surface area.

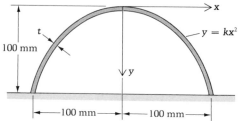

Figure P7.116

7.117 The half cylinder (weight W_C) and the stick (W_s) are attached and in equilibrium as shown in Figure P7.117. Find the x and y coordinates of the combined mass center, and use them to determine the "lean angle" ϕ.

7.118 When a structure is very thin (thickness much smaller than other dimensions) and flat, it is called a plate. If it is very thin but curved, it is called a shell. These are very important structural elements. Give an argument proving that if a plate or a shell is homogeneous, its center of mass is at its centroid of volume, which in turn is approximately located at the centroid of its (flat or curved) surface area.

* **7.119** Referring to Problem 7.118, find the center of mass of the thin conical shell of constant density shown in Figure P7.119.

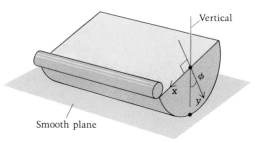

Figure P7.117

Figure P7.119

7.5 The Theorems of Pappus

There are two very old theorems of value when we are calculating either:

 a. The surface area generated by revolving a plane curve around an axis (which doesn't cross the curve) in the plane, or

 b. The volume generated by revolving a plane area around an axis (which doesn't cross the area) in the plane.

The two theorems are due to a Greek geometer of the third century, Pappus of Alexandria, and they are developed below.

 ① The area of A of the surface $\mathcal{A}$ generated by revolving curve $\mathcal{C}$ about the x axis (see Figure 7.5) is equal to $\int 2\pi y \, ds$, or

$$A = 2\pi \int y \, ds$$

So from our knowledge of centroids,

$$A = 2\pi \bar{y} s$$

where s is the arclength of the generating curve $\mathcal{C}$ and $\bar{y}$ is the y coordinate of the centroid of $\mathcal{C}$. Thus the generated area is the arclength (s) of the

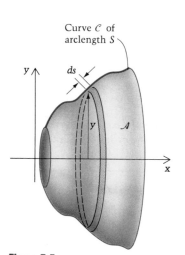

Figure 7.5

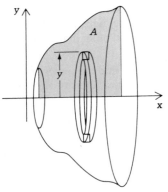

Figure 7.6

curve times the distance $(2\pi\bar{y})$ traveled during the revolving by the centroid of the curve. If the surface is formed by less than a full revolution, say by angle ϕ between 0 and 2π radians, then, of course, $A = \phi\bar{y}s$, still the arclength times the distance traveled by C.

Ⅱ   The volume V generated by revolving the shaded area A shown in Figure 7.6 about the x axis is

$$V = \int 2\pi y \, dA = 2\pi \int y \, dA = 2\pi\bar{y}A$$

where $\bar{y}$ is this time the centroid of the shaded *area*. Thus the generated *volume* is this area multiplied by the distance traveled by its centroid during the revolving. Again, for less than a full revolution, we would simply have $V = \phi\bar{y}A$ with $0 < \phi < 2\pi$.

We now consider five examples of the use of the theorems of Pappus.

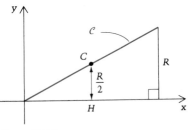

Figure E7.17

EXAMPLE 7.17

Determine the surface area of a right circular cone of radius R and height H.

Solution

Using the first theorem of Pappus, we avoid integration completely. The desired area will be generated by revolving the line $\mathcal{C}$ in Figure E7.17 about the x axis:

$$A = 2\pi\bar{y}s = 2\pi \frac{R}{2} \sqrt{R^2 + H^2}$$

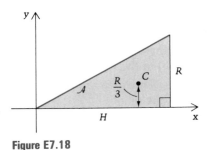

Figure E7.18

EXAMPLE 7.18

Find the volume of a right circular cone of radius R and height H.

Solution

The volume will be generated if we revolve the shaded area $\mathcal{A}$ of Figure E7.18 around the x axis. Then, by the second theorem of Pappus,

$$V = 2\pi\bar{y}A = 2\pi \frac{R}{3} \left(\frac{1}{2} HR \right) = \frac{\pi R^2 H}{3}$$

as we have seen in Example 7.7. Note that the volume of a cone is one-third that of the smallest cylinder that will enclose it — that is, the cylinder with the same height and base radius.

EXAMPLE 7.19

Find the surface area of a complete torus (doughnut).

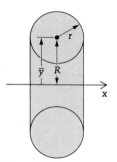

Figure E7.19

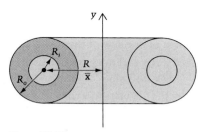

Figure E7.20

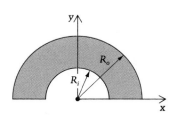

Figure E7.21a

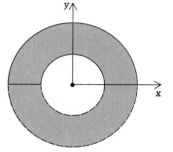

Figure E7.21b

Solution

We shall generate the surface area by revolving the circle about the x axis (see Figure E7.19) and using the first theorem of Pappus:

$$A = 2\pi\bar{y}s$$
$$= 2\pi R(2\pi r)$$
$$A = 4\pi^2 Rr$$

EXAMPLE 7.20

Find the volume of a hollowed torus.

Solution

We shall revolve the darker shaded area in Figure E7.20 about the y axis to generate the required volume. By the second theorem of Pappus,

$$V = 2\pi\bar{x}A = 2\pi R\pi(R_o^2 - R_i^2)$$
$$V = 2\pi^2 R(R_o^2 - R_i^2)$$

The theorems of Pappus are also very useful in finding centroids when volumes and areas (or areas and arcs) are known. We illustrate this with a final example.

EXAMPLE 7.21

Find the centroid of a semicircular annulus.

Solution

Revolving the area in Figure E7.21a about x, we obtain a hollow sphere (with the cross-section shown in Figure E7.21b), and from the second theorem of Pappus,

$$V = 2\pi\bar{y}A$$

$$\underbrace{\frac{4}{3}\pi(R_o^3 - R_i^3)}_{\substack{\text{Volume of} \\ \text{generated} \\ \text{hollow sphere}}} = 2\pi\bar{y}\underbrace{\frac{\pi(R_o^2 - R_i^2)}{2}}_{\text{Generating area}}$$

Thus,

$$\bar{y} = \frac{4(R_o - R_i)(R_o^2 + R_o R_i + R_i^2)}{3\pi(R_o + R_i)(R_o - R_i)}$$

or

$$\bar{y} = \frac{4(R_o^2 + R_o R_i + R_i^2)}{3\pi(R_o + R_i)}$$

Question 7.13   Does the answer check out in the two limiting cases $R_i \to 0$ and $R_i \to R_o$?

Answer 7.13   Yes. As $R_i \to 0$, $\bar{y} \to 4R_o/3\pi$ and as $R_i \to R_o$, $\bar{y} \to 4(3R_o^2)/[3\pi(2R_o)]$ $= 2R_o/\pi$. We obtained these results in Examples 7.5 and 7.2, respectively.

PROBLEMS ▶ Section 7.5

7.120 Find the surface area (exterior plus interior) of the object of revolution shown in Figure P7.120. It has a rectangular cross section with a circle removed.

7.121 Find the volume of the object in Problem 7.120.

7.122 In Figure P7.122 find the surface area of the body of revolution.

7.123 Find the volume of the body in Problem 7.122.

7.124 Find the surface area (exterior plus interior) of the tube, which spans 100° of a circle as shown in Figure P7.124.

7.125 Find the volume of the tube in Problem 7.124.

7.126 The 10-m diameter parabolic antenna in Figure P7.126 has an f/D ratio (focal distance to diameter) of 0.28. Find the surface area of one side of the thin shell.

7.127 Find the volume of water that the parabolic shell of revolution in Problem 7.126 could hold.

7.128 The shaft in Figure P7.128 ends in a conical thrust bearing. Find the surface area of the truncated cone if its upper and lower radii are 2 in. and 1 in. and its height is 1.5 in.

7.129 Find the volume of the truncated cone in Problem 7.128.

7.130 Find the centroid of the semicircular area in Figure P7.130 by revolving it about axis x to generate a spherical volume, using the second theorem of Pappus.

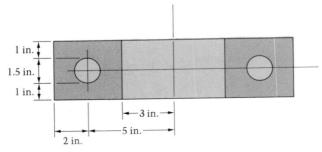

Figure P7.120

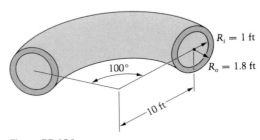

Figure P7.124

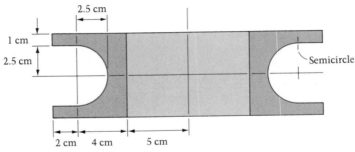

Figure P7.122

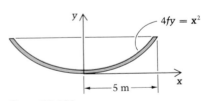

Figure P7.126

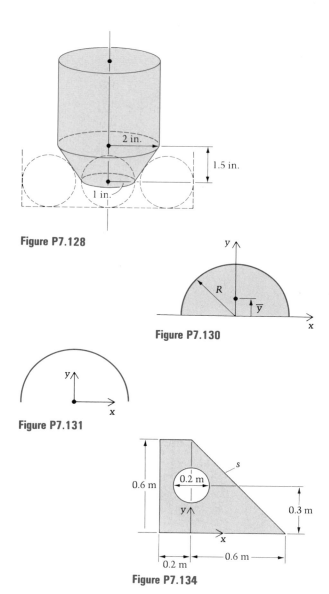

Figure P7.128

Figure P7.130

Figure P7.131

Figure P7.134

7.135 Repeat Problem 7.134 but with the revolution about the line $y = 0$.

7.136 Find the area generated by the slanted line segment s during the revolution of Problem 7.134.

7.137 Find the area generated by the slanted line segment s during the revolution of Problem 7.135.

7.138 Find the approximate volume of the thin, half-toroidal shell shown in Figure P7.138.

7.139 Refer to Problem 7.116. Find $s\bar{x}$ for the parabolic segment between the points $(0, 0)$ and $(100, 100)$ mm. Then use the first theorem of Pappus to determine the outer surface area of the shell.

7.140 In the preceding problem, find $A\bar{x}$ for the area bounded by $x = 0$, $y = 100$ mm, and the segment of the curve between $(0, 0)$ and $(100, 100)$ mm. Use this with the second theorem of Pappus to determine the volume enclosed within the inner surface of the shell.

7.141 Find the shaded surface area of the hollow axisymmetric shell section shown in Figure P7.141.

7.142 In Problem 7.141 find the volume bounded by the inside of the shell and the planes $x = 0$ and $x = 3$ in.

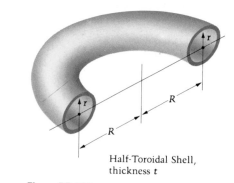

Half-Toroidal Shell, thickness t

Figure P7.138

7.131 Find the centroid of the semicircle (arc) in Figure P7.131 by revolving it about axis x to generate the surface area of a sphere.

7.132 Find the area of the surface formed by revolving an equilateral triangle of side s about one of its sides.

7.133 Find the volume generated by the revolution described in Problem 7.132.

7.134 Find the volume of the three-dimensional figure generated by revolving the area about the line $x = 0.8$ m. See Figure P7.134.

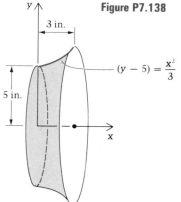

$(y - 5) = \dfrac{x^2}{3}$

Figure P7.141

Problems 7.143–7.147 deal with various parts of a thin half-toroidal shell.

7.143 Find the shaded surface area (one side) of the shell shown in Figure P7.143.

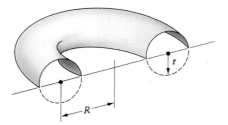

Figure P7.143

7.144 Find the shaded surface area (one side) of the shell shown in Figure P7.144.

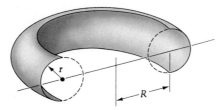

Figure P7.144

7.145 Find the shaded surface area (one side) of the shell shown in Figure P7.145.

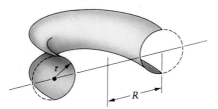

Figure P7.145

7.146 Find the shaded surface area (one side) of the shell shown in Figure P7.146.

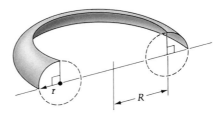

Figure P7.146

7.147 Find the shaded surface area (one side) of the shell shown in Figure P7.147.

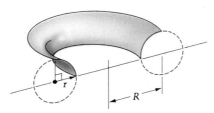

Figure P7.147

7.148–7.152 Find the volumes enclosed by the shells of Figures P7.148–P7.152, respectively.

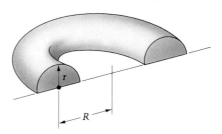

Figure P7.148

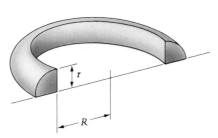

Figure P7.149

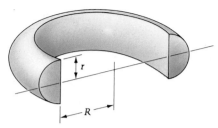

Figure P7.150

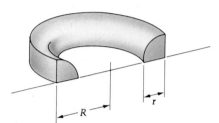

Figure P7.151

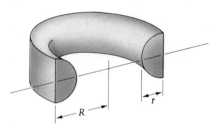

Figure P7.152

7.153 The dam spans 90°, as shown in Figure P7.153. Find the volume of concrete that was required to build it.

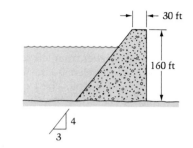

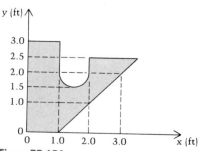

Figure P7.153

COMPUTER PROBLEMS ▶ Chapter 7

7.154 Write a computer program that will accept any number of triplets of numbers A_i, $\bar{x}_i$, $\bar{y}_i$, where A_i is a plane area and $(\bar{x}_i, \bar{y}_i)$ are the x and y coordinates of its centroid. The program is to read N (the number of areas making up a composite area), then read N triplets $(A_i, \bar{x}_i, \bar{y}_i)$, then compute and print the total area A and the coordinates of its centroid. Use the program to find the centroid of the area in Figure P7.154.

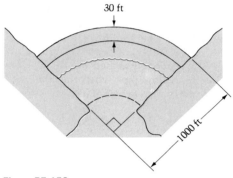

Figure P7.154

* **7.155** A crossed log-periodic dipole array (antenna) looks like a Christmas tree (see Figure P7.155a) with tubes extending in four directions from a mast. One of the four sets of tubes is shown in Figure P7.155b. Not shown are identical sets downward and both into and out of the paper. The figure shows a simplified model of such an antenna; the tubes actually get smaller from left to right and the mast is in sections that also decrease in size.

  Find the mass center of the antenna. The x coordinates of the elements (in inches) are the numbers below the mast. Assume the elements are attached to the outside of the mast. Would the mass center be the same point if the entire antenna was covered with an inch of ice?

* **7.156** A man built a basketball goal for his children out of a 15-ft steel pipe weighing 3.03 lb/ft, and having an inside diameter of 3.33 in. The goal vibrated too much when the basketball hit the backboard, so the man filled the pipe with concrete, which increased the bending stiffness of the pole. The concrete weighed 144 lb/ft^3. How full (percent) was the hollow pipe when the combined mass center of pipe and concrete was at its lowest point?

Figure P7.155a

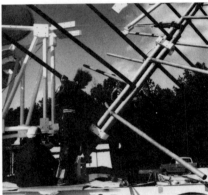

Assume each of these eight elements to be a tube with outside diameter (OD) 1.40 in. and wall thickness 0.085 in. . . .

Mast = tube with OD = 1.81 and wall thickness = 0.091 in.

. . . and these twenty have OD = 0.380 in. and wall thickness 0.050 in.

61.6 in.

3.4 in.

x

10.1 in.
25.3
38.9
51.2
62.2
72.2
81.1
89.2
96.4
102.9
108.8
114.1
118.8
123.1
127.0
130.4
133.6
136.4
138.9
141.2
143.2
145.1
146.7
148.2
149.5
150.8
151.8
152.8

───────── 157.7 in. ─────────

Figure P7.155b

(See Figure P7.156.) *Hint:* Write a program that will list or plot the mass center height versus concrete height, and pick off the answer from the list or plot. Check your answer by using calculus to obtain the exact solution.

* **7.157** The density of the earth varies greatly through its depth. It can be fairly well approximated by a graph consisting of straight lines through the following points:

Depth from Surface of Earth (miles)	Density (g / cc)
0	3.50
400	4.30
1000	5.20
1800	5.70
1800 (jump)	9.60
2500	10.8
3100	11.7
3300	14.0
3500	15.0
3960	16.3

Assuming spherical symmetry, write a computer program to calculate the mass of the earth, and use that result to calculate the average density of the earth.

Figure P7.156

SUMMARY ► **Chapter 7**

In this chapter we learned (in Section 7.2) that the centroid of arclength s, area A, or volume V, as well as the center of mass m, when defined over a region $\mathcal{R}$ can all be given by the same formula:

$$\mathbf{r}_{OC} = \frac{\int \mathbf{r} dQ}{\int dQ} = \frac{\int \mathbf{r} dQ}{Q}$$

in which Q represents s, A, V, or m; **r** is a vector from an origin O to a differential element of Q, and C is the "center of Q," meaning the centroid of s, A, or V or the mass center, as the case may be.

We usually work with scalar components of the above equation, such as $A\bar{x} = \int x dA$, where $\bar{x}$ is the x-component of the vector $\mathbf{r}_{OC}$ from the origin O to the centroid (of area in this case) C.

We studied the Method of Composite Parts in Section 7.3. This extremely useful procedure allows us to find the centroid of a composite area* when we know the areas and centroids of two or more parts that constitute the entire (or "overall") area. For example, the x-component of the centroid of an area made up of two areas A_1 and A_2 is given by

---

* or arclength or volume or mass, as the case may be.

$$\bar{x} = \frac{A_1\bar{x}_1 + A_2\bar{x}_2}{A_1 + A_2},$$

where $\bar{x}_1$ and $\bar{x}_2$ are the respective centroids of the areas A_1 and A_2.

The center of mass, covered in Section 7.4, differs in complexity from centers (centroids) of arclength, area, and volume because of the presence of the density function $\rho = \rho(x, y, z)$. The mass center is defined by

$$\mathbf{r}_{OC} = \frac{\int \mathbf{r} dm}{\int dm} = \frac{\int \mathbf{r}\rho dV}{\int \rho dv}$$

so that if ρ is constant ("uniform density"), it cancels and the center of mass is the same point as the centroid of volume. If ρ is not constant, however, we must include it in the integrand of the numerator and denominator, and the calculation is ordinarily harder.

The Theorems of Pappus were studied in Section 7.5. They are very useful when calculating either (a) the surface area generated by revolving a plane curve around an axis (which doesn't cross the curve) in the plane, or (b) the volume generated by revolving a plane area around an axis (which doesn't cross the area) in the plane. The formulae which correspond to (a) and (b) are, respectively:

$$(a) \qquad A = 2\pi \int y \, ds = 2\pi \bar{y}s$$

and

$$(b) \qquad V = 2\pi \int y \, dA = 2\pi \bar{y}A$$

These formulae allow us to use known centroidal locations to compute surface areas and volumes of objects of (full or partial) revolution about an axis. If the revolution is less than complete, say through ϕ radians, then we simply replace the 2π in the above formulae (Theorems of Pappus) by ϕ.

REVIEW QUESTIONS ▶ Chapter 7

True or False?

1. The centroid of an area depends upon which reference point is used to locate it in the defining equation $A\mathbf{r}_{OC} = \int \mathbf{r} \, dA$.
2. The centroid of a quantity Q is the point where all of Q could be *concentrated* with the same resulting first moment as has the actual distribution of Q.
3. All centroid calculations for areas require two integrations; centroid calculations for volumes require three integrations.
4. Any line through the centroid of an area divides the area into two equal parts.
5. The centroid of a semicircle is farther from the circle center than is the centroid of the semicircular area contained within it.

6. Suppose that in an arbitrary triangle we drop a perpendicular from a vertex V to the opposite base (line l, extended if necessary) and find that the distance from V to l is 9 in. The centroid of the area of the triangle is then 3 in. from l.

7. If an area A is made up of two parts A_1 and A_2 such that $A_1 + A_2 = A$, then the x coordinate of its centroid is given by $\bar{x} = (A_1\bar{x}_1 + A_2\bar{x}_2) / (A_1 + A_2)$.

8. Finding centroids of areas with cutouts is often simplified by the method of composite parts.

9. The center of mass of a body has to be a material (physical) point of the body.

10. The mass center of a rigid body is always the same point of the body (or of a rigid extension of the body) regardless of the position of the body.

11. The centroid of the volume of a body coincides with its center of mass only if its density is constant.

12. The integral $\int_{\mathcal{B}} \mathbf{r} \, dm$ equals the mass of $\mathcal{B}$ multiplied by the vector from the origin of $\mathbf{r}$ to the mass center of $\mathcal{B}$.

13. The use of the Theorem of Pappus always involves the revolution of a curve or an area about a given axis in the same plane.

Answers: 1. F   2. T   3. F   4. F   5. T   6. T   7. T   8. T   9. F   10. T   11. F   12. T   13. T

8

INERTIA PROPERTIES OF PLANE AREAS

8.1 Introduction

In this chapter we study the inertia properties of plane areas. One reason for studying this topic in statics is that these properties arise in the formulas for locating the resultant of hydrostatic pressure forces on a submerged body (which we shall examine in Section 9.2). A more important reason for this study is that it is sometimes considered a prerequisite for courses in strength of materials (or deformable bodies), which follow statics. In these later courses the student will find that stresses in a transversely loaded beam are, under special but important circumstances, inversely proportional to a moment of inertia of the cross-sectional area of the beam. Deflections of the beam will likewise be inversely proportional to this moment of inertia, which forms part of the resistance to bending of the beam. Similarly, the "polar moment of inertia" is a factor in the resistance of a shaft to torsion, or twisting.

The four sections 8.2–8.5 of this chapter can be read by a student who is familiar with no more than single integration. These are the sections normally needed in the first course in the mechanics of deformable solids. The last three sections, however, utilize double integrals when dealing with products of inertia.

Moments and products of inertia of *mass* are needed in dynamics; we shall cover this related topic in our second volume at the point where the subject arises naturally.

8.2 Moments of Inertia of a Plane Area

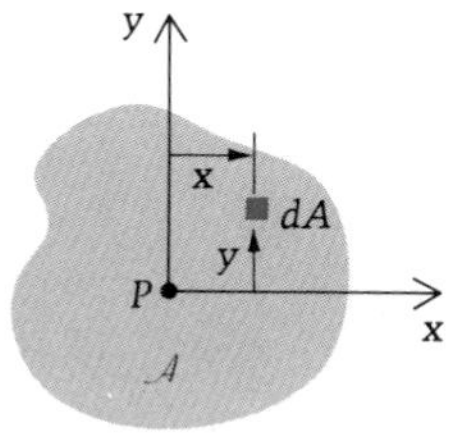

For the plane area shown in Figure 8.1, the moments of inertia* with respect to the x and y axes are defined to be

$$I_x = \int_A y^2 \, dA \tag{8.1}$$

and

$$I_y = \int_A x^2 \, dA \tag{8.2}$$

Figure 8.1

These definitions illustrate why a moment of inertia is sometimes called a "*second* moment" — because of the square of the distance from the x axis for I_x (and from the y axis for I_y). We have seen "first moments" in Chapter 7 in relation to the concept of centroid. Because a moment of inertia is made up of areas multiplied by squares of distances, it has the dimension (length)⁴.

---

* Many prefer to use the term "second moments of area," feeling that the word "inertia" suggests mass and should be reserved for similar integrals that reflect the mass distribution of a body. Nevertheless, the terms "area moment of inertia" and "moment of inertia of area" are widely used in texts on the mechanics of deformable solids.

Equations (8.1) and (8.2) also tell us that a moment of inertia is always positive and is a measure of "how much area is located how far" from a line. If we wish to be specific about the origin of the x and y coordinates, we may write, for example, I_{x_c} if the origin is the centroid, or I_{x_p} if the origin is some other point P.

We now proceed to use the above definitions to find moments of inertia of several common shapes in the following examples.

EXAMPLE 8.1

Find the moments of inertia of the rectangular area of Figure E8.1a about the centroidal x and y axes.

Solution

To find I_{x_c}, we need the integral $\int y^2 \, dA$. Using the horizontal strip shown in Figure E8.1b for our differential area dA, and noting that the y coordinate is the same for all parts of the strip, we get

$$
I_{x_c} = \int y^2 \, dA = \int_{-h/2}^{h/2} y^2 (b \, dy)
$$

$$
= \frac{by^3}{3} \Big|_{-h/2}^{h/2}
$$

$$
= \frac{bh^3}{12}
$$

A similar integration with $dA = h \, dx$ as shown in Figure E8.1c gives I_{y_c}:

$$
I_{y_c} = \int_{x=-b/2}^{b/2} x^2 \underbrace{(h \, dx)}_{dA} = h \frac{x^3}{3} \Big|_{-b/2}^{b/2}
$$

$$
= \frac{hb^3}{12}
$$

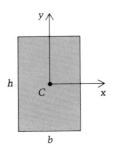

Figure E8.1a

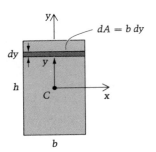

Figure E8.1b

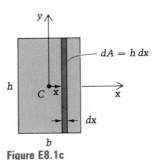

Figure E8.1c

Question 8.1   Was the I_{y_c} calculation really necessary? That is, could the answer for I_{y_c} have been deduced from the result for I_{x_c} obtained first?

For those familiar with double integrals, we use them below to reproduce the result for I_{x_c}:

$$
I_{x_c} = \int_{y=-h/2}^{h/2} \int_{x=-b/2}^{b/2} y^2 \underbrace{dx \, dy}_{dA}
$$

Answer 8.1   No, it was not necessary, because if we change the names of x and y, and of b and h, then the same integral as before yields $I_{y_c} = hb^3 / 12$ without having to integrate again.

$$= \int_{y=-h/2}^{h/2} x \Big|_{-b/2}^{b/2} y^2 \, dy = \int_{-h/2}^{h/2} y^2 \, b \, dy = b \frac{y^3}{3} \Big|_{-h/2}^{h/2} = \frac{bh^3}{12}$$

Note that the first integration (on x) *produces* the strip $b \, dy$ used earlier.

Question 8.2   Could the strip $h \, dx$ have been used for "dA" in the I_{x_c} calculation?

Answer 8.2   No, because then y is not the same for every element of the differential strip.

EXAMPLE 8.2

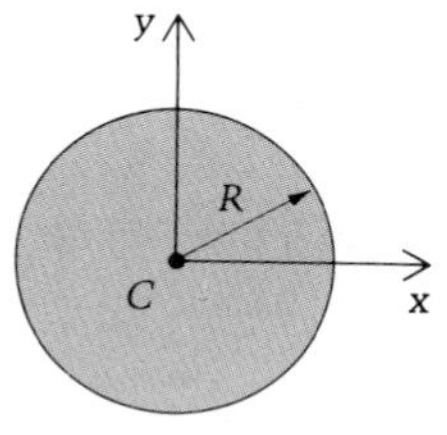

Find the moment of inertia of a circular area (see Figure E8.2a) about any diameter.

Solution

Since $x^2 + y^2 = R^2$ on the boundary, our dA in Figure E8.2b is given by

$$dA = 2\sqrt{R^2 - y^2} \, dy$$

so that

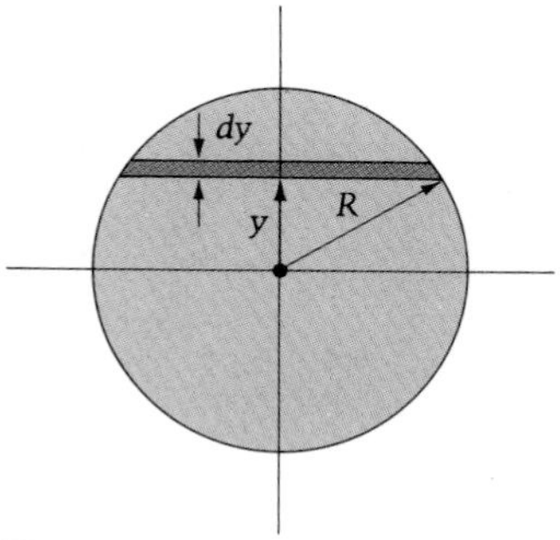

$$I_{x_c} = \int y^2 \, dA = \int_{-R}^{R} 2y^2 \sqrt{R^2 - y^2} \, dy$$

$$= 2 \int_{0}^{R} 2y^2 \sqrt{R^2 - y^2} \, dy$$

$$= 4R^4 \int_{0}^{1} \frac{y^2}{R^2} \sqrt{1 - \left(\frac{y}{R}\right)^2} \, d\left(\frac{y}{R}\right)$$

Substituting $\sin\theta$ for y/R, and noting that $d(y/R) = \cos\theta \, d\theta$,

$$I_{x_c} = 4R^4 \int_{0}^{\pi/2} \sin^2\theta \cos\theta(\cos\theta \, d\theta)$$

where for the integral limits, $y/R = 0$ when $\theta = 0$, and $y/R = 1$ when $\theta = \pi/2$. Continuing,

$$I_{x_c} = 4R^4 \int_{0}^{\pi/2} \frac{\sin^2 2\theta}{4} \, d\theta$$

$$= R^2 \int_{0}^{\pi/2} \frac{1 - \cos 4\theta}{2} \, d\theta$$

$$= \frac{R^4}{2}\left(\theta - \frac{\sin 4\theta}{4}\right)\Big|_{0}^{\pi/2} = \frac{\pi R^4}{4}$$

For the reader acquainted with double integrals, we can obtain the above result with somewhat less effort using polar coordinates (see Figure E8.2c) as follows:

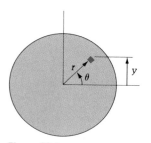

Figure E8.2a

Figure E8.2b

Figure E8.2c

$$I_{x_c} = \int y^2 \, dA = \int_{\theta=0}^{2\pi} \int_{r=0}^{R} \underbrace{(r \sin \theta)^2 r \, dr \, d\theta}_{dA}$$

$$= \int_0^{2\pi} \frac{r^4}{4} \Big|_0^R \sin^2 \theta \, d\theta$$

$$= \frac{R^4}{4} \int_0^{2\pi} \left(\frac{1 - \cos 2\theta}{2} \right) d\theta = \frac{R^4}{8} \left[\theta - \frac{\sin 2\theta}{2} \right] \Big|_0^{2\pi}$$

$$I_{x_c} = \frac{\pi R^4}{4}$$

This is, of course, also I_{y_c} or I about any other diameter of the circle.

EXAMPLE 8.3

Find the moment of inertia of the triangular area in Figure E8.3 about the y axis.

Solution

For our differential area, we shall use the shaded strip in the figure; thus, using Y to locate the lowest (boundary) point of the strip,

$$dA = (h - Y) \, dx$$

But $Y = \dfrac{2h}{b} x$ for the side of the triangle in the first quadrant, so that

$$dA = \left(h - \frac{2hx}{b} \right) dx$$

Therefore,

$$I_y = \int x^2 \, dA = 2 \int_0^{b/2} x^2 \left(h - \frac{2h}{b} x \right) dx$$

$$= 2h \int_0^{b/2} \left(x^2 - \frac{2x^3}{b} \right) dx = 2h \left[\frac{x^3}{3} - \frac{x^4}{2b} \right] \Big|_0^{b/2} = \frac{hb^3}{48}$$

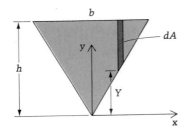

Figure E8.3

PROBLEMS ▶ Section 8.2

8.1 In Figure P8.1:

   a. Find the moment of inertia of the shaded area about the x axis.

   b. Tell why I_y is identical to I_{y_c} of Example 8.1.

8.2 Find I_x for the shaded area in Figure P8.2.

8.3 In Problem 8.2 find I_y for the same area.

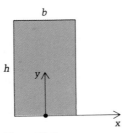

Figure P8.1

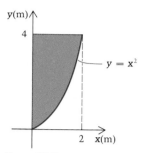

Figure P8.2

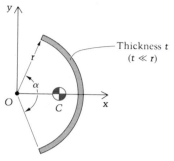

Figure P8.7

8.4   Review Examples 8.1 and 8.3.

  a. Find the moment of inertia of the shaded area in Figure P8.4 about the y axis.

  b. Add the result to that of Example 8.3, and note that the answer is, as it should be, that for the combined, rectangular area.

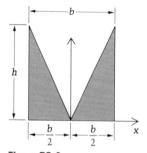

Figure P8.4

8.5   Determine the moment of inertia of the shaded area in Figure P8.5 with respect to the x axis.

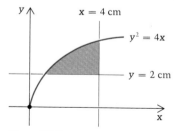

Figure P8.5

8.6   Repeat Problem 8.5 for the moment of inertia about the y axis.

8.7   Find I_x for the thin circular section shown in Figure P8.7. *Hint*: Use polar coordinates, with $dA = tr\, d\theta$.

8.8   In Problem 8.7, find I_y.

8.9   Find the moment of inertia about the x axis of the area under the nth-degree curve shown in Figure P8.9. (n is an integer, $n \geq 1$.)

8.10   In Problem 8.9, find I_y.

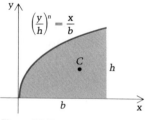

Figure P8.9

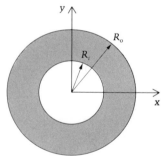

Figure P8.11

8.11   Show that the moment of inertia I_x of the shaded, annular area in Figure P8.11 is $\pi(R_o^4 - R_i^4)/4$ by:

  a. Integrating as in Example 8.2 but with limits of $\int_{R_i}^{R_o}$ on the integration with respect to r.

  b. Subtracting the moments of inertia of the two circles. Why does this work?

8.12   In Problem 8.11, let the figure depict the cross section of a round tube. If the tube is very thin ($R_i \rightarrow R_o$), show that I_x is approximately $\pi R^3 t$, where $R = R_i \approx R_o$ and $t =$ thickness of tube $= R_o - R_i$.

8.3 The Polar Moment of Inertia of a Plane Area

In the study of deformable solids, the "torsion problem" refers to what happens to a shaft when it is twisted. In the same way that the moment of inertia forms part of the resistance of a beam to bending, the *polar moment of inertia* forms part of its resistance to twisting. For this reason, then, we shall discuss the polar moment of inertia in this section.

The polar moment of inertia of an area about a point P is defined to be (see Figure 8.2).

$$J_P = \int (x^2 + y^2)\, dA$$

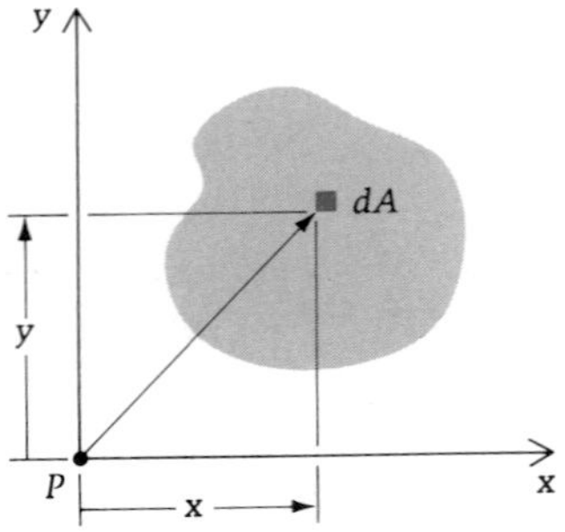

where the axes (x, y) have origin at P. Since the polar coordinate r is given by* $x^2 + y^2 = r^2$, this simplifies to

$$J_P = \int r^2\, dA$$

If we recall that both I_x and I_y are the sums (in integration) of (differential areas) times (squares of their distances from the axis), then we see that J_P is the same type of quantity, this time with respect to the z axis (*normal* to the area) through a point P. Also,

$$J_P = \int y^2\, dA + \int x^2\, dA$$

$$= I_{x_P} + I_{y_P} \tag{8.3}$$

so that for an area moment of inertia, the x and y (in-plane) inertias add up to the z (out-of-plane) inertia.

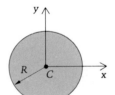

Figure 8.2

EXAMPLE 8.4

Find the polar moment of inertia of the circular cross section of Figure E8.4a with respect to point C.

Solution

By $J_C = I_{x_C} + I_{y_C}$, we obtain immediately from the results of Example 8.2:

$$J_C = \frac{\pi R^4}{4} + \frac{\pi R^4}{4} = \frac{\pi R^4}{2}$$

Alternatively, the shaded "circular strip" in Figure E8.4b has the feature that all its elements are the same distance r from the z axis. Thus with $dA = 2\pi r\, dr$,

$$J_C = \int r^2\, dA = \int_0^R r^2(2\pi r\, dr)$$

$$= 2\pi \left.\frac{r^4}{4}\right|_0^R = \frac{\pi R^4}{2}$$

And if double integrals are familiar to the reader, we obtain the same result a

Figure E8.4a

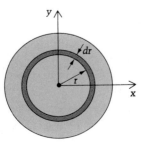

Figure E8.4b

---

* Hence the name "polar moment of inertia."

third time using polar coordinates:

$$J_C = \int r^2 \, dA = \int_{\theta=0}^{2\pi} \int_{r=0}^{R} r^2 (r \, dr \, d\theta)$$

$$= \int_{\theta=0}^{2\pi} \frac{r^4}{4} \Big|_0^R \, d\theta = \frac{R^4}{4} \theta \Big|_0^{2\pi} = \frac{\pi R^4}{2}$$

EXAMPLE 8.5

Find the polar moment of inertia of the rectangular cross section of Figure E8.5, at point C.

Solution

Because $J_C = I_{x_c} + I_{y_c}$, we use previous results and get

$$J_C = \frac{bh^3}{12} + \frac{hb^3}{12} = \frac{bh(b^2 + h^2)}{12}$$

or

$$J_C = \frac{A(b^2 + h^2)}{12}$$

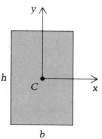

Figure E8.5

PROBLEMS ▶ Section 8.3

In Problems 8.13–8.19 find the polar moment of inertia with respect to the origin for each of the shaded areas.

8.13

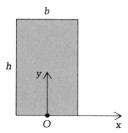

Figure P8.13

8.14

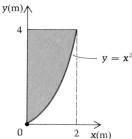

Figure P8.14

8.15

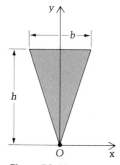

Figure P8.15

8.16

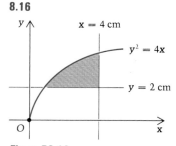

Figure P8.16

8.17

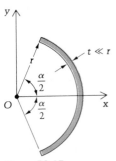

Figure P8.17

8.18

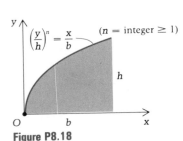

Figure P8.18

8.19

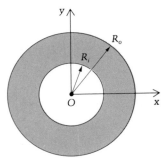

Figure P8.19

8.4 The Parallel-Axis Theorem (or Transfer Theorem) for Moments of Inertia/The Radius of Gyration

Development of the Parallel Axis Theorem

Fortunately, if we wish to determine the moment of inertia I_{x_P} about an axis through a point P other than the centroid, we don't have to integrate if we already know I_{x_C}. All we need do is "transfer" this result, adding the area A times the square of the distance d between the parallel x axes through C and P (see Figure 8.3):

$$I_{x_P} = I_{x_C} + Ad^2 \tag{8.4}$$

We now proceed to prove this very helpful theorem. We have

$$I_{x_P} = \int_A y^2 \, dA = \int_A (\bar{y} + y_1)^2 \, dA$$

$$I_{x_P} = \underbrace{\int_A y_1^2 \, dA}_{I_{x_C}} + \bar{y}^2 \int_A dA + 2\bar{y} \int_A y_1 \, dA$$

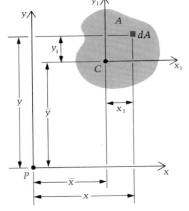

Figure 8.3

where from Figure 8.1 we see that the y coordinate of dA is $\bar{y} + y_1$, where $\bar{y}$ is the y coordinate of C.

Question 8.3 Why may the $\bar{y}$ be brought outside the integrals above?

The last integral, $\int y_1 \, dA$, is zero by the definition of the centroid, and therefore

$$I_{x_P} = I_{x_C} + A\bar{y}^2 = I_{x_C} + Ad^2$$

where d is the distance between the two parallel x axes through P and C.

Answer 8.3 It's constant.

This is the parallel-axis theorem that was stated in Equation (8.4). Note from Equation (8.4) that the smallest value for the moment of inertia of an area is with respect to an axis through the body's centroid. For y axes, we have

$$I_{y_P} = I_{y_C} + A\bar{x}^2 = I_{y_C} + Ad^2 \tag{8.5}$$

where *this* time d is the absolute value of $\bar{x}$, and is again the distance between the parallel axes through P and C.

For the polar moment of inertia, we also have a parallel-axis theorem:

$$\begin{aligned}
J_P \text{ (or } I_{z_P}) &= \int (x^2 + y^2)\, dA = \int [(\bar{x} + x_1)^2 + (\bar{y} + y_1)^2]\, dA \\
&= \int (x_1^2 + y_1^2)\, dA + \bar{y} \int x_1\, dA + \bar{x} \int y_1\, dA + \int (\bar{x}^2 + \bar{y}^2)\, dA \\
&= J_C + 0 + 0 + Ad^2 \\
J_P &= J_C + Ad^2 \tag{8.6}
\end{aligned}$$

where once again d is the distance between the parallel axes, this time z axes through P and C, of interest. It is also the distance between the points P and C in the plane of the area.

> **Question 8.4**   Give an alternative proof of Equation (8.6) using Equation (8.3).

Answer 8.4 $J_P = I_{x_P} + I_{y_P} = I_{x_C} + A\bar{y}^2 + I_{y_C} + A\bar{x}^2 = (I_{x_C} + I_{y_C}) + A(\bar{x}^2 + \bar{y}^2) = J_C + Ad^2$

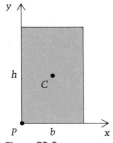

y

h

C

P b

x

Figure E8.6

EXAMPLE 8.6

Find the moments of inertia of the rectangular area about the indicated axes through the lower left corner P in Figure E8.6.

Solution

Using the parallel-axis theorem together with the results of Example 7.1,

$$I_x \text{ (or } I_{x_P}) = I_{x_C} + Ad^2 = \frac{bh^3}{12} + bh\left(\frac{h}{2}\right)^2 = \frac{bh^3}{3}$$

$$I_y \text{ (or } I_{y_P}) = I_{y_C} + Ad^2 = \frac{hb^3}{12} + hb\left(\frac{b}{2}\right)^2 = \frac{hb^3}{3}$$

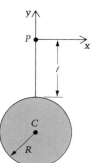

y

P

x

ℓ

C

R

Figure E8.7

EXAMPLE 8.7

Find the moments of inertia of the circular area about the axes x and y shown in Figure E8.7.

Solution

Using the parallel-axis theorem, and the results of Example 8.2,

$$I_x = I_{x_c} + Ad^2 = \frac{\pi R^4}{4} + (\pi R^2)(\ell + R)^2 = \frac{\pi R^4}{4}\left(5 + \frac{8\ell}{R} + \frac{4\ell^2}{R^2}\right)$$

$$I_y = I_{y_c} + Ad^2 = \frac{\pi R^4}{4} + 0$$

[because the distance between the y axes through the two origins vanishes. (They are the same line!)]

EXAMPLE 8.8

(a) Find the moment of inertia of the triangular area in Figure E8.8a about the x axis through O. (b) Then use the transfer theorem to determine I_{x_c}.

Solution

(a) By the definition

$$I_{x_o} = \int_A y^2 \, dA$$

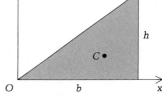

Figure E8.8a

Noting that the hypotenuse has the equation $y = \left(\dfrac{h}{b}\right)x$, we have (see Figure E8.8b):

$$I_{x_o} = \int_0^h y^2\left(b - \frac{b}{h}y\right) dy = \frac{bh^3}{3} - \frac{bh^3}{4} = \frac{bh^3}{12}$$

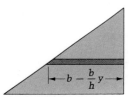

Figure E8.8b

Question 8.5   Show how to obtain this answer without integrating by using Example 8.3.

(b) The transfer theorem gives

$$I_{x_o} = I_{x_c} + Ad^2$$

$$\frac{bh^3}{12} = I_{x_c} + \frac{bh}{2}\left(\frac{h}{3}\right)^2 = I_{x_c} + \frac{bh^3}{18}$$

so that

$$I_{x_c} = \frac{bh^3}{36}$$

Notice that we don't always transfer *from* C to other points, in this case we knew the answer for I_x at O, and we used it "in reverse" to *find* I_{x_c}.

Answer 8.5   Working from the answer to Example 8.3, to make the notations and geometry agree, we must do three things: (1) swap b and h; (2) double the old h; and (3) use half the answer. Thus I_x in Example 8.8 is $\left[\dfrac{b(2h)^3}{48}\right]\Big/ 2 = \dfrac{bh^3}{12}$, which checks.

Answer 8.6   No! We can only transfer to or from an axis through the centroid.

EXAMPLE 8.9

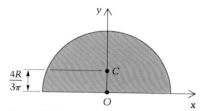

Figure E8.9

Find the polar moment of inertia of the semicircular area of Figure E8.9 with respect to its centroid C.

Solution

At point O we have

$$J_O = \int r^2 \, dA = \frac{\pi R^4}{4}$$

because (see Example 8.4) we must get half of the answer for the full circular area. Using the parallel-axis theorem,

$$J_O = J_C + Ad^2$$

$$\frac{\pi R^4}{4} = J_C + \frac{\pi R^2}{2}\left(\frac{4R}{3\pi}\right)^2$$

$$J_C = \pi R^4 \left(\frac{1}{4} - \frac{8}{9\pi^2}\right)$$

$$= 0.160 \pi R^4 = 0.503 R^4$$

EXAMPLE 8.10

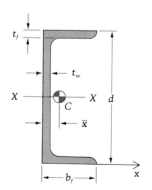

d = depth
b_f = width of flange
t_f = average thickness of flange
t_w = web thickness
$\bar{x}$ = distance locating centroid of section

Figure E8.10

For the C15 × 50 channel section,* (see Figure E8.10) the manual of the American Institute of Steel Construction lists the area of the cross section as 14.7 in.^{2} and the centroidal moment of inertia about line X-X as 404 in.^{4} If $d = 15.0$ in., find the moment of inertia about the baseline axis x.

Solution

We simply use the parallel-axis theorem, and obtain, without integration,

$$I_x = I_{x_C} + Ad^2$$

$$= 404 + 14.7\,(7.5)^2$$

$$= 1230 \text{ in.}^4$$

* Meaning "C" for channel and that the nominal depth is 15 in. and the weight per lineal foot is 50 lb/ft. See the AISC Manual of Steel Construction, American Institute of Steel Construction, Inc., 101 Park Avenue, New York, N.Y. 10017.

Radius of Gyration

This is a good place to introduce the reader to a distance called the *radius of gyration*.* The radius of gyration is a concept associated with the moment of inertia of an area about an axis. The radius of gyration is simply the length that, when squared and multiplied by the area, gives the moment of inertia about the axis. Thus, for example,

$$I_x = A k_x^2$$

For a physical interpretation of the radius of gyration, consider the area A in Figure 8.4(a) having the moment of inertia I_x about the x axis. If

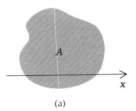

(a)

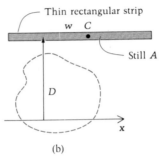

(b)

Figure 8.4

we squeeze the same area into a very thin strip and place it a distance D from the x axis (and parallel to it as shown in Figure 8.4(b), then the moment of inertia of the strip about the x axis is

$$I_x^{\text{STRIP}} = I_{x_c}^{\text{STRIP}} + AD^2 \approx 0 + AD^2$$

Clearly, then, there is a distance D for which I_x^{STRIP} equals I_x of the actual area. This value of D is the radius of gyration k_x and is therefore $\sqrt{I_x / A}$, as in the definition above.

Question 8.7   Why is $I_{x_c}^{\text{STRIP}} \ll AD^2$ in the above equation?

---

* This parameter arises naturally in the analysis of buckling of columns.

Answer 8.7   Because $I_{x_c}^{\text{STRIP}} = \dfrac{wt^3}{12} = \dfrac{At^2}{12}$, much smaller than AD^2 because the strip is given to be very thin.

As examples, the radii of gyration of the areas in Figure 8.5 about the indicated lines are shown beneath the various figures:

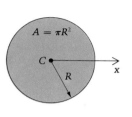

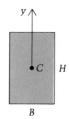

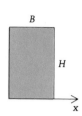

(a) $k_x = \dfrac{R}{2}$

since $I_x = \dfrac{\pi R^4}{4}$

(b) $k_y = \dfrac{\sqrt{3B}}{6}$

since $I_y = \dfrac{HB^3}{12}$

(c) $k_x = \dfrac{\sqrt{3H}}{3}$

since $I_x = \dfrac{BH^3}{3}$

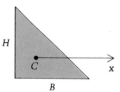

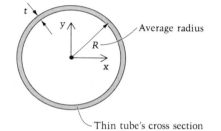

(d) $k_x = \dfrac{\sqrt{2H}}{6}$

since $I_x = \dfrac{BH^3}{36}$

(e) $k_z \approx R$

since $J \approx 2\pi R^3 t$
(and $A = 2\pi Rt$)

Figure 8.5

PROBLEMS ▶ Section 8.4

For each of the areas shown in Problems 8.20–8.23 find the moment of inertia about a line through the centroid parallel to x.

8.20 (See Problem 8.2.)

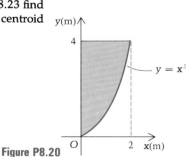

Figure P8.20

8.21

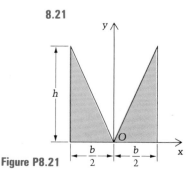

Figure P8.21

8.22 (See Problem 8.5.)

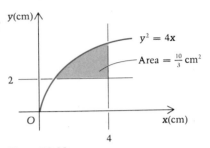

Figure P8.22

8.23 (See Problems 8.9 and 7.23)

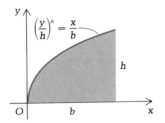

Figure P8.23

In Problems 8.24–8.26, for the problem referred to, find the moment of inertia about a line through the centroid parallel to y.

8.24 Problem 8.20.

8.25 Problem 8.22.

8.26 Problem 8.23.

8.27 In Problem 8.8, the distance from the origin to the centroid of the very thin section is approximately $[r \sin (\alpha / 2)] / (\alpha / 2)$. Use the parallel-axis theorem to find I_{yc}.

8.28 Use the parallel-axis theorem to find the moment of inertia of the semicircular area about an axis through C parallel to x. (See Figure P8.28.)

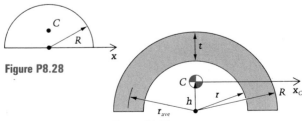

Figure P8.28

Figure P8.29

8.29 Show that the centroidal moment of inertia I_{x_c} of the annular area shown in Figure P8.29 is

$$0.110(R^4 - r^4) - \frac{0.283R^2r^2(R - r)}{R + r}$$

Show that this result becomes $I_{x_c} \approx 0.3tr^3$ if $r \approx R$. Note that

$$h = \frac{4}{3\pi} \left(\frac{R^2 + Rr + r^2}{R + r} \right)$$

8.30 The shaded region $\mathcal{A}$ in Figure P8.30 has an area of 10 ft². If the moment of inertia of the area about the y axis is known to be 600 ft⁴, what is the moment of inertia of the area about the η axis (i.e., about the line $x = 3$ ft)? The centroid C of this area is located at coordinates (7, 3) ft.

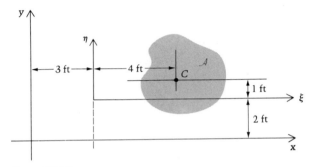

Figure P8.30

8.31 The same shaded region in Problem 8.30 has a radius of gyration with respect to the ζ axis of 2 ft. What is its moment of inertia with respect to the x axis?

8.32 In Figure P8.32:

a. By integration, determine the moment of inertia of the elliptical area about the x axis.

b. What is the radius of gyration of this area with respect to the x axis?

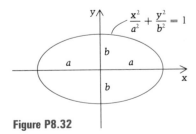

Figure P8.32

8.33 In Figure P8.33:

a. Find the polar moment of inertia J_C of the equilateral triangular area with respect to its centroid C.

b. Compare the result of part (a) with J_C of a square cross section having the same area. Think about which should be larger before making the calculation.

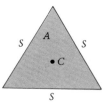

Figure P8.33

8.34 Find the polar moment of inertia of the triangular area of Problem 8.33 about one of its vertices.

For Problems 8.35–8.37 find the polar moments of inertia of the areas about their respective centroids in the problems referred to.

8.35 Problem 8.20.

8.36 Problem 8.22.

8.37 Problem 8.23.

8.38 In Problem 8.20, find the radius of gyration k_x of the shaded area.

8.39 In Problem 8.20, find the radius of gyration k_y of the shaded area.

8.40 Find the radius of gyration k_x for the shaded area of Problem 8.21.

8.5 The Method of Composite Areas

We learned in Chapter 7 that in finding the centroid of a composite area A, the integration could be divided into separate integrals over the various areas comprising A. Thus, for example, if $A_1 + A_2 + A_3 = A$, then

$$A\bar{y} = \int_A y\, dA = \int_{A_1} y\, dA + \int_{A_2} y\, dA + \int_{A_3} y\, dA$$
$$= A_1\bar{y}_1 + A_2\bar{y}_2 + A_3\bar{y}_3$$

or

$$\bar{y} = \frac{A_1\bar{y}_1 + A_2\bar{y}_2 + A_3\bar{y}_3}{A_1 + A_2 + A_3}$$

This same idea, or method of "composite areas," may be used in calculating moments and polar moments of inertia:

$$I_x = \int_A y^2\, dA = \int_{A_1} y^2\, dA + \int_{A_2} y^2\, dA + \int_{A_3} y^2\, dA$$
$$= I_x^{A_1} + I_x^{A_2} + I_x^{A_3}$$

where $I_x^{A_1}$ is the moment of inertia of the area A_1 about the x axis, and similarly for A_2 and A_3.

We consider some examples made simpler by the use of composite areas.

EXAMPLE 8.11

Calculate the polar moment of inertia of the ten identical small bolt cross sections (studied previously in Example 7.10 and shown in Figure E8.11) about both the origin and point C.

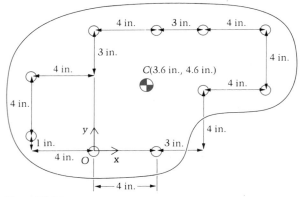

Figure E8.11

Solution

If, as the figure suggests, each bolt diameter is small in comparison with the spacings between bolts, then all points in one of the areas are approximately the same distance from the point with respect to which we desire to calculate the polar moment of inertia. Then

$$J_O = \int r^2 \, dA = \int_{A_1} r^2 \, dA + \cdots + \int_{A_{10}} r^2 \, dA$$

$$\approx r_1^2 \int_{A_1} dA + \cdots + r_{10}^2 \int_{A_{10}} dA = \sum_{i=1}^{10} A_i r_i^2$$

where r_i is the distance from O to the center of the ith area. The expression above corresponds to thinking of each bolt's area as "concentrated" at its center.

With identical areas ($A_i = A$), we obtain

$$J_O = A \sum_{i=1}^{10} r_i^2$$

Starting at O and proceeding counterclockwise we have

$$J_O = A \left[0 + 4^2 + \sqrt{7^2 + 4^2}^2 + \sqrt{11^2 + 4^2}^2 + \sqrt{11^2 + 8^2}^2 \right.$$

$$\left. + \sqrt{7^2 + 8^2}^2 + \sqrt{4^2 + 8^2}^2 + 8^2 + \sqrt{(-4)^2 + 5^2}^2 + \sqrt{(-4)^2 + 1^2}^2 \right]$$

$$= A[4^2 + 7^2 + 4^2 + 11^2 + 4^2 + 11^2 + 8^2 + 7^2 + 8^2 + 4^2 + 8^2$$

$$+ 8^2 + 4^2 + 5^2 + 4^2 + 1^2]$$

$$= 718A \text{ in.}^4$$

where A is in in.^{2} Note, in the above steps, that it is a waste of time to take the square root to get r_i because we are going to then immediately square it to get r_i^2.

Now let us back up and put this analysis on a more rigorous basis. Let J_{C_i} be the polar moment of inertia of the ith area with respect to its own centroid. Using the fact that polar moments of inertia of composite parts add to yield the polar moment of inertia of the whole area, and also using the parallel-axis theorem for each of the parts,

$$J_O = \sum_{i=1}^{10} (J_{C_i} + A_i r_i^2)$$

where r_i is the distance from O to C_i. Thus

$$J_O = \sum_{i=1}^{10} J_{C_i} + \sum_{i=1}^{10} A_i r_i^2$$

and the second sum is precisely what was calculated above. For circular areas (see Example 7.4)

$$J_{C_i} = \frac{A_i R_i^2}{2}$$

where R_i is the radius of the ith bolt's area. With identical areas, we have $R_i = R$, $A_i = A$, and:

$$\begin{aligned} J_O &= \frac{10AR^2}{2} + A \sum_{i=1}^{10} r_i^2 \\ &= 5R^2A + 718A \\ &= (5R^2 + 718)A \text{ in.}^4 \end{aligned}$$

and the smaller is R the better is the concentrated-area approximation. For example, if $R = \frac{1}{4}$ in., then $5R^2A = \frac{5}{16}A$ and the error associated with the concentrated-area approximation is about 0.04%.

To get J_C (the polar moment of inertia about the centroid), we shall use the transfer (or parallel-axis) theorem. The centroid C was computed in Example 7.10 to be at $(\bar{x}, \bar{y}) = (3.6, 4.6)$ in. Therefore,

$$J_O = J_C + A_{\text{Total}}\, d^2$$

where $d = |\mathbf{r}_{OC}|$. Substituting,

$$\begin{aligned} J_C &= 718A - (10A)(3.6^2 + 4.6^2) \\ &= 718A - 341A \\ &= 377\, A \text{ in.}^4 \end{aligned}$$

Note that the transfer theorem can be used to *find* J_C if J is known with respect to some other point such as O in this example.

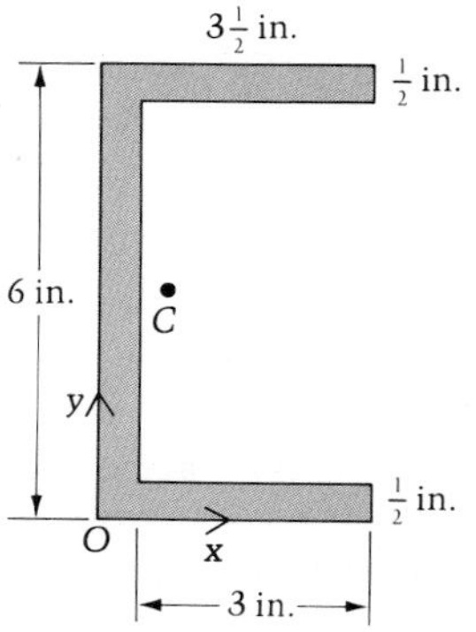

Figure E8.12a

Example 8.12

Find the moment of inertia I_{x_C} for the cross-sectional area of the channel beam shown in Figure E8.12a.

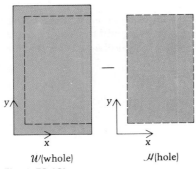

Figure E8.12b

$\mathcal{W}$(whole) $\mathcal{H}$(hole)

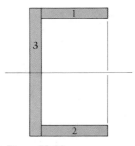

Figure E8.12c

Solution

We shall solve the problem by finding I_x and then using the transfer theorem. We shall find I_x in two different ways. First, using Figure E8.12b,

$$I_x = \overbrace{I_x^{\mathcal{W}}} - \overbrace{I_x^{\mathcal{H}}}$$

$$= \frac{3.5(6^3)}{3} - \left[\frac{3 \times 5^3}{12} + (3 \times 5)(2.5 + 0.5)^2\right]$$

$$= 252 - 166$$

$$= 86 \text{ in.}^4$$

Alternatively, using Figure E8.12c,

$$I_x = I_x^{A_1} + I_x^{A_2} + I_x^{A_3}$$

$$= \left[\frac{3(0.5^3)}{12} + (3 \times 0.5)(5.75^2)\right] + \left[\frac{3(0.5^3)}{12} + (3 \times 0.5)(0.25^2)\right]$$

$$+ \left[\frac{0.5(6^3)}{12} + (0.5 \times 6)(3^2)\right]$$

$$= 49.6 + 0.125 + 36.0 = 85.7 \approx 86 \text{ in.}^4 \qquad \text{(as before)}$$

Using the parallel-axis theorem,

$$I_x = I_{x_c} + Ad^2$$

$$85.7 = I_{x_c} + [6 \times 3.5 - 5 \times 3](3)^2$$

$$I_{x_c} = 85.7 - 6(3^2) = 31.7 \text{ in.}^4$$

The calculation of I_{y_c} will be given as an exercise. (See Problem 8.73.)

EXAMPLE 8.13

Show that even if a triangle, such as the one shown in Figure E8.13, has no right angle, its moment of inertia about a centroidal axis parallel to a base B is still $BH^3/36$, where H is the height of the triangle in the direction normal to that base.

Solution

Let us call the shaded triangle AED of interest $\mathcal{I}$, and denote the two right triangles AEF and DEF by $\mathcal{W}$ and $\mathcal{H}$, respectively. Note that $A_{\mathcal{I}} = A_{\mathcal{W}} - A_{\mathcal{H}}$, so that

$$I_{x_c}^{\mathcal{I}} = \int_{\mathcal{I}} y^2 \, dA = \int_{\mathcal{W}} y^2 \, dA - \int_{\mathcal{H}} y^2 \, dA$$

$$= \frac{(B + Q)H^3}{36} + \frac{(B + Q)H}{2}(d_{\mathcal{W}\mathcal{I}}^2) - \left[\frac{QH^3}{36} + \frac{QH}{2}(d_{\mathcal{H}\mathcal{I}}^2)\right]$$

where $d_{\mathcal{W}\mathcal{I}}$ and $d_{\mathcal{H}\mathcal{I}}$ are the respective distances between the x axis in the figure and the x axes through the centroids of $\mathcal{W}$ and $\mathcal{H}$. But all three of these x axes are coincident, a distance $H/3$ above the base line ADF, because of earlier general results for centroids of triangles. Therefore, $d_{\mathcal{W}\mathcal{I}} = d_{\mathcal{H}\mathcal{I}} = 0$, and we obtain

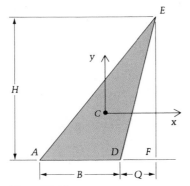

Figure E8.13

$$I_{x_c}^? = \frac{BH^3}{36}$$

the same result that we obtained for the right triangle in Example 8.8.

EXAMPLE 8.14

From previous examples, we have the moments of inertia shown in Figure E8.14. Show that I_x for the rectangle follows from the indicated moments of inertia for the two right triangles, using the transfer theorem and the method of composite areas.

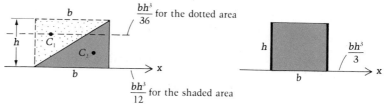

Figure E8.14

Solution

Transferring the moment of inertia of the dotted area to the x axis gives

$$I_x^{dotted} = \underbrace{\frac{bh^3}{36}}_{I_{x_{C_1}}^{dotted}} + \underbrace{\frac{1}{2}bh}_{A}\underbrace{\left(\frac{2}{3}h\right)^2}_{d^2} = \frac{bh^3}{4}$$

Then,

$$I_x^{rectangle} = I_x^{dotted} + I_x^{shaded}$$

$$= \frac{bh^3}{4} + \frac{bh^3}{12} = \frac{bh^3}{3}$$

which agrees with the previously obtained answer for the rectangle.

EXAMPLE 8.15

Find the moments of inertia about the x and y axes through the centroid of the cross-sectional area shown in Figure E8.15. This cross section of a wide-flange beam is called a W14 × 84 because its nominal depth is 14 in. and it weighs 84 lb per ft.

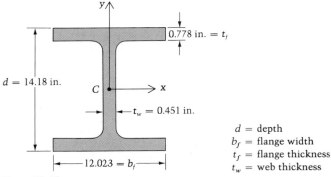

d = depth
b_f = flange width
t_f = flange thickness
t_w = web thickness

Figure E8.15

Solution

$$I_{x_c} = I_{x_c}^{\text{web}} + I_{x_c}^{\text{flange}}$$

$$= \underbrace{\frac{0.451[14.18 - 2(0.778)]^3}{12}}_{\text{web}}$$

$$+ \underbrace{2\left[\frac{12.023(0.778)^3}{12} + 12.023(0.778)\left(\frac{14.18}{2} - \frac{0.778}{2}\right)^2\right]}_{\text{flanges}}$$

Thus

$$I_{x_c} = 75.6 + 2[0.472 + 420] = 916 \text{ in.}^4$$

Notice that the flanges contribute most of the inertia (91.7%!) because of the large "transfer term." Continuing,

$$I_{y_c} = \underbrace{2\left[\frac{0.778(12.023)^3}{12}\right]}_{\text{flanges}} + \underbrace{\frac{[14.18 - 2(0.778)](0.451)^3}{12}}_{\text{web}}$$

$$= 225 + 0.0965 = 225 \text{ in.}^4$$

Notice how tiny the web contribution is here. (Its area is all very close to the y axis!)

The American Institute of Steel Construction manual lists I_{x_c} and I_{y_c} to be 928 in.⁴ and 225 in.⁴ The table values will generally be higher than the results of calculations such as we have made above because of the additional material due to the rounded fillets where the web and flanges are joined. To further illustrate this point, note that we would compute the area of the section to be

$$A = 2(12.023)(0.778) + [14.18 - 2(0.778)](0.451)$$

$$= 18.7 + 5.69 = 24.4 \text{ in.}^2$$

whereas the true are (again from the steel construction manual) is 24.7 in.²

Question 8.8 Why do you suppose I_{x_c} is higher in the manual than we have calculated it to be, while the I_{y_c} values are the same?

Answer 8.8 Because the fillets are so close to the y axis and so relatively far away from the x axis, they will contribute, relatively, a lot to I_{x_c} and very little to I_{y_c}.

PROBLEMS ▸ Section 8.5

8.41 Compute I_x for the I section shown in Figure P8.41.

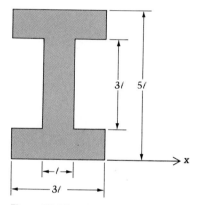

Figure P8.41

8.42 Find the moment of inertia I_{x_c} of the hollow rectangular area shown in Figure P8.42.

Figure P8.42

8.43 Find I_x for the area shown in Figure P8.43.

8.44 In Problem 8.43 find I_y.

In Problems 8.45–8.68 find the moment of inertia about the x axis (odd numbers) and y axis (even numbers) for the areas.

8.45, 8.46

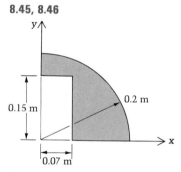

Figure P8.45, P8.46

8.47, 8.48

Figure P8.47, P8.48

8.49, 8.50

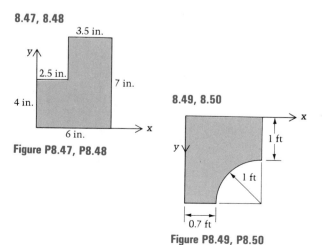

Figure P8.49, P8.50

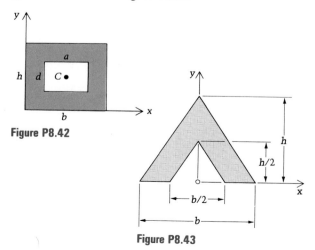

Figure P8.43

For Problems 8.51–8.68, see the instructions above Problem 8.45.

8.51, 8.52

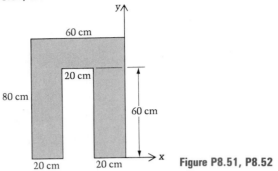

Figure P8.51, P8.52

8.53, 8.54

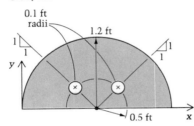

Figure P8.53, P8.54

8.55, 8.56

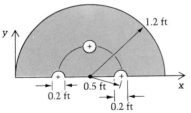

Figure P8.55, P8.56

8.57, 8.58

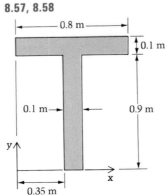

Figure P8.57, P8.58

8.59, 8.60

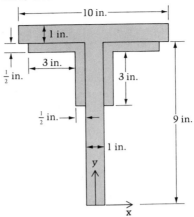

Figure P8.59, P8.60

Problem 8.61 is actually an S24 × 120 I beam, for which the published centroidal moment of inertia is $I_x = 3030$ in.[4]

8.61, 8.62

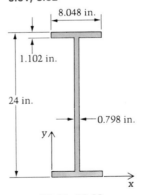

Figure P8.61, P8.62

8.63, 8.64

Hexagon, side length s

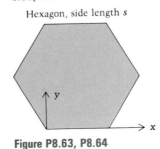

Figure P8.63, P8.64

8.65, 8.66

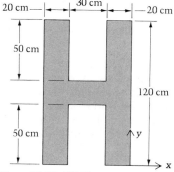

Figure P8.65, P8.66

* 8.67, 8.68

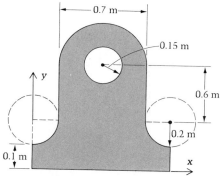

Figure P8.67, P8.68

8.69   Show that the moment of inertia I_{x_C} of the hexagonal area in Figure P8.69 has the value indicated.

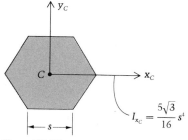

$$I_{x_C} = \frac{5\sqrt{3}}{16} s^4$$

Figure P8.69

8.70   In Problem 8.69 find I_{y_C} and then note that it is the same as I_{x_C}.

---

* Asterisks identify the more difficult problems.

8.71   Find the moment of inertia I_{x_C} of the cross-sectional area of a steel angle beam. Assume a constant thickness of $\frac{5}{8}$ in., and compare your answer with the actual (considering rounded corners and fillets) value of 21.1 in.[4] The x and y coordinates of C in Figure P8.71 are (1.03, 2.03) in.

8.72   Find I_{y_C} in Problem 8.71. Compare with the table value of 7.5 in.[4]

8.73   Find I_{y_C} for the cross-sectional area of the channel beam of Example 8.12. Do it in two ways as was done in the example in finding I_{x_C}. (See Figure P8.73.)

8.74   Find I_{x_C} for the shaded Z section in Figure P8.74.

8.75   Find I_{y_C} for the area of Problem 8.74.

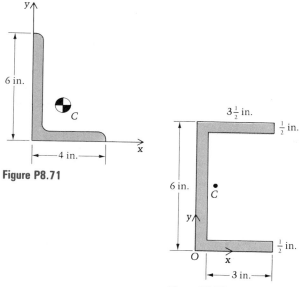

Figure P8.71

Figure P8.73

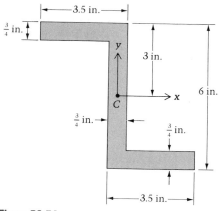

Figure P8.74

8.76 Find the moment of inertia of the composite area about the x axis through the centroid.

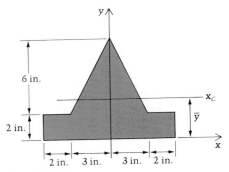

Figure P8.76

8.77 Find the moment of inertia of the composite area in Problem 8.76 about the y axis.

8.78 Prove that the moments of inertia of the triangular area about the two indicated lines have the values shown in Figure P8.78.

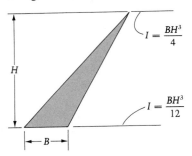

Figure P8.78

8.79 In Figure P8.79 find the moment of inertia of the shaded area about (a) the x axis; (b) the y axis.

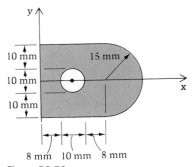

Figure P8.79

8.80 In Figure P8.80 find the polar moment of inertia with respect to C of the annular area by (a) integration; (b) composite areas. (c) Show that your answer approaches $2\pi R_0^3 t$ as $R_i \rightarrow R_o$, where t is the thickness $(R_o - R_i)$.

8.81 A box beam is formed by welding together two steel C15 × 50 channels and two plates as suggested by Figure P8.81(a). The properties of an individual channel are shown in Figure P8.81(b). Find I_{x_c} and I_{y_c} for the area of the box beam. (*Note:* The designation C15 × 50 means: C = channel section; 15 = nominal depth in inches; 50 = weight in lb/ft.)

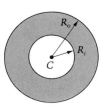

Figure P8.80

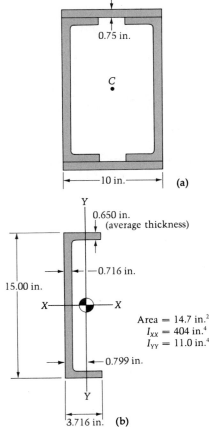

Figure P8.81

8.82 The W12 × 65 steel beam cross section shown in Figure P8.82 has the centroidal moments of inertia I_{XX} − 533 in.^{4} and I_{YY} = 175 in.^{4} Find the centroidal moments of inertia if two of these beams are welded together to form the combined cross section.

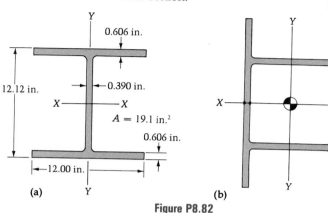

0.606 in.

12.12 in.

0.390 in.

X——X

A = 19.1 in.^{2}

0.606 in.

12.00 in.

(a)   Y

(b)

Figure P8.82

8.83 (a) Find I_x and I_y for the section shown in Figure P8.83 in which $t \ll b, h$. (Neglect all powers of t higher than the first.) (b) Find the ratio b/h for which $I_x = I_y$. *Hint:* You'll have to solve a cubic polynomial equation! Use trial and error with a calculator, and note that there is only one positive root.

8.84 The two 6 × 4 × $\frac{7}{8}$ angles are welded to the shaded steel plate as shown in Figure P8.84. Find the moment of inertia and radius of gyration of the combined area about the x axis.

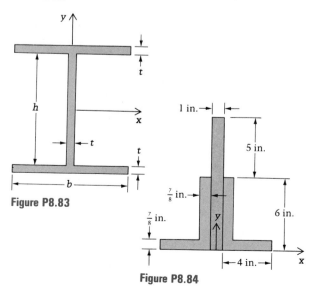

y

t

h

t

x

t

b

Figure P8.83

1 in.

$\frac{7}{8}$ in.

5 in.

$\frac{7}{8}$ in.

y

6 in.

4 in.

x

Figure P8.84

8.85 Repeat Problem 8.84 for the moment of inertia about the y axis.

8.86 In Figure P8.86 show that the moment of inertia of the trapezoidal area about the x axis is $(3b + a)H^3/12$, independent of the angles α_1 and α_2.

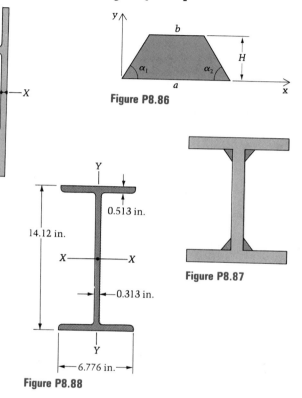

y

b

H

α_1 α_2

a

x

Figure P8.86

Figure P8.87

Y

0.513 in.

14.12 in.

X——X

0.313 in.

Y

6.776 in.

Figure P8.88

8.87 In Example 8.15 assume that four isosceles triangles form the fillets that give the difference between the true area, 24.7 in.^{2}, and the 24.4-in.^{2} area of the web and flanges. (See Figure P8.87.) Show that these triangles (a) contribute the difference between I_{x_c} in the AISC manual (928 in.^{4}) and in the example (916 in.^{4}) and that (b) the triangles contribute negligibly to I_{y_c} so that the manual and example values (225 in.^{4}) should in fact agree as they do.

8.88 A W14 × 38 (meaning the depth is about 14 in. and the weight per ft is 38 lb) wide-flange beam has the cross section shown in Figure P8.88. Calculate the moment of inertia about the centroidal axis X-X and the radius of gyration about this axis. Compare your answers with the table (American Institute of Steel Construction Manual) values of 386 in.^{4} and 5.88 in. (They will differ slightly due to fillets.)

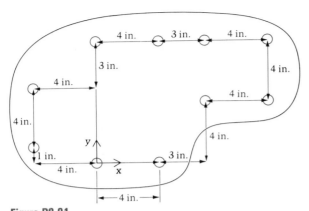

Figure P8.91

8.89 Repeat Problem 8.88 for axis Y-Y. The table results this time are 26.6 in.^{4} and 1.54 in.

8.90 Find the polar moment of inertia of the area of Problem 8.41 about its centroid.

8.91 Find the centroidal polar moment of mertia of the bolt pattern in Example 8.11 by direct calculation using $\Sigma A_i r_i^2$, where r_i = distance from A_i to C. Your answer should, of course, agree with J_C from that example. See Figure P8.91.

8.92 Find the polar moment of inertia of the area of Problem 8.79 about the center of the hole.

8.6 Products of Inertia of Plane Areas

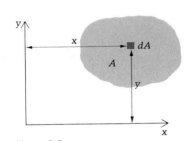

Figure 8.6

Suppose our area A of interest is again in the xy plane, as suggested by Figure 8.6. Then the product of inertia with respect to the axes x and y is defined to be*

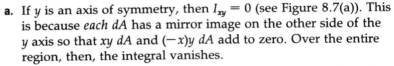

$$I_{xy} = -\int_A xy \, dA$$

Thus the product of inertia is a measure of the imbalance of the area with respect to the two axes. Two important special cases in which the product of inertia vanishes are:

a. If y is an axis of symmetry, then $I_{xy} = 0$ (see Figure 8.7(a)). This is because *each dA* has a mirror image on the other side of the y axis so that $xy \, dA$ and $(-x)y \, dA$ add to zero. Over the entire region, then, the integral vanishes.

b. If x is an axis of symmetry, then I_{xy} again $= 0$ (see Figure 8.7(b)). This time, *each dA* has a mirror image on the other

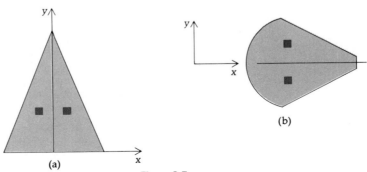

Figure 8.7

---

* Some authors define I_{xy} without the minus sign. The reason for our choice here is noted at the end of Section 8.7.

side of the x axis, and $xy\, dA$ plus $x(-y)\, dA$ add to zero. I_{xy} thus again equals zero.

Frequently, however, we have to deal with *unsymmetrical* areas for which I_{xy} might not be zero. Four such examples follow.

EXAMPLE 8.16

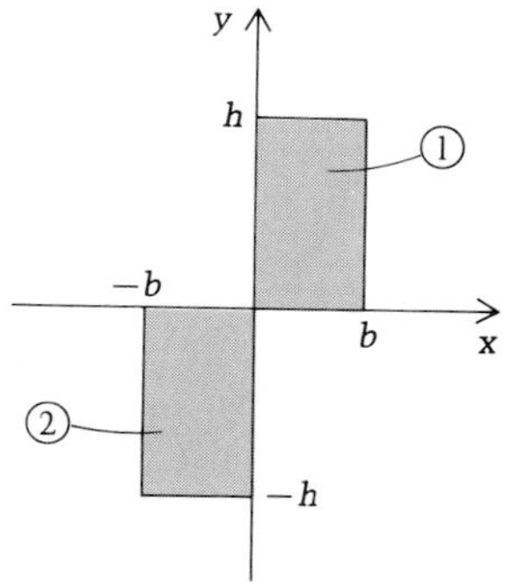

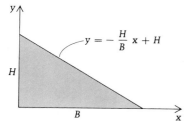

y
h
①
$-b$
b x
②
$-h$

Figure E8.16

Find I_{xy} for the shaded area in Figure E8.16.

Solution

Letting b and h be the base and height of each rectangle, we have

$$I_{xy} = -\int_{A_1} xy\, dA - \int_{A_2} xy\, dA$$

$$= -\int_{y=0}^{h}\int_{x=0}^{b} xy\, dx\, dy - \int_{y=-h}^{0}\int_{x=-b}^{0} xy\, dx\, dy$$

$$= \int_{y=0}^{h} -\frac{x^2}{2}\Big|_0^b\, y\, dy - \int_{y=-h}^{0} \frac{x^2}{2}\Big|_{-b}^0\, y\, dy$$

$$= \frac{-b^2}{2}\frac{y^2}{2}\Big|_0^h + \frac{b^2}{2}\frac{y^2}{2}\Big|_{-h}^0 = \frac{-b^2 h^2}{4} - \frac{b^2 h^2}{4}$$

or

$$I_{xy} = \frac{-b^2 h^2}{2}$$

EXAMPLE 8.17

y
$-y = -\dfrac{H}{B}x + H$
H
B
x

Figure E8.17

Find I_{xy} for the shaded area in Figure E8.17.

Solution

$$I_{xy} = -\int xy\, dA$$

$$I_{xy} = -\int_{y=0}^{H}\int_{x=0}^{\frac{B}{H}(H-y)} xy\, dx\, dy = -\int_{y=0}^{H} y\,\frac{x^2}{2}\Big|_0^{\frac{B}{H}(H-y)}\, dy$$

$$= -\int_0^H \frac{B^2 y}{2H^2}(H-y)^2\, dy = \frac{-B^2}{2H^2}\int_0^H (H^2 y - 2Hy^2 + y^3)\, dy$$

$$= \frac{-B^2}{2H^2} H^4\left(\frac{1}{2} - \frac{2}{3} + \frac{1}{4}\right) = \frac{-B^2 H^2}{24}$$

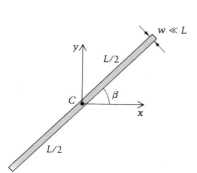

$w \ll L$
y
$L/2$
C
β
x
$L/2$

Figure E8.18a

EXAMPLE 8.18

Find the centroidal product of inertia I_{xyc} for the *thin* strip shown in Figure E8.18a. The strip lies in the xy plane.

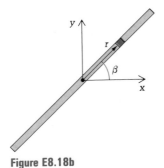

Figure E8.18b

Solution

We shall use the polar coordinate r as shown in Figure E8.18b. The differential area will be $w\,dr$:

$$I_{xy_c} = -\int xy\,dA = -\int_{r=-\frac{L}{2}}^{L/2} (r\cos\beta)(r\sin\beta)w\,dr$$

$$= -w\sin\beta\cos\beta \int_{-L/2}^{L/2} r^2\,dr$$

$$= \frac{-wL^3\sin\beta\cos\beta}{12}$$

$$= \frac{-wL^3\sin 2\beta}{24}$$

Question 8.9   Why is this result only approximately correct?

Answer 8.9   Because $r\cos\beta$ and $r\sin\beta$ are not the (x, y) coordinates of all the points in the "dA." The smaller is w relative to L, the better the approximation.

EXAMPLE 8.19

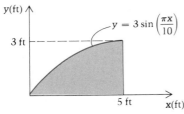

Figure E8.19

Find I_{xy} for the shaded area shown in Figure E8.19.

Solution

$$I_{xy} = -\int xy\,dA = -\int_{x=0}^{x=5}\int_{y=0}^{y=3\sin\frac{\pi x}{10}} xy\,dy\,dx$$

$$= -\int_{x=0}^{5} \frac{xy^2}{2}\Big|_{y=0}^{3\sin\frac{\pi x}{10}} dx$$

$$= -\int_{x=0}^{5} \frac{x}{2}\,9\sin^2\frac{\pi x}{10}\,dx$$

$$= -4.5\int_{x=0}^{5} x\left(\frac{1-\cos(\pi x/5)}{2}\right)dx$$

and using integration by parts,

$$I_{xy} = -4.5\frac{x^2}{4}\Big|_0^5 + 2.25\left(x\frac{\sin(\pi x/5)}{\pi/5}\right)\Big|_0^5 - 2.25\int_{x=0}^{5}\frac{\sin(\pi x/5)}{\pi/5}\,dx$$

$$= -28.1 + 0 + 2.25\left(\frac{5}{\pi}\right)^2\cos(\pi x/5)\Big|_0^5$$

$$= -28.1 - 11.4 = -39.5\text{ ft}^4$$

PROBLEMS ▶ Section 8.6

8.93 Find the product of inertia I_{xy} of the shaded area in Figure P8.93.

8.94 In Figure P8.94:

  a. Find the product of inertia I_{xy} of the thin strip bent into two quarter circles as shown.

  b. Compare your result with that of Example 8.18 for the case in which $\beta = \pi/4$ and when $L/2 = 2\pi R/4$ — that is, when the lengths of the strips in the two problems are the same.

8.95 Find the moments and product of inertia I_x, I_y, and I_{xy} of the shaded area shown in Figure P8.95.

8.96 In Figure P8.96 show that for the shaded triangular area,

$$I_{xy} = -\frac{bH^2(2d + b)}{24}$$

8.97 Show that $I_x = I_y = 2b^4/3$ and $I_{xy} = b^4/2$ for the shaded area in Figure P8.97.

8.98 Find I_{xy} for the shaded area in Figure P8.98.

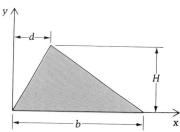

Figure P8.95

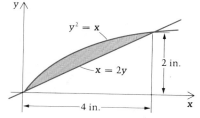

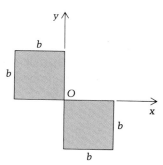

Figure P8.96

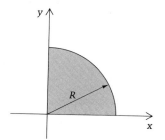

Figure P8.93

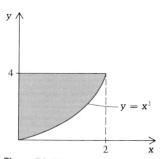

Figure P8.97

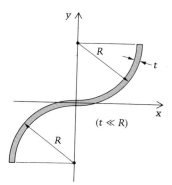

Figure P8.94

Figure P8.98

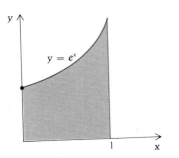

Figure P8.99

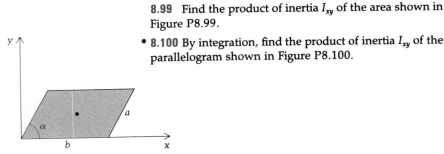

8.99 Find the product of inertia I_{xy} of the area shown in Figure P8.99.

* 8.100 By integration, find the product of inertia I_{xy} of the parallelogram shown in Figure P8.100.

Figure P8.100

8.7 The Parallel-Axis Theorem for Products of Inertia

There is a parallel-axis theorem, or transfer theorem, for products of inertia just as there is for moments of inertia. It is

$$I_{xy_p} = I_{xy_C} - A\bar{x}\bar{y}$$

The proof goes as follows: Referring to Figure 8.8,

$$I_{xy_p} = -\int x\, y\, dA = -\int (\bar{x} + x_1)(\bar{y} + y_1)dA$$

$$I_{xy_p} = -\bar{x}\int y_1\, dA - \bar{y}\int x_1\, dA - \int x_1 y_1\, dA - \bar{x}\bar{y}\int dA$$

The first two integrals vanish by the definition of the centroid, leaving the desired theorem:

$$I_{xy_p} = I_{xy_C} - A\bar{x}\bar{y}$$

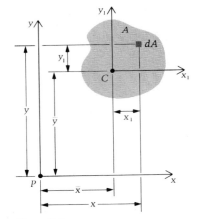

Figure 8.8

> **Question 8.10** Do we need to memorize whether the coordinates $(\bar{x}, \bar{y})$ locate "C with respect to P" as compared to "P with respect to C"?

Answer 8.10 No, because *both* signs change if we go the other way!

EXAMPLE 8.20

Obtain the product of inertia I_{xy} for the area of Example 8.16, shown here as Figure E8.20, by using composite areas and the transfer theorem.

Solution

$$I_{xy} = (I_{x_1y_1}^{A_1} - A_1\bar{x}_1\bar{y}_1) + (I_{x_2y_2}^{A_2} - A_2\bar{x}_2\bar{y}_2)$$

$$= \left[0 - bh\left(\frac{b}{2}\right)\left(\frac{h}{2}\right)\right] + \left[0 - bh\left(-\frac{b}{2}\right)\left(-\frac{h}{2}\right)\right]$$

$$= -\frac{b^2h^2}{4} - \frac{b^2h^2}{4} = -\frac{b^2h^2}{2} \qquad \text{(as before)}$$

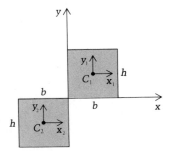

Figure E8.20

Question 8.11   Do we have to have one of the symmetries described at the beginning of Section 8.5 in order for I_{xy} to be zero? If not, think of a case in which $I_{xy} = 0$ even though the area is not symmetric with respect to the x or y axes.

Answer 8.11   No. I_{xy} is zero for this area:

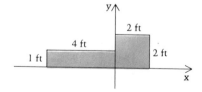

EXAMPLE 8.21

Find I_{xy_c} for the shaded right triangular area shown in Figure E8.21.

Solution

From Example 8.17 we had $I_{xy} = \dfrac{-B^2H^2}{24}$. Using the parallel-axis theorem,

$$I_{xy} = I_{xy_c} - A\overline{x}\overline{y}$$

$$I_{xy_c} = \frac{-B^2H^2}{24} + \frac{BH}{2}\frac{H}{3}\frac{B}{3} = \frac{B^2H^2}{72}$$

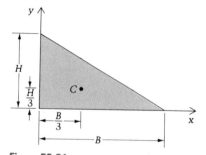

Figure E8.21

EXAMPLE 8.22

Determine the centroidal product of inertia I_{xy} for the "zee" section shown in Figure E8.22.

Solution

We shall label the three parts of the "zee" as indicated. Using composite areas together with the parallel-axis theorem,

$$I_{xy} = I_{xy}^{①} + I_{xy}^{②} + I_{xy}^{③}$$

$$= \left[0 - \underbrace{0.5(3.5)}_{A_①} \underbrace{\left(\frac{3.5 + 0.5}{2}\right)}_{\overline{x}_①} \underbrace{(4)}_{\overline{y}_①} \right] + 0 + \left[0 - \underbrace{0.5(3.5)}_{A_③} \underbrace{\left(\frac{-3.5 - 0.5}{2}\right)}_{\overline{x}_③} \underbrace{(-4)}_{\overline{y}_③} \right]$$

$$= -14.0 - 14.0$$

$$I_{xy} = -28.0 \text{ in.}^4$$

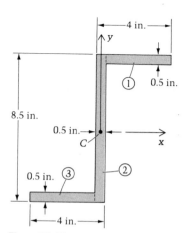

Figure E8.22

   Note that the centroidal axes of ①, ②, and ③, which are parallel to x and y through C are axes of symmetry for the respective parts. Thus each centroidal product of inertia is zero, leaving only the transfer terms. Furthermore, even the transfer term vanishes for ② because its centroid is the same as C so that $\bar{x}_2$ and $\bar{y}_2$ are zero. As expected, then, the only contributions to I_{xyc} are due to the areas ① and ③ being in the "opposite quadrants" (first and third) so that their transfer terms have the same sign.

PROBLEMS ▶ Section 8.7

8.101 What is the product of inertia I_{xyc} of the Z section shown in Figure P8.101?

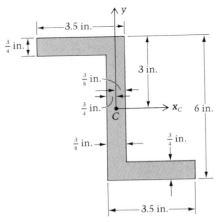

Figure P8.101

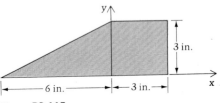

Figure P8.115

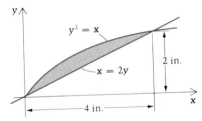

Figure P8.117

8.102 Find the product of inertia I_{xy} for the hollow rectangular area of Problem 8.42.

8.103–8.114 Find the products of inertia I_{xy} for the 12 areas in Problems 8.45–8.68, respectively.

8.115 Find the centroidal moments of inertia I_{xc} and I_{yc} and the product of inertia I_{xyc} of the composite area shown in Figure P8.115.

8.116 Repeat Problem 8.100, this time using the transfer theorem and the fact that $I_{xyc} = -\dfrac{a^3 b}{14} \sin^2 \alpha \cos \alpha$.

8.117 Use the results of Problems 7.23 and 8.95 to find the moments and product of inertia I_{xc}, I_{yc}, and I_{xyc} for the shaded area in Figure P8.117.

8.118 Find I_{xy} for the shaded area in Figure P8.118. (Make use of the parallel-axis theorem!)

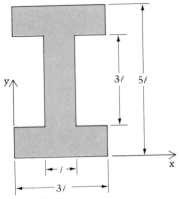

Figure P8.118

8.8 Moments and Products of Inertia with Respect to Rotated Axes Through a Point / Mohr's Circle

Developing the Equation of Mohr's Circle

Suppose we know the moments of inertia I_x and I_y with respect to the orthogonal axes x and y in the plane of an area A, and also the product of inertia I_{xy}. It is possible to derive a pair of equations that will allow us to then find the moments and products of inertia associated with *arbitrarily oriented* orthogonal lines through the point. We shall find these equations in this section, and we shall also see that they may be combined to form the equation of a circle (called **Mohr's circle**), which is a handy tool in finding inertia properties with respect to rotated axes.

We begin by using Figure 8.9 to derive the formula for $I_{x'}$ (the moment of inertia about the arbitrary axis x'). It is assumed that I_x, I_y, and I_{xy} are known for the x and y axes at P.

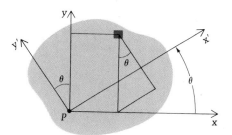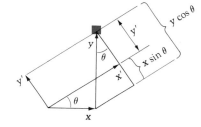

Figure 8.9

Using the definition of the moment of inertia, we obtain

$$I_{x'} = \int y'^2 \, dA = \int (y \cos \theta - x \sin \theta)^2 \, dA$$

in which y' is seen in Figure 8.9 to be $(y \cos \theta - x \sin \theta)$. Expanding,

$$I_{x'} = \cos^2 \theta \int y^2 \, dA - 2 \sin \theta \cos \theta \int xy \, dA + \sin^2 \theta \int x^2 \, dA$$

or, recognizing the definitions of I_x, I_y, and I_{xy},

$$I_{x'} = I_x \cos^2 \theta + I_y \sin^2 \theta + 2I_{xy} \sin \theta \cos \theta \qquad (8.7)$$

Using the following trigonometric identities in Equation (8.7), $I_{x'}$ can be expressed as a function of the angle 2θ:

$$\cos^2 \theta = \frac{1 + \cos 2\theta}{2}$$

$$\sin^2 \theta = \frac{1 - \cos 2\theta}{2}$$

$$\sin \theta \cos \theta = \frac{\sin 2\theta}{2}$$

We thus obtain

$$I_{x'} - \frac{I_x + I_y}{2} = \frac{I_x - I_y}{2} \cos 2\theta + I_{xy} \sin 2\theta \tag{8.8}$$

We will also need $I_{x'y'}$ in terms of I_x, I_y, and I_{xy}:

$$
\begin{aligned}
I_{x'y'} &= -\int x'y' \, dA \\
&= -\int (y \sin \theta + x \cos \theta)(y \cos \theta - x \sin \theta) \, dA \\
&= \sin \theta \cos \theta \left[-\int (y^2 - x^2) \, dA \right] + \underbrace{(\cos^2 \theta - \sin^2 \theta)}_{\cos 2\theta} \left[-\int xy \, dA \right]
\end{aligned}
$$

or

$$I_{x'y'} = -\left(\frac{I_x - I_y}{2} \right) \sin 2\theta + I_{xy} \cos 2\theta \tag{8.9}$$

Equations (8.8) and (8.9) are the transformation equations that give the inertia properties for rotated axes. They can be manipulated into the equation of a circle by squaring both sides of each and adding:

$$\left[I_{x'} - \left(\frac{I_x + I_y}{2} \right) \right]^2 + I_{x'y'}^2 = \left(\frac{I_x - I_y}{2} \right)^2 + I_{xy}^2 \tag{8.10}$$

This is the equation of a circle, which the reader may recognize more easily in the form

$$(x - x_0)^2 + y^2 = R^2$$

The abscissa (x) is seen to be the "moments of inertia axis," while the ordinate is the "products of inertia axis." Thus a point on the circle has the coordinates $(I_{x'}, I_{x'y'})$. The center C of the circle is located at

$$\left(\frac{I_x + I_y}{2}, 0 \right)$$

while the radius R is given by

$$R = \sqrt{\left(\frac{I_x - I_y}{2} \right)^2 + I_{xy}^2}$$

Therefore, Mohr's circle appears as shown in Figure 8.10.

Using Mohr's Circle

Now that we know how to *draw* Mohr's circle, let us proceed to demonstrate how to *use* it. Suppose we wish to know (a) the moment of inertia of an area about an axis x', located angle θ ahead of x as in Figure 8.9, and / or (b) the product of inertia $I_{x'y'}$ with respect to x' and y' (again refer to Figure 8.9). We could, of course, use Equations (8.8) and (8.9) to find these inertia properties directly. But some find it easier to remember, and to make use of, the circle that was derived from these two equations.

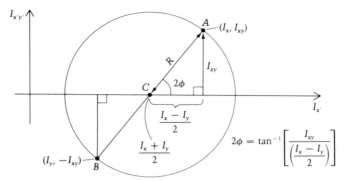

Figure 8.10   Mohr's circle.

Point A in Figure 8.10 has coordinates (I_x, I_{xy}), which correspond to x' lining up with x, and to y' with $y(\theta = 0)$. Point B has the coordinates $(I_y, -I_{xy})$, which corresponds to x' lining up with y, and hence y' with $-x$ $(\theta = 90°)$. These points are 180° apart on the circle because of the "2θ" in the equations — *all the angles on the circle are double what they are on the actual area.* To find $I_{x'}$ for an arbitrary axis x', we rotate, from CA, through 2θ in the opposite direction (i.e., ⟳) on the circle.* We then reach the point P (see Figure 8.11) whose coordinates are $(I_{x'}, I_{x'y'})$. This is all there is to it, but before using this simple procedure, we need to *prove* that the point so identified gives the correct answers for $I_{x'}$ and $I_{x'y'}$ — that is, *that the coordinates of the point P, thus identified, agree with Equations (8.8) and (8.9).*

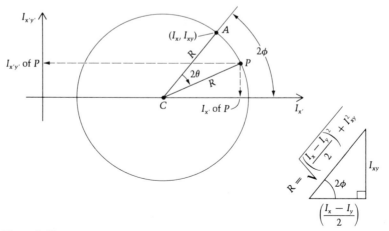

Figure 8.11

---

* If the product of inertia axis is plotted downward instead of upward, the rotation directions on the circle and the actual area will then be the same.

To do this, we first write the $I_{x'}$ coordinate of point P, using Figure 8.11:

$$I_{x'} = \text{center} + (\text{radius}) \cdot [\cos(2\phi - 2\theta)]$$

$$I_{x'} = \left(\frac{I_x + I_y}{2}\right) + R(\cos 2\phi \cos 2\theta + \sin 2\phi \sin 2\theta)$$

$$I_{x'} = \frac{I_x + I_y}{2} + R\frac{\left(\frac{I_x - I_y}{2}\right)}{R}\cos 2\theta + R\frac{I_{xy}}{R}\sin 2\theta$$

$$I_{x'} = \frac{I_x + I_y}{2} + \frac{I_x - I_y}{2}\cos 2\theta + I_{xy}\sin 2\theta$$

This is the same equation as (8.8); therefore, the procedure has yielded the correct answer, in general, for $I_{x'}$. Next we check $I_{x'y'}$. From Figure 8.11 again,

$$I_{x'y'} = (\text{radius})[\sin(2\phi - 2\theta)] = R(\sin 2\phi \cos 2\theta - \cos 2\phi \sin 2\theta)$$

$$= R\frac{I_{xy}}{R}\cos 2\theta - R\frac{\left(\frac{I_x - I_y}{2}\right)}{R}\sin 2\theta$$

$$I_{x'y'} = I_{xy}\cos 2\theta - \left(\frac{I_x - I_y}{2}\right)\sin 2\theta$$

which agrees with Equation (8.9). Therefore, reiterating, once the circle is drawn we can find the values of $I_{x'}$ and $I_{x'y'}$ at any angle $\angle\theta$ with the x axis by rotating on the circle through 2θ ⊃ from the reference line CA and then writing down the coordinates of the point thus located.

We now work an example problem using Mohr's circle, which was named for a German engineer, Otto Mohr (1835–1918), who discovered its relationship to Equations (8.8) and (8.9).

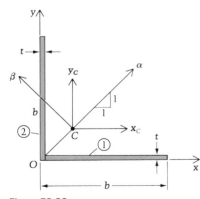

Figure E8.23a

EXAMPLE 8.23

Use Mohr's circle to find the moments of inertia of the cross-sectional area shown in Figure E8.23a about the α and β axes through its centroid C. The thickness t is very small compared to b.

Solution

The steps in this solution will be:

1. Find the centroid C.
2. Find I_x, I_y, and I_{xy}.
3. Use the transfer theorems to get I_{x_c}, I_{y_c}, and $I_{x_cy_c}$.
4. Draw Mohr's circle and use it to obtain I_α and I_β.

(1) The centroid: Let the horizontal rectangle have area A_1 and the vertical one A_2:

$$A\bar{x} = A_1\bar{x}_1 + A_2\bar{x}_2 = (b-t)t\left(\frac{b-t}{2}+t\right) + bt\frac{t}{2}$$

$$\bar{x} = \frac{b^2t/2 + bt^2/2 - t^3/2}{2bt - t^2} \approx \frac{b}{4} \qquad \text{(since } t \ll b\text{)}$$

and

$$\bar{y} \approx \frac{b}{4} \qquad \text{(by symmetry)}$$

(2) The inertia properties at O:

$$I_x = I_{\bar{x}}^{①} + I_{\bar{x}}^{②} = \underbrace{\frac{(b-t)t^3}{12} + (b-t)t\left(\frac{t}{2}\right)^2}_{①} + \underbrace{\frac{tb^3}{12} + tb\left(\frac{b}{2}\right)^2}_{②}$$

$$= \frac{(b-t)t^3}{3} + \frac{tb^3}{3}$$

$$\approx \frac{b^3t}{3}$$

And

$$I_y \approx \frac{b^3t}{3} \qquad \text{(by symmetry)}$$

Also,

$$I_{xy} = \underbrace{0 - (b-t)t\left(\frac{b-t}{2}+t\right)\frac{t}{2}}_{①} + \underbrace{0 - (bt)\frac{t}{2}\frac{b}{2}}_{②}$$

which we shall neglect since $t \ll b$ and the largest term is a "t^2 term."

(3) The inertia properties at C:

$$I_x = I_{x_C} + Ad^2$$

so that

$$I_{x_C} = I_x - Ad^2 \approx \frac{b^3t}{3} - (2bt)\left(\frac{b}{4}\right)^2 = \frac{5}{24}b^3t \Rightarrow I_{x_C} \approx \frac{5}{24}b^3t$$

$$I_{y_C} \approx \frac{5}{24}b^3t \qquad \text{(by symmetry)}$$

$$I_{xy} = I_{xy_C} - A\bar{x}\bar{y}$$

$$I_{xy_C} \approx 0 + (2bt)\frac{b}{4}\frac{b}{4} = \frac{b^3t}{8} \qquad \text{or} \qquad \frac{3b^3t}{24}$$

The reader may wish to check these values by transferring the inertia properties of ① and ② *directly* to C.

(4) We now use Mohr's circle to get the inertias about the α and β axes at C. Factoring $\dfrac{b^3t}{24}$, the circle is drawn as shown in Figure E8.23b in which:

a. Center $= \dfrac{I_{x_c} + I_{y_c}}{2} = 5\left(\dfrac{b^3t}{24}\right)$

b. Radius $= R = \sqrt{\left(\dfrac{I_{x_c} - I_{y_c}}{2}\right)^2 + I_{xy_c}^2} = 3\left(\dfrac{b^3t}{24}\right)$

Point A represents the values (I_{x_c}, I_{xy_c}), while point B represents $(I_{y_c}, -I_{xy_c})$.

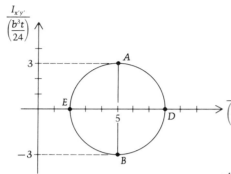

Figure E8.23b

To get from x to α on the actual area, we rotate $45°\circlearrowleft$. Thus we go $90°\circlearrowright$ on the circle, and reach point D of Figure E8.23b, where the moment of inertia is maximum, equaling $I_{x'} = I_\alpha =$ center $+$ radius $= 8\dfrac{b^3t}{24} = \dfrac{b^3t}{3}$, and where $I_{x'y'} = I_{\alpha\beta} = 0$.

> **Question 8.12** Does it make sense that α should be the axis of maximum moment of inertia? That $I_{\alpha\beta}$ should $= 0$? Explain.

Next, to get to β, we continue $90°\circlearrowleft$ *more* on the actual area, therefore $180°$ more $\circlearrowright$ on the circle. This puts us at point E, where $I_{x'} = I_\beta =$ center $-$ radius $= 2\dfrac{b^3t}{24} = \dfrac{b^3t}{12}$, and where again $I_{x'y'} = 0$. Note that the area is distributed fairly close to the β axis through point C, so that it is not surprising that $I_{x'}$ is minimum (see the circle) for this line.

In general, I_x will not equal I_y, so that calculation of the desired $I_{x'}$'s will be slightly more complicated than in the preceding example. However, that example has served to ease us into the use of the circle, and has given us a feel for the axes of maximum and minimum moments of inertia through a point. For these two axes, called the **principal axes,** the product of inertia always vanishes. The associated points on the $I_{x'}$ axis (abscissa) that are farthest to the left and right on the circle give us the **principal moments of inertia.** The circle clearly shows that they are the minimum and maximum moments of inertia at the point and that, for their associated axes, the products of inertia vanish. This has very important implications in the mechanics of deformable solids.

Answer 8.12 Yes; α appears to be the axis for which the area is distributed farthest away, in an overall "$r^2\, dA$" sense. And by symmetry, $I_{\alpha\beta}$ should vanish.

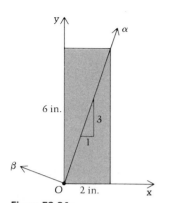

Figure E8.24a

EXAMPLE 8.24

For the rectangular area shown in Figure E8.24a, find the values of I_α, I_β, and $I_{\alpha\beta}$ at the origin (corner O).*

Solution

First we find I_x, I_y, and I_{xy} while reviewing the use of the transfer theorem. We recall that

$$I_{y_c} = \frac{hb^3}{12} = \frac{6(2^3)}{12} = 4 \text{ in.}^4$$

$$I_{x_c} = \frac{bh^3}{12} = \frac{2(6^3)}{12} = 36 \text{ in.}^4$$

$$I_{xy_c} = 0 \quad \text{(by symmetry)}$$

Transferring these centroidal inertia properties to the corner O, we obtain

$$I_{x_O} = I_{x_c} + Ad^2 = 36 + (12)3^2 = 144 \text{ in.}^4$$

$$I_{y_O} = I_{y_c} + Ad^2 = 4 + (12)1^2 = 16 \text{ in.}^4$$

$$I_{xy_O} = I_{xy_c} - A\overline{xy} = 0 - (12)(+1)(+3) = -36 \text{ in.}^4$$

Now we are ready to construct Mohr's circle. We need:

1. The center; it is at

$$\left(\frac{I_x + I_y}{2}, 0\right) = \left(\frac{144 + 16}{2}, 0\right) = (80, 0) \text{ in.}^4$$

2. The reference point A, corresponding to $\theta = 0$, is at $(I_x, I_{xy}) = (144, -36)$ in.^{4}

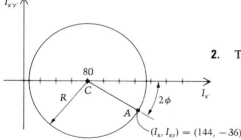

Figure E8.24b

Therefore the circle is constructed as shown in Figure E8.24b. We see from the circle that

$$R = \sqrt{(144 - 80)^2 + (36)^2}$$
$$= \sqrt{(64)^2 + (36)^2} = 73.430 \text{ in.}^4$$

and

$$2\phi = \tan^{-1}\left(\frac{36}{64}\right) = 29.358°$$

Next, the line α is seen in Figure E8.24a to make an angle with the x axis of:

$$\theta = \tan^{-1}\left(\frac{6}{2}\right) = 71.565°$$

---

* We shall start with five-digit accuracy in this example so as not to obscure certain results with roundoff error.

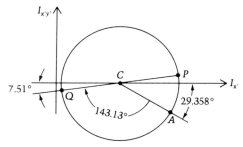

Figure E8.24c

Thus $2\theta = 143.13°$, and we turn on the circle through $2\theta \,\rotatebox[origin=c]{180}{$\circlearrowleft$}$ (opposite to the θ direction ($\circlearrowleft$) on the actual area). We reach the point Q on the circle whose $(I_{x'}, I_{x'y'})$ coordinates correspond to I_α and $I_{\alpha\beta}$ (see Figure E8.24c). The answers for I_α and $I_{\alpha\beta}$ are readily and simply written down with the help of Mohr's circle:

$$I_{x'} = I_\alpha = \text{center} - R \cos 7.51° = 80 - 73.4(0.991) = 7.26 \text{ in.}^4$$

$$I_{x'y'} = I_{\alpha\beta} = -R \sin 7.51° = -73.4(0.131) = -9.59 \text{ in.}^4$$

The value of I_β corresponds to the abscissa of the point P on the circle 180° away (or diametrically opposite) from Q:

$$I_{x'} = I_\beta = \text{center} + R \cos 7.51° = 80 + 73.4(0.991) = 153 \text{ in.}^4$$

It is interesting to note that α (the diagonal of the rectangle) is not the line of minimum moment of inertia. That line is located $143.13 + 7.51 = 151°$ clockwise from CA on the circle; hence $75.5°$ counterclockwise from x on the actual area (see Figure E8.24d).

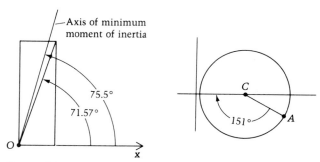

Figure E8.24d

Similarly, to get to the axis of maximum moment of inertia, we turn either $29.36° \circlearrowleft$ from CA on the circle or $151 + 180 = 331° \,\rotatebox[origin=c]{180}{$\circlearrowleft$}$ from CA on the circle. Both these rotations (which, rounded, add to 360°) will reach the point farthest to the right on Mohr's circle — clearly the point with the largest $I_{x'}$, as we have seen before. It has the value of $I_{x'} = \text{center} + \text{radius} = 80 + 73.43 = 153.4 \text{ in.}^4$, and is located 90° away from the previously found axis of minimum moment of inertia. (See Figure E8.24e.)

Note that θ_1 and θ_2 in Figure E8.24e add to 180°, and thus they locate the same line.

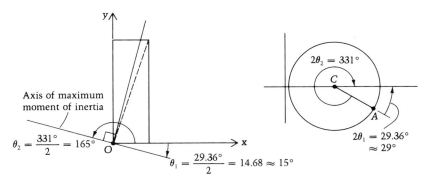

Figure E8.24e

The reader may also come across Mohr's circle again, but in a different context, when studying the mechanics of deformable solids. In that subject, two quantities (stress and strain) each have transformation formulas mathematically identical to Equations (8.8) and (8.9). Therefore, they too have Mohr's circles, by the same derivation as we have presented for inertia properties. Actually, the equations are the transformation formulas for a mathematical entity called a second-order tensor; inertia properties, stress, and strain are thus three examples of second-order tensors. It is a fact that without the minus sign in the definition of the product of inertia ($I_{xy} = -\int xy \, dA$), the inertia properties do not obey the tensor transformation equations, and this is why we have included it in the definition.

PROBLEMS ▶ Section 8.8

8.119 In Figure P8.119 find the principal moments and axes of inertia through the centroid of the shaded area by using Mohr's circle and the results of Problem 8.117.

8.120 Shown in Figure P8.120 is the cross section of a length of $3 \times 3 \times \frac{1}{4}$ "angle iron." Find the exact values of I_α and I_β and compare the results with those of Example 8.23 in which the thickness was neglected.

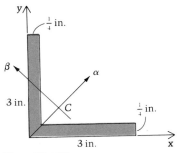

Figure P8.120

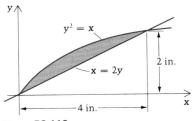

Figure P8.119

8.121 From Examples 8.8 and 8.21 we have these results for the isosceles right triangle shown in Figure P8.121(a):

$$I_{x_c} = I_{y_c} = \frac{h^4}{36} \quad \text{and} \quad I_{xy_c} = \frac{h^4}{72}$$

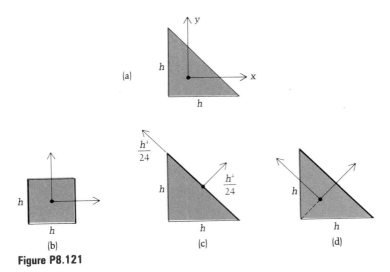

(a)

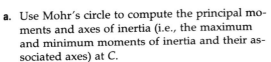

(b)   (c)   (d)

Figure P8.121

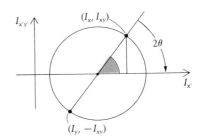

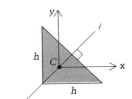

Figure P8.122

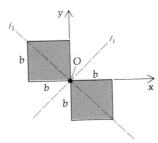

Figure P8.123

a. Use Mohr's circle to compute the principal moments and axes of inertia (i.e., the maximum and minimum moments of inertia and their associated axes) at C.

b. Give an argument that at the center C of the square in Figure P8.121(b), the Mohr's circle is just a point, and that therefore the moment of inertia about all lines through C is $h^4/12$, with the product of inertia vanishing for all pairs of axes there 90° apart.

c. Using Part (b), argue that the moments of inertia of the isosceles right triangle shown in Figure P8.121(c) about the indicated axes are as given in the figure, with the product of inertia associated with these two axes vanishing.

d. Transfer the results of Part (c) to the centroid C of the triangular area [Figure P8.121(d)], and show that the answers agree with those of part (a).

8.122 In Figure P8.122, using Mohr's circle, derive a formula for the angle θ through which we must turn from the x axis in order to reach the axis of maximum moment of inertia. The angle should be in terms of I_x, I_y, and I_{xy}. Test your equation on the bottom sketch shown in Figure P8.122 for which $I_{x_c} = I_{y_c} = h^4/36$ and $I_{xy_c} = h^4/72$, and for which line ℓ is the axis of maximum moment of inertia.

8.123 In an earlier problem, it was shown that $I_x = I_y = \frac{2}{3}b^4$ and $I_{xy} = b^4/2$ for the shaded area shown in Figure P8.123. Now use Mohr's circle to find the maximum and minimum moments of inertia at O and their asso-

ciated axes. You should find that these axes are ℓ_1 and ℓ_2 in the figure. Explain why these lines make sense. Also note that your results are larger and smaller than $\frac{2}{3}b^4$ as they must be.

8.124 Prove with the help of Mohr's circle that $I_x + I_y = I_{x'} + I_{y'}$ for the plane area shown in Figure P8.124.

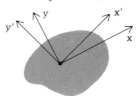

Figure P8.124

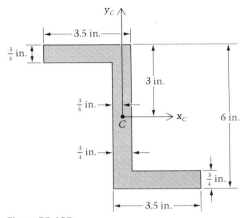

Figure P8.125

8.125 Using Mohr's circle and the results of Problems 8.74, 8.75, and 8.101, find the principal moments and axes of inertia at C for the Z section shown in Figure P8.125.

8.126 Find the smallest moment of inertia for any line drawn through C in Problems 8.71 and 8.72. Compare with the table value of 4.37 in.[4]

COMPUTER PROBLEMS ▶ Chapter 8

8.127 Add to the computer program written for Problem 7.131 the following feature: The program is to also read the centroidal moments of inertia $I_{x_{i_c}}$ and $I_{y_{i_c}}$ of each area A_i. It is then to use these two moments of inertia to *directly* calculate and print (a) the moments of inertia I_x and I_y of the total area about the x and y axes; and (b) the moments of inertia I_{x_c} and I_{y_c} of the total area about axes parallel to x and y through the (calculated) centroid. Also to be printed are the results $I_{x_c} + A\bar{y}^2$ and $I_{y_c} + A\bar{x}^2$, where $\bar{x}$ and $\bar{y}$ are the coordinates of the centroid C. These results should match I_x and I_y if everything is done right. Test the program with the example from Problem 7.131.

8.128 The first three examples in Chapter 8 resulted in centroidal moments of inertia for rectangular, circular, and triangular areas as depicted in Figure P8.128.

Write a program that will use these results, plus the parallel-axis theorem, to place these three areas side by side in the six possible ways shown in the figure and compute I_{y_o} about the left vertical edge. Take $h = 2R = b = 10$ in.

The program should contain an algorithm that does all six problems; i.e., you must teach the computer to permute the bodies and do the transferring of moments of inertia for each permutation.

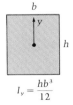

$$I_y = \frac{hb^3}{12}$$

$$I_y = \frac{\pi R^4}{4}$$

Figure P8.128

$$I_y = \frac{hb^3}{48}$$

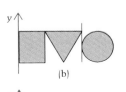

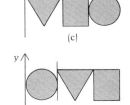

SUMMARY ▶ ## Chapter 8

In this chapter we learned the meaning of the moment and product of inertia of a plane area. With the basic formulas

$$I_x = \int_A y^2 \, dA \qquad \text{and} \qquad I_y = \int_A x^2 \, dA,$$

we derived in Section 8.2 values for the moments of inertia of a number of common shapes, including rectangular, circular, and triangular areas (a number of others are found in Appendix C). The quantities I_x and I_y are important concepts in determining bending and shearing stresses, and also deflections, of laterally loaded beams in studies of the mechanics of deformable bodies.

The polar moment of inertia of a plane area was defined in Section 8.3 to be

$$J_p = \int_A (x^2 + y^2) \, dA \qquad \text{or} \qquad \int_A r^2 \, dA$$

The quantity J_p is crucial in deformable-bodies calculations of shear stress and of the angle of twist in shafts subjected to torsion, or twisting.

In Section 8.4 we discovered the Parallel-Axis Theorem, or Transfer Theorem, for moments of inertia; it is

$$I_{x_p} = I_{x_c} + Ad^2$$

The Transfer Theorem allows us to find I_x about an x-axis through P by adding, to I_x for an x-axis through C, the area multiplied by the square of the distance between the two axes. This means that we don't need to re-integrate to find I_x every time we need the moment of inertia about a different parallel line. The theorem applies as well for y-axes and I_y, and also for the polar moment of inertia.

We also learned in Section 8.4 the meaning of the radius of gyration k_x associated with an x-axis and an area A. It is defined by

$$I_x = Ak_x^2,$$

which shows that the radius of gyration is a number which, when squared and multiplied by the area, gives the moment of inertia of the area about the x-axis. There are also radii of gyration associated with I_y and J_p for an area, defined in the same way.

In Section 8.5 we learned that the idea of composite areas (introduced in conjunction with centroids in Section 7.3) may also be used in inertia calculations by

$$I_x = I_x^{A_1} + I_x^{A_2} + \cdots + I_x^{A_n}$$

where there can be any number of divisions of an area A into $A = A_1 + A_2 + \cdots + A_n$, and where $I_x^{A_i}$ is the moment of inertia of area A_i about the x-axis. It is here that the transfer theorem has its greatest use, for we generally know $I_{x_i}^{A_i}$, the moment of inertia of A_i about the x_i-axis through its own centroid C_i and parallel to x. Thus we merely add $A_i d_i^2$ to $I_{x_i}^{A_i}$ to form $I_x^{A_i}$ and then easily sum all such terms on the right-hand side to form I_x. This idea of course also applies to I_y and to J_p.

Sections 8.6 and 8.7 deal respectively with the product of inertia

$$I_{xy} = -\int_A xy\, dA$$

of a plane area, and with the parallel-axis theorem for products of inertia:

$$I_{xy_p} = I_{xy_c} - A\overline{xy}$$

These concepts become important in deformable bodies and fluid statics whenever x or y are not axes of symmetry for the area in question.

In Section 8.8 we developed formulas (8.8, 8.9) which allow us to find the moments and products of inertia associated with arbitrarily oriented lines through any point at which we already know I_x, I_y and I_{xy}. We then presented the simplifying graphical interpretation of the use of these formulas known as Mohr's circle.

REVIEW QUESTIONS ▶ Chapter 8

True or False?

1. The moment of inertia of a plane area A, about any axis, is always a positive number.

2. The dimension of a moment of inertia of an area is L^3, where L denotes length.

3. The polar moment of inertia of an area A about any axis (z) normal to the plane (xy) of A through a point P in this plane is larger than each of I_{x_p} and I_{y_p}, and in fact equals their sum.

4. Let the moment of inertia of an area A about the x axis through its centroid be called I_{x_c}. Consider the moments of inertia of A about all other axes parallel to x. All these moments of inertia are less than I_{x_c}.

5. The moment of inertia of a circular area about any line in its plane tangent to the circle is $\frac{5}{4}\pi R^4$.

6. If we know the moment of inertia of a plane area A about an axis x in the plane, then the moment of inertia about the parallel axis through the centroid of A is $I_x + Ad^2$, where d is the distance between the axes.

7. The utility of the method of composite areas is significantly enhanced by the parallel-axis theorem.

8. The product of inertia of an area can be positive or negative, but never zero.

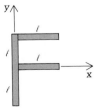

9. The product of inertia I_{xy} of the given letter "F" in the drawing at the left is $\approx -\dfrac{l^3 t}{2}$ where t is the thickness of the letter and $t \ll l$.

10. Mohr's circle is useful in finding the axes and values of the largest and smallest moments of inertia of a plane area with respect to axes through any point.

11. It is possible for Mohr's circle (for area moments of inertia) to cross the ordinate (product of inertia axis).

Answers: 1. T   2. F   3. T   4. F   5. T   6. F   7. T   8. F   9. T   10. T   11. F

9

SPECIAL TOPICS

9.1   Introduction

In this chapter, we introduce the interested reader to two additional subjects in Statics. The first of these (in Section 9.2) is the very old and very interesting principle of virtual work. An alternative to Newton's and Euler's laws, this principle is at the heart of more advanced energy principles of mechanics. But, of course, it can also be used to solve simple statics problems, and, more importantly, to render certain very complicated statics problems simple.

The second special topic in the chapter, examined in Section 9.3, is that of hydrostatic pressure on submerged bodies. In this section, principles of fluids such as Pascal's law and Archimedes' principle are coupled with what we have learned about equilibrium in Chapters 2–4 to solve problems in which some of the forces on a body are caused by pressure from a fluid at rest.

9.2   The Principle of Virtual Work

In Section 4.4 we encountered bodies or structures composed of a number of rigid (or near-rigid) parts with special connections such as pins. When there is a large number of such elements, many free-body diagrams and associated sets of equilibrium equations may be required in the analysis. Among the unknowns appearing in these equations will be all the forces of interaction between adjacent elements or members. It sometimes occurs that most, if not all, of these interactions are not of interest, and we simply have to contend with them in order to get to the results that are important. In such situations there is an equivalent and economical form of analysis that can be used, and it is based upon what is called the *Principle of Virtual Work*. Before stating the principle we must establish some preliminary concepts. For simplicity we restrict our attention here to two-dimensional (plane) problems or situations.

Kinematics of a Rigid Body in Two Dimensions

In two dimensions the general change in configuration possible for a rigid body is shown in Figure 9.1. The vector $\mathbf{r}_{OA}$ is called a position vector for material point A if point O is a fixed point in the frame of reference; $\mathbf{r}_{OA_i}$ denotes that position vector for the initial location of A and $\mathbf{r}_{OA_f}$ the position vector for the final location of A. The vector $\mathbf{u}_A$ is called the displacement that A undergoes when A moves from its initial to final position. So, using the laws of vector addition, we see that

$$\mathbf{u}_A = \mathbf{r}_{OA_f} - \mathbf{r}_{OA_i} \tag{9.1}$$

The rigid-body change in configuration may be accomplished by first moving point A to its final position without changing the orientation of the body (this is called translation), and then rotating the body about an axis (perpendicular to the plane) through A. The angle of rotation is φ.

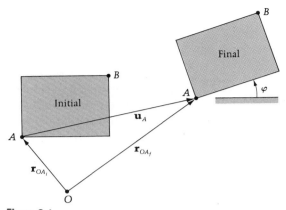

Figure 9.1

The student should use sketches to verify that the same displacement $\mathbf{u}_A$ and angle φ yield the final configuration if the order of the operations is reversed; that is, the body is first rotated about A and then translated to the final configuration. Furthermore, the student should verify that the same change in configuration could be accomplished by a translation taking B to its final position followed by a rotation about B of the *same* angle φ. Thus the most general change in configuration is characterized by the displacement of some point plus a rotation about an axis through that point.

Referring still to Figure 9.1, we note that

$$\mathbf{r}_{OB_f} = \mathbf{r}_{OA_f} + \mathbf{r}_{AB_f}$$

and

$$\mathbf{r}_{OB_i} = \mathbf{r}_{OA_i} + \mathbf{r}_{AB_i}$$

Subtracting, and using Equation (9.1),

$$\mathbf{u}_B = \mathbf{u}_A + (\mathbf{r}_{AB_f} - \mathbf{r}_{AB_i}) \tag{9.2}$$

From the preceding discussion we see that the term $\mathbf{r}_{AB_f} - \mathbf{r}_{AB_i}$ is the displacement that B undergoes when the body is rotated through angle φ about the axis through A. This vector can be put into a particularly useful form if the angle of rotation is infinitesimal; in that case $\mathbf{r}_{AB_f} - \mathbf{r}_{AB_i}$ is infinitesimal and may be written as the differential $d\mathbf{r}_{AB}$. To evaluate it we refer to Figure 9.2 (where $\hat{\mathbf{i}}$ and $\hat{\mathbf{j}}$ are fixed in the frame of reference). The angle θ describes the orientation of the rigid body in which A and B reside, and a change in θ thus represents a rotation of the body. We see from Figure 9.2 that

$$\mathbf{r}_{AB} = |\mathbf{r}_{AB}|(\cos \theta \hat{\mathbf{i}} + \sin \theta \hat{\mathbf{j}}) \tag{9.3}$$

where we note that $|\mathbf{r}_{AB}|$ can't change since the body is rigid. Thus with a differential change in θ we find

$$d\mathbf{r}_{AB} = |\mathbf{r}_{AB}|(-\sin \theta \hat{\mathbf{i}} + \cos \theta \hat{\mathbf{j}})\, d\theta$$

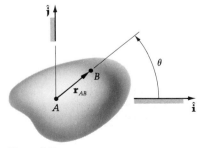

Figure 9.2

which may be expressed as*

$$= |\mathbf{r}_{AB}| (d\theta)\hat{\mathbf{k}} \times (\cos\theta\hat{\mathbf{i}} + \sin\theta\hat{\mathbf{j}})$$
$$= (d\theta)\hat{\mathbf{k}} \times \mathbf{r}_{AB} \qquad (9.4)$$

The vector $(d\theta)\hat{\mathbf{k}}$ is called the infinitesimal rotation vector and it can be seen to conform to a right-hand rule. That is, if the fingers curl in the direction of rotation, the thumb points in the direction of the vector. It is important to recognize that $d\mathbf{r}_{AB}$ is perpendicular to $\mathbf{r}_{AB}$.

We may now return to Equation (9.2) to write

$$\mathbf{u}_B = \mathbf{u}_A + (d\theta)\hat{\mathbf{k}} \times \mathbf{r}_{AB} \qquad (9.5)$$

when the body undergoes an infinitesimal rotation $d\theta$. If $\mathbf{u}_B$ and $\mathbf{u}_A$ are infinitesimal, then they are differentials of $\mathbf{r}_{OB}$ and $\mathbf{r}_{OA}$ and

$$d\mathbf{r}_{OB} = d\mathbf{r}_{OA} + (d\theta)\hat{\mathbf{k}} \times \mathbf{r}_{AB} \qquad (9.6)$$

This equation will be of considerable importance to us shortly, but first we must examine the concept of work.

Work of a Force and Work of a Couple for Infinitesimal Displacements

The increment of work done by a force $\mathbf{F}$ acting on point P of body $\mathcal{B}$ during the infinitesimal displacement $d\mathbf{r}_{OP}$ (see Figure 9.3) is defined to be the scalar

$$dW = \mathbf{F} \cdot d\mathbf{r}_{OP} \qquad (9.7)$$

in which $\mathbf{r}_{OP}$ is a position vector of the point P in the inertial reference frame $\mathcal{I}$, and where it must be stipulated that $\mathbf{F}$ is constant (or at most changes infinitesimally) during the displacement. Note that the definition [Equation (9.7)] tells us that only the component of $\mathbf{F}$ that lies along the displacement $d\mathbf{r}_{OP}$ will contribute to dW; thus this mechanical definition of work is different from our ordinary, everyday perception.†

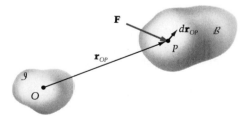

Figure 9.3

---

* An expression of similar form holds in three dimensions, but it is more difficult to derive and to visualize. It is for this reason that the development of this section is restricted to two dimensions.

† Try telling a farmer who has carried a 100-lb sack of feed across a field that he hasn't done any work!

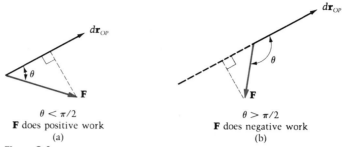

Figure 9.4

Also, note that if the smaller of the two angles between **F** and $d\mathbf{r}_{OP}$ is less than 90°, **F** does positive work during the displacement $d\mathbf{r}_{OP}$ [see Figure 9.4(a)]. If the smaller angle is greater than 90°, however [see Figure 9.4(b)], then the work of **F** during the displacement $d\mathbf{r}_{OP}$ will be negative. We see then that, speaking loosely, we may say that a force does positive work when it "gets to move in the direction it wants to."

> **Question 9.1** What work is done during $d\mathbf{r}_{OP}$ if the angle between **F** and $d\mathbf{r}_{OP}$ is 90°?

Note too that because, if $F = |\mathbf{F}|$,

$$dW = \mathbf{F} \cdot d\mathbf{r}_{OP} = F(|d\mathbf{r}_{OP}|\cos\theta) = (F\cos\theta)|d\mathbf{r}_{OP}| \qquad (9.8)$$

we can view the work increment as either (a) the magnitude of the force times the component of the displacement of P in the direction of the force; or as (b) the magnitude of the displacement of P times the component of **F** in the direction of the displacement.

If we write the two vectors in Equation (9.7) in terms of rectangular Cartesian components parallel to directions fixed in $\mathcal{I}$, then

$$\mathbf{F} = F_x\hat{\mathbf{i}} + F_y\hat{\mathbf{j}}$$
$$\mathbf{r}_{OP} = x\hat{\mathbf{i}} + y\hat{\mathbf{j}}$$

so that

$$d\mathbf{r}_{OP} = dx\hat{\mathbf{i}} + dy\hat{\mathbf{j}}$$

The following equation is another way of writing the work, this time in terms of the rectangular components of the force (see equation 9.7):

$$dW = F_x dx + F_y dy \qquad (9.9)$$

As an example, the work of the force of gravity (or weight) is, with $\hat{\mathbf{j}}$ and positive y downward,

---

Answer 9.1 Zero work.

$$dW = mg\hat{\mathbf{j}} \cdot d\mathbf{r}_{OC} = mg\hat{\mathbf{j}} \cdot (dx\hat{\mathbf{i}} + dy\hat{\mathbf{j}})$$

$$dW = mg\, dy \qquad\qquad (9.10)$$

which is the weight times the downward component of the displacement of the mass center.

The units of work or energy are seen to be the same as are the moments of forces. In the SI system the N · m is used for moment, and the joule (J), which is 1 N · m, is used for work so as to identify the nature of the quantity. Often this distinction is made in the U.S. system by using lb-ft for moment and ft-lb for work.

The increment of work done by a constant couple (of moment $C\hat{\mathbf{k}}$) acting on a body during an infinitesimal rigid change in configuration is defined by

$$dW = C\hat{\mathbf{k}} \cdot [(d\theta)\hat{\mathbf{k}}]$$

$$= C\, d\theta \qquad\qquad (9.11)$$

We see that the work is positive if the sense of turning of the moment of the couple is the same as the direction of the infinitesimal rotation, and negative otherwise.

For this definition to be useful, we surely expect the sum of the increments of work of a pair of equal-in-magnitude but opposite forces to equal the increment of work of the corresponding moment of the couple. To establish this connection, consider the situation depicted in Figure 9.5. The net increment of work done by the forces $\mathbf{F}$ and $-\mathbf{F}$ is

$$dW = \mathbf{F} \cdot d\mathbf{r}_{OB} + (-\mathbf{F}) \cdot d\mathbf{r}_{OA}$$

$$= \mathbf{F} \cdot (d\mathbf{r}_{OB} - d\mathbf{r}_{OA})$$

and using Equation (9.6),

$$dW = \mathbf{F} \cdot [(d\theta)\hat{\mathbf{k}} \times \mathbf{r}_{AB}]$$

$$= (d\theta)\hat{\mathbf{k}} \cdot (\mathbf{r}_{AB} \times \mathbf{F})$$

$$= (d\theta)\hat{\mathbf{k}} \cdot C\hat{\mathbf{k}} = C\, d\theta$$

where we have recognized $C\hat{\mathbf{k}} = \mathbf{r}_{AB} \times \mathbf{F}$ as the moment of the couple composed of the pair of forces.

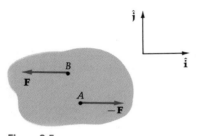

Figure 9.5

Virtual Displacement and Virtual Work

A virtual displacement is an infinitesimal displacement that a material point might be imagined to undergo. The word *virtual* is used to convey that this displacement is imaginary or fictitious; the point in question might be "tied down" (fixed) so that no actual change in configuration of the body would cause the point to move. However, a virtual displacement is a displacement that could take place if the constraint were removed. The symbol δ is used to communicate "virtual" (or imaginary) increments, so that $\delta\mathbf{r}_{OP}$ stands for a virtual displacement and $\delta\theta$ stands for a virtual rotation. Thus, in two dimensions, the most general virtual change of configuration for a rigid body is characterized by a virtual displacement of a point together with a virtual rotation. The virtual displacements of two points A and B are related by

$$\delta \mathbf{r}_{OB} = \delta \mathbf{r}_{OA} + (\delta\theta)\hat{\mathbf{k}} \times \mathbf{r}_{AB} \qquad (9.12)$$

where we have simply substituted δ's for d's in Equation (9.6).

The virtual work of a force is the work that would be done if the point of application of the force were to undergo a virtual displacement *and* the force were not to change during that displacement. Similarly the virtual work of a couple acting on a rigid body is the work that would be done if the body were to undergo a virtual rotation and the couple were not to change during that rotation.

Principle of Virtual Work

The Principle of Virtual Work may be stated as follows. *If a body in equilibrium in an inertial frame is given a virtual rigid change in configuration, starting from the equilibrium configuration, the net virtual work of all of the external forces and couples acting on the body in its equilibrium state vanishes.** We use the symbol δW to stand for the sum of the virtual works of all of the external (equilibrium) forces and couples. Thus the principle is

$$\delta W = 0$$

Before proceeding to a proof of the equivalence of the principle of virtual work and the equilibrium equations, we illustrate this equivalence by a simple example. A coffee cup $\mathcal{B}$ of mass m is in equilibrium on a horizontal table top as shown in Figure 9.6. In the free-body diagram (two-dimensionality has been assumed) we show the weight of the cup and the reaction of the table; this reaction is assumed to have a line of action through B whose distance from the line of action of mg is the unknown a. The problem, of course, is to find the unknown scalars f, N, and a.

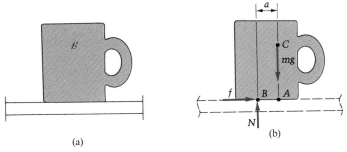

(a)   (b)

Figure 9.6

---

* When the body is caused to deform by a virtual change in configuration, the appropriate generalization of this statement is that the sum of the virtual work of the external forces and the virtual work of the internal forces vanishes. Nontrivial applications of this form are beyond the scope of this book.

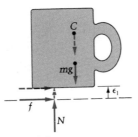

Figure 9.7

First let us give the cup an upward virtual translation, each point in the cup having an upward virtual displacement ϵ_1 (Figure 9.7). We note the virtual work of f is zero since that friction component of reaction is perpendicular to the virtual displacement ϵ_1 of $\mathcal{B}$. The virtual work of N is $N\epsilon_1$, and the virtual work of the weight is $-mg\epsilon_1$ because the mass center C, like every other point, moves upward. Thus

$$\delta W = N\epsilon_1 - mg\epsilon_1 = 0$$
$$(N - mg)\epsilon_1 = 0$$

Because ϵ_1 is arbitrary, we obtain

$$N - mg = 0$$
$$N = mg$$

This, of course, is what we alternatively obtain from the equilibrium equation $\Sigma F_y = 0$. This exercise is instructive with regard to the concept of virtual work. The force N will vanish as soon as the cup breaks contact with the table top, and so an *actual* upward displacement of the cup would be accompanied by zero *actual* work done by N.

Question 9.2 In what way, if at all, does the virtual work calculation above depend upon ϵ_1 being infinitesimal?

As an alternative to the above let us give the cup a downward virtual displacement ϵ_2; in this case (see Figure 9.8 at the left)

$$\delta W = -N\epsilon_2 + mg\epsilon_2 = 0$$

or

$$N = mg \quad \text{(as before)}$$

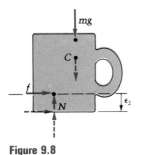

Figure 9.8

This instructive aspect of this part of the exercise is that the virtual displacement ϵ_2 represents a change in configuration that is impossible for the cup to undergo in actuality because it requires the cup to penetrate the table top. Thus in forming virtual changes in configurations, it is not necessary that we conform to actual constraints imposed upon the body.

A horizontal translation of the cup by a displacement ϵ_3 to the right yields

$$\delta W = f\epsilon_3 = 0$$

or

$$f = 0$$

which we recognize as the result of applying the equilibrium equation $\Sigma F_x = 0$.

Not at all.

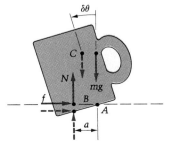

Figure 9.9

Finally let us give the cup a virtual rotation ($\delta\theta$ counterclockwise) about A (see Figure 9.9). We note from the figure that point C will be displaced horizontally, so no virtual work will be done by the weight mg. Point B will move downward a $\delta\theta$, so no virtual work would be done by f even if f were not already known to vanish. Therefore,

$$\delta W = -N(a\,\delta\theta) = 0$$

or

$$a = 0$$

which we recognize as the result that would have been obtained by applying the equilibrium equation $\Sigma M_A = 0$.

> **Question 9.3**   In what way, if at all, does this virtual work-calculation depend upon $\delta\theta$ being infinitesimal?

This example has illustrated the way in which virtual displacements differ from real displacements and the way in which virtual work differs from real work. And we see that the application of the principle for three independent rigid changes in configuration has produced the same results that would be obtained from the three independent equations of equilibrium:

$$\Sigma F_x = 0 \qquad \text{(gives } f = 0\text{)}$$
$$\Sigma F_y = 0 \qquad \text{(gives } N = mg\text{)}$$
$$\Sigma M_A = 0 \qquad \text{(gives } a = 0\text{)}$$

General Equivalence of the Principle of Virtual Work and the Equilibrium Equations

We are now in a position to demonstrate the generality of this equivalence.

Consider the body $\mathcal{B}$ to be acted upon by the external forces $\mathbf{F}_1$, $\mathbf{F}_2, \ldots$ and the couples $C_1\hat{\mathbf{k}}, C_2\hat{\mathbf{k}}, \ldots$, as seen in Figure 9.10.

Now let the body undergo a virtual rigid change in configuration producing changes in the position vectors ($\mathbf{r}_{OP_1}, \mathbf{r}_{OP_2}, \ldots$) of the points of application of the forces (call these virtual changes $\delta\mathbf{r}_{OP_1}, \delta\mathbf{r}_{OP_2}, \ldots$) and a change in the orientation of the body $(\delta\theta)\hat{\mathbf{k}}$. Then the virtual work of the $\mathbf{F}_i$'s and $\mathbf{C}$'s is

$$\delta W = \text{Virtual Work}$$
$$= \mathbf{F}_1 \cdot \delta\mathbf{r}_{OP_1} + \mathbf{F}_2 \cdot \delta\mathbf{r}_{OP_2} + \cdots \qquad (9.13)$$
$$+ C_1\hat{\mathbf{k}} \cdot (\delta\theta)\hat{\mathbf{k}} + C_2\hat{\mathbf{k}} \cdot (\delta\theta)\hat{\mathbf{k}} + \cdots$$

Answer 9.3   Only for infinitesimal $\delta\theta$ does C move horizontally and does the point of application of N and f move vertically.

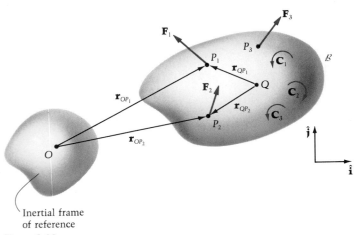

Figure 9.10

Relating the infinitesimal virtual displacements of $P_1, P_2, \ldots$ to that of an arbitrary point of the body, say Q,

$$\delta W = \mathbf{F}_1 \cdot (\delta \mathbf{r}_{OQ} + \delta \theta \hat{\mathbf{k}} \times \mathbf{r}_{QP_1}) + \mathbf{F}_2 \cdot (\delta \mathbf{r}_{OQ} + \delta \theta \hat{\mathbf{k}} \times \mathbf{r}_{QP_2})$$
$$+ \cdots + (C_1 \hat{\mathbf{k}} + C_2 \hat{\mathbf{k}} + \cdots) \cdot \delta \theta \hat{\mathbf{k}}$$

but because

$$\mathbf{F}_1 \cdot (\delta \theta \hat{\mathbf{k}} \times \mathbf{r}_{QP_1}) = (\delta \theta \hat{\mathbf{k}} \times \mathbf{r}_{QP_1}) \cdot \mathbf{F}_1$$
$$= \delta \theta \hat{\mathbf{k}} \cdot (\mathbf{r}_{QP_1} \times \mathbf{F}_1)$$

we obtain

$$\delta W = (\mathbf{F}_1 + \mathbf{F}_2 + \cdots) \cdot \delta \mathbf{r}_{OQ} + \delta \theta \hat{\mathbf{k}} \cdot (\mathbf{r}_{QP_1} \times \mathbf{F}_1$$
$$+ \mathbf{r}_{QP_2} \times \mathbf{F}_2 + \cdots + C_1 \hat{\mathbf{k}} + C_2 \hat{\mathbf{k}} + \cdots)$$
$$\delta W = \Sigma \mathbf{F} \cdot \delta \mathbf{r}_{OQ} + \delta \theta \hat{\mathbf{k}} \cdot \Sigma \mathbf{M}_Q \qquad (9.14)$$

Since the virtual displacement $\delta \mathbf{r}_{OQ}$ of point Q is independent of the virtual angular displacement (rotation) $\delta \theta \hat{\mathbf{k}}$, each can be taken zero while the other is not.* Thus $\delta W = 0$ implies

$$\Sigma \mathbf{F} = \mathbf{0} \qquad \text{and} \qquad \Sigma \mathbf{M}_Q = \mathbf{0} \qquad (9.15)$$

which are the equilibrium equations. Conversely, if $\Sigma \mathbf{F} = \mathbf{0}$ and $\Sigma \mathbf{M}_Q = \mathbf{0}$, it follows from Equation (9.14) that $\delta W = 0$. This establishes (for two dimensions) the equivalence of the principle of virtual work and the equations of equilibrium. This equivalence also holds in three dimensions, but we have not attempted to establish it here because of the more complicated kinematics of rigid bodies in three dimensions.

We now consider an example of the use of the principle of virtual work to establish an equilibrium configuration of a (near-) rigid body.

---

* The virtual work vanishes for *any and all* virtual displacements.

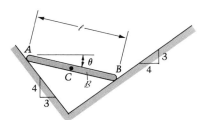

Figure E9.1a

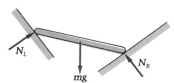

Figure E9.1b

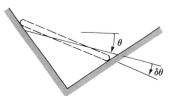

Figure E9.1c

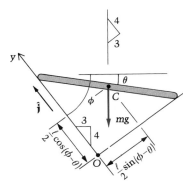

Figure E9.1d

EXAMPLE 9.1

Find the angle θ for equilibrium of the uniform bar $\mathscr{B}$ shown in Figure E9.1a. Both planes are smooth.

Solution

This problem illustrates the way in which we commonly use the principle of virtual work. We can frequently give the body a virtual displacement in which one or more of the unwanted unknowns in the problem do no work and thus do not appear in the equation $\delta W = 0$. We do this here with a virtual displacement in which A and B move along their respective planes.

> **Question 9.4** Why then must B move down its plane if A moves up its plane?

Our sought-after angle increases by $\delta\theta$, as seen in Figure E9.1c. The normal forces N_L and N_R do no virtual work because each is normal to the virtual displacement of the point on which it acts (see Figures E9.1b and E9.1c). Hence only gravity will contribute to the virtual work, which vanishes:

$$\delta W = \mathbf{F}_g \cdot \delta\mathbf{r}_{OC} = 0$$

With the help of Figure E9.1d, we find $\mathbf{r}_{OC}$ and then $\delta\mathbf{r}_{OC}$

$$\mathbf{r}_{OC} = \frac{l}{2}[\sin(\phi - \theta)\hat{\mathbf{i}} + \cos(\phi - \theta)\hat{\mathbf{j}}]$$

$$\delta\mathbf{r}_{OC} = \frac{l}{2}[\cos(\phi - \theta)\hat{\mathbf{i}} - \sin(\phi - \theta)\hat{\mathbf{j}}]\underbrace{\delta(\phi - \theta)}_{-\delta\theta}$$

$$= \frac{l}{2}[\cos\phi\cos\theta + \sin\phi\sin\theta)\hat{\mathbf{i}} - (\sin\phi\cos\theta - \cos\phi\sin\theta)\hat{\mathbf{j}}](-\delta\theta)$$

$$= \frac{-l}{2}\delta\theta[(\tfrac{3}{5}\cos\theta + \tfrac{4}{5}\sin\theta)\hat{\mathbf{i}} - (\tfrac{4}{5}\cos\theta - \tfrac{3}{5}\sin\theta)\hat{\mathbf{j}}]$$

Also, the gravity force, or weight, is

$$\mathbf{F}_g = mg(-\tfrac{3}{5}\hat{\mathbf{i}} - \tfrac{4}{5}\hat{\mathbf{j}})$$

so that, substituting into Equation (1),

$$\delta W = mg\frac{l\,\delta\theta}{2}[\tfrac{3}{5}(\tfrac{3}{5}\cos\theta + \tfrac{4}{5}\sin\theta) - \tfrac{4}{5}(\tfrac{4}{5}\cos\theta - \tfrac{3}{5}\sin\theta)] = 0$$

and since $\delta\theta$ is arbitrary, the bracketed expression vanishes:

$$-\tfrac{7}{25}\cos\theta + \tfrac{24}{25}\sin\theta = 0$$

$$\theta = \tan^{-1}(\tfrac{7}{24}) = 16.3°$$

Answer 9.4 The distance, l, from A to B is constant during the rigid change in configuration.

Virtual Work of Various Common Types of Connections

The real advantage of virtual work is not in analyzing a single rigid body,* but rather in studying problems with systems of rigid bodies, joined together in frictionless ways such as by (a) smooth pins or ball joints, (b) inextensible, taut cables, or (c) linear springs. With (a) and (b), we now proceed to show that in a virtual displacement that is consistent with the connections, no net work is done on the two connected bodies by the pins, ball points, or cables.

If $\mathcal{A}$ and $\mathcal{B}$ are joined by a smooth pin or ball joint, no net work will be done by the forces $\mathbf{F}$ (exerted on $\mathcal{A}$ at A) and $-\mathbf{F}$ (exerted on $\mathcal{B}$ at B), as shown in Figure 9.11. This is because the virtual displacements of the points A and B are the same (they are pinned together) while the forces are equal in magnitude but opposite in direction (in accordance with the principle of action and reaction). Thus,

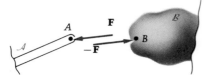

Figure 9.11

$$\delta W = \underbrace{\mathbf{F} \cdot \delta \mathbf{r}_{OA}}_{\text{Work of } \mathbf{F} \text{ on } \mathcal{A}} + \underbrace{-\mathbf{F} \cdot \delta \mathbf{r}_{OB}}_{\text{Work of } -\mathbf{F} \text{ on } \mathcal{B}} = \underbrace{\mathbf{F} \cdot (\delta \mathbf{r}_{OA} - \delta \mathbf{r}_{OB})}_{\text{Zero}} = 0$$

We next let bodies $\mathcal{A}$ and $\mathcal{B}$ be joined by a flexible but inextensible, taut cable, with or without the pulley in between. (See Figure 9.12.) Let the unit vectors $\hat{\mathbf{u}}_{\parallel}$ and $\hat{\mathbf{v}}_{\parallel}$ be along the cable[†] as shown, with $\hat{\mathbf{u}}_{\perp}$ and $\hat{\mathbf{v}}_{\perp}$ being normal to the cable, in the respective directions of the normal (to the cable) components of the virtual displacements $\delta \mathbf{r}_{OA}$ of A and $\delta \mathbf{r}_{OB}$ of B. Then the virtual work done by the cable on the two bodies is

$$\delta W = -T\hat{\mathbf{u}}_{\parallel} \cdot (\delta \mathbf{r}_{OA}\, \hat{\mathbf{u}}_{\parallel} + \delta \mathbf{r}_{OA_{\perp}}\hat{\mathbf{u}}_{\perp}) + T\hat{\mathbf{v}}_{\parallel} \cdot (\delta \mathbf{r}_{OB_{\parallel}}\hat{\mathbf{v}}_{\parallel} + \delta \mathbf{r}_{OB_{\perp}}\hat{\mathbf{v}}_{\perp})$$
$$= -T\delta \mathbf{r}_{OA_{\parallel}} + T\delta \mathbf{r}_{OB_{\parallel}} \tag{9.16}$$
$$= T(\delta \mathbf{r}_{OB_{\parallel}} - \delta \mathbf{r}_{OA_{\parallel}}) = 0$$

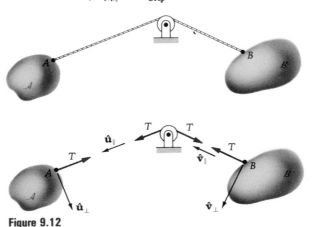

Figure 9.12

---

* For the remainder of this section we shall use the term ''rigid body'' as a shorthand expression to specify that the body is near rigid *and* that it is to be subjected only to rigid virtual changes in configuration.

† Note that $\hat{\mathbf{u}}_{\parallel} = \hat{\mathbf{v}}_{\parallel}$ and $\hat{\mathbf{u}}_{\perp} = \hat{\mathbf{v}}_{\perp}$ if there is no pulley.

because the components of virtual displacement of A and B *along the cable* must be equal if it is inextensible (and, of course, taut).

Question 9.5   Does this result, $\delta W = 0$, depend upon the pulley being frictionless?

Figure 9.13

If we place a spring in the cable as shown above in Figure 9.13, then

$$\delta W = T(\delta \mathbf{r}_{OB_\parallel} - \delta \mathbf{r}_{OA_\parallel})$$

need not vanish, because the spring can stretch. In fact, if we let S be the stretch of the spring, we obtain

$$\delta \mathbf{r}_{OA_\parallel} - \delta \mathbf{r}_{OB_\parallel} = \delta S,$$

the "virtual stretch" of the spring. For a linear spring, the tension in the spring (and cable) is kS where k is the spring modulus, or stiffness, and Equation (9.16) yields

$$\delta W = -kS \, \delta S \qquad (9.17)$$

This means that:

(a) Assuming $S > 0$ (spring is stretched), then if

        (i) $\delta S > 0$, $\delta W < 0$

       (ii) $\delta S = 0$, $\delta W = 0$

      (iii) $\delta S < 0$, $\delta W > 0$

(b) Assuming $S < 0$ (spring is compressed),* then if

        (i) $\delta S > 0$, $\delta W > 0$

       (ii) $\delta S = 0$, $\delta W = 0$

      (iii) $\delta S < 0$, $\delta W < 0$

Of course, if $S = 0$—that is, the spring is unstretched—then it will do *no*

Answer 9.5   Yes, otherwise the cable tensions on the two sides of the pulley would not have to be the same.

* Of course, a transversely flexible cable (or cord) won't support compression. However, a special case of the above analysis is that where the cable and pulley are removed and the ends of the spring are attached directly to bodies $\mathcal{A}$ and $\mathcal{B}$. In that circumstance we can have a compression-supporting spring connection.

virtual work. In the various cases above, the reader is urged to note carefully that the spring does positive virtual work during the virtual displacement if it gets to return *toward* its unstretched length, and negative virtual work if it is forced *farther away* from it.

When we consider two or more rigid bodies connected by frictionless pins or ball joints, and / or by flexible, inextensible cables, we see that the virtual work done by the external forces and couples on the *combined* system of bodies is zero (as was the case for a single rigid body). This is because $\delta W = 0$ on each body and thus it also adds to zero on the combination. Moreover, the virtual work done by a pin (or ball joint or cable) on one body of the connection cancels its work on the other provided the virtual displacements are consistent with the connections, as we have shown. Thus such an internal interaction need not even be considered in the equation $\delta W = 0$! This is the huge advantage in using virtual work: *the non-working forces can be ignored.*

For a linear spring performing non-zero virtual work on a system of rigid bodies, we have seen that its virtual work is $-kS\ \delta S$. If the spring connects the rigid bodies $\mathcal{A}$ and $\mathcal{B}$, then we get the result shown in the following equation:

$$\underbrace{\delta W}_{\substack{\text{(on } \mathcal{A} \text{ except} \\ \text{for spring)} \\ 0}} + \underbrace{\delta W}_{\text{(of spring on } \mathcal{A})} + \underbrace{\delta W}_{\substack{\text{(of spring} \\ \text{on } \mathcal{B})}} + \underbrace{\delta W}_{\substack{\text{(on } \mathcal{B} \text{ except} \\ \text{for spring)} \\ 0}} = 0$$

$$\underbrace{\hspace{3cm}}_{-kS\ \delta S}$$

This way of computing the virtual work done by the spring, as well as the ignoring of non-working internal and reactive forces in computing δW, will now be illustrated by some examples. The reader is urged to mentally consider the difficulty of solving these problems with a set of equilibrium equations. *And* to remember that, if *any* of the forces acting on a body was omitted from the free-body diagram and thence from the equations of equilibrium in Chapters 2, 4, 5, and 6, our solution was doomed!

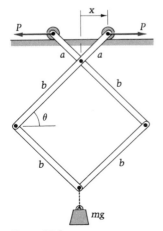

Figure E9.2a

EXAMPLE 9.2

Find the relationship between the forces P and mg and the angle θ, for equilibrium. Neglect the weights of the four bars and the sizes of the two rollers. (See Figure E9.2a.)

Solution

For a virtual displacement (which will be described below), the virtual work will vanish:

$$\delta W = 0 \tag{1}$$

From the equilibrium configuration, we imagine a virtual displacement δx, a slight movement outward of the two points where the forces P are applied. Then the virtual work of these forces is $2[P\ \delta x]$.

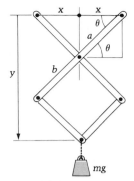

Figure E9.2b

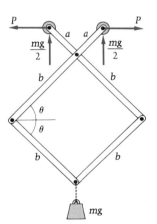

Figure 9.14

The only other force that will perform virtual work is the weight mg. We must determine how much mg "moves" in response to the imagined δx. From Figure E9.2b, we see that $x = a \cos \theta$ so that

$$\delta x = -a \sin \theta \, \delta \theta \qquad (2)$$

which relates δx and the corresponding virtual change $\delta \theta$ in θ according to the rules for differentiation and the formation of differentials. Similarly,

$$y = (a + b) \sin \theta + b \sin \theta = (a + 2b) \sin \theta$$

so that

$$\delta y = (a + 2b) \cos \theta \, \delta \theta$$

which is the virtual displacement of the mass center of the weight. The virtual work of gravity is then $mg \, \delta y$. Thus Equation (1) becomes

$$2P \delta x + mg(a + 2b) \cos \theta \, \delta \theta = 0$$

and using equation (2),

$$2P(-a \sin \theta \, \delta \theta) + mg(a + 2b) \cos \theta \, \delta \theta = 0$$

or

$$[-2Pa \sin \theta + mg(a + 2b) \cos \theta] \, \delta \theta = 0$$

Since $\delta \theta$ is arbitrary, the term in brackets must vanish, giving

$$P = \frac{mg(a + 2b) \cot \theta}{2a}$$

We see from this answer that as $\theta \to 90°$, $P \to 0$; this is because if the rods are vertical, there is no need for any force P to maintain equilibrium.

We now work the problem of the preceding example by the usual method of free-body diagrams accompanied by equilibrium equations of various parts of the structure (see Figure 9.14). From the overall free-body diagram, the roller reactions are each $mg/2$.

The forces exerted on the two lower (two-force!) members shown in Figure 9.15 below are obtained from

$$+\!\uparrow \qquad \Sigma F_y = 0 = 2F_1 \sin \theta - mg$$

$$F_1 = \frac{mg}{2 \sin \theta}$$

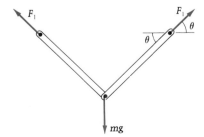

Figure 9.15

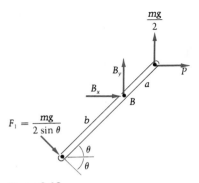

Figure 9.16

Moments about B will now give the value of P for equilibrium. Using the free-body diagram in Figure 9.16,

$$\curvearrowleft_{+} \qquad \Sigma M_B = 0$$

$$= \frac{mg}{2} a \cos \theta + \frac{mg}{2 \sin \theta} \sin(\pi - 2\theta)b - Pa \sin \theta$$

$$P = \left[\frac{mga}{2} \cos \theta + \frac{mgb}{2 \sin \theta} \overbrace{(2 \sin \theta \cos \theta)}^{\sin 2\theta} \right] \Big/ (a \sin \theta)$$

or

$$P = \frac{mg(a + 2b) \cot \theta}{2a}$$

as we obtained earlier using virtual work. Note now that as $\theta \to 0$, $\cot \theta$ (and therefore P) get very large. This can be seen most easily from the figure, where if $\theta \to 0$ only a small fraction ($\sin \theta$) of each F_1 is available to hold up half of mg.

Comparing the two approaches, we see that virtual work is in a sense a "cleaner" approach because forces (such as F_1 above) that do not do any net virtual work on the overall structure do not have to be considered at all in the solution.

EXAMPLE 9.3

Six bars, four of length 2ℓ and two of length ℓ, are pinned together to form a gate as shown in Figure E9.3a. The springs are attached at the midpoints of the indicated members, and the modulus of each is k. The springs can be compressed, and they are unstretched when the gate is fully retracted. Find the angle θ for equilibrium.

Figure E9.3a

Solution

Provided the actual constraints are not violated, the horizontal reaction of the slot onto the roller does no work during a virtual displacement $\delta\theta$ from the equilibrium position, nor do the pin reactions at A. Only the springs and P will contribute to the virtual work δW, which vanishes.

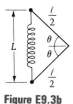

Figure E9.3b

The length L of each spring in the equilibrium configuration is seen in Figure E9.3b to be

$$L = 2\left(\frac{\ell}{2} \sin \theta\right)$$

$$= \ell \sin \theta$$

Therefore, the stretch in the spring is given by

$$S = \ell \sin \theta - \underbrace{2\left(\frac{\ell}{2}\right)}_{\substack{\text{Unstretched} \\ \text{length}}} = \ell(\sin \theta - 1)$$

which is negative because the spring is compressed. The virtual change in S, or "virtual stretch," is

$$\delta S = \ell \cos \theta \, \delta\theta$$

and the virtual work of two of them (with equal S and δS) is

$$\delta W_{\text{springs}} = -2kS \, \delta S = -2k\ell(\sin \theta - 1)\ell \cos \theta \, \delta\theta$$

$$= 2k\ell^2 \cos \theta(1 - \sin \theta) \, \delta\theta$$

As seen in Figure E9.3c, the force P is located at $x = 5\ell \cos \theta$ so that

$$\delta x = -5\ell \sin \theta \, \delta\theta$$

and thus the virtual work of P is

$$\delta W_{\text{by } P} = P \, \delta x = -5P\ell \sin \theta \, \delta\theta$$

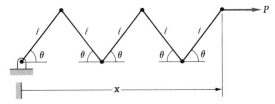

Figure E9.3c

Using $\delta W = 0$, we find

$$\delta W = 2k\ell^2 \cos \theta(1 - \sin \theta)\delta\theta - 5P\ell \sin \theta \, \delta\theta = 0$$

or, since $\delta\theta$ is arbitrary,

$$1 - \sin \theta - 2.5\left(\frac{P}{k\ell}\right) \tan \theta = 0 \tag{1}$$

The solution to this equation may be obtained by trial and error using a calculator or a computer. If the calculator is not programmable, we could proceed as follows for the case $P/k\ell = 1$.

A rough sketch of the functions $1 - \sin \theta$ and $2.5 \tan \theta$ is shown in Figure E9.3d. It appears that they are equal [thus satisfying Equation (1)] at

$\theta \approx 0.4$. Using this as an initial guess and proceeding to compute the left-hand side of Equation (1), call it $f(\theta)$ (with the calculator computing in radians!), we obtain:

θ	$f(\theta)$
0.4	−0.446401
0.3	−0.068861
0.29	−0.031984
0.28	0.004759
0.281	0.001090
0.282	−0.002580

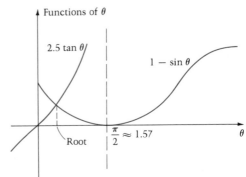

Figure E9.3d

Therefore θ for equilibrium is about 0.281 rad or about 16.1°.

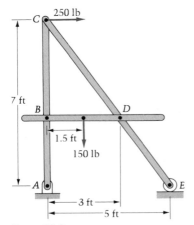

Figure E9.4a

EXAMPLE 9.4

In Figure E9.4a find the force exerted by the pin at D on the member BD in the rigid frame of Example 4.19.

Solution

Referring to the free-body diagram of BD, Figure E9.4b, we see that a virtual rotation with B fixed would cause D_y to be the only unknown component to do virtual work. Remember that the angle of rotation is to be infinitesimal so the displacement of D is essentially vertical. Letting $\delta\theta$ be the virtual rotation, then (see Figures E9.4b,c)

$$\delta W = 0$$
$$D_y(3\ \delta\theta) - 150(1.5\ \delta\theta) = 0$$
$$D_y = 75\ \text{lb}$$

Of course, this is the same result obtained from applying $\Sigma M_B = 0$.

Now giving the bar a virtual horizontal translation, ϵ, as shown in Figure E9.4d, then

$$\delta W = 0$$
$$D_x\epsilon - B_x\epsilon = 0$$
$$D_x = B_x$$

which, of course, is nothing more than the equivalent of applying $\Sigma F_x = 0$.

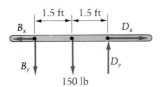

Figure E9.4b

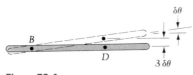

Figure E9.4c

Figure E9.4d

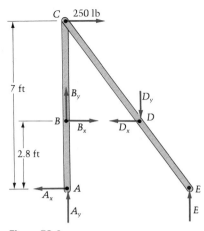

Figure E9.4e

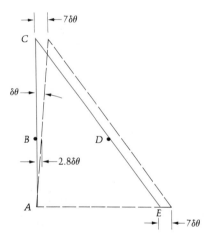

Figure E9.4f

To complete the solution of the problem, we now apply the method of virtual work in a manner that does not provide such an easily recognizable correspondence with the equilibrium equations. We focus our attention on the composite of the bars ABC and CDE for which we have the free-body diagram shown in Figure E9.4e. Now we choose the virtual change in configuration to be given by an infinitesimal rotation, $\delta\theta$, of ABC about A and a horizontal translation, $7\,\delta\theta$, of CDE so that the bars remain connected at C, and E moves horizontally as shown in Figure E9.4f. Thus, point B moves $2.8\,\delta\theta$ to the right and points C and D each move $7\,\delta\theta$ to the right. Applying the principle of virtual work to this assembly and noting that there is no net work done by the forces of interaction at C,

$$\delta W = 0$$

$$B_x(2.8\ \delta\theta) + 250(7\ \delta\theta) - D_x(7\ \delta\theta) = 0$$

or

$$2.8B_x + 7(250) - 7D_x = 0$$

and since

$$B_x = D_x$$

$$4.2D_x = 7(250)$$

or

$$D_x = 417 \text{ lb}$$

as was found before in Example 4.19.

EXAMPLE 9.5

Twelve light rods of length R are connected to six equally spaced points A, B, C, D, E, and F on a circular base as shown in Figure E9.5a. They are joined by ball and socket connections to the base and to each other. The six bar-to-bar connections

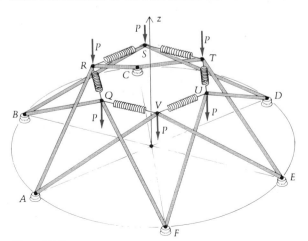

Figure E9.5a

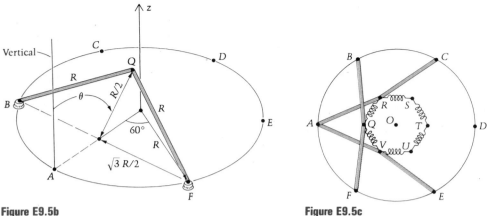

Figure E9.5b

Figure E9.5c

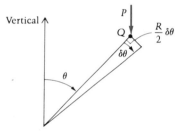

Figure E9.5d

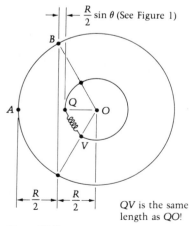

QV is the same length as QO!

Figure E9.5e

at the top are joined by six stiff springs of unstretched lengths $0.4R$. The springs, which can carry compression, form a hexagon in a horizontal plane. If the spring moduli are each equal to k, find the angle(s) θ, between the vertical through A and the plane BQF (see Figure E9.5b), for which the system is in equilibrium under six downward loads P applied at Q, R, S, T, U, and V. Consider only symmetrical displacements—that is, those in which the six top connections Q, R, S, T, U, and V (Figures E9.5a and E9.5c) each move inward and downward identical amounts.

Solution

The sum of the virtual work done by the forces P and the springs will add to zero:

$$\delta W = 0$$

From symmetrical equilibrium positions, the forces P will each move inward toward the z axis (see Figure E9.5a) and downward, during a virtual increase in θ of amount $\delta\theta$. The virtual work done by P is seen in Figure E9.5d to be equal to $P(R/2)\,\delta\theta \sin\theta$, and there are six of these.

Next we consider the virtual work done by the six springs. During the virtual rotation $\delta\theta$ of triangle BQF (Figure E9.5b) about the line BF, the length of each spring will change slightly. In the equilibrium position of the structure at θ, the length of each spring (see Figure E9.5e and consider spring QV) is seen to equal $R - (R/2) - (R/2)\sin\theta$. Thus the stretch in each spring is $0.1R - (R/2)\sin\theta$. This means the "virtual stretch" occurring during the virtual rotation is $-(R/2)\cos\theta\,\delta\theta$, and the virtual work of each spring is

$$-kS\,\delta S = -k\left(0.1R - \frac{R}{2}\sin\theta\right)\left(-\frac{R}{2}\cos\theta\,\delta\theta\right)$$

The total virtual work may now be equated to zero:

$$\delta W = 0 = 6\left[P\frac{R}{2}\,\delta\theta\sin\theta - kS\,\delta S\right]$$

$$0 = \left[\frac{PR}{2}\sin\theta - k\left(0.1R - \frac{R}{2}\sin\theta\right)\left(-\frac{R}{2}\cos\theta\right)\right]\delta\theta$$

or

$$\frac{4P}{kR} \sin \theta + (0.4 - 2 \sin \theta) \cos \theta = 0$$

For $P = kR/4$, the only roots to this equation for $0 \le \theta \le \pi/2$ are $\theta = 27.2°$ and $46.3°$. It is interesting to note that some equilibrium positions are unstable, such as a marble balanced on an inverted spherical bowl. It can be shown using more advanced mechanics that the $27.2°$ equilibrium position that we have found above is stable, while the $46.3°$ position is unstable.

PROBLEMS ▶ Section 9.1

Use the principle of virtual work to solve:

9.1   Problem 4.72

9.2   Problem 4.92

9.3   Problem 4.93

9.4   Problem 4.95

9.5   Problem 4.222

9.6   Problem 4.168

9.7   Problem 4.188

9.8   Problem 4.177

9.9   Problem 6.142

9.10 Problem 6.114

9.11 Find the relationship between P and F for equilibrium of the linkage shown in Figure P9.11.

9.12 Find the force P that must be exerted for equilibrium of the double lever system shown in Figure P9.12.

9.13 In Figure P9.13 find the force F that will hold the weight W in equilibrium.

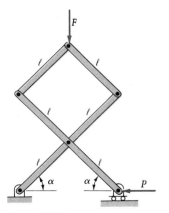

Figure P9.11

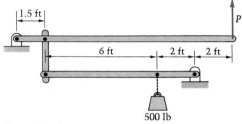

Figure P9.12

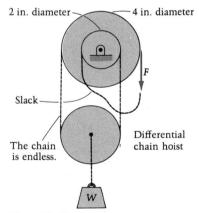

Figure P9.13

9.14 In Figure P9.14 the light, homogeneous, 1-m bars $\mathcal{B}_1$ and $\mathcal{B}_2$ are pinned together at D and are also connected by the spring AB of modulus 200 N/m; the unstretched length of the spring is 0.15 m. What is force P if the system is in equilibrium when triangle DEF is equilateral?

9.15 Repeat Problem 9.14 if $\mathcal{B}_1$ and $\mathcal{B}_2$ each have a mass of 2 kg with respective mass centers at C_1 and C_2.

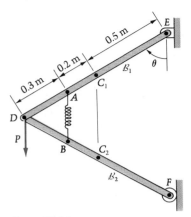

Figure P9.14

9.16 For the data of Problem 9.14, find the angle θ for equilibrium if the force P has a magnitude of 20 newtons.

9.17 Find P as a function of a, b, c, W, and θ for equilibrium of the system shown in Figure P9.17.

9.18 Find the angles θ and φ for equilibrium of the system of two light bars and two heavy weights shown in Figure P9.18.

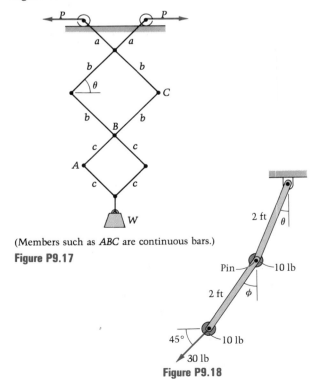

(Members such as *ABC* are continuous bars.)

Figure P9.17

Figure P9.18

•9.19 In Figure P9.19 find the force exerted on the cargo lift by the hydraulic cylinder C if it is pinned to the lift at (a) point A; (b) point B. (c) Use the result of (a) to compute the reaction of pin C onto the member CDE at C. Is this reaction *along CD*?

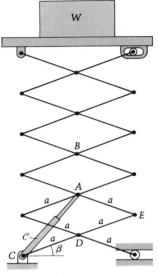

Figure P9.19

9.20 If the system in Figure P9.20 is in equilibrium, find the couple M being applied to crank $\mathcal{A}$ as a function of the angle θ and the force P on the piston. The crank, piston, and connecting rod R are light, and friction is negligible.

9.21 The linkage is made up of 10 light bars of length 2 ft (such as AB) and two bars of length 1 ft (on the extreme right), pinned together as indicated in Figure P9.21. The 300-lb force is to be applied as shown, and it is desired to have the system in equilibrium at $\theta = 60°$. Find the upward force at C on bar CD that will accomplish this.

9.22 In Problem 9.21, suppose that instead of the force at C, the equilibrium is to be maintained by applying a counterclockwise couple onto bar AB. Find the magnitude of the couple.

9.23 In a problem similar to 9.22, there are just four bars, with all other data remaining the same. See Figure P9.23.

   a. Determine the couple M acting on AB for equilibrium using the principle of virtual work.

---

* Asterisks identify the more difficult problems.

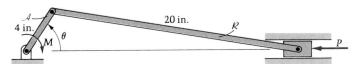

Figure P9.20

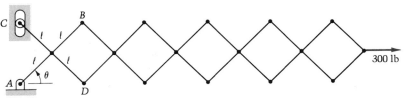

Figure P9.21

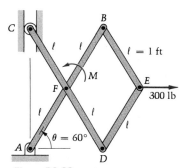

Figure P9.23

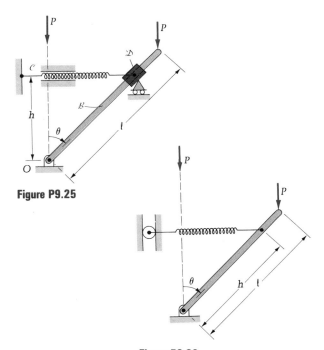

Figure P9.25

Figure P9.26

b. Find the force exerted on the roller at C by the slot, using equilibrium equations. Then take the four bars apart, and, on free-body diagrams of each, show all forces acting on the separate bars.

c. Show (with both virtual work and with equilibrium equations) that with the couple omitted the linkage cannot be in equilibrium.

9.24 In the preceding problem, delete the couple M and add a vertically attached stiff spring joining the midpoints of segments CF and AF with natural length 1 ft. Find the spring modulus if equilibrium exists at $\theta = 60°$.

9.25 The smooth collars $\mathcal{C}$ and $\mathcal{D}$ in Figure P9.25 ensure that the spring remains on the same horizontal level. The spring is unstretched at $\theta = 0$. Show that there is an equilibrium position of bar $\mathcal{B}$ (besides the obvious one at $\theta = 0$) given by

$$\theta = \cos^{-1}\left[\left(\frac{kh^2}{P_\ell}\right)^{1/3}\right]$$

9.26 This problem is similar to Problem 9.25 except that now the spring changes its level while remaining horizontal. Find the "not-obvious" equilibrium angle θ. (See Figure P9.26, where the spring is unstretched at $\theta = 0$.)

9.27 Use virtual work to solve Problem 4.205.

9.28 Verify the result of Example 9.5 by using the equilibrium equations.

In Problems 9.29–9.32 find the relationship between P and W using the principle of virtual work.

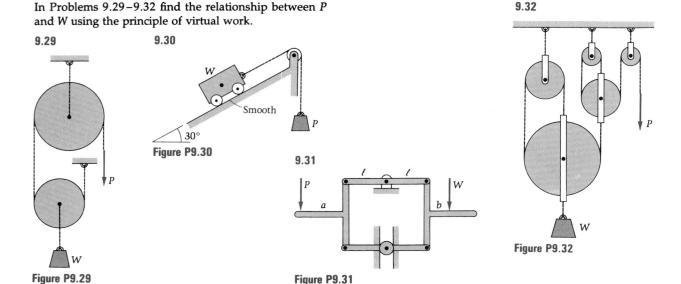

9.29

Figure P9.29

9.30

Smooth

$30°$

Figure P9.30

9.31

Figure P9.31

9.32

Figure P9.32

9.3 Hydrostatic Pressure on Submerged Bodies

In Example 3.35 we recalled that in a fluid at rest:

a. The pressure is equal in all directions at a point (Pascal's law).

b. The pressure is constant through the fluid in each horizontal plane.

c. The pressure causes a force that is normal to every differential area of surface on which it acts.

d. The pressure is equal to γh, where γ is the specific weight of the fluid* (taken constant here) and h is the depth below the free surface.

Also, in Section 7.2 we found that when pressure acts on a plane surface, the resultant of the pressure forces has:

a. A magnitude F_r equal to the volume beneath the loading surface.

b. A line of action normal to the plane and directed toward it, and passing through the centroid of this volume.

In this section we shall first broaden our treatment to allow a *hydrostatic* pressure to act on *non-rectangular* plane areas. It will turn out that the formulas we obtain for F_r and its line of action will be intimately tied to the very same properties of areas we have been studying in detail in Chapters 7 and 8: the amount of area, its centroid, and its inertia properties.

---

* γ is its weight per unit volume, related to the mass density ρ by $\gamma = \rho g$.

We shall look at two results that allow us to obtain the resultant of fluid pressures on *curved* surfaces, often without the need for integration. Finally, we shall complete our look at fluid statics by briefly discussing Archimedes' principle of buoyancy.

Hydrostatic Pressure on Plane Surfaces

We begin by developing some results for hydrostatic pressure on *arbitrary* plane areas. At the depth $x \cos \theta$ (see Figure 9.17), the pressure* on all elements of the line ℓ is given by

$$p = \gamma x \cos \theta \tag{9.18}$$

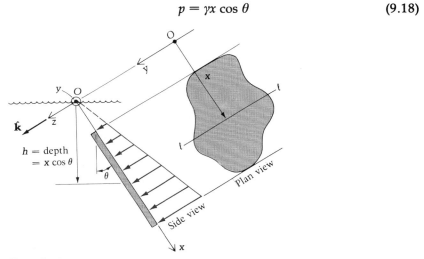

Figure 9.17

where θ is the inclination of the plane area A with the vertical. Noting that all the differential forces on area A due to the pressure ($dF = p \, dA$) are parallel, we know that there will be a "force-alone resultant" in this direction, and we may write

$$d\mathbf{F} = p \, dA\hat{\mathbf{k}}$$

Integrating, we then obtain the resultant:

$$\mathbf{F}_r = \hat{\mathbf{k}} \int \underbrace{\gamma x \cos \theta}_{p} \, dA$$

$$\mathbf{F}_r = \hat{\mathbf{k}}\gamma \cos \theta \int x \, dA \tag{9.19}$$

---

* Throughout this section, we are using gauge pressure — that is, the pressure above atmospheric. If the pressure *behind* one of the submerged surfaces we shall be studying is atmospheric, then the actual front-to-back pressure difference is indeed the gauge pressure we are using. If this is not the case, however, then both should be expressed as absolute pressures or both as gauge pressures so that correct differences will be used in determining resultant fluid forces on the surfaces.

But from our knowledge of centroids,

$$\bar{x}A = \int x \, dA \tag{9.20}$$

so that

$$\mathbf{F}_r = \hat{\mathbf{k}} \, \gamma(\underbrace{\bar{x} \cos \theta}) \, A = p_c A \hat{\mathbf{k}}$$

Depth of
centroid
of area

Pressure at
centroid $= p_c$

Thus the resultant force, normal to the plane of the area, has a magnitude equal to the *pressure at the centroid times the area*. We now seek the *location* of this force-alone resultant which is equipollent to the fluid pressure forces. It will *not* be at the centroid of the plane area unless $\theta = 90°$.

We are looking for the x and y coordinates of a point P where $\mathbf{F}_r$ (above) may be placed and have the same moment about any point (we use the origin O) as does the original system of fluid forces. This point is called the center of pressure.

Thus, we wish to find the vector $\mathbf{r}_{OP}$ in the equation

$$\mathbf{r}_{OP} \times \mathbf{F}_r = \int \mathbf{r} \times d\mathbf{F} \tag{9.21}$$

We represent the sought vector $\mathbf{r}_{OP}$ as

$$\mathbf{r}_{OP} = x_P \hat{\mathbf{i}} + y_P \hat{\mathbf{j}} \tag{9.22}$$

and we write

$$(x_P \hat{\mathbf{i}} + y_P \hat{\mathbf{j}}) \times (p_c A \hat{\mathbf{k}}) = \int (x \hat{\mathbf{i}} + y \hat{\mathbf{j}}) \times (p \, dA \hat{\mathbf{k}}) \tag{9.23}$$

$$-x_P p_c A \hat{\mathbf{j}} + y_P p_c A \hat{\mathbf{i}} = -\hat{\mathbf{j}} \int xp \, dA + \hat{\mathbf{i}} \int yp \, dA \tag{9.24}$$

Equating the $\hat{\mathbf{j}}$ coefficients of Equation (9.24) will give us the x coordinate of the center of pressure:

$$x_p = \frac{\displaystyle\int xp \, dA}{p_c A} = \frac{\displaystyle\int x(\gamma x \cos \theta) \, dA}{\gamma \cos \theta \, \bar{x} A}$$

$$x_p = \frac{\cancel{\gamma \cos \theta} \displaystyle\int x^2 \, dA}{\cancel{\gamma \cos \theta} \, \bar{x} A}$$

and recognizing the moment of inertia (see Section 8.1),

$$x_P = \frac{I_{y_O}}{\bar{x}A}$$ (9.25)

Now, using the parallel-axis (transfer) theorem (see Section 8.3),

$$x_P = \frac{I_{y_c} + A\bar{x}^2}{\bar{x}A} = \bar{x} + \frac{I_{y_c}}{\bar{x}A}$$ (9.26)

We see that, since I_{y_c} is always positive, the center of pressure is always lower than the centroid of the area unless $\theta = 90°$, in which case the pressure is the same at all points of A, and then P and C are at the same depth.

Question 9.6   Why doesn't Equation (9.26) apply for this case?

We next equate the $\hat{\mathbf{i}}$ coefficients of Equation (9.24) and find the lateral position (y_P) of the center of pressure:

$$y_P = \frac{\int yp \, dA}{p_c A} = \frac{\int y\gamma x \cos\theta \, dA}{(\gamma \cos\theta \, \bar{x})A}$$

$$y_P = \frac{\int xy \, dA}{\bar{x}A} = \frac{-I_{xy_O}}{\bar{x}A}$$ (9.27)

where $I_{xy_O} = -\int xy \, dA$ is the product of inertia of the area with respect to the axes x and at y at point O. Using the parallel-axis (or transfer) theorem (see Section 7.6),

$$y_P = \frac{-(I_{xy_c} - A\bar{y}\bar{x})}{\bar{x}A} = \bar{y} - \frac{I_{xy_c}}{\bar{x}A}$$ (9.28)

If either of the axes through C, parallel to x or y, is an axis of symmetry of the plane area, then $I_{xy_c} = 0$ and then $y_P = \bar{y}$. This means that the lateral distance from the x axis to the centroid in this case is the same as the distance from the x axis to the center of pressure* (see Figure 9.18). If, however, $I_{xy_c} \neq 0$, then the center of pressure will be displaced laterally from C; y_P will not equal $\bar{y}$. We shall now consider three examples. In the first, the area is rectangular; in the others, the width varies with depth.

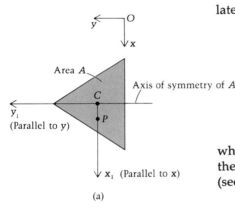

(a)

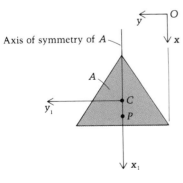

(b)

Figure 9.18   (a) In this case, y_1 is an axis of symmetry of area A, so that $I_{xy_c} = 0$ and thus $y_P = \bar{y}$. (b) This time, x_1 is an axis of symmetry of A, so again $I_{xy_c} = 0$ and $y_P = \bar{y}$.

Answer 9.6   If $\theta = 90°$, we have divided by zero in deriving Equation (9.26).
* For the case when the axis through C parallel to x is an axis of symmetry, the result $y_P = \bar{y}$ becomes obvious when we recall that the line of action of $\mathbf{F}_r$ passes through the centroid of the volume beneath the loading (pressure) surface; the other case's (same) result does not, however, follow from this knowledge about $\mathbf{F}_r$'s line of action.

EXAMPLE 9.6

Show that the results of Example 3.35 follow from Equations (9.20) and (9.26) of this section.

Solution

We have seen that the resultant force on any plane area A equals the centroidal pressure times A; therefore, using Figure E9.6a,

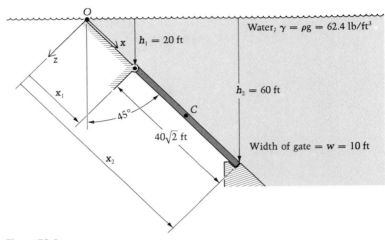

Figure E9.6a

$$|\mathbf{F}_r| = \overbrace{\gamma\left(\frac{x_1 + x_2}{2}\right)(\cos\phi)}^{\text{Pressure at }C}\overbrace{(x_2 - x_1)w}^{\substack{\text{Area of}\\\text{gate}}} = 62.4(x_2^2 - x_1^2)\frac{1/\sqrt{2}}{2} \;(10)$$

$$= 624[(60\sqrt{2})^2 - (20\sqrt{2})^2]/(2\sqrt{2}) = 1.41 \times 10^6 \text{ lb}$$

which is the same answer as we obtained by integration in Example 3.35. For the distance d to the center of pressure P, we use Equation (9.26):

$$x_P = \bar{x} + \frac{I_{y_C}}{\bar{x}A}$$

Using Figure E9.6b, the terms become

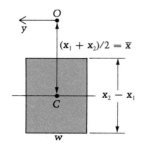

$$I_{y_C} = \frac{(\text{base})(\text{height})^3}{12}$$

Figure E9.6b

$$x_P = \underbrace{\frac{x_1 + x_2}{2}}_{\bar{x}} + \frac{\overbrace{\frac{w(x_2 - x_1)^3}{12}}^{I_{y_C}}}{\underbrace{\frac{(x_1 + x_2)}{2}}_{\bar{x}}\underbrace{[(x_2 - x_1)w]}_{A}}$$

After simplifying, this results in

$$d = \frac{2}{3} \frac{x_2^2 + x_1 x_2 + x_1^2}{x_1 + x_2} = 61.3 \text{ ft}$$

which is again the same answer we obtained in Example 3.35.

EXAMPLE 9.7

Find the center of pressure of the triangular fluid gate, shaped and situated as shown in Figure E9.7

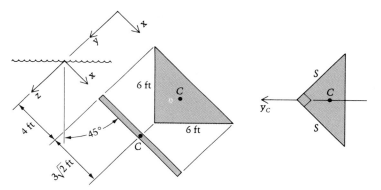

Figure E9.7

Solution

Using Example 8.3, we can obtain

$$I_{y_c} = \frac{S^4}{24}$$

We then obtain, for our triangle,

$$I_{y_c} = \frac{6^4}{24} = 54 \text{ ft}^4$$

Therefore,

$$x_p = \bar{x} + \frac{I_{y_c}}{\bar{x}A}$$

$$= (4 + 3\sqrt{2}) + \frac{54}{\underbrace{(4 + 3\sqrt{2})\frac{1}{2}(6)6}_{A}}$$

$$= 8.61 \text{ ft}$$

For the y coordinate of P,

$$y_p = \bar{y} - \overset{\text{0 by symmetry}}{\frac{I_{\overcancel{xy_c}}}{\bar{x}A}}$$

$$y_p = \bar{y} - 0 = \tfrac{1}{3}(3\sqrt{2})$$

$$= \sqrt{2} = 1.41 \text{ ft}$$

EXAMPLE 9.8

If the same gate as in Example 9.7 is reoriented as shown in Figure E9.8, find the new position of the center of pressure.

Solution

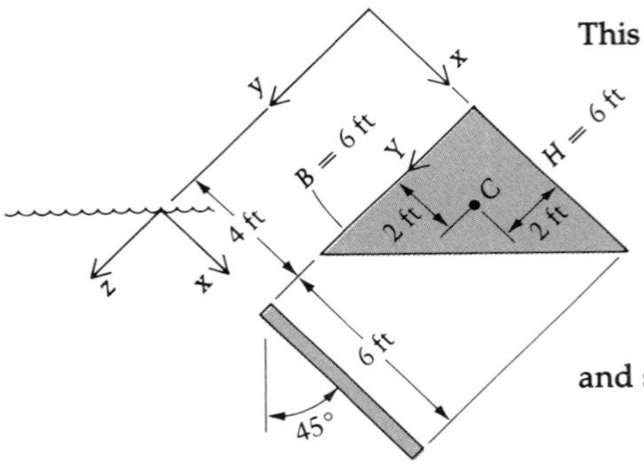

This time, I_Y (see the figure) $= \dfrac{BH^3}{12} = \dfrac{6(6^3)}{12} = 108 \text{ ft}^4$ so that

$$I_{y_c} = 108 - Ad^2$$

$$= 108 - \tfrac{1}{2}(6)6(2^2) = 36 \text{ ft}^{4*}$$

Figure E9.8

and so

$$x_P = \bar{x} + \frac{I_{y_c}}{\bar{x}A}$$

$$= (4 + 2) + \frac{36}{(4 + 2)18}$$

$$= 6 + (\tfrac{1}{3}) = 6.33 \text{ ft}$$

The product of inertia I_{xy_c} is needed for y_P. From Example 8.21, it is

$$I_{xy_c} = \frac{(\text{base} \times \text{height})^2}{72}$$

$$= \frac{(6 \times 6)^2}{72} = 18 \text{ ft}^4$$

Therefore,

$$y_P = \bar{y} - \frac{I_{xy_c}}{\bar{x}A}$$

$$= 2 - \frac{18}{(6)(18)} = 1.83 \text{ ft}$$

---

* Or we could alternatively use $I_{y_c} = \dfrac{BH^3}{36} = \dfrac{6(6^3)}{36} = 36 \text{ ft}^4$.

Hydrostatic Pressure on Curved Surfaces

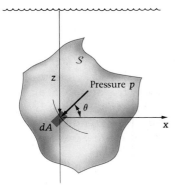

Figure 9.19

If the submerged surface S is not necessarily flat, then formulas such as (9.26) and (9.28) do not apply. Instead we may use the following two results: Let the resultant force due to hydrostatic fluid pressure on one side of a curved surface* be called $\mathbf{F}_r$. Then:

Result (I) The component of $\mathbf{F}_r$ in a horizontal direction x is the force caused by the same pressure acting on the vertical projection of the area onto a plane normal to x. The proof is as follows, with θ being the angle between the normal to the differential area dA and the x direction (see Figure 9.19):

$$\mathbf{F}_{r_x} = \underbrace{\int (p\,dA)}_{dF} \cos\theta = \int p\underbrace{(dA\cos\theta)}$$

dF

Differential force
in x direction

Vertical protection
of dA, normal to x

Differential force on
projected area normal to x

Question 9.7   What is the difference between this result and the first of the three special properties of resultants on page 141 in Section 3.9?

Result (II) The vertical component of $\mathbf{F}_r$ is the weight of the column of fluid above S.

Question 9.8   Why can't the liquid surrounding the column help hold it up?

In this second result, we see that there is no confusion if the fluid above the surface all extends vertically up to the free surface, as shown on the left in Figure 9.20(a). A surface S made up of the curve AB and a

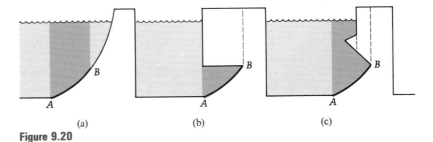

(a)   (b)   (c)

Figure 9.20

---

* With no vertical outward normals except possibly along part of its edge.

Answer 9.7   In the prior result (in Section 3.9) the pressure was constant, and so *any* projection could be used. In this chapter, the (hydrostatic) pressure varies linearly with depth. Thus the vertical projection of the area is the only one it makes sense to use in the theorem.

Answer 9.8   It is only able to exert normal (horizontal, here) forces on the column.

uniform width w normal to the page is being considered, and the vertical component of the fluid force resultant on $\mathscr{S}$ is obviously the weight of the shaded fluid. In Figure 9.20(b), however, the fluid above the same surface $\mathscr{S}$ does not extend up to the free surface. And in Figure 9.20(c), it does so over some of the area elements, on others it does so after interruption, and on still others it doesn't extend upward to any free surface at all. The point is that none of this matters. The vertical force on $\mathscr{S}$ acts as though the fluid existed in a column up to the free surface (extended, if necessary) in all cases because *it is the pressure on the surface that counts and the pressure in the fluid is the same across any horizontal level.*

We now present two related examples. In the second, we shall consider the pressure on a curved surface; in both, we shall use results I and II.

EXAMPLE 9.9

If the dam in Figure E9.9a is 200 ft wide, determine and locate the resultant force exerted on it by the water pressure.

Solution

We note first that in this case, all differential forces exerted on the dam by the water are parallel; therefore, there is a "force-alone" resultant that is normal to the slanted surface.

The horizontal component of the resultant force equals the pressure at the centroid of the projected (shaded) vertical area (see Figure E9.9b) multiplied by this area:

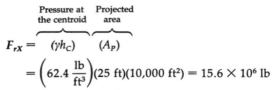

$$F_{rX} = \overbrace{(\gamma h_C)}^{\substack{\text{Pressure at} \\ \text{the centroid}}} \overbrace{(A_P)}^{\substack{\text{Projected} \\ \text{area}}}$$

$$= \left(62.4 \frac{\text{lb}}{\text{ft}^3}\right)(25 \text{ ft})(10{,}000 \text{ ft}^2) = 15.6 \times 10^6 \text{ lb}$$

The vertical component of the resultant force equals the weight of water (see Figure E9.9c) above the wetted surface:

$$F_{rY} = \gamma(\text{Vol}) = 62.4 \frac{\text{lb}}{\text{ft}^3} \left(\tfrac{1}{2} \times 10 \text{ ft} \times 50 \text{ ft} \times 200 \text{ ft}\right) = 3.12 \times 10^6 \text{ lb}$$

Thus the resultant force has the magnitude

$$F_r = \sqrt{15.6^2 + 3.12^2} \times 10^6 = 15.9(10^6) \text{ lb}$$

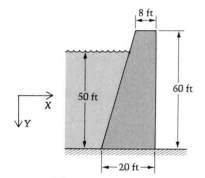

Figure E9.9a

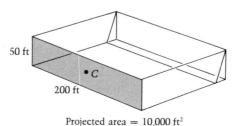

50 ft

200 ft

$\bullet C$

Projected area $= 10{,}000 \text{ ft}^2$

Figure E9.9b

Figure E9.9c

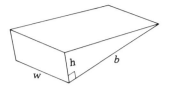

Figure E9.9d

This magnitude of $\mathbf{F}_r$ may be checked. As we have seen, it is also the volume (of the triangular prism in Figure E9.9d) under the loading (or pressure) surface:

$$F_r = \tfrac{1}{2}bhw$$

$$= \tfrac{1}{2}(51)(\gamma 50)(200) \text{ lb}$$

$$= 25.5(62.4)50(200) \text{ lb}$$

$$= 15.9(10^6) \text{ lb} \qquad \text{(as above)}$$

As for the *direction* of $\mathbf{F}_r$, it is, as previously mentioned, normal to the surface because it comprises of infinitely many parallel differential forces, *each one* normal to the surface with the sense

We note that the theorems used to obtain F_{rX} and F_{rY} also yield the correct direction for the resultant:

$$\theta = \tan^{-1}\left(\frac{15.6}{3.12}\right) = \tan^{-1} 5 \qquad \text{(as above and as seen in Figure E9.9e)}$$

where

$$\phi = \tan^{-1}\frac{50}{10} = \tan^{-1} 5 \qquad \text{(also)}$$

$\therefore$ $\mathbf{F}_r$ is seen again to be normal to the slanted surface of the dam.

The place on the dam where $\mathbf{F}_r$ acts (known as the center of pressure) is at the centroid of the distributed loading, which is in this case a triangular prism. Thus the load acts normal to the dam a distance $\tfrac{2}{3}(51) = 34$ ft along the slant from the water surface (see Figure E9.9f), and, of course, halfway across the dam (100 feet from each end).

We may check this result in two ways: First, we use the components F_{rX} and F_{rY} obtained earlier, and let M_{rO} be the resultant moment about O of the forces due to hydrostatic fluid pressure. Then (see Figure E9.9g),

$$M_{rO} = F_{rX}[\tfrac{2}{3}(50) \text{ ft}] + F_{rY}[\tfrac{2}{3}(10) \text{ ft}]$$

$$= 15.6(10^6)\left[\frac{100}{3}\right] + 3.12(10^6)\left[\frac{20}{3}\right] \text{ lb-ft}$$

$$= 520(10^6) + 20.8(10^6) = 541(10^6) \text{ lb-ft}$$

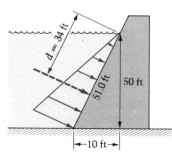

Figure E9.9e   **Figure E9.9f**

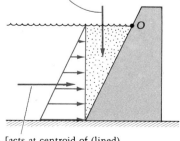

F_{rY}[acts at centroid of (dotted) volume of water above the wetted surface]

F_{rX}[acts at centroid of (lined) pressure curve on vertical projection]

Figure E9.9g

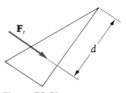

Figure E9.9h

If this is to be equipollent to $\mathbf{F}_r$ at distance d (see Figure E9.9h), then

$$15.9(10^6)d = 541(10^6)$$

$$d = 34.0 \text{ ft} \qquad \text{(as before)}$$

Alternatively, we may determine the distance d by using Equation (9.26):

$$d \text{ or } x_P = \frac{I_{yc}}{\bar{x}A} + \bar{x}$$

$$d = \frac{\left[\dfrac{200(51^3)}{12}\right]}{25.5[200 \times 51]} + 25.5 = 8.5 + 25.5$$

$$d = 34 \text{ ft}$$

EXAMPLE 9.10

Repeat Example 9.9 if the dam is parabolic instead of linear. (See Figure E9.10a, noting the different coordinates.)

Solution

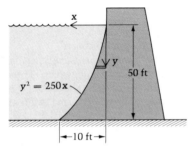

Figure E9.10a

The horizontal component of $\mathbf{F}_r$ is the same because the projected area is unchanged:

$$F_{rX} = -15.6 \times 10^6 \text{ lb}$$

For F_{rY}, we need the specific weight times the volume of the water above the wetted surface of the dam:

$$F_{rY} = \gamma \frac{\text{lb}}{\text{ft}^3}\left[50(10) - \int_{x=0}^{50} \overbrace{(y^2/250)\,dy}^{dA}\right] \text{ft}^2 (200 \text{ ft})$$

$$= \left(62.4\,\frac{\text{lb}}{\text{ft}^3}\right)\left(500 - \frac{50^3}{750}\right) \text{ft}^2 (200 \text{ ft})$$

$$F_{rY} = 4.16 \times 10^6 \text{ lb}$$

which is a bigger force than we found for the triangular dam because there is more water above this one.

The above components of course add vectorially to form the resultant force $\mathbf{F}_r$:

$$\mathbf{F}_r = F_{rX}\hat{\mathbf{i}} + F_{rY}\hat{\mathbf{j}} = -15.6(10^6)\hat{\mathbf{i}} + 4.16(10^6)\hat{\mathbf{j}} \text{ lb}$$

The components act along the lines shown in Figure E9.10b.

We know that there is a force-alone resultant of all the infinitely many differential forces acting on the dam due to water pressure because these forces form a coplanar distributed system once the pressures are multiplied by the 200-ft width. The location of the line along which this force-alone resultant acts is shown in Figure E9.10c.

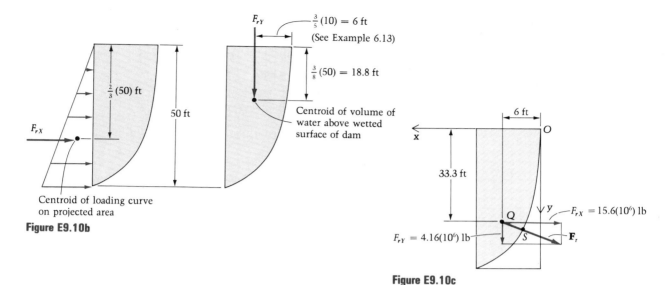

Figure E9.10b

Figure E9.10c

Question 9.9   Why is there no couple accompanying F_r at Q?

An exercise problem will be to locate the spot S where F_r intersects the surface of the dam, and to show that F_r is not in this case normal to the surface, as would *always* be the case for a *plane* area as we have seen in the previous example.

To check the location of the resultant, we shall integrate to get M_{rO}:

$$M_{rO} = \int_{\substack{\text{Wetted} \\ \text{surface}}} r \times dF$$

$$M_{rO} = \int [\underbrace{(x\hat{i} + y\hat{j})}_{r} \times \underbrace{\gamma y}_{\text{Pressure}} \underbrace{w\, ds}_{dA} \underbrace{(-\cos\theta\hat{i} + \sin\theta\hat{j})}_{\substack{\text{Unit vector in} \\ \text{direction of} \\ \text{differential force}}}]^*$$

$$\underbrace{\hspace{5cm}}_{\text{Differential force } dF}$$

$$M_{rO} = \gamma w \hat{k} \int (xy \sin\theta + y^2 \cos\theta)\, ds$$

But as can be seen in Figure E9.10d,

$$\sin\theta = \frac{dx}{ds} \quad \text{and} \quad \cos\theta = \frac{dy}{ds}$$

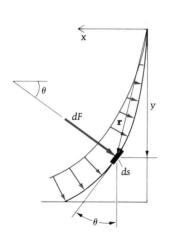

Figure E9.10d

Answer 9.9 Q is the point of intersection of the components constituting F_r.

* Note that $\theta = \tan^{-1}\left(\dfrac{dx}{dy}\right) = \tan^{-1}\left(\dfrac{y}{125}\right)$, which is not needed in this example.

so that

$$\mathbf{M}_{rO} = 62.4(200)\hat{\mathbf{k}}\left[\int_{x=0}^{10} x\underbrace{y}_{\sqrt{250x}}\, dx + \int_{y=0}^{50} y^2\, dy\right]$$

$$\mathbf{M}_{ro} = 12500\hat{\mathbf{k}}\left\{\left[\sqrt{250}\,\frac{x^{5/2}}{5/2}\right]\Bigg|_0^{10} + \left[\frac{y^3}{3}\right]\Bigg|_0^{50}\right\}$$

$$= 12500\hat{\mathbf{k}}[2000 + 41700]$$

$$= 546(10^6)\hat{\mathbf{k}}\ \text{lb-ft}$$

The resultant $\mathbf{F}_r$ has the same moment about O as do the above distributed forces:

$$\overset{15.6 \times 10^6\ \text{lb}}{} \qquad \overset{4.16 \times 10^6\ \text{lb}}{}$$

$$\mathbf{M}_{rO} = F_{rx}(\tfrac{2}{3} \times 50) + F_{ry}(\tfrac{3}{5} \times 10)$$

$$= 545(10^6)\ \text{lb-ft},$$

off by about $\tfrac{1}{5}$ of 1% due to roundoff.

Sometimes we need to compute fluid pressure forces *beneath* a surface. The force due to such pressure is the negative of the force produced by an imaginary column of fluid *above* the same surface and extending up to the free surface level. Thus we may use our previous two results *(with all forces reversed)* for such problems, if we wish. We illustrate this idea with an example worked two ways: first with straightforward integration, and then using the idea of "imaginary fluid *above* the surface."

EXAMPLE 9.11

Find the resultant of the fluid pressure forces acting on the sluice gate AB in the left diagram of Figure E9.11a, expressed as a force at the pin A and a couple. The gate is a thick, quarter-cylindrical surface with radius R and length L perpendicular to the paper.

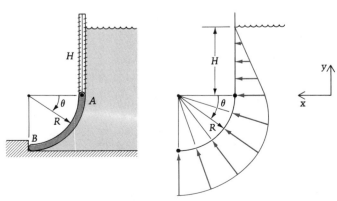

Figure E911.a

Solution

Since the pressure varies linearly with distance from the free surface, as suggested in the right diagram of Figure E9.11a, and since this distance is $H + R \sin \theta$, the resultant force $\mathbf{F}_r$ has the components:

$$F_{rx} = \int_{\theta=0}^{\pi/2} \underbrace{\underbrace{\gamma(H + R \sin \theta)}_{\text{Pressure}} \underbrace{LR \, d\theta}_{dA}}_{\underbrace{\text{Differential force } dF}_{dF_x}} \cos \theta \tag{1}$$

and

$$F_{ry} = \int_{\theta=0}^{\pi/2} \underbrace{\underbrace{\gamma(H + R \sin \theta)}_{\text{Pressure}} \underbrace{LR \, d\theta}_{dA}}_{\underbrace{\text{Differential force } dF}_{dF_y}} \sin \theta \tag{2}$$

Carrying out the integrations leads to

$$\mathbf{F}_r = F_{rx}\hat{\mathbf{i}} + F_{ry}\hat{\mathbf{j}}$$

$$= \gamma RL \left(H + \frac{R}{2} \right) \hat{\mathbf{i}} + \gamma RL \left(H + \frac{\pi}{4} R \right) \hat{\mathbf{j}} \tag{3}$$

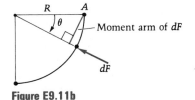

R   A
θ
— Moment arm of dF
dF

Figure E9.11b

For the moment about the pin A, which together with $\mathbf{F}_r$ completes the resultant at A of the fluid forces on the sluice gate, we integrate one more time (see Figure E9.11b):

$$M_{rA} = \int_{\theta=0}^{\pi/2} \underbrace{\gamma(H + R \sin \theta)LR \, d\theta}_{dF \text{ (as above)}} \underbrace{R \sin \theta}_{\substack{\text{Moment} \\ \text{arm}}}$$

$$= \gamma R^2 L \left(H + \frac{\pi}{4} R \right) \tag{4}$$

Thus the resultant at A is the force $\mathbf{F}_r$ together with a clockwise couple of strength M_{rA}.

Consider now the forces resulting from water *imagined* to be *above* the sluice gate (see Figure E9.11c on following page).

For the horizontal component of the resultant force, we add the forces on the vertical projection CB of the gate:

$$\underbrace{\gamma LHR}_{\substack{\text{Volume of rectangular} \\ \text{part of load surface} \\ \text{(shaded, Figure E9.11c) (acts} \\ \text{at } z = H + R/2, \\ \text{the centroid of} \\ \text{this volume)}}} + \underbrace{\tfrac{1}{2}(\gamma LR)R}_{\substack{\text{Volume of triangular} \\ \text{part of load surface} \\ \text{(dotted, Figure E9.11c) (acts} \\ \text{at } z = H + (2/3)R, \\ \text{the centroid of} \\ \text{this volume)}}}$$

It is seen that this force, to the right, agrees with the actual F_{rx} of Equation (3).

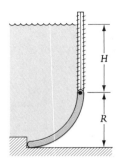

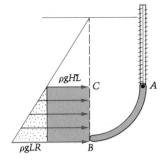

Figure E9.11c

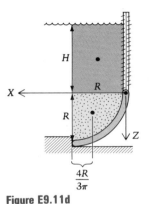

Figure E9.11d

For the y component, we see in Figure E9.11d that the weight of the water shown shaded is $\gamma(HRL)$, acting at the center of the block, and the weight of the water shown dotted is $\gamma\left(\dfrac{\pi R^2}{4} L\right)$, acting at $X = Z = R - \dfrac{4R}{3\pi}$ and $Y = \dfrac{L}{2}$.

It is also seen that the (downward) total y force, the sum of the weights of the two volumes of water in Figure E9.11d, is the same as F_{ry} in Equation (3).

Finally, the above four contributors to $-\mathbf{F}_r$ may also be used, each in reverse, to check M_{rA}:

$$
\circlearrowleft_{+} \quad M_{rA} = \overbrace{(\gamma HRL)\dfrac{R}{2}}^{\text{Rectangle}} + \overbrace{\left(\gamma \dfrac{\pi R^2}{4} L\right)\left(R - \dfrac{4R}{3\pi}\right)}^{\text{Quarter-circle}} + \overbrace{(\gamma LHR)\left(\dfrac{R}{2}\right)}^{\text{Rectangle}} + \overbrace{\left(\dfrac{\gamma LR^2}{2}\right)\left(\dfrac{2}{3}R\right)}^{\text{Triangle}}
$$

$$\underbrace{\hspace{4.5cm}}_{\text{Moments of } y \text{ forces}} \qquad \underbrace{\hspace{4.5cm}}_{\text{Moments of } x \text{ forces}}$$

$$
= \gamma LHR^2 + \dfrac{\pi}{4}\gamma R^3 L
$$

This is the same result that we obtained (Equation 4) much more directly by integration.

Buoyant Force

Archimedes' principle (220 B.C.!) is that the buoyant force on a submerged body equals the weight of the fluid displaced by the submerged portion of the body. The buoyant force acts through the center of mass of the displaced fluid, a point known as the center of buoyancy.

For the proof of the principle, let the *submerged portion* of the body be replaced by an identical volume V_F, in amount and shape, of the fluid itself as suggested by Figure 9.21. Such a volume is obviously in equilibrium, so that the buoyant force B on the fluid is the weight of the fluid volume, acting upward through the mass center C_F of this fluid volume. In Figure 9.22, ρ_F is the mass density of the fluid and B is the single-force resultant of all the differential forces caused by fluid pressure on the fluid volume.

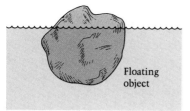

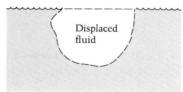

Figure 9.21

Figure 9.22

Question 9.10   How do we know that there is a single-force resultant in this case?

Now, returning to the actual body, the fluid pressure on its submerged portion is everywhere the same as we had on the substituted volume of fluid because this pressure, a function of depth and fluid density, is unaffected by what really comprises the volume V_F. Hence the resultant B of the forces caused by this pressure is the same. The free-body diagram of the actual floating body $\mathcal{B}$ is seen in Figure 9.23. The body is thus buoyed upward by a force equal to the weight of displaced fluid. In Figure 9.23(a), ρ_T and ρ_S are the average mass densities of the parts of $\mathcal{B}$ above and below the free surface of the fluid. Note that the mass center of the part of $\mathcal{B}$ occupying V_F need not be at C_F, or even on the same vertical line as C_F. But the mass center C of *all* of $\mathcal{B}$ is on this line as

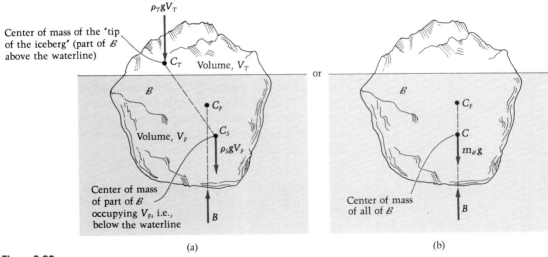

Figure 9.23

Answer 9.10   The resultant of the forces caused by water pressure has to be the negative of the weight of the displaced fluid ($\rho_F g V_F$). Thus if the resultant of these water pressure forces at C_F contains a moment, then ΣM_{C_F} cannot vanish there.

shown in Figure 9.23(b), because otherwise the moment equilibrium equation could not be satisfied.

We now consider briefly the importance of the position of the buoyant force on the rolling motions of a ship in the sea. Consider the sketch of a floating ship's profile in Figure 9.24 with the mass center C_F representing the mass center of the displaced fluid volume V_F.

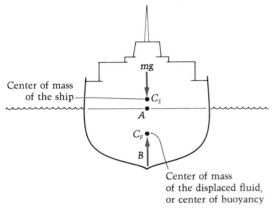

Center of mass of the ship

mg

C_S

A

C_F

B

Center of mass of the displaced fluid, or center of buoyancy

Figure 9.24

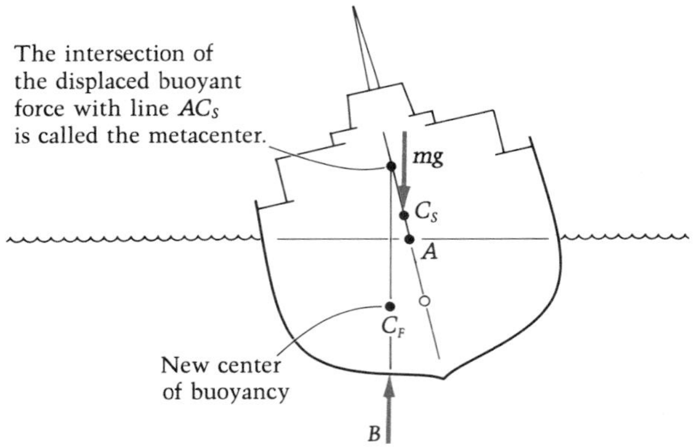

The intersection of the displaced buoyant force with line AC_S is called the metacenter.

mg

C_S

A

C_F

New center of buoyancy

B

Figure 9.25

Now let the ship roll about the z axis normal to the paper through, say, point A in a motion caused by wind and water forces (see Figure 9.25). The buoyant force will, in general, change. There is now more fluid displaced on the left and less on the right. Depending upon the profile of the ship's hull, the weight of this displaced fluid may or may not even equal the weight of the ship. In any case, there will be a definite change in the line of action of B drawn on the ship's profile, and a possible change in its intersection with the line AC_s (which line, relative to the ship, hasn't changed). This intersection point is called the "meta-center." Figure 9.25 shows us that if the "metacenter" lies above the mass center, then the moment formed by B and mg is "restoring" — that is, tends to turn the ship back toward the vertical. In this case, the ship would be stable.

Question 9.11   What would happen if the metacenter were *below* the mass center at a certain angle of roll?

In ship design, the hull profile and mass distributions are carefully computed so as to keep the metacenter above the mass center and also to keep it close to the same spot for fairly large roll angles of the ship, up to 20° and more in some cases.

Answer 9.11   Disaster! The moment of B and mg would then be in a direction that would overturn the ship further.

EXAMPLE 9.12

Compute the largest (uniform) thickness t so that a tub hollowed out of a steel block will float (see Figure E9.12). Let $l = 7$ ft, $w = 3$ ft, $h = 2$ ft, and $\gamma_w = 62.4$ lb/ft³, $\gamma_s = 490$ lb/ft³.

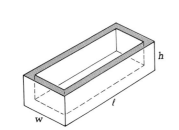

Figure E9.12

Solution

Using Archimedes' principle, we shall equate the weight of the tub (in terms of t) with the buoyant force (weight of water displaced) when the water line is up to the top of the tub:

$$\text{Weight of tub} = \gamma_s \text{ (Volume of steel in tub)}.$$

$$W_T = 490[w l h - (h - t)(w - 2t)(l - 2t)]$$

Substituting and simplifying,

$$W_T = 490(4t^3 - 28t^2 + 61t)$$

The buoyant force is

$$B = \gamma_w(\text{Volume of displaced water})$$

$$= 62.4(w l h) = 62.4(42)$$

$$= 2620 \text{ lb}$$

For equilibrium,

$$+\!\downarrow \quad \Sigma F_y = 0 = W_T - B$$

which gives the following equation in the thickness t:

$$f(t) = t^3 - 7t^2 + 15.3t - 1.34 = 0 \tag{1}$$

This equation has but one real root, which is found below by trial and error to be $t = 0.0913$ ft or about 1.1 inch:

t	$f(t)$
0	−1.34
1	7.96
0.1	0.121
0.09	−0.0190
0.091	−0.0049
0.0912	−0.0021
0.0913	−0.0007
0.0914	+0.0007

PROBLEMS ▶ Section 9.3

9.33 Show that the pressure in a fluid at rest (a) is constant through the fluid in each horizontal plane; (b) obeys the equation $dp = \gamma \, dz$, where z is measured downward. Thus if z is zero at the free surface of a constant density fluid, then $p = \gamma z$.

9.34 Explain why the fluid force is normal to every differential area on which it acts.

9.35 Prove Pascal's law, that the pressure in a fluid at rest is constant in all directions. (See Figure P9.35.) *Hint:* Sum forces on the infinitesimal element of fluid in the x

and z directions, obtaining $p_s = p_x$ and $p_s = p_z$. Then note that θ and the orientation of the slanted surface around the vertical are arbitrary.

9.36–9.40 Find the force-alone resultant of the water pressure forces on the shaded side of the submerged surfaces in Figures P9.36–P9.40.

9.41 For this third orientation of the gate of Examples 9.7 and 9.8, find the center of pressure. See Figure P9.41.

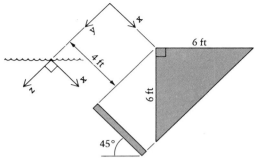

Figure P9.41

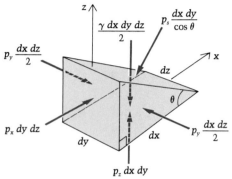

Figure P9.35

9.42 Show that the x coordinate of the center of pressure of fluid force acting on the equilateral triangular area $\mathcal{A}$ in Figure P9.42 is unchanged by any rotation of $\mathcal{A}$ about an axis through the centroid normal to the area. What is this force if the fluid is water with specific weight 62.4 lb / ft³ and if $x_A = 30$ ft and $s = 3$ ft?

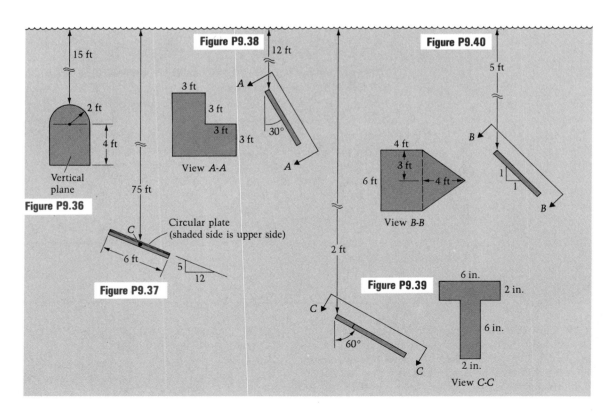

Figure P9.38

Figure P9.40

View A-A

View B-B

Circular plate
(shaded side is upper side)

Figure P9.36

Figure P9.37

Figure P9.39

View C-C

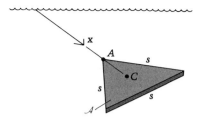

Figure P9.42

9.43 The gate in Figure P9.43 holds back a reservoir of water. The pressure at any depth z is γz, where γ = specific weight of the water. The gate is 5 ft wide perpendicular to the paper. Thus the pressure load may be considered as a distributed line load of intensity $5\gamma z$ lb / ft acting on the "beam" shown. Draw the shear and moment diagrams as suggested by the axes to the left of the figure.

9.44 Find the resultant force caused by fluid pressure on the shaded door in Figure P9.44 if the door is rectangular with a width of w and a height of h.

9.45 Work Problem 9.44 if the door is circular with diameter h.

9.46 Work Problem 9.44 if the door's shape is an equilateral triangle as shown in Figure P9.46.

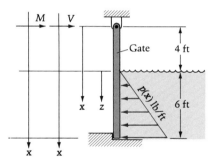

Figure P9.43

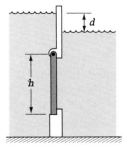

Figure P9.44

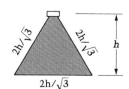

Figure P9.46

9.47 Find the force exerted on the rectangular gate at A by the stop. (See Figure P9.47.)

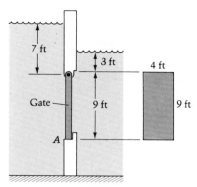

Figure P9.47

9.48 The circular gate in Figure P9.48 has a diameter of 6 ft. It swings about a horizontal pin located 4 in. below its center. At what depth h of water will the gate be in equilibrium?

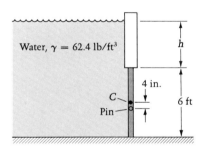

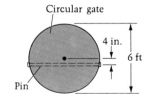

Figure P9.48

9.49 Find the force due to fluid pressure acting on the gate in Example 9.7.

9.50 Find the force due to fluid pressure acting on the gate in Example 9.8.

9.51 In Example 9.10 verify the resultant force by integration. See Figure P9.51. *Hint:* Use:

$$d\mathbf{F} = (\text{pressure})(dA)(-\cos\theta\hat{\mathbf{i}} + \sin\theta\hat{\mathbf{j}})$$

with pressure $= \gamma y$, and $dA = (\text{width})\,ds = 200\,ds$. Then use:

$$ds = \sqrt{dx^2 + dy^2} = dy\sqrt{1 + \left(\frac{dx}{dy}\right)^2}$$

or

$$ds = \sqrt{1 + \left(\frac{y}{125}\right)^2}\,dy$$

Further,

$$\cos\theta = \frac{dy}{ds} \quad \text{and} \quad \sin\theta = \frac{dx}{ds}$$

and substitution gives

$$d\mathbf{F} = 200\gamma y[-dy\hat{\mathbf{i}} + \underbrace{(y\,dy\,/\,125)\hat{\mathbf{j}}}_{dx}]$$

which can be integrated to give $\mathbf{F}_r$.

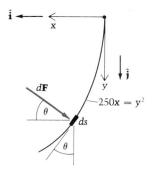

Figure P9.51

★ **9.52** In Example 9.10, find the coordinates of the point S where $\mathbf{F}_r$ intersects the parabolic surface of the dam. Show that $\mathbf{F}_r$ is not normal to the surface at S.

9.53 The dam in Figure P9.53, similar to that of Example 9.9, is holding back water, this time against a vertical (instead of slanted) plane. The dam is made of concrete with average specific weight 155 lb / ft³. There are forces beneath the dam due to two sources: (a) the reaction of the earth ("intergranular," i.e., earth particles onto concrete), and (b) the "uplift" force due to water pressure from fluid that has seeped underneath the dam. The first of these will have a horizontal and a vertical component and act somewhat between B and D; the second is the

vertical force-alone resultant of the upward fluid pressure forces (assumed to act over approximately the entire bottom surface of the dam). Assume the uplift pressure to vary linearly from the full 50γ at B to zero at D, and determine the magnitude and direction of the earth's (force-alone resultant) reaction, and the point where it intersects line BD.

9.54 In Problem 9.53 find the factor of safety against overturning of the dam about D, defined as the ratio of the counterclockwise moment about D due to the weight of the dam, to the clockwise moment caused by the forces from the water pressure on AB and BD. The numerator resists overturning and the denominator promotes it.

9.55 Let $H = 2R$ and find the magnitude, direction, and line of action of the single-force resultant of the fluid forces on the gate of Example 9.11.

9.56 Find the force at B in Figure P9.56 required to hold the gate AB of Example 9.11 in equilibrium.

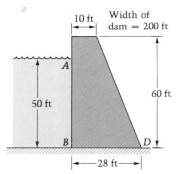

Figure P9.53

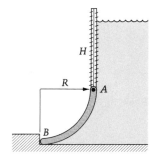

Figure P9.56

9.57 In Example 9.11 compute the line of action of the horizontal component F_{rx} of the fluid force on the sluice gate in two ways, and check one against the other:

a. Using the rectangular and triangular distributions in the example

b. Using Equation (9.26).

9.58 The steel gate has the cross section shown in Figure P9.58. It is 10 ft long, and has a specific weight of 490 lb/ft³. Determine the depth H of water that will cause the gate to open (i.e., pivot clockwise, letting water flow under).

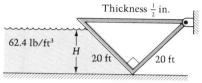

Figure P9.58

9.59 In Figure P9.59 find the water level H at which the hinged gate will open and let water spill out along the line PQ. Neglect the weight of the gate.

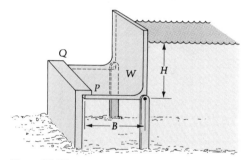

Figure P9.59

9.60 A bowl with the dimensions shown in Figure P9.60 is filled with a liquid of density ρ. By integration, find the resultant force exerted by the liquid on the bowl.

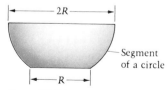

Figure P9.60

9.61 In Figure P9.61:

a. Find the single-force resultant acting on the gate $\mathcal{G}$ caused by hydrostatic pressure, for the case $\theta = 0$. The width of the gate is 5 m and the mass density of the water $= 1$ g/cm³.

b. Find the reactions at the pin A and floor B.

9.62 Repeat Problem 9.61 when $\theta = 20°$.

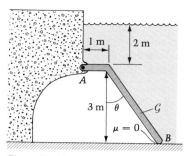

Figure P9.61

9.63 In Figure P9.63 find the magnitude, direction, and line of action of the force-alone resultant caused by hydrostatic pressure acting on a 1-m width of the approximately parabolic surface.

9.64 Consider the law of buoyancy as it relates to ships floating in the sea. For the simplified "ship" in Figure P9.64, assume that the center of mass is at C on the waterline in calm sea. Assume that the ship then rolls as indicated. Locate the centroid of the displaced fluid (which used to be at F) and show that the same buoyant force has now shifted to the right sufficiently to cause the ship to begin to roll back counterclockwise.

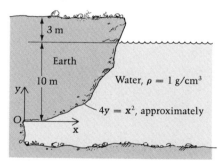

Figure P9.63

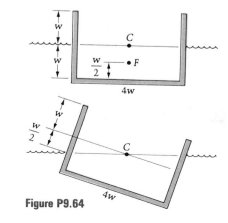

Figure P9.64

9.65 In Problem 9.64 show that if the simplified ship is only 2w wide, then it will continue to roll clockwise toward capsizing if it has rolled to the point where the same 3w/2 is above the waterline on the left side. (This angle of roll is 26.6°, which is *very* large.) See Figure P9.65.

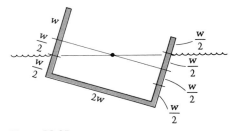

Figure P9.65

9.66 The hollowed sphere in Figure P9.66 is sealed at its base and is being used as a tank. How full is it $(H - z)$ at the instant it lifts off the ground? (Assume the seal exerts no downward force on the tank.)

9.67 Huck Finn has gone to sleep with one end of his fishing pole resting between his toes. If the density of the pole is half that of water, find the fraction of the pole that is under water. See Figure P9.67.

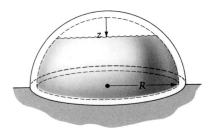

Figure P9.66

Figure P9.67

COMPUTER PROBLEMS ▶ Chapter 9

9.68 Write a Newton-Raphson iterative algorithm (see Appendix D) for solving equations of the form $f(x) = 0$. Use the algorithm to solve the equation

$$f(\theta) = 1 - \sin\theta - \frac{2.5P}{k\ell}\tan\theta = 0$$

from Example 9.3, for a wide range of values of $P/k\ell$ from zero to very large. Compare with the text when $P/(k\ell) = 1$.

9.69 Use the algorithm in the preceding problem to solve the equation

$$f(t) = t^3 - 7t^2 + 15.3t - 1.34$$
$$= 0$$

for the critical tub thickness in Example 9.12; compare with the result found in the text.

* **9.70** The hollowed cone in Figure P9.70 is sealed at its base and is being used as a tank. How full is it $(H - z)$ at the instant it lifts off the ground? (Assume the seal cannot exert a downward force on the tank.)

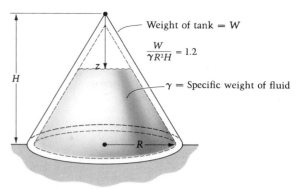

Weight of tank $= W$

$$\frac{W}{\gamma R^2 H} = 1.2$$

$\gamma =$ Specific weight of fluid

Figure P9.70

REVIEW QUESTIONS ▶ Chapter 9

True or False?

1. A virtual displacement is an imagined one that constraints on the body may prevent from occurring in reality.

2. An advantage of virtual work is that forces not of interest may be suppressed in the analysis by appropriate choices of virtual displacements.

3. In using virtual work, we need not worry about the body being in equilibrium in an *inertial* frame; any old frame will do.

4. Virtual displacements really need not be restricted to infinitesimals provided rotations are infinitesimal or zero.

5. The method of virtual work cannot be used to find the force of interaction where two parts of a body are pinned together.

6. The Principle of Virtual Work is restricted to situations where there is no friction.

7. The resultant force due to hydrostatic pressure on a submerged plane surface always equals the product of the projected area on a vertical plane and the pressure at the centroid of this projection.

8. For hydrostatic pressure on a submerged plane area, the center of pressure is closer to the free surface of the liquid than is the centroid of the area.

9. Consider the vertical plane that (a) contains the centroid of a plane area submerged in a liquid; and (b) is perpendicular to the plane area. The center of pressure has to lie in this vertical plane.

10. The magnitude of the resultant force on a curved surface equals the pressure at the centroid of the curved area multiplied by this area.

11. The buoyant force is always vertical.

Answers: 1. T   2. T   3. F   4. T   5. F   6. F   7. F   8. F   9. F   10. F   11. T

▶
▶
▶

Appendix Contents

A $\blacktriangleright$ $\blacktriangleright$ $\blacktriangleright$ Vectors

A.1 Vectors: Addition, Subtraction, and Multiplication by a Scalar*

Vectors are mathematical entities possessing the qualities of magnitude and direction and obeying certain algebraic rules. Although the concepts involved may be extended to spaces of higher dimensions, our concern shall be with vectors in ordinary three-dimensional space. In this circumstance it is possible to represent a vector by an "arrow" — that is, a line segment whose length is proportional to the magnitude of the vector, with directionality along the line segment indicated by the arrowhead.

Many of the features of the algebra of vectors may be either deduced or readily seen from the *parallelogram law* by which vector addition is defined (Figure A.1). Using boldfaced type to denote vectors, we write

$$\mathbf{C} = \mathbf{A} + \mathbf{B}$$

Clearly from the parallelogram law there is no significance attributable to order, so that also

$$\mathbf{C} = \mathbf{B} + \mathbf{A}$$

which is to say that vector addition is commutative. We also note that the parallelogram law suggests a slightly different way of depicting vector addition; that is, the same result is obtained by a head-to-tail arrangement as shown in Figure A.2.

We are now in a position to illustrate (at least for two dimensions) the fact that vector addition is associative. In Figure A.3 we see that

$$\mathbf{A} + (\mathbf{B} + \mathbf{C}) = (\mathbf{A} + \mathbf{B}) + \mathbf{C}$$

Thus, in vector addition, order is irrelevant.

The negative of a vector **A** is the vector having the same magnitude as **A** but the opposite direction. Writing $(-\mathbf{A})$ for the negative of **A**, the parallelogram law dictates that

$$\mathbf{A} + (-\mathbf{A}) = 0$$

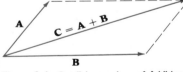

Figure A.1 Parallelogram Law of Addition

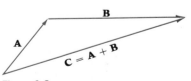

Figure A.2

---

* Examples pertinent to Sections A.1–A.3 may be found in Section 2.2 of the text.

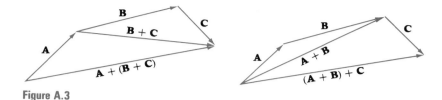

Figure A.3

Furthermore, letting **D** be the sum of a vector **B** and $(-\mathbf{A})$ we have

$$\mathbf{D} = \mathbf{B} + (-\mathbf{A})$$

or more simply

$$\mathbf{D} = \mathbf{B} - \mathbf{A}$$

Thus we have the process of *subtraction*.

The simplest of the processes of multiplication involving a vector is multiplication by a scalar. A scalar is an entity whose measure is a real number. If α is a scalar, then by $\alpha\mathbf{A}$ we mean the vector whose magnitude is the product of the absolute value of α and the magnitude of **A**, and $\alpha\mathbf{A}$ has the same direction as **A** if α is positive and the opposite direction if α is negative. The parallelogram law shows that multiplication by a scalar is distributive — that is

$$\alpha(\mathbf{A} + \mathbf{B}) = \alpha\mathbf{A} + \alpha\mathbf{B}$$

A.2   Unit Vectors and Orthogonal Components

The vector $\hat{\mathbf{e}} = \mathbf{A}/|\mathbf{A}|$, where $|\mathbf{A}|$ is the magnitude of **A**, is dimensionless, has a magnitude of unity, and has the same direction as **A**. Thus $\hat{\mathbf{e}}$ is called the *unit vector* in the direction of **A**.

Unit vectors of preassigned directions provide the mechanism by which vectors are usually expressed. Suppose we let (as in Figure A.4) x, y, and z be mutually perpendicular axes, or reference directions, and we let $\hat{\mathbf{i}}$, $\hat{\mathbf{j}}$, and $\hat{\mathbf{k}}$ be unit vectors in those directions. The parallelogram law allows us to decompose a vector **A** into three mutually perpendicular parts written $A_x\hat{\mathbf{i}}$, $A_y\hat{\mathbf{j}}$, and $A_z\hat{\mathbf{k}}$ so that

$$\mathbf{A} = A_x\hat{\mathbf{i}} + A_y\hat{\mathbf{j}} + A_z\hat{\mathbf{k}}$$

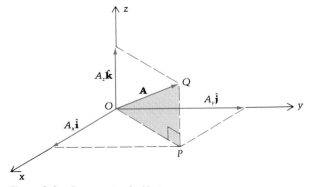

Figure A.4   Components of a Vector

If we write $\mathbf{A} = (A_x\hat{\mathbf{i}} + A_y\hat{\mathbf{j}}) + A_z\hat{\mathbf{k}}$, then the vector in the parentheses extends from the origin O to the projection point P of $\mathbf{A}$ in the xy plane. This vector, by the Pythagorean theorem, has magnitude $\sqrt{A_x^2 + A_y^2}$. If to this vector we then add $A_z\hat{\mathbf{k}}$ (which extends from P to the tip Q of the arrowhead of $\mathbf{A}$), then we have succeeded in building up the vector $\mathbf{A}$ from its three components. The magnitude of $\mathbf{A}$, using the (shaded) right triangle OPQ, is then given in terms of the components of $\mathbf{A}$ as

$$|\mathbf{A}| = \sqrt{\sqrt{A_x^2 + A_y^2}^2 + A_z^2}$$
$$= \sqrt{A_x^2 + A_y^2 + A_z^2} \tag{A.1}$$

We can see that the vectors $A_x\hat{\mathbf{i}}$, $A_y\hat{\mathbf{j}}$, and $A_z\hat{\mathbf{k}}$ are the (orthogonal) projections of $\mathbf{A}$ onto the x, y, and z directions. They are often called orthogonal components of $\mathbf{A}$; the same terminology is also used for the scalars A_x, A_y, and A_z. When context does not make clear which components are intended, we distinguish them by calling the former "vector components" and the latter "scalar components." If θ_x, θ_y, and θ_z are the angles between vector $\mathbf{A}$ and x, y, and z, respectively, then from Figure A.5 we see that

$$A_x = |\mathbf{A}| \cos \theta_x$$
$$A_y = |\mathbf{A}| \cos \theta_y$$
$$A_z = |\mathbf{A}| \cos \theta_z \tag{A.2}$$

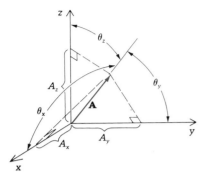

Figure A.5 Angles Whose Cosines Are the Direction Cosines of $\mathbf{A}$

These cosines are called the direction cosines of $\mathbf{A}$. We note that since a unit vector has a magnitude of unity, Equations (A.2) show that the direction cosines of a unit vector *are* its components.

Squaring both sides in each of Equations (A.2), adding, and using Equation (A.1), we confirm an important property of the direction cosines of a line — that is,

$$\cos^2 \theta_x + \cos^2 \theta_y + \cos^2 \theta_z = 1 \tag{A.3}$$

Finally, it is of great practical importance to realize that, from the properties of addition and multiplication by a scalar enumerated in Section A.1,

$$\alpha\mathbf{A} = \alpha A_x\hat{\mathbf{i}} + \alpha A_y\hat{\mathbf{j}} + \alpha A_z\hat{\mathbf{k}} \tag{A.4}$$

α being any scalar. Moreover, if

$$\mathbf{C} = \mathbf{A} + \mathbf{B}$$

then

$$C_x = A_x + B_x$$
$$C_y = A_y + B_y$$
$$C_z = A_z + B_z \tag{A.5}$$

A.3 Scalar (Dot) Product

The scalar product of two vectors $\mathbf{A}$ and $\mathbf{B}$ is denoted $\mathbf{A} \cdot \mathbf{B}$ (from whence comes the name "dot" product) and is defined by

$$\mathbf{A} \cdot \mathbf{B} = |\mathbf{A}||\mathbf{B}| \cos \theta$$

where $\cos \theta$ is the cosine of the angle between the two vectors. This product, as its name communicates, is a scalar, and from the definition it is clear that scalar multiplication is commutative — that is,

$$\mathbf{A} \cdot \mathbf{B} = \mathbf{B} \cdot \mathbf{A}$$

For a unit vector $\hat{\mathbf{e}}$,

$$\mathbf{A} \cdot \hat{\mathbf{e}} = |\mathbf{A}|(1) \cos \theta$$

so that $\mathbf{A} \cdot \hat{\mathbf{e}}$ is the scalar component of $\mathbf{A}$ associated with the direction of $\hat{\mathbf{e}}$. Thus if, as in the preceding section, we express $\mathbf{A}$ by

$$\mathbf{A} = A_x\hat{\mathbf{i}} + A_y\hat{\mathbf{j}} + A_z\hat{\mathbf{k}}$$

then

$$A_x = \mathbf{A} \cdot \hat{\mathbf{i}}$$
$$A_y = \mathbf{A} \cdot \hat{\mathbf{j}}$$
$$A_z = \mathbf{A} \cdot \hat{\mathbf{k}}$$

A very important property is that the scalar product is distributive over addition — that is,

$$\mathbf{A} \cdot (\mathbf{B} + \mathbf{C}) = \mathbf{A} \cdot \mathbf{B} + \mathbf{A} \cdot \mathbf{C}$$

We may obtain this property easily if we now define the unit vector $\hat{\mathbf{e}}$ to be $\mathbf{A}/|\mathbf{A}|$, so that

$$\mathbf{A} \cdot (\mathbf{B} + \mathbf{C}) = |\mathbf{A}|\hat{\mathbf{e}} \cdot (\mathbf{B} + \mathbf{C})$$

From Figure A.6 it is clear that

$$\hat{\mathbf{e}} \cdot (\mathbf{B} + \mathbf{C}) = \hat{\mathbf{e}} \cdot \mathbf{B} + \hat{\mathbf{e}} \cdot \mathbf{C})$$

so that

$$\mathbf{A} \cdot (\mathbf{B} + \mathbf{C}) = |\mathbf{A}|\hat{\mathbf{e}} \cdot (\mathbf{B} + \mathbf{C})$$
$$= |\mathbf{A}|(\hat{\mathbf{e}} \cdot \mathbf{B} + \hat{\mathbf{e}} \cdot \mathbf{C})$$
$$= |\mathbf{A}|\hat{\mathbf{e}} \cdot \mathbf{B} + |\mathbf{A}|\hat{\mathbf{e}} \cdot \mathbf{C}$$
$$= \mathbf{A} \cdot \mathbf{B} + \mathbf{A} \cdot \mathbf{C}$$

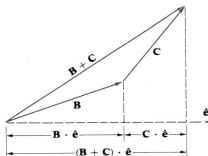

Figure A.6   Distributive Property of the Scalar Product

The commutative and distributive properties of the scalar product allows us to obtain a result of great practical importance. Suppose we express the vectors $\mathbf{A}$ and $\mathbf{B}$ as

$$\mathbf{A} = A_x\hat{\mathbf{i}} + A_y\hat{\mathbf{j}} + A_z\hat{\mathbf{k}}$$
$$\mathbf{B} = B_x\hat{\mathbf{i}} + B_y\hat{\mathbf{j}} + B_z\hat{\mathbf{k}}$$

Then

$$\mathbf{A} \cdot \mathbf{B} = (A_x\hat{\mathbf{i}} + A_y\hat{\mathbf{j}} + A_z\hat{\mathbf{k}}) \cdot (B_x\hat{\mathbf{i}} + B_y\hat{\mathbf{j}} + B_z\hat{\mathbf{k}})$$
$$= A_xB_x(\hat{\mathbf{i}} \cdot \hat{\mathbf{i}}) + A_xB_y(\hat{\mathbf{i}} \cdot \hat{\mathbf{j}}) + A_xB_z(\hat{\mathbf{i}} \cdot \hat{\mathbf{k}})$$
$$+ A_yB_x(\hat{\mathbf{j}} \cdot \hat{\mathbf{i}}) + A_yB_y(\hat{\mathbf{j}} \cdot \hat{\mathbf{j}}) + A_yB_z(\hat{\mathbf{j}} \cdot \hat{\mathbf{k}})$$
$$+ A_zB_x(\hat{\mathbf{k}} \cdot \hat{\mathbf{i}}) + A_zB_y(\hat{\mathbf{k}} \cdot \hat{\mathbf{j}}) + A_zB_z(\hat{\mathbf{k}} \cdot \hat{\mathbf{k}})$$

But, since $\hat{\mathbf{i}}$, $\hat{\mathbf{j}}$, and $\hat{\mathbf{k}}$ are mutually perpendicular unit vectors, then by the definition of the scalar product we obtain

$$\hat{\mathbf{i}} \cdot \hat{\mathbf{i}} = \hat{\mathbf{j}} \cdot \hat{\mathbf{j}} = \hat{\mathbf{k}} \cdot \hat{\mathbf{k}} = 1$$

$$\hat{\mathbf{i}} \cdot \hat{\mathbf{j}} = \hat{\mathbf{i}} \cdot \hat{\mathbf{k}} = \hat{\mathbf{j}} \cdot \hat{\mathbf{k}} = 0$$

Substituting above, then, we obtain

$$\mathbf{A} \cdot \mathbf{B} = A_x B_x + A_y B_y + A_z B_z \qquad (A.6)$$

This important result tells us how to compute the scalar product of two vectors in terms of its scalar components. When we apply it to $\mathbf{B} = \mathbf{A}$ we obtain

$$\mathbf{A} \cdot \mathbf{A} = A_x^2 + A_y^2 + A_z^2$$

$$= |\mathbf{A}|^2$$

which was already available by the definition — that is,

$$\mathbf{A} \cdot \mathbf{A} = |\mathbf{A}||\mathbf{A}| \cos(0)$$

$$= |\mathbf{A}|^2$$

Finally, before leaving this section we point out that the scalar product gives us an elegant and practical way to express the angle between any two vectors, and, of particular importance in mechanics, it gives us a compact way to compute the orthogonal component of a vector in any given direction.

A.4 Vector (Cross) Product

A second multiplicative operation between two vectors is called the vector product and is usually denoted $\mathbf{A} \times \mathbf{B}$ from whence comes the term "cross" product. The vector product is defined to be a vector so that, if $\mathbf{C} = \mathbf{A} \times \mathbf{B}$, then $|\mathbf{C}| = |\mathbf{A}||\mathbf{B}| \sin \theta$, where θ is the angle (not greater than 180°) between $\mathbf{A}$ and $\mathbf{B}$, and $\mathbf{C}$ is perpendicular to both $\mathbf{A}$ and $\mathbf{B}$ with its direction determined by the "right hand rule" as indicated in Figure A.7. The assignment of the direction of $\mathbf{C}$ may be visualized as in the figure by first giving $\mathbf{A}$ and $\mathbf{B}$ a common point of origin or "tail," and then by rotating $\mathbf{A}$ about its tail so as to line it up with $\mathbf{B}$. If the sense of turning of $\mathbf{A}$ (through angle θ) to line it up with $\mathbf{B}$ is mimicked by curling the fingers of the right hand, then the extended thumb points in the direction of $\mathbf{C} = \mathbf{A} \times \mathbf{B}$. This is also the correspondence between the turning and advance of a screw or bolt with the customary right-hand threads. A sometimes useful geometric interpretation of the cross product is that its magnitude is the area of a parallelogram whose sides have as their lengths the magnitudes of $\mathbf{A}$ and $\mathbf{B}$. (See Problem A.13 at the end of this appendix.)

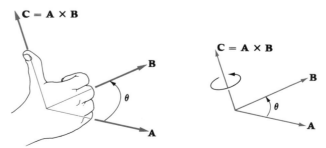

Figure A.7 Direction of the Vector Product

From its definition, vector multiplication is not commutative, and, in fact,

$$\mathbf{B} \times \mathbf{A} = -\mathbf{A} \times \mathbf{B}$$

Vector multiplication is, however, distributive, and this is an important property of which we shall take advantage many times. The property is that

$$\mathbf{A} \times (\mathbf{B} + \mathbf{C}) = \mathbf{A} \times \mathbf{B} + \mathbf{A} \times \mathbf{C} \tag{A.7}$$

To establish this property we begin with a special case. Let **B** and **C** be vectors, each perpendicular to **A** as shown in Figure A.8. Because **A** is perpendic-

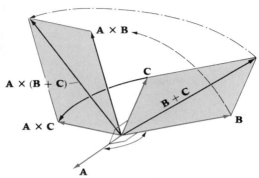

Figure A.8   Distributive Property of the Vector Product

ular to **B**, to **C** and to **B** + **C**, then the vector product of **A** with each of these is obtained by rotating the original vector through 90° about **A** as shown, and multiplying its magnitude by | **A** |. Thus by the parallelogram law of addition,

$$\mathbf{A} \times (\mathbf{B} + \mathbf{C}) = \mathbf{A} \times \mathbf{B} + \mathbf{A} \times \mathbf{C}$$

We now need to relax the restriction that **B** and **C** are perpendicular to **A**. To that end we decompose the vector **B** into two parts: one part, $\mathbf{B}_1$, in the direction of **A** and the remainder, $\mathbf{B}_2$, necessarily perpendicular to **A**. Thus

$$\mathbf{B}_1 + \mathbf{B}_2 = \mathbf{B}$$

We now need the result that

$$\mathbf{A} \times \mathbf{B} = \mathbf{A} \times \mathbf{B}_2$$

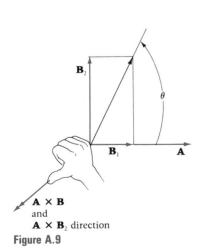

Figure A.9

Figure A.9 shows us that the direction of $\mathbf{A} \times \mathbf{B}_2$ will be the same as that of $\mathbf{A} \times \mathbf{B}$. As for the magnitudes, we have

$$|\mathbf{A} \times \mathbf{B}| = |\mathbf{A}||\mathbf{B}| \sin \theta$$

and this equals $|\mathbf{A} \times \mathbf{B}_2|$, since

$$|\mathbf{A} \times \mathbf{B}_2| = |\mathbf{A}||\mathbf{B}_2| \sin 90°$$

$$= |\mathbf{A}|(|\mathbf{B}| \sin \theta)1$$

Now let us call **D** the sum of **B** and **C** and similarly decompose **C** and **D**. Thus

$$\mathbf{D} = \mathbf{B} + \mathbf{C}$$

and

$$D_1 = B_1 + C_1$$

and thus

$$D_2 = B_2 + C_2$$

Because B_2 and C_2 (and hence D_2) are each perpendicular to A, we use our restricted distributive law, which is that if B_2 and C_2 are normal to A,

$$A \times (B_2 + C_2) = A \times B_2 + A \times C_2$$

or

$$A \times D_2 = A \times B_2 + A \times C_2$$

But

$$A \times B_2 = A \times B$$

and similarly,

$$A \times C_2 = A \times C$$

and

$$A \times D_2 = A \times D$$

so that $A \times (B + C) = A \times B + A \times C$ without any restriction, which completes our proof.

We close this section with a formula for the cross product in terms of components of the multiplying vectors. As usual let $\hat{i}, \hat{j},$ and $\hat{k}$ be mutually perpendicular unit vectors in the x, y, and z directions, but let us also specify that this is a *right-handed system*; that is

$$\hat{i} \times \hat{j} = \hat{k}$$
$$\hat{k} \times \hat{i} = \hat{j}$$

and

$$\hat{j} \times \hat{k} = \hat{i}$$

It then follows that

$$\hat{j} \times \hat{i} = -\hat{k}$$
$$\hat{i} \times \hat{k} = -\hat{j}$$

and

$$\hat{k} \times \hat{j} = -\hat{i}$$

and of course

$$\hat{i} \times \hat{i} = \hat{j} \times \hat{j}$$
$$= \hat{k} \times \hat{k} = 0$$

Expressing A and B in terms of components along x, y, and z,

$$A \times B = (A_x\hat{i} + A_y\hat{j} + A_z\hat{k}) \times (B_x\hat{i} + B_y\hat{j} + B_z\hat{k})$$
$$= A_xB_y(\hat{i} \times \hat{j}) + A_xB_z(\hat{i} \times \hat{k}) + A_yB_x(\hat{j} \times \hat{i})$$
$$\quad + A_yB_z(\hat{j} \times \hat{k}) + A_zB_x(\hat{k} \times \hat{i}) + A_zB_y(\hat{k} \times \hat{j})$$
$$= (A_yB_z - A_zB_y)\hat{i} + (A_zB_x - A_xB_z)\hat{j} + (A_xB_y - A_yB_x)\hat{k} \qquad \text{(A.8)}$$

Sometimes it is useful to express this result in the form of a determinant,

$$\begin{vmatrix} \hat{\mathbf{i}} & \hat{\mathbf{j}} & \hat{\mathbf{k}} \\ A_x & A_y & A_z \\ B_x & B_y & B_z \end{vmatrix}$$

which will be recognized, when expanded, as the right-hand side of Equation (A.8).

It is necessary for the reader to become thoroughly familiar with the vector, or cross, product, because it is this operation that lies at the heart of the concept of the *moment of a force* (see Section 3.2).

A.5   Scalar and Vector Triple Products

Two different multiplicative operations, each involving three vectors, naturally arise in mechanics. The first of these is called the *scalar triple product.* It is, as the name implies, a scalar, and it results from the scalar (dot) product of a vector with the vector (cross) product of two others, as in $\mathbf{C} \cdot (\mathbf{A} \times \mathbf{B})$. A sometimes useful geometric interpretation of this product is that its magnitude is the volume of a parallelepiped whose sides have as their lengths the magnitudes of $\mathbf{A}$, $\mathbf{B}$, and $\mathbf{C}$. Of greater importance to us here is that the value of the scalar triple product is not changed by certain exchanges of elements; in particular;

$$\mathbf{C} \cdot (\mathbf{A} \times \mathbf{B}) = \mathbf{A} \cdot (\mathbf{B} \times \mathbf{C}) = \mathbf{B} \cdot (\mathbf{C} \times \mathbf{A})^* \qquad (A.9)$$

Equation (A.9) may be established by returning to the expression for $\mathbf{A} \times \mathbf{B}$ given by Equation (A.8):

$$\mathbf{A} \times \mathbf{B} = (A_y B_z - A_z B_y)\hat{\mathbf{i}} + (A_z B_x - A_x B_z)\hat{\mathbf{j}} + (A_x B_y - A_y B_x)\hat{\mathbf{k}}$$

If we similarly express $\mathbf{C}$ by

$$\mathbf{C} = C_x \hat{\mathbf{i}} + C_y \hat{\mathbf{j}} + C_z \hat{\mathbf{k}}$$

then

$$\mathbf{C} \cdot (\mathbf{A} \times \mathbf{B}) = (A_y B_z - A_z B_y)C_x + (A_z B_x - A_x B_z)C_y$$
$$+ (A_x B_y - A_y B_x)C_z \qquad (A.10)$$

The right-hand side of Equation (A.10) may be expressed as a determinant so that

$$\mathbf{C} \cdot (\mathbf{A} \times \mathbf{B}) = \begin{vmatrix} C_x & C_y & C_z \\ A_x & A_y & A_z \\ B_x & B_y & B_z \end{vmatrix} \qquad (A.11)$$

Because the value of a determinant is not altered by two interchanges of its rows, the right-hand side of Equation (A.11) has the same value as

$$\begin{vmatrix} A_x & A_y & A_z \\ B_x & B_y & B_z \\ C_x & C_y & C_z \end{vmatrix} \quad \text{or} \quad \begin{vmatrix} B_x & B_y & B_z \\ C_x & C_y & C_z \\ A_x & A_y & A_z \end{vmatrix} \qquad (A.12)$$

---

* This third expression may also be written $(\mathbf{C} \times \mathbf{A}) \cdot \mathbf{B}$ so that Equation (A.9) expresses that an identity follows from "interchanging" the "dot" and "cross." It is of course understood that the parentheses, indicating the order of the operations, always group the elements of the cross product.

The first of these may now be recognized as $\mathbf{A} \cdot (\mathbf{B} \times \mathbf{C})$ and the second as $\mathbf{B} \cdot (\mathbf{C} \times \mathbf{A})$, which establishes the identity (A.9). We note that, because $\mathbf{B} \times \mathbf{A} = -\mathbf{A} \times \mathbf{B}$, then $\mathbf{C} \cdot (\mathbf{B} \times \mathbf{A}) = -\mathbf{C} \cdot (\mathbf{A} \times \mathbf{B})$. This result is also immediately available from the determinant form because one interchange of rows changes the sign of a determinant.

Finally, we point out a feature of the scalar triple product that the reader may already have deduced. That is, if any two of the three vectors are proportional,* the scalar triple product vanishes. This follows from the fact that the cross product vanishes if the two vectors are proportional, since one of the forms of Equation (A.9) will have the two proportional vectors in the cross product. The result may also be deduced from the determinant form since a determinant vanishes if two rows are proportional. It is also seen from the fact that, in this case, the parallelepiped mentioned in the opening paragraph has zero volume.

EXAMPLE A.1

For the vectors:

$$\mathbf{A} = 2\hat{\mathbf{i}} - 3\hat{\mathbf{j}} + \hat{\mathbf{k}}$$
$$\mathbf{B} = 5\hat{\mathbf{i}} + \hat{\mathbf{j}} - 4\hat{\mathbf{k}}$$
$$\mathbf{C} = 3\hat{\mathbf{i}} + \hat{\mathbf{k}}$$

find the scalar triple product $\mathbf{C} \cdot (\mathbf{A} \times \mathbf{B})$.

Solution

$$\mathbf{C} \cdot (\mathbf{A} \times \mathbf{B}) = \mathbf{C} \cdot [(2\hat{\mathbf{i}} - 3\hat{\mathbf{j}} + \hat{\mathbf{k}}) \times (5\hat{\mathbf{i}} + \hat{\mathbf{j}} - 4\hat{\mathbf{k}})]$$

and omitting the zero cross products such as $\hat{\mathbf{i}} \times \hat{\mathbf{i}}$,

$$= \mathbf{C} \cdot [2(1)(\hat{\mathbf{i}} \times \hat{\mathbf{j}}) + 2(-4)(\hat{\mathbf{i}} \times \hat{\mathbf{k}}) + (-3)5(\hat{\mathbf{j}} \times \hat{\mathbf{i}}) + (-3)(-4)(\hat{\mathbf{j}} \times \hat{\mathbf{k}})$$
$$+ 1(5)(\hat{\mathbf{k}} \times \hat{\mathbf{i}}) + 1(1)(\hat{\mathbf{k}} \times \hat{\mathbf{j}})]$$
$$= (3\hat{\mathbf{i}} + \hat{\mathbf{k}}) \cdot [2\hat{\mathbf{k}} + 8\hat{\mathbf{j}} + 15\hat{\mathbf{k}} + 12\hat{\mathbf{i}} + 5\hat{\mathbf{j}} - \hat{\mathbf{i}}]$$
$$= (3\hat{\mathbf{i}} + \hat{\mathbf{k}}) \cdot [11\hat{\mathbf{i}} + 13\hat{\mathbf{j}} + 17\hat{\mathbf{k}}]$$
$$= 3(11) + 0(13) + 1(17) = 50$$

Interchanging the dot and cross and recomputing as a check,

$$(\mathbf{C} \times \mathbf{A}) \cdot \mathbf{B} = [3(-3)(\hat{\mathbf{i}} \times \hat{\mathbf{j}}) + 3(1)(\hat{\mathbf{i}} \times \hat{\mathbf{k}}) + 1(2)(\hat{\mathbf{k}} \times \hat{\mathbf{i}})$$
$$+ 1(-3)(\hat{\mathbf{k}} \times \hat{\mathbf{j}})] \cdot \mathbf{B}$$
$$= [-9\hat{\mathbf{k}} - 3\hat{\mathbf{j}} + 2\hat{\mathbf{j}} + 3\hat{\mathbf{i}}] \cdot \mathbf{B}$$
$$= (3\hat{\mathbf{i}} - \hat{\mathbf{j}} - 9\hat{\mathbf{k}}) \cdot (5\hat{\mathbf{i}} + \hat{\mathbf{j}} - 4\hat{\mathbf{k}})$$
$$= 3(5) + (-1)(1) + (-9)(-4) = 50$$

The reader is encouraged to confirm this result by using the determinant form, Equation (A.11).

---

* For example, $\mathbf{A} = \alpha\mathbf{B}$, where α is a scalar.

A *vector triple product* is the result of the cross product of a vector with the result of a preceding cross product as in $\mathbf{A} \times (\mathbf{B} \times \mathbf{C})$. The parentheses, which denote the order of multiplication, were really unnecessary in the case of the scalar triple product because the result of a dot product is a scalar and the cross product of a scalar and a vector is without meaning. Here, however, the parentheses *are* needed because $(\mathbf{A} \times \mathbf{B}) \times \mathbf{C}$ is itself a legitimate vector triple product. Moreover, in general

$$\mathbf{A} \times (\mathbf{B} \times \mathbf{C}) \neq (\mathbf{A} \times \mathbf{B}) \times \mathbf{C}$$

We may easily illustrate this fact by an example: let $\hat{\mathbf{i}}$, $\hat{\mathbf{j}}$, and $\hat{\mathbf{k}}$ be the usual right-handed system of mutually perpendicular unit vectors; then

$$\hat{\mathbf{i}} \times (\hat{\mathbf{i}} \times \hat{\mathbf{j}}) = \hat{\mathbf{i}} \times \hat{\mathbf{k}} = -\hat{\mathbf{j}}$$

But

$$(\hat{\mathbf{i}} \times \hat{\mathbf{i}}) \times \hat{\mathbf{j}} = 0 \times \hat{\mathbf{j}} = 0$$

which establishes the importance of the order of multiplication.

Identities involving vector triple products that will be particularly useful in dynamics are

$$\mathbf{A} \times (\mathbf{B} \times \mathbf{C}) = (\mathbf{A} \cdot \mathbf{C})\mathbf{B} - (\mathbf{A} \cdot \mathbf{B})\mathbf{C} \tag{A.13}$$

and

$$(\mathbf{A} \times \mathbf{B}) \times \mathbf{C} = (\mathbf{A} \cdot \mathbf{C})\mathbf{B} - (\mathbf{B} \cdot \mathbf{C})\mathbf{A} \tag{A.14}$$

Both (A.13) and (A.14) can perhaps be remembered by noting that in both formulas, the answer is "the vector in the middle times the dot product of the other two, *minus* the other one in parentheses times the dot product of the other two."

To prove the identity (A.13), we apply the component form (A.7) of the cross product to $\mathbf{B} \times \mathbf{C}$:

$$\mathbf{B} \times \mathbf{C} = (B_y C_z - B_z C_y)\hat{\mathbf{i}} + (B_z C_x - B_x C_z)\hat{\mathbf{j}} + (B_x C_y - B_y C_x)\hat{\mathbf{k}}$$

Now letting $\mathbf{B} \times \mathbf{C}$ play the role of $\mathbf{B}$ in Equation (A.7),

$$\begin{aligned}
\mathbf{A} \times (\mathbf{B} \times \mathbf{C}) &= [A_y(B_x C_y - B_y C_x) - A_z(B_z C_x - B_x C_z)]\hat{\mathbf{i}} \\
&\quad + [A_z(B_y C_z - B_z C_y) - A_x(B_x C_y - B_y C_x)]\hat{\mathbf{j}} \\
&\quad + [A_x(B_z C_x - B_x C_z) - A_y(B_y C_z - B_z C_y)]\hat{\mathbf{k}} \\
&= (A_y C_y + A_z C_z)B_x\hat{\mathbf{i}} + (A_z C_z + A_x C_x)B_y\hat{\mathbf{j}} \\
&\quad + (A_x C_x + A_y C_y)B_z\hat{\mathbf{k}} - [(A_y B_y + A_z B_z)C_x\hat{\mathbf{i}} \\
&\quad + (A_z B_z + A_x B_x)C_y\hat{\mathbf{j}} + (A_x B_x + A_y B_y)C_z\hat{\mathbf{k}}]
\end{aligned}$$

If on the right-hand side we add and subtract the vector

$$A_x B_x C_x\hat{\mathbf{i}} + A_y B_y C_y\hat{\mathbf{j}} + A_z B_z C_z\hat{\mathbf{k}}$$

then we have, after regrouping,

$$\begin{aligned}
\mathbf{A} \times (\mathbf{B} \times \mathbf{C}) &= (A_x C_x + A_y C_y + A_z C_z)(B_x\hat{\mathbf{i}} + B_y\hat{\mathbf{j}} + B_z\hat{\mathbf{k}}) \\
&\quad - (A_x B_x + A_y B_y + A_z B_z)(C_x\hat{\mathbf{i}} + C_y\hat{\mathbf{j}} + C_z\hat{\mathbf{k}})
\end{aligned}$$

But by Equation (A.6),

$$A_x C_x + A_y C_y + A_z C_z = \mathbf{A} \cdot \mathbf{C}$$

and

$$A_xB_x + A_yB_y + A_zB_z = \mathbf{A} \cdot \mathbf{B}$$

Therefore,

$$\mathbf{A} \times (\mathbf{B} \times \mathbf{C}) = (\mathbf{A} \cdot \mathbf{C})\mathbf{B} - (\mathbf{A} \cdot \mathbf{B})\mathbf{C}$$

which is the desired result. Identity (A.14) may be similarly established.

EXAMPLE A.2

Determine $\mathbf{A} \times (\mathbf{B} \times \mathbf{C})$ for the vectors

$$\mathbf{A} = 2\hat{\imath} - 3\hat{\jmath} + \hat{k}$$
$$\mathbf{B} = 5\hat{\imath} + \hat{\jmath} - 4\hat{k}$$
$$\mathbf{C} = 3\hat{\imath} + \hat{k}$$

Solution

$$\begin{aligned}
\mathbf{A} \times (\mathbf{B} \times \mathbf{C}) &= (\mathbf{A} \cdot \mathbf{C})\mathbf{B} - (\mathbf{A} \cdot \mathbf{B})\mathbf{C} \\
&= [2(3) + (-3)0 + 1(1)](5\hat{\imath} + \hat{\jmath} - 4\hat{k}) \\
&\quad - [2(5) + (-3)1 + 1(-4)](3\hat{\imath} + \hat{k}) \\
&= 7(5\hat{\imath} + \hat{\jmath} - 4\hat{k}) - 3(3\hat{\imath} + \hat{k}) \\
&= 26\hat{\imath} + 7\hat{\jmath} - 31\hat{k}
\end{aligned}$$

The reader is encouraged to confirm this result by first computing $\mathbf{B} \times \mathbf{C}$ and then "crossing" $\mathbf{A}$ into it.

PROBLEMS ▶ Appendix A

A.1 Find $\mathbf{A} + \mathbf{B}$ and $\mathbf{A} - \mathbf{B}$ if $\mathbf{A} = 2\hat{\imath} + 6\hat{\jmath} - 4\hat{k}$ and $\mathbf{B} = -3\hat{\imath} - \hat{\jmath} + 10\hat{k}$.

A.2 What is the unit vector in the direction of the vector $(3\hat{\imath} + \hat{\jmath} - \hat{k})$?

A.3 Find the dot product of the vectors:

$$\mathbf{F} = 2\hat{\imath} + 3\hat{\jmath} + 6\hat{k}$$
$$\mathbf{Q} = -\hat{\imath} - 6\hat{\jmath} - 3\hat{k}$$

A.4 Make up a rule about the direction of $\mathbf{A} \times \mathbf{B}$ that relates to a right-handed screw turning from $\mathbf{A}$ into $\mathbf{B}$. (See Figure PA.4.)

A.5 Given:

$$\mathbf{A} = 6\hat{\imath} + 2\hat{\jmath} + 9\hat{k}, \quad \mathbf{B} = 10\hat{\imath} + 6\hat{k},$$
$$\mathbf{C} = 2\hat{\imath} + 4\hat{\jmath} + 6\hat{k}$$

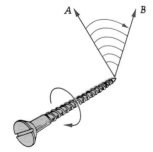

Figure PA.4

Find:

$$(\mathbf{A} \cdot \mathbf{B})\mathbf{C}, \quad \mathbf{A} \cdot \mathbf{B} \times \mathbf{C},$$
$$(\mathbf{A} \times \mathbf{B}) \times \mathbf{C}, \quad \mathbf{A} \times (\mathbf{B} \times \mathbf{C})$$

For the following sets of vectors in Problems A.6–A.9, find: (a) $\mathbf{A} + \mathbf{B}$; (b) $\mathbf{B} - \mathbf{C}$; (c) $\mathbf{A} \cdot \mathbf{B}$; (d) $\mathbf{B} \times \mathbf{C}$; (e) $\mathbf{A} \cdot (\mathbf{B} \times \mathbf{C})$; (f) $\mathbf{A} \times (\mathbf{B} \times \mathbf{C})$; (g) the unit vector in the direction of $\mathbf{C}$; (h) the direction cosines of $\mathbf{B}$.

A.6 $\mathbf{A} = \hat{\mathbf{i}} + \hat{\mathbf{j}}; \mathbf{B} = 2\hat{\mathbf{i}} - 3\hat{\mathbf{j}} + 4\hat{\mathbf{k}}; \mathbf{C} = -5\hat{\mathbf{i}} + 4\hat{\mathbf{j}} - 2\hat{\mathbf{k}}$

A.7 $\mathbf{A} = 2.4\hat{\mathbf{i}} - 6.3\hat{\mathbf{k}}; \mathbf{B} = 4.1\hat{\mathbf{i}} - 20.5\hat{\mathbf{j}} + 6.0\hat{\mathbf{k}};$
$\mathbf{C} = 5.1\hat{\mathbf{i}} + 8.7\hat{\mathbf{j}}$

A.8 $\mathbf{A} = 3\hat{\mathbf{i}}; \mathbf{B} = -7\hat{\mathbf{j}} + \hat{\mathbf{k}}; \mathbf{C} = \hat{\mathbf{i}} + \hat{\mathbf{j}} + \hat{\mathbf{k}}$

A.9 $\mathbf{A} = 2\hat{\mathbf{i}} - \hat{\mathbf{j}} + \hat{\mathbf{k}}; \mathbf{B} = 15\hat{\mathbf{i}} - 20\hat{\mathbf{j}} + 18\hat{\mathbf{k}};$
$\mathbf{C} = \hat{\mathbf{i}} + 7\hat{\mathbf{k}}$

A.10   Find the cross product of the vectors

$$\mathbf{P} = 2\hat{\mathbf{i}} + 3\hat{\mathbf{j}} + 6\hat{\mathbf{k}} \quad \text{and} \quad \mathbf{Q} = -\hat{\mathbf{i}} - 6\hat{\mathbf{j}} - 3\hat{\mathbf{k}}$$

A.11   Find the dot and cross products of

$$\mathbf{A} = 2\hat{\mathbf{i}} - \hat{\mathbf{j}} + 3\hat{\mathbf{k}} \quad \text{and} \quad \mathbf{B} = 3\hat{\mathbf{i}} + 2\hat{\mathbf{j}} - 5\hat{\mathbf{k}}$$

A.12   Find the angle between $\hat{\mathbf{k}}$ and the sum of the vectors in Problem A.11.

A.13   Show that the magnitude of $\mathbf{A} \times \mathbf{B}$ equals the "area" of a parallelogram having side lengths equal to $|\mathbf{A}|$ and $|\mathbf{B}|$.

A.14   Find the angle between the forces $\mathbf{F}_1$ and $\mathbf{F}_2$ in Figure PA.14.

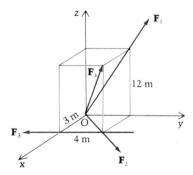

Figure PA.14

* **A.15**   In Problem A.14, find the distance between forces $\mathbf{F}_4$ and $\mathbf{F}_3$ (i.e., the length of the line segment intersecting the two and perpendicular to them both).

A.16   Show that $\mathbf{A} \cdot (\mathbf{B} \times \mathbf{C})$ is the volume of the parallelepiped in Figure PA.16.

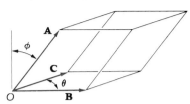

Figure PA.16

A.17   The three vectors $\mathbf{A} = 2\hat{\mathbf{i}} + 3\hat{\mathbf{j}}$, $\mathbf{B} = 5\hat{\mathbf{j}} + \hat{\mathbf{k}}$, and $\mathbf{C} = \hat{\mathbf{i}} + \hat{\mathbf{j}} + \hat{\mathbf{k}}$ define a paralellepiped with the origin O as one vertex. Find the length of the diagonal from O to the opposite corner.

A.18   Show that any vector $\mathbf{C}$ perpendicular to both $\mathbf{A}$ and $\mathbf{B}$ has the form

$$\mathbf{C} = \pm |\mathbf{C}| \frac{\mathbf{A} \times \mathbf{B}}{|\mathbf{A} \times \mathbf{B}|}$$

A.19   Given $\mathbf{A} = 3\hat{\mathbf{i}} + 4\hat{\mathbf{j}}$ and $\mathbf{B} = 4\hat{\mathbf{j}} + \hat{\mathbf{k}}$, write unit vectors (a) having the same direction as $\mathbf{A}$; (b) perpendicular to $\mathbf{B}$ and lying in the xy plane; (c) perpendicular to both $\mathbf{A}$ and $\mathbf{B}$.

A.20   The vector $\mathbf{D}$ in Figure PA.20 has a magnitude of 60. Resolve the vector into two nonorthogonal components, one along AB and another along BC.

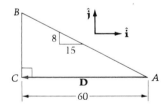

Figure PA.20

---

* Asterisks identify the more difficult problems.

B

▶
▶
▶

TABLES RELATING TO UNITS

The tables in this appendix are useful in identifying and converting units between the SI and U.S. systems.

Table B.1   Units Commonly Used in Engineering Mechanics

Quantity	SI (Standard International or "Metric") Unit	U.S. Unit
force	newton (N)	pound (lb)
mass	kilogram (kg)	slug
length	meter (m)	foot (ft)
time	second (s)	second (sec)
moment of force	N · m	lb-ft
work or energy	joule (J)(= N · m)	ft-lb
pressure or stress	pascal (Pa)(= N/m^2) —	lb/ft^2
velocity	m/s	ft/sec
angular velocity	rad/s	rad/sec
acceleration	m/s^2	ft/sec^2
angular acceleration	rad/s^2	rad/sec^2
mass moment of inertia	$kg · m^2$	$slug\text{-}ft^2$
moment of inertia of area	m^4	ft^4
momentum	kg · m/s	slug-ft/sec
moment of momentum	$kg · m^2/s$	$slug\text{-}ft^2/sec$
impulse	N · s	lb-sec
angular impulse	N · m · s	lb-ft-sec
mass density	kg/m^3	$slug/ft^3$
specific weight	N/m^3	lb/ft^3
power	watt (W)(= J/s)	ft-lb/sec
frequency	hertz (Hz)(= 1 cycle/s)	Hz (same)

Table B.2   Standard Prefixes Used in the SI System of Units

tera	T	10^{12}	centi	c	10^{-2}
giga	G	10^{9}	milli	m	10^{-3}
mega	M	10^{6}	micro	μ	10^{-6}
kilo	k	10^{3}	nano	n	10^{-9}
hecto	h	10^{2}	pico	p	10^{-12}
deka	da	10^{1}	femto	f	10^{-15}
deci	d	10^{-1}	atto	a	10^{-18}

Table B.3   Conversion Factors for SI and U.S. Units*

To Convert From	To	Multiply By	Reciprocal (to Get from SI to U.S. Units)
Length, area, volume			
foot (ft)	meter (m)	0.30480	3.2808
inch (in.)	m	0.025400	39.370
statute mile (mi)	m	1609.3	6.2137×10^{-4}
foot^2 (ft^2)	meter^2 (m^2)	0.092903	10.764
inch^2 (in.^{2})	m^2	6.4516×10^{-4}	1550.0
foot^3 (ft^3)	meter^3 (m^3)	0.028317	35.315
inch^3 (in.^{3})	m^3	1.6387×10^{-5}	61024
Velocity			
feet/second (ft/sec)	meter/second (m/s)	0.30480	3.2808
feet/minute (ft/min)	m/s	0.0050800	196.85
knot (nautical mi/hr)	m/s	0.51444	1.9438
mile/hour (mi/hr)	m/s	0.44704	2.2369
mile/hour (mi/hr)	kilometer/hour (km/h)	1.6093	0.62137
Acceleration			
feet/second^2 (ft/sec^2)	meter/second^2 (m/s^2)	0.30480	3.2808
inch/second^2 (in./sec^2)	m/s^2	0.025400	39.370
Mass			
slug (lb-sec^2/ft)	kg	14.594	0.068522
Force			
pound (lb) or pound-force (lbf)	newton (N)	4.4482	0.22481
Density			
slug/foot^3 (slug/ft^3)	kg/m^3	515.38	0.0019403
Energy, work, or moment of force			
foot-pound or pound-foot (ft-lb) (lb-ft)	joule (J) or newton · meter (N · m)	1.3558	0.73757

(Continued)

* Rounded to the five digits cited. Note, for example, that 1 ft = 0.30480 m, so that

$$(\text{Number of feet}) \times \left(\frac{0.30480 \text{ m}}{1 \text{ ft}} \right) = \text{Number of meters}$$

Table B.3   continued

To Convert From	To	Multiply By	Reciprocal (to Get from SI to U.S. Units)
Power			
foot-pound/second (ft-lb/sec)	watt (W)	1.3558	0.73756
horsepower (hp) (550 ft-lb/sec)	W	745.70	0.0013410
Stress, pressure			
pound/inch^2 (lb/in.^{2} or psi)	N/m^2 (or Pa)	6894.8	1.4504×10^{-4}
pound/foot^2 (lb/ft^2)	N/m^2 (or Pa)	47.880	0.020886
Mass moment of inertia			
slug-foot^2 (slug-ft^2 or lb-ft-sec^2)	kg · m^2	1.3558	0.73756
Momentum			
(or linear momentum)			
slug-foot/second (slug-ft/sec)	kg · m/s	4.4482	0.22481
Impulse			
(or linear impulse)			
pound-second (lb-sec)	N · s (or kg · m/s)	4.4482	0.22481
Moment of momentum			
(or angular momentum)			
slug-foot^2/second (slug-ft^2/sec)	kg · m^2/s	1.3558	0.73756
Angular impulse			
pound-foot-second (lb-ft-sec)	N · m · s (or kg · m^2/s)	1.3558	0.73756

C

MOMENTS AND PRODUCTS OF INERTIA OF AREAS

	Shape	Area	Moment of Inertia
Parallelogram		bh	$I_x = \dfrac{bh^3}{3}$ $I_y = \dfrac{bh}{3}(b + h \cot \alpha)^2 - \dfrac{b^2 h^2}{6} \cot \alpha$ $I_{xy} = -\dfrac{bh^2}{12}(3b + 4h \cot \alpha)$
Rectangle		bh	$I_x = \dfrac{bh^3}{3}$ $I_y = \dfrac{bh^3}{3}$ $I_{xy} = -\dfrac{b^2 h^2}{4}$ $I_{x_C} = \dfrac{bh^3}{12}$ $I_{y_C} = \dfrac{hb^3}{12}$ $I_{xy_C} = 0$

	Shape	Area	Moment of Inertia
Triangle		$\frac{1}{2}bh$	$I_x = \dfrac{bh^3}{12}$ $I_y = \dfrac{bh}{12}(a^2 + ab + b^2)$ $I_{xy} = -\dfrac{bh^2}{24}(2a + b)$
Right Triangle		$\frac{1}{2}bh$	$I_x = \dfrac{bh^3}{12}$ $I_y = \dfrac{hb^3}{12}$ $I_{xy} = -\dfrac{b^2h^2}{8}$ $I_{x_C} = \dfrac{bh^3}{36}$ $I_{y_C} = \dfrac{hb^3}{36}$ $I_{xy_C} = -\dfrac{b^2h^2}{72}$
Circle		πR^2	$I_{x_C} = I_{y_C} = \dfrac{\pi R^4}{4}$ $I_{xy_C} = 0$
Semicircular Area		$\dfrac{\pi R^2}{2}$	$I_x = I_y = \dfrac{\pi R^4}{8}$ $I_{xy} = 0$

	Shape	Area	Moment of Inertia
Quarter Circle		$\dfrac{\pi R^2}{4}$	$I_x = I_y = \dfrac{\pi R^4}{16}$ $I_{xy} = -\dfrac{R^4}{8}$
Annulus		$\pi(R_o^2 - R_i^2)$	$I_{x_C} = I_{y_C} = \dfrac{\pi(R_o^4 - R_i^4)}{4}$ $I_{xy_C} = 0$
Thin Annulus		$2\pi R t \quad (t \ll R)$	$I_{x_C} = I_{y_C} \approx \pi R^3 t$ $I_{xy_C} = 0$
Thin Annular Sector		$tR\alpha \quad (t \ll R)$	$I_x \approx \dfrac{tR^3}{2}(\alpha - \sin \alpha)$ $I_y \approx \dfrac{tR^3}{2}(\alpha + \sin \alpha)$ $I_{xy} = 0$

	Shape	Area	Moment of Inertia
Circular Sector	$\dfrac{4R\sin\frac{\alpha}{2}}{3\alpha}$	$\dfrac{\alpha R^2}{2}$	$I_x = \dfrac{\alpha R^4}{8}\left(1 - \dfrac{\sin\alpha}{\alpha}\right)$ $I_y = \dfrac{\alpha R^4}{8}\left(1 + \dfrac{\sin\alpha}{\alpha}\right)$ $I_{xy} = 0$
Segment of a Circle	$\dfrac{4R\sin^3\frac{\alpha}{2}}{3(\alpha - \sin\alpha)}$	$\dfrac{R^2}{2}(\alpha - \sin\alpha)$	$I_x = \dfrac{R^4}{24}(3\alpha - 4\sin\alpha + \sin\alpha\cos\alpha)$ $I_y = \dfrac{R^4}{8}(\alpha - \sin\alpha\cos\alpha)$ $I_{xy} = 0$
Ellipse		πab	$I_{x_C} = \dfrac{\pi ab^3}{4}$ $I_{y_C} = \dfrac{\pi ba^3}{4}$ $I_{xy_C} = 0$
Semiellipse		$\dfrac{\pi ab}{2}$	$I_x = \dfrac{\pi ab^3}{8}$ $I_y = \dfrac{\pi a^3 b}{8}$ $I_{xy} = 0$

	Shape	**Area**	**Moment of Inertia**
Segment of a Parabola		$\frac{4}{3}ab$	$I_x = \dfrac{4}{15}\,ab^3$ $I_y = \dfrac{4}{7}\,a^3b$ $I_{xy} = 0$
Bent Strip		$2lt \quad (t \ll l)$	$I_x \approx \dfrac{2\,l^3t\,\cos^2\alpha}{3}$ $I_y \approx \dfrac{2\,l^3t\,\sin^2\alpha}{3}$ $I_{xy} = 0$

D

EXAMPLES OF NUMERICAL ANALYSIS/THE NEWTON-RAPHSON METHOD

There are a few places in the book where equations arise whose solutions cannot easily be found by elementary algebra. These equations are either polynomials of degree higher than two, or else transcendental equations.

In this appendix, we explain in brief the fundamental idea behind the Newton-Raphson numerical method for solving such equations and others of the form $f(\theta) = 0$. We shall do this for two equations. The first is

$$f(\theta) = 1 - \sin \theta - 2.5 \tan \theta \qquad (D.1)$$

which occurs in Section 9.2.

We are looking for the smallest positive root θ of Equation (D.1). We see that the function has the value unity at $\theta = 0$ and its slope, $-\cos \theta - 2.5 \sec^2 \theta$, is -3.5 there. Furthermore (again by inspection), $f(\theta)$ is negative at $\theta = \pi/4$. Thus (see Figure D.1) we expect a root between 0 and $\pi/4$. The Newton-Raphson algorithm, found in more detail in any book on numerical analysis, works as follows: If θ_0 is an initial estimate of a root θ, then a better approximation is

$$\theta_1 = \theta_0 - \frac{f(\theta_0)}{f'(\theta_0)}$$

Figure D.2 shows what is happening. The quantity $f(\theta_0)/f'(\theta_0)$ causes a backup, from the initial θ_0 approximation, toward the actual root we seek.*

Let us use the initial guess $\theta_0 = \pi/4$ (see Example 9.3). Then $f(\pi/4) = -2.207106781$ and $f'(\pi/4) = -5.707106786$ so that the improved estimate is

$$\theta_1 = 0.785398164 - \frac{-2.207106781}{-5.707106786} = 0.398668624$$

Repeating the algorithm, $f(\theta_1) = -0.441253558$, $f'(\theta_1) = -3.865154335$, and so

$$\theta_2 = 0.398668624 - \frac{-0.441253558}{-3.865154335} = 0.284506673$$

Further applications of the Newton-Raphson algorithm yield

$$\theta_3 = 0.281298920$$
$$\left.\begin{array}{l} \theta_4 = 0.281297097 \\ \theta_5 = 0.281297097 \\ \theta_6 = 0.281297097 \end{array}\right\} \text{convergence!}$$

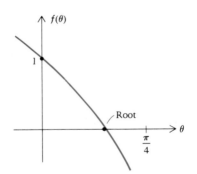

Figure D.1

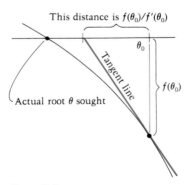

This distance is $f(\theta_0)/f'(\theta_0)$

θ_0

Tangent line

Actual root θ sought

$f(\theta_0)$

Figure D.2

---

* The reader may wish to show that the idea works as well for the other three possible sign combinations of f and f' (i.e., $++$, $+-$, and $-+$).

We have nine good digits after just four iterations, and the approximation $\theta \approx 0.281$ agrees with the solution in the text.

As a second example, we consider the cubic polynomial equation from Example 9.12:

$$f(t) = t^3 - 7t^2 + 15.3t - 1.34 = 0 \qquad (D.2)$$

We begin with the estimate $t_0 = 1$ in. and quickly obtain, using $f'(t) = 3t^2 - 14t + 15.3$,

$$t_1 = t_0 - \frac{f(t_0)}{f'(t_0)} = 1 - \frac{7.96000000}{4.30000000} = -0.851162791$$

Continuing, the results are:

$$t_1 = 1.000000000$$
$$t_2 = -0.851162791$$
$$t_3 = -0.168924556$$
$$t_4 = 0.063694406$$
$$t_5 = 0.090990090$$
$$t_6 = 0.091349686$$
$$t_7 = 0.091349748$$
$$t_8 = 0.091349748 \quad \Bigg\} \quad \text{convergence!}$$
$$t_9 = 0.091349748$$

This time we have eight significant digits after only seven iterations. Note that we have in fact found the *only* root of Equation (D.2), as the reader may wish to show by examining the values of f at the two roots of $f' = 0$. (Note that f is negative at $t = 0$ and positive for large t, and that by Descartes' rule of signs, the *maximum number* of positive real roots is three, and of negative ones is zero.)

E ▶▶▶ EQUILIBRIUM: A BODY AND ITS PARTS

In this appendix, we present a general proof of the ideas given by example in Section 4.4.

Suppose we decompose a body $\mathcal{B}$ into two parts $\mathcal{B}_1$ and $\mathcal{B}_2$ as shown in Figure E.1:

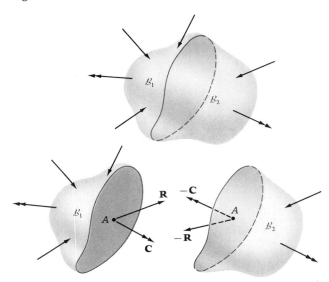

Figure E.1

Furthermore:

1. Let $\mathbf{F}_r$ be the sum of all the external forces on $\mathcal{B}$, the combined body.
2. Let $\mathbf{M}_{rP}$ be the sum of the moments, about an arbitrary point P, of all the external forces and couples on $\mathcal{B}$.
3. Let $\mathbf{F}_r'$ be the sum of all the forces external to $\mathcal{B}$ that act on $\mathcal{B}_1$, and $\mathbf{M}_{rP}'$ be the sum of the moments of those forces and couples about P.
4. Let $\mathbf{F}_r''$ be the sum of all the forces external to $\mathcal{B}$ that act on $\mathcal{B}_2$, and $\mathbf{M}_{rP}''$ be the sum of the moments of those forces and couples about P.
5. Let force $\mathbf{R}$ and couple $\mathbf{C}$ constitute the resultant at A of all the forces that $\mathcal{B}_2$ exerts on $\mathcal{B}_1$; these are *internal* to $\mathcal{B}$.

From the definitions just given we see that

$$\mathbf{F}_r = \mathbf{F}_r' + \mathbf{F}_r''$$

and

$$\mathbf{M}_{rP} = \mathbf{M}_{rP}' + \mathbf{M}_{rP}''$$

We now apply the equilibrium equations to $\mathcal{B}_1$ and obtain

$$\mathbf{F}_r' + \mathbf{R} = 0 \qquad\qquad (E.1)$$

$$\mathbf{M}_{rP}' + \mathbf{r}_{PA} \times \mathbf{R} + \mathbf{C} = 0 \qquad\qquad (E.2)$$

Applying the equilibrium equations to $\mathcal{B}_2$ and using the action-reaction principle as indicated on the free-body diagram,

$$\mathbf{F}_r'' + (-\mathbf{R}) = 0 \qquad\qquad (E.3)$$

$$\mathbf{M}_{rP}'' + \mathbf{r}_{PA} \times (-\mathbf{R}) + (-\mathbf{C}) = 0 \qquad\qquad (E.4)$$

Adding Equations (E.1) and (E.3),

$$\mathbf{F}_r' + \mathbf{F}_r'' = 0 \qquad \text{or} \qquad \mathbf{F}_r = 0 \qquad\qquad (E.5)$$

which is of course the force-equation of equilibrium for the combined body $\mathcal{B}$. Adding Equations (E.2) and (E.4),

$$\mathbf{M}_{rP}' + \mathbf{M}_{rP}'' = 0$$

or

$$\mathbf{M}_{rP} = 0 \qquad\qquad (E.6)$$

which is the moment equation of equilibrium for $\mathcal{B}$.

Two important conclusions can be drawn from this analysis. First, we note that the equilibrium equations for body $\mathcal{B}$ are the sums of the equilibrium equations for its constituents $\mathcal{B}_1$ and $\mathcal{B}_2$. Therefore, any two of the sets will be independent, but not all three. Thus while there may be profit to be gained from writing the equilibrium equations for two, say $\mathcal{B}$ and $\mathcal{B}_1$, nothing additional can be obtained from the equilibrium equations for $\mathcal{B}_2$.

Second, if we know, or can determine, all of the external forces on $\mathcal{B}$, then the resultant interaction between $\mathcal{B}_1$ and $\mathcal{B}_2$ can be determined from the equations of equilibrium for $\mathcal{B}_1$ (or $\mathcal{B}_2$). This is the basis for much of stress analysis in the mechanics of deformable solids [for example, the determination of shear and bending moments in beams (Chapter 5)].

F ▶▶ ANSWERS TO ODD-NUMBERED PROBLEMS
▶

Unless suggested otherwise by the problem statement or figure, $\hat{\mathbf{i}}, \hat{\mathbf{j}}$, and $\hat{\mathbf{k}}$ are unit vectors in the respective directions $\rightarrow$, $\uparrow$, and out of the page.

CHAPTER 1

1.1 An automobile is parked on a hill. Find the forces exerted on the tires by the road.
1.3 Velocity is defined from the concepts of space and time. The dimension of velocity is L/T. Energy is defined from the concepts of force and space; the dimension of energy is $L \cdot F$. **1.5** 1.609 km
1.7 1060 J **1.9** Yes; both sides have dimension L/T.
1.11 3.44×10^{-8} ft³/(slug-sec²) [or ft⁴/(lb-sec⁴)]
1.13 2.07×10^{11} Pa; 2.07×10^{8} kPa; 2.07×10^{5} MPa; 207 GPa **1.15** 1% **1.17** 0.006%; 0.000006%

CHAPTER 2

2.1 $\mathbf{F}_2, |\mathbf{F}_2| = 9.00$ N
2.3 Result is apparent from Figure 2.3.
2.5 (a) 87.8 N (b) 0.456, 0.569, −0.684
2.7 $0.545\hat{\mathbf{i}} - 0.545\hat{\mathbf{j}} + 0.636\hat{\mathbf{k}}$
2.9 From 2.6, $\mathbf{F} = 70\hat{\mathbf{i}} - 20\hat{\mathbf{j}} \pm 68.6\hat{\mathbf{k}}$ lb
For +, $(\mathbf{F} \cdot \hat{\mathbf{e}})\hat{\mathbf{e}} = 42.1(-0.3\hat{\mathbf{i}} + 0.1\hat{\mathbf{j}} + 0.949\hat{\mathbf{k}})$ lb
For −, $(\mathbf{F} \cdot \hat{\mathbf{e}})\hat{\mathbf{e}} = -88.1(-0.3\hat{\mathbf{i}} + 0.1\hat{\mathbf{j}} + 0.949\hat{\mathbf{k}})$ lb
2.11 $18\hat{\mathbf{i}} + 3\hat{\mathbf{j}} - 48\hat{\mathbf{k}}$ lb **2.13** $0.8\hat{\mathbf{i}} + 0.6\hat{\mathbf{j}}$ **2.15** 33 N
2.17 (a) $25.7\hat{\mathbf{i}} - 10.7\hat{\mathbf{j}} + 16\hat{\mathbf{k}}$ lb
(b) $55.1\hat{\mathbf{i}} - 37.1\hat{\mathbf{j}} + 16\hat{\mathbf{k}}$ lb
2.19 12.9 kN; $-18.8\hat{\mathbf{j}}$ kN **2.21** $11.5\hat{\mathbf{i}} - 16.4\hat{\mathbf{j}}$ kN
2.23 $(\mathbf{A} \cdot \mathbf{B})^2 = |\mathbf{A}|^2 |\mathbf{B}|^2 \cos^2 \theta \le |\mathbf{A}|^2 |\mathbf{B}|^2$

2.25 50.2 $\uparrow$ N and 39.8 ⟋35° N
2.27 (a) $21\left(\dfrac{-2\hat{\mathbf{i}} + 6\hat{\mathbf{j}} - 3\hat{\mathbf{k}}}{7}\right)$ lb
(b) $-6\hat{\mathbf{i}} + 18\hat{\mathbf{j}} - 9\hat{\mathbf{k}}$ lb
2.29 Answer given in problem.

2.31 $\mathbf{F}_{PR} = 202$ ⟋45° N; $\mathbf{F}_{QR} = 418$ ⟋20° N

2.33 $97.5\left(\dfrac{3}{13}\hat{\mathbf{i}} - \dfrac{4}{13}\hat{\mathbf{j}} + \dfrac{12}{13}\hat{\mathbf{k}}\right)$ lb;
$122\left(\dfrac{8}{17}\hat{\mathbf{i}} + \dfrac{15}{17}\hat{\mathbf{j}}\right)$ lb; $42.2\hat{\mathbf{j}}$ lb; no
2.35 $20\hat{\mathbf{i}} + 77.5\hat{\mathbf{j}}$ lb and $20\hat{\mathbf{i}} - 77.5\hat{\mathbf{j}}$ lb
2.37 $\pm(0.0683\hat{\mathbf{i}} + 0.820\hat{\mathbf{j}} - 0.569\hat{\mathbf{k}})$
2.39 (a) $|\mathbf{F}_1| = 12.5$ lb; (b) $|\mathbf{F}_1| = 1.08$ N
2.41 The angle is greater in (b) between the sections of the sling above the block and the vertical. The vertical components of force in these sections add to the weight, so the larger angle means the tension must be larger in (b). **2.43** 825 lb **2.45** 30°
2.47 4.57 lb each **2.49** 609 lb; 7.5° **2.51** 41.7 lb

2.53 491 ⟋30° N; 850 ⟍60° N
2.55 $N_1 = W \sin \theta_2 / \sin (\theta_2 - \theta_1)$;
$N_2 = W \sin \theta_1 / \sin (\theta_2 - \theta_1)$;
As $\theta_1 \rightarrow 0$, $N_1 \rightarrow W$ and $N_2 \rightarrow 0$ ✓. **2.57** 592 N
2.59 $F_{AD} = 294$ N; $F_{BD} = 208$ N; $F_{CD} = 339$ N
2.61 $F_1 = W$; $F_2 = 0.530\,W$; $F_3 = 0.385\,W$;
$F_4 = 0.335\,W$; $F_5 = 0.362\,W$; $F_6 \rightarrow \infty$ (They are horizontal, so cannot support the vertical load!)
2.63 $\Sigma \mathbf{F} = 10\hat{\mathbf{j}}$ N $\ne \mathbf{0}$ so no! Need $\mathbf{F}_4 = -10\hat{\mathbf{j}}$ N
2.65 In R_1, 373 N; R_2, 275 N; R_3, 78.5 N **2.67** $W/8 \uparrow$
2.69 (a) Answer given in problem; (b) 30 lb

2.71 $W/8$ **2.73** $\dfrac{21}{64} W$

2.75 (a) A: 450 lb, B: 225 lb; (b) A: 900 lb, B: 450 lb;
(c) A: 600 lb, B: 300 lb
2.77 Tension $= 157$ N; His force onto scaffold $= 530 \downarrow$ N **2.79** 1 in.
2.81 Outside sections: 1651 lb; center section: 1650 lb
2.83 392 N
2.85 From a FBD of the pulley, as θ goes up so does

the tension; $T = 125 / \cos \theta$ N, which is 125 N @ $\theta = 0$ and 483 N @ $\theta = 75°$. When $T =$ man's weight, $\theta = 81.0°$. **2.87** 230 N

2.89 From ground, 280 $\uparrow$ N;

from incline, 127 $\overset{30°}{\diagup}$ N;

from gravity, 196 $\downarrow$ N; from $\mathscr{A}$, 184 $\overset{4}{\diagup}_{3}$ N

2.91 (a) $\dfrac{Qr}{\cdot 2\sqrt{R^2 + 2Rr}}$ (b) $P + Q/2$ (c) $\dfrac{Q(r + R)}{2\sqrt{R^2 + 2Rr}}$

2.93 8 **2.95** 0.383 in. **2.97** $W/\sqrt{6}$; $W\sqrt{2}/6$

CHAPTER 3

3.1 $-60\hat{k}$ N · cm **3.3** $36\hat{k}$ lb-ft
3.5 $4700\hat{k}$ lb-ft; 4.7 ft **3.7** $5210 \curvearrowright$ lb-ft
3.9 (a) $-9\hat{i} + 12\hat{j} + 16\hat{k}$ N
(b) $8.4\hat{i} + 4.8\hat{j} - 7.2\hat{k}$ N · m
3.11 $400\hat{i} - 800\hat{j} + 300\hat{k}$ lb-ft **3.13** 824 N
3.15 117° **3.17** $44.4\hat{e}_{CD}$ N · m **3.19** $-800\hat{k}$ lb-ft
3.21 $10\hat{k}$ N · m **3.23** 1.63 ft
3.25 (a) $-71\hat{i} - 105\hat{j} - 71\hat{k}$ N · m (b) $-105\hat{j}$ N · m
3.27 $-560\hat{k}$ lb-ft **3.29** $360\hat{i} + 360\hat{j}$ lb-ft
3.31 (a) $1080\hat{i} + 1200\hat{k}$ lb-in.
(b) $900(0.667\hat{j} + 0.750\hat{k})$ lb-in.
(c) $-810\hat{i} - 1350\hat{j} - 2850\hat{k}$ lb-in.
(d) $923(0.543\hat{i} + 0.466\hat{j} - 0.699\hat{k})$ lb-in.
3.33 $18\hat{k}$ lb-ft
3.35 (a) $-0.231, 0.923, 0.308$ (b) $-9\hat{i} + 36\hat{j} + 12\hat{k}$ lb
(c) $96\hat{i} + 84\hat{j} - 180\hat{k}$ lb-ft (d) $84\hat{j}$ lb-ft (e) **0**
3.37 (a) $-12\hat{i} + 16\hat{j} - 10\hat{k}$ lb-ft (b) $-10\hat{k}$ lb-ft
(c) $(-0.537, 0.716, -0.447)$; $-12\hat{i} + 16\hat{j} - 10\hat{k}$ lb-ft
3.39 (a) $15\hat{i} + 5\hat{j} + 5\hat{k}$ N · m
(b) $\dfrac{5}{3}\hat{i} + \dfrac{5}{3}\hat{j} - \dfrac{10}{3}\hat{k}$ N · m **3.41** $-2400\hat{k}$ lb-ft
3.43 (a) $355\hat{k}$ lb-ft (b) same about any point!
3.45 (a) $-258\hat{i} - 388\hat{j} - 774\hat{k}$ lb-ft
(b) same as (a) (c) $-388\hat{j}$ lb-ft (d) $-388\hat{j}$ lb-ft
3.47 $180\hat{i} + 240\hat{k}$ lb-in., same about all points.
3.49 $-74\hat{k}$ lb-ft **3.51** $-52.0\hat{k}$ lb-ft
3.53 (a) $-20\hat{i} + 20\hat{k}$ N · m (b) same as (a);
(c) $\dfrac{80}{17}\left(\dfrac{8\hat{i} - 9\hat{j} + 12\hat{k}}{17}\right)$ N · m **3.55** $-176\hat{u}_{\ell}$ N · m
3.57 (a) $3\hat{i} + 9\hat{j}$ lb (b) $-36\hat{k}$ lb-ft (c) $18\hat{k}$ lb-ft
(d) $18\hat{k}$ lb-ft
3.59 (a) $\hat{i}$ N; $20\hat{k}$ N · m
(b) Both answers to (a) must be zero for equilibrium
(c) $a\hat{i} + \dfrac{20}{3}\hat{j}$ N and $(-a - 1)\hat{i} - \dfrac{20}{3}\hat{j}$ N,
respectively, where "a" is arbitrary.

3.61 $1,730\hat{i}$ lb; $1,000\hat{j}$ lb; $-10,700\hat{k}$ lb-ft
3.63 Second force is $-180\hat{j}$ N, along $x = -70$ cm.
3.65 $5\hat{i} + 8\hat{j}$ N; $-0.9\hat{k}$ N · m **3.67** (c) and (d)
3.69 Forces with largest x are 1540 lb;
others are 639 lb. **3.71** $40\hat{i} + 40\hat{j} + 40\hat{k}$ N, 0 N · m
3.73 $P\hat{j} + P\hat{k}$; $-Pa(2\hat{i} + \hat{k})$
3.75 $-500\hat{k}$ N; $-650\hat{i} + 462\hat{j} + 28\hat{k}$ N · m
3.77 $2130\hat{i} - 66\hat{j}$ lb
3.79 $\mathbf{F}_r = 0.866\hat{i} - 1.50\hat{j}$ lb; $\mathbf{M}_{rC} \neq 0$
3.81 130 N · m; $\theta = 67.4°$
3.83 $\mathbf{F} = 40\hat{i} - 20\hat{j} - 30\hat{k}$ lb;
$\mathbf{M}_{rO} = 40\hat{i} + 180\hat{j} - 120\hat{k}$ lb-ft
3.85 $\mathbf{F}_r = -9\hat{i} + 12\hat{j} + 16\hat{k}$ lb;
$\mathbf{M}_{rA} = -36\hat{i} - 18\hat{j} - 90\hat{k}$ lb-ft
3.87 Resultant at C: $3.4\hat{i} + 67.8\hat{j} + 31.2\hat{k}$ lb and
$-271\hat{i} + 80\hat{j} + 45\hat{k}$ lb-ft
3.89 (a) $24\hat{i} - 24\hat{j} - 32\hat{k}$ lb-ft
(b) $36\hat{i} - 36\hat{j} + 48\hat{k}$ lb-ft
(c) $6\hat{i} - 6\hat{j} + 8\hat{k}$ lb-ft (d) **0** (e) $24\hat{k}$ lb-ft
(f) $\Sigma\mathbf{M}_0$, less its 2 $\perp$ components *in* the plane EOG,
leaves its component $\perp$ EOG.

3.91 $591\hat{i} + 82.1\hat{j}$ N **3.93** $33.7 \overset{60.8°}{\diagup}$ lb

3.95 $-823\hat{i} + 1215\hat{j}$ lb

3.97 $145.4 \overset{35.8°}{\diagup}$ kips; 9.24 ft above B.
3.99 $850\hat{i} - 1110\hat{j}$ lb; $y + 1.35x = 1.11$; at $x = 0.82$ ft
3.101 $\mathbf{F}_r = -2\hat{i}$ lb, along the line $y = 9$ in.
(doesn't cross x axis!)
3.103 $\mathbf{F}_r = 3\hat{i} + 3\hat{j}$ lb, crossing at $x = 5$ in. Equation is
$y = x - 5$ **3.105** $-90\hat{j}$ lb, 10.3 ft to the right of O.
3.107 $-220\hat{k}$ N, passing through (x, y)
$= (3.18, 3.55)$ m **3.109** $18\gamma\hat{i}$ lb; $y = -4.02$ ft
3.111 Yes; $|\mathbf{M}_{rP}|/|\mathbf{F}_r|$
3.113 $30\hat{i}$ lb; 25 in. above 80-lb force
3.115 (a) $-2200\hat{i} + 1200\hat{j}$ lb,
with intercepts $x = -2.08$ ft, $y = -1.14$ ft
(b) $(x, y) = (4.33, -3.50)$ ft
3.117 (a) $10\hat{k}$ lb, $-80\hat{i} + 105\hat{j}$ lb-ft
(b) $10\hat{k}$ lb at $(-10.5, -8)$ ft
3.119 (a) $3P\hat{i} - P\hat{j} + 2P\hat{k}$, $Pb\hat{i} - 2Pc\hat{j} + 3Pa\hat{k}$
(b) $6a + 3b + 2c = 0$ (c) No, $\Sigma\mathbf{F} \neq 0$
3.121 (a) $100\hat{i} - 200\hat{j} - 300\hat{k}$ N;
$(-200\ell - 10)\hat{i} + (200\ell - 20)\hat{j} + (-300\ell + 30)\hat{k}$ N · m
(b) $\ell = 0.200$ m
3.123 (a) $8W\hat{k}$ at $(x, y) = (0.884R, 1.06R)$
(b) $3.16W\hat{k}$ at $(1.15R, 1.60R)$ producing $-4.84W\hat{k}$
at middle of plate; or $-3.16W\hat{k}$ at $(0.260R, -0.187R)$
producing $-11.2W\hat{k}$ at middle of plate.
3.125 $6\hat{i} + 8\hat{j} - 8\hat{k}$ lb, piercing xy plane at $(3, 1.4)$ ft

$\hat{i}$ lb at A; original's screwdriver:

    low A

    along $x = 0.667$ m, $y = 1.5$ m

3.131 $-1.73\hat{i} - 3\hat{j}$ lb through $x = -2.5$ ft, $y = 0$

3.133 $3\hat{k}$ lb through the origin

3.135 $\mathbf{F}_r = 3P\hat{i} - P\hat{j} + 2P\hat{k}$ along $28y + 14z = 9a$, $28x$

$-42z = 73a$; $\mathbf{C} = Pa\left(\dfrac{75}{14}\hat{i} - \dfrac{25}{14}\hat{j} + \dfrac{50}{14}\hat{k}\right)$

3.137 $-50\hat{i} - 120\hat{j}$ N intersecting a horizontal line through A 4.33 m to the right of A

3.139 $\mathbf{F}_r = 9\hat{i} + 5\hat{j}$ lb along $z = 9$ ft, $5x - 9y = 0$; $\mathbf{C} = 45\hat{i} + 25\hat{j}$ lb-ft

3.141 (a) Replace system by $\mathbf{F}_r$ and $\mathbf{M}_{rP}$ at P; (b) Replace $\mathbf{M}_{rP}$ by its components; (c) Replace $\mathbf{M}_{rP_\perp}$ by $\mathbf{F}_r$ and $-\mathbf{F}_r$; (d) Cancel $\mathbf{F}_r$ and $-\mathbf{F}_r$ at P; slide $\mathbf{M}_{rP_\parallel}$ over to Q to form,

with $\mathbf{F}_r$, the screwdriver there.   **3.143** $\dfrac{q_0 L^2}{2\pi}\hat{k}$

3.145 $F_r = \dfrac{1}{2}(6\gamma)6 = 18\gamma$ lb at $z_r = 2$ ft. In 3.109, lines of action of concentrated loads are not properly placed for equipollence.   **3.147** 33,900 N at $x_G = 4$ m

3.149 $25 \uparrow$ N at $x_r = 0.8$ m

3.151 $200\hat{j}$ N and $200\hat{k}$ N · m

Let x be measured from the left end of the beam in the next three solutions:

3.153 1500 N down at $x = 1.33$ m, 3000 N down at $x = 3$ m, and 1500 N down at $x = 4.67$ m

3.155 Letting x be measured from left end of beam, 300 lb down at $x = 1$ ft, 700 lb down at $x = 3.5$ ft, and 600 lb up at $x = 8.33$ ft.

3.157 95.0 N down at $x = 1.1$ m and 38.5 N up at $x = 2.9$ m

3.159 $450\hat{i} - 290\hat{j}$ lb; $y = -0.644x + 6.69$

CHAPTER 4

4.1

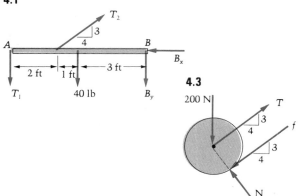

4.3

4.5

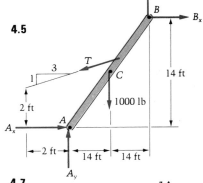

4.7

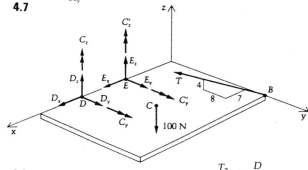

4.9

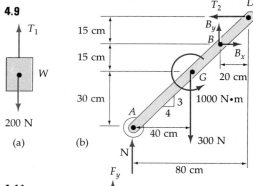

4.11

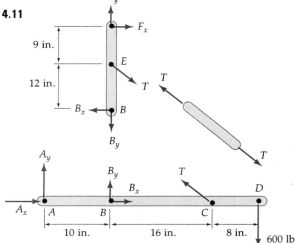

4.13 190 ↓ lb; 31.6 → lb

4.15 Answer given in problem

4.17 0.289 mg **4.19** 792 N

4.21 2.45 ft; 158 lb and 186 lb, respectively, normal to the planes

4.23 $(4 \tan \theta)/(3\pi)$; $T = 0$ at $\theta = 0$, $T \to \infty$ as $\theta \to \pi/2$

4.25 B is not on a line parallel to y drawn through A

4.27 At A, 50 ← N; at B, 100 ↑ N

4.29 222 ↑ lb; 0.919 ft   **4.31** 290 N; 220 ↑ N

4.33 $707\hat{i} + 1040\hat{j}$, 1670 ⟳ N · m

4.35 On top face, 1263 lb ↓, one-third of distance from left end; on bottom face, 1233 lb ↑, one-third of distance from right end; smaller distance between resultant forces.   **4.37** $WL/(2\ell)$; the other tension is <0 if $\ell < L/2$, and "you can't push a rope."

4.39 (a) 229 ⟍ 37° lb; 242 lb; 120 ⟲ lb-in.

4.41 At A, 190 ↑ N; at B, 303 ⟋ 60° N; at C, 85.6 N

4.43 At A, 250 ↑ N; at B, 50 ↑ N

4.45 $wL/4$ ↑ at each end

4.47 At A, 800 ↑ lb; at B, 1000 ↑ lb

4.49 At B, 3500 ↑ lb; at D, 500 ↑ lb

4.51 At A, $-216\hat{i} - 178\hat{j}$ lb; at F, 178 ↑ lb

4.53 $150\hat{i} + 1500\hat{j}$ lb and 9000 ↳ lb-ft

4.55 7110 lb; $4570\hat{i} + 7450\hat{j}$ lb

4.57 86.2 ↳ N · m   **4.59** 6.58 ft

4.61 4790 lb (compression)

4.63 89,600 lb (tension); $-17,800\hat{i} + 112,000\hat{j}$ lb

4.65 At C, 220 ↑ N; at D, 1060 ↑ N   **4.67** 0.897 W

4.69 12.1 → N   **4.71** 1710 → N

4.73 Tension is $mg/2$

4.75 $40\hat{i} + 80\hat{j}$ lb   **4.77** $-1.80H\hat{i} - 2.40H\hat{j}$

4.79 3.03 ft and 48.5 tons; $54.2 \le W_c \le 100$ tons

4.81 $13W/11$

4.83 $\theta \ge 7.18°$ (considered only θ's between 0 and 90°)

4.85 (a) It blows over; the moment imbalance about B is 3450 lb-ft.   (b) 0.51 ft deep

4.87 $300\hat{i} + 4800\hat{j}$ lb; 6000 ⟳ lb-ft   **4.89** 24.8 cm

4.91 89.6 lb   **4.93** 5260 lb compression

4.95 At A, $10,900\hat{i} - 14,500\hat{j}$ N; at B, 18,200 N

4.97 $88\hat{i} - 108\hat{j}$ lb

4.99 On $\mathcal{A}$, $-112\hat{i} + 1360\hat{j}$ lb; on $\mathcal{B}$, 328 ⟍ 20° lb

4.101 167 ↑ N   **4.103** 0.309ℓ **4.105** 23.2°

4.107 3250 N; $750\hat{i} + 3000\hat{j} + 1000\hat{k}$ N; $1800\hat{k}$ N · m   **4.109** $117\hat{i} + 2070\hat{j}$ N; 884 N

4.111 BC: 30800 N, BE: 50800 N

4.113 $149\hat{i}$ lb   **4.115** 333 ↑ N; $667\hat{k}$ N, $1000\hat{i}$ N · m

4.117 $0.214\hat{i}$ mg; all forces pass through the bar, so that moments about its centerline add to zero automatically.

4.119 Answer given in problem.

4.121 $F_{BF} = 3840$ N; $F_{CE} = 7510$ N

4.123 $202\hat{i} + 122\hat{j} + 82\hat{k}$ lb; $20\hat{k}$ lb-ft

4.125 371 N; $C_{B_y} = 24.9$ N · m; $C_{B_z} = -31.2$ N · m

4.127 At A, $-12,000\hat{j} + 6,000\hat{k}$ N; at B, $-12,000\hat{i} + 12,000\hat{j}$ N; at C, $16,000\hat{i} - 4,000\hat{k}$ N

4.129 144 lb compression each

4.131 42 lb   **4.133** $-10\hat{i} - 3\hat{j} - 100\hat{k}$ lb; $-1060\hat{i} + 3200\hat{j} + 10\hat{k}$ lb-ft

4.135 250 N tension; 0.0500 m

4.137 $\cos^{-1}\left(\dfrac{P\ell}{kh^2}\right)$; cosines can't be greater than 1

4.139 29.0°

4.141 Solution outlined in problem.

4.143 (a) If any one vanished, would have a 3-force, coplanar system with the forces neither concurrent nor parallel; If two vanished, the remaining one isn't "equal and opposite" to P; If all three vanished, $\Sigma F_y = -P \ne 0$. (b) 35.3°

4.145 From left to right: (a) AB (b) AB; BC; EF (c) BD (d) BD **4.147** $-355\hat{i} + 119\hat{j}$ lb; $733\hat{j}$ lb

4.149 1110 ⟋ $\theta/2$ lb; 27.9°

4.151 $25/\sin\alpha$ lb; $25\cot\alpha$ lb (↓ on top; ↑ on bottom); zero   **4.153** 16.7 ← lb   **4.155** 39.2 ⟳ N · m

4.157 49.7 lb   **4.159** 14.1 ← lb

4.161 (a) $2W(D - 2R)/D$ (b) Answer given in problem.

4.163 24,000 lb compression

4.165 (a) At A, $39.0\hat{i} - 353\hat{j}$ N; at D, $-39.0\hat{i} + 549\hat{j}$ N;   (b) at B, $157\hat{i} + 549\hat{j}$ N; at C, $-196\hat{i} - 196\hat{j}$ N

4.167 At F, $436\hat{i} + 390\hat{j}$ N; 1170 N onto BDE at D

4.169 (a) $4000 \sin(\theta/2)$ lb (b) 2000 lb directed from B toward C **4.171** 1.09 W

4.173 1860 N

4, $-120\hat{i} - 147\hat{j}$ lb;
(b) $120\hat{i} + 333\hat{j}$ lb
, at A, $25\hat{i} - 25\hat{j}$ lb; at B, $-25\hat{i} + 225\hat{j}$ lb
(D) on EB at D, 350 ↓ lb   **4.179** $-958\hat{i} + 708\hat{j}$ lb
4.181 120 lb of compression on pecan;
100 lb of tension in link AB

4.183 261 N   **4.185** 189 N
4.187 $16\,W/15$ **4.189** $-1880\hat{i} - 1000\hat{j}$ lb
4.191 (a) $-533\hat{i} - 200\hat{j}$ lb   (b) $-533\hat{i} + 200\hat{j}$ lb
4.193 (a) $-960\hat{i} - 483\hat{j}$ N   (b) $620\hat{i}$ N
(c) $-860\hat{i} + 483\hat{j}$ N
4.195 (a) $-1020\hat{i}$ N   (b) $870\hat{i} + 1200\hat{j}$ N
(c) $150\hat{i}$ N   **4.197** $2\,mg/\pi$
4.199 27.2°   **4.201** $15.0\,P$
4.203 $2(b + c)(c + d)/(ad)$ **4.205** 9140 ← lb
4.207 $\alpha = 39.8°$, $\beta = 68.2°$
4.209 Trailer, 506 lb each; truck rear, 891 lb each;
truck front, 753 lb each; 289 lb downward on ball
4.211 (a) 52 ↑ lb   (b) 14.9 lb Ⓒ (needs a sleeve
over pin to prevent buckling)   (c) 96 → lb
4.213 2210 lb Ⓣ; $1040\hat{i} + 8850\hat{j}$ lb
4.215 451 N Ⓣ; 284 → N
4.217 5250 lb Ⓒ; $-1360\hat{i} - 3980\hat{j}$ lb   **4.219** 177 lb
4.221 (a) 150 ↑ lb   (b) 100 ↑ lb

4.223 At E, $\dfrac{W}{e}\left[d + \dfrac{b(d - e)}{b - 2e}\right]$

$+ \dfrac{w}{e}\left[\dfrac{b - L}{2} + \dfrac{b(b - L - 2e)}{2(b - 2e)}\right]$;

At C, $\dfrac{2W(d - e)}{b - 2e} + w\left(\dfrac{b - L - 2e}{b - 2e}\right)$

4.225 1410 lb compression; 382 → lb by AB;
$-382\hat{i} - 710\hat{j}$ lb by CD **4.227** 480 → N
4.229 $F = W$ for any x **4.231** 828 ← N
4.233 With $\hat{i}$ toward the wall: $T_3 = 200\hat{i}$, $T_2 = -100\hat{i}$,
$T_1 = 400\hat{i}$ lb-ft, each on a FBD including the part of the
shaft farther from the wall.   **4.235** $-27.3\hat{i} + 266\hat{j}$ lb
4.237 63,000 lb   **4.239** At A, $50\hat{j}$ lb; at B, $-50\hat{j}$ lb
4.241 Answer given in problem.   **4.243** $25\,L/24$
4.245 70.8 kips Ⓒ   **4.247** 1500 lb Ⓣ
4.249 Solution outlined in problem.
4.251 ABC: $-167\hat{i} + 320\hat{j}$ lb; EDC: $417\hat{i} - 320\hat{j}$ lb;
The applied load is the difference in the x-components
acting on the pin.   **4.253** 57.6 kN Ⓣ
4.255 245 lb Ⓣ each
4.257 At E: $3600\,\hat{K}$ lb with $-3000\,\hat{J}$ lb-ft;
at O: $3600\,\hat{K}$ lb with the couple $6000\,\hat{J}$ lb
4.259 Answer given in problem.
4.261 50   **4.263** Answers given in problem.

CHAPTER 5

5.1 BC: 1100 lb Ⓒ; CD: 880 lb Ⓣ; BD: 500 lb Ⓣ;
AD: 880 lb Ⓣ; AB: 200 lb Ⓣ
5.3 AB: 4.24 lb Ⓒ; AH and HG: 3 lb Ⓣ; BH, EF, and
FG: 2 lb Ⓣ; BC and CD: 4 lb Ⓒ; CG and DF: 0;
BG: 1.41 lb Ⓣ; DG: 2.83 lb Ⓣ; DE: 2.83 lb Ⓒ
5.5 CD: 800 N Ⓒ; BD: 800 N Ⓣ; AE: 800 N Ⓒ;
BE: 2000 N Ⓒ; AB: 2260 N Ⓣ; BC: 1130 N Ⓣ;
DE: 1130 N Ⓒ
5.7 AG: 150 lb Ⓣ; AF: 175 lb Ⓒ; AB: 90 lb Ⓣ;
EF: 90 lb Ⓒ; BF: 350 lb Ⓒ; CD: 233 lb Ⓒ;
DE: 120 lb Ⓣ; BE: 408 lb Ⓣ; CE: 1100 lb Ⓒ;
BC: 120 lb Ⓒ; $FG = 0$
5.9 CD: 167 N Ⓒ; AC: 208 N Ⓣ; others: 0
5.11 AB: 10 kN Ⓣ; BC: 0; FE: 15 kN Ⓣ; FD: 0;
GF: 15 kN Ⓣ; GD: 11.2 kN Ⓒ; CD: 5 kN Ⓒ;
AC: 11.2 kN Ⓒ; AG: 10 kN Ⓣ; GC: 10 kN Ⓣ
5.13 AB: 667 lb Ⓒ; BE: 1200 lb Ⓒ; BC: 2670 lb Ⓒ;
CD: 2670 lb Ⓒ; BQ: 2670 lb Ⓣ; GQ: 3330 lb Ⓣ;
QC: 0; QD: 0
5.15 OH: 4 k Ⓣ; AB: 3 k Ⓒ; BH: 3.89 k Ⓣ
5.17 AB: 0.415W Ⓒ; AG: 0.915W Ⓒ; BG: 0.215W Ⓣ;
BC: 0.414W Ⓒ; CG: 0.704W Ⓒ; FG: 0.289W Ⓒ;
CD: 1.05W Ⓒ; CF: 0.395W Ⓣ; DF: 0.354W Ⓒ;
DE: 0.317W Ⓒ; EF: 0.183W Ⓣ
5.19 AB: 67 lb Ⓣ; AE: 83 lb Ⓒ; BC: 400 lb Ⓣ;
BE: 50 lb Ⓣ; BF: 417 lb Ⓒ; CD: 400 lb Ⓣ;
CF: 400 lb Ⓒ; DF: 500 lb Ⓒ; EF: 67 lb Ⓒ
5.21 AG: 3.33 kN Ⓣ; AF: 18.3 kN Ⓣ;
AB: 9.0 kN Ⓣ; CD: 3 kN Ⓣ
5.23 CE and CF: 0; DE, EF, and FG: 19.2 k Ⓣ;
BF: 8 k Ⓒ; BC and CD: 20.8 k Ⓒ; AB: 27.7 k Ⓒ;
AG: 10.7 k Ⓣ; BG: 8.33 k Ⓣ
5.25 $F_1 = F_2 = 0$; $F_3 = 3.54$ k Ⓣ; $F_4 = 20$ k Ⓣ;
$F_5 = 63.5$ k Ⓒ
5.27 BG: 0; AG: 50 kN Ⓒ; CG: 39.3 kN Ⓣ
5.29 (a) 6   (b) 354 lb Ⓣ   (c) A_y is same either way,
but A_x is zero for the roller. F_{AB} is same either way,
so F_{AJ} must change with a roller at A.
5.31 Start with a triangle in the middle
and work outward.
5.33 $m = m_1 + m_2 - 1$, $p = p_1 + p_2 - 2$,
$m_1 = 2p_1 - 3$, and $m_2 = 2p_2 - 3$ yields $m = 2p - 3$
5.35 2.77 kN Ⓒ   **5.37** 10.8 kN Ⓒ
5.39 CF: 1.12 kN Ⓒ; BG: 1.41 kN Ⓒ
5.41 BH: 1.58 kN Ⓣ; DF: 1.42 kN Ⓣ; CF: 1.78 kN Ⓒ;
CD: 0.935 kN Ⓣ   **5.43** JM: 0; DG: 8.49 kips Ⓣ
5.45 (a) BC: 1000 lb Ⓒ   (b) CE: 3000 lb Ⓣ
5.47 AF: 0; CF: 3.75 kN Ⓣ
5.49 IE: 300 lb Ⓒ; JC: 486 lb Ⓒ; KC: 0; DE: 901 lb Ⓣ
5.51 BF: 1600 lb Ⓣ; BC: 2260 lb Ⓒ

5.53 *GH*: 5830 lb ©; *CH*: 4170 lb ©; *BC*: 8000 lb Ⓣ
5.55 *BH*: 4.19 kips Ⓣ; *FE*, *DE*, *AJ* are zero-force members. **5.57** 1.40W ©
5.59 *DC*: 3320 lb ©; *DG*: 212 lb ©; *DF*: 0
5.61 *CD*: 29.3 kN ©; *DH*: 4 kN ©; *CH*: 21.6 kN Ⓣ
5.63 *DC*: 893 N Ⓣ; *DE*: 631 N ©
5.65 *CD*: 20.8 kips ©; *KJ*: 3.25 kips ©; *LJ*: 22.9 kips Ⓣ **5.67** *CD*: 473 lb ©; *DJ*: 0
5.69 *EL*: 16.3 kN Ⓣ; *GH*: 8.79 kN ©
5.71 (a) *BG*, *CG*, *CF* are zero-force members
(b) *FG*: 9600 N Ⓣ; *FD*: 389 N Ⓣ; *FE*: 11300 N Ⓣ; *CD*: 19400 N ©
5.73 *AC*: 12.8 kN ©; *OC*: 38.8 kN Ⓣ; *BC*: 29.2 kN ©
5.75 172 lb © **5.77** 52 lb © **5.79** (a) at *A*: $33.3\hat{j}$ lb; at *B*: $-75\hat{i}$ lb; at *C*: $-25\hat{i} - 33.3\hat{j} + 100\hat{k}$ lb
(b) *AD*: 33.3 lb ©; *CD*: 108 lb Ⓣ
(c) *BC*: 0; *AB*: 0; *BD*: 75 lb Ⓣ (d) 0
5.81 *AB*: 0.410*P* Ⓣ; *BC*: 0.273*P* ©
5.83 347 N © each
5.85 $\mathcal{B}_1$: 2080 N ©; $\mathcal{B}_2$: 2080 N Ⓣ; $\mathcal{B}_3$: 0
5.87 $\mathcal{B}_1$: 307 N ©; $\mathcal{B}_2$: 2100 N Ⓣ; $\mathcal{B}_3$: 1190 N ©
5.89 (a) Start with tetrahedron *CBDF*; (b) 6750 lb ©
5.91 In psi: *AB*: 100, tensile; *BD*: 250, tensile; *BC*: 550, compressive; *DC* and *AD*: 440, tensile
5.93 In MPa: *AB*: 0.753 Ⓣ; *BE*: 0.333 ©; *BC*: 0.565 Ⓣ; *BD*: 0.267 Ⓣ; *AE*: 0.160 ©; *ED*: 0.283 ©; *CD*: 0.400 © where Ⓣ means tensile stress and © means compressive.
5.95 $\sigma_{BG} = 7.05$ MPa; $\sigma_{CF} = 4.48$ MPa, both compressive
5.97 *LC*: 398 kPa, *CR*: 99.5 kPa, both tensile
5.99 *DE* (stress is 4.92 MPa, compressive)
5.101 0.812 MPa, compressive
5.103 *AB*: 6000 psi, *BC*: 2990 psi, and *BD*: 11,600 psi, all compressive **5.105** 53.9 kPa
5.107 0.0305 in. elongation
5.109 0.00742 in. elongation
5.111 $\delta_{BG} = 0.0000475\ell$ m, $\delta_{CF} = 0.0000239\ell$ m, both shortened **5.113** 0.0112 in. **5.115** 308 MPa
5.117 3*L* **5.119** $0.111 \times 10^{-4}\hat{i} - 12.5 \times 10^{-4}\hat{j}$ in.
5.121 Answer given in problem.
5.123 $(0.490\hat{i} - 0.128\hat{j}) \times 10^{-3}$ m
5.125 $2.41 \times 10^{-4}L$ in. (with *L* in inches); $F_A = 24,100$ lb; $F_s = 43,300$ lb; $F_C = 32,700$ lb **5.127** 105.8°C
5.129 $F_A = 7600$ lb ©; $F_s = 79,900$ lb Ⓣ; $F_C = 27,700$ lb Ⓣ
5.131 $-10\hat{i} + 5\hat{j}$ kN with the couple $-7.5\hat{k}$ kN · m
5.133

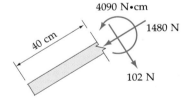
750 N · m
500 N

5.135 On *A-A*, 20 ↓ lb with 20 ↻ lb-in.; on *B-B*, 20 ↑ lb with 40 ↺ lb-in.; on *C-C*, 20 ↓ lb with 40 ↻ lb-in.

5.137 At *AA*,

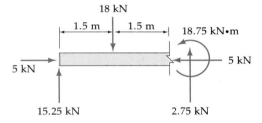

1 k
8 k-ft
; at *BB*,
2 k
8 k-ft

at *CC*,
3 k

5.139 75 ↓ N with 45 ↻ N · m (or *V* = 75 N; *M* = 45 N · m)
5.141 16.7 ↓ N with 8 ↺ N · m (or *V* = 16.7 N; *M* = −8 N · m)
5.143 250 ↑ N with 125 ↺ N · m (or *V* = −250 N; *M* = −125 N · m)
5.145 40 ↑ lb and 260 ↺ lb-ft
5.147 56.6 ← kips with 7.52 ↑ kips with 28.2 ↻ kip-ft
5.149

18 kN
1.5 m   1.5 m
18.75 kN•m
5 kN
5 kN
15.25 kN
2.75 kN

5.151

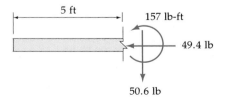

4090 N•cm
1480 N
40 cm
102 N

5.153

5 ft
157 lb-ft
49.4 lb
50.6 lb

5.155 (a) *x* = 8 m
(b) 36 kN for $0 \le x \le 4$ and $12 \le x \le 16$ m
(c) *x* = 0, 16 m (d) 216 kN · m at *x* = 8 ft

at $x = a$

t center; 1800 N $\cdot$ m at the wall
+ $3000\hat{j}$ lb and $18000\hat{k}$ lb-ft

... = 150 lb Ⓒ for $0 \le x < 6$ ft; $N = 150$ lb Ⓣ for
$6 < x \le 12$ ft; $V = 250x - 3000$ lb, $0 \le x \le 12$ ft;
$M = -18000 + 3000x - 125x^2$ lb-ft, $0 \le x \le 12$ ft

5.163 (a) $27000 \uparrow$ N with $48000 \circlearrowleft$ N $\cdot$ m
(b) $N = 0$; $V = -27000 + 9000x - 563x^2$;
$M = -48000 + 27000x - 4500x^2 + 188x^3$

5.165 $V = \dfrac{2P}{3}$ for $\dfrac{L}{3} < x < \dfrac{2L}{3}$; $M = \dfrac{PL}{9}$ at $x = \dfrac{L}{3}$

and $\dfrac{-PL}{9}$ at $x = \dfrac{2}{3}L$

5.167 $V = \dfrac{-3P}{4}$ for $0 < x < \dfrac{L}{2}$; $M = \dfrac{3PL}{8}$ at $x = \dfrac{L}{2}$

5.169 $V = P$ for $0 \le x \le L$; $M = \dfrac{PL}{2}$ at $x = \left(\dfrac{L}{2}\right)^+$;

$M = -\dfrac{PL}{2}$ at $x = \left(\dfrac{L}{2}\right)^-$

5.171 $V = \dfrac{wx^2}{L} - \dfrac{wL}{4}$ and $M = \dfrac{wLx}{4} - \dfrac{wx^3}{3L}$,

for $0 \le x \le \dfrac{L}{2}$. For $\dfrac{L}{2} \le x \le L$,

obtain V and M by symmetry.

5.173 For $0 \le x \le 3$ ft, $V = -800$ lb and
$M = 800x$ lb-ft; for $3 \le x \le 9$ ft,
$V = -50x^2 + 900x - 3050$ lb and

$M = \dfrac{50}{3}x^3 - 450x^2 + 3050x - 3150$ lb-ft.

5.175 In the six intervals, from left to right,
$(V, M) = (-31, 31x)$;

$\left[\dfrac{2000}{6}(x - 0.1)^2 - 31, -\dfrac{1000}{9}(x - 0.1)^3 + 31x\right]$;

$(100x - 41, -50x^2 + 41x + 1)$; $(19, -19x + 19)$;
$(19, -19x - 81)$: $(-200, 200x - 300)$
with units of (N, N $\cdot$ m).

5.177 (a) For $0 < x < 0.6$ m, $V = \dfrac{500}{3}x^2$ and

$M = \dfrac{-500}{9}x^3$; for $0.6 < x < 0.8$, $V = 60$ and

$M = -60x + 24$; for $0.8 < x < 1$, $V = 660$ and
$M = -660x + 504$. Units are N for V, and N $\cdot$ m for M.
(b) 660 N for $0.8 < x \le 1$ (c) -156 N $\cdot$ m at the wall
5.179 $V = 400x$ lb; $M = 1500x - 12500$ lb-ft
5.181 $V_{\text{extremes}} = 13.3$ lb for $6 < x < 12$ ft, and
-46.7 lb at $x = 0$; $M_{\text{extremes}} = 109$ lb-ft at $x = 4.67$ ft,
and 0 at $x = 0$ and 12 ft

5.183 $V = \text{constant} = -\dfrac{C}{L}$ throughout;

$M_{\text{extremes}} = 0$ at $x = 0$ and C at $x = L$

5.185 $V_{\text{extremes}} = 500$ lb for $0 \le x < 3$ ft
and -500 lb for $3 < x \le 6$ ft, $M_{\text{extremes}} = 0$ at $x = 0$,
6 ft and -1500 lb-ft at $x = 3$ ft
5.187 $V_{\text{extremes}} = 1500$ kN at $x = 6^-$ m, and -1200 kN
at $x = 6^+$ m; $M_{\text{extremes}} = \pm1800$ kN $\cdot$ m at $x = 2$ and
6 m, respectively. For $2 < x < 6$ m, $V = 300(x - 1)$ kN
and $M = -150x^2 + 300x + 1800$ kN $\cdot$ m.
5.189 $V_{\text{extremes}} = 20$ kN at $x = 9^-$ m, and -16 kN
for $0 \le x < 3$ m; $M_{\text{extremes}} = 48$ kN $\cdot$ m at $x = 3$ m,
and -18 kN $\cdot$ m at $x = 9$ m

5.191 $V_{\text{extremes}} = \dfrac{q_0L}{\pi}$ at $x = \dfrac{L}{2}$, and 0 at $x = 0$ and L;

$M_{\text{extremes}} = \dfrac{2q_0L^2}{\pi^2}$ at $x = 0$, and 0 at $x = L$

5.193 $V_{\text{extremes}} = 550$ lb for $6 < x < 12$ ft,
and -1000 lb at $x = 12^+$ ft; $M_{\text{extremes}} = 1200$ lb-ft
at $x = 4$ ft, and -2500 lb-ft at $x = 12$ ft
5.195 $V_{\text{extremes}} = 83.3$ N for $6 \le x < 9$ m,
and -100 N for $9 < x \le 12$ m; $M_{\text{extremes}} = 0$ at $x = 0$,
and -500 N $\cdot$ m at $x = 9$ m
5.197 $V_{\text{extremes}} = 14.4$ N for $0.3 < x \le 1.2$ m,
and -5.6 N for $0 \le x < 0.3$ m; $M_{\text{extremes}} = 22.4$ N $\cdot$ m
at $x = 0.6^+$ m, and -2.64 N $\cdot$ m at $x = 0.6^-$ m
5.199 50.7 ft from P; 3310 lb at P

5.201 $y = H\left(1 - \cos\dfrac{\pi x}{L}\right)$

5.203 Answer given in problem.
5.205 214 ft; $T_A = 175,000$ lb, $T_B = 144,000$ lb

5.207 $fL\left[\dfrac{2^{n+2}(x/L)^{n+2} - 2(x/L)}{2^{n+2} - 2}\right]$

5.209 $\sqrt{K - 1}\cosh^{-1}\sqrt{K/(K - 1)}$, where $K = \left(\dfrac{2T_0}{q_0\ell}\right)^2$

5.211 2.3 m **5.213** (a) 1.3% (b) 6.8% (c) 38.3%
5.215 2.95 N/m
5.217 (a) $120 \uparrow$ lb; (b) $57.4 \rightarrow$ lb with line of action
8.3 ft above B **5.219** q_0L; $2.06L$; $0.152L$
5.221 $S_1 = 3.94$ ft; $S_2 = 4.61$ ft; $T_1 = 1240$ lb;
$T_2 = 1110$ lb; $T_3 = 1400$ lb
5.223 Assuming a uniform beam,
(a) $\phi_1 = \phi_2 = 39.9°$ (b) Each $= 156$ lb
(c) 6.42 ft above the horizontal line through A
5.225 (a) 0.436 m (b) $T_1 = 600$ N; $T_2 = 557$ N;
$T_3 = 572$ N **5.227** 60°

CHAPTER 6

6.1 (a) 60 lb; (b) 40 lb **6.3** 31.0°
6.5 $1.38W$, by tipping **6.7** 82.9 lb **6.9** 94.6 N
6.11 $28mg/29$ **6.13** $0.030R$ **6.15** 5.71°
6.17 115 lb; 0.577 **6.19** $-129.0 \le P \le 65.1$ N
6.21 0.289 **6.23** 52.9 N $\cdot$ m **6.25** 0.885

6.27 0.247   **6.29** (a) $mg \cos \phi / [2 \sin (\theta + \phi)]$
(b) 0.577   **6.31** 41.2°; 0.875   **6.33** 166 N · m
6.35 0.625   **6.37** $(x, y) = (0.395, 0.500)$ m
6.39 (a) $2Wr \sin \phi \circlearrowright$ (b)Answer given in problem.
6.41 0.0577; 25.8 lb
6.43 (a) 20 lb;   (b) 20 lb each;   (c) Mr. A; 0.125
6.45 No; 6.56 → lb
6.47 51.8 lb; answer is same if P pushes to the left.
6.49 0.373L; before slip occurs have 5 unknowns
but only three independent equations.   **6.51** 2
6.53 (a) Answer given in problem;
(b) rolling to the right.
6.55 $P = 40$ lb to the right; 40 ← lb by slot on $\mathcal{A}$;
60 ↑ lb by track on $\mathcal{C}$.   **6.57** 0.350   **6.59** 0.511
6.61 100 lb-ft   **6.63** 31.3 N   **6.65** 0.333
6.67 112 N   **6.69** $\mu_T \leq \mu_P / [(1 - \mu_P) \tan 8° + \mu_P]$
6.71 (a) 0.819 cm (in tension)   (b) 0.588
6.73 1270 N   **6.75** 982 N   **6.77** 115 N   **6.79** 0
6.81 $(\cot \alpha) / 2$ **6.83** 0.208   **6.85** 563 N
6.87 3.61°, tips   **6.89** (a) 115 lb   (b) 0.577
6.91 (a) 2.05;   (b) 1.45   **6.93** 85 lb
6.95 (a) $N = 33.3$ lb; $f = 19.2$ lb directed along AB
(b) 0.577

6.97 (a) $8.250 + 0.533P$ (b) $6.925 - 0.858P$ ⟋40°
(c) 3.87 N   (d) 14.6 N (by reversing the direction of f)
6.99 1.55 lb
6.101 If $\tan \alpha > \mu_o$, slips first at $P = \mu_o mg$. If
$\tan \alpha < \mu_o$, tips first at $P = mg \tan \alpha$. If $\tan \alpha = \mu_o$,
slips and tips simultaneously at $P = \mu_o mg$.
6.103 $mg(32\mu_o + 3x) / (32 - 3\mu_o x)$
6.105 2/3   **6.107** 24.0°
6.109 (a) $\Sigma M_{\text{line } OB} = 0$ shows that N is in plane AOB;
(b) $\sqrt{x^2 + y^2} / (2z)$
6.111 (a) 240 lb, slips only on plane and to the right as
$\mathcal{A}$ turns ↺   (b) 480 lb, slips only on $\mathcal{B}$ and to the right
as $\mathcal{A}$ turns ↻
6.113 $\mathcal{B}$ is needed for equilibrium; $W_{\beta_{\min}} = 29.2$ lb;

73.2 ⟋60° lb   **6.115** 12 ft
6.117 $N_{\text{rear}} = 140$ lb; $N_{\text{front}} = 43.2$ lb; $f = 49.2$ lb;
$\mu_{\min} = 0.351$ **6.119** at the wall; $1.17mg$ **6.121** 0.857
6.123 0.791   **6.125** 5/12
6.127 $5W_B / (12W_C + 6W_R)$
6.129 (a) 0.575   (b)1.15 m   **6.131** 5.04 ft
6.133 0.213   **6.135** 5.53 N; 0.237   **6.137** 0.348
6.139 $W\sqrt{\mu^2 \cos^2 \phi - \sin^2 \phi}$ **6.141** 0.318
6.143 $(\tan \alpha) / 2$
6.145 107° < 2α < 116°, where 2α = angle between
bars   **6.147** 90°   **6.149** Answer given in problem.
6.151 Answer given in problem.   **6.153** 623 lb

6.155 3070 N   **6.157** 122 N ≤ $W_{\mathcal{A}}$ ≤ 2060 N
6.159 81.4 lb   **6.161** 422 lb   **6.163** 3.17
6.165 800 lb   **6.167** Yes. (He could go 8.2 ft further
out before it slipped.)   **6.169** 95 lb
6.171 67 → lb and 602 ↻ lb-ft
6.173 2   **6.175** 40.2 lb
6.177 (a) B: 94.4 ↓ lb, D: 94.4 ↑ lb   (b) 5.60 lb
6.179 $(1 - \mu \tan \alpha) \tan \alpha / (\mu + \tan \alpha)$
6.181 14.1 lb-ft   **6.183** 41.2 lb-in.
6.185 The integration yields $M = P\mu R_o / 2$,
which is 25% less than $2P\mu R_o / 3$.
6.187 10.2 lb-in. (95% of the moment in 6.186)
6.189 μPR **6.191** 213 lb-in.   **6.193** 107 N
6.195 Some results: $\theta = 90° \Rightarrow \rho = 0.414$;
$\theta = 45° \Rightarrow \rho = 1.414$
6.197 72.0° (Equation to solve is
$\sin^3 \beta + \dfrac{1}{2} \sin^2 \beta \cos \beta - 1 = 0$)

CHAPTER 7

7.1 Answer given in problem.
7.3 $\bar{x} = [2r \sin (\alpha / 2)] / \alpha$ **7.5** 2.10
7.7 $(0.549a, 0)$ ft   **7.9** (1.00, 0.393) ft   **7.11** 2.73 in.
7.13 4.50 in.   **7.15** 3.60 in.   **7.17** 4 m   **7.19** 0.637R
7.21 3 m   **7.23** 1 in.   **7.25** 2.087 m   **7.27** $\bar{z} = 3$ ft
7.29 Answer given in problem.   **7.31** $(\frac{3}{4}, \frac{3}{4}, \frac{3}{4})$ m
7.33 5 cm   **7.35** Answer given in problem.
7.37 $(-0.05, 0.117)$ m
7.39 lines: (2.5, 1.0) cm; triangle: (2.67, 1.0) cm
7.41 (0, 6.24) in.   **7.43** (2.39, 2.39) in.
7.45 $(0, -0.849)$ cm   **7.47** (33.5, 17.0) cm
7.49 (6.67, 2.50) in.   **7.51** (6.19, 5.77) in.
7.53 $(0, -4r^3 / [3\pi(2R^2 - r^2)])$ **7.55** (0.943, 1.19) in.
7.57 (0, 0.584R)   **7.59** (0, 2.42) in.
7.61 (1.03, 2.03) in.; no difference to 3 digits
7.63 $(0.255, 0.225, -0.0200)$ m   **7.65** $\left(\dfrac{l}{2}, \dfrac{l}{6}, \dfrac{l}{6}\right)$
7.67 (15, 21.6, 28.4) cm; $\bar{x}$
7.69 $(2.21, 4.77, 0.794)$ in.   **7.71** $(0.602, -0.244)$ in.
7.73 (4.40, 1.79) in.   **7.75** 3.46 cm
7.77 (0.397, 0.260) m   **7.79** (3.94, 2.85) in.
7.81 (1.25, 1.25) m   **7.83** Answer given in problem.
7.85 0.675 units of length
7.87 0.494   **7.89** $\dfrac{4}{3\pi} (a + b)$; yes
7.91 $\lim\limits_{\alpha \to 0} \left(\dfrac{2r \sin \alpha}{3\alpha}\right) = \dfrac{2r}{3} = \bar{y}$ for the triangle
7.93 $(0, -0.198R)$ **7.95** 3.00 in.
7.97 $(10.4, -3.81, 35)$ mm   **7.99** 0.707
7.101 1.73R **7.103** $\left(\dfrac{2l}{7}, \dfrac{2l}{7}, \dfrac{l}{2}\right)$ **7.105** $\frac{2}{3}L$

7.107 $(0, -R/\pi)$ **7.109** (a) $1.71r$ (b) $1.29r$
7.111 $\bar{z} = 3R(1 + \cos \alpha)^2 / [4(2 + \cos \alpha)]$
7.113 (a) 16γ; $(\frac{3}{8}, \frac{12}{5}, -3)$ (b) 5.4γ, 8.6γ, 2γ,

respectively **7.115** $14.7°$ **7.117** $\tan^{-1}\left(\dfrac{3\pi W_s}{4W_c}\right)$

7.119 On the axis, $H/3$ from the base toward the vertex **7.121** 384 in.^{3} **7.123** 993 cm^3
7.125 123 ft^3 **7.127** 87.5 m^3 **7.129** 11.0 in.^{3}
7.131 $(0, 2R/\pi)$ **7.133** $0.785 s^3$ **7.135** 0.393 m^3
7.137 1.60 m^2 **7.139** 8,484 mm^2; 53,300 mm^2
7.141 174 in.^{2} **7.143** $\pi^2 Rr$ **7.145** $\pi r(\pi R - 2r)$

7.147 $\pi r \left(\dfrac{\pi R}{2} - r\right)$ **7.149** $\dfrac{\pi r^2}{12}(3\pi R + 4r)$

7.151 $\dfrac{\pi r^2}{12}(3\pi R - 4r)$ **7.153** 21,500,000 ft^3

7.155 $(57.5, 0, 0)$; Because the mast's diameter is much larger, $\bar{x}$ would increase with an inch of ice covering everything.
7.157 6.10×10^{24} kg (Handbook actual mass is 5.98×10^{24} kg, so not bad!); 5.63 g/cc

CHAPTER 8

8.1 (a) $bh^3/3$; (b) y and y_C are the same line!
8.3 4.27 m^4 **8.5** 25.1 cm^4 **8.7** $tr^3(\alpha - \sin \alpha)/2$
8.9 $bh^3n/[3(n + 3)]$ **8.11** Answer given in problem.
8.13 $bh(4h^2 + b^2)/12$ **8.15** $bh(12h^2 + b^2)/48$
8.17 $\alpha t r^3$ **8.19** $\pi(R_0^4 - R_i^4)/2$ **8.21** $bh^3/36$

8.23 $bh^3n\left[\dfrac{n^2 + 4n + 7}{12(n + 3)(n + 2)^2}\right]$ **8.25** 1.82 cm^4

8.27 $r^3t\left(\dfrac{\alpha}{2} + \dfrac{\sin \alpha}{2} - \dfrac{4}{\alpha}\sin^2 \dfrac{\alpha}{2}\right)$

8.29 Answers given in problem. **8.31** 120 ft^4
8.33 $0.0361 s^4$; $0.0313 s^4 = 0.867$ of J_C for the triangle, which has some areas farther from the axis.
8.35 7.15 m^4
8.37

$$bh^3n\left[\dfrac{n^2 + 4n + 7}{12(n + 3)(n + 2)^2}\right] + hb^3n\left[\dfrac{n^2}{(3n + 1)(2n + 1)^2}\right]$$

8.39 0.894 m **8.41** $83.0 \ell^4$ **8.43** $\frac{5}{64}bh^3$
8.45 0.000235 m^4 **8.47** 453 in.^{4} **8.49** 2.53 ft^4
8.51 8,800,000 cm^4 **8.53** 0.806 ft^4
8.55 0.806 ft^4 **8.57** 0.0966 m^4 **8.59** 1560 in.^{4}
8.61 8070 in.^{4} **8.63** $2.49 s^4$ **8.65** 0.252 m^4
8.67 0.326 m^4 **8.69** $5\sqrt{3} s^4/16$
8.71 21.1 in.^{4}, the same as table value **8.73** 6.91 in.^{4}
8.75 15.4 in.^{4} **8.77** 194 in.^{4}
8.79 (a) 77,900 mm^4 (b) 538,000 mm^4
8.81 $I_{x_c} = 1740$ in.^{4}; $I_{y_c} = 666$ in.^{4}

8.83 (a) $I_x = \dfrac{th^3}{12} + \dfrac{tbh^2}{2}$; $I_y = \dfrac{tb^3}{6}$ (b) 1.81

8.85 62.5 in.^{4} **8.87** Answers given in problem.
8.89 $I_{YY} = 26.6$ in.^{4}, right on; $k_Y = 1.55$ in., off by less than 1% **8.91** 377 A **8.93** $-R^4/8$
8.95 $\frac{8}{5}$ in.^{4}; $\frac{32}{7}$ in.^{4}; $\frac{-8}{3}$ in.^{4}
8.97 Answers given in problems. **8.99** -1.05
8.101 19.0 in.^{4} **8.103** -0.000172 m^4
8.105 -1.33 ft^4 **8.107** -1.36 ft^4
8.109 -0.0466 m^4 **8.111** -1700 in.^{4}
8.113 11,300,000 cm^4
8.115 12.4 in.^{4}; 79.9 in.^{4}; -12.4 in.^{4}
8.117 0.267 in.^{4}; 1.16 in.^{4}; -0.533 in.^{4}
8.119 1.41 in.^{4}, $64.9°$ clockwise from x_C; 0.0174 in.^{4}, $25.1°$ counterclockwise from x_C
8.121 (a) $h^4/24$, $h^4/72$ (respectively normal and parallel to the hypotenuse) (b) If $I_x = I_y$ and $I_{xy} = 0$, the circle is a point. (c) The triangle contributes half the inertia of the square in all directions.
(d) Answer given in problem. **8.123** $7b^4/6$; $b^4/6$
8.125 $I_{max} = 52.0$ in.^{4} at $27.5°$ ↶ from x_C; $I_{min} = 5.60$ in.^{4} at $62.5°$ ↷ from x_C.
8.127 (a) $I_x = 14.96$ ft^4, $I_y = 10.51$ ft^4;
(b) $I_{x_c} = 3.091$ ft^4, $I_{y_c} = 3.883$ ft^4; the results check.

CHAPTER 9

9.1 $\dfrac{mg}{2}\tan \theta$; the limits are as stated. **9.3** 5250 lb
9.5 43.7 N **9.7** At D, 2 ↓ kN; at B, 4 ↑ kN
9.9 $\theta_1 = 26.27°$; $\theta_2 = 59.67°$; $T = 0.7116mg$; $N_{left} = 0.8967mg$; $N_{right} = 0.5050mg$

9.11 $P = (1.5 \cot \alpha)F$ **9.13** $\dfrac{W}{2}\left(1 - \dfrac{r}{R}\right) = \dfrac{W}{4}$

9.15 21.3 ↑ N **9.17** $P = \dfrac{(a + 2b + 2c)\cot \theta}{2a} W$

9.19 (a) $\dfrac{5W\sqrt{9 \sin^2 \beta + \cos^2 \beta}}{4 \sin \beta}$

(b) $\dfrac{5W\sqrt{25 \sin^2 \beta + \cos^2 \beta}}{12 \sin \beta}$

(c) $-\dfrac{5}{4}W(\cot \beta)\hat{i} - \dfrac{13}{4}W\hat{j}$; no **9.21** 2860 ↑ lb

9.23 (a) 779 ↶ lb-ft (b) 300 → lb,
(c) Answer given in problem. (See FBDs at top of next page.)
9.25 Answer given in problem.
9.27 9140 lb **9.29** $P = W/2$
9.31 $P = W$ (independent of a and b!)
9.33 Answers given in problem.
(Consider thin horizontal and vertical cylinders of fluid, respectively.)
9.35 Answer given in problem.
9.37 132,000 lb, normal to the area at depth 75.004 ft. (Note from the fifth decimal digit that,

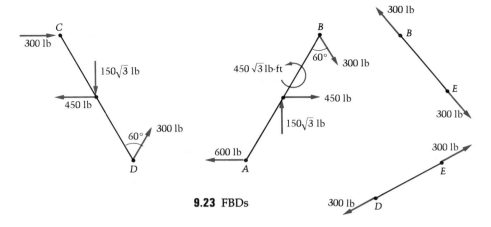

9.23 FBDs

as depth increases, distance between centroid and center of pressure decreases!)
9.39 22.1 lb normal to the area at depth 2.130 ft
9.41 $(x_P, y_P, z_P) = (6.98, 0, 0)$ ft
9.43

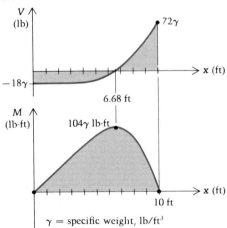

γ = specific weight, lb/ft³

9.45 $\dfrac{\pi\gamma dh^2}{4} \rightarrow$ at center of door   **9.47** $72\gamma \leftarrow$

9.49 105γ ◿ 1 , at the center of pressure of

Example 9.7
9.51 Answer given in problem.
9.53 $(-15.6\hat{i} + 26.6\hat{j}) \times 10^6$ lb, 7.71 ft to the left of D

9.55 $3.74\gamma R^2 L$ ◿ 48.1° through center of quarter-circle
9.57 $\dfrac{2(3H^2 + 3HR + R^2)}{3(2H + R)}$ below the water line

9.59 $\sqrt{3}B$

9.61 (a) $-515,000\hat{i} - 98,100\hat{j}$ N on the line

$y = 0.190x - 1.81$ with ⌐ at A (b) at A,

$515,000\hat{i} - 834,000\hat{j}$ N; at B, 932,000 ↑ N

9.63 642,000 ◺ 40.1° N, along the line
$y = -0.843x + 5.33$ **9.65** Answer given in problem.
9.67 0.293 **9.69** 0.091349748

APPENDIX A

A.1 $-\hat{i} + 5\hat{j} + 6\hat{k}$; $5\hat{i} + 7\hat{j} - 14\hat{k}$ **A.3** -38
A.5 $228\hat{i} + 456\hat{j} + 684\hat{k}$; 120; $404\hat{i} - 112\hat{j} - 60\hat{k}$;
$512\hat{i} - 456\hat{j} - 240\hat{k}$
A.7 (a) $6.5\hat{i} - 20.5\hat{j} - 0.3\hat{k}$ (b) $-\hat{i} - 29.2\hat{j} + 6\hat{k}$
(c) -28.0 (d) $-52.2\hat{i} + 30.6\hat{j} + 140\hat{k}$ (e) -1010
(f) $193\hat{i} - 7.14\hat{j} + 73.4\hat{k}$ (g) $0.506\hat{i} + 0.863\hat{j}$
(h) $(0.189, -0.943, 0.276)$
A.9 (a) $17\hat{i} - 21\hat{j} + 19\hat{k}$ (b) $14\hat{i} - 20\hat{j} + 11\hat{k}$
(c) 68 (d) $-140\hat{i} - 87\hat{j} + 20\hat{k}$ (e) -173
(f) $67\hat{i} - 180\hat{j} - 314\hat{k}$ (g) $0.141\hat{i} + 0.990\hat{k}$
(h) $(0.487, -0.649, 0.584)$
A.11 -11; $A \times B = -\hat{i} + 19\hat{j} + 7\hat{k}$
A.13 Area of shaded parallelogram = Area of dashed rectangle = (Base)(Height) = $|A|(|B|\sin\theta)$ and $|A \times B| = |A||B|\sin\theta$

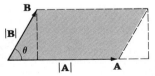

A.15 2.91 m **A.17** 9.70
A.19 (a) $\frac{3}{5}\hat{i} + \frac{4}{5}\hat{j}$ (b) $\hat{i}$ (c) $\frac{4}{13}\hat{i} - \frac{3}{13}\hat{j} + \frac{12}{13}\hat{k}$

INDEX

659